Todd

PATTY'S INDUSTRIAL HYGIENE AND TOXICOLOGY

Fourth Edition

**Volume I, Parts A and B
GENERAL PRINCIPLES**

**Volume II, Parts A, B, and C
TOXICOLOGY**

**Volume III, Parts A and B
THEORY AND RATIONALE
OF INDUSTRIAL HYGIENE
PRACTICE**

PATTY'S INDUSTRIAL HYGIENE AND TOXICOLOGY

Fourth Edition
Volume I, Part B
General Principles

GEORGE D. CLAYTON
FLORENCE E. CLAYTON
Editors

CONTRIBUTORS

R. E. Allan
E. J. Baier
M. C. Battigelli
L. R. Birkner
R. P. Briggs
G. D. Clayton
M. A. Coffman
T. Ferry
J. F. Gamble

J. Grumer
P. D. Halley
A. M. Hyman
S. Nandan
C. F. Phillips
G. R. Rosenblum
M. A. Shapiro
L. K. Simkins

J. Singh
R. D. Soule
J. F. Stockman
R. P. Thompson
E. R. Tichauer
A. Turk
R. L. Vincent
G. M. Wilkening

A Wiley-Interscience Publication

JOHN WILEY & SONS, INC.

New York / Chichester / Brisbane / Toronto / Singapore

Copyright © 1991 by John Wiley & Sons, Inc.

All rights reserved. Published simultaneously in Canada.

Reproduction or translation of any part of this work
beyond that permitted by Section 107 or 108 of the
1976 United States Copyright Act without the permission
of the copyright owner is unlawful. Requests for
permission or further information should be addressed to
the Permissions Department, John Wiley & Sons, Inc.

Library of Congress Cataloging in Publication Data:
Patty, F. A. (Frank Arthur), 1897–1981
  Patty's industrial hygiene and toxicology.

  "A Wiley-Interscience publication."
  Includes bibliographical references.
  Contents: v. 1. General principles (2 v.)

  1. Industrial hygiene. 2. Industrial toxicology. I. Clayton,
George D. II. Clayton, Florence E. III. Allan, R. E. (Ralph E.)
IV. Title.
RC967.P37 1991   613.6′2   90-13080
ISBN 0-471-50196-4 (v. 1B)

Printed in the United States of America

10 9 8 7 6 5 4 3 2 1

Contributors

Ralph E. Allan, C.I.H., J.D., University of California at Irvine, Irvine, California

Edward J. Baier, C.I.H., Consultant, formerly Director of Technical Support, OSHA, Washington, D.C.

Mario C. Battigelli, M.D., West Virginia University, School of Medicine, Occupational Medicine, Institute of Occupational Health and Safety, Morgantown, West Virginia

Lawrence R. Birkner, C.I.H., Manager Corporate Safety and Industrial Hygiene, ARCO, Los Angeles, California

Raymond P. Briggs, Ph.D., Department of Science, ISSM, University of Southern California, Los Angeles, California

George D. Clayton, C.I.H., retired, formerly Chairman of the Board, Clayton Environmental Consultants, Inc., San Luis Rey, California

Michael A. Coffman, C.I.H., Manager Industrial Hygiene, Federal Mogul Corporation, Detroit, Michigan

Ted Ferry, Ed.D., Professor Emeritus, Safety Science, University of Southern California, Los Angeles, California

John F. Gamble, Ph.D., Exxon Biomedical Sciences, Inc., Mettlers Rd., CN2350, East Millstone, New Jersey

Joseph Grumer, (deceased) formerly Supervisory Research Chemist, Fire and Explosion Prevention, Pittsburgh Mining and Research Center, Bureau of Mines, U.S. Department of the Interior, Pittsburgh, Pennsylvania

Paul D. Halley, C.I.H., retired, formerly Director of Industrial Hygiene and Toxicology, Standard Oil Company (Indiana), Chicago, and as Editor in Chief, American Industrial Hygiene Association JOURNAL, Akron, Ohio

Angela M. Hyman, The Rockefeller University, New York, New York

Shri Nandan, P.E., Vice President and Director, Environmental Engineering Services, CLAYTON ENVIRONMENTAL CONSULTANTS, Pleasanton, California

Carolyn F. Phillips, C.I.H., Manager Industrial Hygiene, Health, Safety and Environmental Services, Shell Oil Co., Houston, Texas

Gary R. Rosenblum, Corporate Safety and Industrial Hygiene, ARCO, Los Angeles, California

Maurice A. Shapiro, Professor, Environmental Health Engineering, University of Pittsburgh, Graduate School of Public Health, Pittsburgh, Pennsylvania

Lisa K. Simkins, P.E., C.I.H., Vice President, Director Western Operations, CLAYTON ENVIRONMENTAL CONSULTANTS, Pleasanton, California

Jaswant Singh, Ph.D., C.I.H., Senior Vice President, Director of Pacific Operations, CLAYTON ENVIRONMENTAL CONSULTANTS, Cypress, California

Robert D. Soule, C.I.H., C.S.P., Indiana University, Indiana,

Pennsylvania and Occupational Health Consultant

Judith F. Stockman, R.N., C-A.N.P., C.O.H.N., Certified Workers' Compensation Administrator, The STOCKMAN Group, Whittier, California

Richard P. Thompson, Environmental Engineering Services, CLAYTON ENVIRONMENTAL CONSULTANTS, Pleasanton, California

Erwin R. Tichauer, Sc.D., retired, formerly Professor, New York University Medical Center, Division of Biomechanics, New York, New York

Amos Turk, Ph.D., Consultant, The City College, New York, New York

Richard L. Vincent, Director of Development and Programs, Lighting Research Institute, New York, New York

George M. Wilkening, C.I.H., Science Coordinator, Environmental and Occupational Health Sciences Institute, Piscataway, New Jersey

Preface

The preface of Volume 1A, which precedes this publication, discusses the origin of the name *Industrial Hygiene* and the genesis of the series that emerged in 1948.

An astute reader may wonder why certain topics were selected for inclusion in the first book of this revision, and others for the second book of "General Principles" and why there are two books to this volume.

The previous revision in 1978 (the third) comprised 27 chapters and was contained in one book, with a 3-inch spine. It was very heavy and unwieldy, unsuitable for use on field trips. This fourth revision was expanded from 27 chapters to 43, expanding simultaneously with the new challenges and responsibilities facing the industrial hygienist today. Thus the reason for two books on "General Principles" is obvious. Why certain subjects are in a particular book, the answer is "accommodation." Although we had hoped to organize the two books by topics, this became unfeasible. Some chapters were submitted earlier than others. The chapters of the second book of Volume 1 are as salient as those of the first book, and the reader is urged to consider these two books as a unit, as one without the other leaves a wide gap in coverage of subjects that are important to the practicing industrial hygienist.

GEORGE D. CLAYTON
FLORENCE E. CLAYTON

San Luis Rey, California
May 1991

Contents

43 Industrial Hygiene Aspects of Hazardous Material Emergencies and

    Gary R. Rosenblum and Lawrence R. Birkner, C.I.H.

USEFUL EQUIVALENTS AND CONVERSION FACTORS

1 kilometer = 0.6214 mile
1 meter = 3.281 feet
1 centimeter = 0.3937 inch
1 micrometer = 1/25,4000 inch = 40 microinches
    = 10,000 Angstrom units
1 foot = 30.48 centimeters
1 inch = 25.40 millimeters
1 square kilometer = 0.3861 square mile (U.S.)
1 square foot = 0.0929 square meter
1 square inch = 6.452 square centimeters
1 square mile (U.S.) = 2,589,998 square meters
    = 640 acres
1 acre = 43,560 square feet = 4047 square
    meters
1 cubic meter = 35.315 cubic feet
1 cubic centimeter = 0.0610 cubic inch
1 cubic foot = 28.32 liters = 0.0283 cubic meter
    = 7.481 gallons (U.S.)
1 cubic inch = 16.39 cubic centimeters
1 U.S. gallon = 3.7853 liters = 231 cubic inches
    = 0.13368 cubic foot
1 liter = 0.9081 quart (dry), 1.057 quarts (U.S.,
    liquid)
1 cubic foot of water = 62.43 pounds (4°C)
1 U.S. gallon of water = 8.345 pounds (4°C)
1 kilogram = 2.205 pounds

1 gram = 15.43 grains
1 pound = 453.59 grams
1 ounce (avoir.) = 28.35 grams
1 gram mole of a perfect gas ⇌ 24.45 liters (at
    25°C and 760 mm Hg barometric pressure)
1 atmosphere = 14.7 pounds per square inch
1 foot of water pressure = 0.4335 pound per
    square inch
1 inch of mercury pressure = 0.4912 pound per
    square inch
1 dyne per square centimeter = 0.0021 pound
    per square foot
1 gram-calorie = 0.00397 Btu
1 Btu = 778 foot-pounds
1 Btu per minute = 12.96 foot-pounds per
    second
1 hp = 0.707 Btu per second = 550 foot-pounds
    per second
1 centimeter per second = 1.97 feet per minute
    = 0.0224 mile per hour
1 footcandle = 1 lumen incident per square foot
    = 10.764 lumens incident per square meter
1 grain per cubic foot = 2.29 grams per cubic
    meter
1 milligram per cubic meter = 0.000437 grain per
    cubic foot

To convert degrees Celsius to degrees Fahrenheit: °C (9/5) + 32 = °F
To convert degrees Fahrenheit to degrees Celsius: (5/9) (°F − 32) = °C
For solutes in water: 1 mg/liter ⇌ 1 ppm (by weight)
Atmospheric contamination: 1 mg/liter ⇌ 1 oz/1000 cu ft (approx)
For gases or vapors in air at 25°C and 760 mm Hg pressure:
    To convert mg/liter to ppm (by volume): mg/liter (24,450/mol. wt.) = ppm
    To convert ppm to mg/liter: ppm (mol. wt./24,450) = mg/liter

CONVERSION TABLE FOR GASES AND VAPORS[a]

(Milligrams per liter to parts per million, and vice versa; 25°C and 760 mm Hg barometric pressure)

Molecular Weight	1 mg/liter ppm	1 ppm mg/liter	Molecular Weight	1 mg/liter ppm	1 ppm mg/liter	Molecular Weight	1 mg/liter ppm	1 ppm mg/liter
1	24,450	0.0000409	39	627	0.001595	77	318	0.00315
2	12,230	0.0000818	40	611	0.001636	78	313	0.00319
3	8,150	0.0001227	41	596	0.001677	79	309	0.00323
4	6,113	0.0001636	42	582	0.001718	80	306	0.00327
5	4,890	0.0002045	43	569	0.001759	81	302	0.00331
6	4,075	0.0002454	44	556	0.001800	82	298	0.00335
7	3,493	0.0002863	45	543	0.001840	83	295	0.00339
8	3,056	0.000327	46	532	0.001881	84	291	0.00344
9	2,717	0.000368	47	520	0.001922	85	288	0.00348
10	2,445	0.000409	48	509	0.001963	86	284	0.00352
11	2,223	0.000450	49	499	0.002004	87	281	0.00356
12	2,038	0.000491	50	489	0.002045	88	278	0.00360
13	1,881	0.000532	51	479	0.002086	89	275	0.00364
14	1,746	0.000573	52	470	0.002127	90	272	0.00368
15	1,630	0.000614	53	461	0.002168	91	269	0.00372
16	1,528	0.000654	54	453	0.002209	92	266	0.00376
17	1,438	0.000695	55	445	0.002250	93	263	0.00380
18	1,358	0.000736	56	437	0.002290	94	260	0.00384
19	1,287	0.000777	57	429	0.002331	95	257	0.00389
20	1,223	0.000818	58	422	0.002372	96	255	0.00393
21	1,164	0.000859	59	414	0.002413	97	252	0.00397
22	1,111	0.000900	60	408	0.002554	98	249.5	0.00401
23	1,063	0.000941	61	401	0.002495	99	247.0	0.00405
24	1,019	0.000982	62	394	0.00254	100	244.5	0.00409
25	978	0.001022	63	388	0.00258	101	242.1	0.00413
26	940	0.001063	64	382	0.00262	102	239.7	0.00417
27	906	0.001104	65	376	0.00266	103	237.4	0.00421
28	873	0.001145	66	370	0.00270	104	235.1	0.00425
29	843	0.001186	67	365	0.00274	105	232.9	0.00429
30	815	0.001227	68	360	0.00278	106	230.7	0.00434
31	789	0.001268	69	354	0.00282	107	228.5	0.00438
32	764	0.001309	70	349	0.00286	108	226.4	0.00442
33	741	0.001350	71	344	0.00290	109	224.3	0.00446
34	719	0.001391	72	340	0.00294	110	222.3	0.00450
35	699	0.001432	73	335	0.00299	111	220.3	0.00454
36	679	0.001472	74	330	0.00303	112	218.3	0.00458
37	661	0.001513	75	326	0.00307	113	216.4	0.00462
38	643	0.001554	76	322	0.00311	114	214.5	0.00466

CONVERSION TABLE FOR GASES AND VAPORS (*Continued*)

(Milligrams per liter to parts per million, and vice versa; 25°C and 760 mm Hg barometric pressure)

Molecular Weight	1 mg/liter ppm	1 ppm mg/liter	Molecular Weight	1 mg/liter ppm	1 ppm mg/liter	Molecular Weight	1 mg/liter ppm	1 ppm mg/liter
115	212.6	0.00470	153	159.8	0.00626	191	128.0	0.00781
116	210.8	0.00474	154	158.8	0.00630	192	127.3	0.00785
117	209.0	0.00479	155	157.7	0.00634	193	126.7	0.00789
118	207.2	0.00483	156	156.7	0.00638	194	126.0	0.00793
119	205.5	0.00487	157	155.7	0.00642	195	125.4	0.00798
120	203.8	0.00491	158	154.7	0.00646	196	124.7	0.00802
121	202.1	0.00495	159	153.7	0.00650	197	124.1	0.00806
122	200.4	0.00499	160	152.8	0.00654	198	123.5	0.00810
123	198.8	0.00503	161	151.9	0.00658	199	122.9	0.00814
124	197.2	0.00507	162	150.9	0.00663	200	122.3	0.00818
125	195.6	0.00511	163	150.0	0.00667	201	121.6	0.00822
126	194.0	0.00515	164	149.1	0.00671	202	121.0	0.00826
127	192.5	0.00519	165	148.2	0.00675	203	120.4	0.00830
128	191.0	0.00524	166	147.3	0.00679	204	119.9	0.00834
129	189.5	0.00528	167	146.4	0.00683	205	119.3	0.00838
130	188.1	0.00532	168	145.5	0.00687	206	118.7	0.00843
131	186.6	0.00536	169	144.7	0.00691	207	118.1	0.00847
132	185.2	0.00540	170	143.8	0.00695	208	117.5	0.00851
133	183.8	0.00544	171	143.0	0.00699	209	117.0	0.00855
134	182.5	0.00548	172	142.2	0.00703	210	116.4	0.00859
135	181.1	0.00552	173	141.3	0.00708	211	115.9	0.00863
136	179.8	0.00556	174	140.5	0.00712	212	115.3	0.00867
137	178.5	0.00560	175	139.7	0.00716	213	114.8	0.00871
138	177.2	0.00564	176	138.9	0.00720	214	114.3	0.00875
139	175.9	0.00569	177	138.1	0.00724	215	113.7	0.00879
140	174.6	0.00573	178	137.4	0.00728	216	113.2	0.00883
141	173.4	0.00577	179	136.6	0.00732	217	112.7	0.00888
142	172.2	0.00581	180	135.8	0.00736	218	112.2	0.00892
143	171.0	0.00585	181	135.1	0.00740	219	111.6	0.00896
144	169.8	0.00589	182	134.3	0.00744	220	111.1	0.00900
145	168.6	0.00593	183	133.6	0.00748	221	110.6	0.00904
146	167.5	0.00597	184	132.9	0.00753	222	110.1	0.00908
147	166.3	0.00601	185	132.2	0.00757	223	109.6	0.00912
148	165.2	0.00605	186	131.5	0.00761	224	109.2	0.00916
149	164.1	0.00609	187	130.7	0.00765	225	108.7	0.00920
150	163.0	0.00613	188	130.1	0.00769	226	108.2	0.00924
151	161.9	0.00618	189	129.4	0.00773	227	107.7	0.00928
152	160.9	0.00622	190	128.7	0.00777	228	107.2	0.00933

CONVERSION TABLE FOR GASES AND VAPORS (*Continued*)

(*Milligrams per liter to parts per million, and vice versa; 25°C and 760 mm Hg barometric pressure*)

Molec- ular Weight	1 mg/liter ppm	1 ppm mg/liter	Molec- ular Weight	1 mg/liter ppm	1 ppm mg/liter	Molec- ular Weight	1 mg/liter ppm	1 ppm mg/liter
229	106.8	0.00937	253	96.6	0.01035	277	88.3	0.01133
230	106.3	0.00941	254	96.3	0.01039	278	87.9	0.01137
231	105.8	0.00945	255	95.9	0.01043	279	87.6	0.01141
232	105.4	0.00949	256	95.5	0.01047	280	87.3	0.01145
233	104.9	0.00953	257	95.1	0.01051	281	87.0	0.01149
234	104.5	0.00957	258	94.8	0.01055	282	86.7	0.01153
235	104.0	0.00961	259	94.4	0.01059	283	86.4	0.01157
236	103.6	0.00965	260	94.0	0.01063	284	86.1	0.01162
237	103.2	0.00969	261	93.7	0.01067	285	85.8	0.01166
238	102.7	0.00973	262	93.3	0.01072	286	85.5	0.01170
239	102.3	0.00978	263	93.0	0.01076	287	85.2	0.01174
240	101.9	0.00982	264	92.6	0.01080	288	84.9	0.01178
241	101.5	0.00986	265	92.3	0.01084	289	84.6	0.01182
242	101.0	0.00990	266	91.9	0.01088	290	84.3	0.01186
243	100.6	0.00994	267	91.6	0.01092	291	84.0	0.01190
244	100.2	0.00998	268	91.2	0.01096	292	83.7	0.01194
245	99.8	0.01002	269	90.9	0.01100	293	83.4	0.01198
246	99.4	0.01006	270	90.6	0.01104	294	83.2	0.01202
247	99.0	0.01010	271	90.2	0.01108	295	82.9	0.01207
248	98.6	0.01014	272	89.9	0.01112	296	82.6	0.01211
249	98.2	0.01018	273	89.6	0.01117	297	82.3	0.01215
250	97.8	0.01022	274	89.2	0.01121	298	82.0	0.01219
251	97.4	0.01027	275	88.9	0.01125	299	81.8	0.01223
252	97.0	0.01031	276	88.6	0.01129	300	81.5	0.01227

[a] A. C. Fieldner, S. H. Katz, and S. P. Kinney, "Gas Masks for Gases Met in Fighting Fires," U.S. Bureau of Mines, Technical Paper No. 248, 1921.

PATTY'S INDUSTRIAL HYGIENE AND TOXICOLOGY

Fourth Edition

Volume I, Part B
GENERAL PRINCIPLES

CHAPTER TWENTY-FOUR

Emerging Industrial Hygiene Concerns

Edward J. Baier, C.I.H.

There are many issues evolving in the industrial hygiene field. Some are part of a continuum and some are current.

A major concern for practitioners of this art and science has been the ability to market their skills adequately. Failure to market is currently threatening the very existence of the profession of industrial hygiene after five decades of development.

Over the years the profession has allowed its comprehensive scope to be diluted and usurped. It spawned many environmental programs, but as each new program developed, such as air pollution control and radiological health, the basic concept of industrial hygiene diminished.

Following passage of the Occupational Safety and Health Act of 1970, much of industrial hygiene program emphasis shifted from one of surveillance at all types of workplaces to one of simply monitoring exposures of workers for compliance with standards. Over the past two decades a cadre of persons have entered the field of industrial hygiene with this singular purpose. Because the legal process for standards development takes several years, the number of standards generated represents only a small fraction of the stresses to which workers are exposed. Persons engaged solely in compliance activities have become overspecialized in this one aspect of industrial hygiene, and growth of their careers in the profession has been severely limited.

The general public is unaware of "industrial hygiene" and what "industrial hygienists" do. The public, through its legislative representatives, has demonstrated

Patty's Industrial Hygiene and Toxicology, Fourth Edition, Volume 1, Part B, Edited by George D. Clayton and Florence E. Clayton.
ISBN 0-471-50196-4 © 1991 John Wiley & Sons, Inc.

that it is concerned only with addressing a specific need at a point in time. When community air pollution became a problem, the public policy approach was to create an "air pollution expert." When ionizing radiation became a problem, a "radiation safety expert" was created. When asbestos became a problem, an "asbestos expert" was created. When concern for radon was expressed, it resulted in the creation of a "radon expert." Now, public awareness of indoor air quality is an emerging problem that must be addressed in a new public policy.

Perhaps the field of industrial hygiene is too abstract to convey its comprehensive concept and too complex in its practice to market. The profession may be better understood in the so-called "smokestack" industries, because there is a more specific risk identification of acute effects and their control. Emerging highly technical enterprises, on the other hand, are perceived as being different from the conventional workplace. The public and the workers in these environments believe that the hazards are more subtle and that there are no skills currently available to cope with these hazards. There is a perceived fear of long-term chronic effects from very low levels of exposure to various unidentified substances. Lack of risk assessment data and failure to communicate what data are available contribute to the anxiety.

The practice of industrial hygiene has been marked by change. Modifications have occurred over time in products manufactured, in process development, in sophistication of technical and scientific instrumentation, in the introduction of new chemicals into the marketplace, and in the manner in which public interest in the environment has increased. As the profession matured, the word "anticipate" was added to the traditional scope of industrial hygiene, to "recognize, evaluate, and control."

During the 1980s dramatic changes occurred which had and will have a significant impact on the profession into the next century. These changes include the demography of the work force, completely new and different types of workplaces and associated hazards, and significant changes in public policy. The industrial hygienist must audit each new technology as it develops and assure that worker exposures are controlled. The professional industrial hygienist needs to foster the sophisticated practice of industrial hygiene by studying processes and correcting stress, rather than simply adhering to and following the strict regulatory compliance function as it developed in the past two decades. Further, public policy seems to act only with a single purpose at a point in time as awareness emerges. The professional can no longer be an observer in the public policy arena. The professional must provide technical expertise to bring rational order to the public policy development system.

The traditional manufacturing industries have given way to service-type and "high-technology" enterprises. These new workplaces are smaller and have few workers, making an effective delivery of industrial hygiene services difficult. A recent study of the populations of current workplaces indicated 14 percent of all employees work in establishments with 10 or fewer employees. Approximately 39 percent of all employees work in places with less than 50 employees.

The demography of the work force also has changed dramatically. Those on payrolls today represent a smaller percentage of males, who are getting older, and a higher percentage of females, who cover a wide range of age groups. The female work population has almost doubled in the past two decades, from about 30 percent to about 57 percent. Fewer younger workers of both sexes are entering heavy industries. Automation is replacing older workers.

Persons entering the work force today are better educated than their predecessors and have been sensitized by the media to the existence of a myriad of "environmental dangers." Dire consequences to health are often believed to exist merely from the presence of a substance or material in the work environment, whether or not there has been any real exposure. This information is frequently translated on the job not only as real stress, but as perceived job stress. Whether or not the effect of the job stress is real or perceived, it is real to the individual who is affected.

The organization of these smaller, but more highly skilled, places of employment is in transformation. The newer technologies are reducing what had been termed "middle management." Managers are becoming working specialists with work teams brought together as needed to carry out particular tasks. Reorganizations of many corporations have further reduced the number of managers.

Many high-technology and service-type industries operate in clean rooms and in office-type settings.

Energy conservation became a slogan of the 1970s. To minimize energy loss in buildings, construction became tight, with controlled environments. Air contaminants generated within these structures are recirculated through the heating and ventilating system. The result was "sick building syndrome." Carbon dioxide together with combustion products from heating systems and appliances tend to concentrate. Organic chemicals from carpets and furniture, glues, duplicators, paints and coatings, plywood and particle board, and paneling, along with insecticides and related materials, contribute to the normal indoor air pollution of a constricted occupied working space. Microorganisms and biological contaminants including viruses, fungi, molds, bacteria, nematodes, pollen, dander and mites from cooling towers, damp organic materials, plants, animals, and insects, together with high humidity areas that favor the growth of biological organisms, add to the contaminated air space.

Heating and cooling systems designed for the "controlled environments" recirculate the envelope of air within the structure, and very little, if any, outside makeup air is supplied. The resultant effects on the occupants are hypersensitive or allergic reactions; skin rashes; eye, respiratory, and mucous membrane irritation; objectionable odors; exacerbation of chronic respiratory diseases; and a host of physiological reactions, such as fever, chills, headache, sore throat, nausea, and general malaise.

Evaluation and control of complaints in clean-room and office-type environments are complex, because of the many factors that contribute to the environment. The emotional aspects of the situation further complicate the mystery. Because chemical concentrations in the air in these structures are almost always less than accepted standards, the people affected have little confidence in the investigation or the investigator. Presenting the results of the findings in a meaningful way is a great challenge to the industrial hygienist's skill in communication.

The developing field of biotechnology or genetic engineering presents new opportunities to enhance the profession. This science was spawned with the identification of deoxyribonucleic acid or DNA, the material that contains the genetic information of living organisms. It has been estimated that the biotechnology industry could contribute 10 billion dollars to the economy by the year 1995 and in excess of 100 billion dollars by the year 2000.

Applications of this emerging science have already been achieved. These include hepatitis B vaccine without contamination and hybridoma cells which produce monoclonal antibodies that fuse with cancer cells, growth hormones, and immune regulators. Pest-resistant, temperature-resistant, and drought-resistant plants have been produced. Enzymes have been produced that can literally digest oil spills and toxic materials. Microorganisms can degrade lignin, fix nitrogen, and leach valuable metals from mine tailings.

Because development of this technology deals with manipulation of living organisms, polarization of the public has all but eliminated a middle ground for discussion. A better understanding of popular perceptions is required to deal logically with the issues. Public policy, the protection of intellectual property, foreign trade and balance of trade, patent rights, and related concepts are in a developmental stage. The industrial hygienist must become familiar with all of the technical and social aspects of biotechnology in order to harmonize a solution.

Another developing technology that will become economically significant by the year 2000 is the semiconductor industry. It, as with biotechnology, typifies an evolving workplace with a work force of highly skilled labor and highly technical management. The workplace is typically a clean room, but many of the chemicals used in semiconductor processes are highly toxic. These substances include arsine, phosphine, gallium, and silane, plus many different acids and alkalies. Although exposure evaluations of the workers have not indicated the presence of a serious risk, an emotional issue of miscarriages among pregnant women employed in this industry has emerged. An epidemiologic study by researchers at the University of Massachusetts prompted this industry to begin a comprehensive health status evaluation of all workers. The study design and protocol for this evaluation should provide a model for future epidemiologic studies of this and future technologies.

Superconductor development is also emerging as a new industry. The goal is to provide conductivity more efficiently with reduced electrical resistance at temperatures as far above zero degrees Kelvin as possible. Metallic ceramics are under study to replace the metals, primarily copper and aluminum, that are currently in use. The toxicity of the components of the ceramics, superconducting magnets, and other potential hazards to workers in this industry have not been fully assessed and characterized and will have to be studied during development.

Although ergonomics, or human engineering, has been part of the industrial hygiene regime for many years, it is now receiving public attention and political emphasis. Increased use of computers in offices has prompted reevaluation of work stations. Employees at checkout counters where bar codes are used to price merchandise in supermarkets have experienced cumulative trauma disorders. Carpal tunnel syndrome is experienced by workers in many industries, including meatpacking and automobile manufacturing. Each work site must be evaluated to assure that the work station and task are designed to protect the worker from muscular and skeletal disorders.

Public focus is now being centered on the entire health care industry. Toxic substances used in the hospital environment, such as anesthetic waste gases, sterilizing agents, and ionizing radiation have been prime targets for industrial hygiene audits. On occasion, studies of health care facilities were conducted to determine the cause of cases of hepatitis B in order to develop preventive measures.

Since the recognition of acquired immune deficiency syndrome (AIDS) in the United States in 1981, cases of this disease have been reported in every state. Any worker exposed to blood and body fluids or to contaminated syringe needles has the potential to develop AIDS and/or hepatitis B. Health care workers are at primary risk, but those engaged in other unrelated occupations may also be infected. These include law enforcement personnel, firefighters, paramedics, morticians, lifeguards, housekeepers, laundry workers, and others who may come in contact with body fluids. Education, workplace practices, and personal protective equipment are the principal methods for preventing exposure to AIDS and hepatitis B. These are tools common to the profession of industrial hygiene.

Another emerging concern is the handling of hazardous waste. For years, the dumping of hazardous waste has been a threat to environmental quality and to humans and wildlife who may be exposed through direct contact with the materials, either at ground surface runoff or by leaching of the waste into drinking water. Under the Resource Conservation and Recovery Act, the Environmental Protection Agency (EPA) was charged with identifying and licensing treatment, storage, and disposal of hazardous waste. Many sites had been used and abandoned or failed to meet EPA standards. In some cases, the materials had to be relocated to approved sites. This required workers, and safety and health standards had to be developed to protect them. There are concerns as to the most cost-effective worker protection program, and an opportunity for industrial hygienists to contribute their skills currently exists.

A continuing controversy is whether or not exposure to electromagnetic energy causes effects in humans and in livestock. High-tension electrical transmission wires, radio and television transmission, magnetic resonance imaging, electric transformers, electric motors, and even electric toasters, electric alarm clocks, and electric blankets have been purported to be linked to various health effects. The literature, both technical and mass media, provides conflicting reports. This is a fertile area for further and definitive study.

The profession of industrial hygiene is well positioned to address the emerging issues in the workplace. Its basic tenets—anticipation, recognition, evaluation, and control—place it in readiness to cope with the science and technology changes in the workplace. Industrial hygienists must continue to remain vigilant, to evaluate the impact of technological ingenuity on human health, and to assure control of all stress that may result.

The profession of industrial hygiene, on the other hand, is lacking in its ability to market its skills and the ability to raise the level of public awareness. It must modify the manner in which it has conducted its business and spread its doctrine. It must act quickly to spread its message or public policy will rely on others to carry out the industrial hygiene mission.

The profession grew and evolved as a scientific body, dedicated to protecting the health and well-being of people, both in the workplace and in the community. It is "people" oriented. It struggled to maintain high standards. Qualified practitioners of industrial hygiene were recognized by certification and a strict code of ethics was established for their conduct. The profession is most familiar with risk assessment and has the tools required to conduct scientific evaluation of stress. It has demonstrated the skills required to manage risk. It has the education and background to manage available resources and obtain the greatest benefit from those available resources.

Public consciousness, on the other hand, has not shown an interest in the technical aspects of emerging problems. The public perceives or recognizes health threats such as "cancer" and "reproductive effects" among workers or persons in the community and demands that action be taken. Legislatures at federal, state, or local levels act in response to the demand. Often the legislative action may not adequately address the concern. Rather, it satisfies the outcry.

When a concern is acted upon, such as asbestos or radon, a number of persons, often entrepreneur types with little or no experience, enter the scene. When a question of credentials arises, the government affected may establish minimum criteria and training, and license or establish lists of persons considered to be approved practitioners. This practice has resulted in board-certified industrial hygienists being required to take short courses and examinations in order to qualify to carry out a miniscule part of the practice of the industrial hygiene profession.

The industrial hygienists have been slow to achieve management level positions in those places where they were employed. They have failed to communicate their abilities to manage anything other than risk. The time is ripe, and the opportunity exists to be heard in the political arena. The industrial hygienist is thoroughly skilled in carrying out professional duties. The industrial hygienist must now communicate skills in management and in public policy. This opportunity must not be squandered!

CHAPTER TWENTY-FIVE

Industrial Hygiene Records and Reports

Carolyn F. Phillips, C.I.H., and Paul D. Halley, C.I.H.

1 INTRODUCTION

An integral and essential part of any good occupational health and industrial hygiene program is suitable and useful records. Industrial hygienists collect, maintain, and use data for recognizing, evaluating, and controlling health hazards. Hygiene records include, but are not limited to, assessment of hazards, exposure measurements, development and maintenance of controls, training and education, and auditing. The industrial hygienist will need information to document which employee has worked at which job for which time period, including job descriptions from an exposure viewpoint.

Record keeping serves as an essential tool for the industrial hygienist in managing, monitoring, and documenting efforts in evaluating and controlling employee exposure. This, however, is not the only function of the industrial hygienist's records. Physicians or epidemiologists will require employee exposure data if a causal relationship is suspected between a specific or mixed exposure and illness. Corporate legal and employee relations groups may consider the industrial hygienist's records essential for health-related grievance or arbitration cases, and for compensation claims or litigation. Exposure records serve as the primary source for statistical evaluations and epidemiologic studies, and as a basis for further research. Properly documented exposure–illness records can assist in determining potentially unhealthy working conditions and developing new limits of exposure.

Patty's Industrial Hygiene and Toxicology, Fourth Edition, Volume 1, Part B, Edited by George D. Clayton and Florence E. Clayton.
ISBN 0-471-50196-4 © 1991 John Wiley & Sons, Inc.

In addition, many of today's regulations require that certain records be kept. Records and reports are also useful in meeting employee and community right-to-know information requests.

Recognizing that record keeping is an essential part of the industrial hygienist's job, it is natural to ask what constitutes "adequate" records. The novice industrial hygienist soon discovers that there is no one "adequate" record keeping system for all-purpose use. However, the various systems judged to be most effective all have certain characteristics in common. Records should be as detailed as possible for the data required (yet physically manageable), and they should be appropriately structured to relate to other available pertinent data (i.e., medical, personnel, weather). Care must be taken not to collect data for their own sake but with a specific program need in mind.

Records are only as good as the measurements or activity they document. It is of primary importance not only to record data accurately, but also to document the methods used to obtain the results. This allows the user of the records to evaluate and compare historical sample results.

Record keeping in itself serves no useful purpose unless certain objectives have been defined and the records are translated into some form of report. The industrial hygienist must present data in a report format that is readily understandable, and in sufficient detail to permit the user to make adequate decisions. The report should reflect the special expertise of the industrial hygienist to interrelate all facets of the worker and the worker's environment in evaluating the potential impact on the worker's health. For example, inadequate assessment by the industrial hygienist could result either in the impairment of the health of the worker or in excessive costs for unnecessary control measures. This chapter reviews the scope and contents of record keeping and reports, and gives a few examples of workable forms. It also covers regulatory requirements and the role of the computer in record-keeping systems.

2 INDUSTRIAL HYGIENE RECORDS

Any practical system for documenting industrial hygiene surveys and activities must be comprehensive, flexible, and simple. A system of storage and retrieval should be developed that will accomplish the following functions:

1. Allow the user to retrieve information in the form required
2. Cover all foreseeable areas of current and future interest
3. Exclude extraneous data not expected to be required
4. Minimize cost
5. Maximize efficiency

A major purpose of an exposure data base is tracking people, chemicals and other potential stresses, locations, and personal exposures.

Industrial hygiene studies often produce large amounts of data of various types. Consideration must be given to the use of the data by others for different purposes. The data that constitute the original industrial hygiene records can be stored easily in a uniform or constant format. It must also be recognized that there will be many users of the data and different outputs will be required. The permanent record from which the industrial hygiene report is developed should include notes logged in a field notebook or on a specialized survey form. The practice of jotting down random bits of information on pieces of scrap paper that may be lost is poor practice. Details not recorded on the spot are often forgotten and so do not get into the report. Reports written from memory, though they may be sufficiently accurate in some instances, tend to be incomplete and are usually inadequate as a legal record. For example, unless the original notes from the survey can be produced, any record of conditions in a plant offered as evidence in court may be open to question.

It is just as important to record all data pertinent to the sample as it is to collect the sample. The experienced industrial hygienist will later compare his or her field observations and assessments with analytic results and judge whether the data appear to be valid. Not every data record is as valid as some others. Some are strictly factual, others are subjective judgments or approximations by the hygienist. A method is needed to indicate the validity of the data so that more credibility is not given to the data than they warrant. Proper validation also reduces the likelihood of misuse of the data in the future.

Records should be sufficiently detailed to permit another individual to duplicate the previous survey without the assistance of the person who originally recorded the data. Last, the speed and accuracy of preparing a final report from survey records of an investigation depend largely on information recorded in the form of notes. Photographs tell stories with a minimum of explanation and serve as a permanent record of the specific conditions. "Before and after" pictures can be particularly effective. They may also provide information that was not recorded at the work site. Plot plans or engineering drawings of the facility showing location of equipment can also be useful, particularly for sound level surveys or noise dosimetry.

Evaluation of a potential health hazard may involve calibration of equipment, air sampling, laboratory analysis, evaluation of physical stresses (noise, light, radiation, heat), and biological monitoring (e.g., urinary phenol for benzene exposure). Each of these may require its own form or set of forms. When an industrial hygienist is called on to make many similar exposure assessment surveys, it is often advisable to design a specific form to be used at the time of sampling.

Emphasis is on time-weighted average exposure sampling data, with less consideration for area and short-term monitoring data. It is, however, important that all three types of data be obtained and recorded. Then if adverse health effects are found in employees and personal time-weighted sampling data are found to be quite low, a review of short-term personal sample data and area data may point to the intermittent exposures as the causative factors. Of course, hitherto unsuspected effects of low exposure may also be discovered.

In any event, recording data obtained in any survey activity requires the use of suitable forms and their orderly completion in a neat and efficient manner. Forms with spaces labeled for essential data help avoid the common failure to ensure needed information has been obtained at the time of the survey. The use of a poorly designed or incomplete form will lead to incomplete reports or will require repetitive follow-up with field personnel or perhaps even a resurvey. In addition, the use of standardized forms within an organization assists in maintaining a consistent data collection and documentation system for more than one user.

Recognizing potential health hazards involves an inventory of all materials and processes likely to create a health hazard by job or occupation or area. If the site is covered by the Occupational Safety and Health Administration (OSHA) Hazard Communication Rule (CFR 1920.1200), then the area chemical records developed for that purpose can be used as a checklist during a preliminary survey as well as a record of potential exposure. Typically an initial "walk-through" appraisal is made, during which forms or at least a list of items can be reviewed and observations noted. Material Safety Data Sheets (MSDS) from the suppliers should be available in the workplace for each material used in that workplace. The MSDS, another "record," lists health and safety hazards, composition of mixtures, special precautions, toxicology data, medical precautions, and related information.

Table 25.1 lists the basic data elements that should be part of an industrial hygienist's record on exposure assessment. Examples of sample forms for industrial hygiene air samples are given in Figures 25.1a and b and 25.2. There is no single standardized format. Many companies have designed their own forms to fit their own record-keeping systems. Records for the industrial hygiene laboratory should include adequate numbering and laboratory identification schemes for samples, as well as established calibration and quality-control standards. Calibration data for both field and laboratory equipment must be recorded and kept. This can be done on the air sample forms or the lab analysis form, or a separate record can be developed. The efficiency of the system depends on who does the calibrations and where they are done. The laboratory doing the analytical work should have documented records of its quality-assurance procedures and follow a strict quality-control program and good laboratory practice.

Physical stress data include noise, light, radiation, and heat. Separate forms may be developed or the basic air sampling form can be amended to fit the specific needs. Figure 25.3 is an example of a noise exposure form. A biological monitoring form may also be appropriate. Records on inspection of ventilation systems may be recorded on forms such as in Figures 25.4 and 25.5.

3   REPORT OF SURVEYS AND STUDIES

An effective industrial hygiene survey report conveys, accurately and efficiently, both the data and pertinent observations and recommendations developed by the evaluation. These serve as a base for solving immediate problems and also document information for future reference. Management needs carefully defined and timely information in order to initiate effective control action. The report should present

Table 25.1. Industrial Hygiene Exposure Data

Employee name
Employee code number (Social Security and/or company no.)
Company name
Site and/or location name
Work area or unit name
Job title and/or job code
Operation condition (normal, shutdown, upset)

Sample date
Sample code number
Sample type (area, personal, source, bulk)
Substance(s) sampled
Sample length (time-weighted average, peak, task, grab)
Task description (work activity during sample)
Shift length
Shift time (morning, night, etc.)
Skin contact (potential or actual)
Personal protective equipment used (gloves, respirator, etc.)
Weather conditions (as applicable)

Sample results
Validity code (how good are the data?)
Sample and analytic method
Potentially interfering chemicals
Analytic lab (name and credentials)
Calibration data (lab and field)

the facts, analyze and interpret the findings, and develop conclusions and recommendations. The writer must try to anticipate and answer questions, because feedback from the written communication may not be received. The report must be organized and written for the needs and understanding of a specific reader or group of readers. The content, approach, style of writing, and choice of words depend on the varied backgrounds of the intended readers. The same data may be presented differently depending on whether the report is for management, engineering or medical personnel, or the employees. This does not mean that the content of the facts or conclusions and recommendations is different, but that it may be phrased differently. Often the report is distributed to more than one of the above groups and then a great deal of attention must be paid to the way the data are presented to be sure the report will be understood clearly by all readers. A system must be established to retain these reports for future reference.

In report writing, the following five steps are recommended:

1. **Plan**. Outline purpose of the study, define the problem, establish the scope

0-2801 (REV 3-85)

INDUSTRIAL HYGIENE
SAMPLING FORM

SENT TO
HEAD OFFICE

IH. NO.

NAME EMPLOYEE SAMPLED

SOCIAL SECURITY NUMBER

EMPLOYEE NUMBER

JEP OR AREA CODE

JOB TITLE

EMPLOYEE'S COMPANY

LOCATION (SHELL, FACILITY OR OTHER)

AREA/UNIT DESCRIPTION

OPERATING CONDITON/ OPERATION

SAMPLE TYPE

1 ☐ PERSONAL-TWA
2 ☐ PERSONAL-PEAK
3 ☐ AREA
4 ☐ BLANK
5 ☐ BULK
6 ☐ LAB CONTROL
7 ☐ SOURCE
8 ☐ PERSONAL-TASK

POSSIBLE INTERFERENCES/INSTRUCTIONS

COLLECTOR

1 ☐ CHARCOAL TUBE
2 ☐ FILTER (RESP.)
3 ☐ FILTER (TOTAL)
4 ☐ IMPINGER
5 ☐ PASSIVE DOSIMETER
6 ☐ SILICA GEL
7 ☐ PACKED TUBE
8 ☐ _____

BAROMETER ____ IN. HG   TEMPERATURE ____ °F

WIND ____ M.P.H.   HUMIDITY ____ % R.H.

SAMPLE NO.

DESCRIPTION OF COLLECTOR (Mfg., Lot No., Etc.)

ANALYSIS REQUIRED

CODE	NAME	≤	RESULT	UNIT
				1 ☐ PPMV 5 ☐ MG
				2 ☐ MG/M³ 6 ☐ FIBER
				3 ☐ PPBV 7 ☐ % WT
				4 ☐ FIBER/CC 8 ☐ % VOL
				9 ☐ µ G/m³

DATE SAMPLED

SHIFT   1 2 3
         ☐ ☐ ☐

SAMPLED BY _____   PHONE _____

RETURN RESULTS TO

WORK TASK

PPE

TIME

START TIME

INSTRUMENT TYPE   **SERIAL NO.**

CALIBRATION DATA

	BY	REFERENCE/METHOD	DATE
INITIAL			
FINAL			

COUNTER READING	VOLUME PER COUNT	FLOW RATE
FINAL:	FINAL:	INITIAL:
INITIAL:	INITIAL:	AVG.
NET:	AVG.	

SAMPLING VOLUME _____ LITERS

SAMPLING TIME _____ MIN

INFORMATION FOR TWA DETERMINATION

- LENGTH OF SHIFT _____ MIN.
- DOES THIS SAMPLE REPRESENT EXPOSURE OVER THE ENTIRE SHIFT   ☐ YES   ☐ NO
- IF NO, COMPLETE THE FOLLOWING:
- INCLUDE RESULTS FROM _____ ADDITIONAL SAMPLES (ATTACHED)
- ASSIGN ZERO EXPOSURE FOR _____ MINUTES OF SHIFT DURING PERIOD NOT SAMPLED
  ☐ TWA CANNOT BE CALCULATED FROM THESE SAMPLES

WAS A RESPIRATOR WORN DURING SAMPLE TO CONTROL EXPOSURE?   ☐ YES   ☐ NO

DID ANY SIGNIFICANT SKIN CONTACT OCCUR?   ☐ YES   ☐ NO

SKETCH AREA (And/Or Remarks)

COMMENTS

CLASSIFIED BY

REVIEWED BY

DATA CLASSIFICATION

Figure 25.1a. Industrial hygiene sampling form.

ANALYTICAL REPORT

CONDITION OF SAMPLE	DATE RECEIVED	DATE REPORTED	METHOD CODE	NO. OF ANALYSES	LH. NO.
☐ OK ☐ SEE BELOW					

ANALYZED BY	DATE ANALYZED	LAB RECORD NO.	ACCOUNT NO.

METHOD/REFERENCE

COMMENT

RESULTS ▸ ☐ PPMV ☐ MGM³ ☐ MG USE < LIMIT OF DETECTION (NOT N.D.) APPROVALS ▸
☐ PPBV ☐ μG/M³ ☐ μG

OPERATIONAL CODES

OPERATING CONDITION CODE (01-15)
(For TWA Samples)

Normal	01
Startup	02
Upset/spill/leak	03
Shutdown/turnaround	05
Maintenance	06
Other (specify in written Comments section)	08

OPERATION CODE (16-60)
(For Peak or Task Samples)

Operations

Handling/disposal of product/process sample(s)	
Working around spill/leak	
Gauging tank(s)	
Analyzing product/process sample(s)	
Collecting product/process sample(s)	
Other Operating Tasks (specify in Comments section)	36

Maintenance

Entry/work in confined space	16
Removing insulation	21
Applying insulation	54
Breaking into process line	23
Changing product/process filters	26

Maintenance (Continued)

Installing/removing blinds	01
Cleaning equipment (draining, purging, decontaminating)	02
Tank/equipment inspection (involving entry)	03
Degreasing/solvent washing	05
Welding/grinding/cutting	06
Painting	08
Abrasive blasting	
Carpentry	
Other mechanical work (specify in Comment section)	

Loading/Shipping

Filling drum(s)	53
Emptying drum(s)	34
Filling container(s) (sacks, bins, etc.)	31
Emptying container(s) (sacks, bins, etc.)	24
Loading tank truck(s)	17
Unloading tank truck(s)	
Loading tank car(s)	
Unloading tank cars(s)	16
Loading barge(s)	21
Unloading barge(s)	54
Loading ship	23
Unloading ship	26

COMMENT CODES

Monitoring Problems

Pump/dosimeter malfunctioned	02
Initial and final flow rates differ by > 10%	55
Pump may have not run for entire sampling period	14

Monitoring Conditions

Monitoring in conjunction with government agency	03
Local ventilation system in work area	04
Major engineering changes made	05
Random sample	08
Worst case situation	09
Duplicate sample	43

Analysis

Sample lost in shipment	28
Sample lost in analysis	20
Sample arrived contaminated/damaged	23
Incorrect sample collector used	35
Interference by other compound suspected	21
Sample shipped cold arrived ambient	61
Reported result is minimum value due to possible sample loss/breakthrough	56
Result corrected for recovery/desorption efficiency of < 75%	60

PPE (specify types in **Comments** section)

Respirator worn	57
Gloves worn	58
Protective clothing worn	59

Figure 25.1b. Analytical report.

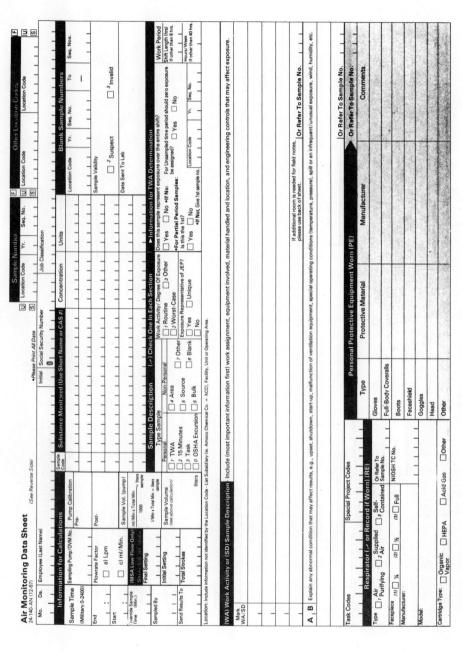

Figure 25.2. Air monitoring data sheet.

14

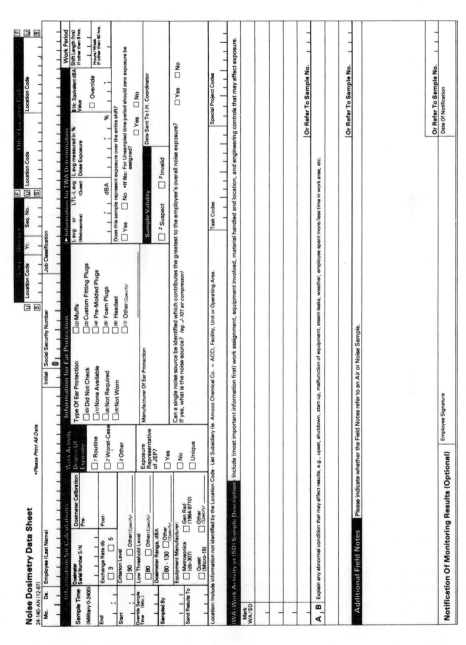

Figure 25.3. Noise dosimetry data sheet.

15

PLANT _____ DEPT. _____ DATE _____
OPERATION EXHAUSTED _____ BY _____

LINE SKETCH SHOWING POINTS OF MEASUREMENT

DATE SYSTEM INSTALLED _____

HOOD AND TRANSPORT VELOCITY

POINT	DUCT		VP	SP	FPM	CFM	REMARKS
	D	AREA (Tbl. 5-5)	IN. H₂O	IN. H₂O	(Tbl. 9-1)	Q =VA	

PITOT TRAVERSE
PITOT READINGS— SEE TABLES 9-1 TO 9-4

FAN
TYPE _____
SIZE _____

POINTS	VP	VEL.	VP	VEL.	VP	VEL.
1						
2						
3						
4						
5						
6						
7						
8						
9						
10						
TOTAL VEL.						
AVERAGE VEL.						
CFM						

POINT	DIA.	SP	VP	TP	CFM
INLET					
OUTLET					

FAN SP_____ (SEE SECTION 6)
MOTOR
  NAME_____ SIZE_____
  HP_____ E___ I___ W___
COLLECTOR
  TYPE & SIZE_____

POINT	DIA.	SP	Δ SP
INLET			
OUTLET			

NOTES _____

AMERICAN CONFERENCE OF GOVERNMENTAL INDUSTRIAL HYGIENISTS	SURVEY FORM	
	DATE 1-88	FIGURE 9-17

Figure 25.4. Survey form—American Conference of Governmental Industrial Hygienists.

of the report, and consider the way in which information and materials will be presented.

2. **Organize**. Collect, tabulate, analyze, and interpret data, to permit development of conclusions and recommendations.
3. **Outline**. Present information in a form that will allow written discussion.
4. **Write**. Prepare a rough draft.
5. **Revise**. Rewrite and correct the first draft after an interval of time to allow a fresh viewpoint.

No standard format exists for all cases, but some technical writing books and

LABORATORY HOOD VENTILATION TEST FORM

AREA/UNIT _____   DATE TESTED _____

BLDG./ROOM _____   TESTED BY _____

HOOD NO. _____   TEST INSTRUMENT:
                                            MODEL NO. _____
                                            SERIAL NO. _____

HOOD DOOR FULLY OPEN[1]

	1	2	3
A	X	X	X
B			
	X	X	X
C			
	X	X	X

TEMPERATURE (°F) _____

BAROMETRIC PRESSURE (IN. HG) _____

HOOD USE (CHECK ONE):
LOW TOXICITY MATERIALS (TLV>10 PPM)

_____

TOXIC MATERIALS (TLV≤ 10 PPM)

_____

VELOCITY MEASUREMENTS

LOCATION	VELOCITY (FPM)
A-1	_____
A-2	_____
A-3	_____
B-1	_____
B-2	_____
B-3	_____
C-1	_____
C-2	_____
C-3	_____
SUM OF MEASUREMENTS	_____

HOOD DIMENSIONS:
HEIGHT[1] (IN.) _____
WIDTH (IN.) _____

HOOD EXHAUST VOLUME (Q):
Q = AREA (FT.^{2}) X AVERAGE
Q = _____ CFM

HOOD STATIC PRESSURE (IN. H_2O) ____

AVERAGE VELOCITY (FPM) _____

MINIMUM VELOCITY (FPM) _____

COMMENTS:

$$\text{AVG. VELOCITY} = \frac{\text{SUM OF MEASUREMENTS}}{\text{NO. OF MEASUREMENTS}}$$

[1]The hood door should be fully open or open to the designated operating
level during measurements.

Figure 25.5. Laboratory hood ventilation test form.

company style manuals do prescribe various report structures. Books on writing
style and on preparing technical reports are indispensable and should be part of a
reference library. Writing skills improve through the use of such references, accompanied by practice. References 1–4 are examples of useful resources. In all
cases, the final report format used by the writer should fit the needs of the orga-

nization, as well as those of the reader and the writer. One commonly used report format contains the following sections:

A. A summary that concisely presents the work reported, including a statement of why the work was done, an abridgment of background information, conclusions, and recommendations

B. Recommendations that list all proposed changes, supported by brief comments on the reasons for the suggested courses of action

C. Discussion that presents findings at length and makes and evaluates conclusions

D. An attachment that contains result data and material too detailed for inclusion in the discussion, providing necessary support information

For a long report on a major study, a title page and a table of contents are useful. As an aid to the reader, headings and subdivisions should be used. Illustrations, tables, graphs, diagrams, and photographs reduce verbosity. The report should tell what was done, why it was done, what was seen, and what data were available to the industrial hygienist. Appropriate sections of the report should contain a brief description of the plant, department, or operation, and any significant changes that may have taken place since any previous survey; a discussion of control measures already implemented; potential health effects resulting from exposure to the stresses surveyed; regulatory requirements, samples, measurements, and test results, and an interpretation of these results. Sampling, measuring, and analytical procedures; documentation of equipment calibration; and findings from other pertinent studies may be included or referenced as appropriate. The written report should be aimed at the intended readers, logically directing their attention to the facts in the shortest possible time. Forceful writing accomplishes this by using plain words and proper grammar for clarity and by omitting needless words and sentences. These characteristics will make the report more readable and will not distort the communication.

Adequate follow-through is a prime factor in gaining acceptance of recommendations with a minimum of delay after a report has been issued. Follow-through methods include:

1. Offering assistance in the report transmittal letter

2. Presenting the report in person and reviewing contents with appropriate plant personnel, including management

3. Providing assistance in carrying out recommendations

4. Reviewing designs

5. Conducting a follow-up survey

4   LINKING INDUSTRIAL HYGIENE DATA TO HEALTH RECORDS

The health experience of workers in relation to exposure must be followed closely to achieve a complete occupational health program. Both the industrial hygienist and the physician are concerned with monitoring. One monitors the work environment, the other, the human body. The monitoring systems used are personal, environmental or area, biological, and medical. (Note: in this context, biological monitoring is restricted to measuring for the material or its metabolites in the body, usually via blood or urine analysis.) Personal and area monitoring provide the exposure information necessary for designing effective engineering controls and work practices. Biological and medical monitoring provide information on exposure only after absorption takes place. For adequate evaluation of the effect of the work environment, it may be necessary to use all four monitoring systems (5,6,7).

Industrial hygienists must keep in mind the value of reciprocal information. Employee exposure assessments are useful in diagnosing occupational illness, or they may indicate areas for medical surveillance. Medical findings may indicate areas for industrial hygiene study. Biological monitoring data may reveal exposure trends before illness symptoms. Although medical surveillance should of course never be utilized as the primary means for evaluating employee exposure, it can be a supplementary tool to evaluate the effectiveness of a control program involving engineering or other control techniques, and/or personal protective controls.

In cases where significant skin absorption of a chemical may occur, personal and area monitoring may not provide sufficient information. In such cases, medical and biological monitoring may be required to evaluate the exposure fully. Accordingly, there must be a close and ongoing relationship between industrial hygiene and medicine to determine what exposure limits are needed for potentially hazardous materials. Figure 25.6 indicates the data relationships involved in a combined industrial hygiene and medical assessment of the worker.

5   OTHER RECORDS

Coordinating record-keeping activities from many sources, including personnel records, medical data, environmental data, and chemical audits, is essential if linking of exposure data to individual employees is to be accomplished. Personnel records should contain the cumulative summary of jobs held, with dates and department. If personnel records are not kept for a sufficient length of time the industrial hygienist will need to set up a method to retain the information for future use. Job lists will then identify potentially exposed workers. Identifying the chemical, biological, and physical agents is essential, as well as any process changes that could affect job activity or exposure levels.

Fundamental to the process is a review to identify the chemical, biological, and physical agents in the work environment. Any process changes that could affect exposure conditions should be reported promptly to the industrial hygienist for assessment. However, the real world often makes this a very difficult condition to

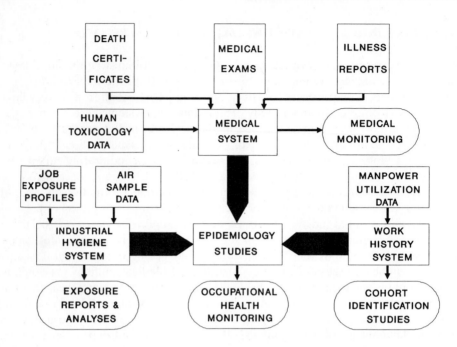

Figure 25.6.   Health surveillance system.

meet. With the introduction of new processes, elimination of departments, and modifications of jobs, business is dynamic. Linking exposure measurement data to employees through the personnel record system can be extremely complicated unless there is careful cross-referencing. Exposure records for a job may be considered representative for all employees who held that job during the given time interval. In such cases it may be easier to retrieve the needed information if exposure data are cross-referenced to job and department designations, rather than to process and location in a department. Employees may frequently change jobs within a department, so information regarding an exposure assessment for all jobs should be available. Another reason to assess the exposure for all jobs is that the levels of exposure that are currently believed to be safe may later show a potential risk. This will often include monitoring, but professional judgment can be used to document estimated exposures based on objective data on the job and process.

Many factors in addition to work exposure can contribute to the cause of diseases. Among these are individual habits such as smoking or the use of alcohol, dietary preferences, air pollution or polluted water supplies, exposure to noise, solvents, and other conditions or materials, or exposure through an avocation or second job. Unfortunately the employer generally does not always have access to much of this type of data, so there will be gaps in a total exposure record.

Obtaining appropriate industrial hygiene data is especially difficult in dealing with health effects that develop slowly or after a long latent period, yet acquiring such data is necessary. It may be necessary to obtain and retain such data for as long as 30 to 40 years for correlation with chronic health effects (see Section 6). Cross-correlation of industrial hygiene data, medical history, and periodic medical surveillance examinations (e.g., audiometric examinations, pulmonary function tests) provides human exposure data of great value, particularly if large populations can be studied. Computerized data storage retrieval is an asset for carrying out such important epidemiologic studies. Currently the weakest link in an epidemiology study is the historical exposure record.

As indicated in Section 1, an important part of an industrial hygiene survey and report is a review and evaluation of potential health effects from exposure to environmental stresses. The industrial hygienist will need input from toxicologists and product safety and medical personnel. The toxicologist's records contain information and data on the new product, often varying from a preliminary evaluation based on professional judgment through short-term and long-term chronic testing of the product in experimental animals. These data will not allow us to define the probable risk to humans with certainty, but they do constitute a valuable guide to the industrial hygienist in evaluating potential chemical stresses in the workplace.

Using permissible exposure limits (PELs) and/or threshold limit values (TLVs), as well as toxicology and epidemiology data, the industrial hygienist evaluates the work environment in terms of exposure. The usual practice is to allow for a safety factor for human exposure as compared to animals or to conduct risk assessment on the material. The safety factor should be based on an evaluation of the type and degree of toxic effects. The industrial hygienist, in addition to using such data for his or her own evaluations, can also be expected to provide operating management with brief and concise information on toxicity, hazards, and controls. After the industrial hygienist has evaluated the magnitude of the worker exposure, the data may have to be coordinated with medical records available on these same workers. Working with medical staffs to recommend any special biological tests that may be pertinent, the industrial hygienist can determine whether available yardsticks (e.g., PELs, TLVs) are adequate for evaluating a "safe" environment or whether they should be further refined.

Workers' compensation or claims records may provide information on employee exposures. Morbidity records may provide clues to developing trouble. For example, a high incidence of absenteeism for respiratory problems in a particular department or plant, as compared to other such locations, may be an indication of excessive environmental stress from a particular chemical or condition. Such situations should be investigated thoroughly by the physician to determine whether a common medical problem exists, and then, if these suspicions are confirmed, by the industrial hygienist.

The indication of the presence or absence of similar local effects, signs, symptoms, or disease in co-workers sharing essentially the same work environment may provide significant information in assessing potentially hazardous locations or processes.

Personnel records can prove invaluable especially in investigating causative factors in demonstrated or suspected occupational illness. For example, if medical examination reveals conditions such as silicosis or hearing threshold shift, a review of the worker's job assignment history may tell whether the worker has ever had related exposures. A check of personnel records might provide some assistance in determining whether the worker's condition may have resulted from exposure at a previous job.

6   OSHA RECORD-KEEPING REQUIREMENTS

This section is primarily concerned with the OSHA aspects of record-keeping requirements. Governmental requirements to maintain occupational records are standard in many countries. The OSH Act, Section (c)(1), states that:

> Each employer shall make, keep and preserve, and make available to the Secretary or the Secretary of Health, Education and Welfare, such records regarding his activities relating to this Act as the Secretary may prescribe by regulation as necessary or appropriate to the enforcement of this Act or for developing information regarding the causes and prevention of occupational accidents and illnesses. In order to carry out the provisions of this paragraph such regulations may include provisions requiring employers to conduct periodic inspections. The Secretary shall also issue regulations requiring that employers, through posting of notices or other appropriate means, keep their employees informed of their protection and obligations under this Act, including the provisions of applicable standards (8).

OSHA has established numerous requirements for record keeping including such items as training, certification, exposure data, programs, and injury/illness data. These requirements are contained in the Code of Federal Regulations Title 29, Labor Chapter XVII—Occupational Safety and Health Administration, Parts 1903 to 1920.1500 and 29 CFR, 1926 (9). Some examples of particular interest to industrial hygienists follow.

6.1   Part 1903—Inspections, Citations, and Proposed Penalties

OSHA is authorized to inspect, investigate, and to review records required by the act and other records that are related to the purpose of the inspection.

6.2   Part 1904—Recording and Reporting of Occupational Injuries and Illnesses

This part requires each employer to maintain a log and summary of all recordable injuries and illnesses no later than six days after receiving information of an occurrence. It also requires the posting of an annual summary of occupational injuries and illnesses for the previous calendar year (by February 1 for at least a one month period). Records shall be kept for five years and be available to employees or their representative. Failure to maintain these records may result in a citation and penalties.

The criteria for deciding what injuries and illnesses are recordable are established by the Bureau of Labor Statistics (BLS). The BLS guidelines are supplemental instructions to the mandatory OSHA record-keeping forms and have significant impact on a record program. All people responsible for injury and illness record-keeping decisions must understand the BLS guideline.

6.3 Part 1910—General Industry Standards

6.3.1 Subpart C—General Safety and Health Provisions

Section 1910.20—Access to Employee Exposure and Medical Records This section gives employees and designated representatives the right of access to relevant exposure and medical records. It also gives OSHA the right of access. These records include those taken both for compliance with other OSHA standards and for the companies' own purposes. Medical records are to be kept for length of employment plus 30 years. Employee exposure records shall be maintained at least 30 years. Any analyses using exposure or medical records shall be kept for at least 30 years. The section also requires the employer to notify employees annually of their right to access these records.

6.3.2 Subpart G—Occupational Health and Environmental Controls

Section 1910.95—Occupational Noise Exposure Employers shall have a hearing conservation program. The record keeping will involve exposure measurements and audiometric test data. Noise exposure data must be kept for two years and audiometric data for the duration of employment.

Section 1910.96—Ionizing Radiation The employer must maintain adequate past and current exposure records on exposure to ionizing radiation for individuals in restricted areas. The employer must advise the employee of the exposures on at least an annual basis.

6.3.3 Subpart I—Personal Protective Equipment

Section 1910.134—Respiratory Protection When engineering controls are not feasible, respiratory protection may be used to protect employees from exposure. If respirators are used, the employer shall have a written respirator program covering safe use of respirators in normal and emergency situations. These written procedures shall include selection, fit, use, and maintenance and storage. Records shall be kept of inspections. Substance specific standards may include more specific requirements including fit testing and training records.

6.3.4 Subpart Z—Toxic and Hazardous Substances

In addition to the PEL list (1910.100), Subpart Z contains a number of substance specific standards. The following ethylene oxide example is typical of record-keeping requirements in these standards.

Section 1910.1047—Ethylene Oxide This section requires that records of objective data used to exempt certain operations be kept during the period the employer relies on the data. An accurate record of all measurements to monitor employee exposure to ethylene oxide must be kept at least 30 years. Also to be retained are medical records for each employee required to be in the medical surveillance program.

Section 1910.1200—Hazard Communication This section requires developing and using a hazard communication program. Manufacturers or importers of chemicals shall have a written hazard determination procedure. The written programs shall be available to employees and to OSHA. The standard requires area chemical lists that can be very useful to an industrial hygiene program and when maintained efficiently, an asset to epidemiology studies.

OSHA is not unique in its record-keeping requirements. In 1988, a new law was passed in the United Kingdom called Control of Substances Hazardous to Health Regulations (COSHH). This regulation is administered by the Health and Safety Executive. It introduced a new legal framework for controlling exposure to hazardous substances during work activities. The regulation requires both an assessment of health risk and measures to protect workers' health.

Part of the regulation requires records of the assessments. This requirement is to record why decisions about risks were made and the precautions taken. The records are to reflect the details with which the assessment is carried out, be useful to those who need it now and in the future, and also indicate under what conditions the assessment must be reviewed. The regulations also indicate that the style and presentation of the assessment is to be influenced by the recipient. The records also have to be available to authorized inspectors. The regulations are similar to OSHA's requirement of maintaining monitoring records for at least 30 years if they are representative of personal exposure. It also requires that a record be kept for at least five years of tests and repairs to control measures.

7 CONFIDENTIALITY OF RECORDS

Between the employee and the company physician there is a patient–physician relationship dictating that all medical information be considered and retained in confidence by the physician. This confidentiality of medical records follows recognized standards of law and medical ethics. The role of a company-employed physician is further clarified by the provisions regarding proper conduct of an industrial physician, including codes of ethics of the American Medical Association, the American College of Occupational Medicine, and most state medical societies.

Management does not have a need to know specific diagnoses to fill the management function of assigning employees to specific tasks and operations. However, management does need to know what medically based restriction on work duties must be considered in assigning employees to jobs. Because management is not privy to an employee's medical history and specific disabilities, the physician must have an adequate knowledge of the various work environments and their potential adverse effects on employees, and must advise management of any limitations in the placement of its employees. As an example, an employee with certain respiratory disabilities should not be assigned to operations presenting potential exposure to chemicals that are respiratory tract irritants. The industrial hygienist has a key role in working with the physician to review the work environments and the limitations for placement of employees.

Biological monitoring records of the employee, such as results of blood-lead analyses, are not quite as clear-cut. The results are important to both the physician and the industrial hygienist. Such data are needed by the industrial hygienist for a complete exposure assessment, for the results may indicate exposure not found during exposure monitoring of the employee. Management use of such data should be limited to implementing engineering or other control measures and to temporary assignment of the employee, if needed, to another area without that specific exposure.

When medical and exposure data are computerized, as discussed in the following section, it is important that the confidentiality of these restricted medical data be maintained. Only those privileged to see medical data should be able to obtain this information from the computer. The industrial hygienist will have a continuing need for exposure and biological monitoring data and should have free access to them.

8 COMPUTERIZATION OF RECORDS

In the past, and even at present in many cases, numerous data on employee exposures were compiled in the form of narrative-style, written reports. These are used primarily to document employee exposure evaluation as measured by some yardstick such as threshold limit values and to recommend control measures where indicated. Medical records as confidential files, on the other hand, are largely kept on an individual employee basis. Under such circumstances, epidemiologic studies and other correlations between industrial hygiene and medical data are often difficult and time consuming. One need only consider the practice of attempting to correlate hearing tests (involving 14 measurements on each employee) with noise level measurements, each being repeated as often as yearly, to recognize the complexity of manually correlating industrial hygiene and medical data.

With the use of the computer for industrial hygiene data, there is now a much better opportunity and likelihood that adequate evaluation of the various occupational health parameters will be accomplished. This, in the final analysis, means better control of the work environment and early detection of any adverse health effects. A significant benefit of a computer data base for industrial hygiene records is in usable output from the system (10,11).

Table 25.2 shows some of the activities that can be efficiently supported by a computerized health surveillance system. Table 25.3 lists the attributes required for such a system to be effective.

With an adequately designed computer program and input of pertinent data,

Table 25.2. Activities to be Supported by Health Surveillance System

Epidemiology studies
Evaluation of workplace health conditions
Medical surveillance programs
Compliance with record-keeping and reporting mandates
Providing data for litigation purposes
Public understanding programs
Response to proposed legislation or regulatory activities

the industrial hygienist can evaluate employee exposure to workplace contaminants more effectively than before. For example, the input data needed for benzene exposure would include those listed in Table 25.1. Urinary phenol measurements would also be included if available. In addition to personal monitoring data, area and source monitoring data would be included. Other data sets could include a record of all job assignments for an individual and a list of all chemicals used in the job over time. From such data bases, a computer printout can be obtained of all employees potentially exposed to benzene, indicating which have measured exposure above or below the allowed limit.

Because the capability of computers is constantly evolving and because each company has different needs, this section can be no more than a general discussion of the considerations for developing a system. These include but are not limited to data to be stored, reports to be generated, statistical analysis, coordination with medical records, and work history for epidemiology studies. Other aspects to be considered include number of users, number of locations, size of data base, financial resources, and staff capabilities and needs. Only after evaluation of this type of information can a decision be made as to use of a personal computer, mainframe or network, or whatever the latest technology allows.

Table 25.3. Required Attributes of Health Surveillance System

Large storage capacity
Confidentiality
Operability (by noncomputer experts)
Quality assurance
Flexibility
Service levels
   Readily available
   Reliable
   Easily maintained
Cost effective
Minimum impact on user locations
Easy data retrieval and display

In a large company with numerous locations, the use of a company-wide system, based on either mainframe or personal computers, has many benefits. These include:

1. Response to ad hoc queries that involve information on more than one facility
2. Standardizing data and record keeping
3. Ability to include data validity checks and error warnings in the base program
4. Potential for earlier acknowledgment and interpretation of trends related to exposure
5. Central repository of information for regulatory response
6. Reduction of clerical errors and data handling time

Finally, a summarized report of the employee's exposure potential and biological monitoring data could be made available to the examining physician at the time of the employee's periodic or special medical examination. This allows the physician to review the individual's medical findings in the light of environmental stresses.

In employing computer techniques, the industrial hygienist, working with the physician(s) and operations personnel, must first establish goals and define accurately what will be needed in the way of computer reports. Without these two factors carefully thought out in advance, the data recorded may not be retrievable, or if retrievable, may not be usable. The actual users of the system should be included in the development process in order to gain acceptance of the system and meet their needs. It is technically easy to enter data that will yield the usual statistical or other relatively simple information. What must be remembered is that the computer will not provide answers if the data input has not supplied the correct base for such answers, or if adequate formatting has not been provided to permit meaningful retrieval programming.

Another significant goal in computerizing a record-keeping system is assistance in complying with regulatory and legal requirements.

There are many commercial computer systems available that try to meet the industrial hygienists' needs as described above. A number of companies have chosen to design and develop their own systems tailored to interface with other corporate systems.

Another fast developing area is the direct interface of monitoring equipment with the computer data base. New sampling and analysis devices do or will include smart samplers, data logging, and electronic dosimeters. In some cases, the data, part or all, can be fed directly into the computer for processing. If a dosimeter can provide minute-by-minute exposure increments for the workday, the industrial hygienist will face the critical decision—what data are to be kept? Some statistical summary of the time-weighted average and peak data or the entire record? In order for industrial hygienists to avail themselves of technologic advances, they must stay abreast of progress in microprocessor technology.

9 RELATED GOVERNMENTAL RECORD-KEEPING SYSTEMS

OSHA is implementing a computerized data base for field inspection reports as a reference for compliance officers. In addition to being a resource for OSHA, it could be used to compare different facilities of the same company in various geographic locations so that citations in one location can be referenced for similar operations and added citations given in future visits if the same problem has not been corrected.

From a different perspective, the United Kingdom Health and Safety Executive (HSE) has initiated a National Exposure Data Base (NEDB) that aims to bring together exposure data on a wide range of workplaces in the United Kingdom (12). Input includes all airborne sampling data plus details on control measures gathered during visits to workplaces by HSE occupational hygiene inspectors. In the long term, the HSE also hopes to enter data from industry so that the data base becomes a focal point for exposure information. The data will be used in developing new regulations and may offer a sounder base for future epidemiology studies.

Figure 25.7 is an example of the form used to collect the data for entry into the NEDB. The data includes substances monitored, industry, TWA values, control measures, job tasks, and other data. The data base will be useful in identifying particular industries with problems or specific processes or tasks that may need improved controls.

Figure 25.8 shows a similar data base concept for noise data developed by the Canadian Centre for Occupational Health and Safety (13). The contents have been compiled from data reported in journals, health and safety reports, and surveys by various industries and agencies. The objective is to share noise data by making it available either on compact disk or on-line.

10 CONCLUSIONS

It should be clear that for industrial hygiene to fulfill its obligation and goal of protecting the health of workers, adequate records and reports are essential and must be employed to the fullest. The extent and sophistication of such records and reports, of course, vary with the individual situation. The industrial hygiene report is the industrial hygienist's most forceful and enduring communication with management and employees. Done properly, in clear and concise language and understandable to those who are not industrial hygienists, it can effect the changes and improvements necessary to protect the health of the worker. If the report is done poorly, the effort that went into the original industrial hygiene survey record and exposure assessment may have been useless.

Industrial hygiene records, though recognized as a necessary and valuable part of a report, can go far beyond the immediate intent of verifying environmental stresses that exist for the moment. Careful and full use of such records and data can assist the industrial hygienist and other health professionals to accomplish the following:

Health and Safety Executive

Environmental monitoring results

	FCG file reference			Date of visit	Substance								
					Units								
	Occupier			Total number of people on site	Monitoring procedure								
					In-house atmos. monitoring								
					In-house biol. monitoring								
					Males exposed								
					Females exposed								
	Reference number	Sample type	Sample description (eg name/process/job)		Sample period	Duration (minutes)	Result	TWA	Result	TWA	Result	TWA	Result	TWA
1													
2													
3													
4													
5													
6													
7													
8													
			Exposure limits		8 hour								
					10 min								

DEC1 (4/85)

Figure 25.7a. Environmental monitoring results—Part 1.

	Exposure details		Control measures	
HSE Area no.				
SHIELD ID no.				
Industry				
Type of visit				
OHVR reference no.				
	Type	Pattern	RPE	LEV
1				
2				
3				
4				
5				
6				
7				
8				

Figure 25.7b.   Environmental monitoring results—Part 2.

1. Evaluate the adequacy of current acceptable levels of exposure.
2. Establish acceptable levels of exposure for chemicals (or other hazards) for which none currently exist.
3. Provide a basis for input to the regulatory process.
4. Show trends in exposure and controls.
5. Conduct retrospective studies if occupational illnesses occur.

NOISE LEVEL DATA BASE FIELDS (+ searchable)

******* ORIGIN OF DATA *******

Author+   The person(s) responsible for measurements.

Publication   The published source of measurements.

Title+   The title of the study (and language, if other than
                 English).

Reference Data   Bibliographic details of the publication.

Year of Study+   The year of measurement.

******* WORKPLACE *******

Industry+   Standard Industrial Classification Code and corresponding
                 term, plus CIS facet code(s) and term(s) which describe a
                 particular industry.

Operation+   Description of work-related human actions causing the
                 noise.

Occupation+   Canadian Classification and Dictionary of Occupations
                 description of the worker group affected by the noise.

Source+   The source of the noise (eg. piece of equipment).

Type of Noise+   The duration characteristics of the noise measured (eg.
                 continuous, intermittent, impulse).

Exposure Duration Per Day   The number of hours (per day) the workers are
                            exposed to the measured noise levels.
Engineering Control   Devices or structures installed to reduce
                            noise.
Protection   Ear protection worn by workers.

******* MEASUREMENT(S) *******

Location of Measurement   Detailed information concerning the position
                            of measuring device from source of noise
                            and/or worker.
Measuring Device Used   Brand name of the equipment and settings used
                            to record the measurement.
Duration of Measurement   The amount of time the measuring devices were
                            operating.

Noise Level   Sound Pressure Level (SPL) in dB(A) or peak SPL in dB.
                 Minimum, maximum, and average sound pressure levels
                 in dB(A).
                 Equivalent Continuous Noise Level;
                 Time-Weighted Average (TWA) sound level in dB(A) and
                 exchange rate used (5 dB or 3 dB).
                 Octave Band Analysis (31.5-8000 Hz) in dB.

Figure 25.8. Noise level data base fields.

6. Confirm or refute the validity of any compensation claims.
7. Compare data for like operations or workplaces, especially for the purpose of evaluating the effectiveness of differing approaches to control of the work environment.
8. In cooperation with medical staff, detect any early evidence of developing occupational health problems and effect controls at an early stage.
9. Provide support for legal staff or the industrial hygienist as an expert witness during litigation.

Finally, industrial hygiene records and reports must be retained long enough to ensure that they have fulfilled their intended purpose. Industrial hygiene engineering control recommendations, for example, may not be needed again after they have served their original purpose of effecting control of the work environment. However, the fact that controls were installed on a specific date (and the impact on the exposures) could be critical to epidemiology studies or future liability issues. Environmental measurements of work stresses, on the other hand, may be useful 20, 30, or even 40 years after serving their initial purpose. In some instances, retention of such records for a given period is required by law. In the final analysis, except for legal requirements, the industrial hygienist, with his or her background of training and experience, can best determine the point at which such records and data are no longer of value in protecting and promoting the health of the worker. The use of a carefully developed, flexible computer system can facilitate handling complex industrial hygiene data in an organized fashion and provide important information to the managers, medical staff, and employees.

REFERENCES

1. R. R. Rathbone and J. B. Stone, *A Writer's Guide for Engineers and Scientists*, Prentice-Hall, Englewood Cliffs, NJ, 1962.
2. W. Strunk, Jr., *The Elements of Style*, 3rd ed., revised by E. B. White, Macmillan, New York, 1979.
3. M. Stiertzer, *The Elements of Grammar*, Collier Books, Macmillan, 1986.
4. A. Plotnik, *The Elements of Editing*, Macmillan, 1982.
5. J. J. Bloomfield, "Industrial Health Records: The Industrial Hygiene Survey," *Am. J. Pub. Health*, **35**, 559 (1945).
6. J. T. Siedlecki, "How Medicine and Industrial Hygiene Interact," *Int. J. Occup. Health Saf.* (Sept.–Oct. 1974).
7. M. G. Ott, H. R. Hoyle, R. R. Langner, and H. C. Scharnweber, "Linking Industrial Hygiene and Health Records," *Am. Ind. Hyg. Assoc. J.* **36:**760–766 (1975).
8. Occupational Safety and Health Act of 1970. Public Law 91-596, Dec. 1970.
9. U.S. Dept. of Labor, "Occupational Safety and Health Standards for General Industry, 1989."
10. W. B. Austin and C. F. Phillips, "Development and Implementation of a Health Surveillance System," *Am. Ind. Hyg. Assoc. J.* **44:**638–642 (1983).

11. L. T. Daigle and R. H. Cohen, "Computerized Occupational Health and Records System," *Appl. Ind. Hyg.* (May 1985, preview issue).

12. D. K. Burns and P. L. Beaumont, "The HSE National Exposure Database (NEDB)," *Ann. Occup. Hyg.* **33**(1), 1–14 (1989).

13. B. Pathak, K. Marha, and W. J. Louch, "An Industrial Noise Levels Database," *Ann. Occup. Hyg.* **33**(2), 269–274 (1989).

CHAPTER TWENTY-SIX

Occupational Epidemiology: Some Guideposts

John F. Gamble, Ph.D., and Mario C. Battigelli, M.D.

Epidemiology may be defined as the study of health and illness in human populations. It is the study of the distribution of determinants of health-related states and events in populations, and the application of this study to the control of health problems. It is a

> Method of reasoning about disease that deals with biological inferences derived from observations of disease phenomenon in population groups (1).

1 INTRODUCTION

Although epidemiologists do not always agree on a definition of epidemiology, or on the scope of activities included in this scientific discipline, there are features of epidemiology that can serve to characterize it as a field of study. We shall list a number of statements and viewpoints that should help to orient one as to what epidemiology is about.

1. Findings should relate to a defined population, a population at risk. A population is either a finite or infinite universe of people about whose health a statement is to be made. Conventionally, in a study of a population of concern, three groups

Patty's Industrial Hygiene and Toxicology, Fourth Edition, Volume 1, Part B, Edited by George D. Clayton and Florence E. Clayton.
ISBN 0-471-50196-4   © 1991 John Wiley & Sons, Inc.

are considered in succession: Target population, available study population, and study sample.

In general, the examiner evaluates the status of the study sample and then proceeds to make statistically justifiable generalizations based on findings, provided that the sample is representative, that is, that there are (a) an unambiguous definition of the study population composition (example: all factory employees hired between 1950–1960); (b) an enumeration of all the members of the study population (denominator); and (c) adequate definition of the sampling process to ensure an even probability for all members to be selected into the sample. The study sample should provide, in other words, an adequate insight on the target population (2).

2. Epidemiology deals with biological inferences about the cause and/or natural history of a disease. Clues to etiology come from comparing disease rates in groups with different levels of exposure, and from quantities defining the relationship in the form of exposure–response curves. Rules of causation generally also include criteria such as consistency of the association, strength of association (both magnitude and gradient), temporal association (cause must precede or coincide with effect), and biological plausibility [see Hill (3), Rothman (4), and (5) for discussion].

3. Association is a term meaning the quantitative relationship between two variables such as, for example, smoking and lung cancer. In epidemiology there are two main measures of association and these can also be measures of effect. We compare disease frequency (e.g., in experimental and control groups) by assessing their differences (attributable risk) or their ratio. The best measures of the latter parameter are incidence and prevalence. *Incidence rate* is the proportion of new cases of disease in a population at risk during a specified period of time. In practice this becomes the total population. *Prevalence rate* is the total number of cases divided by total population at a given point in time or over a short time interval.

4. Often in epidemiology, one investigates associations between exposure (independent variable) and disease (dependent variable). When the comparison is between exposed and nonexposed populations, the paradigm is what is called a natural experiment. In the occupational experience, epidemiology deals with the real world and can not control the composition of the exposure groups. Unlike traditional experiments, comparability between exposed and referent groups is often affected by factors other than mere exposure. Therefore, in all epidemiology studies one must be concerned about bias, that is, an effect tending to produce results that depart systematically from the results of a given exposure.

5. Epidemiology deals with the evaluation of scientific hypotheses. A common practice is to assume that hypothesis testing involves simply accepting the null hypothesis or rejecting the hypothesis in favor of an alternative. Statistical testing quantifies the likelihood of falsely rejecting the null hypothesis. It is not uncommon to think of the *p* value as the decision point for the truth or falsity of the hypothesis.

In epidemiology, however, many if not most so-called hypothesis-testing situations are actually estimation exercises. Every epidemiologic study is an exercise in measurement for the purpose of obtaining estimates of the disease occurrence (such as prevalence and incidence rates) or some derivatives of these measures (i.e., means). It is an attempt to quantify the association between exposure and effect.

6. We can define epidemiology by what epidemiologists do; what are the hypotheses or questions they ask? In general terms these questions relate to the occurrence of disease by time, place, and person, such as the following:

- Has there been an increase or decrease in disease over the years?
- Does one geographic area (industry, plant, work area, job title) have a higher frequency of disease than another?
- What are the characteristics of persons with a particular disease or condition that distinguish them from those without the disease or condition? These characteristics may be risk factors such as age, sex, race, physiological function, biochemical status, socioeconomic status, occupation, and personal habits.

In occupational epidemiology, specific studies address questions relating, for instance, the association between low-level ionizing radiation and leukemia or between asbestos and cancer (or more specifically, chrysotile exposure and mesothelioma or non-asbestiform amphibole exposure and lung cancer).

In the next section we describe some methods epidemiologists use to answer these types of questions.

2 TYPES OF EPIDEMIOLOGIC RESEARCH

There are basically two types of research: (1) experimental and (2) nonexperimental or observational.

2.1 Experimental Research

Experiments in epidemiology consist of clinical trials or field trials where the objective is to test etiologic hypotheses and to detect "exposure" effects. The questions are, for example, of the form: does drug A work better than drug B? If so, at what dose? Analogous logic can be followed for assessing an adverse effect (i.e., reduction of it) by rearranging work schedule, intensity of work, and so on.

The paradigm of this scientific process is an experiment or controlled observation, for example, an animal experiment. In a clinical epidemiologic study (or animal experiment) designed to estimate the effect of a particular exposure (such as a drug), the procedure is the following:

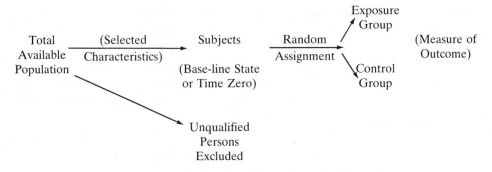

Out of a total population available for study one might select subjects of given sex and age. Subjects would then be randomly assigned, respectively, to exposed and nonexposed groups and followed over a specified time period adequate to determine the incidence of a particular condition. The measure of effect in this case is a relative risk (risk ratio, rate ratio) which is the ratio of the incidence rate of the condition among exposed to the incidence rate among the nonexposed.

Variations on this theme include the following:

- Observing a continuous semiquantifiable variable, such as fibrosis, instead of a discrete variable (such as the occurrence of a tumor); the outcome, measured by severity ranks, can be contrasted between the exposed and unexposed groups.
- Several exposure groups (none, low, medium, high) may allow the computation of an exposure–response relationship. Thus one gains evidence not only of etiology but also of potency or risk at various levels of exposure or dose.

2.2   Nonexperimental or Observational ("Real World") Research

The goal of an observational study is to simulate an experiment had one been possible. This is the common kind of research in occupational epidemiology because one cannot assign workers to industries and jobs, nor require that they stay on the job. Because of the observational rather than experimental nature of occupational epidemiologic research, there are a number of potential problems one must be aware of in the conduct and interpretation of epidemiologic studies.

2.2.1   Potential Problems

Experimental and observational research often appear to have a similar structure, and their analysis is parallel. There are, however, scientific standards in experimental research that cannot be met in observational studies. These are (a) random assignment to an exposure group and (b) control of exposure. Other scientific standards that are routinely applied in laboratory experiments but may be absent in observational research include (c) a stipulated research hypothesis; (d) a well-specified cohort or population at risk (PAR); (e) high-quality data, and (f) analysis of attributable actions (6).

Conclusions about cause–effect relationships drawn from these studies may be fraught with difficulties introduced by unknown or undetected biases, which are quite common in observational studies. Bias usually results when required standards of quality are not applied. We now discuss scientific standards and bias as they relate to epidemiology.

2.2.1.1 *Randomization*. When the members of a population selected for an experiment are similar in relevant and measurable characteristics (such as age, sex, species, or health status), each individual is randomly assigned to an exposed or unexposed group. In this way one hopes that the random exposure assignments will prevent most bias. If there are undetermined characteristics that can influence the outcome, one hopes that such characteristics will be more or less equally distributed in all exposure groups and therefore have no significant effect on the ultimate analysis. Indeed, if it is equally distributed, this effect will be the same for all exposure groups, and it becomes irrelevant for the results of interest.

It is a major function of randomization to achieve "approximate comparability with all variables, whether known or not" (7). If exposure and nonexposure groups differ with respect to the outcome, we may indeed attribute that difference to the exposure. Obviously, if the exposed and comparison (referent) groups differ with respect to variables other than the exposure, a cause–effect relationship is weakened.

An often overlooked aspect of the randomization principle is the fact that "in order to insure generalizability it is necessary that samples be random. Random samples imply that each member of the population should have the same probability of being included in the sample" (8). This idea has been stated most emphatically with the warning to the observer that "in the absence of random sampling, the whole apparatus of inference from sample to population falls to the ground, leaving the sample without a scientific basis for the inferences which he wishes to make" (quoted in Reference 7).

However, randomization is not a cure-all because it does not always assure comparability and prevent bias. Randomization is neither necessary nor sufficient for rigorous statistical inferences. However, with or without randomization one should try to control for base-line inequalities or biases, for example, by stratification (7).

2.2.1.2 *Control of Exposure*. A second important characteristic of an experiment is the ability to measure exposure with adequate precision. In a bench experiment one can and does precisely control both the duration and intensity of the exposure of interest and eliminate all unwanted or confounding exposures. In epidemiologic research, comparability must be assured by more complex arrangements. A common procedure, for instance, is to exclude from the experiment all subjects presenting the outcome at the beginning of the study period. Each study group can begin with the same base-line demographic characteristics and health status. One can further control the experiment to include sufficient time (latency) for the exposure to have an effect. Weiss (9) discusses two biases that can arise because of differences in the onset of exposure. In his examples these biases may account for some apparent elevated risks observed in occupational cohorts. A *chronology bias* can occur if cohort members have a different length of exposure at the beginning of the study period. Because exposure intensity is generally greater in earlier eras, workers with a longer duration of exposure and who had worked during earlier time periods will probably also have a greater intensity of exposure than workers hired more recently. This bias can be avoided if comparisons are made between subjects whose exposure begins at about the same time.

Selection bias also occurs when exposure begins before the study period begins. The more remote the onset of exposure before the beginning of the study period, the greater the selection pressures and therefore the greater the reduction in the study population to a fraction of the original. The remedy is to include all subjects beginning work at about the same time.

Occupational exposures are generally erratic in nature because of changes in the work schedule and in the intensity of the environmental characteristics. Workers are often exposed to a variety of nonoccupational agents, such as tobacco, dietary, and other risks of everyday life, that can have effects as significant as the occupational exposure singled out for a study. Problems in attributing to a given outcome an exposure in these situations may depend on the quantitative characteristics of an exposure, the eligibility of workers for inclusion in the "exposed" category, the identification of independent secular trends as a factor of pathology, and so on. Thus indicating an exposure history as causing an outcome is a somewhat subjective judgment, based upon considerations of the intensity and duration of exposure, classification of exposure status, and whether there is a progressive increase in the outcome variable that parallels with increased exposure.

2.2.1.3 A Priori Hypothesis. In an experiment or clinical trial there is a research hypothesis that identifies the cause–effect comparison that will be tested. Similarly, in an epidemiologic study there is usually an *a priori* hypothesis. Because of the wide availability of computing equipment, our ability to collect large amounts of diverse information has vastly increased, and consequently many different associations can be explored in "data dredging" activities. In this process a new hypothesis may be suggested by newly collected data. Setting statistical significance at $p < .05$, one might expect a positive computer-generated conjecture by chance alone for every 20 associations explored. And voilà, a new cause–effect relationship is discovered. Beware!

2.2.1.4 Well-specified Cohort. In experiments, the characteristics of the cohort members are determined at the beginning of the experiment, that is, before exposure has occurred. During the course of the experiment periodic checks must monitor the occurrence of the outcome. This process assures (1) that only eligible subjects are included in the cohort, and (2) all subjects are accounted for statistically.

In epidemiology the actual base-line conditions are often unknown and unknowable. In retrospective studies one relies on data collected for other purposes.

In retrospective cohort studies and case-control studies, for example, work history and study population are determined from personnel records and from exposure estimates collected for other purposes; often smoking information is not available at all.

The "healthy worker effect" is a phenomenon where the overall health experience (based on both mortality and morbidity) of the employed population is usually more favorable than that of the general population, often the actual comparison population in a mortality study. The employed population usually appears healthier than the general population because of two main factors, selection bias and higher socioeconomic status (10).

The most important factor is probably selection bias, that is, selection by the employer for healthy "employable" workers, and self-selection by the employee out of the work force. The selecting out process does not necessarily lead to a healthier work force, because short-term employees are often less healthy than longer-term workers (11, 12).

Wang and Miettinen (13) describe the healthy worker effect as a matter of *confounding*. Owing to different requirements and incentives for job entry and exit between the study population and the control or reference population, these two populations may not be comparable for many risk factors. Therefore there will be interference (confounding) when their results are compared.

2.2.1.5 High-Quality Data. In a bench experiment base-line data are obtained directly with calibrated methods worked out before the onset of the experiment. These measurements are specific for the outcome defined in the hypothesis. In epidemiology, outcome information often must be obtained from second-hand sources and therefore errors are less readily screened out. For instance, in mortality studies the outcome variable, death, is determined from death certificates. The determination of cause of death leaves plenty of room for inaccuracies and omissions, even deception (14). Patients may die at home of chronic illness in which the immediate cause of death may be an intervening complication rather than the underlying pathological process. Even hospital deaths with high autopsy rates show discrepancies between the autopsy results and the cause of death (COD) originally recorded on the death certificate. The current trend for fewer autopsies and the differential rates of autopsies for ethnic and racial groups increases further the potential for misclassification. An example, taken from asbestos workers, is provided by Newhouse and Wagner (15). When comparing the COD recorded from autopsy and/or hospital records with the respective COD recorded on the death certificate, they found underestimates of mesotheliomas and overestimates of gastrointestinal and other tumors (Table 26.1).

The accuracy of the COD varies depending on disease category (>90 percent for ischemic heart disease, for example). Mortality studies are quite useful for some causes of death (e.g., lung cancer), when the COD used is broad and when very elderly decedents are not used (14).

Exposure information can come from a variety of sources. In retrospective cohort studies, they may come from personnel records that are accurate in terms of time spent (tenure) in given jobs at that facility. Actual environmental conditions are

Table 26.1. Revision of Cause of Death in Light of Autopsy Findings (15)

Disease	Death Certificate	Added	Removed	Revised Number
1. Cancer of lung and asbestosis	29	3 (from 2)	5	27
2. Asbestosis without lung cancer	15	0	3	12
3. Gastrointestinal tumors	14	0	7	7
4. Other tumors	10	0	4	6
5. Mesotheliomas	4	15 (from 1, 3, 4)	0	19
6. Other diseases	12	0	0	12

often unmeasured or sparsely documented. As a result, duration of exposure is often used as a surrogate of dose in the determination of exposure–response relationship. To minimize this source of error, exposed subjects included in the exposure–response analysis should belong to uniform exposure categories, with similar exposure durations (16). Unfortunately such remedies are not always feasible.

In case-control studies information on exposure may come from surrogate sources, in both cases and controls (such as friends and relatives). Often the information is obtained long after the occurrence of the event(s) in question, and problems therefore exist in verifying the accuracy of such information. For example, in attempting to control confounding from tobacco smoking the only source of information may be the spouse and/or children. In general, measurements of work history, job exposure, and exposure to toxic substances are less accurate than smoking. Table 26.2 reports examples of these discrepancies.

2.2.1.6 Analysis of Attributable Actions. An ideal experiment allows one to determine if there is a statistical difference between the exposed and nonexposed groups. If there is such a difference (or no difference) one may conclude the exposure is (or is not) the responsible causative agent. Such a conclusion is the accepted outcome of an experiment, assuming that scientific standards have been maintained and the number of observations is adequate to achieve sufficient power.

In epidemiology the exposure is often erratic, variable in intensity and duration, uneven among different subjects, and often contaminated by exposures to agents other than those characteristic of the workplace, such as personal habits (hobbies, smoking), diet, and similar risks of life. These risk factors for the disease of interest are often unmeasured or are often not amenable to statistical tests or adequate analysis.

To reiterate, scientific standards common in experimental science are difficult to achieve in all instances in observational science. Therefore bias is a constant concern in cause–effect research. Feinstein (25) has suggested a basic strategy for cause–effect studies, with emphasis on the location of biases (Figure 26.1).

Susceptibility bias occurs when the base-line states of the contrasted groups are not comparable. The differences can be quite varied and might include factors such

Table 26.2. Percent Agreement for Smoking Status and Work History

Reference	Respondent	Units	Agreement (%)
17	Husband and wife	Smoking—yes/no	96
		Age Started	28
		Cigarettes/day	36
18	Surrogates	Smoking—yes/no	
		Smoker	91
		Nonsmoker	85
19	Self-report	Work history	82
20	Spouse	Exposure to solvents—yes/no	58
21	Self-report	Fact of Employment Date of Hire	95
		Exact	53
		±One Year	73
		Tenure—Exact	48
22	Spouse	Number of work area assignments	48
		Usual work area	72
		Type of Chemical or physical exposure	3
23	Next of kin	Smoking status	87
24	Wife	Industry	51
		Occupation	48
		Exposure to asbestos	50
		Exposure to radiation	86
		Exposure to arsenic	70
		Exposure to formaldehyde	70
		Smoking—yes/no	100

as smoking, alcohol consumption, socioeconomic status, and health status. Perhaps the best-known bias at this stage is the healthy-worker effect or selection bias.

Performance bias occurs when there are differences in the acquisition of information on the "maneuvers" or exposures of the compared groups. Enterline (26) discusses two pitfalls that fit in this category. The first case occurs when an occupational cohort experiences competing causes of death so the probability of death from the cause(s) of interest may be underestimated. A second pitfall can occur when there are variations in latency periods. Because susceptibilities to disease is on a continuum and is influenced by the intensity of exposure, observations should be collected by time since first exposure. For an effect having a long latency, the inclusion of large numbers of workers with a short time since first exposure will reduce the computed relative risk for that effect. Similarly, the evaluation of an exposure–response relationship for a chronic disease with a long latency should be done on workers observed through adequate length of time. For example, to evaluate lung cancer in its relationships with asbestos exposure, the effect should be evaluated in workers with 20 or more years since exposure inception.

Detection bias occurs in the determination of outcome and arises because of

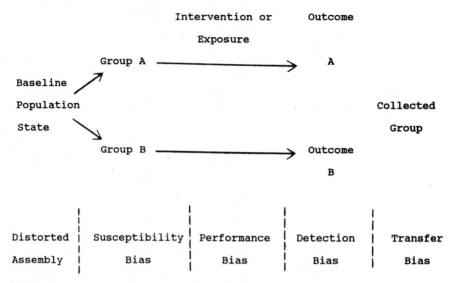

Figure 26.1. Basic architecture of cause–effect research in epidemiology and location of major problems (25).

inconsistencies of diagnostic criteria and/or exposure. Examples include the following:

- A greater effort to get smoking or other exposure information from members of a case than from those of a control group can bias estimates of exposure.
- If a radiologist interpreting chest X-rays for pneumoconiosis knows the origin of the population, he may interpret nonspecific changes as consistent with a specific effect in the exposed group. The cohort and exposed population should be read "blind," without distinction evident to the reader between "exposed" and "control" or "referent" group. In addition, more than one reader should be used.
- Correcting COD on death certificates from hospital and/or autopsy records for exposed (but not control) populations invalidates both comparisons and the calculated relative risks (26).

Transfer bias occurs after the events relating to base-line state, exposure, and outcome have occurred. It is an incomplete collection of data or an uneven loss of data when transferred for analysis. This bias occurs in cohort studies when subjects are lost to follow-up because of unknown status and/or incomplete information such as sex, race, date of birth, or work history. In a cross-sectional morbidity study this bias occurs when study subjects do not participate for reasons unknown, which may include being late, or absent, or coincidence of vacation schedule with examination time.

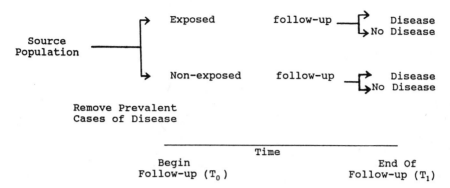

Figure 26.2. Basic design of a cohort study.

In summary, occupational epidemiologic studies may appear to have the same structure as experiments. But unlike experiments the exposed and comparison groups are often not perfectly comparable, random apportionment to exposures and control groups is not possible, and exposure cannot be controlled. In order to reduce bias and increase accuracy the sources of possible errors must be diligently reviewed and when possible controlled, lest the results and interpretations be distorted.

2.2.2   Study Designs

Let us now look at specific characteristics of study design and biases in occupational epidemiology.

In epidemiology there are three main study designs: *longitudinal studies*, composed of cohort and case-control studies, and *cross-sectional* or *prevalence studies*. Because there is no *a priori* control of exposure, and members of exposure groups are not randomly assigned, a primary device to minimize bias is in the objective selection of subjects for comparison.

2.2.2.1 Cohort (Follow-up, Incidence, Prospective) Study. For recent reviews of cohort studies see Checkoway et al. (27, 28) and Liddell (29). A prospective or cohort study usually analyzes a group of persons (cohort) defined by some exposure; the cohort is followed forward from the initial time of exposure. The basic design of cohort study includes the following characteristics (also see Figure 26.2):

1. Enumeration of an exposed cohort (population at risk)
2. Identification of a comparison nonexposed population
3. Follow-up of each individual in the cohort to determine the incidence of disease
4. Comparison rates of disease between cohort and reference population

We discuss mainly the historical cohort using mortality data, which is the most common type of occupational cohort study.

Some examples of typical occupational cohorts include all workers ever employed at a single location (mine, plant, company, etc.), or all workers from several plants, factories, or mines from the same or different companies. The important distinction is that the industrial process (exposure) is similar. A cohort might also comprise the members of a particular union or association with common occupational exposures. A historical cohort mortality study is often based on company employment or union records. The minimum amount of information required includes name, social security number, date of birth, date of hire, and date of termination with enough detail to identify where and for how long each individual worked. The records must not only contain this information but must also be complete; that is, they must include all persons ever employed (30).

There are basically two kinds of reference or comparison populations: *external* and *internal*. The most common is an external reference population, usually provided by the national population experience. Sometimes, state or even county populations are similarly used to compare the rates of disease. Disease rates from these sources are available; can be stratified by gender, race, year, and age at death; and are large enough for the disease rates to be stable. The standardized mortality rate or ratio (SMR) is the measure of effect. Indeed, the SMR may be thought of as the ratio of observed number of deaths to expected number of deaths. Observed deaths are from the exposed population and are stratified by age, sex, and race. Expected deaths are derived from the reference population and are the number of deaths one would expect in a nonexposed population with the same age, race, and sex distribution as the exposed population.

The choice of a comparison population can produce different results and conclusions, so the actual SMR in any given study has no absolute meaning. Although an increased SMR is often used as the driving force for deriving interpretations of risk, there are dangers and pitfalls in using this information in an unqualified and absolute fashion [see Tsai and Wen (31) for a discussion of the use of the SMR].

The use of the national population in these comparisons has a variety of justifications:

a. Rates of disease are stable for even rare conditions.
b. These rates are available on computer programs, making the related calculation convenient and prompt.
c. The national population is the reference group in common use and makes comparisons with other studies possible.

An internal reference group comprises all eligible workers in the cohort with little or no exposure. The internal comparison group is advantageous in that selective forces and quality of data are similar, if not identical, to the exposed workers. Except for exposure, the internal controls should be quite similar to the exposed group, thus reducing the bias originating from the healthy worker effect. The internal comparison group also has obvious disadvantages, because it is usually

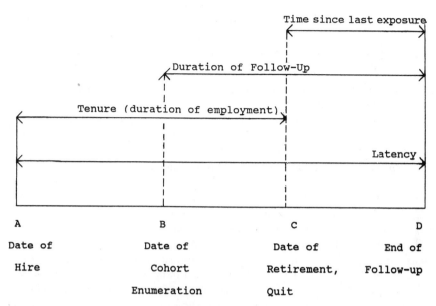

Figure 26.3. Time-related factors in a cohort study.

small and often presents less stable disease rates. Sometimes there is no "nonexposed" group but only a minimal or less exposed group, for reference.

The base for cohort studies is the person-time experience of their members. Stratification according to time-related factors may be necessary in addition to the specification of age at first exposure as well as calendar year of first exposure. Risk of disease varies with age and often calendar year. Other time-related variables such as latency (time since exposure onset), cumulative exposure, and tenure (a surrogate assessment of exposure) contribute to the estimates of exposure.

The mutual relationships of these time-related factors are illustrated in Figure 26.3.

The essence of a cohort study consists of comparing rates between the exposed and reference populations. When the comparison is obtained with an external reference population the results may indicate exposure effects but should not be considered, in any way, the final proof of adverse effects (27–29). To establish causation, rather than just association, the demonstration of a graded exposure–response relationship cannot be overemphasized (3, 29, 32).

Measures of exposure can be quite varied. The most common is tenure (length of exposure), a surrogate of cumulative exposure (33). Measurements of exposure obviously are often subject to error, and the quality and quantity of past exposure estimates vary widely. Important factors to consider are *duration* of exposure (with or without gaps), *intensity* of exposure, age at first exposure, and time since last exposure. Stratification by cumulative exposure and by latency is also important for diseases of gradual evolution (i.e., chronic).

The method of evaluating exposure–response trends across multiple levels of

Table 26.3. Criteria of Quality in Evaluating Cohort Studies

Enumeration and Verification: Is 100% of the population at risk enumerated and has the enumeration been verified, such as through Social Security records? (41).

Follow-up: Is the loss to follow-up less than 5 percent?

Latency: Is the period between first exposure and end of follow-up longer than the latency of the disease of interest?

A priori hypothesis: Is the hypothesis being investigated clearly stated and hypothesized prior to data analysis? (Other significant findings should be tested in an independent study, otherwise there is the possibility of "data-dredging bias") (42).

Control Populations: The national population is a minimum requirement. State or local reference populations may be more desirable. An internal reference population is seldom available, but often is the ideal reference.

Stratification: Are results stratified by latency and exposure, or is this adjustment accounted for in another manner (e.g., regression techniques)?

Exposure: Is there an exposure–response analysis? Is it stratified by latency? Are there adequate data to assess intensity and duration of exposure? Are cumulative exposure and/or tenure used as exposure variables? Is there a trend analysis? Is exposure time long enough and intensity great enough to evaluate an exposure–response relationship?

Confidence Intervals: Are there confidence intervals on SMRs and on the slopes of the exposure–response regression?

Bias: Are confounding biases such as selection, smoking, and other occupational exposures considered?

Design: Is the appropriate study design used to answer the hypothesis of interest? Please note that case-referent studies are better for rare diseases.

Interpretation: Is there a discussion of possible sources of error and their potential impact? Are the descriptions of methods, material, and data in sufficient detail that the reader has the opportunity to reach his own conclusions? Are criteria relating to causality (consistency, plausibility, strength of association) considered?

---

exposure is an extension of the process comparing incidence rate between exposed and reference populations. In this type of analysis the low or nonexposure category is set at unity, and the higher exposure categories are gauged by the lowest exposure category.

A general form of an exposure–response model is: $RR = a + b(x)$, where RR = rate ratio or relative risk; x = exposure value such as years exposed or mg/m^3 years-dust or ppm-years; b = slope of the exposure-response trend; and a = intercept, usually set to 1.0.

In a SMR analysis using an external reference, a relative slope b/a can be calculated that adjusts for the departure of the intercept from 1 and allows comparisons between different cohorts (34).

Table 26.3 provides a checklist of factors to be considered in interpreting the results of individual cohort studies. Methodological issues relating to retrospective cohort studies are discussed in detail in References 35 through 38.

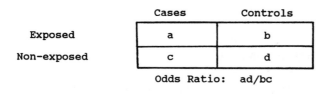

	Cases	Controls
Exposed	a	b
Non-exposed	c	d

Odds Ratio: ad/bc

Example 1.

	Lung Cancer Cancer	Controls	Odds Ratio
Exposed	60	160	2.25
Non-exposed	40	240	1.0
	100	400	

Odds Ratio: (60 x 240)/(40 x 160) = 2.25

Example 2.

	Smokers			Non-smokers		
	Cases	Controls	OR	Cases	Controls	OR
Exposed	60	80	4.5	10	80	1.5
Non-exposed	20	120	1.0	10	120	1.0
	80	200		20	200	

OR = (60 x 120)/(20 x 80) = 4.5   OR = (10 x 120)/(80 x 10) = 1.5

Figure 26.4. Calculation of odds ratio in case-control study.

2.2.2.2 Case-Control Study.

2.2.2.2 Case-Control Study. The case-control study (also called a case-referent or a "trohoc" study) is, like the cohort design, a prospective or longitudinal study. The most basic difference is that the cohort or prospective study identifies exposure and looks for disease, whereas the case-control study identifies disease and looks for exposure. As a result the case-control study involves only a sample and not the total of the population at risk (39, 40). In a "nested" case-control (or cohort-based) study, for example, the cases would be all those in the cohort who had the disease of interest and the controls would be a sample of those in the cohort without the disease.

In the case-control design, cases are selected on the basis of disease. Indexes of effect may be computed by obtaining from the same cohort and source population the ratio $(a/c)/(b/d)$, or ad/bc (Figure 26.4). For example, suppose there were 100 lung cancer cases, with 60 percent of these exposed to a particular chemical, and 400 non-lung cancer controls with 40 percent exposed. The controls are selected from those free of disease. The measure of association is the odds ratio, that is, the ratio of the odds of exposure among the cases to the odds of exposure among the controls. This computation indicates that the relative risk (odds ratio) for lung cancer is 2.25 times greater among workers exposed to the chemical than among those not exposed.

To control for possible confounding from smoking, one could also obtain smoking histories, stratify on smoking, and thus calculate odds ratios for smokers and nonsmokers.

If this were a nested case-control study, the cohort might comprise 10,000 workers. It would be costly and time-consuming to examine such a large study and to code each work history and determine who was exposed. It is essentially impossible to get smoking histories on this many workers. It is relatively cheap and quick to do so for 500 workers, thus making case-control studies attractive alternatives.

If there is less than 95 percent follow-up, a cohort analysis should include an appropriate scrutiny of the nonresponders, in order to ascertain potential biases.

If diagnosis of disease (or death) is incorrect, the inclusion of a misclassified case into either a cohort or case-control study may bias the SMR and odds ratio toward the null. Collection of additional diagnostic data may create an ascertainment bias if this addition is not applied equally among exposed and controls in a cohort study or among cases and controls in a case-control study.

A disadvantage of a case-control design is said to be that often there is less familiarity with this design than the cohort method, making findings more difficult to qualify and interpret. The problem of unfamiliarity is compounded by the incorrect idea that a case-control study is a fundamentally different type of study. However, the cohort and case-control study designs are architecturally the same in that both are prospective, following workers from some base-line state forward in time to some end point (25). In cohort studies the population studied is defined on the basis of exposure to a particular substance. Data on risk factors and incidence are then collected on all cohort members. The association is a comparison of the incidence of disease in exposed and nonexposed members. In a case-control study the cases may be obtained from the same population base, but only a subsample of the population at risk comprises the controls. The contrast is obtained by the comparison of the exposure parameter in diseased and nondiseased persons (43).

In practice, case-control studies tend to have more potential sources of bias than cohort studies. The potential pitfalls have been distinguished into the following categories: (1) selection of exposed subjects, (2) exposure estimates, (3) selection of controls, and (4) confounding factors (44–46).

Bias in the selection of subjects might occur when a particular exposure requires medical surveillance and the diagnosis of symptoms, signs, or cause of death are influenced by the knowledge of that exposure. This has been called the "diagnostic suspicion" bias by Sackett (47) and "publicity bias" by Feinstein (25). For example, working in the rubber industry might encourage the diagnosis of bladder cancer (48). Such a suspicion could result in a detection bias as well. Wells and Feinstein (49) showed that male smokers with a chronic cough were 22 times more likely to be given a sputum test in the diagnostic pursuit of lung cancer than nonsmoking women without a cough.

Information on a surrogate of exposure such as job tenure should not be a problem in a nested case-control study when company personnel records are used to confirm work history. If interviews are used to ascertain work exposure and/or nonoccupational exposure, an exposure suspicion bias could result (47). This could occur if the interviewer has knowledge of a subject's disease status, and such knowledge may in turn influence the intensity and outcome of the search for exposure. The importance of this bias is shown in the Wells and Feinstein study (49) summarized above, and in two studies of thyroid cancer among children described by Sackett (47). Upon routine questioning about exposure to irradiation, the evidence of exposure was, respectively, 28 and 0 percent. Upon intensive questioning and search of records, the evidence of previous irradiation was 47 and 50 percent, respectively. Such bias can be reduced by using standardized methods and employing blinded interviewers.

Recall bias can also be a problem if the subject (or surrogate) attributes the disease to a particular exposure, and/or if the exposure is rare. The former problem may be alleviated by using diseased controls that are apt to answer questions in a manner similar to cases.

Selection of controls can be a knotty problem in case-control studies. Controls should reflect the frequency of exposure in the source population. If diseased controls are selected, the diagnosis should be unrelated to exposure. If the disease in controls is related to the exposure, the frequency of exposure among controls will be falsely elevated and the effects of exposure underestimated. Two important principles are involved in selecting controls. Controls (1) should be *comparable* to the cases in all respects except the presence of disease and (2) should be as representative as possible of the nondiseased population.

Thus in the literature there are discussions such as that initiated by Gordis (50) about using dead controls matched to dead cases. McLaughlin et al. (51, 52) argue against using cancer patients as controls in case-control studies, but Smith et al. (53) argue in favor of the procedure. Similarly Linet and Brookmeyer (54) are generally in favor. The discussion about controls will go on indefinitely, for "it is unlikely any formula for control selection in studies involving occupational diseases will hold for all circumstances" (55).

Confounding is a potential problem in all study designs in nonexperimental research. The confounding factor is important when (a) it is predictive of the response and (b) it is associated with the exposure. For example, smoking is often said to be a confounder in a lung cancer study because smoking is predictive of the response, lung cancer. However, if the proportion of smokers in the case and control groups is the same, smoking is not a confounder because it is not associated with exposure.

Confounding factors must be identified in an epidemiologic study to avoid misinterpretation of the results. Some methods of controlling confounding are:

- Matching: select one or more controls with the same confounding factor(s) as the cases; that is, match a smoking case with a smoking control.
- Stratification: group cases and controls in similar categories of the confounding factors; that is, compare the risk of cases who smoke using controls who have the same smoking status.

• Restriction: include only subjects within a specified range of the confounder.

Confounding is a problem one must always consider. An appropriate comment is offered by Feinstein (56):

> The term *confounding factors* (or *confounding variables*) is sometimes used as a general name for problems that can distort or bias the result of an analytic study. The exact sources of the confounding variables are seldom specified, however, and their discovery often seems to occur via intuition rather than by direct attention to their hiding places. . . .

> For . . . statistical differences to be attributed to the casual agent, . . . scientific standards for performing and interpreting trohoc research . . . should reflect the same rigorous criteria that pertain when cause–effect reasoning is applied in other scientific activities; and the standards should contain the same careful attention to the prevention or removal of susceptibility bias, performance bias, and detection bias (56).

2.2.2.3 Cross-sectional (Survey, Prevalence) Study. Cross-sectional studies are among the most common of epidemiologic studies (at least when studying morbidity). The distinguishing characteristic of a cross-sectional study is that each person in the study population is examined only once. When quotients are calculated to summarize the *prevalence* of disease, the conditions in the numerator and denominator are determined at the time of the examination. Ideally, these parameters should be determined at a midpoint of the time interval studied.

For the longitudinal study design, in contrast to the cross-sectional design, each person is examined on at least two separate occasions. When quotients are calculated to summarize the incidence of disease, the first examination delineates the base-line conditions (denominator), the second examination delineates the change in state (occurrence of new cases), and the latter appears in the numerator.

A cross-sectional study is distinctive in that the data collected at one point can be analyzed and interpreted (1) for the current situation and (2) looking backward or retrospectively, as well as (3) forward or prospectively. For example, suppose we examine all underground coal miners in 10 bituminous coal mines and a control group of surface workers at these same mines, collecting the following observations on each person: chest X-ray for determination of pneumoconiosis, pulmonary function, age, height, smoking history, and work history.

In analyzing the data we look *backward*, or retrospectively, by comparing the prevalence of pneumoconiosis and reduced pulmonary function in both the underground miners and controls. We assess the *current* situation by describing the prevalence of pneumoconiosis (cases/total), and the related pulmonary function indexes. A *forward* look at the data occurs when we stratify by smoking category, grouping the miners into discrete exposure categories (<5, 5–10, 10–20, and >20 years tenure), and reporting the prevalence of pneumoconiosis and indexes of pulmonary function (adjusted for age, race, sex, and height). Again it should be stressed that observations obtained within the context of cross-sectional studies may not be used as definite estimates of pertinent risks. The presentation of these

Table 26.4. Sample Data Layout for Cross-sectional Study: Prevalence of Pneumoconiosis

Tenure (years)	Underground Miners (%)		Controls (Surface Workers) (%)	
	Smokers	Nonsmokers	Smokers	Nonsmokers
<5	1	0	1	0
5–15	5	2	2	0
>15	10	6	3	1
Total	8	2.5	2	0.5

results for all temporal directions can be arranged in a single table. Table 26.4 is an example for the computation of prevalence of pneumoconiosis in men 40 to 50 years of age. These hypothetical data show that the prevalence of pneumoconiosis among smokers is four times higher among underground miners, compared to surface workers. The current overall prevalence of pneumoconiosis is 8 percent among smokers and 2.5 percent among nonsmokers. The prevalence of pneumoconiosis increases in both smokers and nonsmokers.

A similar format could be used if the outcome variable was lung function. In that instance average values could be used.

The interpretation becomes more difficult in regard to a reduced pulmonary function, because this is a nonspecific effect, often related to a number of different exposures.

According to Table 26.5 both smoking and exposure to coal mine dust appear to affect lung performance. However, in a study of this type, it may not be possible to differentiate completely the causes of the reduced lung function. One reason is the intercorrelation existing between age, smoking, and exposure, thereby making separate analysis of each variable difficult if not impossible. A second problem has to do with the study design. Everything that has happened to the exposed and control group occurred before the data collection began. Did the reduced pulmonary function happen as a result of exposure, had it already occurred before

Table 26.5. Sample Data Layout for Cross-sectional Study Expressed as Percent of Predicted Pulmonary Function

Tenure (years)	Underground Miners (%)[a]		Controls (Surface Workers) (%)[a]	
	Smokers	Nonsmokers	Smokers	Nonsmokers
<5	100 (3.5)	95 (3.4)	97 (3.2)	96 (3.3)
5–15	90 (3.3)	93 (3.3)	92 (3.0)	94 (3.2)
>15	80 (3.0)	90 (3.2)	85 (2.9)	92 (3.2)

[a]Figures in parentheses are mean FEV_1 in liters.

Table 26.6. Lung Cancer Mortality—Comparison by Study Design (9)

	SMR for Lung Cancer	
Occupation	Cohort	Cross-sectional
New Jersey insulation workers	444	867
Asbestos manufacturing workers	111	151
Homestake gold miners	124	321

exposure, or is the relationship observed in the data largely the result of healthier workers selectively leaving employment? The use of cross-sectional study design does not allow the observer to address these questions directly.

Cross-sectional studies are relatively easy to conduct and are useful for conditions that are measured quantitatively and vary over time, such as lung function or pneumoconiosis. They are also useful for relatively frequent conditions of a chronic nature (i.e., chronic bronchitis). It is quite easy in cross-sectional studies to collect information for many different hypotheses. Because of the usually abundant data, cross-sectional studies can result in new etiologic hypotheses regarding risk factors and/or disease. A major value of the cross-sectional study is descriptive, that is, the simple characterization of a target population.

A survey is also the initial stage of a cohort study. If a cross-sectional study is used to substitute for a longitudinal study, the occurrence rates in the two types of studies will probably be different. For example, Glindmeyer et al. (57) found the age coefficients for lung function derived from a five-year longitudinal study were lower than those from cross-sectional analysis. Weiss (9) discusses three examples that show differences in lung cancer mortality between cohort and cross-sectional analysis. These results are summarized in Table 26.6. All workers included in these analyses had 20 or more years latency. Clearly the cross-sectional design shows the higher risk.

Both kinds of studies must consider errors due to differential losses or selection. In a cross-sectional study one is investigating a residual cohort consisting of the remains of an original cohort affected by attrition. Often the causes of attrition, and most important, the health status of those leaving are unknown. Even the magnitude of attrition is not known. Alternatively, in a cohort study one knows the number and original health status of all members and so can better estimate the effort of attrition.

2.3   The Rules of Evidence in Epidemiology

Experiments are an attempt to determine a cause–effect relationship and provide the model for epidemiologic studies. In animal experiments, for example, the question of whether the rate of disease in the exposed group differs from the rate of disease in the nonexposed group is answered (generally) by either *chance* or *identifiable cause*:

	Disease		
Exposure	Yes	No	
Yes	a	b	$a + b$
No	c	d	$c + d$

Is the difference between $a/a + b$ and $c/c + d$ due to chance or to exposure? The answer to this question is based largely on tests of statistical significance that provide reasonable assurance of the probability that the results are (or are not) due to chance alone. Statistics play an important role because the exposed and nonexposed groups are *comparable* in essentially every way but their exposure. This is so because (*a*) the assignment to exposure groups is determined by the investigator by random selection among similar subjects; and (*b*) the investigator has continuous control and observation during the course of the entire experiment.

Epidemiologic studies are also cause–effect studies. They are investigations of the association between "exposure" and "disease." Because they are based on observations rather than experiments, *noncomparability* between exposed and nonexposed groups is practically guaranteed. The investigator has no control over the content of these groups and no control of exposure. Therefore, in epidemiology, one must be concerned about *bias*—an effect tending to produce results that depart systematically from the true values (A partial list of biases is presented in Appendix 1). As a result the decision about whether an association is causal or not cannot be a mere statistical judgment.

Epidemiology is concerned with determining causation and estimating the magnitude of the effect of the causative factor. The science of epidemiology is largely observational because the makeup and exposure of the groups to be compared cannot be directly controlled by the investigator. There is no single study or critical point that "decides" or determines the etiologic significance of an association. Biases can occur at all stages of a study. There are, however, well-known and widely used criteria or rules of evidence that are applied when determining whether an association is causal or not. One of the best-known applications is recognized in the association of smoking and lung cancer presented by the first Surgeon General's report on smoking (58).

The following guidelines provide a framework for the most difficult part of epidemiology, the interpretation of the meaning of the data.

3   CAUSATION IN EPIDEMIOLOGY

In cause–effect research a decision ultimately has to be made as to whether a given disease is likely to result from a given circumstance. How does one arrive at a judgment for or against causality?

Much of scientific knowledge is a collection of causal statements. The ability to understand a phenomenon, in general, is often equated with the success of predicting its occurrence (4). Pitfalls in judgment of this process abound. Among the classic errors affecting the validity of epidemiological interpretations and beliefs are the following:

(*a*) Method of consensus: A causal inference is drawn on the basis of a consensus stipulated among knowledgeable experts.

(*b*) Appeal to authority: Many inferences are made on the basis of approval by a noted authority. The flip side of this argument occurs when one rejects an inferential argument because the proponent doesn't have the necessary credentials. Research is sometimes discounted because it is funded or done by interested parties and is therefore biased. For example, certain publishers require disclosure of funding sources. Sacks et al. (59) have criticized a number of studies because the source of support was not provided. Thus prejudice sometimes appears to be used rather than judgment on the findings and arguments.

(*c*) "After this therefore on account of this" (post hoc ergo propter hoc): Chronological sequence of coupled phenomena does not prove a causative relationship.

As summarized by Susser (60), criteria for "making inferences about cause depend . . . on subjective yet reasoned judgment. Such judgments are reached by weighing the available evidence; there are no absolute rules, and different workers often come to conflicting conclusions."

What is needed to establish causality from epidemiologic evidence is a positive association between exposure and disease in groups of individuals with known exposure, provided that bias (or biases) is (are) minimized. In one of the seminal papers in epidemiology, Hill (3) summarized rules of evidence to use in differentiating causation and association. The most important of the rules are mentioned herewith. The reader of epidemiologic studies obviously has the responsibility of considering the merits of a study, of applying "common sense" to the methods and numbers given by the study, and of not accepting figures "at their face value without considering closely the various factors influencing them." The careful reader must assess whether the observations were "well and fairly made" and whether the differences observed had such biological significance and consistency that it is a matter of "indifference whether it is technically [statistically] significant or not" (61).

The second responsibility of the reader (and writer) in the field of epidemiology is to consider the evidence for causality in light of the following criteria.

3.1 Temporality

A criterion reliable under all circumstances is that the cause must precede the effect. Not only must the cause precede the effect, but for chronic diseases that require a lengthy latent period for the disease to develop, there must be adequate time from first exposure for causation to be affirmed. Temporality is generally an indeterminate criterion in cross-sectional studies because the time relationship between cause and effect is not known.

3.2 Strength of Association

The strength of association is the magnitude of the outcome, measured in the exposed group compared to the nonexposed. How large is the rate ratio (SMR in cohort studies, *odds ratio* in case–control studies)? The stronger the association the more likely it is a causative link, because strong associations are less likely to be due to bias than weak associations. A weak association could be due to undetected biases, or it could be due, for example, to low exposure, small exposure group, or young cohort with short latency. The strength of an association is also a function of other causes. A rare but effective causal factor should have a strong association if competing causes are common.

3.3 Biological Gradient

The presence of an exposure–response gradient is generally considered strong evidence of a causal association. An exposure–response gradient could be non-causal if there is confounding. For example, one might observe an association of increased SMRs for lung cancer with increasing tenure. In fact, increasing age is associated with both lung cancer risk and tenure, and so the gradient could be in part or totally due to age. The stronger the gradient, the less likely the effect is caused by a confounder. The use of a cumulative exposure index sometimes reduces the interference of age on exposure. If the study is large enough one can stratify on age and compare risk of the outcome among different exposure groups of the same approximate age (and smoking status).

The lack of a biological gradient does not by itself disprove a causative association. Misclassification of exposure or too narrow an exposure range to show an effect will reduce a gradient toward the null. Variants on the idea of a narrow exposure range include exposure of such magnitude that it produces a maximum effect so there is no gradient, as well as exposures too low to produce a detectable effect.

3.4 Consistency

Consistency is the repeated and converging observations of an association in different populations under different circumstances. Exact replication is impossible in epidemiology, but even though separate situations differ, a consistent health effect related to similar exposure provides strong evidence of a causal association.

Consistency does not apply to an updated analysis in the same population and with similar results. A hypothesis based on a study of one population should be confirmed by study of a different population. An updated study (if the later study is better) can change the conclusions. One should take only the best study information from each population.

In the 1964 Surgeon General's report on smoking and health (58), 28 out of 29 case-control studies and 7 of 7 cohort studies of different populations all showed similar relative risks of smoking. When this happens, it is unlikely that constant error or fallacy is present in every study. Different results from a different inquiry (such as 1 of 29 case-control studies) do not refute the causal association.

Consistency can also mean parallelism in the strength of association and in the biological gradient and parallelism with the results of animal studies. If there are studies where association is not detected, but a plausible reason exists for this absence, the argument is strengthened. In this sense consistency can be thought of as a coherent and verified pattern whenever all relevant factors are taken into account.

3.5 Plausibility

Are the findings plausible in terms of biological knowledge? Do the facts fit the theory? Do findings from experiments in nonhuman species verify the hypothesis?

If an association appears plausible it is easier to accept. Conversely, if the association is not plausible, acceptance may not occur no matter how strong the evidence appears to be.

Results that are plausible in terms of current theory affirm the hypothesis; results that contradict the theory suggest that either the study results are in error or the theory needs revision. For example, if cigarette smokers who inhale are at no greater risk than those who do not inhale (or women are at less risk than men) then perhaps one should either reject the study results or modify the hypothesis.

A consistent negative finding in studies of a range of animal species (say 5 or more) casts doubt on a positive finding in human studies—unless the finding is replicated and/or the findings are so compelling as to leave little doubt. On the other hand, a positive finding in a single animal species may verify a finding in a human study.

Are the epidemiologic findings compatible with the facts? For formaldehyde to cause brain tumors there should be evidence that inhaled formaldehyde is transported to the brain or in some way affects metabolism in the nervous system. The presence of such a mechanism affirms the hypothesis; the lack of such a mechanism may make the hypothesis less plausible. A similar question relates to asbestos and intestinal cancer. The presence of asbestos in intestinal tissue supports the hypothesis. The lack of fibers in a pathological specimen makes the hypothesis less plausible.

3.6 Summary

Conclusiveness in inferring causality is a goal, but it is not always an accomplishment. The rules of evidence must be applied to the question of causality. They overlap with each other, can be mutually reinforcing, and have different levels of conclusiveness (60). A suggested framework to assess the relative importance of each criterion in inferring causality is presented in Table 26.7.

Table 26.7. Effect of Rules of Evidence on a Causal Hypothesis

Criterion	Effect on Hypothesis		
	Support	Detract	Indeterminate
Time order			
Compatible			+
Incompatible		+ +	
Uncertain			+
Biological gradient			
Yes	+ +		
Questionable or no			+
Consistency			
High	+ +		
Positive	+ +		
Negative		+ +	
Low			+
Plausibility			
Consistent with theory?			
Yes	+		
No		+	
Consistent with fact?			
Yes	+		
No		+ +	
Consistent with animal studies?			
Yes	+		
No		+	
Probability			
Statistically significant association	+		
Not statistically significant			
Adequate power		+	
Lacks power			+

Source: Modified from Reference 60.

It seems appropriate to conclude this discussion with a quote from Feinstein (25):

> If industrial workers and the public are to receive suitable protection against occu-
> pational and environmental hazards, and if everyone is to be protected against irra-
> tional fears that may lead to a needless loss of income, jobs, and homes, the first step
> in the process is to obtain accurate and unbiased information about what the hazards
> are, where they occur, and how relatively great they may be. The task of acquiring
> credible information is difficult but not impossible. Until this task is approached with
> rigorous scientific methods, workers and the public may be victimized far more often
> by the products of defective research than by the toxins of occupation or the pollutions
> of environment.

4   CONCLUSION

We have tried to present some guideposts and points of view that will assist in the reading and interpretation of epidemiologic studies. To conclude, we reiterate some of the viewpoints discussed herein but in the words of others.

Although there is long historical precedent for using "tests of significance" as a primary means with which to interpret epidemiologic data, there is no compelling reason to do so. To describe relations between exposure and disease, there is no need for a commitment to reject or to fail to reject null hypotheses. Furthermore, dichotomization of the p value range into regions labeled "significant" and "nonsignificant" is highly arbitrary and can result in misleading interpretation (62).

Formal statistical tests are framed to give mathematical answers to structural questions leading to judgments, whereas in . . . [epidemiology one] must give answers to unstructured questions leading from judgment to decision implementation. These questions of decision generally hinge around judgments about causality and prediction (63).

Full epidemiologic analysis assesses bias, confounding, causation and chance. Of these, chance is least important but still receives the most attention (64).

The results of epidemiological studies are often difficult to interpret, and claims and counter-claims of the existence, or absence, of a risk are sometimes made on the basis of fragmentary and contradictory data. It is understandable, therefore, that those responsible for protecting the health of individual workers have sometimes sought to lay down rules that could ease their task. Unfortunately, epidemiology cannot be regulated in the same way as laboratory science. A laboratory investigator may properly be told to use so many animals, test so many species, test at so many levels of dose, and observe for an optimum length of time, as all these are under his personal control—subject only to the constraints of finance. It is unproductive, however, to lay down similar rules for the epidemiologist, as unintended experiments on Man cannot be repeated at will and the conditions of the experiment are not under the observer's control. All we can do is to require that the observations that are made should be relevant and not biased and that consideration should be given to the possibility of confounding as an explanation for them. So far as numbers are concerned, we can deal with this only by combining the results from different series, for this purpose, all data that are relevant and unbiased must be grist to the mill. Indeed, if there is one general rule in the assessment of epidemiological evidence, it is that no conclusion can be reached until the totality of the evidence is taken into account (65).

Sound epidemiologic evidence offers a robust and often unique resource in shaping our beliefs and driving our conclusions. By the same token, erroneous epidemiologic data and observations may be the prescription for waste and unjustifiable decisions. The task for any responsible practitioner is to examine the evidence with patience, care, and an open mind. This should be an admonition and a rule.

Appendix 1

BIASES THAT CAN OCCUR IN EPIDEMIOLOGIC RESEARCH*

1. In reading-up on the field
   (a) Positive results bias (publication or file drawer bias): Positive results are more likely to get published than negative results (66).
   (b) One-sided reference bias: References may be restricted to those supporting the authors' results. Or only the results that support the authors' view or a governmental or other policy are presented.
2. In specifying and selecting the study sample (selection bias)
   (a) Diagnostic suspicion bias: Knowledge of occupation or exposure may influence the diagnosis and/or the search for a putative cause. This is similar to what Feinstein (25) calls the diagnostic review bias. The elimination of this bias can be accomplished by the examiners being blind to the other information.
   (b) Sample size bias: Too small a sample may prove nothing; too large a sample can make any difference statistically significant (67).
   (c) Nonrespondent bias: Nonrespondents (or "latecomers") in a survey may have different exposures or health status from respondents or "early comers." A mortality follow-up of gold miners examined in a cross-sectional morbidity study showed a positive association of lung cancer to exposure (68). Because this was not a complete cohort and subsequent mortality studies of the complete cohort (69, 70) were negative, this original result may be incorrect because of nonrespondent bias. Williams (71) suggests that in suveys the effects of nonrespondent bias can be subtle, can change relationships, do not react to increasing response rates in a desirable way, and can be serious in magnitude. For example, a 4 percent bias in estimating unemployment rates can occur with a nearly 98 percent response.
   (d) Volunteer bias: Volunteers or "early comers" may have different exposures or health status from nonvolunteers or "latecomers." In fact, volunteers are in many ways different from nonvolunteers.
   (e) Healthy worker bias: An employed population of active workers is generally healthier than the general population, often with a mortality risk 60 to 90 percent of that of the general population. There is a large literature on the healthy worker effect. Early and recent comments can be found in References 72 and 73.
   (f) Chronology bias: This is a selection bias caused by variation in exposure over time. The effects of this bias were observed in a cross-sectional survey where the risk of respiratory cancer was overestimated by about 30 percent compared to a cohort study (74).
3. In measuring and classifying exposures and outcomes (information bias)

---

*By stage of research [see Sackett (47) for additional references and discussion].

(a) Detection bias: This occurs if the outcome (response) event is detected unequally in the exposed and nonexposed groups. For example, Enterline (75) showed that correction in the cause of death of asbestos-exposed workers in three studies incorrectly increased respiratory cancer risk by about 13 percent and mesothelioma risk by about 67 percent (75; see also 25 for other examples).

(b) Instrumental error: Faulty calibration and inaccurate measuring instruments result in bad data and wrong results. Graham et al. (76) reasoned that a leaky spirometer resulted in a reported decrease in pulmonary function in a cohort of granite shed workers. The decrease was incorrectly attributed to exposure (77) because the instrument error was undetected during its use.

(c) Interviewer bias: This is a systematic error due to interviewers gathering selective data. The data may be biased if the interviewers are aware of the research hypothesis and the health status of the individual. For example, an interviewer who believes asbestos causes lung cancer may readily accept no exposure from a control but may inquire more intensively of a case. Such knowledge may influence both an interviewer's attitude and mode of interrogation. This bias can be eliminated by blinded interviewers and a standard questionnaire (25). Gould (78) discusses an interesting example of "unconscious manipulation of data," which argues for the necessity of blindness as a minimum, and perhaps even supports the suggestion of Feinstein (25) that the investigator should not be the one both to collect the data and later to test the hypothesis.

(d) Recall bias: This bias is the underreporting of true exposures in controls or overreporting of true exposures in cases. The potential for its occurrence exists whenever past self-report information is required by participants. Although this bias is given a lot of attention in epidemiologic textbooks, the research literature on this subject is sparse (79).

(e) Measurement bias: This is the inaccurate or misclassification of subjects on a study variable such as smoking category or exposure. Bias due to misclassification of exposure is generally considered to reduce risk toward the null (80). Greenberg et al. (81) showed that misclassification occurring in a nuclear worker noncancer group produced a substantial bias away from the null hypothesis of no association between cancer and nuclear work.

4. In analyzing the data

(a) Data dredging bias: If there is no hypothesis, significant associations are suitable for hypothesis-forming activities but should be confirmed in specific studies of different cohorts to test the hypothesis. So-called "hypothesis-generating research" presents a hazard of interpreting a statistical association as causative when it is not. Such an association may be due to factors overlooked during the data dredging or may be simply accidental because of the numerous associations examined. (See References 25 and 42 for more discussion of this bias.)

(b) Managing unknown data: Unknown data should not be used as evidence of nonexposure, and the report should indicate how unknown data are handled. The treatment of lost-to-follow-up in a cohort study can affect the apparent forces of mortality. Methods of dealing with those lost-to-follow-up include counting them as lost at the time of loss, assuming they are alive till the end of follow-up, and assuming them dead at loss or at the end of follow-up. The last two methods result in abnormally high SMRs. The second method is considered the most conservative and is a common method in many cohort studies. However, considering subjects lost to follow-up at the time of loss may be the least biased estimate of expected mortality (82, 83).

(c) Retroactive demarcation of exposure: Odds ratios may show an eightfold change in the same data set by changing the exposure criteria. To avoid this, bias exposure should be set before analysis of the data, and there should be no changes in exposure categories after collection of the data (25).

5. In interpreting the analysis

(a) Significance bias: This bias occurs when there is a confusion about the statistical (or stochastic) significance of a result and the quantitative (or biological) significance. If n is too small, a biologically important difference may not be statistically significant. If n is quite large, a biologically insignificant difference may appear to be statistically significant. For whatever reason, statistical significance is commonly set at $p < .05$. There are no accepted standards for biologically significant differences (25).

(b) Correlation bias: Correlation should not be equated with causation. One might avoid being misled by correlation coefficients (r) by examining a scattergraph of the data, considering r^2, which indicates the proportion of the variance that has been reduced by the linear model, and remembering that correlations from relatively small sample sizes can easily achieve statistical significance (25).

(c) Bias of interpretation: This error arises from failure to consider all interpretations consistent with the facts and failure to assess the credentials of each. One of the major advances of scientific method in the nineteenth and twentieth centuries is the obligation of investigators to rule out alternative hypotheses that might explain the observed results.

Appendix 2

GLOSSARY*

Association: An association is when events occur together more frequently than would be expected to occur by chance. For example, a *statistical association* exists when the occurrence of disease in two groups is statistically significant. A statistical association does not necessarily imply a causal relationship.

*See also Last (84).

An *artificial* or *spurious association* occurs when a statistically significant relationship between two events is due to chance, creating a type I error. One implication of probability is that a certain proportion of outcomes will be declared statistically significant even though due to random fluctuation. Artificial associations can also be a result of bias, flaws in methods, flaw in study design, and/or nonrepresentative selection of study groups.

A *noncausal* (indirect) *association* is when there is a statistically significant relationship between an exposure and health outcome, but the association occurs because of a common relationship to an underlying condition. For example, the miasma theory hypothesized that cholera death was due to bad air and was supported by the inverse relationship between altitude and cholera death. However, bad air and low altitudes were also associated with contaminated water, and the true association with contaminated water was missed.

A *causal association* is when A causes B and can occur only if (1) A occurs before B, (2) a change in A is correlated with a change in B, and (3) the correlation is not due to confounding; that is, the correlation is not due to both A and B being correlated to some factor C. A number of criteria are widely used to help evaluate the likelihood of a causal association. These include strength of association, exposure–response relationship, a consistent association in different studies, a correct temporal sequence of events (A occurs before B), and biological plausibility.

Attributable Risk or Risk Difference: This is a measure to estimate the magnitude of a given cause (referred to in epidemiology as an exposure). This is an absolute effect, namely, differences in incidence or prevalence rates between exposed and nonexposed populations. It should be used only when there is a cause–effect relationship between exposure and disease. It is a useful measure to estimate the magnitude of the public health problem caused by the exposure.

Bias: A bias is an effect that can occur at any stage of an investigation or inference. The effect of the bias is to produce results that depart systematically from the truth. We have described some biases in Appendix 1.

Confounding: Confounding is a central issue in nonexperimental research and is a mixing of the effect of exposure with the effect of some extraneous factor, thereby distorting the effect one is trying to estimate. For example, if lung cancer is the effect or response and asbestos the exposure, a confounder could be smoking if the exposed asbestos workers have a higher (or lower) prevalence of smokers than the nonexposed control group. For an extraneous factor to be a confounder, two criteria must be met:

1. The extraneous factor must be associated with the exposure under study. That is, the occurrence of the factor must be "different" in exposed and control groups.
2. The extraneous factor must be associated with the disease under study. The factor must be predictive of the occurrence of the disease but not necessarily causal. Thus a lower social class is predictive of increased cancer risk, not

because it causes cancer, but because it is correlated with other factors that do cause cancer.

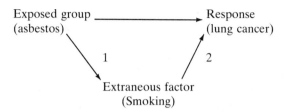

One could remove the confounding effect of smoking by stratifying exposed and nonexposed groups, that is, comparing asbestos workers who smoked with controls who smoked. Then smoking would not be associated with exposure.

Dose–response Relationship: More correctly termed *exposure–response*, this is a relationship in which a change in exposure is associated with a change in the risk of a specified outcome (response).

Exposure is an estimate of dose. In occupational epidemiology a common surrogate of dose is years worked (tenure). If quantitative estimates of exposure are available (such as mg/m^3, dust-years, fibers/cm^3-years, million particles per cubic foot-years or mppcf-years, and ppm-years) then cumulative exposure can be estimated. Cumulative exposure is defined as (concentration × time) and so cumulative exposure is estimated by multiplying the estimated exposure in each job times the time worked in that job (such as mg/m^3 × years exposed) and adding the results over the life work history of each worker. This is not a measure of "dose" because the relationship between the external level of contact with the exposure agent(s) and contact at the critical site for each individual may be quite different from one individual to another. "Dose" also varies depending on properties of the agent, circumstance of exposure, performance characteristic of portal of entry, pathways of transport to the critical site, and sampling method (85).

The main point of an epidemiologic analysis is to estimate the magnitude of the response as a function of exposure status (32). This cannot be accomplished by determining if average exposures of, say, cases and controls are statistically different. The most desirable analysis is to determine response rates at several levels of exposure. Estimation of effects at every level is possible, but such a separate analysis at each exposure level results in lost information because the continuity of the underlying variables is not considered and because of a potential loss in precision due to fewer data in each exposure strata. If there is an exposure–response relationship, the estimates of effect in bordering exposure categories provide additional information such that the effect may be characterized in an exposure category that has few data.

Exposure–response relationships are among the most important criteria for inferring causality. Statistical hypothesis testing using central measures of exposure are not the preferred measure of exposure–response because this method suffers from loss of information and is more susceptible to confounding. It is more desirable to estimate an overall measure of trend in response at various levels of exposure rather than a separate estimate of overall effect.

Two types of exposure–response curves are postulated (Figure 26.5). Curve A shows a sigmoid relationship that implies a *threshold*; below some low exposure level there is no detectable response. Curve B shows a straight-line relationship between exposure 2 response and implies *no threshold*. This is the popular one-hit model for exposure–response relationships of carcinogens, where even one molecule of a substance results in a response.

Latent Period (Latency, Induction Period): This is the time between exposure to a disease-causing agent and the appearance of the disease. For chronic disease this can be years. For example, the latent period for lung cancer due to occupational exposure is generally considered to be at least 20 years.

Null Hypothesis: This is a statement about causation phrased in the negative, such as "smoking is not a cause of lung cancer." If the p value is statistically significant (say, $p < .05$), then we reject the hypothesis that exposure does not cause the effect.

Odds Ratio: Commonly used in connection with case-control data, the odds ratio (OR) is the ratio of the odds in favor of exposure among cases (a/b) to the odds in favor of exposure among controls or noncases (c/d):

$$OR = \frac{a/b}{c/d} = \frac{a/d}{b/c}$$

	Exposed	Not Exposed	OR
Cases	$a = 30$	$b = 20$	1.5
Controls	$c = 20$	$d = 20$	1

$$OR = \frac{ad}{bc} = \frac{20 \cdot 30}{20 \cdot 20} = 1.5$$

Thus the cases are at 1.5 times greater risk of exposure than the controls.

Proportional Mortality Ratio (PMR): Number of deaths from a given cause during a given time period divided by the total number of deaths during the same time period times 100; [(number of deaths)/(total deaths)] $\times$ 100.

Significance Testing (Statistical Inference versus Scientific Inference): Feinstein (6) has called "significance" the "single greatest intellectual pathogen in both biological and statistical domains today." There are two types of significance; the one used almost exclusively is "statistical significance." This is a measure of the probability that an observed difference is due to chance. It is an assessment of random variability and involves primarily a test to reject a *null hypothesis* of no association between two variables and, by exclusion, accept the alternative hypothesis. The p value is a quantification of error in the decision process. If p is set by the investigator at .05, then when $p < .05$ the observation is "significant" and the null hypothesis

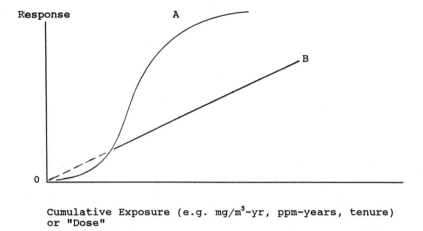

Cumulative Exposure (e.g. mg/m³-yr, ppm-years, tenure) or "Dose"

Figure 26.5. Hypothetical exposure–response curves. *A*, Threshold, sigmoid exposure–response curve. *B*, No threshold, linear exposure–response curve.

is rejected; about 5 percent of the time the hypothesis will be rejected when true. This kind of "decision" is both misleading and counterproductive because:

- The results of a single study virtually never provide the only basis for a decision in epidemiology.
- Statistical testing is derived from experiments set up to provide a choice between two or more alternatives of action. Observational studies are not experiments formed to give "mathematical answers to structured questions" (63).
- Statistical testing does not account for biases, quality of data, lack of follow-up, and all other inherent problems in observational studies. The *p* value addresses only the likelihood of falsely rejecting the null hypothesis, and has nothing to do with the chain of reasoning that must precede a conclusion.
- Statistical significance is a function of the number of observations and the variability of the observations. One can make any difference "statistically significant" by making *n* large enough. It is only when *"scientific significance"* (or *"biological significance"*) is apparent that statistical significance becomes important (86, 87).
- "Chance as an explanation of an apparent hazard is most effectively eliminated by repetition of the study on another population or over a further period. It can seldom be eliminated on the basis of a test of significance alone, for such tests are only guides to interpretation and never absolute arbiters" (65).

Scientific (or biological) *significance* refers to the quantitative difference in the observations. For example, a mean difference of 1 in. in height between smokers and nonsmokers may be statistically significant, but its quantitative significance is trivial. (See also Reference 32.)

Table 26.8. Circumstances when Type I and Type II Errors Occur

	True Biological Significance	Observed Biological Significance	Calculated Statistical Significance ($p < .05$)	Agreement of Biological and Statistical Significance	Agreement of Statistical Conclusion and "Truth"
1	Yes	Yes	No	No	Type II error
2	Yes	No	No	Yes	Type II error
3	No	Yes	Yes	Yes	Type I error
4	No	No	Yes	No	Type I error

STANDARDIZATION: the weighted averaging of the stratum-specific rates according to an artificial or standard distribution for the factor of interest. The artificial distribution (known as the standard) migh be the exposed population (indirect method), comparison population (direct method), or the combined populations. For example, the standardized mortality ratio (SMR) is the ratio of observed deaths in the study population (stratified by age, race, and sex) to the expected number of deaths based on mortality rates in the comparison population stratified in the same manner.

Stratification: To control for possible effects of confounding, the comparison is between samples of exposed and controls separated into subsamples according to specified criteria such as age, race, sex, and smoking status. By stratifying in the analysis of results one compares "like with like." After stratifying the data into selected strata to compare strata-specific rates and thereby reducing the effects of unfair comparisons, a standardized rate can be calculated.

TYPE I: Alpha errors occur when the null hypothesis is rejected even though it is true. When the p value is set at .05, the likelihood of this error will be about 5 percent.

TYPE II: Beta errors occur when the null hypothesis is erroneously accepted. As the criteria for statistical significance is made more stringent (i.e., as the "significant" p value is made smaller) the probability of making a type II error increases. The circumstances when type I and II errors occur can be visualized as shown in Table 26.8. In the first instance there is a true biologically significant effect but not a statistically significant effect. The reason for this error may be that the sample size is too small.

The fear in the second instance is that one may miss a truly significant difference. The finding of no statistical and no biological significance could be due to a small sample size. The lack of quantitative difference could be due to a flaw in the research design or to a small sample size that did not detect a true effect because of the low precision.

In the third instance sample size is not the problem for there is statistical significance. The error is caused by distorted results such as bias or confounding, or because of the 1 out of 20 times the probabilities are wrong.

In the fourth instance statistical significance is achieved despite the lack of biological significance. This may occur because of the huge sample size. (See References 25 and 32 for further discussion.)

REFERENCES

1. D. E. Lilienfeld, *Am. J. Epidemiol.*, **107**, 87–90 (1978).
2. G. Rose and D. J. P. Barker, *Br. Med. J.*, **2**, 803–804 (1978).
3. A. B. Hill, *Proc. Roy. Soc. Med.*, **58**, 295–300 (1965).
4. K. J. Rothman, Ed., *Causal Inference*, Epidemiology Resources, Inc., Chestnut Hill, MA, 1988.
5. M. J. Gardner and D. G. Altman, *Br. Med. J.*, **292**, 746–750 (1986).
6. A. R. Feinstein, *Science*, **242**, 1257–1263 (1988).
7. R. M. Royall, *Am. J. Epidemiol.*, **104**, 463–474 (1976).
8. J. R. Jamison, *S. Afr. Med. J.*, **57**, 783–785 (1980).
9. W. Weiss, *J. Occup. Med.*, **25**, 290–294 (1983).
10. C. P. Wen and S. P. Tsai, *Scand. J. Work. Environ. Health*, **8**, Suppl. 1, 48–52 (1982).
11. E. S. Gilbert, *Am. J. Epidemiol.*, **116**, 177–188 (1982).
12. J. C. McDonald, F. D. K. Liddell, G. W. Gibbs, G. E. Eyssen, and A. C. McDonald, *Br. J. Ind. Med.*, **7**, 11–24 (1980).
13. J. D. Wang and O. S. Miettinen, *Scand. J. Work Environ. Health*, **8**, 153–158 (1982).
14. J. M. Harrington, *Scand. J. Work. Environ. Health*, **10**, 347–352 (1984).
15. M. L. Newhouse and J. C. Wagner, *Br. J. Ind. Med.*, **26**, 302–307 (1969).
16. E. S. Johnson, *Br. J. Ind. Med.*, **43**, 427–429 (1986).
17. L. N. Kolonel, T. Hirohata, and Nomura, *Am. J. Epidemiol.*, **106**, 476–484 (1977).
18. J. K. McLaughlin, M. S. Dietz, E. S. Mehl, and N. J. Blot, *Am. J. Epidemiol.*, **126**, 144–146 (1987).
19. M. Baumgarten, J. Siemiatycki, and G. W. Gibbs, *Am. J. Epidemiol.*, **118**, 583–591 (1983).
20. S. L. Shalat, D. C. Christiani, and E. L. Baker, *Scand. J. Work. Environ. Health*, **13**, 67–69 (1987).
21. W. F. Stewart, J. A. Tonascia, and G. M. Matanoski, *J. Occup. Med.*, **29**, 795–800 (1987).
22. G. G. Bond, K. M. Bodner, W. Sobel, R. J. Shellenberger, and G. H. Flores, *Am. J. Epidemiol.*, **128**, 343–351 (1988).
23. K. Steenland and T. Schnorr, *J. Occup. Med.*, **30**, 348–353 (1988).
24. M. L. Lerchen and J. M. Samet, *Am. J. Epidemiol.*, **123**, 481–489 (1986).
25. A. R. Feinstein, *Clinical Epidemiology*, Saunders, Philadelphia, 1985.
26. P. E. Enterline, *J. Occup. Med.*, **18**, 150–156 (1976).
27. H. Checkoway, N. Pearce, and J. M. Dement, *Am. J. Ind. Med.*, **15**, 363–373 (1989).

28. H. Checkoway, N. Pearce, and J. M. Dement, *Am. J. Ind. Med.*, **15**, 375–394 (1989).
29. F. D. K. Liddell, *J. Clin. Epid.*, **41**, 1217–1237 (1988).
30. K. Steenland, L. Staynes, and A. Griefe, *Am. J. Ind. Med.*, **12**, 419–430 (1987).
31. S. P. Tsai and C. P. Wen, *Int. J. Epidemiol.*, **15**, 8–21 (1986).
32. K. J. Rothman, *Modern Epidemiology*, Little, Brown, Boston, 1986.
33. E. S. Johnson, *Br. J. Ind. Med.*, **43**, 427–429 (1986).
34. J. Hanley and F. D. F. Liddell, *J. Occup. Med.*, **27**, 555–560 (1985).
35. G. M. H. Swaen and J. M. M. Meijers, *Br. J. Ind. Med.*, **45**, 624–629 (1988).
36. J. Wang and O. Miettinen, *Scand. J. Work. Environ. Health*, **8**, 153–158 (1982).
37. C. M. J. Bell and D. A. Coleman, *Stat. Med.*, **2**, 363–371 (1983).
38. S. Hernberg, *Scand. J. Work. Environ. Health*, **7**, Suppl. 4, 121–126 (1981).
39. N. Pearce, H. Checkoway, and J. Dement, *Am. J. Ind. Med.*, **15**, 395–402 (1989).
40. N. Pearce, H. Checkoway, and J. Dement, *Am. J. Ind. Med.*, **15**, 403–16 (1989).
41. G. M. Marsh and P. E. Enterline, *J. Occup. Med.*, **21**, 665–670 (1979).
42. D. C. Thomas, J. Siemiatycki, R. Dewar, J. Robins, M. Goldberg, and B. G. Armstrong, *Am. J. Epidemiol.*, **122**, 1080–1095 (1985).
43. S. Greenland and H. Morgenstern, *J. Clin. Epidemiol.*, **41**, 715–716 (1988).
44. O. Axelson, *Scand. J. Work. Environ. Health*, **5**, 91–99 (1979).
45. R. I. Horwitz and A. R. Feinstein, *Am. J. Med.*, **66**, 556–564 (1979).
46. A. R. Feinstein, *J. Chronic Dis.*, **38**, 127–133 (1985).
47. P. L. Sackett, *J. Chronic Dis.*, **32**, 51–63 (1979).
48. A. J. Fox and G. C. White, *Lancet*, **1**, 1009–1010 (1976).
49. C. K. Wells and A. R. Feinstein, *Am. J. Epidemiol.*, **128**, 1016–1026 (1988).
50. L. Gordis, *Am. J. Epidemiol.*, **115**, 1–5 (1982).
51. J. K. McLaughlin, W. J. Blot, E. S. Mehl et al., *Am. J. Epidemiol.*, **121**, 131–139 (1985).
52. J. K. McLaughlin, W. J. Blot, E. S. Mehl, and J. S. Mandel, *Am. J. Epidemiol.*, **122**, 485–494 (1985).
53. A. H. Smith, N. E. Pearce, and P. W. Callas, *Am. J. Epidemiol.*, **17**, 298–306 (1988).
54. M. S. Linet and R. Brookmeyer, *Am. J. Epidemiol.*, **125**, 1–11 (1987).
55. P. A. Hessel and G. K. Sluis-Cremer, *Int. Arch. Occup. Environ. Health*, **59**, 97–105 (1987).
56. A. R. Feinstein, *J. Chronic Dis.*, **32**, 35–41 (1979).
57. H. W. Glindmeyer, J. E. Diem, R. N. Jones et al., *Am. Rev. Resp. Dis.*, **125**, 544–548 (1982).
58. *Smoking and Health: Report of the Advisory Committee to the Surgeon General*, Public Health Service, Washington, DC, 1964.
59. H. S. Sacks, J. Berrier, D. Reitman et al., *New Engl. J. Med.*, **316**, 450–455 (1987).
60. M. Susser, "Falsification, Verification and Casual Inference in Epidemiology: Reconsiderations in the Light of Sir Karl Popper's Philosophy," in Reference 32.
61. A. B. Hill, *Principles of Medical Statistics*, 9th ed., Oxford University Press, New York, 1971.
62. S. F. Lanes and C. Poole, *J. Occup. Med.*, **26**, 571–574 (1984).

63. M. Susser, *Am. J. Epidemiol.*, **105**, 1–15 (1977).

64. P. Cole, *J. Chronic Dis.*, **32**, 15–28 (1979).

65. R. Doll, *Ann. Occup. Hyg.*, **28**, 291–305 (1984).

66. C. B. Begg and J. A. Berlin, *J. Roy. Stat. Soc., Ser.* A **151**, 419–463 (1988).

67. J. A. Freiman, T. C. Chalmers, H. Smith, and R. R. Kuebler, *N. Engl. J. Med.*, **299**, 690–694 (1978).

68. J. D. Gillam, J. M. Dement, R. A. Lemen et al., *Ann. NY Acad. Sci.*, **271**, 336–352 (1976).

69. J. C. McDonald, G. W. Gibbs, F. D. K. Liddell et al., *Am. Rev. Resp. Dis.*, **118**, 271–277 (1978).

70. D. P. Brown, S. D. Kaplan, R. D. Zumwalde et al., "Retrospective Cohort Mortality Study of Underground Gold Mine Workers," in D. F. Goldsmith, D. M. Winn, and C. M. Shy, Ed., *Silica, Silicosis, and Cancer*, Praegle, New York, 1986.

71. B. Williams, "How Bad Can "Good" Data Really Be?," presented at American Statistical Association Meeting, Atlanta, GA, August, 1975.

72. A. J. McMichael, *J. Occup. Med.*, **18**, 165–168 (1976).

73. L. M. Carpenter, *Br. J. Ind. Med.*, **44**, 289–291 (1987).

74. W. Weiss, *J. Occup. Med.*, **31**, 102–105 (1989).

75. P. E. Enterline, *Am. Rev. Resp. Dis.*, **113**, 175–180 (1976).

76. W. G. B. Graham, R. V. O'Grady, and B. Dubuc, *Am. Rev. Resp. Dis.*, **123**, 25–28 (1981).

77. G. P. Theriault, J. M. Peters, and W. M. Johnson, *Arch. Environ. Health*, **28**, 23–27 (1974).

78. S. J. Gould, *Science*, **200**, 503–509 (1978).

79. K. Raphael, *Int. J. Epidemiol.*, **16**, 167–170 (1987).

80. K. T. Copeland, H. Checkoway, and A. J. McMichael, *Am. J. Epidemiol.*, **105**, 488–495 (1977).

81. E. R. Greenberg, B. Rosner, C. Hennekens, R. Rinsky, and T. Colton, *Am. J. Epidemiol.*, **121**, 301–308 (1985).

82. J. E. Vena, H. A. Sultz, G. L. Carlo, R. C. Fiedler, and R. E. Barnes, *J. Occup. Med.*, **29**, 256–261 (1988).

83. E. S. Johnson, *J. Occup. Med.*, **30**, 60–62 (1988).

84. J. M. Last, *A Dictionary of Epidemiology*, Oxford University Press, New York, 1983.

85. T. F. Hatch, *Arch. Env. Health*, **16**, 571–578 (1968).

86. C. Poole, *Am. J. Public Health*, **77**, 195–199 (1987).

87. D. R. Cox, *Scand. J. Stat.*, **4**, 49–70 (1977).

CHAPTER TWENTY-SEVEN

Industrial Hygiene Sampling and Analysis

Robert D. Soule, C.I.H., C.S.P.

1 INTRODUCTION

In the decade that has passed between the third and fourth editions of this book, several changes have taken place, quite subtly, that will have long-lasting effects on the practice of industrial hygiene. Perhaps the two most significant of these changes have been the infusion of computer assistance and the trend toward "generic" government regulations.

An integral part of industrial hygiene sampling and analysis is the proper utilization of data. The use of computers has greatly facilitated the handling of such data. Chapter 25, Industrial Hygiene Records and Reports, provides valuable information on this subject. In Volume 1A, Chapter 6, entitled Hazard Communication and Worker Right-To-Know Programs, presents information on the requirements of supplying employees with data obtained through sampling and analysis.

Reasons for conducting industrial hygiene sampling include (1) identification and quantification of specific contaminants present in the environment, (2) the determination of exposures of workers in response to complaints, (3) assessment of compliance status with respect to various occupational health standards, and (4) evaluation of the effectiveness of engineering controls installed to minimize workers' exposures. The reasons for sampling dictate to some extent the sampling strategy that should be used. Sampling can be conducted in a grid pattern throughout a plant to document the environmental characteristics of the workplace, or

Patty's Industrial Hygiene and Toxicology, Fourth Edition, Volume 1, Part B, Edited by George D. Clayton and Florence E. Clayton.
ISBN 0-471-50196-4 © 1991 John Wiley & Sons, Inc.

personal breathing zone samples can be obtained to document actual exposure conditions. The substances being evaluated determine the type of sampling devices to be used, and the analytical requirements will specify time and perhaps flow rate of sampling. The occupational health standard will indicate whether continuous or grab sampling is required. In short, consideration must be given to a number of questions pertaining to the fundamental purpose of the sampling.

Many analytical methods available to the industrial hygienist have been so standardized and simplified that they require relatively little experience. On the other hand, many seemingly simple tests call for a basic understanding of solubility and gas laws, partial pressures, and chemical reactions. In many instances questions arising from such considerations can be answered only by qualified specialists. The ultimate methods of analysis to be used depend on the problem at hand rather than mere application of a "standard method." The trend in recent years has been toward development of methods that give relatively prompt results with a high degree of accuracy. The latter aspect has increased in importance because of legal significance given to occupational health standards, particularly as promulgated under authority of the Occupational Safety and Health Act of 1970. The National Institute for Occupational Safety and Health (NIOSH) and the Occupational Safety and Health Administration (OSHA) have developed specific methods for sampling and analyzing many atmospheric contaminants in the workplace. The objective of these procedures typically is an accuracy of at least ± 25 percent with 95 percent confidence at the permissible exposure limit.

This chapter presents a development of proper strategy for sampling in the workplace, statistical bases for industrial hygiene sampling, sampling techniques for gaseous and particulate contaminants, summaries of techniques available for analyzing atmospheric samples, and a discussion of biological monitoring as it relates to industrial hygiene sampling. Relatively complete discussions of detailed methods of analysis for specific contaminants are presented elsewhere in this series (1). It is interesting to note that many of the early descriptions of industrial hygiene sampling/analysis (2, 3) still serve as a foundation for modern, more "sophisticated" methods.

2   GENERAL CONSIDERATIONS

The magnitude of chemical and physical stresses can be evaluated in various ways. One form of evaluation is qualitative, that is, using one or more of the human senses without taking any actual measurements. This kind of inspection and evaluation of a work situation is very beneficial, particularly when done by an experienced industrial hygienist. Another form of evaluation is quantitative, that is, involving collection and analysis of samples representative of actual workers' exposures. Generally, this type of evaluation is most desirable and necessary in many cases, particularly when the purpose of the sampling is to determine compliance with occupational health standards or to form the basis for designing engineering controls.

2.1 Preliminary Survey

An experienced, professional industrial hygienist often can evaluate, quite accurately and in some detail, the magnitude of chemical and physical stresses associated with an operation without benefit of any instrumentation. In fact, professionals use this qualitative evaluation every time they make a survey, whether it is intended to be the total effort of their work or a preliminary inspection prior to actual sampling and analysis of potential stress. Qualitative evaluation can be applied by anyone familiar with an operation, from the worker to the professional investigator, to ascertain some of the potential problems associated with work activities.

The first step in evaluating the occupational environment is to become as familiar as possible with particular operations. The person evaluating the operation should be aware of the types of industrial process and the chemical raw materials, by-products, and contaminants encountered. The evaluator should also know what protective measures are provided, how engineering controls are being used, and how many workers are exposed to contaminants generated by specific job activities.

The number of chemical and physical agents capable of producing occupational illnesses continues to increase. New products that require the use of new raw materials or new combinations of familiar substances are continually being introduced. This is particularly true in the chemical industries, where new chemicals and products and the operations for their processing are being developed. It is important that responsible industrial hygienists establish and maintain a list of the chemical and physical agents encountered in their particular area of jurisdiction. In fact, OSHA's hazard communication standard, and many state "right-to-know" laws, make such inventorying a legal obligation. The composition of the products and by-products and as many as possible of the associated contaminants and "undesirables" should be known. This means that the industrial hygienist must obtain complete information on the composition of various commercial products. In most instances the desired information can be obtained from descriptive material provided by the suppliers in the form of a Material Safety Data Sheet. Similarly, the labels on the containers of the material should be read carefully. Although there are explicit requirements under hazard communication and right-to-know regulations, labels still do not always give complete information, and further investigation of the composition of the materials is necessary.

After the inventory is obtained, it is necessary to determine the toxicity and other hazardous properties of the chemicals. Information of this type can be found in several excellent reference texts on toxicology and industrial hygiene (4–8).

During a qualitative walk-through evaluation, many potentially hazardous operations can be detected visually. Operations that produce large amounts of dusts and fumes can be spotted. However, "visible" does not necessarily mean "hazardous"; airborne dust particles that *cannot* be seen by the unaided eye normally are more hazardous, because they are more likely to be inhaled into the lungs. Concentrations of dust of respirable size usually must reach extremely high levels before they are visible. Thus the absence of a visible cloud of dust is not a guarantee that a "dust-free" atmosphere exists. However, operations where activities generate dust that can be spotted visually are likely to warrant implementation of additional controls.

In addition to sight, the sense of smell can be used to detect the presence of many vapors and gases. Trained observers are able to estimate rather accurately the concentrations of various gases and solvent vapors present in the workroom air. For many substances the odor threshold concentration, that is, the lowest concentration that can be detected by smell, is greater than the permissible exposure level. In these cases, if the substances can be detected by their odors, excessive levels are indicated. However, many substances, notably hydrogen sulfide, can cause olfactory fatigue, that is, anesthesia of the olfactory nerve endings, to the extent that even dangerously high concentrations cannot be detected by odor. A detailed presentation of evaluation and control of "odors" is contained elsewhere in this book (see Chapter 41).

Although it is usually possible to determine the presence or absence of potentially hazardous physical agents at the time of the qualitative evaluation, rarely can the potential hazard be evaluated without the aid of special instruments. As a minimum, however, the sources of physical agents such as radiant heat, abnormal temperatures and humidities, excessive noise, improper or inadequate illumination, ultraviolet radiation, microwaves, and various other forms of radiation can be noted.

An important aspect of the qualitative evaluation is an inspection of the types of control measure in use at a particular operation. In general, the control measures include such features as shielding from radiant or ultraviolet energy, local exhaust and general ventilation provision, respiratory protection devices, and other personal protective measures. General indexes of the relative effectiveness of these controls are the presence or absence of accumulated dust on floors, ledges, and other work surfaces, the condition of ductwork for the ventilation systems, that is, whether there are holes or badly damaged sections of ductwork, whether the fans for ventilation systems appear to provide adequate control of contaminants generated by the process, and the manner in which personal protective measures are accepted and used by the workers.

2.2   Representative Quantitative Surveys

Although the information obtained during a qualitative evaluation or walk-through inspection of a facility is important and always useful, only by measurement can the hygienist document the actual level of chemical or physical agent associated with a given operation. Of course, the strategy used for any given air sampling program depends to a great extent on the purpose of the study. The specific objectives of any sampling program may include one or more of the following:

a. Provide a basis on which unsatisfactory or unsafe conditions can be detected and the sources identified
b. Assist in designing controls
c. Provide a chronicle of changes in operational conditions

d. Provide a basis for correlating disease or injury with exposure to specific stresses
e. Verify and assess the suppression of contaminants by methods designed to do so
f. Document compliance with health and safety regulations

These objectives can be condensed into two major categories:

a. Sampling for industrial health engineering surveillance, testing, or control
b. Sampling for health research or epidemiological purposes

A sampling program for engineering purposes should be designed to yield the specific information desired. For example, one might need only single samples before and after a change in ventilation to determine whether the change has had the desired effect. On the other hand, industrial hygiene primarily is directed at predicting the health effects of an exposure by comparing sampling results with hygienic guides, determining compliance with health codes or regulations, or defining as precisely as possible environmental factors for comparison with observed medical efforts.

Regardless of the objective or objectives of the sampling program, the investigating industrial hygienist must answer the following questions to be able to implement the correct strategy.

1. Where should samples be obtained?
2. Whose work area should be sampled?
3. For how long should the samples be taken?
4. How many samples are needed?
5. Over what period of work activity should the samples be taken?
6. How should the samples be obtained?

In answering these questions, the importance of adequate field notes must be emphasized. While the sample is being obtained, notes should be made of the time, duration, location, operations underway, and all other factors pertinent to the sample, and the exposure or condition it is intended to define. Printed forms with labeled spaces for essential data help to avoid the common failure to record needed information. Industrial hygiene record keeping is discussed in detail elsewhere; see Chapter 25.

2.2.1 Where to Sample

Three general locations are used for collection of air samples: at a specific operation, in the general workroom air, and in a worker's breathing zone. The choice of sampling location is dictated by the type of information desired, and combination of the three types of sampling may be necessary. Most frequently these days, the sampling is intended to determine the level of exposure of a worker or group of workers to a given contaminant throughout a work day. To obtain this type of information it is necessary to collect samples at the worker's breathing zone as well as in the areas adjacent to his particular activities. When the purpose of the survey is to determine sources of contamination, or to evaluate engineering controls, a strategic network of area sampling would be more appropriate.

2.2.2 Whom to Sample

Logically, samples should be collected in the vicinity of workers directly exposed to contaminants generated by their own activities. In addition, however, samples should be taken from the breathing zone of workers in nearby work areas not directly involved in the activities that generate the contaminant, and from those of workers remote from the exposure who have either complained or have reason to suspect that the contaminants have been drawn into their work areas.

2.2.3 Sample Duration

In most cases, minimum sampling time is determined by the time necessary to obtain an amount of the material sufficient for accurate analysis. The duration of the sampling period therefore is based on the sensitivity of the analytical procedure, the acceptable concentration of the particular contaminant in air, and the anticipated concentration of the contaminant in the air being sampled.

Preferably, the sampling period should represent some identifiable period of time of the worker's exposure, usually a minimum of one complete cycle of activity. This is particularly important in studying nonroutine or batch-type activities, which are characteristic of many industrial operations. Exceptions include operations that are highly automated and enclosed operations where the processing is done automatically and the operator's exposure is relatively uniform throughout the workday. In many cases it is desirable to sample the worker's breathing zone for the duration of the full shift. This is particularly important if sampling is being done to determine compliance status relative to occupational health standards.

Evaluation of a worker's daily time-weighted average exposure is best accomplished, when analytical methods permit, by allowing the person to work a full shift with a personal breathing zone sampler attached. The concept of full-shift integrated personal sampling is much preferred to that of short-term or general area sampling, if the results are to be compared to standards based on time-weighted average concentrations. When methods that permit full-shift integrated sampling are not applicable, time-weighted average exposures can be calculated from alternative short-term or general area sampling methods.

The first step in calculating the daily, time-weighted average exposure of a worker or a group of workers is to study the job descriptions obtained for the persons under consideration and to ascertain how much time during the day they spend at various tasks. Such information usually is available from the plant personnel office or foreman on the job. In many situations, the investigator must make time studies to obtain the correct information. Information obtained from plant personnel should be checked by the investigator because in many situations job activities as observed by the investigator do not fit official job descriptions. From this information and the results of the environmental survey, a daily, 8-hr, time-weighted average exposure can be calculated, assuming that a sufficient number of samples have been collected, or measurements obtained with direct-reading instruments under various plant operating conditions to represent accurately the "exposure profiles."

When sampling is done for the purpose of comparing results with airborne contaminants whose toxicologic properties warrant short-term and ceiling limit values, it is necessary to use short-term or grab sampling techniques to define peak concentrations and estimate peak excursion durations. For purposes of further comparison, the 8-hr, time-weighted average exposures can be calculated using the values obtained by short-term sampling.

2.2.4 Number of Samples

The number of samples needed depends to a great extent on the purpose of sampling. Two samples may be sufficient to estimate the relative efficiency of control methods, one sample being taken while the control method is in operation and the other while it is off. On the other hand, several dozen samples may be necessary to define accurately the average daily exposure of a worker who performs a variety of tasks. The number of samples also depends to some extent on the concentrations encountered. If the concentration is quite high, a single sample may be sufficient to warrant further action. If the concentration is somewhat near the acceptable level, a minimum of three to five samples usually is desirable for each operation being studied.

There are no set rules regarding the duration of sampling or the number of samples to be collected. These decisions usually can be reached quickly and reliably only after much experience in conducting such studies. A primer on the statistical bases of industrial hygiene sampling for compliance purposes is presented in Section 3.

2.2.5 When to Sample

The type of information desired and the particular operations under study determine when sampling should be done. If, for example, the operation continues for more than one shift, it usually is desirable to collect air samples during each shift. The airborne concentrations of chemicals or exposure to physical agents may be quite different for each shift. It usually is desirable to obtain samples during both summer and winter months, particularly in plants located in areas where large temperature variations occur during different seasons of the year. For one thing, there generally is more natural ventilation provided to dilute the airborne contaminants during the summer months than during the winter.

2.2.6   Choosing a Method

In general, the choice of instrumentation when sampling for a particular substance depends on a number of factors, including:

a. The portability and ease of use
b. The efficiency of the instrument or method
c. The reliability of the equipment under various conditions of field use
d. The type of analysis or information desired
e. The availability of the instrument
f. The "personal choice" of the industrial hygienist based on past experience and other factors

No single, universal air sampling instrument is available today, and it is doubtful that such an instrument will ever be developed. In fact, the present trend in the profession is toward a greater number of specialized instruments. The sampling instruments used in the field of industrial hygiene generally can be classified by type as follows: (*a*) direct-reading; (*b*) those that remove the contaminant from a measured quantity of air; and (*c*) those that collect a fixed volume of air. The three methods are listed in order of their general application and preference in use today.

Industrial hygienists must consider a proposed sampling program in relation to their own familiarity with the sampling and analytical method. As a rule, a method should never be relied on unless and until the hygienist has personally evaluated it under controlled conditions, such as by the following:

a. Sampling a synthetic atmosphere from a proportioning apparatus or gastight impervious chamber of sufficient size to permit creating and sampling mixtures without introducing significant errors
b. Introducing measured amounts of contaminants into a device attached to the sampling arrangement in a manner that utilizes the entire amount deposited
c. Comparing performance of the method with a device of proven performance by sampling from a common manifold over the same period of time

In other words, the sampling program must be preceded by appropriate calibration of all equipment to be used. This subject is discussed in greater detail in Chapter 16, Volume 1A.

Regardless of the sampling instrumentation selected for use in conducting industrial hygiene surveys, it is critical and imperative that the actual performance characteristics be known. This requires that various types of calibration be done periodically, or at any time that the performance of the device is questioned.

3   STATISTICAL BASES FOR SAMPLING

Since the implementation of the Occupational Safety and Health Act, statistical tests for noncompliance are being applied more consistently when environmental

Table 27.1. Decision Options for Compliance versus Noncompliance

TRUE STATE / ACTION	COMPLIANCE WITH STANDARD	NONCOMPLIANCE WITH STANDARD
DECLARE COMPLIANCE	No Error	Type II Error
DECLARE NONCOMPLIANCE	Type I Error	No Error

data are used to make a decision concerning a worker's exposure to a specific contaminant. The decision options concerning compliance and noncompliance are given in Table 27.1. In statistical terms, these options correspond to the null hypothesis that the worker is in compliance with the industrial hygiene standard. The detailed discussion of procedures is presented from the compliance officer's viewpoint. In this approach, samples are collected to see whether it is possible to reject the hypothesis that compliance exists with appropriate certainty. The type I error is the probability of declaring noncompliance given that the true state is compliance. This probability is a measure of any uncertainty that noncompliance does exist, and it should be kept small to ensure that noncompliance decisions are correct.

The employer's responsibility, on the other hand, is the protection of the employees. The employer's goal is to keep the type II error, which is the probability of declaring compliance when the true state is noncompliance, as small as possible. In statistical decision terms, therefore, the employer wants to assume that a given worker is in a state of noncompliance (null hypothesis). Then data are collected to show that the hypothesis can be rejected, with a goal of keeping as small as possible the probability of wrongly rejecting the null hypothesis.

Three questions are of primary importance when considering a statistical basis for industrial hygiene sampling. Over what time span should each sample be taken? How many samples should be obtained? At what times during the workday should the samples be obtained?

3.1 Duration of Sampling

For some types of sampling unit, such as the common colorimetric detector tubes, the sampling period is predetermined. In other cases, such as sampling for asbestos, the sampling period is defined fairly closely by the requirements of the analytical procedure. However, in most cases, the industrial hygienist has a choice over a wide range of sampling times from a few seconds to a complete work shift. The industrial hygienist's first and intuitive reaction might be to maximize the sampling period, expecting that this would increase significantly the reliability of the data, that is, that it would ultimately provide a "better answer." This is not true for short-term or "grab" samples, however. In such cases the primary consideration in arriving at sampling duration should be the requirements of the analytical method. Each analytical procedure requires some minimum amount of material for reliable analysis. This should be known in advance by the industrial hygienist, and the sampling period should be selected accordingly. Any increase in the duration of the sampling past this minimum time required to collect an adequate amount of material is both unnecessary and unproductive.

When attempting to make a decision on a possible noncompliance situation, as the statistical discussion in this chapter is directed, it is better to take shorter samples because this allows more samples to be taken in a given time period. It is much more important to collect several samples of short duration than to collect one medium-length (partial-shift) sample covering the same total sampling period. The random sampling and analytical errors can be averaged out, along with the longer-term environmental fluctuations during the sampling period, by taking a mean of random independent short-term samples. If sampling can be conducted over essentially 100 percent of the work period, either by a single sample or several consecutive samples, a better estimate of the true average exposure of the employee can be obtained. The sampling and analytical error must be contended with, and these are typically much smaller than the environmental fluctuations that affect short-term samples.

Thus there is a marked advantage in using a single, full-period sample (or several consecutive samples over the full work period) when attempting to demonstrate noncompliance with an occupational health standard. It is much more difficult to demonstrate noncompliance using the mean of several grab samples because the additional variability due to environmental fluctuations lowers the lower confidence limit (LCL) of the sampling result. This is demonstrated in Figure 27.1, which presents the effect of number of grab samples on statistical requirements for demonstrating noncompliance, using three levels of data variability, as expressed by the geometric standard deviation (GSD). For much industrial hygiene data the geometric standard deviations range between 1.5 and 2.5. Within this range, and for sample sizes of 3 to 10, it generally is necessary to obtain a measured mean exposure of 150 to 250 percent of the occupational health standard to demonstrate noncompliance.

If a full-period sample, or series of several consecutive samples, is used in attempting to demonstrate noncompliance, the degree to which the observed mean must exceed the occupational health standard is much lower. Figure 27.2 illustrates the effect of sample size for consecutive samples covering a full period on requirements for demonstrating noncompliance. The curves represent different sampling and analytical coefficients of variation (CVs) ranging from 0.05 to 0.25. Table 27.2 summarizes typical coefficients of variation. As Figure 27.2 indicates, for sample sizes of one to four it is necessary only that the average concentration exceed the occupational health standard by 5 to 14 percent to demonstrate noncompliance.

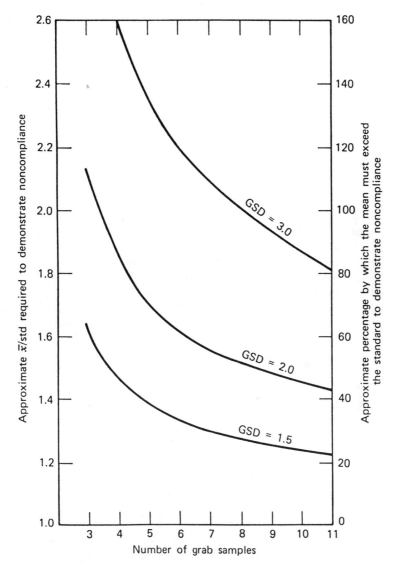

Figure 27.1. Effect of small grab sample sizes on requirements for demonstration of noncompliance; the three GSDs reflect the amount of variability in the environment.

3.2 Number of Samples

The second question of importance to the industrial hygienist is, "How many samples should be obtained?" This question is vital because it relates directly to the confidence that can be placed in the resulting estimate of the airborne con-

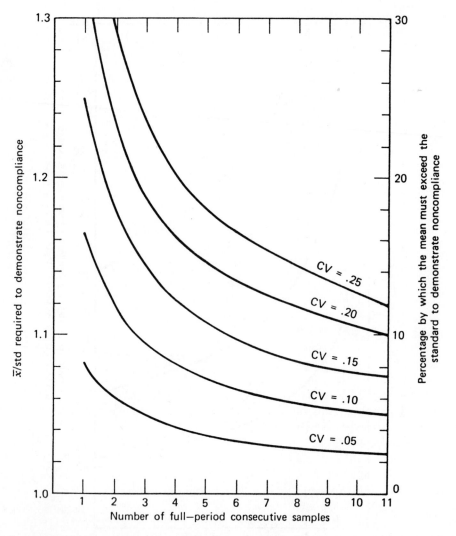

Figure 27.2. Effect of full-period consecutive sample size on requirements for demonstration of noncompliance.

centration of the contaminant in question, and subsequently the employee exposure. The effects of sample size on requirements for demonstrating noncompliance are illustrated in Figures 27.1 and 27.2. As these figures demonstrate, the curves for geometric standard deviations of 1.5 to 2.5 change relatively slowly after sample sizes of seven or eight (Figure 27.1), and a similar leveling off occurs in the curves relating the value of the "measured mean-to-standard" ratio necessary to demonstrate noncompliance versus sample size.

For full-period consecutive samples, Figure 27.2 shows that an appropriate num-

Table 27.2. Coefficients of Variation (CV) for Some Sampling–Analytical Procedures

SAMPLING / ANALYTICAL METHODS	C.V.
Asbestos (sampling and counting)	0.22
Charcoal Tubes (sampling and analytical)	0.10
Colorimetric Detector Tubes	0.14
Gross Dust (sampling and weighing)	0.05
Respirable Dust (sampling and weighing)	0.09

ber of samples is between four and seven. Practical considerations include costs of sampling and analyses, and the impossibility of running some long-duration sampling methods for longer than about 4 hr per sample. Thus most full-period consecutive sampling strategies result in at least two samples when an 8-hr, time-weighted average standard is being applied. As the curves demonstrate for a sampling and analytical technique with a coefficient of variation of 10 percent, the degree to which the estimate of the mean concentration must exceed the standard drops from 12 to 6 percent with an increase from two to seven samples. The relatively small decrease in percentage, with sample size exceeding seven, normally cannot be justified when compared to the time and effort required to obtain the additional samples. Thus on a cost–benefit basis, it can be concluded that two consecutive samples covering a full work period, for example, two consecutive 4-hr measurements, is the best number of samples to be obtained when comparison is made to an 8-hr, time-weighted average standard.

For grab samples, Figure 27.1 shows that the estimate of the mean exposure concentration must exceed the standard by unreasonably large amounts to demonstrate noncompliance when less than four grab samples are obtained. As discussed earlier, there is a point beyond which little is gained in attempting to reduce errors in the mean by taking more than seven grab samples. Because the level of variability in the mean of grab samples usually is much higher than for the same number of full-period samples, however, it might be necessary to take more than seven grab samples to attain the same level of precision afforded by even four or fewer full-period samples. Thus on the basis of the statistical criterion that can lead to economies in sampling by permitting reduction in the sampling effort with a calculable degree of confidence, it can be concluded that the optimal number of grab samples to be taken over the time period appropriate to the standard is between four and seven.

Figure 27.3 represents the procedure to be used when consecutive samples are obtained in a series over only a portion of the total period for which the standard applies; that is, the full work period is not included in the consecutive sampling

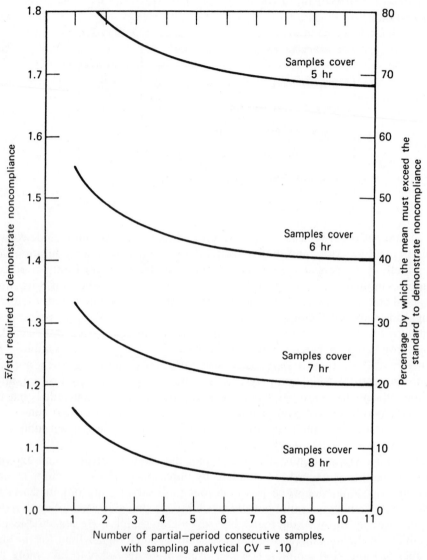

Figure 27.3. Effect of partial-period consecutive sample size on requirements for demonstration of noncompliance.

period. The effect of sample size and total time covered by all samples on the requirement for demonstrating noncompliance is shown, using a typical sampling and analytical coefficient of variation of 10 percent.

Obviously, the taking of partial-period consecutive samples is a compromise between the preferred full-period sampling and the less desirable approach with grab samples. Comparison of the curves in Figures 27.2 and 27.3 reveals that a geometric standard deviation of 2.5 (Figure 27.2) would be comparable to a curve of approximately 5.5 hr in Figure 27.3 for the same ratios of estimates of the mean to the standard necessary to judge noncompliance. Thus in general, if it is not possible to sample for at least 70 percent of the time period appropriate to the standard, for example, 5.5 hr for an 8-hr standard, it is better to use the grab sampling strategy.

3.3 "Grab" Sampling

The last of the three statistical questions to be answered concerns the periods of exposure during which grab sampling should be conducted. The accuracy of the probability level for the test depends on assumptions of the log-normality and independence of sample results that are averaged. These assumptions are not highly restrictive if precautions are taken to avoid any bias when selecting the sampling times over the period for which the standard is defined. Thus it usually is preferable to choose the sampling period in a statistically random manner. For a standard that is defined as a time-weighted average concentration over a period significantly longer than the sampling interval, an unbiased estimate of the true mean can be assured by taking samples at random. On the other hand, it is valid to sample at regular intervals if the contaminant level varies randomly about a constant mean, and any fluctuations are of short duration relative to the length of the sampling interval. If means and their confidence limits are to be calculated from samples taken at regularly spaced intervals, however, biased results could occur if the industrial operation being monitored were cyclic and in phase with the sampling periods. Results from random sampling would be valid nevertheless, even when cycles or trends occurred during the period of the standard. In this context, "random" refers to the manner of selecting the sample, and any particular sample could be the outcome of a random sampling procedure. A practical definition of random sampling is "a strategy by which any portion of the work shift has the same chance of being sampled as any other."

Strictly speaking, sampling results are valid only for the portion of the work period during which measurements were obtained. However, if it is not possible to sample during the entire workday or the entire length of a particular operation, professional judgment may permit inferences to be made about concentrations during other unsampled portions of the day. Reliable knowledge concerning the operation obviously is required to make these types of extrapolation.

3.4 Statistical Procedures

For those interested in, or required to use, statistical procedures, the following should be used to compare sampling results with the applicable occupational health standard. As is the case with other material in this section, the statistics have been oriented toward determining whether noncompliance with the time-weighted average, ceiling, or excursion standard exists.

3.4.1 Full-Period, Single Sample

The following procedure can be used to determine noncompliance with either a time-weighted average or a ceiling standard. It is used when only one sample is being tested. For a time-weighted average standard, the sample must have been taken for the entire period for which the standard is defined (usually 8 hr). The variability of the sampling, expressed either as a standard deviation or as a coefficient of variation, and the analytical methods used to collect and analyze the sample must be well known from previous measurements. The statistical test given is the "one-sided comparison-of-means" test using the normal distribution at the 95 percent confidence level.

Only if the lower confidence limit of the sample exceeds the standard is there 95 percent confidence that the true average concentration exceeds the standard, and thus that a condition of noncompliance exists.

The lower confidence limit can be expressed as

$$LCL = x - 1.645\sigma$$

where x = measurement being tested
    1.645 = critical standard normal deviate for 95 percent confidence
    σ = standard deviation of sampling/analytical method

If the coefficient of variation (CV) is known, the LCL can be computed from

$$LCL = x - [(1.645)(CV)(standard)]$$

The nomogram in Figure 27.4 can be used to aid this calculation. Some coefficients of variation are presented in Table 27.2.

3.4.2 Full-Period, Consecutive Samples

The procedure involving consecutive samples for a full period should be used to determine noncompliance with either a time-weighted average or a ceiling standard. That is, several consecutive samples are taken for the entire time period for which the standard is defined. If the samples do not cover the entire time period of the standard, refer to the procedure described in Section 3.4.3. The variability (standard deviation or coefficient of variation) of the sampling and analytical methods used to collect and analyze the samples must be well known from previous measurements. The statistical test given is the "one-sided comparison-of-means" test using the normal distribution at the 95 percent confidence level.

Only if the lower confidence limit of the mean of the consecutive samples exceeds the standard is there 95 percent confidence that the true average concentration exceeds the standard and that a condition of noncompliance exists.

$$LCL = \bar{x} - 1.645\sigma_x$$

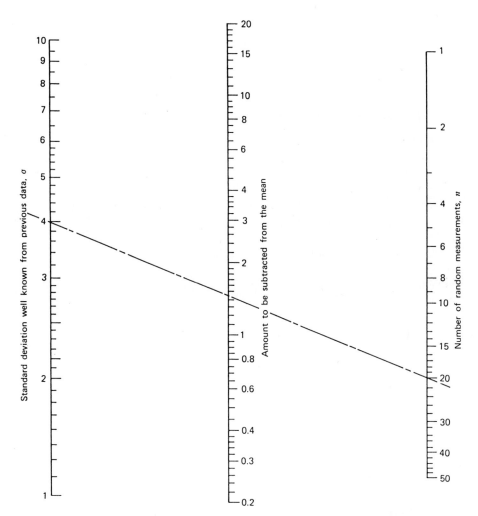

Figure 27.4. Nomogram for full-period or partial-period procedures using a well-known standard deviation; both σ and the amount can be multiplied by the same power of 10 (7).

where $\bar{x}$ = time-weighted average of n samples
$\qquad = (T_1x_1 + T_2x_2 + \cdots + T_nx_n)/T$
$\quad T_i$ = duration of ith sample
$\quad x_i$ = measurement of concentration in ith sample
$\quad T = T_1 + T_2 + \cdots + T_n$ = total duration for n consecutive samples

$$\sigma_x = \frac{\sigma}{T}(T_1^2 + T_2^2 + \cdots + T_n^2)^{1/2}$$

$$\quad\;\; = \frac{\sigma}{(n)1/2} \text{ if } T_1 = T_2 = \cdots = T_n$$

If the coefficient of variation is known, σ can be computed from

$$\sigma = (CV)(\text{standard})$$

Again, Figure 27.4 can be used to aid this calculation for the case of n equal-duration samples. Some coefficients of variation are given in Table 27.2.

3.4.3 Partial-Period, Consecutive Samples

One sample or a series of consecutive samples collected over less than the period for which a standard is defined is referred to as a "partial-period" sample. Because the concentration during the period not covered by the sample could not be less than zero, the full-period standard can be multiplied by a factor to obtain a conservative "partial-period standard." The lower confidence limit then can be calculated as in the preceding section and compared to the "partial-period standard."

Factor = actual time of sample/time period of standard

Thus for an 8-hr, time-weighted average standard, typical factors would be as follows:

Total Time of Sample(s) (hr)	Factor
8.00	1.000
7.75	1.032
7.50	1.067
7.25	1.103
7.00	1.143
6.75	1.185
6.50	1.231
6.25	1.280
6.00	1.333

3.4.4 "Grab Sampling" Data

If full-period samples of industrial contaminant concentrations are available, the best method of modeling the uncertainties of the result is with a normal distribution. When only a set of grab samples is available, however, the log-normal distribution best describes the uncertainties of the process. One procedure for analyzing grab samples by estimating the average concentration of a contaminant and making a decision on the level of the contaminant is that developed with NIOSH funding (9, 10). Advantages offered by the technique are listed here:

a. It is contaminant independent. Thus it becomes possible to use only a single decision chart for any contaminant.
b. It is capable of both the noncompliance and the "no-action" decision. Each

of these decisions is subject to a predetermined probability of type I or type II error. A type I error is said to occur if the noncompliance decision is wrongly asserted. A type II error is said to occur if the no-action decision is wrongly asserted.

c. The estimation and decision procedures are implemented by a single, straight-forward nomographic method. For estimation, the procedure yields the best estimate of the actual average contaminant level.

One disadvantage of the procedure is that the lower confidence limit is not directly computed. However, the simplicity of the calculations required and the plotting of a single point on the decision chart far outweigh any advantages that direct calculation of a lower confidence limit would yield. Modifications of this basic approach are being used by industrial hygienists today in a variety of employment settings.

4 SAMPLING FOR GASES AND VAPORS

For purposes of definition, a substance is considered to be a gas if at 70°F and atmospheric pressure, the normal physical state is gaseous. A vapor is the gaseous state of a substance in equilibrium with the liquid or solid state of the substance at the given environmental conditions. This equilibrium results from the vapor pressure of the substance causing volatilization or sublimation into the atmosphere. The sampling techniques discussed in this section are applicable to a substance in "gaseous" form, regardless of whether it is technically a gas or a vapor.

4.1 General Requirements

Particulate substances can be readily scrubbed or filtered from sampled airstreams because of the larger relative physical dimension of the contaminant and the operation and interaction of agglomerative, gravitational, and inertial effects. Gases and vapors, however, form true solutions in the atmosphere, thus requiring either sampling of the total atmosphere using a gas collector or the use of a more vigorous scrubbing mechanism to separate the gas or vapor from the surrounding air. Sampling reagents can be chosen to react chemically with the contaminants in the airstream, thus enhancing the collection efficiency of the sampling procedure. In the development of an integrated sampling scheme it is necessary to consider the following basic requirements:

a. The method must have an acceptably high efficiency of collecting the contaminant of interest.

b. A rate of airflow that can provide a sufficient sample for the required analytical procedure, maintain the acceptable collection efficiency, and be accomplished in a reasonable time period must be available.

c. The collected gas or vapor must be kept in the chemical form in which it exists in the atmosphere under conditions that maintain the stability of the sample before analysis.

d. The sample must be submitted for analysis in a suitable form and medium.

e. A very minimal amount of analytical procedure in the field must be associated with the overall method.

f. To the extent practicable, the use of corrosive (e.g., acidic or alkaline) or relatively toxic (e.g., benzene) sampling media should be avoided.

Of these general requirements, perhaps the most important is the first, that is, knowing the collection efficiency of the sampling system chosen or anticipated. This efficiency information can be obtained either from published data describing the method or as a result of independent evaluations as an essential part of planning the industrial hygiene survey. In making such evaluations, known concentrations of the gases must be prepared by means of either dynamic or static test systems. Once the known atmosphere is generated, the sampling device should be used as anticipated or intended and the efficiency defined in terms of such variables as characteristic of the sampler, rate of airflow, stability of the sampling during collection, and apparent losses by adsorption on walls on the sampling device.

The collection of sufficient sample for the subsequent analytical procedure is a matter that must be clearly understood not only by the laboratory analyst but by the field investigator as well. Understanding can best be promoted by discussions between the two professionals before the industrial hygiene field survey. The field investigator must discuss as fully as possible with the chemist the nature of the process involved in the survey so that together they may select the best combination of sampling and analytical methods to satisfy the sensitivity requirements of the analytical method, minimize effects of potential interferences, and complete the sampling within a time frame consistent with process conditions or potential exposure.

4.2   Collection Techniques

There are two basic methods for collecting samples of gaseous contaminants: instantaneous or "grab" sampling, and integrated or long-term sampling. The first involves the use of a gas-collecting device, such as an evacuated flask or bottle, to obtain a fixed volume of a contaminant-in-air mixture at known temperature and pressure. This is called "grab" sampling because the contaminant is collected almost instantaneously, that is, within a few seconds or minutes at most; thus the sample is representative of atmospheric conditions at the sampling site at a given point in time. This method commonly is used when atmospheric analyses are limited to such gross contaminants as mine gases, sewer gases, carbon dioxide, or carbon monoxide, or when the concentrations of contaminants likely to be found are sufficiently high to permit analysis of a relatively small sample. However, with the increased sensitivity of modern instrumental techniques such as gas chromatography and infrared spectrometry, instantaneous sampling of relatively low concentrations of atmospheric contaminants is becoming feasible.

The second method for collection of gaseous samples involves passage of a known volume of air through an absorbing or adsorbing medium to remove the contaminants of interest from the sample airstream. This technique makes it possible to sample the atmosphere over an extended period of time, thus "integrating" the sample. The contaminant that is removed from the airstream becomes concentrated in or on the collection medium; therefore it is important to establish a sampling period long enough to permit collection of a quantity of contaminant sufficient for subsequent analysis.

4.2.1 Instantaneous Sampling

Various devices can be used to obtain instantaneous or grab samples. These include vacuum flasks or bottles, gas or liquid displacement collectors, metallic collectors, glass bottles, syringes, and plastic bags. The temperature and pressure at which the samples are collected must be known, to permit reporting of the analyzed components in terms of standard conditions, normally 25°C and 760 mm Hg for industrial hygiene purposes.

Grab samples are usually collected when analysis is to be performed on gross amounts of gases in air (e.g., methane, carbon monoxide, oxygen, and carbon dioxide). The samplers should not be used for collecting reactive gases such as hydrogen sulfide, oxides of nitrogen, and sulfur dioxide unless the analyses can be made directly in the field. Such gases may react with dust particles, moisture, wax sealing compound, or glass, altering the composition of the sample.

The introduction of highly sensitive and sophisticated instrumentation has extended the applications of grab sampling to low levels of contaminants. In areas where the atmosphere remains constant, the grab sample may be representative of the average as well as the momentary concentration of the components; thus it may truly represent an integrated equivalent. Where the atmospheric composition varies, numerous samples must be taken to determine the average concentration of a specific component. The chief advantage of grab sampling methods is that collection efficiency is essentially 100 percent, assuming no losses due to leakage or chemical reaction preceding analysis. The more common types of grab sampling devices are discussed in detail below.

4.2.1.1 Evacuated Containers.

Evacuated flasks are heavy-walled containers of glass or other suitable material, usually of 250- or 300-ml capacity, but frequently holding as much as 500 or 1000 ml, from which essentially all (99.97 percent or more) of the air has been removed by a heavy-duty vacuum pump. The internal pressure after the evacuation is practically zero. The container is closed; for example, for glass units, the neck is sealed by heating and drawing during the final stages of evacuation. These units are simple to use because no metering devices or pressure measurements are required. The pressure of the sample is taken as the barometric pressure reading at the site. After the sample has been collected by breaking the

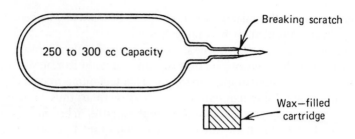

Figure 27.5. Evacuated flask.

heat-sealed end of the glass unit or opening the container, the flask is resealed and transported to the laboratory for analysis. A typical evacuated container is depicted in Figure 27.5.

A variation of this procedure with evacuated flasks is to add a liquid absorbent to the flask before it is evacuated and sealed to preserve the sample in a desirable form following collection.

Partially evacuated containers or vacuum bottles are prepared with a suction pump just before sampling is performed, although frequently they are evacuated in the laboratory the day before a field visit. No attempt is made to bring the internal pressure to zero, but temperature readings and pressure measurements with a manometer are recorded after the evacuation, and again after the sample has been collected. Examples of this type of collector are heavy-walled glass bottles and metal or heavy plastic containers with tubing connectors closed with screw clamps or stopcocks.

4.2.1.2 Displacement Collectors. Gas or liquid displacement collectors include 250- to 300-ml glass bulbs fitted with end tubes that can be closed with greased stopcocks or with rubber tubing and screw clamps. They are used widely in collecting samples containing oxygen, carbon dioxide, carbon monoxide, nitrogen, hydrogen, or other combustible gases for analysis by an Orsat or similar analyzer. Another device operating on the liquid displacement principle is the aspirator bottle, which has exit openings at the bottom through which the liquid is drained during sampling. Figure 27.6 shows a typical displacement collector.

In applying the gas displacement technique, the samplers are purged conveniently with a bulb aspirator, hand pump, small vacuum pump, or other suitable

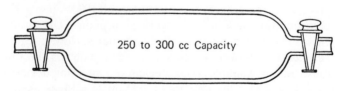

Figure 27.6. Gas or liquid displacement collector.

source of suction. Usually satisfactory purging can be achieved by drawing a minimum of 10 air changes of the test atmosphere through the gas collector.

The devices used for gas displacement collectors can be filled by liquid displacement, the most frequently employed liquid being water. In sampling, liquid in the container is drained or poured out slowly in the test area and replaced by air to be sampled. Of course, this method is limited to gases that are insoluble in and nonreactive with the displaced liquid. The solubility problem can be minimized by using mercury or water conditioned with the gas to be collected. Mercury, however, must be used with caution, because it may create an exposure problem if handled carelessly.

4.2.1.3 Flexible Plastic Bags. Evacuated flasks and displacement containers are being used less frequently because of the introduction of flexible containers. Flexible plastic bags can be used to collect air and breath samples containing organic and inorganic vapors and gases in concentrations ranging from parts per billion to more than 10 percent volume in air. They also are convenient for preparing known concentrations of gases and vapors for equipment calibration. The bags are available commercially in a variety of sizes, up to 9 ft^3, and can be made easily in the laboratory.

Sampling bags are manufactured from various plastic materials, most of which can be purchased in rolls or sheets cut to the desired size. Some materials, such as Mylar, may be sealed with a hot iron using a Mylar tape around the edges. Others, such as Teflon, require high temperature and controlled pressure in sealing. Certain plastics, including Mylar and Scotchpak, may be laminated with aluminum to seal the pores and reduce the permeability of the inner walls to sample gases and the outer walls to sample moisture. Sampling ports may consist of a sampling tube molded into the fabricated bag and provided with a closing device or a clamp-on air valve.

Plastic bags have the advantages of being light, unbreakable, and inexpensive, and they permit the entire sample to be withdrawn without the difficulty associated with dilution by replacement air, as is the case with rigid containers. However, they must be used with caution because generalization of recovery characteristics of a given plastic cannot be extended to a broad range of gases and vapors. Important factors to be considered in using these collectors are absorption and diffusion characteristics of the plastic material, concentration of the gas or vapor, and reactive characteristics of the gas or vapor with moisture and with other constituents in the sample. Information on the storage properties of gases and vapors in plastic containers has been published (11, 12).

The bags must be leak tested and preconditioned for 24 hr before they are used for sampling. Preconditioning consists of flushing the bag three to six times with the test gas, the number of times depending on the nature of the bag material and the gas. In some cases it is recommended that the final refill remain in the bag overnight before the bag is used for sampling. Such preconditioning usually is helpful in minimizing the rate of decay of a collected gas. At the sampling site the air to be sampled is allowed to stand in the bag for several minutes, if possible, before removal and subsequent refilling of the bag with a sample. Once collected, the interval between sampling and analysis should be as short as practicable.

4.2.2   Integrated Sampling

Integrated sampling of the workroom atmosphere is necessary when the composition of the air is not uniform, when the sensitivity requirements of the method of analysis necessitate sampling over an extended period, or when compliance or noncompliance with an 8-hr, time-weighted average standard must be established. Thus the professional observations and judgment of the industrial hygienist are called on in devising the strategy for the procurement of representative samples to meet the requirements of an environmental survey of the workplace. This deliberation was discussed earlier in this chapter.

4.2.2.1 Sampling Pumps. Integrated air sampling requires a relatively constant source of suction as an air-moving device. A vacuum line, if available, may be satisfactory. However, the most practical source for prolonged periods of sampling is a pump powered by electricity. These pumps come in various sizes and types, and must be chosen for the sampling devices with which they will be used.

If electricity is not available or if flammable vapors present a fire hazard, aspirator bulbs, hand pumps, portable units operated by compressed gas, or battery-operated pumps are suitable for sampling at rates up to 3 liters/min. The latter have become the workhouses of the industrial hygiene profession, particularly in judging compliance with health standards. For higher sampling requirements, ejectors using compressed air or a water aspirator may be employed.

When compressed air or batteries are to be used as the driving force for a pump, the duration of the sampling period is important in relation to the supply of compressed air or the life of the rechargeable battery. These units must be allowed to run unattended, and periodic checks on the airflow must be made.

The common practice in the field is to sample for a measured period of time at a constant, known rate of airflow. Direct measurements are made with rate meters such as rotameters and orifice or capillary flowmeters. These units are small and convenient to use, and have become quite accurate even at the very low flow rates common with modern industrial hygiene sampling, that is, 10 to 20 cm^3/min. The sampling period must be timed carefully with a stopwatch.

Many pumps have inlet vacuum gauges or outlet pressure gauges attached. These gauges, when properly calibrated with a wet or dry gas meter, can be used to determine the flow rate through the pump. The gauge may be calibrated in terms of cubic feet per minute or liters per minute. If the sample absorber does not have enough resistance to produce a pressure drop, a simple procedure is to introduce a capillary tube or other resistance into the train behind the sampling unit.

Samplers are always used in assembly with an air-moving device (source of suction) and an air metering unit. Frequently, however, the sampling train consists of filter, probe, absorber (or adsorber), flowmeter, flow regulator, and air mover. The filter serves to remove any particulate matter that may interfere in the analysis.

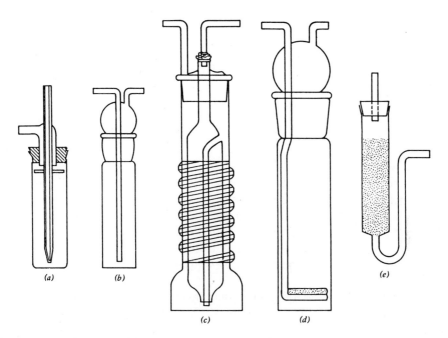

Figure 27.7. Absorbers: (*a*) impinger; (*b*) wash bottle; (*c*) spiral absorber; (*d*) fritted glass bubbler; (*e*) packed column.

It should be ascertained that the filter does not also remove the gaseous contaminant of interest. The probe or sampling line is extended beyond the sampler to reach a desired location. It also must be checked to determine that it does not collect a portion of the sample. The meter that follows the sampler indicates the flow rate of air passing through the system. The flow regulator controls the airflow. Finally at the end of the train the air mover provides the driving force.

4.2.2.2 Absorbers. Four basic types of absorber are employed for collecting gases and vapors: simple gas washing bottles, spiral and helical absorbers, fritted bubblers, and glass-bead columns. Such units are used more in research applications than in routine exposure monitoring. The absorbers provide sufficient contact between the sampled air and the liquid surface to ensure complete absorption of the gaseous contaminant. In general, the lower the sampling rate, the more complete is the absorption (Figure 27.7).

Simple gas washing bottles include Drechsel types, standard Greenburg-Smith devices, and midget impingers. The air is bubbled through the liquid absorber without special effort to secure intimate mixing of air and liquid, and the length of travel of the gas through the collecting medium is equivalent to the height of the absorbing liquid. These scrubbers are suitable for gases and vapors that are readily soluble in the absorbing liquid or react with it. One or two units may be enough for efficient collection, but in some cases several may be required to attain the efficiency of a single fritted glass bubbler. Advantages of these devices are simplicity in construction, ease of rinsing, and the small volume of liquid required.

Spiral and helical absorbers provide longer contact between the sampled air and the absorbing solution. The sample is forced to travel a spiral or helical path through the liquid 5 to 10 times longer than that in the simpler units. In fritted glass bubblers air passes through porous glass plates and enters the liquid in the form of small bubbles. The size of the air bubbles depends on the liquid and on the diameter of the orifices from which the bubbles emerge. Frits are classified as fine, coarse, or extra coarse, depending on the number of openings per unit area. The extra-coarse frit is used when a more rapid flow is desired. The heavier froth generated by some liquids increases the time of contact of gas and liquid. These devices are more efficient collectors than the simple gas washing bottles and can be used for the majority of gases and vapors that are soluble in the reagent or react rapidly with it. Flow rates between 0.5 and 1.0 liter/min are used commonly. These absorbers are relatively sturdy, but the fritted glass is difficult when used for contaminants that form a precipitate with the reagent, in which cases a simple gas washing bottle would be preferable.

Packed columns are used for special situations calling for a concentrated solution. Packing material, such as glass pearl beads, wetted with the absorbing solution provides a large surface area for the collection of a sample and is especially useful when a viscous liquid is required. The rate of sampling is low, usually 0.25 to 0.5 liter of air per minute.

4.2.2.3 Adsorption Media. When it is desired to collect insoluble or nonreactive vapors, an adsorption technique frequently is the method of choice. Activated charcoal and silica gel are common adsorbents (Figures 27.8 and 27.9). Solid adsorbents require less manipulative care than do liquid adsorbents; they can provide high collection efficiencies, and with improved adsorption tube design and a better definition of desorption requirements, they are becoming increasingly popular in industrial hygiene surveys.

Activated charcoal is an excellent adsorbent for most vapors boiling above 0°C; it is moderately effective for low-boiling gaseous substances (between -100 and 0°C), particularly if the carbon bed is refrigerated, but a poor collector of gases having boiling points below -150°C. Its retentivity for adsorbed vapor is several times that of silica gel. Because of their nonpolar characteristics, organic gases and vapors are adsorbed in preference to moisture, and sampling can be performed for long periods of time.

Silica gel has been used widely as an adsorbent for gaseous contaminants in air samples. Because of its polar character it tends to attract polar or readily polarizable substances preferentially. The general order of decreasing polarizability or attraction is: water, alcohols, aldehydes, ketones, esters, aromatic compounds, olefins, and paraffins. Organic solvents are relatively nonpolar in comparison with water, which is strongly adsorbed onto silica gel; such compounds are displaced by water in the entering airstream. Consequently the volume of air sampled under humid

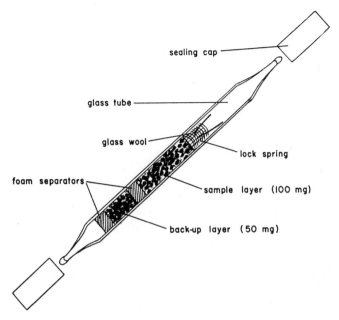

Figure 27.8. Activated charcoal sampling tube.

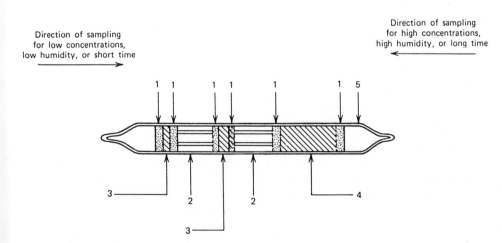

Figure 27.9. Silica gel sampling tube for aromatic amines: 1, 100-mesh stainless steel screen plugs; 2, 12-mm glass tube separator; 3, 150-mg silica gel section, 45/60 mesh; 4, 700-mg silica gel section, 45/60 mesh: 5, 8-mm ID glass tube.

conditions may have to be restricted. Despite this limitation, silica gel is a very useful adsorbent.

4.2.2.4 Condensation. In condensation methods, vapors or gases are separated from sampled air by passing the air through a coil immersed in a cooling medium, dry ice and acetone, liquid air, or liquid nitrogen. The device is not considered to be a portable field technique ordinarily. It may be necessary to use this method when the gas or vapor may be altered by collecting in liquid or when it is difficult to collect by other techniques. A feature of this method is that the contaminating material is obtained in concentrated form. The partial pressure of the vapor can be measured when the system is brought back to room temperature.

4.2.2.5 Collection Efficiency. The collection efficiency (the ratio of the amount of contaminant retained by the absorbing or adsorbing medium to that entering it) need not be 100 percent as long as it is known, constant, and reproducible. The minimum acceptable collection performance in a sampling system is usually 90 percent, but higher efficiency is certainly desirable. When the efficiency falls below the acceptable minimum, sampling may be carried out at a lower rate, or in the case of liquid absorbers, at a reduced temperature by immersing the absorber in a cold bath to reduce the volatility of both the solute and solvent.

Frequently, the relative efficiency of a single absorber can be estimated by placing another in series with it. Any leakage is carried over into the second collector. The absence of any carryover is not in itself an absolute indication of the efficiency of the test absorber, because the contaminant may be stopped effectively by either absorber. Analysis of the various sections of silica gel or activated charcoal tubes used in sampling a contaminant is a useful check on the collection efficiency of the first section of the tube. Another valuable technique is the operation of the test absorber in parallel or in series with a different type of collector having a known high collection efficiency (an absolute collector if one is available) for the contaminant of interest. By running the test absorber at different rates of low, the maximum permissible rate of flow for the device can be ascertained.

4.3 Direct-Reading Techniques

Various direct-reading techniques that can be used to evaluate airborne concentrations of gases and vapors are available to the industrial hygienist. These include instruments capable of direct response to airborne contaminants, various reagent kits that can be used for certain substances, colorimetric indicator (detector) tubes, and passive dosimeters.

4.3.1 Instruments

A direct-reading instrument, for purposes of this discussion, is "an integrated system capable of sampling a volume of air, making a quantitative analysis and displaying the result." Direct-reading instruments can be portable devices or fixed-site monitors. Generally these devices are characterized by disadvantages that limit their application for measuring the low concentrations of significance to the industrial hygienist.

Direct-reading instruments are used commonly for on-site evaluations for a variety of reasons, depending primarily on the understood purpose of the survey. Direct-reading instruments are useful in the following applications:

a. To find sources of emission of hazardous substances "on the spot."
b. To determine the performance characteristics of specific operations or control devices, usually by comparing results of "before and after" surveys.
c. As a qualitative industrial hygiene monitoring instrument to ascertain whether specific air quality standards are being complied with.
d. As continuous monitoring devices, by establishment of a network of sensors at fixed locations throughout a plant. Readout from such a system can be used to activate either an alarm or an auxiliary control system in the event of process upsets, or to obtain permanent recorded documentation of concentrations of contaminants in the workroom atmosphere.

The advantages of having direct-reading instruments available for industrial hygiene surveys are obvious. Such on-site evaluations of atmospheric concentrations of hazardous substances make possible the immediate assessment of unacceptable conditions and enable industrial hygienists to initiate immediate corrective action in accordance with their judgment of the seriousness of the situation without causing further risk of injury to the workers. It cannot be overemphasized that great caution must be employed in the use of direct-reading instruments and, more important, in the interpretation of their results. Most of these instruments are nonspecific, and before recommending any action industrial hygienists often must verify their on-site findings by supplemental sampling and laboratory analyses to characterize adequately the chemical composition of the contaminants in a workroom area and to develop the supporting quantitative data with more specific methods of greater accuracy. Such precautions become mandatory if the industrial hygienist or other professional investigator conducting the sampling has not had extensive experience with the process in question, or when a change in the process or substitution of chemical substances may have occurred.

The calibration of any direct-reading instrument is an absolute necessity if the data are to have any meaning. Considering this to be axiomatic, it must also be recognized that the frequency of calibration depends on the type of instrument. Certain classes of instruments, because of their design and complexity, require more frequent calibration than others. It is also recognized that peculiar "quirks" in an individual instrument produce greater variations in its response and general performance, thus requiring a greater amount of attention and more frequent calibration than other instruments of the same design. Direct personal experience with a given instrument serves as the best guide in this matter.

Another unknown factor that can be evaluated only by experience is the environmental variability of sampling sites. For example, when locating a particular fixed-station monitor at a specific site, consideration must be given to the presence of interfering chemical substances, the corrosive nature of contaminants, vibration, voltage fluctuations, and other disturbing influences that may affect the response of the instrument.

Finally, the required accuracy of the measurements must be determined initially. If an accuracy of ±3 percent is needed, more frequent calibration must be made than if ±25 percent accuracy is adequate in the solution of a particular problem.

As indicated earlier, direct-reading instruments for atmospheric contaminants are classified as devices that provide an immediate indication of the concentration of contaminants by a dial reading, a strip-chart recording, or a tape printout. When properly calibrated and used with full cognizance of their performance characteristics and limitations, these services can be extremely helpful to industrial hygienists who are engaged in on-site evaluations of potentially hazardous conditions. Many types of instruments depend on certain physical or chemical principles for their operation. They are discussed briefly below and are described in detail in other publications (13, 14). In general, the advantages of direct-reading instruments include the following:

a. Rapid estimation of the concentration of a contaminant, permitting on-site evaluations and implementation of appropriate measures

b. Provision of permanent 24-hr records of contaminant concentrations when used as continuous monitors

c. Easy incorporation of alarm systems onto continuous monitoring instruments to warn workers of buildup of hazardous conditions

d. Reduction of the number of manual tests needed to accomplish an equivalent amount of sampling

e. Similar reduction of the number of laboratory analyses

f. Provision of evidence of monitoring environmental conditions for presentation in litigation proceedings

g. Reduced cost per sample of obtaining data

The disadvantages of direct-reading instruments usually include at least one of the following:

a. High initial cost of instrumentation and, if used as a continuous monitor, installation of the sensing network

b. Need for frequent calibration

c. General lack of adequate calibration facilities

d. Lack of portability

e. Lack of specificity, the most critical negative factor

Descriptions of several types of direct-reading instrument follow; they can be used if necessary precautions are taken and the limitations of the devices are understood. In the following discussions, when the lower end of the working range is indicated as zero, the reader should recognize that "0" is a relative term, related to the sensitivity of the instrument.

4.3.1.1 Colorimetry. Colorimetry is the measurement of the relative power of a beam of radiant energy in the visible, ultraviolet, or infrared region of the electromagnetic spectrum, which has been attenuated as a result of passing a suspension of solid or liquid particulates in air or other gaseous medium, or a photographic image of a spectral line or an X-ray diffraction pattern on a photographic film or plate. Such photometers contain a lamp or other generating source of energy, an optical filter arrangement to limit the bandwidth of the incident beam of radiation, and an optical system to collimate the filtered beam. The beam then is passed through the sample system contained in a cuvette or gas cell to a photocell, bolometer, thermocouple, or pressure-sensor type of detector, where the signal is amplified and fed to a readout meter or to a strip-chart recorder. The more sophisticated technique, termed spectrophotometry, employs prisms made of glass (visible region), quartz (ultraviolet), and sodium chloride or potassium bromide (infrared), or diffraction gratings, instead of optical filters, to provide essentially monochromatic radiation as the source of energy.

Although spectrophotometers are used mostly in laboratories for highly specific and precise analytical determinations, field-type colorimetric analyzers have been designed to function primarily as fixed-station monitors for active gases such as oxides of nitrogen, sulfur dioxide, "total oxidant," ammonia, aldehydes, chlorine, hydrogen fluoride, and hydrogen sulfide. These instruments require frequent calibration with zero and span gases at the sampling site to assure generation of reliable data. However, built-in automated calibration systems, which regularly standardize zero and span controls against pure air and a calibrated optical filter, are now available for several gases including nitrogen oxides, sulfur dioxide, and aldehydes.

4.3.1.2 Heat of Combustion. With these instruments a combustible gas or vapor mixture is passed over a filament heated above the ignition temperature of the substance of analytical interest. If the filament is part of a Wheatstone bridge circuit, the resulting heat of combustion changes the resistance of the filament, and the measurement of the imbalance is related to the concentration of the gas or vapor in the sample mixture. The method is basically nonspecific, but it may be made more selective by choosing appropriate filament temperatures for individual gases or vapors or by using an oxidation catalyst for a desired reaction, such as Hopcalite for carbon monoxide.

Combustible gas indicators must be calibrated for their response to the anticipated individual test gases and vapors. These instruments are definitely portable and they are valuable survey meters in the industrial hygienist's collection of field instruments. Readings are in terms of 0 to 1000 ppm or 0 to 100 percent of the lower explosive limit (LEL). However, industrial atmospheres rarely contain a single gaseous contaminant, and these indicators will respond to all combustible gases present. Hence supplementary sampling and analytical techniques should be used for a complete definition of environmental conditions.

4.3.1.3 Electrical Conductivity. Electrical conductivity instruments function by drawing a gas–air mixture through an aqueous solution. Gases that form electrolytes, such as vinyl chloride, produce a change in the electroconductivity as a summation of the effects of all ions thus produced. Hence the method is nonspecific. If the concentrations of all other ionizable gases are either constant or insignificant, the resulting changes in conductivity may be related to the gaseous substances of interest. Temperature control is extremely critical in conductance measurements; if thermostated units are not used, electrical compensation must enter into the measurements to allow for the 2 percent/°C conductivity temperature coefficient average for many gases.

The electrical conductivity method has found its greatest application in the continuous monitoring of sulfur dioxide in ambient atmospheres. However, a lightweight portable analyzer that uses a peroxide absorber to convert SO_2 to H_2SO_4 is available; this battery-operated instrument can provide an integrated reading of the SO_2 concentration over the 0 to 1 ppm range within 1 min. A larger portable model that may be operated from a 12-V automobile battery is also available for the high concentration ranges of SO_2 encountered in field sampling.

4.3.1.4 Thermal Conductivity. The specific heat of conductance of a gas or vapor is a measure of its concentration in a carrier gas such as air, argon, helium, hydrogen, or nitrogen. However, thermal conductivity measurements are nonspecific, and the method has had only limited application as a primary detector. It has found its greatest usefulness in estimating the concentrations of separately eluted components from a gas chromatographic column. This method operates by virtue of the loss of heat from a hot filament to a single component of a flowing gas stream, the loss being registered as a decrease in electrical resistance measured by a Wheatstone bridge circuit. This method is applied mainly to binary gas mixtures, and uses are based on the electrical imbalance produced in the bridge circuit by the difference in the filament resistances of the sample and reference gases passed through separate cavities in the thermal conductivity cell.

4.3.1.5 Flame Ionization. The hydrogen flame ionization detector typically consists of a stainless steel burner in which hydrogen is mixed with the sample gas stream in the base of the unit. Combustion air or oxygen is fed axially and diffused around the jet through which the hydrogen–gas mixture flows to the cathode tip, where ignition occurs. A loop of platinum, set about 6 mm above the tip of the burner, serves as the collector electrode. The current carried across the electrode gap is proportional to the number of ions generated during the burning of the sample. The detector responds to essentially all organic compounds, but its response is greatest with simple hydrocarbons and diminishes with increasing substitution of other elements, notably oxygen, sulfur, and chlorine. A low noise level of 10^{-12}A provides a high sensitivity of detection and the detector is capable of the wide linear dynamic range of 10^7. Its usefulness is enhanced by its insensitivity to water, the permanent gases, and most inorganic compounds, thus simplifying the analysis of aqueous solutions and atmospheric samples. It serves to great advantage in both laboratory and field models of gas chromatographs (an application discussed in detail subsequently), as well as in hydrocarbon analyzers that are set up as fixed station monitors of ambient atmospheres in the laboratory or field.

Hydrocarbon analyzers, operating with an FID detector, are literally carbon counters; their response to a given quantity of a typical C_6 hydrocarbon is six times that to methane, at a fixed flow rate of the sample stream. Thus the instrument's characteristics, such as sensitivity, usually are given as a methane equivalent. In addition to hydrocarbons, these analyzers respond to alcohols, aldehydes, amines, and other compounds, including vinyl chloride, which produces an ionized carbon atom in the hydrogen flame.

4.3.1.6 Gas Chromatography. Gas chromatography, a physical process for separating components of complex mixtures, is used routinely as a portable technique for in-plant studies. A gas chromatograph has the following components:

a. A carrier gas supply complete with a pressure regulator and flowmeter
b. An injection system for the introduction of a gas or vaporizable sample into a port at the front end of the separation column
c. A stainless steel, copper, or glass separation column containing a stationary phase consisting of an inert material such as diatomaceous earth, used alone as in gas–solid chromatography or as a support for a thin layer of a liquid substrate, such as silicone oils, in gas–liquid chromatography
d. A heater and oven assembly to control the temperature of the column(s) injection port, and detector unit
e. A detector
f. A recorder for the chromatograms produced during the separations

The separations are based on the varied affinities of the sample components for the packing materials of a particular column, the rate of carrier gas flow, and the operating temperature of the column. Improved separations are made possible by the use of temperature programming. The sample components, as a consequence of their varied affinities for a given column, are eluted sequentially; thus they evoke separate responses by the detection system, from which the signal is amplified to produce a peak on the strip chart or other output. The height and area of the peak are proportional to the concentration of the eluted sample components. Calibrations can be made using known mixtures of the pure substance in a gas–air mixture prepared in a flexible plastic bag or other suitable container. The time of retention on the column and supporting analytical techniques (e.g., infrared spectrophotometry) can be used in qualitative analysis of the individual peaks of a chromatogram. The method is capable of providing extremely clean-cut separations and is one of the most useful techniques in the field of organic analysis. It is sensitive to fractional part per billion concentrations of many organic substances. The most commonly used detectors include flame ionization, thermal conductivity, and electron capture.

Rugged, battery-operated, portable gas chromatographs have been refined sufficiently to be practical for many field study applications. These instruments may be obtained with a choice of thermistor, thermal conductivity, flame ionization, and electron capture detectors. Complete with gas sampling valve, rechargeable batteries, appropriate columns, and self-contained supplies of gas, these chromatographs have much to offer the industrial hygienist engaged in on-site analyses of trace quantities of organic compounds and the permanent gases. Gas lecture bottles provide 8 to 20 hr of operation, depending on the flow rates, and they must be recharged using high pressure gas regulators. The retention times of the compounds of analytical interest must be determined in the laboratory for a given type of column, as is true for the laboratory type chromatographs.

4.3.1.7 Other Principles of Operation. Various additional types of direct-reading instrument are available, although their range of applicability is more limited. The principles of operation include chemiluminescence, coulometry, polarography, potentiometry, and radioactivity. Table 27.3 summarizes the types of direct-reading instrument available, along with typical operating characteristics and examples of gases and vapors for which the instruments have been used successfully.

4.3.2 Reagent Kits

Direct-reading colorimetric techniques, which utilize the chemical reaction of an atmospheric contaminant with a color-producing reagent, are available in a variety of forms. Detector kit reagents may be in either liquid or solid phase or supplied in the form of chemically treated papers. The liquid and solid reagents are generally supported in sampling devices through which a measured amount of contaminated air is drawn. Chemically treated papers are usually exposed to the atmosphere, and the reaction time for a color change to occur is noted.

Liquid reagents come in sealed ampules or in tubes for field use. Such preparations are provided in concentrated or solid form for easy dilution or dissolution at the sampling site. Representative of this type of reagent are the *o*-tolidine and the Griess–Ilsovay kits for chlorine and nitrogen dioxide, respectively. Although the glassware needed for these applications may be somewhat inconvenient to transport to the field, methods based on the use of liquid reagents are more accurate than those that use solid reactants. This is due to the inherently greater reproducibility and accuracy of color measurements made in a liquid system.

Papers impregnated with chemical reagents have found wide applications for many years for the detection of toxic substances in air. Examples include the use of mercuric bromide papers for the detection of arsine, and lead acetate for hydrogen cyanide. When a specific paper is exposed to an atmosphere containing the contaminant in question, the observed time of reaction provides an indication of the concentration of that substance. For example, a 5-sec response time by the *o*-

tolidine–cupric acetate paper indicates a concentration of 10 ppm HCN in the tested atmosphere.

Similarly, sensitive detector crayons have been devised for the preparation of a reagent smear on a test paper for which response to a specific toxic substance in a suspect atmosphere may then be timed to obtain an estimation of the atmospheric concentration of a contaminant. Crayons for phosgene, hydrogen cyanide, cyanogen chloride, and lewisite (ethyl dichloroarsine) have been developed.

4.3.3 Colorimetric Indicators

Colorimetric indicating tubes containing solid reagent chemicals provide compact direct-reading devices that are convenient to use for the detection and semiquantitative estimation of gases and vapors in atmospheric environments. Presently there are tubes for more than 200 atmospheric contaminants on the market; five major companies manufacture and/or distribute these devices.

Whereas it is true that the operating procedures for colorimetric indicator tubes are simple, rapid, and convenient, there are distinct limitations and potential errors inherent in this method of assessing atmospheric concentrations of toxic gases and vapors. Therefore dangerously misleading results may be obtained with these devices unless they are used under the supervision of an adequately trained industrial hygienist who (1) enforces rigidly the periodic (as required) calibration of individual batches of each specific type of tube for its response to known concentrations of the contaminant, as well as the refrigerated storage of all tubes, to minimize their rate of deterioration; (2) informs the staff of the physical and chemical nature and extent of interferences to which a given type of tube is subject and limits the tube's usage accordingly; and (3) stipulates how and when other independent sampling and analytical procedures will be employed to derive needed quantitative data. A manual describing recommended practice for colorimetric indication tubes (15) discusses in detail the principles of operation, applications, and limitations of these devices. A brief summary is given below.

Colorimetric indicating tubes are filled with a solid granular material, such as silica gel or aluminum oxide, which has been impregnated with an appropriate chemical reagent. The ends of the glass tubes are sealed during manufacture. When a tube is to be used, its end tips are broken off, the tube is placed in the manufacturer's holder, and the recommended volume of air is drawn through the tube by means of the air-moving device provided by the manufacturer. This device may be one of several types, such as a positive displacement pump, a simple squeeze bulb, or a small electrically operated pump with an attached flowmeter. Each air-moving device must be calibrated frequently, for example, after sampling 100 tubes as an arbitrary rule, or more often if there are reasons to suspect changes due to effects of corrosive action from contaminants in the tested atmospheres. An acceptable pump should be correct to within ±5 percent by volume; with use, its flow characteristics may change. It should also be checked for leakage and plugging of the inlet after every 10 samples.

In most cases a fixed volume of air is drawn through the detector tube, although

Table 27.3. Direct-Reading Physical Instruments

Principle of Operation	Applications and Remarks	Code[a]	Range	Repeatability (precision)	Sensitivity	Response Time
Aerosol photometry	Measures, records, and controls particulates continuously in areas requiring sensitive detection of aerosol levels; detection of 0.05–40 μm diameter particles. Computer interface equipment is available	A and B	10^{-3} to 10^2 μg/liter	Not given	10^{-3} μg/liter (for 0.3 μm DOP)	Not given
Chemiluminescence	Measurement of NO in ambient air selectively and NO_2 after conversion to NO by hot catalyst. Specific measurement of O_3. No atmospheric interferences	B	0–10,000 ppm	±0.5–±3%	Varies: 0.1 ppb to 0.1 ppm	ca 0.7 sec, NO mode and 1 sec, NO_x mode; longer period when switching ranges
Colorimetry	Measurement and separate recording of NO_2-NO_x, SO_2, total oxidants, H_2S, HF, NH_3, Cl_2 and aldehydes in ambient air	A and B	ppb and ppm	±1–5%	0.01 ppm (NO_2, SO_2)	30 sec to 90% of full scale
Combustion	Detects and analyzes combustible gases in terms of percent LEL on graduated scale. Available with alarm set at ⅓ LEL	A	ppm to 100%	—	ppm	<30 sec
Conductivity, electrical	Records SO_2 concentrations in ambient air. Some operate off a 12-V car battery. Operates unattended for periods up to 30 days	A and B	0–2 ppm	<±1–±10%	0.01 ppm	1–15 sec (lag)
Coulometry	Continuous monitoring of NO, NO_2, O_3 and SO_2 in ambient air. Provided with stripchart recorders. Some require attention only once a month	A and B	Selective: 0–1.0 ppm overall, or scale to 100 ppm (optional)	±4% of full scale	varies: 4–100 ppb dependent on instrument range setting	<10 min to 90% of full scale

Instrument	Description	Class	Range	Accuracy	Sensitivity	Response time
Flame ionization (with gas chromatograph)	Continuous determination and recording of methane, total hydrocarbons, and carbon monoxide in air. Catalytic conversion of CO to CH_4. Operates up to 3 days unattended	B	Selective: 0–1 ppm; 0–100 ppm	±1% of full scale	Not given	5 min (cycle time)
	Separate model for continuous monitoring of SO_2, H_2S, and total sulfur in air. Unattended operation up to 3 days	B	0–20 ppm	±4% of full scale	0.005 ppm (H_2S); 0.01 ppm (SO_2)	5 min (cycle time)
Flame ionization (hydrocarbon analyzer)	Continuous monitoring of total hydrocarbons in ambient air; potentiometric or optional current outputs compatible with any recorder. Electronic stability from 32 to 110°F	B	0–1 ppm as CH_4; ×1, ×10, ×100, ×1000 with continuous span adjustment	±1% of full scale	1 ppm to 2% full scale as CH_4; 4 ppm to 10% as mixed fuel	<0.5 sec to 90% of full scale
Gas chromatograph, portable	On-site determination of fixed gases, solvent vapors, nitro and halogenated compounds, and light hydrocarbons. Instruments available with choice of flame ionization, electron capture, or thermal conductivity detectors and appropriate columns for desired analyses. Rechargeable batteries	A	Depends on detector	Not given	<1 ppb (SF_6) with electron capture detector; <1 ppm (HCs)	—
Infrared analyzer (photometry)	Continuous determination of a given component in a gaseous or liquid stream by measuring amount of infrared energy absorbed by component of interest using pressure sensor technique. Wide variety of applications include CO, CO_2, Freons, hydrocarbons, nitrous oxide, NH_3, SO_2, and water vapor	B	From ppm to 100% depending on application	±1% of full scale	0.5% of full scale	0.5 sec to 90% of full scale

continued on next page

Table 27.3. (*Continued*)

Principle of Operation	Applications and Remarks	Code[a]	Range	Repeatability (precision)	Sensitivity	Response Time
Photometry, ultraviolet (tuned to 253.7 nm)	Direct readout of mercury vapor; calibration filter is built into the meter. Other gases or vapors that interfere include acetone, aniline, benzene, ozone, and others that absorb radiation at 253.7 mµ	A	0.005–0.1 and 0.03–1 mg/m³	±10% of meter reading or ± minimum scale division, whichever is larger	0.005 mg/m³	Not given
Photometry, visible (narrow-centered 394 nm band pass)	Continuous monitoring of SO_2, SO_3, H_2S, mercaptans, and total sulfur compounds in ambient air. Operates more than 3 days unattended	B	1–3000 ppm (with airflow dilution)	±2%	0.01–10 ppm	<30 sec to 90% of full scale
Particle counting (near forward scattering)	Reads and prints directly particle concentrations at 1 of 3 preset time intervals of 100, 1000 or 10,000 seconds, corresponding to 0.01, 0.1, and 1 ft³ of sampled air	B	Preset (by selector switch); particle size ranges:	±0.05% (probability of coincidence) 0.3, 0.5, 1.0, 2.0, 3.0, 5.0, and 10.0, µm; counts up to 10^7 particles per ft³ (35×10^3/liter)	—	Not given
Polarography	Monitor gaseous oxygen in flue gases, auto exhausts, hazardous environments, and in food storage atmospheres and dissolved oxygen in wastewater samples. Battery operated, portable, sample temperature 32 to 110F, up to 95% relative humidity. Potentiometric recorder output. Maximum distance between sensor and amplifier is 1000 ft	A	0–5 and 0–25%	±1% of reading at constant sample temperature	Not given	20 sec to 90% of full scale
Radioactivity	Continuous monitoring of ambient gamma and X-radiation by measurement of ion chamber currents, aver-	B	0.1–10^7 mR/hr	±10% (decade accuracy)	—	<1 sec

aging or integrating over a constant recycling time interval, sample time interval, sample temperature limits 32 to 120°F; 0 to 95% relative humidity (weatherproof detector); up to 1000 ft remote sensing capability. Recorder and computer outputs. Complete with alert, scram, and failure alarm systems. All solid-state circuitry

			Range	Accuracy	Sensitivity	Response time
Radioactivity	Continuous monitoring of beta- or gamma-emitting radioactive materials within gaseous or liquid effluents; either a thin-wall Geiger-Müller tube or a gamma scintillation crystal detector is selected depending on the isotope of interest; gaseous effluent flow, 4 cfm; effluent sample temperature limits 32 to 120°F using scintillation detector and 65 to 165°F using G-M detector. Complete with high radiation, alert and failure alarms	B	10–10^6 counts/min	±2% full scale (rate meter accuracy)	$<10^{-7}$ μCi of ^{131}I per cm^3 of air and 10^{-7} μCi of ^{137}Cs per cm^3 of water using a scintillation detector	0.2 sec at 10^6 counts/min (rate meter)
Radioactivity	Continuous monitoring of radioactive airborne particulates collected on a filter tape transport system; rate of airflow, 10 scfm; scintillation and G-M detectors, optional but a beta-sensitive plastic scintillator is provided to reduce shielding requirements and offer greater sensitivity. Air sample temperature limits 32 to 120°F; weight 550 lb. Complete with high and low flow alarm and a filter failure alarm	B	10 to 10^6 counts/min	±2% of full-scale (rate-meter accuracy)	10^{-12} μCi of ^{137}Cs per cm^3 of air using a scintillation detector	0.2 sec at 10^6 counts/min (rate meter)

[a]Codes: A, portable instruments; B, fixed monitor "transportable" instruments.

in some systems varied amounts of air may be sampled. The operator compares either an absolute length of stain produced in the column of the indicator gel or a ratio of the length of stain to the total gel length against a calibration chart, to obtain an indication of the atmospheric concentration of the contaminant that reacted with the reagent. To make estimates using another type of tube, a progressive change in color intensity is compared with a chart of color tints. In a third type of detector, the volume of sampled air required to produce an immediate color change is noted; it is intended that this air volume be inversely proportional to the concentration of the atmospheric contaminant. Basic mathematical analyses of the relationships among the variables affecting the length of stain (i.e., the concentration of test gas, volume of air sample, sampling flow rate, grain size of gel, tube diameter, and other variables) have been published (15, 16). These sources should be consulted for a full appreciation of the complex interrelationships among the factors affecting the kinetics of indicator tube reactions. It is sufficient to point out here that the length of stain is proportional to the logarithm of the product of gas concentration and air sample volume as follows:

$$\frac{L}{H} = \ln(CV) + \ln\left(\frac{K}{H}\right)$$

where L = length of stain (cm)
      C = gas concentration (ppm)
      V = volume of air sampled (cm^3)
      K = a constant for a given type of indicator tube and test gas
      H = mass transfer proportionality factor (cm); height of a mass transfer unit

If this mathematical model is correct for a given indicator tube, a linear plot of L versus the logarithm of the CV product, for a fixed constant flow rate, will yield a straight line with slope equal to H. The significance of this equation is the implication that it is important to control the flow rate, which may produce a greater effect on the length of stain than does the concentration of the test gas. In the optimal design of an indicator tube, therefore, it is desirable that the reaction rate be rapid enough to permit the establishment of equilibrium between the indicating gel and the test gas, thus producing a stoichiometric relationship between the volume of stained indicating gel and the quantity of the absorbed test gas. Such equilibrium conditions may be assumed to exist when stain lengths are directly proportional to the volume of sampled air and are not affected by the sampling flow rate. With this situation a "log–log" plot of stain length versus concentrations for a fixed sample volume may be prepared in the calibration of a given batch of tubes.

From the preceding discussion of the complexity of the heterogeneous phase kinetics of indicator tube reactions, the quality-control problems associated with their manufacture and storage, and the difficulties posed by interfering substances, it is obvious that frequent, periodic calibration of these devices should be made by the user. Dynamic dilution systems for the reliable preparation of low concentrations of a test gas or vapor are recommended for this purpose.

4.3.4 Passive Monitors

Passive dosimetry has become very popular for personal monitoring in recent years, taking advantage of gaseous diffusion. The dosimeters, or monitors, utilize Brownian motion to control the sampling process into a collection medium. This technology is particularly well suited for personal monitoring devices, resulting in lightweight, low-cost monitors that require no power source.

The monitors rely on a concentration gradient across a static or placid layer of air to induce a mass transfer. The following equation, based on Fick's law, gives the steady-state relationship for the rate of mass transfer:

$$W = D\left(\frac{L}{A}\right)(C_1 - C_0)$$

where W = mass transfer rate
D = diffusion coefficient
A = frontal area of static layer
L = length or depth of static layer
C_1 = ambient concentration
C_0 = concentration at collection surface

It can be seen that by choosing an effective collection surface, such that C_0 is essentially zero, the mass transfer or collection rate is proportional to the ambient vapor concentration C_1. It may also be noted that the units of $D(A/L)$ are volume per unit time, the same as for volumetric flow in a pump monitoring system. The rate of sampling of the contaminant is then the product of the $D(A/L)$ term and the average ambient concentration.

Figure 27.10 illustrates the construction of a monitor for mercury vapor in which the collection surface is a gold layer (19). A similar configuration (Figure 27.11) has been developed for nitrogen dioxide (20). In the case of the mercury monitor, mercury vapor in the atmosphere passes through the microporous barrier film and continues through the static air column according to Flick's law. The average time required to reach the collection surface is

$$T = \frac{L^2}{2D} \text{ in seconds}$$

For example, the average time for a mercury atom to progress to the collection surface when the depth L is 0.65 cm can be shown, using $D = 0.12$ cm^2/sec, to be 1.75 sec. Thus the sampling occurs rapidly when thin static air columns are used.

The accuracy and precision of the sampling process are functions of the measured exposure time, velocity effects, and temperature effects. The accuracy and precision

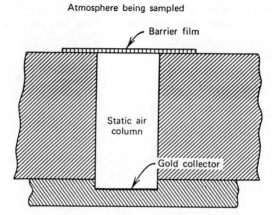

Atmosphere being sampled

Barrier film

Static air
column

Gold collector

Figure 27.10. Cross-sectional view of mercury detector.

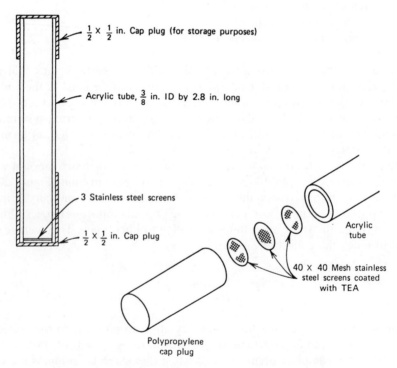

$\frac{1}{2} \times \frac{1}{2}$ in. Cap plug (for storage purposes)

Acrylic tube, $\frac{3}{8}$ in. ID by 2.8 in. long

3 Stainless steel screens

$\frac{1}{2} \times \frac{1}{2}$ in. Cap plug

Acrylic
tube

40 × 40 Mesh stainless
steel screens coated
with TEA

Polypropylene
cap plug

Figure 27.11. Exploded view of nitrogen dioxide monitor.

of the reported concentration are functions of the calibration standards, collection media, and analytical method used. Of these factors, the potential velocity and temperature effects distinguish this type of monitoring device from the conventional dynamic or flow monitor. All other factors are common to both methods.

Regarding velocity effects, the thickness of the attached boundary layer on the surface of the barrier film is a function of the velocity of air movement over the face of the monitor. Sampling that is independent of this air velocity can be achieved when the L term is large compared to the average boundary layer thickness. For temperatures between 50°F and 88°F, the temperature factor is constrained to ±1.8 percent. For use at higher or lower temperatures, the temperature data may be rerecorded to allow corrections to be made.

This type of personal monitor is attached to the worker in his breathing zone. The total exposure time is noted, and analysis results give the amount of vapor collected. These data provide an average mass collection rate, which can then be used to calculate the time-weighted average concentration. The physical parameters of the sampler design are chosen according to desired exposure time and the substance to be monitored.

Corroborative testing of both the mercury and nitrogen dioxide monitors and comparison of results to reference methods have shown excellent agreement. With the increasing emphasis on development of specific methods for monitoring workers' exposures to contaminants, and the extension of the concept to colorimetric indicator tubes, it is likely that the passive monitor concept will become the vital basis for a new generation of industrial hygiene monitoring equipment.

5   SAMPLING FOR PARTICULATES

In classifying airborne particulates, the term "aerosol" normally refers to any system of liquid droplets or solid particles dispersed in a stable aerial suspension. This requires that the particles or droplets remain suspended for significant periods of time.

Liquid particulates usually are classified into two subgroups, mists and fogs, depending on particle size. The larger particles generally are referred to as mists, whereas small particle sizes result in fogs. Liquid droplets normally are produced by such processes as condensation, atomization, and entrainment of liquid by gases.

Solid particulates usually are subdivided into three categories: dusts, fumes, and smoke, the distinction among them being primarily related to the processes by which they are produced. Dusts are formed from solid organic or inorganic materials by reducing their size through some mechanical process such as crushing, drilling, or grinding. Dusts vary in size from the visible to the submicroscopic, but their composition is the same as the material from which they were formed. Fumes are formed by such processes as combustion, sublimation, and condensation. The term is generally applied to the oxides of zinc, magnesium, iron, lead, and other metals, although solid organic materials such as waxes and some polymers may form fumes by the same methods. These particles are very small, ranging in size from 1 μm to as small as 0.001 μm in diameter. Smoke is a term generally used to refer to airborne particulate resulting from the combustion of organic materials (e.g., wood, coal, or tobacco). The resulting smoke particles are all usually less then 0.5 μm in diameter.

The nature of the airborne particulate dictates to a great extent the manner in which sampling of the environment must be accomplished. Sampling is performed by drawing a measured volume of air through a filter, impingement device, electrostatic or thermal precipitator, cyclone, or other instrument for collecting particulates. The concentration of particulate matter in air is denoted by the weight or the number of particles collected per unit volume of sampled air. The weight of collected material is determined by direct weighing or by appropriate chemical analysis. The number of particles collected is determined by counting the particles in a known portion or aliquot of the sample and extrapolating to the whole sample.

5.1 General Requirements

The general requirements discussed earlier for gases and vapors apply, for the most part, to particulate contaminants as well. There are several aspects of sampling that apply only for particulates, however, because of the wide range of particle sizes of airborne particulates confronting the industrial hygienist in most industrial settings.

The sampling train for particulates consists of the following components: air inlet orifice, particle collection medium (and preselector, if classification of the total particulate is being done), flowmeter, flow rate control device, and air mover or pump. Of these, the most important by far is the particle collection medium, used to separate the particles from the sampled air stream. Both the efficiency of the device and its reliability must be high. The pressure drop across the medium should be low, to keep the size of the required pump to a minimum. The medium may consist of a single element, such as a filter or impinger, or there may be two or more elements in series, to classify the particulate into two different size ranges. Proper selection, care, and calibration of the other components of the sampling train, particularly the flowmeter and flow control mechanism, are discussed elsewhere in this book.

Usually it is important that the sampling method not alter chemical or physical characteristics of the particles collected. For example, if the material is soluble, it cannot be collected in a medium capable of dissolving it. If the particles have a tendency to agglomerate, and it is important to be able to distinguish individual particles, deep-section collection on a filter should not be used.

An additional consideration, which is of much greater concern with particulate than with gases and vapors, is the variation of concentration in space. Many cases have been reported of significant differences in concentrations being documented with sampling units placed equidistant from a source of contaminant generation. Similarly, with personal breathing zone sampling, it is not uncommon for simultaneous samples obtained on both shoulders of a worker to indicate substantially different concentrations. The importance of these observations lies in the understanding that particulate sampling results by themselves indicate conditions within a short distance of the sampling unit and should be augmented with additional information, such as studies of airflow patterns within the workroom.

5.2 Collection Techniques

The concept of grab sampling for particulates is not as valuable as for gases or vapors; however, there are methods for collecting instantaneously a sample of airborne particles. One such device is the konimeter which, although of perhaps only historical interest today, served a very useful purpose in industrial hygiene studies. The konimeter draws a small, measured volume of air into the instrument and literally blasts it at high velocity against a glass plate on which the particles deposit. The particles then can be examined microscopically and counted or defined in other terms. Because there are millions of particles per cubic foot of most industrial workrooms, a very small volume of air is needed for this technique.

Another early means of obtaining an instantaneous sample for particulates was the settling chamber. With this device, a chamber is opened in the atmosphere being tested and is closed rapidly to trap the sample. The particles in the air then are allowed to settle by gravitational forces and are collected on a glass plate for subsequent microscopic analysis.

Today, the most meaningful sampling for particulate is done over extended periods of time with various collection techniques, depending on the material being collected and the availability of sampling equipment. The most common collection techniques are filtration, impingement, impaction, elutriation, electrostatic precipitation, and thermal precipitation.

5.2.1 Filtration

Undoubtedly the most common method of collecting airborne particulate is filtration. The fibrous type of filter matrices consists of irregular meshes of fibers, usually about 20 μm or less in diameter. Air passing through the filter changes direction around the fibers and the particles impinge against the filter, where they are retained. The largest particles (30 μm and greater) deposit to some extent by sieving action; the smaller particles (submicrometer sizes) also deposit through their Brownian motion, which carries them into the filter material. Efficiency of collection generally increases with airstream velocity, density, and particle size for particles greater than 0.5 μm in diameter. Deposition by diffusion dominates for the smallest particle sizes and decreases as the diameter of the particle increases. Thus there is a size at which the combined efficiency by impingement and diffusion is at a minimum; this is usually 0.1 μm diameter. Because the total weight of particles less than 0.1 μm diameter usually is less than 2 percent of the total collected dust, deposition by diffusion can practically be ignored. Of course, there are exceptions, such as sampling for freshly formed metal fumes, where diffusion deposition is very significant.

Filters are available in a wide variety of matrices including cellulose, glass,

Table 27.4. Common Applications of Filters

FILTER MATRIX	COMMON APPLICATIONS
Cellulose Ester	asbestos counting, particle sizing, metallic fumes, acid mists
Fibrous Glass	total particulate, oil mists, coal tar pitch volatiles
Paper	total particulate, metals, pesticides
Polycarbonate	total particulate, crystalline silica
Polyvinyl chloride	total particulate, crystalline silica, oil mists
Silver	total particulate, coal tar pitch volatiles, crystalline silica
Teflon	special applications (high temp.)

asbestos, ceramic, carbon, metallic, polystyrene, and other polymeric materials (see Table 27.4). Filters made of these fibrous materials consist of thickly matted fine fibers and are small in mass per unit face area, making them useful for gravimetric determinations. Of these, cellulose fiber filters are the least expensive, are available in a wide range of sizes, have high tensile strengths, and are relatively low in ash content. Their greatest disadvantage is their hygroscopicity, which can present problems during weighing procedures. The filters made of synthetic fibers, particularly glass and polyvinyl chloride, are more common, partly because stable tare weights can be determined easily.

Membrane filters, microporous plastic films made by precipitation of a resin under controlled conditions, are used to collect samples that are to be examined microscopically, although they can also serve for gravimetric sampling and for specific determinations using instrumentation. Thus the cellulose ester membrane filters are the most commonly used filters for sampling for such substances as asbestos (analyzed microscopically for fiber count) and metal fumes (analyzed for specific elements by atomic absorption techniques).

5.2.2 Impaction and Impingement

Impactors take advantage of a sudden change in direction in airflow and the momentum of the dust particles to cause the particles to impact against a flat surface. Usually impactors are constructed in several stages, to separate dust by size fractions. The particles adhere to the plate, which may be dry or coated with an adhesive, and the material on each plate is weighed or analyzed at the conclusion of sampling. It is imperative that the impactors be calibrated for the particular material of interest, because the manufacturer's calibration typically is based on a uniformly sized particle that may not give accurate results for the substance of interest.

Impingers also utilize inertial properties of particles to collect samples. Although the interest in, and application of, impingers is waning, they still play an important role in industrial hygiene sampling. The impinger consists of a glass nozzle or jet submerged in a liquid, frequently water. Air is drawn through the nozzle at high velocity, and the particles impinge on a flat plate, lose their velocity, are wetted by the liquid, and become trapped. Gases that are soluble in the liquid also are collected. Usually the contents of the impinger samples are analyzed microscopically, gravimetrically, or in a few cases, by specific methods.

The principles of collection for impaction and impingement are quite similar. The primary distinction between them is that with impaction, the particles are directed against a dry or coated surface, whereas a liquid collecting medium is used in impingement.

5.2.3   Elutriation

Elutriators have been essential elements in the sampling trains used to characterize dust levels in many mineral dust surveys, usually as preselectors at the front of the sampling train. Elutriators can have either a horizontal or a vertical orientation.

Elutriators are quite similar to inertial separators in the theoretical basis of operation. The primary difference is that elutriators operate at normal gravitational conditions, whereas the inertial collectors induce very high momentum forces to achieve collection of particles. Horizontal elutriators make use of the fact that as air moves across a horizontal channel with laminar flow, particles of greatest mass tend to cross the streamlines and settle because of gravity. The smaller particles remain airborne by resistance forces of the air for longer times and distances, depending on their size and mass. It is imperative that elutriators be used and operated exactly as described by the manufacturer to avoid disturbances at the inlet, to ensure laminar flow along the elutriator, and avoid risk of redispersion of settled particles.

Verticle elutriators utilize the same dependence on gravitational forces to separate the dust into fractions, except that with the vertical elutriator the natural force works in a direction opposite to the induced airflow instead of normal to it. The vertical elutriator is recognized as a practical device to sample for cotton dust when it is desired to avoid collecting the lint or fly.

5.2.4   Electrostatic Precipitation

Electrostatic precipitators have been used for many decades for industrial air analysis in workrooms. These systems have the advantage of negligible flow resistance, no clogging, and precipitation of the dust onto a metal cylinder or foil liner whose weight is unaffected by humidity. In most units, the "wire and tube" system is used; a stiff wire, supported at one end, is aligned along the center of the tube and serves as the charged electrode. The tip of the wire is sharpened to a point, and a high voltage (10 to 25 kV dc) is applied to the electrode. The corona discharge from the tip charges particles suspended in the air that is drawn through the tube. The electrical gradient between the wire and the wall of the cylinder (or foil) causes the charged particles to migrate to the inside surface of the tube.

The migration velocity of the charged particles greater than 1 μm in diameter increases in proportion to particle diameter, although migration velocity is almost independent of particle diameter for particles smaller than about 1 μm. Therefore very high separation efficiencies are attainable with electrostatic precipitators, and they have become particularly well suited for particles of submicrometer size, such as those in metal fumes.

5.2.5 Thermal Precipitation

A particle in a thermal gradient in air is directed away from a high-temperature source by the differential bombardment from gas molecules around it. This action is utilized in the design of thermal precipitation units. Air is drawn past a hot wire or plate, and the dust collects on a cold glass or metal surface opposite the hot element. Because a high thermal gradient is needed, the gap between the wire or plate and the deposition surface is kept very small, typically less than 2 mm. The migration velocity induced by the thermal gradient is small and is very nearly independent of particle diameter. Because of the severe limitations on maximum flow rate possible with high deposition efficiency, however, these units have been used only for collecting sufficient particulate matter for examination under a microscope. An additional limitation is the inappropriateness of these devices for sampling mists or other liquid particulates, unless their boiling points are high enough to ensure that the liquid is not volatilized by the operating temperatures of the instrument.

5.2.6 Centrifugal Collection

With the increasing emphasis on personal monitoring, there has been increasing use of preselectors in conjunction with industrial hygiene sampling to document concentrations of "respirable" dust. This is most commonly done with centrifugal separators, such as the cyclone. Air enters the cyclone tangentially through an opening in the side of a cylindrical or inverted cone-shaped unit. The larger particles are thrown against the side of the cyclone and fall into the base of the assembly. The smaller particles are drawn toward the center of the unit, where they swirl upward along the axis of a tube extending down from the top. The air in the cyclone rotates several times before leaving, and consequently the dust deposits as it would in a horizontal elutriator having an area several times that of the cyclone's outer surface. Thus the volume of a cyclone is much smaller than a horizontal elutriator or other inertial collector with the same flow rate and efficiency.

Cyclones used to sample for respirable dust, such as in the course of determining compliance with the respirable mass standard for silica-containing dusts, should have performance characteristics meeting the criteria of the American Conference of Governmental Industrial Hygienists (ACGIH) (Figure 27.12). The orientation

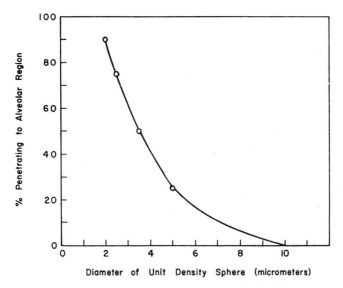

Figure 27.12. ACGIH respirable dust criteria.

of the cyclone is not as critical as for the elutriators, and small 10-mm diameter cyclones have become commonplace for personal breathing zone sampling.

5.3 Direct-Reading Techniques

The development of methods for continuous monitoring of particles in air has been more substantial in the field of air pollution than in industrial hygiene. In general, these instruments have limitations on their sensitivity from two primary sources: the random property fluctuations of the accompanying gas molecules, and the noise level of the electronic circuitry, which converts the physical change to a measurable signal. Instruments that read out particle "sizes" are often calibrated using well-characterized aerosols not necessarily representative of the particles sampled; thus the accuracy of these instruments may be highly variable. The response of the instruments is nonspecific; that is, the devices respond to a property of the substance (size, shape, or mass), rather than to the material itself. However, there are needs of the industrial hygienist that can be met by direct-reading instruments for particulates, and continuing application of these devices is anticipated.

5.3.1 Aerosol Photometry (Light Scattering)

Instruments using light scattering techniques are based on generation of an electrical pulse by a photocell that detects the light scattered by a particle. A pulse height analyzer estimates the effective diameter. The number of pulses is related to the number of particles counted per unit flowrate of the sampled medium. Instruments that give a size analysis based on the measurement of total particle concentration in a large illuminated volume are used in monitoring particulate concentrations in experimental rooms or exposure chambers.

Aerosol photometry can usually provide only an approximate analysis of particulates, classified according to particle size, in industrial surveys. Calibrating the instrument with each type of particulate to be measured is not practicable. The great variations in shape, size, agglomerative effects, and refractive indexes of the various components in a given dust suspension make such calibrations virtually impossible. Whereas aerosol photometry can indicate the particulate concentration in the different size ranges of interest, it usually is necessary to perform size distribution analyses by microsieving or microscopic procedures.

5.3.2   Respirable Mass Monitors

The first direct-reading respirable mass monitor to be introduced to the industrial hygiene field was based on the attenuation of beta radiation resulting from collection of a sample of dust by taking comparative readings of a beta-radiation source. The instrument uses a two-stage collection system. The first stage consists of a cyclone precollector for the retention of the nonrespirable fraction of the dust. The precollector retains the larger particles and allows those of respirable size to pass to the second stage, which consists of a circular nozzle impactor and beta absorption assembly with an impaction disk. The dust collected by impaction on the thin plastic film increasingly absorbs the beta radiation from a ^{14}C source being monitored by a Geiger–Müller detector. The penetration of low-energy beta radiation depends almost exclusively on the mass per unit area of the absorber and the maximum beta energy of the impinging electrons; it is independent of the chemical composition or the physical characteristics of the collected, absorbing matter. This unit, and modifications thereof, incorporate a digital readout of the respirable mass concentration.

A second type of unit is based on the change in resonant frequency of a piezoelectric quartz crystal accompanying deposition of particulate on the face of the crystal. This unit uses an impactor inlet to separate the dust into respirable and nonrespirable fractions. Modifications have utilized electrostatic precipitation for particle collection as well as multistage impactors. The mass concentration-to-frequency relationship is linear, and the instrument has sufficient sensitivity to measure ambient particulate mass concentration in a few seconds sampling at 1 to 3 liters/min.

5.3.3   Other Developments

These monitoring devices, like all direct-reading instruments, are not personal monitoring devices. Because the readings are instantaneous, they cannot be used directly to determine compliance with 8-hr, time-weighted average standards. Thus such units cannot be used as compliance instruments. However, the usefulness of direct-reading equipment for analyzing environmental conditions "on the spot" has been great, and continued development and refinement is likely.

6 ANALYSIS OF SAMPLES

Continuing advances and improvements in analytical capabilities have made it possible to measure minute quantities of specific compounds, ions, or elements. The industrial hygienist can process a very small sample of air and accurately determine the presence of suspected contaminants. As presented earlier in this chapter, the field industrial hygienist should work closely with the industrial hygiene analyst, to become familiar with the limitations of the analytical equipment of interest, thus being able to plan a sampling strategy with maximum efficiency.

It is beyond the scope of this chapter to outline specific procedures for industrial contaminants. Instead, brief descriptions of the various analytical methods and techniques that have been applied to industrial hygiene samples are presented, with the expectation that the reader will consult more detailed sources for complete understanding of analytical requirements for particular substances of interest. Of particular note are the descriptions of analytical methods published by NIOSH and OSHA (17, 18).

6.1 Gravimetric Techniques

The most frequently employed analytical method for industrial hygiene samples is gravimetric analysis of filters or other collection media to determine the weight gain. This requires careful handling and processing of the media before collecting the sample, as well as the conditioning of the media after collection of the sample in the exact manner as used to obtain tare weights. In so doing, any necessary correction for the "blank" can be incorporated into the analysis. Often gravimetric analysis is done as a "gross" analysis, or general indicator of conditions, with subsequent analyses performed for specific constituents of the sample.

Another type of gravimetric technique involves the formation of a precipitate by combining a sample solution with a precipitating agent, with subsequent weighing of the solid precipitate formed.

6.2 Titrimetric Methods

Acid–base and oxidation–reduction volumetric procedures are outstanding examples of simple but useful analytical methods still employed in the analysis of industrial hygiene samples. Hydrogen chloride gas and sulfuric acid mist can be collected in an impinger or bubbler containing a standard sodium hydroxide solution and quantified by back-titration with a standard acid. Ammonia and caustic particulate matter can be collected in acid solution with similar apparatus, and the airborne concentration determined by titration with a standard base. Oxidation–reduction titrations, principally iodometric, are useful for measuring sulfur dioxide, hydrogen sulfide, and ozone. Improved volumetric methods utilize electrodes to indicate acid–base null points, and amperometric methods are available for oxidation–reduction titrations. These electrical techniques increase analytical precision and speed up the analyses but do not affect the sensitivity appreciably.

6.3   Optical Methods

Much of the sampling for dust requires analysis by microscopic techniques. The use of light microscopy for "dust counting" is decreasing, being replaced by more specific, and more reproducible, mass sampling techniques. However, it is frequently necessary to determine the particle size distribution of airborne particulate, and optical methods offer an effective way of doing this.

In addition to the more classic counting applications of microscopy, the present sampling and analysis method for asbestos is based on actual fiber counting at 400 to 450× magnification using phase contrast illumination. As with all optical methods, the analytical results (i.e., actual counts) are somewhat analyst-dependent, because much of the technique requires subjective analysis by the microscopist.

6.4   Colorimetric Procedures

Changes of color intensity or tone have been the bases of many useful industrial hygiene analytical methods. For example, the use of Saltzman's reagent in a fritted glass bubbler to determine the airborne concentrations of nitrogen dioxide is a classic application of such methods. Under controlled conditions of sampling, the concentration of NO_2 in the air is inversely proportional to the time required to produce the color change. Titrations employing acid–base and iodometric reactions with color indicators are conducted in similar fashion.

Usually such titrations of air samples lack the accuracy and precision obtainable with careful laboratory procedures, but they are adequate for most field studies and have the great advantage of giving a direct and immediate indication of the environmental concentrations. Relatively sensitive and specific analyses of many contaminants can be made, using the spectrophotometers available as both laboratory and field instruments.

Colorimetric methods involve analytical reactions to produce a color in proportion to the quantity of the contaminant of interest in the sample. For example, in the determination of metals, the dithizone extraction method is able to determine selectively the various metallic elements depending on the pH of the solution.

6.5   Spectrophotometric Methods

In addition to the colorimetric procedures, which take advantage of spectrophotometry operating in the visible range, infrared and ultraviolet spectrophotometers have considerable application in the industrial hygiene area. The interaction of electromagnetic radiation with matter is the basis for such analytical techniques. Principles of operation extend from the infrared radiation spectra, to the ultraviolet and, in fact, to the X-ray region. The latter can be used to provide information on elemental composition (fluorescence) and crystal structure (diffraction).

In most cases the sample, whether gas, liquid, or solid, is exposed to radiation of known characteristics and specific wavelengths (fluorescence), and the fractions transmitted or scattered are determined and quantified. Color production, turbidity, and fluorescence are examples of properties determined by electromagnetic radiations that are widely used for quantifying industrial hygiene air samples.

6.6 Spectrographic Techniques

Because the smallest trace of materials can be detected by the spectrograph, spectrographic procedures may be employed for small amounts of metallic ions and elements when other procedures cannot be used. The chief limitations are the high cost and the need for a highly trained technician who has access to a rather complete spectrographic laboratory to do quantitative work. Generally the degree of sensitivity afforded by these units is not required by the industrial hygienist, although it frequently is desirable to obtain a complete elemental analysis of a sample of unknown composition as a starting point for an elaborate analytical program.

In applying emission spectroscopy, a solid sample is vaporized in a carbon arc, causing the formation of characteristic radiation, which is dispersed by a grating or a prism, and the resulting spectrum is photographed. Each metallic or metal-like element can be identified from the spectra that are formed. Elemental analyses of body tissues, dust, ash, and air samples can be qualitatively analyzed by this technique.

With mass spectroscopy gases, liquids, or solids are ionized by passage through an electron beam. The ions thus formed are projected through the analyzer by means of an electromagnetic or electrostatic field, or simply by the time necessary for the ions to travel from the gun to the collector. Each compound has a characteristic ionization pattern that can be used to identify the substance. This analytical tool, in conjunction with gas chromatography, has become a powerful technique for separating and identifying a wide range of trace contaminants in industrial hygiene and ambient air samples.

6.7 Chromatographic Methods

The development of chromatographic methods of analysis has given the industrial hygienist an extremely versatile means of quantifying low concentrations of airborne contaminants, particularly organic compounds. Gas chromatography utilizes the selective absorption and elution provided by appropriately chosen packings for the columns to separate mixtures of substances in an air sample or in a desorption solution. The various compounds in the air sample have different affinities for the material in the column, thus "slowing" some of the constituents more than others, with the result that as the individual compounds reach the detector associated with the chromatograph, they can be quantified by running standards of known concentration of the various substances along with the unknowns. This separation is achieved without any appreciable change in the entities; thus the chromatograph can serve as an analytical technique in its own right by attaching an appropriate detector. Thermal conductivity, flame ionization, and electron capture detectors are commonly used for this purpose. The chromatograph can also serve to "purify" a sample by separating the constituents and selecting a narrow portion of the eluted sample. This portion then can be subjected to other more sophisticated types of analysis, such as mass spectroscopy.

6.8 Atomic Absorption Spectrophotometry

With atomic absorption spectrophotometry (flame photometry), monochromatic radiation from a discharge lamp containing the vapor of a specific element, such as lead, passes through a flame into which the sample is aspirated. The absorption of the monochromatic radiation is measured by a double-beam method, and the concentration is determined. This technique permits rapid determination of almost all metallic elements. Solutions of the metals are aspirated into the high-temperature flame, where they are reduced to free atoms. The absorption generally obeys Beer's law in the parts per million range, where quantitative determinations can be made. The characteristic absorption gives this technique high selectivity. Most interferences can be overcome by proper pretreatment of the samples. Atomic absorption methods have found substantial use in industrial hygiene in the determination of both major and trace metals in industrial hygiene samples, as well as in blood, urine, and other body fluids and tissues.

6.9 Other Techniques

Many additional, specialized analytical methods are available to the industrial hygienist and analytical chemist for application to specific qualitative and quantitative needs. For detailed information on particular methods or procedures, the reader is advised to consult the references listed at the end of this chapter and, more important, to keep abreast of current developments in the industrial hygiene analytical field by subscribing to journals or routinely reviewing the wealth of new information constantly coming forth.

As a general guide to the industrial hygienist, a summary of sampling and analytical techniques appropriate for a variety of contaminants commonly encountered in industry is presented in Table 27.5.

7 BIOLOGICAL MONITORING

In the final analysis, the degree of success associated with attempts to provide a safe and healthful work environment must be determined by assessing the amount of the contaminant of concern that has been actually absorbed by the workers. Regardless of the degree of sophistication applied to environmental and personnel monitoring, the extent to which workers have absorbed the contaminant should be determined by some clinical measurement on the individual. In the industrial hygiene profession, this can be derived from direct quantitative analysis of body fluids, tissue, or expired air for the presence of the substance or metabolite. An indirect determination of the effect of the substance on the body can be made by measurements on the functioning of the target organ or tissue. With the possible

Table 27.5. Sampling–Analytical Methods

Substance	Sampling Medium					Analytical Method				
	charcoal/silica gel	direct-reading unit	filter	solvent/reagent	std. acid/base	atomic absorption	gas chromatography	gravimetric/colorim.	ion-selective electrode	X-ray diffraction
Acid mists					X				X	
Alcohols	X						X			
Ammonia					X				X	
Asbestos			X							X
Carbon mon-oxide		X								
Coal tar pitch volatiles			X					X		
Hydrocarbon	X						X			
Hydrogen sulfide				X				X		
Metals			X			X				
Oil mists			X					X		
Organic va-pors	X						X			
Pesticides			X	X			X			
Phenol					X		X			
Silica			X							X
Solvents	X						X			

exception of carcinogenic substances, even the most hazardous materials have some "no effect" level below which exposure can be tolerated by most workers for a working lifetime without incurring any significant physiological injury.

An ideal approach to establishment of no-effect or tolerance levels would require a classical chemical engineering materials balance applied to humans and their environment. Simply stated, the amount of any substance entering the body must equal the products and by-products leaving the system, plus any accumulation within the body. A material balance such as this was established for lead by and through the efforts of Dr. Robert Kehoe and his co-workers (19). The study involved quite elaborate test facilities for analyzing food, beverages, air intake, and urinary, fecal, and expired breath outputs, as well as a closely controlled environment in which volunteer subjects were willing to put in 40-hr weeks under conditions that closely simulated actual work environments. The studies indicated that lead did not undergo metabolism to other forms within the body. This made the determination of the overall material balance relatively straightforward. Unfortunately, such is not the case with most occupational contaminants of interest today, particularly the organic materials. Material balances are much more difficult to establish for compounds that undergo metabolic change to other chemical structures within, and during passage through, the human body.

The following general presentation of potential applications for biological monitoring should be a useful supplement to an industrial hygiene monitoring program.

7.1   Urine Analysis

Perhaps the most common biological fluid analyzed in attempts to determine the extent to which individuals have been exposed to contaminants is urine. The following procedure for collection and analysis of urine samples for phenol as an indication of exposure to benzene is illustrative of procedure, because it encompasses many of the considerations that must be made in processing such samples (21). Deviations or alternatives available for urine samples for other contaminants are indicated in parentheses.

For urinary phenol samples, "spot" urine specimens of about 10 ml are collected as close to the end of the workday as possible. (For other contaminants, it is preferable to collect the specimen at the beginning of the work shift or perhaps composite a 24-hr sampling.) If any worker's urine phenol level exceeds 75 mg/liter, procedures should be instituted immediately to determine the cause of the elevated levels and to reduce the exposure of the worker to benzene. Weekly specimens are collected as described until three consecutive determinations indicate that the urinary phenol levels are below 75 mg/liter.

To collect the sample, the workers should be instructed to wash their hands thoroughly with soap and water and to provide a urine sample from a single voiding into a clean, dry specimen container having a tight closure and at least 120-ml capacity. The containers may be of glass or of wax-coated paper or other disposable materials. After collection of the specimens, 1 ml of a 10 percent copper sulfate solution should be added to each sample as a preservative. (Other contaminants will require the addition of specific preservatives to maintain the stability of the sample.) The samples should be stored immediately under refrigeration, preferably below 4°C. Such refrigerated specimens will remain stable for approximately 90 days. (The stability of specimens for other metabolites or contaminants will vary; this factor should be ascertained before establishing a biological monitoring program.) If shipment of samples is necessary, the most rapid method available should be employed, using acceptable packing procedures specified by the carrier. Each specimen must include proper identification, including the worker's name, the date, and time of collection.

The urine samples are analyzed by treating the specimen with perchloric acid at 95°C to hydrolyze the phenol conjugates, phenyl sulfate, and phenyl glucuronide formed as detoxification products following absorption of benzene. Total phenol is determined by gas chromatographic analysis of the extract.

Although the specific requirements of methods for other compounds may differ significantly from those outlined above, it should be emphasized that the specific concentrations indicating excessive buildup of contaminants or metabolites within the body are not known for most compounds with any certainty. The biological

Table 27.6. Urine Biological Limit Values

SUBSTANCE	BLV, mg/l	
	Warning	Intervention
Aniline	10	20
Arsenic	0.3	0.6
Cadmium	0.05	0.1
Lead	0.10	0.15
Mercury	0.05	0.1
Nitrobenzene	25	50

threshold limit values, that is, concentrations indicative of excessive exposure for some compounds, have been determined; biological exposure indexes are published annually by the ACGIH (22). Table 27.6 lists representative organic and metallic compounds for which biological threshold limit values (BLVs) in urine have been determined. The substances in this table are analyzed directly in the urine; that is, metabolic products are not evaluated, as was the case for exposure to benzene and consequent analysis for phenol in the urine. The first concentration presented in the table is intended to serve as a warning that exposures to the indicated compounds are approaching the limits for acceptable absorption of the substance into the body; the second concentration is a level at which medical intervention and/or treatment of the individual is advised.

7.2 Blood Analysis

Collection of samples of workers' blood and subsequent analysis for specific indications of exposure, either for the contaminant in question directly or for key metabolites, is another useful biological monitoring technique. The following procedure for collecting and analyzing blood samples for lead is presented as an illustration of the technique for this method (23).

A 10-ml sample of whole blood is collected using a vacutainer and a sterilized, stainless steel needle. (It is important that the vacutainers have been determined to be free of the contaminant of interest, because any leaching of the material from the vacutainers into the blood samples would distort the analytical result. The possibility of such distortion has been of particular concern in the analysis of blood samples for lead.) In the laboratory, the sample is transferred to a tared, leadfree beaker; no aliquoting of the sample is permissible, because most of the lead is present in the clotted portion. The weight of the blood sample is determined to the nearest 0.01 g, weighing rapidly to minimize evaporation of the sample. A system for ashing the sample is employed to permit the analyst to handle the sample

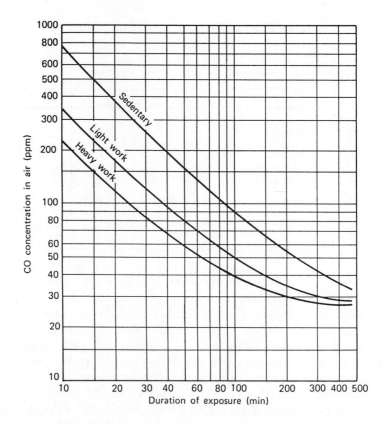

Figure 27.13. Length of time to achieve 5 percent carboxyhemoglobin at various concentrations of carbon monoxide (25)

easily, for the blood clot will break up readily and smoothly. The sample then is evaporated just to dryness and, after cooling, additional acid is added and ashing continued until the residue fails to darken upon additional heating. The residue then is put into solution and analyzed directly by atomic absorption spectrophotometry or other suitable method.

Another useful application of blood analysis as an index of exposure to contaminants is the determination of carboxyhemoglobin in the blood as an indication of exposure to carbon monoxide. Extensive studies of the concentration of carbon monoxide in the air and consequent level of carboxyhemoglobin in the blood have been made (24). Figure 27.13 presents a family of curves giving the length of time of exposure to various concentrations of carbon monoxide in air necessary to achieve 5 percent carboxyhemoglobin levels in the blood as a function of the relative activity of the workers, that is, sedentary, light work, or heavy work. Table 27.7 summarizes typical biological threshold limits for compounds or their metabolic indicators in blood.

Table 27.7. Blood Biological Limit Values

SUBSTANCE	BLV
Carboxyhemoglobin, as CO index	~10 %
Ethanol	0.08 %
Lead	0.08 mg/100 g blood
Methemoglobin, as nitro-amino index	~7 %
Methyl bromide	10 mg/100 g blood
Organophosphates	15 % inhibition of cho-linesterase activity

7.3 Expired Breath Analysis

A biological monitoring technique that is increasing in application is analysis of expired air for contaminants for which equilibrium between the body and respired air can be used as an indication of the concentration of the contaminant in the workroom air to which the individual has been exposed. For the most part, this type of analysis has been limited to the chlorinated hydrocarbon solvents such as methylene chloride, carbon tetrachloride, vinyl chloride, 1,1,1-trichloroethane, trichloroethylene, tetrachloroethylene, and some of the Freons (25).

In collecting the sample, the worker is instructed to take several deep breaths and direct the expired air into a flexible bag or to pass it through a glass tube after flushing the contents several times with expired air. In either case, the container is tightly sealed and the contents subsequently analyzed directly for the substance, usually using gas chromatographic techniques. The analytical result is compared to "breath decay curves," such as in Figure 27.14, to determine the concentration of the contaminant to which the worker had been exposed recently (25).

7.4 Future Considerations

It is likely that, with the increasing interest in providing safe work conditions for persons exposed to a tremendously wide range of contaminants, development of cause-and-effect relationships between the level of exposure to a particular contaminant and the amount of the substance remaining in the body will become a more integral part of the total occupational health monitoring procedure. Accordingly, greater emphasis will be placed on the medical aspects of the total monitoring effort in attempts to document, as an end result, that workers have not been exposed to excessive levels of contaminants. Where the results of biological monitoring indicate excessive exposures, and the environmental monitoring aspects indicate

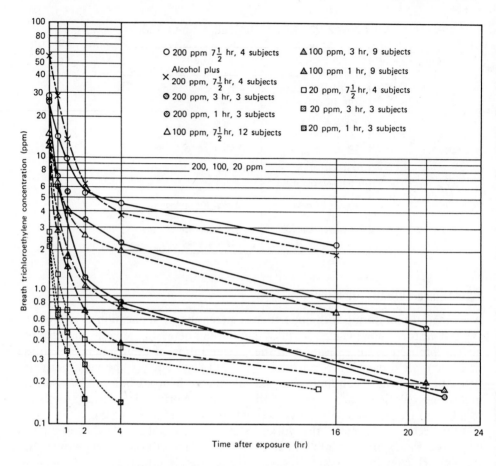

Figure 27.14. Trichloroethylene postexposure breath decay curves.

that concentrations are within acceptable limits, analysis of individuals' work practices, or at least specific analysis of an individual's work activity, may be required
to ascertain the cause of the elevated readings in a given individual. As such, the
biological monitoring program is a viable and extremely useful supplement to the
ongoing environmental and medical surveillance programs for contaminants for
which such coordinated efforts are possible.

8 SUMMARY

Recent laws that have been passed, both federal and state, require accurate, thorough data on the work environment. This requires the industrial hygienist not only
to be thoroughly knowledgeable about the laws affecting the profession, but also
to be technically competent in the methods of sampling, with emphasis on sampling
limitations.

8.1 Computer Assistance

Computers have made their presence felt, and appreciated, in almost every walk of life today. It is logical to expect the computer to be recognized as a real ally of the industrial hygienist. Projects that could only have been talked about decades ago, because of the tremendous volume of data needed to complete the work, are now rather trivial exercises and are accomplished with minimal effort. Calculations of time-weighted average exposures, using air sampling data and analytical results from dozens of individual samples, can be made in seconds; hours of calculations were required prior to computer assistance.

More important than assisting in the more mundane activities of industrial hygiene (equipment calibration, exposure calculations, etc.), the computer has made it possible to *analyze* data in a manner and to a degree virtually impossible without computer assistance. Of particular long-term benefit is the application of computers to "prospective epidemiology." This refers to the continuing maintenance of data bases that document the environmental condition, as compiled by routine industrial hygiene monitoring, and the various indexes of health status, derived from periodic medical surveillance. Information of this nature, referenced to individual workers to whom the data apply, is being compiled by many corporate industrial hygiene functions. Many software programs, incorporating data base management, spreadsheet profiles, rapid search-and/or-match provisions, and other aids, are available to the industrial hygienist (26, 27).

8.2 Regulatory Activity

At the beginning of the 1990s, OSHA was in the midst of what appears to be a move toward development and promulgation of "generic" standards. There was successful completion of the "PEL Project" in which the list of permissible exposure limits was updated, essentially by incorporation of the current list of ACGIH's threshold limit values. In addition, guidelines for occupational safety and health programs had been published, focusing on four key elements: management commitment, hazard surveys, hazard control, and training and education efforts. Both of these activities had an impact on the practicing industrial hygienist, the PEL Project to a much greater extent.

Two items of proposed rule making that, at the time this book went to press, had not been fashioned into final form (either as standards or guidelines) were generic requirements for exposure monitoring and for medical surveillance. In many respects, this activity is a direct continuation of efforts initiated and brought into focus by OSHA's hazard communication standard. In fact, even without passage of generic standards that require implementation of routine monitoring programs, it is logical to expect the emphasis on hazard communication to result in a significant increase in monitoring activity. As employees are informed periodically of the hazards of the materials with which they work, they are likely to begin asking for information regarding their own exposures.

Regardless of the motivation, that is, whether it is to add information to an ongoning data base of corporate exposure/medical information or to satisfy new exposure monitoring regulations, it appears likely that the next decade or two of industrial hygiene practice will see a dramatic increase in emphasis on industrial hygiene sampling and analysis.

REFERENCES

1. G. D. Clayton and F. E. Clayton, Eds., *Patty's Industrial Hygiene and Toxicology*, Vol. 2, 3rd rev. ed., Wiley Interscience; New York, 1981.

2. M. B. Jacobs, *The Analytical Toxicology of Industrial Inorganic Poisons*, Wiley Interscience, New York, 1967.

3. H. B. Elkins, *The Chemistry of Industrial Toxicology*, 2nd ed., Wiley, New York, 1959.

4. A. Hamilton and H. Hardy; *Industrial Toxicology*, 3rd ed., Publishing Sciences Group, Acton, MA, 1974.

5. J. Doull, C. D. Klaassen, and M. O. Amdur, eds., *Casarett and Doull's Toxicology: The Basic Science of Poisons*, 2nd ed., Macmillan, New York, 1980.

6. P. L. Williams and J. L. Burson, *Industrial Toxicology: Safety and Health Applications in the Workplace*, Van Nostrand Reinhold, New York, 1985.

7. E. Hodgson, R. B. Mailman, and J. E. Chambers; *Dictionary of Toxicology*, Van Nostrand Reinhold, New York, 1988.

8. American Conference of Governmental Industrial Hygienists, *Documentation of the Threshold Limit Values*, 3rd ed., ACGIH, Cincinnati, OH, 1971.

9. Y. Bar-Shalom, A. Segall, D. Budenaers, and R. B. Shainker, "Statistical Theory for Sampling of Time-Varying Industrial Atmospheric Contaminant Levels," Systems Control, Inc., Report to NIOSH Contract HSM-99-73-78, Palo Alto, CA, 1974.

10. D. Budenaers, Y. Bar-Shalom, A. Segall, and R. B. Shainker, *Handbook for Decisions on Industrial Atmospheric Contaminant Exposure Levels*, U.S. Department of Health, Education and Welfare, Cincinnati, OH, 1975.

11. F. J. Schuette, "Plastic Bags for Collection of Gas Samples," *Atm. Environ.*, **1**, 515 (1967).

12. G. O. Nelson, *Controlled Test Atmospheres, Principles and Techniques*, Ann Arbor Sciences Publishers, Ann Arbor, MI, 1971.

13. American Conference of Governmental Industrial Hygienists, *Air Sampling Instruments for Evaluation of Atmospheric Contaminants*, 4th ed., ACGIH, Cincinnati, OH, 1972.

14. B. A. Plog, ed., *Fundamentals of Industrial Hygiene*, 3rd ed., National Safety Council, Chicago, IL, 1988.

15. American Industrial Hygiene Association, *Direct Reading Colorimetric Indicator Tubes Manual*, AIHA, Akron, OH, 1976.

16. B. E. Saltzman, "Direct Reading Colorimetric Indicators, Section S," in *Air Sampling Instruments for Evaluation of Atmospheric Contaminants*, 4th ed., American Conference of Governmental Industrial Hygienists, Cincinnati, OH, 1972.

17. P. M. Eller, Ed., *NIOSH Manual of Analytical Methods*, 3rd ed., U.S. Department of Health and Human Services, Cincinnati, OH, 1985.

18. *OSHA Analytical Methods Manual*, OSHA Analytical Laboratory, Salt Lake City, UT, 1985.

19. A. Linch, *Biological Monitoring for Industrial Chemical Exposure Control*, CRC Press, Cleveland, OH, 1974.

20. G. H. Schnakenberg, "A Passive Personal Sampler for Nitrogen Dioxide," U.S. Dept. of the Interior, Bureau of Mines, Technical Progress Report, Pittsburgh, 1976.

21. National Institute for Occupational Safety and Health, "Criteria for a Recommended Standard . . . Benzene," U.S. Department of Health, Education and Welfare; Cincinnati, OH, 1974.

22. American Conference of Governmental Industrial Hygienists, Threshold Limit Values and Biological Exposure Indices for 1989–1990, ACGIH, Cincinnati, OH, 1989.

23. National Institute for Occupational Safety and Health, "Criteria for a Recommended Standard . . . Inorganic Lead," U.S. Department of Health, Education and Welfare, Cincinnati, OH, 1972.

24. National Institute for Occupational Safety and Health, "Criteria for a Recommended Standard . . . Carbon Monoxide," U.S. Department of Health, Education and Welfare, Cincinnati, OH, 1972.

25. R. D. Stewart et al., "Biological Standards for the Industrial Worker by Breath Analysis: Trichloroethylene," U.S. Department of Health, Education and Welfare, Cincinnati, OH, 1974.

26. ACGIH, *Microcomputer Applications in Occupational Health and Safety*, Lewis Publishers, Chelsea, MI, 1987.

27. J. T. Garrett, L. J. Cralley, and L. V. Cralley, *Industrial Hygiene Management*, Wiley-Interscience, New York, 1988.

CHAPTER TWENTY-EIGHT

Industrial Hygiene Engineering Controls

Robert D. Soule, C.I.H., C.S.P.

1 INTRODUCTION

Control of health hazards in the work environment is one of the traditional responsibilities of the industrial hygienist. Although diseases suspected of being related to the work environment were recognized more than 2500 years ago (e.g., lead poisoning of workers engaged in mining operations), the systematic application of industrial hygiene engineering controls is a recent technical development. The "profession" of industrial hygiene is a young one, the leading association in this field having celebrated its fiftieth anniversary in 1989.

Evidence of the importance of engineering control of the work environment among the alternative solutions to industrial hygiene problems is found in every current industrial hygiene text: all list the possible solutions in priority fashion as engineering controls, good work practices, administrative controls, and as a last resort, use of personal protective equipment. In fact, some decisions of the Occupational Safety and Health Review Commission make it clear that engineering control of specific substances must be attempted even when complete control is not feasible just by engineering means. In other words, the use of personal protective equipment alone cannot be justified simply by analyzing a problem and determining that complete engineering control cannot be provided. Instead, the language of federal regulations has been interpreted as mandating that engineering controls be provided to the extent feasible and, in the event these are not sufficient

Patty's Industrial Hygiene and Toxicology, Fourth Edition, Volume 1, Part B, Edited by George D. Clayton and Florence E. Clayton.
ISBN 0-471-50196-4 © 1991 John Wiley & Sons, Inc.

to achieve acceptable limits of exposure, use of personal protection or other corrective measures may be considered.

Thus it is clear that the importance of applying engineering controls, as a component of the total responsibility of the industrial hygienist, will continue to increase tremendously in the foreseeable future. All professionals in the field of industrial hygiene must be cognizant of the principles of engineering controls, and the engineering professional must be familiar with the details of implementing the control concepts. Accordingly, this chapter presents material useful to the industrial hygiene engineer and nonengineer alike in applying their resources to the industrial hygiene problems occurring in their respective settings. The focus here is on control of airborne contaminants; control of other types of stress, for example, noise, heat stress, and radiation, is discussed in the respective chapters elsewhere in this book.

2 PRINCIPLES OF CONTROL

The industrial hygiene engineering control principles are deceptively few: substitution, isolation, and ventilation, both general and localized. In a technological sense, an appropriate combination of these principles can be brought to bear on any industrial hygiene control problem to achieve satisfactory quality of the work environment. It may not be, and usually is not, necessary or appropriate to apply all these principles to any specific potential hazard. A thorough analysis of the control problem must be made, to ensure that a proper choice from among these methods will produce the proper control in a manner that is most compatible with the technical process, is acceptable to the workers in terms of day-to-day operation, and can be accomplished with optimal balance of installation and operating expenses.

With advancing technology and the tendency for acceptable exposure limits to become increasingly more stringent, the industrial hygiene engineer's function must include an ongoing analysis of installed control measures. Aside from problems associated with operation and maintenance of these provisions, the concept of the control provisions itself might become outdated by changes in process or regulations. The engineer then must be able to develop effective control methods, and must have the capability to continue to evaluate the effectivenes of these methods regularly.

2.1 Substitution

Although it can be debated that substitution is not an engineering option, it frequently offers the most effective solution to an industrial hygiene problem. There is a tendency to analyze any problem from the standpoint of correction rather than elimination. For example, the first inclination in considering a vapor exposure problem in a degreasing operation is to provide or increase ventilation rather than substitute a solvent having a much lower degree of hazard associated with its use. In its broadest sense, substitution includes replacing hazardous substances, changing from one type of process equipment to another, or in some cases even changing the process itself. In other words, material, equipment, or an entire process can be substituted to provide effective control of a hazard, often at minimal expense.

There are many examples of substituting materials of lower toxicity; some are classics in the history of industrial hygiene. Most industrial hygienists are familiar with the substitution of red phosphorus for white in the manufacture of matches. Although this was done primarily in reaction to a tax law, the result was a markedly reduced potential hazard. The sequence of substitutions of degreasing solvents is an interesting one: from petroleum naphtha to carbon tetrachloride to chlorinated hydrocarbons to fluorinated hydrocarbons. Each of these substitutions alleviated one problem but resulted in a new one. This underscores a basic problem associated with using substitution as a control method, in that one hazard can be replaced by another inadvertently.

In some processes, there is only limited opportunity to substitute materials; however, it might be possible to substitute, or at least modify, process equipment. This approach almost always is taken as a result of an obvious potential, usually physical, hazard. Applications to counter potential safety hazards are common: substituting safety glass for regular glass in enclosures, replacing unguarded equipment with properly guarded machines, replacing safety gloves or aprons with garments made of materials more impervious to chemicals being handled. Because substitution of equipment frequently is done as an immediate response to an obvious problem, it may not be recognized as an engineering control, even though the end result is every bit as effective.

Substituting one process for another may not be considered except in major modifications of a process. In general, a change in any process from a batch to a continuous type of operation carries with it an inherent reduction in potential hazard. This is true primarily because the frequency and duration of potential contact of workers with the process materials are reduced when the overall process approach becomes one of continuous operation. The substitution of processes can be applied on a fundamental basis. For example, substitution of airless spray for conventional spray equipment can reduce the exposure of a painter to solvent vapors. Substitution of a paint dipping operation for the paint spray operation can reduce the potential hazard even further. In any of these cases the automation of the process can further reduce the potential hazard.

2.2 Isolation

Application of the principle of isolation frequently is envisioned as consisting of installation of a physical barrier between a hazardous operation and the workers. Fundamentally, however, this isolation can be provided without a physical barrier by appropriate use of distance and time.

Perhaps the most common example of isolation as a control measure is associated with storage and use of flammable solvents. Large tank farms with dikes around tanks, underground storage of some solvents, detached solvent sheds, and fireproof solvent storage rooms within buildings all are commonplace in American industry.

Although the primary reason for the isolation of solvents is the risk of fire and explosion, the principle is no less valid as an industrial hygiene measure.

Frequently the application of the principle of isolation maximizes the benefits of additional engineering concepts such as local exhaust ventilation. For example, the charging of mixers is the most significant operation in many processes that use formulated ingredients. When one of the ingredients in the formulation is of relatively high toxicity, it is worthwhile to isolate the mixing operation, that is, install a mixing room, thereby confining the airborne contaminants potentially generated by the operation to a small area rather than having them influence the larger portion of the plant. Thus in ensuring isolation, the application of ventilation principles to control this contaminant at the source, that is, the mixer, is much more effective.

2.3 Ventilation

For purposes of industrial hygiene engineering, ventilation is a method for providing control of an environment by strategic use of airflow. The flow of air may be used to provide either heating or cooling of a work space, to remove a contaminant near its source of release into the environment, to dilute the concentration of a contaminant to acceptable levels, or to replace air exhausted from an enclosure. Ventilation is by far the most important engineering control principle available to the industrial hygienist. Applied either as general or local control, this principle has industrial significance in at least three applications: control of heat and humidity primarily for comfort reasons, prevention of fire and explosions, and most important to the industrial hygienist, maintenance of concentrations of airborne contaminants at acceptable levels in the workplace.

Detailed discussions of the principles and application of general and local exhaust ventilation appear in subsequent sections of this chapter. Application of these principles to industrial problems, whether general or localized, requires a basic understanding of the fundamentals of airflow. The scientific laws that define completely the motion of any fluid, including air, are complex, and except in the relatively simple case of laminar flow, we know relatively little about them. Nevertheless there are fundamental relationships that must be understood and conscientiously applied by the industrial hygiene engineer. These are described briefly in the following sections; more elaborate discussions of these principles are given in several of the general references identified at the end of this chapter.

2.3.1 Conservation of Mass

The most basic principle of airflow is referred to as the "continuity equation" or principle of "conservation of mass." This relationship states that the mass rate of flow remains constant along the path taken by the fluid. For any two points in the fluid stream, therefore, we can express the relationship as

$$A_1 v_1 \delta_1 = A_2 v_2 \delta_2$$

where A = cross-sectional area (ft^2 or m^2)
v = linear velocity (ft/min or m/sec)
δ = specific weight (lb/ft^3 or g/m^3)

For most industrial applications, δ is relatively constant because the absolute pressure within a ventilation system varies over a narrow range and air remains essentially incompressible. Therefore,

$$A_1 v_1 = A_2 v_2 \quad \text{and} \quad Q_1 = Q_2$$

where $Q = Av$, the volumetric rate of airflow (ft^3/min or m^3/sec).

2.3.2 Conservation of Energy

The energy content for steady flow of a frictionless, incompressible fluid along a single streamline can be described by Bernoulli's theorem, which states algebraically:

$$H + \frac{P}{\delta} + \frac{v^2}{2g_c} = C$$

where H = elevation above any arbitrary datum plane (ft or m)
P = absolute pressure (lb/ft^2 or g/m^2)
δ = specific weight (lb/ft^3 or g/m^3)
v = velocity (ft/sec or m/sec)
g_c = gravitational acceleration (ft/sec^2 or m/sec^2)
C = a constant, different for each streamline

The unit for each of the three terms in this expression is "foot-pounds per pound" (gram-meters per gram) of fluid, or "feet" of fluid, frequently referred to as elevation head, pressure head, and velocity head, respectively.

The elevation term H usually is omitted when Bernoulli's equation is applied to industrial exhaust systems, because only relatively small changes in elevation are involved. Because all streamlines originate from the atmosphere, a reservoir of nearly constant energy, the constant C, is the same for all streamlines; the restriction of the equation to a single streamline thus can be removed. Furthermore, because the pressure changes in nearly all exhaust systems are at most only a few percent of the absolute pressure, the assumption of incompressibility may be made with negligible error.

2.3.3 Velocity Pressure

Velocity pressure is that pressure exerted by air in motion; it is analogous to kinetic energy and maintains air velocity. It exists only when air is in motion, it acts in the direction of air flow, and it is always positive in sign. In Bernoulli's equation,

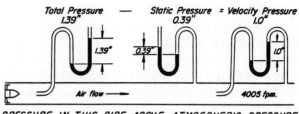

Figure 28.1.   Relationship among velocity pressure, static pressure, and total pressure.

the term $v^2/2g_c$ represents the velocity head. The relationship between the velocity of air and velocity pressure is

$$v = \sqrt{2g_c h}$$

where v = velocity (ft/sec or m/sec)
$\quad g_c$ = gravitational acceleration (ft/sec^2 or m/sec^2)
$\quad h$ = head of air (ft or m)

2.3.4   Static Pressure

Static pressure overcomes the resistance in a system caused by friction of the air against the duct walls, produces initial air velocity, and overcomes turbulence and shock caused by a change in direction or velocity of air movement. Static pressure is analogous to potential energy, and it exists even where there is no air motion. It acts equally in all directions and tends either to collapse the walls of the duct upstream from the fan or to expand the walls of the duct on the downstream side. Static pressure usually is negative in sign upstream from a fan and positive downstream. It is measured as the difference between duct pressure and atmospheric pressure. The most common unit of static pressure is "inches of water" (in H$_2$O).

2.3.5   Total Pressure

The driving force for the airflow is a pressure difference that is required to start and maintain flow. This pressure is called total pressure (*TP*) and has two components, velocity pressure (*VP*) and static pressure (*SP*); the relationship among the three is simply:

$$SP + VP = TP$$

Figure 28.1 shows the relationship among static, velocity, and total pressures at different points in a duct system.

   If gas flowing through a duct system undergoes an increase in velocity, a part of the available static pressure is used to create the additional velocity pressure necessary to accelerate the flowing gas. Conversely, if the velocity is reduced at some point in a duct system, a portion of the kinetic energy or velocity pressure at that point is converted into potential energy or static pressure. Static pressure and velocity pressure are therefore mutually convertible. However, this conversion always is accompanied by a net loss of total pressure due to turbulence and shock; that is, the conversion is never 100 percent efficient.

2.3.6 Energy Losses

Resistance is provided to any air in motion by all surfaces encountered by the flowing volume of air. Consequently, some of the energy of the air is given up in overcoming this friction and is transformed into heat. The rougher the surface confining the flow or the higher the flow rate, the higher the frictional losses.

Frictional loss in a duct varies directly as the length, inversely as the diameter, and directly as the square of the velocity of air flowing through the duct. This loss can be calculated from charts (1) using the Fanning friction factor, which is an empirical function of Reynolds number, duct material, and type of construction.

Another type of energy loss encountered in airflow results from turbulence caused by a change in direction or velocity within a duct. The pressure drop in a duct system due to dynamic losses increases with the number of elbows or angles and the number of velocity changes within the system. The resulting pressure drop from these energy losses is expressed in units of "equivalent length." For example, an elbow of 10-in. diameter and 20-in. centerline radius is said to have an equivalent length of 14, meaning that the loss through the elbow will be the same as through 14 ft of straight pipe with the same diameter operating under the same conditions (2, 3).

Another method of defining the losses due to turbulence and friction is to express the losses in terms of velocity pressure. For example, a loss of $0.28VP$ in a transition or elbow means the incremental pressure drop is equal to 0.28 of the velocity pressure of the airstream at that point (2–4).

Acceleration and hood entrance loss is a drop in pressure caused by turbulence when air is accelerated from rest to enter a duct or opening. Turbulence losses of this type vary with the type of opening.

This entry loss plus the acceleration energy required to move the air at a given velocity (1 VP) make up the hood static pressure SP_h, which is expressed algebraically as

$$SP_h = h_e + VP$$

where h_e is hood entry loss. The SP_h can be measured directly at a short distance downstream from the hood entrance. The calculation of SP_h is the first step in the design or evaluation of a local exhaust system, discussed later.

The coefficient of entry, C_e, is a measure of how efficiently a hood entry is able to convert static pressure to velocity pressure. The coefficient of entry is the ratio of rate of flow by the hood static pressure to the theoretical flow if the hood static pressure were completely converted to velocity pressure.

The result of the friction and dynamic losses to air flowing through ductwork is a pressure drop in the system. Bernoulli's theorem can be restated in a simplified expression of conservation of energy as follows:

$$SP_1 + VP_1 = SP_2 + VP_2 + \text{energy losses}$$

Static pressure plus velocity pressure at a point upstream in the direction of airflow equals the static pressure plus velocity pressure at a point downstream in the direction of airflow plus friction and dynamic losses.

3 GENERAL VENTILATION

For purpose of discussion in this chapter, "general ventilation" is applied to the practice of supplying and exhausting large volumes of air throughout a work space. It is used typically in industry to achieve comfortable work conditions (temperature and humidity control) or to dilute the concentrations of airborne contaminants to acceptable limits throughout the work space. Properly used, general ventilation can be effective in removing large volumes of heated air or relatively low concentrations of low toxicity contaminants from several decentralized sources.

General ventilation can be provided by either natural or mechanical means; often the best overall result is obtained with a combination of mechanical and natural air supply and exhaust.

3.1 Natural Ventilation

Natural ventilation can be provided either by gravitational forces, using thermal forces of convection primarily, or by anemotive forces, created by differences in "wind pressure." These two natural forces operate together in most cases, resulting in the natural displacement and infiltration of air through windows, doors, walls, floors, and other openings in an industrial building. Unfortunately, the wind currents and thermal convection profiles on which natural ventilation depends are erratic and frequently unpredictable. Thus it is perhaps a misnomer to refer to natural ventilation as a "control" method, because to employ this technique requires dependence on, rather than control of, natural forces.

On the other hand, there are applications for general natural ventilation. The pressure exerted on the upwind side and concurrent suction exerted on the downwind side of a building as a result of wind movement can be predicted for flat terrain fairly reliably. Thus the wind forces exerted on an isolated building in a relatively flat area permit the prediction of natural ventilation forces. In the more common complex industrial building, however, the effects of the presence of one building on the others normally cannot be calculated; thus the use of wind pressure models in the development of general ventilation systems is not feasible.

The amount of natural ventilation provided to a building depends both on the wind pressure profiles and on the thermal effects occurring within the structure.

The warmer inside air tends to rise and leave the building through any available openings in the upper structure; cooler air tends to infiltrate the building by the reverse process in the lower structure. These thermal effects, which are much more predictable than the external wind forces, can be useful in design of a general ventilation scheme (5).

Because this chapter is devoted to "engineering control" and because of the lack of control inherent in natural ventilation systems, the remainder of this chapter presents concepts of ventilation and quality control of the work environment in terms of mechanically controlled ventilation systems. Obviously, many of the principles associated with use of mechanical ventilation systems apply equally well to natural ventilation. However, the remainder of this chapter focuses on concepts and design principles for which it is assumed there is indeed control.

3.2   Mechanically Controlled Ventilation

A modern industrial complex characterized by a low profile building structure (i.e., large floor space, low height) as well as multistory buildings of masonry and glass construction present ventilation problems that were not found among older industrial plants. For the most part the industrial facilities constructed 30 or more years ago incorporated by design many features that permitted—in fact, expected—natural ventilation of the work space. By contrast, modern buildings generally defy the exertion of natural ventilation forces, and mechanical ventilation must be relied upon almost completely. Mechanical ventilation systems exhaust contaminated air by mechanical means (exhaust fans) with the concomitant use of appropriate air supply to replace the exhausted air. The best method of achieving this in modern closed buildings is to supply air through a system of ductwork, distributing the air into the work areas in a manner that will provide optimal benefit to the workers for both comfort and control of contaminants.

3.3   Applications of General Ventilation

Normally, either or both of two end results are desired when applying general ventilation: an environment that is comfortable to the workers, or one that is free of harmful concentrations of airborne contaminants.

3.3.1   Comfort Ventilation

General ventilation for comfort, principally heat relief, includes certain aspects of what is commonly considered to be "air conditioning engineering." Here, air is treated to control simultaneously temperature, humidity, and cleanliness, and is distributed to maximize effects in the conditioned space. In the typical residential or office building, these requirements are primarily associated with comfort for the occupants. In many industrial situations, however, comfort conditions are impractical if not impossible to maintain, and the chief function of ventilation for comfort control is to prevent acute discomfort and the accompanying adverse physiological effects.

General exhaust ventilation also may be used to remove heat and humidity if a source of cooler air is available. If it is possible to enclose the heat source, as in the case of ovens or furnaces, a gravity or forced-air stack may be all that is necessary to prevent excessive heat from entering the workroom.

Air Conditioning. Complete, or integrated, air conditioning improves the control of temperature, humidity, radiation, air movements, or drafts, as well as the cleanliness and purity of the air. Under ordinary circumstances of comfort and health the body temperature is kept approximately constant at its normal level of 98.6°F, through a balance between the internal production of heat (metabolism) and thermal equilibrium with the environment. The exchange of heat between humans and the environment is discussed in detail elsewhere in this book. According to the various equations that have been developed to express the thermal exchange between the human body and the environment, four external physical factors are of primary importance: air temperature, radiant temperature, air velocity, and the moisture content of the air. These four environmental factors are independently variable and may be combined in many different ways to produce the same thermal effect on humans. Thus an increasing velocity can offset a rise in air temperature, extreme radiant heat can be compensated somewhat by lowering the air temperature, and a considerable elevation of air temperature can be tolerated if the relative humidity is low. Combinations of thermal conditions that induce the same sensations are called thermoequivalent conditions. The effective temperature charts of the American Society of Heating, Refrigeration and Air-Conditioning Engineers (ASHRAE) give the combinations of equivalent conditions for clothed, sedentary individuals and for workers stripped to the waist. These charts are presented for illustration in Figures 28.2 to 28.4. Although many indexes of heat stress and subjective response to this stress have been developed over the years, the effective temperature charts have been the primary tool of the air conditioning engineer. These scales were derived by experiments with test subjects exposed to a variety of thermal conditions by repeating trials with combinations of temperature, humidity, and air motion. Despite its limited subjective basis, subsequent laboratory and field experience with the effective temperature scale has indicated that it is useful in defining thermal comfort conditions and, to a great extent, it correlates with objective measurements of physiological response to heat.

Any other features of ventilation for purposes of comfort are similar to those associated with control of airborne contaminants and are discussed below. For aspects of engineering control unique to the thermal problem, refer to Chapter 21 on the control of heat stress in Volume 1A.

3.3.2 Dilution Ventilation

Dilution ventilation has as its primary function the maintenance of concentrations of airborne contaminants at or below acceptable exposure limits, in terms of either potential fire and explosion or occupational health considerations. If enough clean air is mixed with contaminated air, the concentration of the contaminant can be

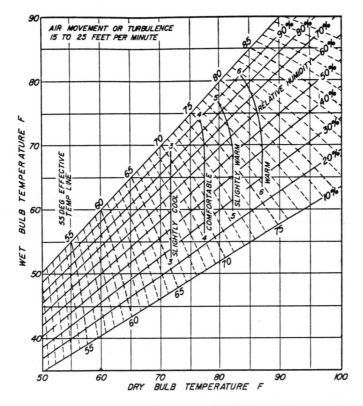

Figure 28.2. ASHRAE comfort chart for normally clothed people at rest in still air. Effective temperature (broken) lines indicate sensation of warmth immediately after entering conditioned space. Solid lines (3 to 6) indicate sensations experienced after 3-hr occupancy.

diluted to any reasonable level. Of course, dilution ventilation for purposes of fire and explosion control should not be employed in areas occupied by workers, because the concentrations of concern from an occupational health standpoint invariably are orders of magnitude below those of concern for explosive limits. Exposures to atmospheres controlled to concentrations below the lower explosive limit or even a fraction thereof could cause narcosis, severe illness, or even death. Therefore it is extremely important not to confuse dilution ventilation requirements for health hazard control with those for fire and explosion prevention.

When considering whether dilution ventilation is appropriate for health hazard control, it should be remembered that dilution ventilation has four limiting factors:

a. The quantity of contaminant generated must not be excessive, or the air volume necessary for dilution would be impractically large.

b. Workers must be far enough away from evolution of the contaminant, or the contaminant must be of sufficiently low concentration, that the workers will not be exposed above acceptable limits.

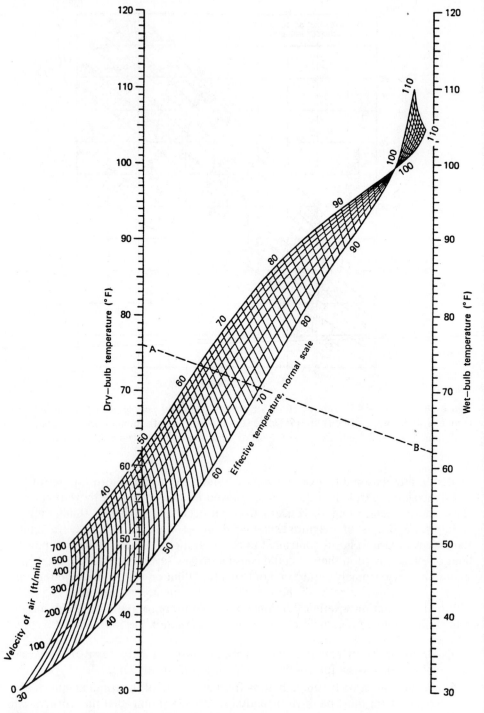

Figure 28.3. ASHRAE effective temperature scale for normally clothed persons at rest.

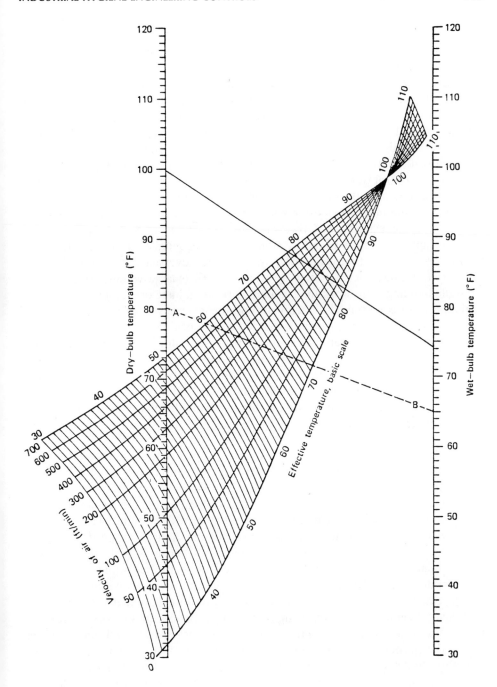

Figure 28.4. ASHRAE effective temperature scale for men stripped to the waist at rest or doing light physical work in rooms heated by convection methods. (From current edition of ASHRAE Handbook.)

c. The toxicity of the contaminant must be relatively low.

d. The evolution or generation of the contaminant must be reasonable, uniform, and consistent.

A review of these factors indicates clearly that dilution ventilation is not normally appropriate for control of fumes and dust, because the high toxicity often encountered requires excessively large quantities of dilution air. Moreover, the velocity and rate of evolution usually are very high, resulting in locally high concentrations, thus rendering the dilution ventilation concept inappropriate.

In general, dilution ventilation is not as satisfactory as local exhaust ventilation for primary control of health hazards. Occasionally, however, dilution ventilation must be used because the operation of the process prohibits local exhaust. Dilution ventilation sometimes provides an adequate amount of control more economically than a local exhaust system. However, this condition is an exception rather than the rule, and it should be kept in mind that the economic considerations of the long-term use of dilution ventilation systems often overshadow the initial cost of the system because such a system invariably exhausts large volumes of heated air from the building. This work load can easily result in huge operating costs because of the need for conditioned makeup air, and the general ventilation scheme would be much more expensive over an extended period of time.

In practice, dilution ventilation for control of health hazards is used to best advantage in controlling the concentration of vapors from organic solvents of relatively low toxicity. To apply the principle of dilution successfully to such a problem, data must be available on the rate of vapor generation or on the rate of liquid evaporation. Usually such data can be obtained from the plant records on material consumption.

Calculating Dilution Ventilation. The volume of a room to be ventilated and the ventilation rate are frequently related by taking the ratio of the ventilation rate to the room volume to yield a "number of air changes per minute" or "number of air changes per hour." These terms, used quite frequently in discussion of ventilation requirements, frequently are employed incorrectly in industrial hygiene.

Ventilation requirements based on room volume alone have no validity. Calculations of the required rate of air changes can be made only on the basis of a material balance for the contaminant of interest. Similar calculations can be made for the rate of concentration increase or decrease; however, they require not only the air change rate but also the rate of generation of contaminant. In the design of industrial ventilation, "*x* air changes" has valid application only very rarely. The term is useful when applied to meeting rooms, offices, schools, and similar spaces, where the purpose of ventilation is simply the control of odor, temperature, or humidity, and the only contamination of air is the result of activity of people. Dilution ventilation requirements should always be expressed in cubic feet per minute or some other absolute unit of airflow, not in "air changes per hour."

The concentration of a gas or vapor at any time can be expressed by a differential material balance which, when integrated, provides a rational basis for relating ventilation to the generation and removal rates of a contaminant. Let C be concentration of gas or vapor at time, t, and take the following additional quantities:

G = rate of generation of contaminant

Q = rate of ventilation

K = design distribution constant, allowing for incomplete mixing

Q' = Q/K = effective rate of ventilation, corrected for incomplete mixing

V = volume of room or enclosure

Starting with a fundamental material balance, and assuming no contaminant in the air supply, the following relationship applies:

rate of accumulation = rate of generation − rate of removal

This basic relationship can be applied in various ways.

Concentration Buildup. Rearranging the differential material balance, the buildup of contaminant can be expressed as follows:

$$\int_{c_1}^{c_2} \frac{dC}{G - Q'C} = \frac{1}{V} \int_{t_1}^{t_2} dt$$

$$\ln\left(\frac{G - Q'C_2}{G - Q'C_1}\right) = -\frac{Q'}{V}(t_2 - t_1)$$

If $C_1 = 0$ at $t_1 = 0$,

$$\ln\left(\frac{G - Q'C}{G}\right) = -\frac{Q'}{V}t$$

or

$$\frac{G - Q'C}{G} = \varepsilon^{-(Q't/V)}$$

Rate of Purging. When a volume of air is contaminated but further contamination or generation has ceased, the rate of decrease of concentration over a period of time is as follows:

$$V \, dC = -Q'C \, dt$$

and

$$\int_{c_1}^{c_2} \frac{dC}{C} = -\frac{Q'}{V} \int_{t_1}^{t_2} dt$$

$$\ln\frac{C_2}{C_1} = -\frac{Q'}{V}(t_2 - t_1)$$

Maintaining Acceptable Concentrations at Steady State. At steady state, $dC = 0$, and

$$G\ dt = Q'C\ dt$$

and

$$\int_{t_1}^{t_2} G\ dt = \int_{t_1}^{t_2} Q'C\ dt$$

At a constant concentration C, and uniform generation rate G,

$$G(t_2 - t_1) = Q'C(t_2 - t_1)$$

$$Q' = \frac{G}{C}$$

$$Q = \frac{KG}{C}$$

Therefore the rate of flow of uncontaminated dilution air required to reduce the atmospheric concentration of a hazardous material to an acceptable level can be easily calculated if the generation rate can be determined. Usually the acceptable concentration is considered to be the acceptable 8-hr time-weighted average (TWA) concentration. For liquid solvents the steady-state dilution ventilation requirement can be conveniently expressed as

$$Q = \frac{(6.71)(10^6)(SG)(ER)(K)}{(MW)(PEL)}$$

where Q = actual ventilation rate (ft^3/min)
  SG = specific gravity of volatile liquid
  ER = evaporation rate of liquid (pints/hour)
  MW = molecular weight of liquid
  K = design safety factor for incomplete mixing
  PEL = permissible exposure limit (ppm)

Specifying Dilution Ventilation. The foregoing discussion introduced the concept of a "design safety factor" K for calculating dilution ventilation requirements. The K factor is based on several considerations:

a. The efficiency of mixing and distribution of makeup air introduced into the room or space being ventilated.

b. The toxicity of the solvent. Although PEL and toxicity are not synonymous, the following guidelines have been suggested for choosing the appropriate *K* value:

Slightly toxic	PEL > 500 ppm
Moderately toxic	PEL = 100–500 ppm
Highly toxic	PEL < 100 ppm

c. A judgment of any other circumstances the industrial hygienist determines to be of importance, based on experience and the individual situation. Included in these criteria are considerations such as (*a*) seasonal changes in the amount of natural ventilation; (*b*) reduction in operation efficiency of mechanical air-moving devices; (*c*) duration of the process, operational cycle, and normal location of workers relative to sources of contamination; (*d*) location and number of points of generation of the contaminant in the workroom or area; and (*e*) other circumstances that may affect the concentration of hazardous material in the breathing zone of the workers. The *K* value selected usually varies from 3 to 10, depending on the foregoing considerations.

Table 28.1 lists the air volumes required to dilute the vapors of common organic solvents to the PEL level based on the liquid volume of solvent evaporated per unit time. These values must be multiplied by a *K* factor to allow for variations in uniformity of air distribution and for other considerations. Hemeon (6) includes a table of recommended dilution rates for 53 organic solvents. The "Ventilation Design Concentrations" in this table are not based on the PEL alone; they also incorporate the effects of odor. All the concentrations in this table are lower than the PELs, but the substances that are especially toxic or have a very disagreeable odor have the greatest safety factors.

It must be emphasized that PELs are subject to revision, and the dilution values estimated from such tables become obsolete if the PELs are changed; therefore, such a table should be used with caution, and with reference to the latest PEL (or threshold limit) values (7, 8).

3.3.3   General Ventilation Airflow Patterns

General ventilation presents some disadvantages, the most significant one being that it permits the occupied space to become, in effect, a large settling chamber for the contaminants, even though the concentration may be within acceptable limits as far as potential health hazard is concerned. In some cases, the settling or separation of contaminants from the air may represent condensation of materials that were vaporized by high-temperature processes and accumulated on surfaces after condensation. The undesirability of the settling of contamination onto surfaces

Table 28.1. Dilution Air Volumes for Vapors

The following values are tabulated using the TLV values shown in parentheses, parts per million. TLV values are subject to revision if further research or experience indicates the need. If the TLV value has changed, the dilution air requirements must be recalculated. The values on the table must be multiplied by the evaporation rate (pts/min) to yield the effective ventilation rate (Q′)

	Ft³ of Air (STP) Required for Dilution to TLV[b]
Liquid (TLV in ppm)[a]	Per Pint Evaporation
Acetone (750)	7,350
n-Amyl acetate (100)	27,200
Benzene (10)	Not recommended
n-Butanol (butyl alcohol) (50)	88,000
n-Butyl acetate (150)	20,400
Butyl Cellosolve	
(2-butoxyethanol) (25)	Not recommended
Carbon disulfide (10)	Not recommended
Carbon tetrachloride (5)	Not recommended
Cellosolve (2-ethoxyethanol) (5)	Not recommended
Cellosolve acetate	
(2-ethoxyethyl acetate) (5)	Not recommended
Chloroform (10)	Not recommended
1-2 Dichloroethane (10)	
(ethylene dichloride)	Not recommended
1-2 Dichloroethylene (200)	26,900
Dioxane (25)	Not recommended
Ethyl acetate (400)	10,300
Ethyl alcohol (1000)	6,900
Ethyl ether (400)	9,360
Gasoline (300)	Requires special consideration
Isoamyl alcohol (10)	37,200
Isopropyl alcohol (400)	13,200
Isopropyl ether (250)	11,400
Methyl acetate (200)	25,000
Methyl alcohol (200)	49,100
Methyl n-butyl ketone (5)	Not recommended
Methyl Cellosolve	
(2-methoxyethanol) (5)	Not recommended
Methyl Cellosolve acetate	
(2-methoxyethyl acetate) (5)	Not recommended
Methyl chloroform (350)	11,390
Methyl ethyl ketone (200)	22,500
Methyl isobutyl ketone (50)	64,600
Methyl propyl ketone (200)	19,900
Naptha (coal tar)	Requires special consideration
Naptha VM & P (300)	Requires special consideration
Nitrobenzene (1)	Not recommended

Table 28.1. (*Continued*)

Liquid (TLV in ppm)[a]	Ft³ of Air (STP) Required for Dilution to TLV[b]
	Per Pint Evaporation
n-Propyl acetate (200)	17,500
Stoddard solvent (100)	30,000–35,000
1,1,2,2-Tetrachloroethane (1)	Not recommended
Tetrachloroethylene (50)	79,200
Toluene (100)	38,000
Trichloroethylene (50)	90,000
Xylene (100)	33,000

Source: American Conference of Governmental Industrial Hygienists, Committee on Ventilation, *Industrial Ventilation—A Manual of Recommended Practice*, 20th ed., ACGIH, Lansing, MI, 1988, pp. 2–3.
[a]See threshold limit values 1988–1989 in Appendix of manual.
[b]The tabulated dilution air quantities must be multiplied by the selected K value.

within the plant has been dramatically demonstrated in plants handling highly combustible dusts where, although the dust concentrations at any one time were not sufficient to constitute a hazard, accumulation over many months resulted in disastrous explosions. It is therefore necessary to maintain an effective ongoing good housekeeping program in conjunction with any broad-based engineering control such as general ventilation.

The design of general ventilation systems for a plant is not complete without consideration of the routes by which the air will enter and leave the work space. The routes are of critical concern not only to the occupants of the building but potentially to the neighbors and the community at large. In general, the relative locations of air inlets and outlets should be considered in the implementation of a general ventilation system.

Air Inlets. Sufficient combined or total inlet area should be planned to accommodate the required volume of makeup air during the heating or cooling season, whichever needs the larger volume. This will prevent excessive inlet velocities that consume power, create drafts on workers, stir up dust, and interfere with performance of local exhaust systems. The chosen inlet velocity is likely to represent an engineering compromise between low rates, which require large inlet areas, and higher rates, which facilitate rapid dilution of contaminants.

If widespread hazardous operations are controlled by dilution or general ventilation, inlets should be well distributed around the building to provide uniform circulation. If contaminants are controlled by localized dilution, or by forceful diffusion into the general room-air reservoir, inlets may be purposely located to give nonuniform air supply distribution.

Inlets should be located to take full advantage of any thermal or convection effects within the building. For this purpose the designer must avoid the location of inlets (and outlets) near the "neutral zone" of inside–outside pressure differentials, which is approximately midway between the floor and roof. Inlets should be located remote from stacks or ventilators discharging contaminants from the same or neighboring structures.

Air Outlets. Outlets should be located as far as possible from air inlets to prevent "short-circuiting." This advice holds for both natural and mechanical ventilation systems.

If widely scattered operations are controlled by general dilution, outlets should be uniformly distributed around the building. Similarly, if contaminants are controlled by localized dilution, air outlets may be purposely located to short-circuit a corner of a large room, with the intent of creating a high rate of air change there without involving the atmosphere of the entire space. This method of localized space ventilation requires a higher rate of exhaust air than supply air within the area to be controlled, to prevent the spread of contaminants throughout the room.

Outlets should be placed to take full advantage of thermal effects. They should be protected against the direct force of prevailing winds, which reduce the capacity of any exhaust fans or destroy the anticipated airflow route planned on the basis of thermal effects inside the building.

In general, dilution air should enter the work space at approximate breathing zone height, pass through the workers' breathing zone, then through the zone of contamination; afterward, if it is not exhausted, it should enter a space of higher relative contamination. A useful concept in the interest of air-handling economy is "progressive ventilation." The air removed from an industrial plant frequently may be directed in a way that will provide both local and general ventilation. In fact, the air may be routed through several areas in succession, always in the direction of increasing air contamination, as long as the exhaust from one area is acceptable as the supply for the next. Progressive ventilation saves energy in the form of airborne heat and horsepower to move the air.

3.3.4 General Ventilation versus Local Exhaust Ventilation

The science and art of industrial process ventilation has been somewhat evolutionary from the natural general ventilation achieved by correct design of industrial buildings for hot, humid, and dusty operations to the highly effective local exhaust hoods in current use. Local ventilation prevents the spread of air contaminants throughout the building atmosphere with surprisingly small quantities of airflow in comparison with the volumes of air required by general ventilating systems. In spite of this characteristic air conservation of local ventilation, there are highly persuasive reasons for the selection by many industrial managements of general ventilation: (1) simplicity and economy of natural general ventilation, (2) relatively low first cost of mechanical general ventilation, (3) absence of interference with manufacturing operations, (4) flexibility in plants with constantly changing layouts, (5) existence of contaminating processes throughout the entire plant, (6) desire for large volumes of air circulation in hot weather, and (7) discovery that local exhaust systems do not eliminate the necessity of supplying large volumes of heated air in the wintertime to replace that escaping by exfiltration from loosely constructed buildings. This volume of air change may be more than ample to control process air contaminants by the method of dilution, if means are provided to disperse contaminants from the immediate vicinity of the workers.

4   LOCAL EXHAUST VENTILATION

Local exhaust ventilation (LEV) incorporates the concept of controlling a contaminant by capturing it at or near the place where it is generated and removing it from the work space. Local ventilation relies more heavily on mechanical methods of controlling airflow than does general ventilation. A local exhaust system usually includes all of the following components: a hood or enclosure, ductwork, an air-cleaning device (where necessary for air pollution abatement purposes), and an air-moving device (usually an exhaust fan) to draw the contaminated air through the exhaust system and discharge it to the outside air. In general, local exhaust systems consist of more individual components than do general exhaust systems; because they also offer more operational parameters that must be controlled within acceptable ranges, they require more maintenance and involve higher operating expenses as well.

When the primary purpose of the ventilation is to provide control of airborne contaminants, local exhaust systems generally are much superior to general ventilation. The advantages of the local exhaust ventilation system over general exhaust for any particular application will include many of the following:

a. If the system is designed properly, the capture and control of a contaminant can be virtually complete. Consequently, the exposure of workers to contaminants at the sources exhausted can be prevented. With general ventilation the contaminant is diluted when the exposure occurs, and at any given workplace this dilution may be highly variable, and therefore inadequate at certain times.

b. The volumetric rate of required exhaust is less with local ventilation; as a result, the volume of makeup air required is less. Local ventilation offers savings in both capital investment and heating costs.

c. The contaminant is contained in a smaller exhausted volume of air. Therefore if air pollution control is needed, it is less expensive because the cost of air pollution control is approximately proportional to the volume of air handled.

d. Local exhaust systems can be designed to capture large, settleable particles or at least confine them within the hood, and thus greatly reduce the effort needed for good housekeeping.

e. Auxiliary equipment in the workroom is better protected from such deleterious effects of the contaminant as corrosion and abrasion.

f. Local exhaust systems usually require a fan of fairly high pressure charac-

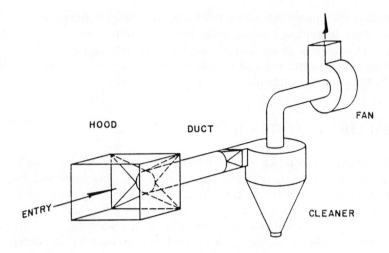

Figure 28.5. Interrelated components of a local exhaust system.

teristics to overcome pressure losses in the system. Therefore the performance of the fan system is not likely to be affected adversely by such influences as wind direction or velocity or inadequate makeup air. This is in contrast to general ventilation, which can be affected greatly by seasonal factors.

4.1 Components of a LEV System

The four components of a simple local exhaust system are illustrated in Figure 28.5: (*a*) a hood, (*b*) ductwork, (*c*) an air-cleaning device (cleaner), and (*d*) an air-moving device (fan). Typically the system is a network of branch ducts connected to several hoods or enclosures, main ducts, air cleaner for separating the contaminants from the airstream, exhaust fan, and discharge stack to the outside atmosphere.

4.1.1 Hoods

A hood is a structure designed to enclose a contaminant-producing operation partially and to guide airflow in an efficient manner to capture the contaminant. The hood is connected to the ventilation system with ductwork that removes the contaminant from the hood. The design and location of the hood is one of the most critical aspects in the successful operation of a local exhaust system.

4.1.2 Ductwork

The ductwork in an exhaust system provides a path for flow of the contaminated air exhausted from the hood to the point of discharge. The following points are important in design of the ductwork:

   a. In the case of dust, the duct velocity must be high enough to prevent the dust from settling out and plugging the ductwork.

   b. In the absence of dust, the duct velocity should strike an economic balance between ductwork cost and fan, motor, and power costs.

   c. The location and construction of the ductwork must furnish sufficient protection against external damage, corrosion, and erosion to maximize the useful life of the local exhaust system.

4.1.3 Air Cleaner

Most exhaust systems installed for contaminant control need an air cleaner. Occasionally the collected material has some economic reuse value, but this is seldom the case. To collect and dispose of the contaminant is usually inconvenient and certainly represents an added expense. Yet the growing concern with air pollution control and the need to comply with legal restrictions on discharges from sources of atmospheric emissions place new importance on the air-cleaning device within a local exhaust system.

4.1.4 Air-Moving Device

The heart of the local exhaust system is the fan, usually of the centrifugal type. Wherever practicable a fan should be placed downstream from the air cleaner so that it will handle uncontaminated air. In such an arrangement, the fan wheel can be the backward curved blade type, which has a relatively high efficiency and lower power cost. For equivalent air handling, the forward-curved blade impellers run at somewhat lower speeds, and this may be important when noise is a factor. If chips and other particulate matter have to pass through the impeller, the straight blade or paddlewheel-type fan is best because it is least likely to clog.

   Fans and motors should be mounted on substantial platforms or bases and isolated by antivibration mounts. At the fan inlet and outlet the main duct should attach through a vibration isolator, that is, a sleeve or band of very flexible material, such as rubber or fabric.

   When the system has several branch connections, consideration should be given to using a belt drive instead of a direct-connected motor. Then if increased airflow is required at a future date, the need can be accommodated, to some degree, by adjusting the fan speed.

   The importance and interaction of the various components of the exhaust system will become evident in the discussion on design of a ventilation system later in this chapter. Because of the importance of the role played by the exhaust hood in the overall success of the ventilation system, and because it is the component of the system that typically permits most innovation in concept, specific discussion of the application of the various types of hood follows.

4.2 Applications of Local Exhaust Hoods

The local exhaust hood is the point at which air first enters the exhaust ventilation system. As such, the term can be used to apply to any opening in the exhaust system regardless of its shape or physical disposition. An open-ended section of ductwork, a canopy-type hood situated above a hot process, and a conventional laboratory booth-type hood all could be called hoods in the context of this discussion.

The hood captures the contaminant generated by a particular process or operation and causes it to be carried through the ductwork to a convenient discharge point. The quantity of air required to capture and convey the air contaminants depends on the size and shape of the hood, its position relative to the point of generation of the contaminant, and the nature and quantity of the air contaminant itself. It should be emphasized that there is not necessarily a "standard hood" that is correct for all applications of a particular operation, because the methods of processing are unique to each operating plant. On the other hand, standard concepts for exhaust hooding have been developed and have been recommended for specific types of operations (2).

Hoods can be classified conveniently into four categories, based on the concept of contaminant capture/control: enclosures, booth-type hoods, receiving hoods, and exterior hoods.

4.2.1 Enclosure Hoods

Enclosure hoods normally surround the points of emission or generation of contaminant as completely as practicable. In essence, they surround the contaminant source to such a degree that all contaminant dispersal action takes place within the confines of the hood itself. Because of this, enclosure hoods generally require the lowest rate of exhaust ventilation, and therefore are economical and quite efficient. They should be used whenever possible, and they deserve particular consideration when a moderately or highly toxic contaminant is involved.

4.2.2 Booth-Type Hoods

The common spray-painting enclosure and other booths are special cases of enclosure-type hoods. These are typified further by the common laboratory hood in which one face of an otherwise complete enclosure is open for ready access. Air contamination takes place within the enclosure, and air is exhausted from it in such a way, and at such a rate, that an average velocity is induced across the face of the opening sufficient to overcome the tendency of the contaminant to escape from within the hood. The three walls of the booth greatly reduce the exhaust air requirements, although not to the extent of a complete enclosure.

4.2.3 Receiving Hoods

The term "receiving" refers to a hood in which a stream of contaminated air from a process is exhausted by a hood located near the source of generation of the contaminants specifically for purposes of control. Two examples of this type of hood are canopies situated above hot processes and hoods attached to grinders, positioned to take advantage of centrifugal and gravitational forces to maximize control of the dust generated by the process. Canopy hoods, frequently located above hot processes, are similar to exterior hoods in that the contaminated air originates beyond the physical boundaries of the hood. The fundamental difference between receiving and exterior hoods is that in the former the hood takes advantage of the natural movement of the released contaminant, whereas in the latter, air is induced to move toward the exterior hood.

In practice, receiving hoods are positioned to be in the pathway of the contaminant as normally released by the operation. If hood space is limited by the process, baffles or shields may be placed across the line of throw of the particles to remove their kinetic energy. Then the particles may be captured and carried into the hood by lower air velocities. Additional examples of receiving hoods include those associated with many hand tool operations (surface grinders, metal polishers, stone cutters, and sanding machines).

4.2.4 Exterior Hoods

Exterior hoods must capture air contaminants generated from a point outside the hood itself, sometimes at quite a distance. Exterior hoods therefore differ from enclosure or receiving hoods in that their sphere of influence must extend beyond their own dimensions in capturing contaminants without the aid of natural forces such as natural drafts, buoyancy, and inertia. In other words, directional air currents must be established adjacent to the suction opening of the hood to provide adequate capture. Thus they are quite sensitive to external sources of air disturbance and may be rendered completely ineffective by even slight lateral movement of air. They also require the most air to control a given process and are the most difficult of the various hoods to design. Examples of exterior hoods include the exhaust slots on the edges of the tanks or surrounding a workbench such as a welding station, exhaust grilles in the floor or workbench below a contaminated process, and the common propeller-type exhaust fans frequently mounted in walls adjacent to a source of contamination.

4.3 Fundamental Concepts of LEV

4.3.1 Capture/Control Velocities

Local exhaust hoods perform their function in one of two ways: capture or control. With the "capture" approach, air movement is created to draw the contaminant into the hood. When the air velocity that accomplishes this objective is created at a point outside a nonenclosing hood, it is called "capture velocity." Some exhaust hoods essentially enclose the contaminant source and create an air movement that prevents the contaminant from escaping from the enclosure. The air velocity created at the openings of such hoods is called the "control velocity."

The successful design of any exhaust hood is based on correct determination of the two quantities control velocity and capture velocity. The air velocity, which must be developed by the exhaust hood at the point or in the area of desired control, is based on the magnitude and direction of the air motion to be overcome

Table 28.2. Range of Recommended Capture Velocities

Condition of Dispersion of Contaminant	Examples	Capture Velocity (fpm)
Released with practically no velocity into quiet air	Evaporation from tanks; degreasing, etc.	50–100
Released at low velocity into moderately still air	Spray booths; intermittent container filling; low speed conveyor transfers; welding; plating; pickling	100–200
Active generation into zone of rapid air motion	Spray painting in shallow booths; barrel filling; conveyor loading; crushers	200–500
Released at high initial velocity into zone of very rapid air motion	Grinding; abrasive blasting, tumbling	500–2000

In each cateory above, a range of capture velocity is shown. The proper choice of values depends on several factors:

Lower End of Range	Upper End of Range
1. Room air currents minimal or favorable to capture	1. Disturbing room air currents
2. Contaminants of low toxicity or of nuisance value only	2. Contaminants of high toxicity
3. Intermittent, low production	3. High production, heavy use
4. Large hood—large air mass in motion	4. Small hood—local control only

American Conference of Governmental Industrial Hygienists, Committee on Ventilation, *Industrial Ventilation—A Manual of Recommended Practice*, 20th ed., ACGIH, Lansing, MI, 1988, pp. 3–6.

and is not subject to direct and exact evaluation (Table 28.2). Many empirical ventilation standards, especially concerning dusty equipment such as screens and conveyor belt transfers, are based on parameters such as "cubic feet per minute per foot of belt width." These so-called exhaust rate standards usually are based on successful experience, are easily applied, and usually give satisfactory results if not extrapolated too far. In addition, they minimize the effort and uncertainty involved in calculating the fan action of falling material, thermal forces within hoods, and external air currents. However, such standards have three major pitfalls:

  a. They are not of a fundamental nature; that is, they do not follow directly from basic "laws."

  b. They presuppose a certain minimum quality of hood or enclosure design, although it may not be possible or practical to achieve the same quality of hood design in a new installation.

  c. They are valid only for circumstances similar to those that led to their development and use. It should be clear that the nature of the process generating

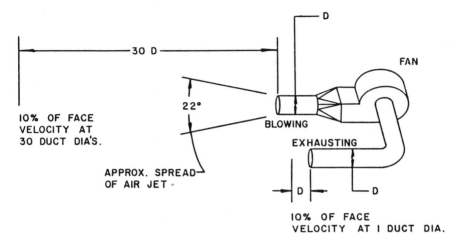

Figure 28.6. Airflow characteristics of blowing and exhausting.

the contaminant will be an important determinant of the required capture velocity.

4.3.2 Airflow Characteristics

The flow characteristics at a suction opening are much different from the flow pattern at a supply or discharge opening. Air blown from an opening maintains its directional effect in a fashion similar to water squirting from a hose and, in fact, is so pronounced that it is often called "throw." However, if the flow of air through the same opening is changed so that the opening operates as an exhaust or intake with the same volumetric rate of airflow, the flow becomes almost completely nondirectional and its range of influence is greatly reduced. As a first approximation, when air is blown from a small opening, the velocity 30 diameters in front of the plane of the opening is about 10 percent of the velocity at the discharge. The same reduction in velocity is achieved at a much smaller distance in the case of exhausted openings, such that the velocity equals 10 percent of the face velocity at a distance of only one diameter from the exhaust opening. (Figure 28.6 illustrates this point.) Therefore local exhaust hoods must not be applied for any operation that cannot be conducted in the immediate vicinity of the hood.

4.3.3 Airflow into Openings

Airflow into round openings was studied extensively by Dalla Valle (9). His theory of airflow into openings is based on a point source of suction that draws air from all directions. The velocity at any point in front, distance X, of such a source is equivalent to the quantity of air Q flowing to the source divided by the effective area of the sphere of the same radius.
Conversely,

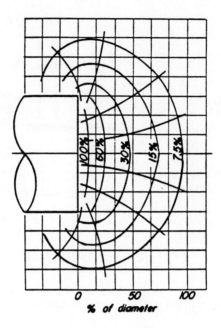

Figure 28.7. Velocity contours for unflanged circular opening.

$$Q = VA$$
$$A = 4\pi X^2$$

Thus

$$Q = V(12.57X^2)$$

where Q = airflow (ft³/min or m³/sec)
$\quad\quad V$ = velocity at point X (ft/min or m/sec)
$\quad\quad X$ = centerline distance (ft or m)
$\quad\quad A$ = duct area (ft² or m²)
$\quad\quad \pi$ = 3.14159 (dimensionless constant)

Postulating that a point source is approximated by the end of an open pipe, Brandt (10) and Dalla Valle determined the actual velocity contours for a circular opening (Figure 28.7). These contours, or line of constant velocity, are best described by the following equation:

$$Q = V(10X^2 + A)$$

4.3.4 Effects of Flanging

Flanges surrounding a hood opening force air to flow mostly from the zone directly in front of the hood. Thus the addition of a flange to an open duct or pipe improves the efficiency of the duct as a hood for a distance of about one diameter, as shown by the following equation:

$$Q = 0.75V(10X^2 + A)$$

For a flanged opening on a table or bench, the equation becomes

$$Q = 0.5V(10X^2 + A)$$

4.3.5 Slots

When the width-to-length ratio (aspect ratio) of an exhaust opening approaches 0.1, caution must be used in applying the generalized continuity equation because the opening behaves more like a slot. Using a line of reasoning similar to that of Dalla Valle, Silverman (11) considered the slot to be a line source of suction. Disregarding the end, the area of influence then approaches a cylinder and the velocity is given by

$$V = \frac{Q}{2\pi XL}$$

where L = length of slot (ft or m)
$\quad$ X = centerline distance (ft or m)
$\quad$ π = 3.14159 (dimensionless constant)

Correcting for empirical versus theoretical considerations, the design equation that best applies for freely suspended slots is

$$V = \frac{Q}{3.7XL}$$

Because flanging the slots will give the same benefits as flanging an open pipe, only 75 percent of the air is required to produce the same velocity at a given point. Therefore, for a flanged slot

$$V = \frac{Q}{2.8XL}$$

4.3.6 Air Distribution in Hoods

To provide efficient capture with a minimum expenditure of energy, the airflow across the face of a hood should be uniform throughout its cross section. For slots and lateral exhaust applications, this can be done by incorporating external baffles. Another method of design is to provide a velocity of 2000 to 2500 ft/min into the slot with a low-velocity plenum or large area chamber behind it. For large, shallow hoods, such as paint spray booths, laboratory hoods, and draft shakeout hoods, the same principle may be used. In these cases unequal flow may occur, with resulting higher velocities near the takeoffs. Baffles provided for the hood improve the air distribution and reduce pressure drop in the hood, giving the plenum effect. When the face velocity over the whole hood is relatively high or the hood or booth is quite deep, baffles may not be required.

4.3.7 Entrance Losses in Hoods

Hood static pressure, SP_h, is the term used to refer to the negative static pressure exhibited in ductwork a short distance downstream from the hood. It represents the energy needed to accelerate the air from ambient velocity (often near zero) to the duct velocity and also to overcome the frictional losses resulting from turbulence of the air upon entering the hood and ductwork. Therefore,

$$SP_h = VP + h_e$$

where VP = velocity pressure in the duct
$\quad\quad\quad h_e$ = hood entry loss

The hood entry loss, h_e, is expressed as a function of the velocity pressure VP. For hoods of most types, $h_e = F_h VP$, where F_h is the hood entry loss factor. For plain hoods, where the hood entry loss is a single expression $F_h VP$, the new VP referred to is the duct velocity pressure. The hood static pressure can be expressed as

$$SP_h = VP_{duct} + h_e$$

or

$$SP_h = VP_{duct} + F_h VP_{duct} = (1 + F_h)VP_{duct}$$

However, for slot, plenum, or compound hoods there are two entry losses, one through the slot and the other into the duct. Thus

$$SP_h = h_{e(slot)} + VP_{duct} + h_{e(duct)}$$

$$= F_{slot} VP_{slot} + VP_{duct} + F_{duct} VP_{duct}$$

The velocity pressure resulting from acceleration through the slot is not lost as long as the slot velocity is less than the duct velocity, and usually this is the case.

Another constant used to define the performance of a hood is "coefficient of entry", C_e. This is defined as the ratio of the actual airflow to the flow that would exist if all the static pressure were present as velocity pressure. Thus

$$C_e = \frac{Q_{actual}}{Q_{VP} = SP_h} = \frac{KA(VP)^{0.5}}{KA(SP_h)^{0.5}} = \left(\frac{VP}{SP_h}\right)^{0.5}$$

This quantity is constant for a given shape of hood and is very useful for determining the flow into a hood by a single hood static pressure reading. The coefficient of entry, C_e, is related to the hood entry loss factor, F_h, by the following equation only where the hood entry loss is a single expression:

$$C_e = \left(\frac{1}{1 + F_h}\right)^{0.5}$$

Listings of the entry loss coefficients, C_e, and the entry losses, h_e, in terms of velocity pressure, VP, are available (12). Most of the more complicated hoods have coefficients that can be obtained by combining some of the more simple shapes illustrated.

4.3.8 Static Suction

The air volume for a hood can be specified by giving the hood static pressure, SP_h, and duct size. For example, the hood static pressure at a typical grinding wheel hood is 2 in. of water. This reflects a conveying velocity of 4500 ft/min and entrance coefficient, C_e, of 0.78. For other types of machinery where the type of exhaust hood is relatively standard, a specification of the static suction and the duct size can be found in various reference sources (2–4). Specification of the static suction without duct size is of course meaningless, because decreased size increases velocity pressure and static suction while actually decreasing the total flow and the degree of control. Therefore static suction measurements for standard hoods or for systems in which the airflow has been measured previously are quite useful to estimate, in a comparative way, the quantity of air flowing through the hood.

4.3.9 Transport Velocities

In order to prevent particles of dust and fumes from settling and plugging ductwork, air velocities for transporting these materials must be relatively high. The minimum velocity, called "transport velocity," is typically 3500 to 4000 linear ft/min. At these velocities, frictional loss from air moving along the surface of the ducts becomes significant; therefore, all fittings, such as elbows and branches, must be wide-swept and gradual, having smooth interior surfaces. The cross-sectional area of the main duct generally equals the sum of the areas of cross sections for all branches upstream, plus a safety factor of approximately 20 percent. When the main duct is enlarged to accommodate an additional branch, the connection should be tapered, not abrupt.

Local exhaust systems for gases and vapors may have lower duct velocities (1500 to 2500 ft/min) because there is little to settle and plug the ducts. Lower velocities reduce markedly the frictional and pressure losses against which the fan must operate, thereby realizing a saving in power cost for the same airflow.

4.4 Principles of Local Exhaust Ventilation

When applying local exhaust ventilation to a specific problem, control of the contaminants is more effective if the following basic principles are followed.

4.4.1 Enclosure of the Source

A process to be exhausted by local ventilation should be enclosed as much as possible. This generally provides better control per unit volume of air exhausted. Nevertheless, the requirement of adequate access to the process must always be considered. An enclosed process may be costly in terms of operating efficiency or capital expenditure, but the savings gained by exhausting smaller air volumes may make the enclosure worthwhile.

4.4.2 Capture of Contaminants

Air velocity through all hood openings must be high enough not only to contain the contaminant but also to remove the contaminant from the hood. The importance of optimal capture and control velocity was discussed in the preceding sections.

4.4.3 Keeping Contaminants Out of Breathing Zone

Exhaust hoods that do not completely enclose the process should be located as near to the point of contaminant generation as possible and should provide airflow in a direction away from the worker toward the contaminant source. This item, drawing upon the characteristics of blowing and exhausting from openings in ductwork, was considered in more detail in the preceding sections.

4.4.4 Providing Makeup Air

All air exhausted from a building or enclosure must be replaced to keep the building from operating under negative pressure. This applies to local exhaust systems as well as general exhaust systems. Additionally, the incoming air must be tempered by a makeup air system before being distributed inside the process area. Without sufficient makeup air, exhaust ventilation systems cannot work as efficiently as intended. Makeup air requirements are discussed below.

4.4.5 Discharge of Exhaust Air

Exhaust air must be discharged in such a way as to prevent its reentry into the work space. It is disconcerting to observe the frequency with which this principle is violated. The beneficial effect of a well-designed local exhaust system can be offset by undesired recirculation of contaminated air back into the work area. Such recirculation can occur if the exhausted air is not discharged away from supply air inlets. The location of the exhaust stack, its height, and the type of stack weather cap all can have a significant effect on the likelihood of contaminated air reentering through nearby windows and supply air intakes.

4.5   Low-Volume High-Velocity Exhaust Systems

A unique application of the fundamentals of local exhaust ventilation incorporating the principles discussed in the preceding section is what is referred to as "low-volume high-velocity" (LVHV) exhaust ventilation. With typical application of this approach, small volumes of air at relatively high velocities are used to control dust generated by portable hand tools and machining operations. Control is achieved by exhausting the air essentially at the point of generation by means of close-fitting, custom-designed hoods. To maximize the flexibility required by operators of the equipment, flexible, small-diameter, lightweight plastic hoses are used, being built into the overall design of the tool. Figure 28.8 illustrates a typical application for control of dust generated by a disk sander. The capture velocities in these situations are high; slot velocities typically are in excess of 20,000 linear ft/min. However, because of the short distances required for capture of the dust and the small diameter tubing (ductwork) associated with the system, the total airflow rate is low.

Although this technique is not common, there are applications for which it has optimized control of dust. Successful use of LVHV exhaust ventilation has been made in such diverse activities as rock drilling, conventional machining, machining of beryllium, wood sanding, metal surface grinding, pneumatic chipping, internal surface grinding, and chiseling. Development of these exhaust systems has been somewhat empirical. However, it is emphasized that proper design of such systems requires the same careful, deliberate consideration of ventilation principles as more conventional exhaust systems, the topic of the following section.

5   DESIGN OF VENTILATION SYSTEMS

With the increasing interest in and concern for environmental conditions in the American workplace, the design of ventilation systems has advanced from the "art" it was in the not-too-distant past to a true science of engineering design. Many of the empirical guidelines or rules of thumb still are applied appropriately through continuing experimentation and satisfactory experience. These rules and guidelines have developed into special forms and charts published for the direct purpose of facilitating the design of ventilation systems.

In the industrial workplace there are two basic types of ventilation system: general and local exhaust. These concepts have been described earlier in this chapter. It is noted here that industrial hygienists' primary concern in the design of ventilation systems is to provide a safe and healthful environment for the worker. Although hygienists usually deal mainly with local exhaust ventilation considerations, general ventilation is an integral part of their concern as well.

5.1   General Ventilation

As described earlier, "general ventilation" refers to the general flushing of a work environment with a supply of fresh air, to maintain desirable temperature and humidity conditions or to dilute a contaminant or contaminants to acceptable levels.

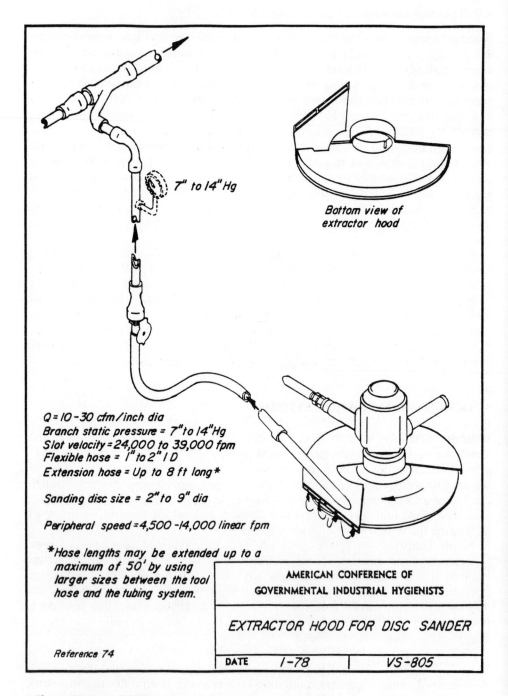

7" to 14" Hg

Bottom view of
extractor hood

Q = 10 -30 cfm/inch dia
Branch static pressure = 7" to 14" Hg
Slot velocity = 24,000 to 39,000 fpm
Flexible hose = 1" to 2" I D
Extension hose = Up to 8 ft long *

Sanding disc size = 2" to 9" dia

Peripheral speed = 4,500 - 14,000 linear fpm

*Hose lengths may be extended up to a
  maximum of 50' by using
  larger sizes between the tool
  hose and the tubing system.

AMERICAN CONFERENCE OF
GOVERNMENTAL INDUSTRIAL HYGIENISTS

EXTRACTOR HOOD FOR DISC SANDER

Reference 74

| DATE | 1-78 | VS-805 |

Figure 28.8. ACGIH-recommended arrangement for low-volume high-velocity hood.

5.1.1 Office Buildings

In nonproduction areas such as office buildings, the principal concern is worker comfort, as contrasted with general ventilation considerations for production areas where the focus is on maintaining contaminant levels at acceptable limits. Accordingly, it is necessary as a preliminary phase of general ventilation design for office space to consider the conditions that may affect worker comfort: air temperature, humidity, radiant heat sources, concentrations of tobacco smoke or other potentially offensive odors, the possibility of body odors, and general airflow patterns desirable in the office space.

Various attempts have been made to develop indexes that describe a "comfort zone," that is, a definition of a combination of parameters (usually temperature, humidity, and air velocity) that results in a "comfortable" environment. The comfort zone defined in the effective temperature charts (Figure 28.2) is one of the most frequently used indexes for this purpose. These comfort conditions can be met by proper selection of air conditioning equipment for office spaces. A more pressing modern-day problem is that referred to commonly as "tight building syndrome," resulting usually from insufficient supplies of fresh air to the interior of modern airtight buildings of masonry and glass. This condition of "stale air" as a unique industrial hygiene problem is treated in Chapter 17, Volume 1A.

Generally the problem of "odors" is affected by the air supply, the space provided per occupant, and the odor-absorbing capacity of the air conditioning process, as well as the temperature and humidity of the air.

In summary, the design of general office space ventilation requires that an analysis be made of the worker comfort zone parameters (primarily temperature and humidity), the tolerable level of odor, the equivalent space allowed per person in the room, and the operating characteristics of the air conditioning unit to be installed. Table 28.3 summarizes the requirements for ventilation under a variety of conditions typical of office-type space.

5.1.2 Industrial Process Buildings

The general ventilation systems typically employed in process operations can be natural or mechanically controlled. As discussed earlier in this chapter, the two forces available for use in natural ventilation are wind pressure and thermal or convective currents produced within the building. Although many of the older production facilities were designed with features such as sawtooth or monitor-type roofs to provide maximum ventilation, these applications are becoming less common, particularly in light of the relatively straightforward application of mechanically controlled ventilation systems. In some industrial buildings, such as warehouses and pump rooms, natural ventilation is sufficient to provide adequate movement of air, particularly because the number of people normally employed in such situations is limited. Despite the economic incentive associated with natural ventilation, however, it is somewhat restricted in its application, whereas ventilation by mechanically controlled devices is virtually without limit, particularly when used in conjunction with ductwork to distribute the air strategically throughout the space.

Table 28.3. Recommended Quantities of Outdoor Air Supplied to Occupied Space

Outdoor Air (ft³/min per person)	Type of Space or Occupancy[a]
5–10	High-ceiling space such as bank, auditorium, church, department store, theater; room with no smoking
10–15	Apartment, barber shop, beauty parlor, hotel room; room with light smoking
15–20	Cafeteria, drug store with lunch counter, general office, hospital room, public dining room, restaurant; room with moderate smoking
20–30	Brokers' boardroom, private office, tavern; room with heavy smoking
30–60	Conference room, night club; crowded room with heavy smoking

[a]Air-cleaning or odor-removing devices not used. (Space not less than 150 ft³ per person or floor area not less than 15 ft² per person.) State and city codes should be consulted to make certain that minimum outdoor air requirements are followed.

A primary reason for applying general ventilation within a process area is to secure the dilution of the contaminant level to acceptable limits. Although this is accompanied by benefits, usually in terms of worker comfort, the chief concern is generally the elimination of a potential health hazard. In that respect, dilution ventilation has application chiefly for vapors or gases, and only under unusual circumstances is it appropriate to consider general ventilation as a control means when the contaminant of concern is a particulate material. In these instances application of local exhaust ventilation would be more appropriate. Table 28.1 summarizes the air volumes necessary to dilute the concentrations of specifically identified vapors to acceptable limits. These values, used in conjunction with the considerations presented earlier in this chapter concerning the airflow patterns created by the general ventilation system, form the basis for the design of dilution ventilation systems.

5.2 Design of Local Exhaust Ventilation

When enclosure or other form of isolation or substitution is not appropriate, or has been used to the extent practicable, the industrial hygiene engineer looks to local exhaust ventilation as the primary control measure. With proper application of local exhaust ventilation, the potential workroom contaminant is controlled at or near the source of generation or release of the substance. It should be kept in mind that local exhaust systems are not necessarily designed to be 100 percent efficient in capturing and controlling contaminants released by the process. It is because of this residual or background level of contamination throughout the workroom that the general ventilation provisions are included as a complementary aspect of the total ventilation within a plant.

Many approaches to the overall design of the local exhaust ventilation system are available to the engineer. However, they all consist of a series of similar steps.

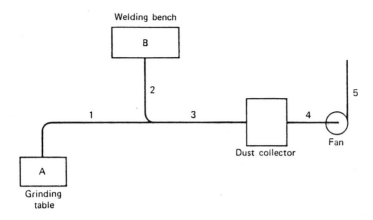

Figure 28.9. Simple sketch of local exhaust system.

The following section presents the sequential procedure to be used in the design of local exhaust systems. For illustrative purposes, an example of design of a local exhaust system is carried through the complete procedure.

a. Prepare a sketch or layout of the workroom and operations within it. This will facilitate preparation of a sketch and/or layout of the contemplated exhaust system, using single-line drawings for ductwork. Changes in elevation need not be shown at this point. Although it is helpful if the layout is approximately to scale, subsequent steps in the procedure, such as calculating proper duct sizes, can be accomplished from even a very simple drawing. For purposes of illustration, consider the design of a local exhaust system for a manual grinding station and a welding operation within the same general work area. The concept for the ventilation system to provide control of contaminants at these operations is illustrated in Figure 28.9.

b. Make a rough design or sketch of the desired hood for each operation, indicating direction and elevation of the outlet for the duct connections. The type of hood should be selected with an eye to performing the job with the least amount of air. This requires some ingenuity in designing enclosures that will not interfere with production; they may be equipped with movable sections. There may not be a "standard hood" available for a particular application. For convenience, the more or less standard designs of local exhaust hood for portable hand grinding and bench welding operations are used in this illustration. Figures 28.10 and 28.11, respectively, depict the configuration and design parameters for these hood arrangements (2).

c. Estimate the required control or capture velocity for each operation and type of hood or enclosure considered. This part of the process involves some judgment, particularly with hoods of nonstandard design. For this illustration, the information presented in Figures 28.10 and 28.11 is used in making preliminary estimates of

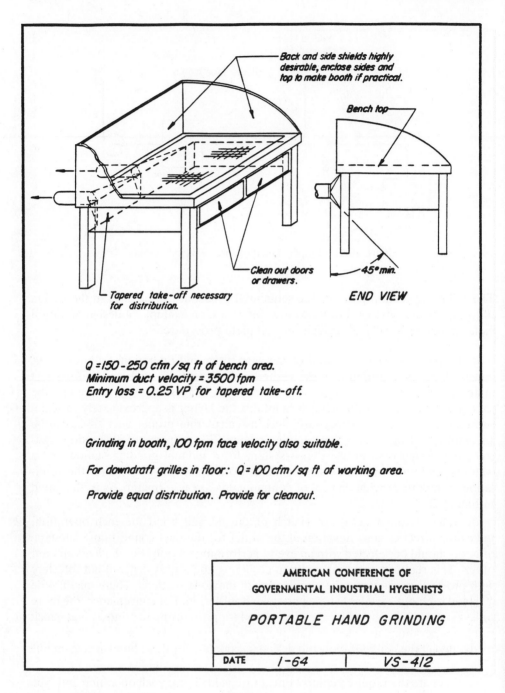

Back and side shields highly
desirable, enclose sides and
top to make booth if practical.

Bench top

END VIEW

45° min.

Clean out doors
or drawers.

Tapered take-off necessary
for distribution.

Q = 150 - 250 cfm / sq ft of bench area.
Minimum duct velocity = 3500 fpm
Entry loss = 0.25 VP for tapered take-off.

Grinding in booth, 100 fpm face velocity also suitable.

For downdraft grilles in floor: Q = 100 cfm / sq ft of working area.

Provide equal distribution. Provide for cleanout.

AMERICAN CONFERENCE OF
GOVERNMENTAL INDUSTRIAL HYGIENISTS

PORTABLE HAND GRINDING

DATE   1-64   VS-412

Figure 28.10. ACGIH-recommended hood for portable hand grinding.

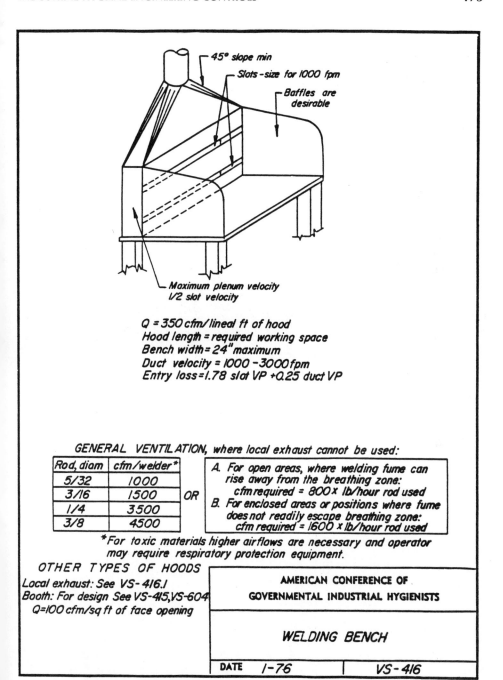

45° slope min

Slots - size for 1000 fpm

Baffles are desirable

Maximum plenum velocity 1/2 slot velocity

$Q = 350$ cfm/lineal ft of hood
Hood length = required working space
Bench width = 24" maximum
Duct velocity = 1000 - 3000 fpm
Entry loss = 1.78 slot VP + 0.25 duct VP

GENERAL VENTILATION, where local exhaust cannot be used:

Rod, diam	cfm/welder*
5/32	1000
3/16	1500
1/4	3500
3/8	4500

OR

A. For open areas, where welding fume can rise away from the breathing zone:
   cfm required = 800 x lb/hour rod used
B. For enclosed areas or positions where fume does not readily escape breathing zone:
   cfm required = 1600 x lb/hour rod used

*For toxic materials higher airflows are necessary and operator may require respiratory protection equipment.

OTHER TYPES OF HOODS
Local exhaust: See VS-416.1
Booth: For design See VS-415, VS-604
Q=100 cfm/sq ft of face opening

AMERICAN CONFERENCE OF
GOVERNMENTAL INDUSTRIAL HYGIENISTS

WELDING BENCH

| DATE | 1-76 | VS-416 |

Figure 28.11. ACGIH-recommended hood for welding bench.

Table 28.4. Typical Recommended Transport Velocities

Material, Operation, or Industry	Minimum Transport Velocity (ft/min)	Material, Operation, or Industry	Minimum Transport Velocity (ft/min)
Abrasive blasting	3500–4000	Lead dust	4000
Aluminum dust, coarse	4000	With small chips	5000
Asbestos carding	3000	Leather dust	3500
Bakelite molding powder dust	2500	Limestone dust	3500
Barrel filling or dumping	3500–4000	Lint	2000
Belt conveyors	3500	Magnesium dust, coarse	4000*
Bins and hoppers	3500	Metal turnings	4000–5000
Brass turnings	4000	Packaging, weighing, etc.	3000
Bucket elevators	3500	Downdraft grille	3500
Buffing and polishing		Pharmaceutical coating pans	3000
Dry	3000–3500	Plastics dust (buffing)	3800
Sticky	3500–4500	Plating	2000
Cast iron boring dust	4000	Rubber dust	
Ceramics, general		Fine	2500
Glaze spraying	2500	Coarse	4000
Brushing	3500	Screens	
Fettling	3500	Cylindrical	3500
Dry pan mixing	3500	Flat deck	3500

Dry press	3500
Sagger filling	3500
Clay dust	3500
Coal (powdered) dust	4000
Cocoa dust	3000
Cork (ground) dust	2500
Cotton dust	3000
Crushers	3000 or higher
Flour dust	2500
Foundry, general	3500
Sand mixer	3500–4000
Shakeout	3500–4000
Swing grinding booth exhaust	3000
Tumbling mills	4000–5000
Grain dust	2500–3000
Grinding, general	3500–4500
Portable hand grinding	3500
Jute	
Dust	2500–3000
Lint	3000
Dust shaker waste	3200
Pickerstock	3000

Silica dust	3500–4500
Soap dust	3000
Soapstone dust	3500
Soldering and tinning	2500
Spray painting	2000
Starch dust	3000
Stone cutting and finishing	3500
Tobacco dust	3500
Woodworking	
Wood flour, light dry sawdust and shavings	2500
Heavy shavings, damp sawdust	3500
Heavy wood chips, waste, green shavings	4000
Hog waste	3000
Wool	3000
Zinc oxide fume	2000

appropriate velocities and volumetric flow rates. Accordingly, an airflow rate of 200 ft³/min per square foot (cfm/ft²) of bench area for the grinding table and a slot velocity of 1000 (linear) ft/min for the welding bench are selected.

d. Estimate or compute the "sphere of influence" over which the selected capture velocities must function. It will be advantageous to consider the hood arrangements in three dimensions. In the illustrated problem, assume that work at both the grinding table and the welding bench will be conducted on a flat surface 2 × 4 ft and that most of the work will be conducted within 2 in. of the table surface.

e. Compute the total volumetric airflow needed for each hood and indicate this on the working sketch of the system or in an appropriate work sheet. From the information supplied in Figures 28.10 and 28.11, the required volumetric airflows for the two hoods can be calculated. For this example, a volumetric flowrate of 200 ft³/min per square foot of work surface and a bench area of 8 ft² have been assumed. Thus a volumetric flow rate of 1600 ft³/min will be required for hood A, the grinding table. For the welding bench, the required volumetric rate will be 350 ft³/min per linear foot of hood times 4 ft (the length of the hood), or a total of 1400 ft³/min.

f. Summarize the volumetric airflow requirements for each section of the branch and main ductwork on the work sheet, or enter the air flows directly on the working sketch of the system. For this simple illustration, the summation of volumetric rate requirements is straightforward:

Duct section 1	1600 ft³/min
Duct section 2	1400 ft³/min
Duct sections 3, 4, and 5	3000 ft³/min

g. Select the minimum effective transport velocities for the contaminant-laden air being moved through the ductwork. For transporting particulate matter, data such as those in Table 28.4 can be used. Velocities in the range of 1000–2000 linear ft/min commonly are used for gases and vapors, although in these cases duct velocities can be as low as is consistent with space and weight limitations. The recommended transport velocities presented in Table 28.4 indicate values of 4000 and 3500 linear ft/min for ductwork servicing the grinding table and welding bench hoods, respectively.

h. Determine the nearest practical duct diameter necessary to carry the required volumetric rate of air at the proper velocity. It is often necessary at this juncture to consider the standard sizes of ducts available from the fabricator, to avoid the unnecessary expense of special fabrication. If the available duct sizes differ greatly from the theoretically required diameters, the actual transport velocity should be computed for use in subsequent design steps. In the example problem, dividing the volumetric flow rates by the desired transport velocities, and assuming that standard ductwork is available in full-inch diameter increments only, the duct diameter nearest to the calculated necessary diameters is 8 in. for both the grinding table and the welding bench. Using 8-in. diameter ductwork for these hoods would produce transport velocities of 4580 and 4010 linear ft/min for the grinding hood and welding hood, respectively.

i. Compute the velocity pressures corresponding to air velocities in each branch duct connected to an exhaust opening. Velocity pressures are calculated using the formula

$$VP = \left(\frac{V}{4005}\right)^2$$

j. Estimate the entry loss for each hood in percentage of velocity pressure in the branch duct. In the example problem, these relationships are given in Figures 28.10 and 28.11.

k. Compute the entrance losses for all hoods, using information developed in items 9 and 10. Using the relationships presented in Figures 28.10 and 28.11, the entrance losses for the grinding table hood and the welding bench hood can be calculated to be 0.327 and 0.357 in H_2O, respectively.

l. Compute the losses in elbows, the branch-to-main connections, or other transition points in the duct system. This can be done in terms of either equivalent straight-duct lengths or velocity pressures, whichever form corresponds to the data being used. The equivalent straight-duct method has the advantage that such transition losses can be added directly to the straight-duct sections for use with friction charts. The velocity-head method may be preferred by those who are especially familiar with the fundamentals of fluid mechanics. For this sample problem, refer to Figure 28.12 for the relative arrangement of equipment in the exhaust system. In computing friction and transition losses assume that branches 1 and 2 contain two 90° elbows, that the centerline radius of the elbow turns is twice the duct diameter, that branch 2 merges with duct section 3 on a 45° angle of entry, and that duct section 4 contains one 90° elbow also with a centerline radius of two diameters. On the basis of these assumptions, and using information such as that presented in Table 28.5, it can be shown that each of the elbows in duct sections 1 and 2 is equivalent to 10 ft of straight duct, that the elbow in duct section 4 is equivalent to 17 ft of straight duct, and that the transition point from duct section 2 to 3 is equivalent to 11 ft of straight duct.

m. Obtain the duct friction losses from an appropriate chart, table, or formula. A friction chart like the one illustrated in Figure 28.13 can be used at this point. For the various sections of ductwork associated with the example problem, the following friction losses can be derived by referring to Figure 28.13 or similar chart:

Duct section 1	4.0 in. H_2O/100 ft of duct
Duct section 2	2.9 in. H_2O/100 ft of duct
Duct sections 3, 4, 5	1.0 in. H_2O/100 ft of duct

n. Summarize the losses for each branch to determine where to increase or

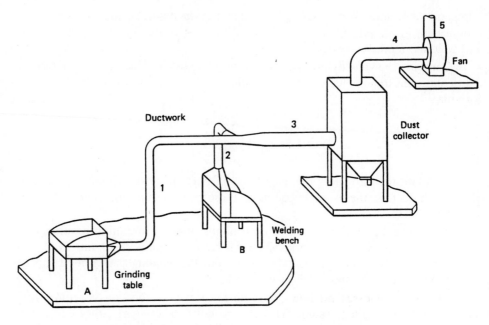

Figure 28.12. Representation of ventilation system.

decrease duct sizes to ensure that a balance of pressure drop will be maintained. The total pressure drop from any points in the system to each upstream opening or exhaust point should be the same if balance is to be achieved. In actual operation, the airflow quantities through all open branches automatically adjust themselves to fulfill this requirement of balance. If the flow through any branch is too high, further closing of the damper increases the friction or energy loss through that branch, thus making it possible for a smaller volume of airflow to create the necessary pressure drop for a balanced condition. In summing the entry losses for the hoods and the friction losses in the ductwork for duct sections 1 and 2, it can be shown that these two sections are balanced by virtue of the total pressure drop at the common point (where duct section 2 merges with the main duct) being the same, namely, 1.60 in. H_2O.

o. After the system has been balanced on paper to a reasonable degree by adjusting round duct sizes, any necessary conversion to rectangular duct sizes can be made using a chart such as that in Figure 28.14. In the sample problem, this step is not applicable.

p. Determine the adjusted total pressure drop of the branch and main duct system by adding the pressure losses, beginning with any one exhaust opening and proceeding toward the fan or dust collector. If the branch duct system has been reasonably balanced (in accordance with item 14), it will be a simple matter to select the starting point for totaling the pressure losses. However, if no balancing has been attempted, the losses must be estimated for the single run of duct that is

Table 28.5. Equivalent Resistance of Elbows and Duct Connections

EQUIVALENT RESISTANCE IN FEET OF STRAIGHT PIPE							

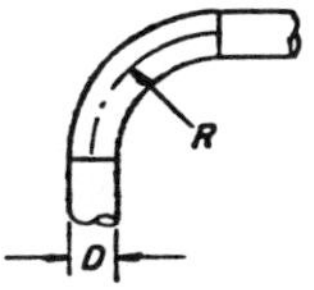

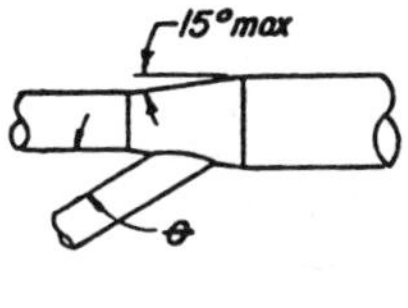

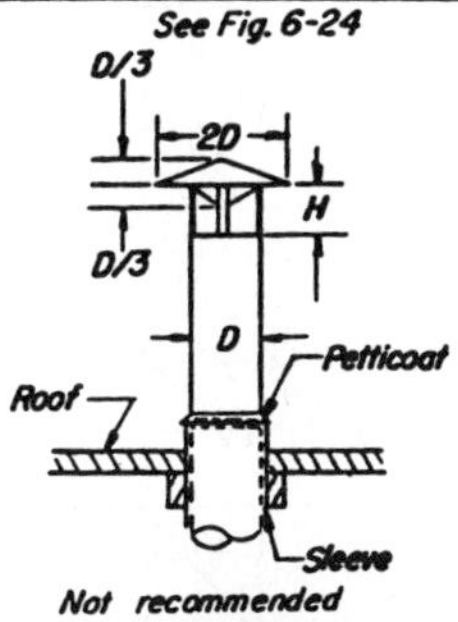

Pipe D	90° Elbow * Centerline Radius			Angle of Entry		H, No of Diameters		
	1.5 D	2.0 D	2.5 D	30°	45°	1. D	.75 D	.5 D
3"	5	3	3	2	3	2	2	9
4"	6	4	4	3	5	2	3	12
5"	9	6	5	4	6	2	4	16
6"	12	7	6	5	7	3	5	20
7"	13	9	7	6	9	3	6	23
8"	15	10	8	7	11	4	7	26
10"	20	14	11	9	14	5	9	36
12"	25	17	14	11	17	6	11	44
14"	30	21	17	13	21	7	13	53
16"	36	24	20	16	25	9	15	62
18"	41	28	23	18	28	10	18	71
20"	46	32	26	20	32	11	20	80
24"	57	40	32			13	24	92
30"	74	51	41			17	31	126
36"	93	64	52			22	39	159
40"	105	72	59					
48"	130	89	73					

* For 60° elbows —— 0.67 x loss for 90°
45° elbows —— 0.5 x loss for 90°
30° elbows —— 0.33 x loss for 90°

AMERICAN CONFERENCE OF GOVERNMENTAL INDUSTRIAL HYGIENISTS	
DUCT DESIGN DATA	
DATE 1-76	Fig. 6-11

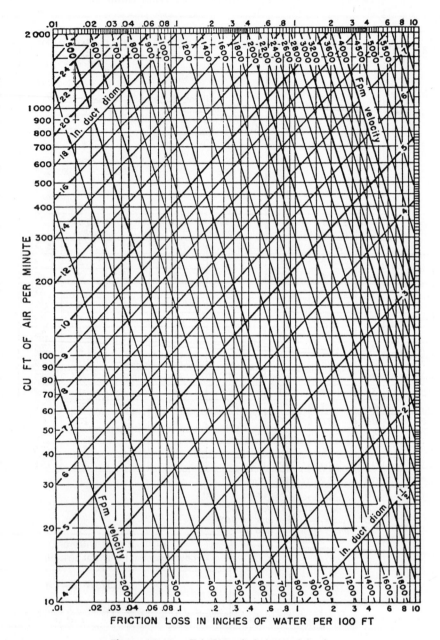

Figure 28.13. Friction of air in straight ducts.

Side Rectangular Duct	4.0	4.5	5.0	5.5	6.0	6.5	7.0	7.5	8.0	8.5	9.0	9.5	10.0	10.5	11.0	11.5	12.0	12.5	13.0	13.5	14.0	14.5	15.0	15.5	16
3.0	3.8	4.0	4.2	4.4	4.6	4.8	4.9	5.1	5.2	5.4	5.5	5.6	5.7	5.9	6.0	6.1	6.2	6.3	6.4	6.5	6.6	6.7	6.8	6.9	7.0
3.5	4.1	4.3	4.6	4.8	5.0	5.2	5.3	5.5	5.7	5.8	6.0	6.1	6.3	6.4	6.5	6.7	6.8	6.9	7.0	7.1	7.2	7.3	7.4	7.5	7.6
4.0	4.4	4.6	4.9	5.1	5.3	5.5	5.7	5.9	6.1	6.3	6.4	6.6	6.8	6.9	7.1	7.2	7.3	7.5	7.6	7.7	7.8	7.9	8.1	8.2	8.3
4.5	4.6	4.9	5.2	5.4	5.6	5.9	6.1	6.3	6.5	6.7	6.9	7.0	7.2	7.4	7.5	7.7	7.8	8.0	8.1	8.2	8.4	8.5	8.6	8.7	8.9
5.0	4.9	5.2	5.5	5.7	6.0	6.2	6.4	6.7	6.9	7.1	7.3	7.4	7.6	7.8	8.0	8.1	8.3	8.4	8.6	8.7	8.9	9.0	9.1	9.3	9.4
5.5	5.1	5.4	5.7	6.0	6.3	6.5	6.8	7.0	7.2	7.4	7.6	7.8	8.0	8.2	8.4	8.6	8.7	8.8	9.0	9.2	9.4	9.5	9.6	9.8	9.8

Side Rectangular Duct	6	7	8	9	10	11	12	13	14	15	16	17	18	19	20
6	6.6														
7	7.1	7.7													
8	7.5	8.2	8.8												
9	8.0	8.6	9.3	9.9											
10	8.4	9.1	9.8	10.4	10.9										
11	8.8	9.5	10.2	10.8	11.4	12.0									
12	9.1	9.9	10.7	11.3	11.9	12.5	13.1								
13	9.5	10.3	11.1	11.8	12.4	13.0	13.6	14.2							
14	9.8	10.7	11.5	12.2	12.9	13.5	14.2	14.7	15.3						
15	10.1	11.0	11.8	12.6	13.3	14.0	14.6	15.3	15.8	16.4					
16	10.4	11.4	12.2	13.0	13.7	14.4	15.1	15.7	16.3	16.9	17.5				
17	10.7	11.7	12.5	13.4	14.1	14.9	15.5	16.1	16.8	17.4	18.0	18.6			
18	11.0	11.9	12.9	13.7	14.5	15.3	16.0	16.6	17.3	17.9	18.5	19.1	19.7		
19	11.2	12.2	13.2	14.1	14.9	15.6	16.4	17.1	17.8	18.4	19.0	19.6	20.2	20.8	
20	11.5	12.5	13.5	14.4	15.2	15.9	16.8	17.5	18.2	18.8	19.5	20.1	20.7	21.3	21.9
22	12.0	13.1	14.1	15.0	15.9	16.7	17.6	18.3	19.1	19.7	20.4	21.0	21.7	22.3	22.9
24	12.4	13.6	14.6	15.6	16.6	17.5	18.3	19.1	19.8	20.6	21.3	21.9	22.6	23.2	23.9
26	12.8	14.1	15.2	16.2	17.2	18.1	19.0	19.8	20.6	21.4	22.1	22.8	23.5	24.1	24.8
28	13.2	14.5	15.6	16.7	17.7	18.7	19.6	20.5	21.3	22.1	22.9	23.6	24.4	25.0	25.7
30	13.6	14.9	16.1	17.2	18.3	19.3	20.2	21.1	22.0	22.9	23.7	24.4	25.2	25.9	26.7
32	14.0	15.3	16.5	17.7	18.8	19.8	20.8	21.8	22.7	23.6	24.4	25.2	26.0	26.7	27.5
34	14.4	15.7	17.0	18.2	19.3	20.4	21.4	22.4	23.3	24.2	25.1	25.9	26.7	27.5	28.3
36	14.7	16.1	17.4	18.6	19.8	20.9	21.9	23.0	23.9	24.8	25.8	26.6	27.4	28.3	29.0
38	15.0	16.4	17.8	19.0	20.3	21.4	22.5	23.5	24.5	25.4	26.4	27.3	28.1	29.0	29.8
40	15.3	16.8	18.2	19.4	20.7	21.9	23.0	24.0	25.1	26.0	27.0	27.9	28.8	29.7	30.5
42	15.6	17.1	18.5	19.8	21.1	22.3	23.4	24.5	25.6	26.6	27.6	28.5	29.4	30.4	31.2
44	15.9	17.5	18.9	20.2	21.5	22.7	23.9	25.0	26.1	27.2	28.2	29.1	30.0	31.0	31.9
46	16.2	17.8	19.2	20.6	21.9	23.2	24.3	25.5	26.7	27.7	28.7	29.7	30.6	31.6	32.5
48	16.5	18.1	19.6	20.9	22.3	23.6	24.8	26.0	27.2	28.2	29.2	30.2	31.2	32.2	33.1
50	16.8	18.4	19.9	21.3	22.7	24.0	25.2	26.4	27.6	28.7	29.8	30.8	31.8	32.8	33.7
52	17.0	18.7	20.2	21.6	23.1	24.4	25.6	26.8	28.1	29.2	30.3	31.4	32.4	33.4	34.3
54	17.3	19.0	20.5	22.0	23.4	24.8	26.1	27.3	28.5	29.7	30.8	31.9	32.9	33.9	34.9
56	17.6	19.3	20.9	22.4	23.8	25.2	26.5	27.7	28.9	30.1	31.2	32.4	33.4	34.5	35.5
58	17.8	19.5	21.1	22.7	24.2	25.5	26.9	28.2	29.3	30.5	31.7	32.9	33.9	35.0	36.0
60	18.1	19.8	21.4	23.0	24.5	25.8	27.3	28.7	29.8	31.0	32.2	33.4	34.5	35.5	36.5
62	18.3	20.1	21.7	23.3	24.8	26.2	27.6	29.0	30.2	31.4	32.6	33.8	35.0	36.0	37.1
64	18.6	20.3	22.0	23.6	25.2	26.5	27.9	29.3	30.6	31.8	33.1	34.2	35.5	36.5	37.6
66	18.8	20.6	22.3	23.9	25.5	26.9	28.3	29.7	31.0	32.2	33.5	34.7	35.9	37.0	38.1
68	19.0	20.8	22.5	23.7	25.8	27.3	28.7	30.1	31.4	32.6	33.9	35.1	36.3	37.5	38.6
70	19.2	21.1	22.8	23.5	26.1	27.6	29.1	30.4	31.8	33.1	34.3	35.6	36.8	37.9	39.1
72															39.6
74															40.0
76															40.5
78															40.9
80															41.3
82															41.8
84															42.2
86															42.6
88															43.0
90															43.4
92															43.8
94															44.2
96															44.6

Side Rectangular Duct	22	24	26	28	30	32	34	36	38	40	42	44	46	48	50	52	56	60	64	68	72	76	80	84	88	Side Rectangular Duct
6																										6
7																										7
8																										8
9																										9
10																										10
11																										11
12																										12
13																										13
14																										14
15																										15
16																										16
17																										17
18																										18
19																										19
20																										20
22	24.1																									22
24	25.1	26.2																								24
26	26.1	27.2	28.4																							26
28	27.1	28.2	29.5	30.6																						28
30	28.0	29.3	30.5	31.6	32.8																					30
32	28.9	30.1	31.4	32.6	33.8	35.0																				32
34	29.7	31.0	32.3	33.6	34.8	36.0	37.2																			34
36	30.5	32.0	33.0	34.6	35.8	37.0	38.2	39.4																		36
38	31.4	32.8	34.2	35.5	36.7	38.0	39.2	40.4	41.6																	38
40	32.1	33.6	35.1	36.4	37.6	39.0	40.2	41.4	42.6	43.8																40
42	32.8	34.4	35.9	37.3	38.6	39.9	41.1	42.4	43.6	44.8	45.9															42
44	33.5	35.2	36.7	38.1	39.5	40.8	42.0	43.4	44.6	45.8	46.9	48.1														44
46	34.2	35.9	37.4	38.9	40.3	41.7	43.0	44.3	45.6	46.8	47.9	49.1	50.3													46
48	34.9	36.6	38.2	39.7	41.2	42.6	43.9	45.2	46.5	47.8	48.9	50.2	51.3	52.6												48
50	35.5	37.3	38.9	40.4	42.0	43.5	44.8	46.1	47.4	48.8	49.8	51.2	52.3	53.6	54.7											50
52	36.2	38.0	39.6	41.2	42.8	44.3	45.7	47.1	48.3	49.7	50.8	52.2	53.3	54.6	55.8	56.9										52
54	36.8	38.7	40.3	42.0	43.6	45.0	46.5	48.0	49.2	50.6	51.8	53.2	54.3	55.6	56.8	57.9										54
56	37.4	39.3	41.0	42.7	44.3	45.8	47.3	48.8	50.1	51.5	52.7	54.1	55.3	56.5	57.8	58.9	61.3									56
58	38.0	39.8	41.7	43.4	45.0	46.6	48.1	49.6	51.0	52.4	53.7	55.0	56.2	57.5	58.8	60.0	62.3									58
60	38.6	40.4	42.3	44.0	45.8	47.3	48.9	50.4	51.8	53.3	54.6	55.9	57.1	58.5	59.8	61.0	63.3	65.7								60
62	39.2	41.0	42.9	44.7	46.5	48.0	49.7	51.2	52.6	54.2	55.5	56.8	58.0	59.4	60.7	62.0	64.3	66.7								62
64	39.7	41.6	43.5	45.4	47.2	48.7	50.4	52.0	53.4	55.0	56.4	57.7	59.0	60.3	61.6	62.9	65.3	67.7	70.0							64
66	40.2	42.2	44.1	46.0	47.8	49.5	51.1	52.8	54.2	55.8	57.2	58.6	59.9	61.2	62.5	63.9	66.3	68.7	71.1							66
68	40.7	42.8	44.7	46.6	48.4	50.2	51.8	53.5	55.0	56.6	58.0	59.5	60.8	62.1	63.4	64.8	67.3	69.7	72.1	74.4						68
70	41.3	43.3	45.3	47.2	49.0	50.9	52.5	54.2	55.8	57.3	58.8	60.3	61.7	63.0	64.3	65.7	68.3	70.7	73.1	75.4						70
72	41.8	43.8	45.9	47.8	49.7	51.5	53.2	54.9	56.5	58.0	59.6	61.1	62.6	63.9	65.2	66.6	69.2	71.7	74.1	76.4	78.8					72
74	42.3	44.4	46.4	48.4	50.3	52.1	53.9	55.6	57.2	58.8	60.4	61.9	63.3	64.8	66.1	67.5	70.1	72.7	75.1	77.4	79.9					74
76	42.8	44.9	47.0	49.0	50.8	52.7	54.6	56.3	57.9	59.5	61.2	62.7	64.1	65.6	67.0	68.4	71.0	73.6	76.1	78.4	80.9	83.2				76
78	43.3	45.5	47.5	49.5	51.5	53.3	55.2	57.0	58.6	60.3	62.0	63.4	64.9	66.4	67.9	69.3	71.8	74.5	77.1	79.4	81.8	84.2				78
80	43.8	46.0	48.0	50.1	52.0	53.9	55.8	57.6	59.3	61.0	62.7	64.1	65.7	67.2	68.7	70.1	72.7	75.4	78.1	80.4	82.8	85.2	87.5			80
82	44.2	46.4	48.6	50.6	52.6	54.5	56.4	58.2	60.0	61.7	63.4	64.9	66.5	68.0	69.5	71.0	73.6	76.3	79.0	81.4	83.8	86.2	88.6			82
84	44.6	46.9	49.2	51.1	53.2	55.1	57.0	58.9	60.7	62.4	64.1	65.7	67.3	68.8	70.3	71.8	74.5	77.2	79.9	82.4	84.8	87.2	89.6	91.9		84
86	45.0	47.4	49.6	51.6	53.7	55.7	57.6	59.5	61.3	63.0	64.8	66.4	68.0	69.5	71.1	72.6	75.4	78.1	80.8	83.3	85.8	88.2	90.6	92.9		86
88	45.4	47.9	50.1	52.2	54.3	56.3	58.2	60.1	62.0	63.7	65.4	67.0	68.7	70.3	71.8	73.4	76.3	79.0	81.6	84.2	86.8	89.2	91.6	93.9	96.3	88
90	45.9	48.3	50.6	52.8	54.8	56.9	58.8	60.7	62.6	64.4	66.0	67.8	69.4	71.1	72.6	74.2	77.1	79.9	82.5	85.1	87.8	90.2	92.6	94.9	97.3	90
92	46.3	48.7	51.1	53.4	55.4	57.4	59.4	61.3	63.2	65.0	66.8	68.5	70.1	71.8	73.3	74.9	77.8	80.8	83.4	86.0	88.7	91.2	93.6	95.9	98.3	92
94	46.7	49.1	51.6	53.9	55.9	57.9	60.0	61.9	63.8	65.6	67.5	69.2	70.8	72.5	74.1	75.6	78.6	81.7	84.3	86.9	89.6	92.1	94.6	96.9	99.3	94
96	47.2	49.5	52.0	54.4	56.3	58.4	60.5	62.4	64.4	66.2	68.2	69.8	71.5	73.2	74.8	76.3	79.4	82.6	85.2	87.8	90.5	93.0	95.6	97.9	100.3	96

Figure 28.14. Chart for converting round to rectangular duct sizes having equivalent friction losses. (Reprinted by permission, "Friction Equivalents for Round, Square and Rectangular Ducts," by R. G. Huebacher, from A.S.H.V.E. Journal Section, *Heating, Piping and Air Conditioning*, December, 1947.

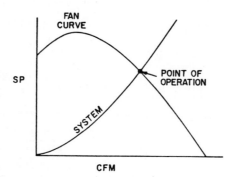

Figure 28.15. Fan and system curves.

believed to have the greatest total resistance. Experienced designers, after a few moments of inspection, usually can determine the longest equivalent run in terms of friction loss—that is, the course that offers the greatest resistance to airflow, not necessarily the longest one in tems of physical length. In summing the total pressure drop of the branch and main duct system, it can be calculated that the total pressure drop up to the point of entry into the dust collector is 1.91 in. H_2O.

q. Add the resistance of accessory equipment such as air cleaners and weather caps. For the most part, this information should be obtained directly from the manufacturers, because equipment varies greatly in resistance to airflow. For this example, assume that the pressure drop across the dust collector is 2.0 in. H_2O.

r. Select a fan of proper type, size, and speed to produce the highest operating efficiency, or lowest horsepower, consistent with other considerations. Use the calculated total airflow in cubic feet per minute and the total system resistances or static pressure for fan selection purposes. In selecting a fan, it is necessary to be familiar with the fan laws and system curves, that is, how the static pressure and volumetric airflow are related. A fan curve or system curve shows graphically the possible combinations of volumetric flow and static pressure for a given application. Because the fan and system each can operate only at a single point on its own curve, the combination can operate only where these curves intersect (Figure 28.15). Thus a fan must be chosen on the basis of its characteristics and the requirements of the system in which it will be applied. Each fan is characterized by the volume of gas flow, the pressure at which the flow is produced, the speed of rotation, the power required, and the efficiency. These quantities are measured by manufacturers with standard testing methods, and the results are plotted to furnish the characteristic "fan curves" of most manufacturers. Tables such as Table 28.6 are available to assist the design engineer in selecting a fan. The fan static pressure is calculated using the following equation:

$$\text{fan } SP = SP(\text{fan inlet}) + SP(\text{fan outlet}) - VP(\text{fan inlet})$$

$$= 4.20 + 0.10 - 0.91$$

$$= 3.39 \text{ in. } H_2O$$

Table 28.6. Typical Multi-Rating Table for Fans

Volume (ft^3/min)	Outlet Velocity (ft/min)	Velocity Pressure (in. water)	Static Pressure, SP (in. H_2O)																	
			1 in.		2 in.		3 in.		4 in.		5 in.		6 in.		7 in.		8 in.		9 in.	
			rpm	bhp	rpm	bhp	rpm	bhp	rpm	bhp	rpm	bhp	rpm	bhp	rpm	bhp	rpm	bhp	rpm	bhp
2,520	1000	0.063	437	0.63	595	1.27	728	2.00	837	2.66										
3,120	1200	0.090	459	0.85	610	1.55	735	2.30	842	3.10										
3,530	1400	0.122	483	1.05	626	1.87	746	2.72	847	3.57	943	4.60								
4,030	1600	0.160	513	1.33	642	2.18	759	3.17	858	4.12	950	5.21	1030	6.29						
4,530	1800	0.202	532	1.61	666	2.56	774	3.63	876	4.63	964	5.82	1040	6.92	1125	8.18				
5,040	2000	0.250	572	2.00	688	2.97	797	4.12	890	5.30	976	6.50	1052	7.75	1134	8.96	1208	10.15	1270	11.67
5,540	2200	0.302	603	2.36	712	3.43	816	4.66	910	5.93	999	7.38	1068	8.60	1145	9.93	1210	11.18	1279	12.82
6,040	2400	0.360	637	2.79	746	3.99	840	5.33	926	6.73	1017	8.17	1088	9.50	1160	10.88	1230	12.25	1288	13.92
6,550	2600	0.422	670	3.27	762	4.62	866	6.05	954	7.83	1032	9.08	1095	10.50	1171	11.98	1245	13.50	1298	15.10
7,060	2800	0.489	708	3.81	795	5.32	892	6.72	963	8.78	1050	9.97	1125	11.60	1188	13.06	1257	14.70	1310	16.48
7,560	3000	0.560	746	4.42	833	6.05	920	7.70	993	9.32	1068	11.00	1142	12.75	1210	14.28	1277	15.98	1328	17.80
8,060	3200	0.638			866	6.96	943	8.71	1020	10.40	1097	12.10	1168	14.02	1228	15.50	1292	17.36	1340	19.15
8,560	3400	0.721			900	7.93	964	9.80	1053	11.48	1120	13.30	1188	15.35	1248	16.93	1310	19.00	1360	20.90
9,070	3600	0.808					1010	11.00	1078	12.70	1148	14.65	1213	16.70	1270	18.42	1335	20.75	1380	22.60
9,570	3800	0.900					1038	12.25	1108	14.15	1170	14.90	1240	18.80	1292	19.46	1355	22.35	1405	24.40
10,080	4000	0.998					1162	13.60	1138	15.40	1200	17.35	1270	19.70	1320	21.70	1380	23.15	1430	26.40
10,580	4200	1.100							1168	16.90	1230	19.05	1283	21.50	1348	23.50	1405	26.10	1450	28.45
11,100	4400	1.210							1198	18.58	1258	20.55	1322	22.50	1373	25.40	1430	27.95	1478	30.60
11,600	4600	1.310							1232	20.30	1290	22.50	1355	23.80	1405	27.40	1450	30.15	1500	32.90
12,100	4800	1.450							1270	21.00	1321	24.40	1383	25.65	1432	29.60	1482	32.40	1528	35.20
12,600	5000	1.570							1301	24.20	1355	26.40	1410	28.80	1462	31.80	1513	34.60	1555	37.80
15,120	6000	2.230													1622	45.90	1670	49.00	1702	51.50

Table 28.7. Industrial Activities Covered by OSHA Standards Specifying
Ventilation Requirements

Industrial Application	Standard Reference
Abrasive blasting	29 CFR 1910.94(a)
Asbestos	29 CFR 1910.1001(c)
Bulk gas plants	29 CFR 1910.106(f)
Confined spaces	29 CFR 1910.146
Dip tanks containing combustible or flammable liquids	29 CFR 1910.108(b)
Electrostatic spraying	29 CFR 1910.107(i)
Flammable or combustible liquids in storage rooms and enclosures	29 CFR 1910.106(d)
Grinding, polishing, and buffing	29 CFR 1910.94(b)
Hydrogen	29 CFR 1910.103(b)
Laundries	29 CFR 1910.262(c)
Open surface tanks	29 CFR 1910.94(d)
Oxygen	29 CFR 1910.104(b)
Powder coatings	29 CFR 1910.107(i)
Sawmills	29 CFR 1910.265(c)
Spray-finishing operations (general)	29 CFR 1910.107(d)
Surface coating	29 CFR 1910.94(c)
Textiles	29 CFR 1910.262(rr)
Welding, cutting, and brazing	29 CFR 1910.252(f)

Thus the fan and motor selected for this system should be capable of handling a
static pressure of 3.5 to 4.0 in. H_2O.

5.3 Ventilation Regulations

Some applications of ventilation, both general and local exhaust, are covered specifically by occupational safety and health regulations. The federal standards as
established by the Occupational Safety and Health Administration (OSHA) now
include specific ventilation requirements for the applications identified in Table
28.7.

In addition to the specific references to ventilation requirements in those standards identified above, most occupational health standards for specific substances,
for example, benzene, ethylene oxide, formaldehyde, and lead, incorporate language clearly meant to include ventilation considerations. In most recent OSHA
health standards there is a section titled "Methods of Compliance" which contains
a statement equivalent to ". . . employer shall institute engineering controls and
work practices to reduce and maintain employee exposure to or below the PEL–
TWA, except to the extent that such controls are not feasible." It is inconceivable
that such "engineering controls" would not rely heavily on ventilation.

For the most part, the ventilation requirements incorporated into OSHA standards are based on consensus-type standards developed by organizations such as the

American National Standards Institute (ANSI) and the National Fire Protection Association (NFPA). In all likelihood, the number and degree of specificity of ventilation standards will increase in the future in the form of regulations promulgated by OSHA and other regulatory agencies.

Many of the occupational health standards currently proposed by OSHA are couched in language that makes it clear that engineering control of the specific contaminants covered by the standard is the primary responsibility of the employer. In fact, engineering controls (chiefly ventilation in the case of airborne contaminants) must be applied to the extent feasible even if they are not sufficient by themselves to reduce the airborne concentration of the contaminant to the specified acceptable level. Thus in addition to the occupational health standards dealing with specific aspects of ventilation, there are indirect references to the requirement for installation of ventilation systems to maintain the work environment within stated acceptable limits. Accordingly, the need for application of sound engineering in the control of workplace contaminants undoubtedly will continue to increase.

5.4 Computer Assistance

The preceding sections have dealt with the process of designing ventilation systems, using calculations based on determination of a "balanced" system. Thus, to a large extent, the calculations use "trial-and-error" solutions to optimize the dimensions and other quantitative aspects of the system. For example, there generally is an interrelationship among the key variables affecting airflow through a section of ductwork: duct diameter, air velocity, volumetric flow rate, and resistance to airflow. These relationships are depicted in graphs or "friction charts" such as that in Figure 28.13. One would expect, therefore, that a means of "number crunching," that is, repetitive calculations using similar sets of data, would facilitate the ventilation design process. Similarly, because once the parameters of the system have been selected the calculations are relatively straightforward, a mechanical calculator would be helpful. For these reasons, then, it is not surprising that computer programs have been suggested for use in the design of ventilation systems. These programs typically follow the same procedure outlined in the preceding sections and therefore assume the user has a basic understanding of ventilation design. Perhaps the most useful of these programs (13, 14) incorporate the "equivalent foot" method in which all elements of the system (elbows, fittings, and transitions) are converted to equivalent feet of straight duct, thus simplifying the analytical process. I have worked with these programs, both as a classroom instructor in ventilation design and as a ventilation consultant, and have found them to be useful and fairly reliable with only minor modifications of the programs. Programming for hand-held computers has been published (15, 16), facilitating even more the design process for the ventilation engineer.

Many large companies and architect/engineering firms have, of course, incorporated ventilation design and specification preparation aspects into their computer-assisted engineering work. Access to computer-generated optimal characteristics of a ventilation system, together with use of auto-CAD (computer-assisted drawing) programs, makes proper ventilation design a relatively simple, rapid process. It is fully anticipated that, with the rapidly advancing computer technology, availability of such assistance to the typical, practicing industrial hygiene engineer will be improved greatly in the very near future.

6 MAKEUP AIR REQUIREMENTS

Adequate supplies of outside air are necessary to ensure proper operation of ventilation systems. Properly designed exhaust systems by themselves will remove toxic contaminants; however, they should not be relied upon to draw air into the building to replace that which is exhausted, because this will result in negative pressure within the building, leading to undesirable effects. Mechanically controlled systems for supplying the air to the work space are preferred and in most modern industrial facilities are necessary, to achieve the overall desired ventilation. Most outside air supply systems currently installed in industrial plants have been placed in operation only to replace the exhausted air, with little or no regard to the overall positive effects that can be achieved by the air supply system in the total environmental control system. There is a definite trend toward consideration of the environmental control potential of air supply systems, that is, ability to affect beneficially the air quality, temperature, and humidity of the workplace. Control of air volume, velocity, and temperature is important for a satisfactory work environment; a proper air supply system can provide both replacement air and some degree of environmental control.

6.1 Need for Adequate Makeup Air

Air enters a work space in an amount equal to the volume actually exhausted, regardless of whether mechanical provisions are made for replacement of the air. Of course in some cases, particularly when the total exhaust volume is small, the replacement air enters the building with no adverse effects, even in the absence of a mechanical makeup air system. When the exhaust volumes are relatively large in comparison to the size of free inlet area of the building, however, undesirable effects will be experienced. Older industrial buildings with large void spaces in the enclosing walls may provide significant openings, and air leakage into the building can be quite pronounced. Modern windowless plants of masonry construction, however, can be essentially airtight, with the result that the building is starved for air if there is a significant volume of exhaust ventilation. In a building that is relatively open, an influx of outside air is particularly undesirable in northern climates, where the incoming air cools the perimeter of the building, workers exposed to the incoming air are subjected to drafts, temperature gradients are produced within the building, and the internal heating system usually is overtaxed. Although the incoming air eventually may be tempered to acceptable conditions by mixing with interior air, it is an ineffective heat transfer process, usually resulting in wasted energy.

Reasons for considering an adequate supply of makeup air in conjunction with a total ventilation system include the following:

a. Lack of adequate makeup air creates a negative pressure within the building; this increases the static pressure that must be overcome by the exhaust fans if they are to function properly. This effect can result in a reduction of the total volume of air exhausted from the enclosure, and it is particularly serious with low-pressure fans such as the common wall fans and roof exhausters of the propeller type.

b. Cross-drafts created by the infiltration of uncontrolled makeup air interfere with proper operation of exhaust hoods and also may disperse contaminated air from one area of the building to another; moreover, they can seriously interfere with proper operation of process equipment. The relatively high-velocity cross-drafts through windows and doors created by the infiltration of air can even dislodge settled material on beams and other horizontal surfaces, resulting in resuspension of solid contaminants in the workroom.

c. If the negative pressure created within the building is significant enough, there may be actual backdraft on flues and other natural draft stacks, posing a potential health hazard from release of combustion products into the workroom. Such backdrafting can occur in natural draft stacks at very low negative pressures. Secondary problems associated with the backdraft phenomenon include difficulty in maintaining pilot lights in burners, erratic operation of temperature controls, and potential corrosion damage in stacks and heat exchangers due to condensation of water vapor from the flue gases.

d. The cold drafts resulting from infiltration of uncontrolled makeup air can produce worker discomfort, leading to reduced work efficiency attributable to lower ambient temperatures and the drafts impinging on the workers.

e. The differential pressures created across internal partitions may make doors difficult to open or close if the amount of infiltrated makeup air is high enough. In some extreme cases, this can pose a safety hazard when doors are moved by the force of the pressure gradient itself in an uncontrolled manner.

f. The conditions of cooler ambient temperatures near the building perimeter often lead to decisions to correct the problem by installing more heating equipment in those areas when installation of a proper makeup air system would eliminate the problem. The heaters warm the air only after it has entered the building, usually too late to be of benefit to the workers in the perimeter. Then the overheated air moves toward the building interior, making those areas often uncomfortabley warm. Often attempts then are made to alleviate this problem by installing more exhaust fans to remove the excess heat, but this further aggravates the problem of temperature gradients within the building. The net result is the unnecessary wasting of heat without solving the problem. The fuel consumption associated with attempts to achieve comfortable temperatures within the building by this approach is much higher than it would be with a properly installed system for providing makeup air.

6.2 Principles of Supplying Air

The volume of makeup air provided to a work space generally should equal the total volume of air removed from the building by exhaust ventilation systems, process venting, and combustion processes. The determination of the actual amount of air removed from the building usually requires no more than a simple inventory of air exhaust locations, with measurements of the amount of air being exhausted and compilation of the exhaust volumes. It frequently is desirable to incorporate any reasonable projections for additional exhaust requirements in the near future, certainly if process changes or additions or modifications to the plant are being contemplated. Often it is desirable to install an air supply system of a capacity slightly larger than actually needed, to allow for future needs. In most cases the drive mechanism on the air supply system can be modified to supply only the desired quantity of air.

The locations of the makeup air units, with the consequent distribution of air and creation of airflow patterns within the building, often do not receive sufficient thought before installation, and in many instances this aspect of the system is neglected altogether. However, successful operation of the total ventilation system depends upon proper distribution of makeup air. The following principles for supplying makeup air should be incorporated into the design of an air supply system.

a. Locate the inlet for the fresh air away from any sources of contamination, such as exhaust stacks or furnace exhausts. It is advisable to filter the fresh air, in order to protect equipment and to ensure maximum efficiency of heat exchange in the tempering systems.

b. Incorporate a mechanically operated air mover, that is, fan or blower, in the air supply system to prevent the development of negative pressure within the room or building.

c. Design the locations of the makeup air entries so as to maximize utilization of the air within the work space. Properly located makeup air supply inlets can be used to provide general dilution ventilation first, after which the air provides makeup for exhaust systems operated within the building. Obviously, this concept does not apply for specialized types of makeup air such as those installed for spot-cooling purposes, where the air is introduced at temperatures significantly below room temperature, and at specific work stations. In general, the air distribution pattern created by the combination of makeup air supply and process exhaust must be engineered carefully to provide effective coverage within the building without creating excessive drafts, which could interfere with process operations or compromise the comfort of workers.

d. Introduce the makeup air into the plant at a height of approximately 8 to 10 ft wherever practicable. In this way, the air is used first by the workers, maximizing the results of general or dilution ventilation possibilities. Such an arrangement for distribution of the incoming air also permits closer control of the ambient workplace temperature.

e. Heat or cool the makeup air, as necessary, to approximately the desired inside temperature. Accordingly, the temperature range for incoming air typically varies between 65 and 80°F, depending on the geographic location of the building and the climatological and meteorological conditions characteristic of the location and time of year.

6.3 Recirculation of Exhaust Air

Typical application of local exhaust ventilation includes discharge of the contaminated air to the atmosphere, with or without benefit of an air-cleaning device. In recent years, industries have begun actively investigating exhaust systems that will clean the air of contaminants to a degree sufficient to permit discharge of the air directly to the workplace. This approach is desirable because air-cleaning or pollution control devices have been employed to substantial degrees to meet air quality or emission standards. Additionally, the heating and/or cooling of makeup air needed to replace exhausted air is an expensive item, particularly because costs of energy are increasing and in some cases energy to operate new installations is not even available.

The acceptability of recirculating air systems obviously depends to a great extent on the health hazard associated with the contaminant being exhausted. It has been a fairly consistent policy of official occupational health agencies not to condone the recirculation of exhausted air if the contaminant may have an adverse effect on the health of the workers, mainly because even though the air cleaner used is efficient enough to clean the air sufficiently for health protection, incorrect operation or poor maintenance of the system would result in return of the contaminated air.

6.3.1 Circumstances Permitting Recirculation of Air

Recirculation of exhausted air is sometimes a feasible method of supplementing ventilation within the workplace. In some situations, the economic and energy conservation factors may be important enough to warrant the additional capital and operating expenditures of assuring the safe provision of recirculated air. In general, recirculation of exhausted air may be permitted under the following circumstances:

a. The ventilation system must be furnished with an air cleaner system efficient enough to provide an exit concentration, that is, concentration of contaminant in the air recirculated to the workroom, not exceeding an allowable value, which may be calculated for equilibrium conditions using the following equation:

$$C_R = \frac{(PEL - C_O)Q_T}{2Q_R K}$$

where C_R = concentration of contaminant in exit air from the collector before mixing

Q_T = total ventilation flow through the affected work space (ft³/min)
Q_R = recirculated air flow (ft/min)
K = a mixing factor, usually varying between 3 and 10 (3 = good mixing conditions)
PEL = permissible exposure limit for contaminant
C_O = concentration of contaminant in workers' breathing zone with local exhaust discharged outside

This air cleaning system is referred to as the primary system.

b. A secondary air-cleaning system of efficiency equal to or greater than the primary system should be installed in series with the primary cleaner. As an alternative to this approach, a reliable monitoring device may be installed to furnish a representative sample of the recirculated air. This monitoring system must be fail–safe with respect to failure of the power supply, environmental contamination, or the obvious results of poor maintenance.

c. A warning signal should be provided to indicate the need for attention to the secondary air-cleaning system, or the air monitor should indicate when the concentrations have exceeded predetermined allowable limits.

d. There should be a mechanism for either immediate bypass of the recirculated air to the outdoors or complete shutdown of the contaminant-generating process to become operative under conditions that activate the system's warning device.

Although the application of recirculated air systems in industry has been quite limited, the potential for and anticipated future applications of this approach are quite extensive. NIOSH (17) indicated that 356 of the 514 compounds constituting the ACGIH list of threshold limit values at the time of the study were potentially recirculatable. The compounds excluded from consideration were primarily those identified as having carcinogenic properties and those for which a ceiling exposure limit had been established. This preliminary analysis of design and operating criteria for recirculation systems emphasized that air monitoring equipment should be on-line, automatic, and specific for the individual contaminant. Although a preliminary effort, this study did indicate that use of recirculated air in industrial workplaces will become much more commonplace than it is today. An excellent treatment of considerations for recirculated air has been published (18).

6.3.2 Design Considerations for Recirculated Air

With the full expectation that interest in and development of refined recirculated air systems will continue, the following factors and considerations should be incorporated into analysis of the appropriateness of recirculated air systems in any particular setting.

a. Usually it is necessary to provide general ventilation air in addition to that recirculated, to ensure continual dilution of the contaminants in the recirculated airstream. When it is proposed or possible that all the supplied ventilation air be that recirculated from exhaust systems, all possible contaminants in the airstream must be evaluated, not just the most significant. Of particular concern would be minor contaminants in the airstream that might pass through the primary and secondary air-cleaning devices. For example, a fabric filter–high efficiency particulate filter combination, although providing essentially complete collection of particulate contaminants, could allow the concentration of gases and vapors to build up during the course of operation from relatively insignificant to potentially hazardous levels.

b. Wherever practicable, the recirculating air system should be designed to permit bypass to the outdoors when weather conditions permit. For example, if the system is intended to conserve heat during winter months, it can discharge outdoors in warmer weather if windows and doors are designed to permit sufficient makeup air when opened. Of course continuous operation of the bypass mode would not be desirable where the work space is conditioned or where mechanically supplied makeup air is required at all times.

c. Air recirculated through wet collectors may pose a problem in that the humid air from such equipment usually causes uncomfortably high humidity within the workplace and possible condensation problems. Excessive humidity may be prevented through the use of auxiliary ventilation equipment.

d. Design data and testing programs preceding installation of recirculating air systems should consider all operational time periods of the system, for it can be reasonably expected that the concentration of contaminants exiting typical collectors will vary over time.

e. As with design and layout of any air supply system, the ductwork from the recirculating air system should provide adequate mixing with other air supplies, to avoid uncomfortable drafts on workers or air currents that could adversely affect the performance of local exhaust hoods.

f. In general, even when a monitoring device is available, the installation of a secondary air-cleaner system is preferable because it usually lends a greater degree of reliability to the system and requires a less sophisticated maintenance program.

g. Although the primary concern for the quality of the recirculated air centers on acceptable exposure limits, odors or nuisance values for contaminants should be considered as well. In fact, in some locations, adequately cleaned recirculated air, provided by a system with appropriate safeguards, can be of better quality than the makeup air supply entering the building from outside.

h. Routine testing, maintenance procedures, and records should be developed for the recirculating air system. In addition, the workroom air should be tested periodically.

The foregoing considerations for recirculating air systems do not appear to impose any insurmountable obstacles to continuing development of proper criteria for design and operation of such systems. With the probable continuing concern for energy conservation and air quality (both within and outside the industrial plant), it is likely that the next generation of industrial hygiene engineers will deal routinely with recirculating air systems.

REFERENCES

1. R. H. Perry and D. Green, Eds., *Perry's Chemical Engineers' Handbook*, 6th ed., McGraw-Hill, New York, 1989.
2. American Conference of Governmental Industrial Hygienists, Committee on Industrial Ventilation, *Industrial Ventilation—A Manual of Recommended Practice*, 20th ed., ACGIH, Lansing, MI, 1988.
3. H. J. McDermott, *Handbook of Ventilation for Contaminant Control*, 2nd ed., Butterworth, Stoneham, MA, 1985.
4. J. L. Alden and J. M. Kane., *Design of Industrial Ventilation Systems*, 5th ed., Industrial Press, New York, 1982.
5. American Society of Heating, Refrigeration and Air-Conditioning Engineers, *ASHRAE Guide and Data Book—Fundamentals and Equipment*; ASHRAE, New York, 1987.
6. W. C. L. Hemeon, *Plant and Process Ventilation*, 2nd ed., Industrial Press, New York, 1963.
7. *OSHA General Industry Safety and Health Regulations*, U.S. Code of Federal Regulations, Title 29, Chapter XVII, Part 1910.1000.
8. American Conference of Governmental Industrial Hygienists, *Threshold Limit Values for Chemical Substances and Physical Agents in the Workroom Environment*, ACGIH, Cincinnati, OH, 1989.
9. J. M. Dalla Valle, *Exhaust Hoods*, Industrial Press, New York, 1944.
10. A. D. Brandt, *Industrial Health Engineering*, Wiley, New York, 1947.
11. L. Silverman, "Velocity Characteristics of Narrow Exhaust Slots," *Ind. Hyg. Toxicol. J.*, **24**, 267 (1942).
12. American National Standards Institute, *Fundamentals Governing the Design and Operation of Local Exhaust Systems*, ANSI Z9.2 Committee, New York, 1979.
13. D. E. Clapp, D. J. Groh, and C. M. Nenadic, "Ventilation Design by Microcomputer," *Am. Ind. Hyg. Assoc. J.*, **43**, 212 (1982).
14. American Conference of Governmental Industrial Hygienists, *Microcomputer Applications in Occupational Health and Safety*, Lewis Publishers, Chelsea, MI, 1987.
15. J. J. Loeffler, "Using Programmable Calculators for Ventilation System Design, II;" *Plant Eng.* (Sept. 14, 1978).
16. H. P. Shotwell, "A Ventilation Design Program for Hand-Held Programmable Computers," *Am. Ind. Hyg. Assoc. J.*, **45**, 749 (1984).
17. National Institute for Occupational Safety and Health, *Recirculation of Exhaust Air*, NIOSH Report No. 76-186, Cincinnati, OH, 1976.
18. L. V. Cralley and L. J. Cralley. *In-Plant Practices for Job Related Health Hazards Control*, Vol. 2, Engineering Aspects, Wiley, New York, 1989.

CHAPTER TWENTY-NINE

Air Pollution

George D. Clayton, C.I.H.

1 INTRODUCTION

Congress has passed numerous legislative acts to monitor and control air pollution in the United States. The two major legislative acts that had, and will have, the greatest influence on the quality of air in the United States are the 1970 Clean Air Act and the 1990 Clean Air Act. This chapter discusses both acts and scrutinizes the major air pollution episodes in which illness and death occurred.

The human species and nature both create air pollution. Because mankind has flourished throughout the centuries, with burgeoning populations throughout the world that live in areas with varying degrees of polluted air, it is logical to assume that individuals have a built-in mechanism to accommodate, to a certain degree, their exposure to air pollutants. When an individual's threshold of tolerance has been exceeded, that person's health is affected, at times even causing death. Thus the need to develop standards to protect the majority of the population from the deleterious effects of air pollution has been of concern for many years. The concept of "zero" pollution is a laudable goal, but impractical, not only because nature itself is a source of air pollution, but because humanity is a source of pollution.

2 HISTORY

Air pollution is not a phenomenon of the twentieth century. It is as old as recorded history, and beyond; only the magnitude of the problem has increased and been

Patty's Industrial Hygiene and Toxicology, Fourth Edition, Volume 1, Part B, Edited by George D. Clayton and Florence E. Clayton.
ISBN 0-471-50196-4   © 1991 John Wiley & Sons, Inc.

recognized more widely, simultaneously with the industrialization and growth of urban populations.

In 361 B.C. Theophrastus wrote that "fossil substance called 'coals' burns for a long time, but the smell is troublesome and disagreeable." Several centuries later, in 65 B.C., Horace, the poet, was lamenting that the shrines of Rome were blackened by smoke. During the reign of Edward I in England, in 1273, the first smoke abatement law was passed in response to people's fears that the air pollutants were detrimental to health. During this period it was believed that food cooked over burning coals would cause illness, and even death. In 1306 the people became so concerned that a royal proclamation was signed, which prohibited burning coal in London. One industry owner who was caught disobeying this royal proclamation was tried, found guilty, and beheaded. This was the first recorded penalty given as a result of violating an air pollution code.

People have been concerned with the effects of air pollution on health since its inchoation, and the first air pollution law was passed in an attempt to curb air pollution because of its detrimental health effects. A number of events we refer to as disasters that have occurred during the twentieth century justified this concern, and they have stimulated additional legislation.

Air pollution disasters may be "natural" in origin or "man-made," emanating from societal activities. The five principal recorded man-made disasters occurred in the Meuse Valley, Belgium in 1930; in Donora, Pennsylvania in 1948; in Poza Rica, Mexico in 1950; in London in 1952; and in Bhopal, India in 1984. A discussion of these episodes follows.

Meuse Valley (1). The first well-known air pollution disaster occurred between Seraning and Hug, in the Meuse Valley, Belgium. During the period December 1–5, 1930 a large number of persons were taken ill, and more than 60 died. Older persons with previously known disease of the heart or lungs accounted for the majority of fatalities. The signs and symptoms, primarily those caused by a respiratory irritant, included chest pain, cough, shortness of breath, and irritation of the mucous membranes and of the eyes. The area of the Meuse Valley where the episode occurred is approximately 15 miles long and $1\frac{1}{2}$ miles wide, and is surrounded by hills 330 ft high. Within this area at the time of the disaster, there were the following types of industry:

1. Four very large steel plants, each having coking installations, blast furnaces, rolling mills, welding furnaces, boilers, and other operations compatible with large steel plants
2. Three large metallurgical works
3. Four electric power generating plants
4. Six glass works, ceramic plants, or brickworks, which were equipped with coal or producer-gas heating furnaces
5. Three groups of lime kilns
6. Three zinc plants, each equipped with reduction furnaces, retort drying ovens, and other operations compatible with zinc plants

7. One coking plant (other than those associated with steel plants)
8. One sulfuric acid plant with roasting ovens
9. One concentrated fertilizer plant, with concentration and drying ovens

In addition to these industries discharging pollutants into the atmosphere, most of the homes and communities burned coal, thus adding considerably to the pollution load. Other sources of pollution at the time of the episode were railroads and automobiles. The investigators considered the sources of pollution and listed the following substances that could have been, and probably were, present in the atmosphere at the time of the disaster:

Hydrogen
Sulfur dioxide
Sulfur trioxide (sulfuric acid)
Hydrochloric acid
Hydrofluoric acid and its salts, ammonium fluoride and zinc fluoride
Carbon monoxide
Carbon dioxide
Hydrosulfuric acid
Nitric oxide
Nitrogen dioxide (NO_2)
Nitrous acid (HNO_2)
Nitric acid (HNO_3)
Ammonia and ammonium salts (thiocyanates, sulfides)
Ammonium sulfide
Saturated and unsaturated hydrocarbons, including natural gas and gasoline vapors

Odor of organic products (not identified) from phosphate plants
Drops of tar, phenol, naphthalene, and so forth
Soot
Cement dust
Lime dust
Metal dust
Zinc oxide
Lead
Arsenous acid anhydride
Arsine
Methyl alcohol
Ethyl alcohol
Formaldehyde
Zinc chloride
Silica from Bessemer steel works

The meteorological investigators found the conditions for Monday, December 1, through Friday, December 5, 1930, to have been anticyclonic, characterized by high atmospheric pressures and mild winds of a general easterly direction. During this period the fog became increasingly worse. There was considerable cooling during the night, often as much as 10°C. The wind was a very mild east wind, not exceeding 5 miles per hour and often dropping during the days of December 2, 3, and 4 to 1.6 miles per hour.

After the investigators had checked all the available data on the medical, meteorological, and environmental factors, efforts were made to determine the causative agent. Many of the contaminants could be eliminated immediately because their known toxicological action was not compatible with the medical findings. Other contaminants were eliminated because it was decided that the concentrations could not have been sufficient to cause the damage found. From all the findings the investigators concluded the contaminants most likely to have caused the symptoms were sulfur compounds; however they added that the synergistic effect of the other contaminants may have played an important part.

***Donora* (2).** The second well-known air pollution disaster occurred in Donora, Pennsylvania during the period of Wednesday, October 27, to Sunday, October 31, 1948. Within this period 20 people died and 6000 became ill.

At the time the investigation was being conducted, there were the following industries in Donora:

1. One large steel plant having blast furnaces, looping rod mills, wire drawing department, and wire finishing department   .
2. One zinc plant equipped with reduction furnaces, retort drying ovens, and other operations compatible with zinc plants
3. One sulfuric acid plant

Other heavy industries in the nearby area included two steel companies and one by-product plant in Monessen, a steel and by-product plant in Clairton, a glass company in Charleroi, and a power company and a railroad in Elrama.

A study of the plant processes and analyses of the raw materials used in those industries led the investigators to analyze the atmosphere for the following constituents:

Total particulate matter	Acid gases	Carbon monoxide
Lead	Hydrogen	Oxygen
Cadmium	Arsine	Oxide of nitrogen
Zinc	Arsenic	Stibine
Iron	Sulfur dioxide	Manganese
Chloride	Total sulfur	Iron carbonyl
Fluoride	Carbon dioxide	

The investigators made the following conclusions (2):

It seems reasonable to state that while no *single* substance was responsible for the October 1948 episode, the syndrome could have been produced by a combination or summation of the action of two or more of the contaminants. Sulfur dioxide and its oxidation products, together with particulate matter, are considered significant contaminants. However, the significance of the other irritants as important adjuvants to the biological effects, cannot be finally estimated on the basis of present knowledge.

Information available on the toxicological effects of mixed irritant gases is meager, and data on possible enhanced action due to adsorption of gases on particulate matter is limited. Further, available toxicological information pertains mainly to adults in relatively good health. Hence, the lack of fundamental data on the physiologic effects of a mixture of gases and particulate matter, over a period of time, is a severe handicap in evaluating the effects of atmospheric pollutants on persons of all ages and in various stages of health.

Although sufficient data were not available to determine the exact causative agent, sufficient environmental data were available to make 10 recommendations which, if fulfilled, would prevent a recurrence of the disaster.

Poza Rica (3). The third much-publicized incident occurred in Poza Rica, Mexico on November 24, 1950, setting a record of 22 persons dead and 320 people hospitalized.

A review of the history of the incident shows that the sulfur-removal unit (Girbotol unit) began operation on November 21, 1950. A few days after the unit was in operation, trouble developed when the amine solution overflowed, partly plugging the gas lines to the pilot lights of the flares. However, the flare appeared normal under the reduced rate of flow.

On November 24, 1950, at approximately 2:00 A.M., efforts were made to increase the rate of flow of gas to the plant's rated capacity of 60 billion ft^3/day. The desired rate of flow was reached at approximately 2:30 A.M. At approximately 4:00 A.M. difficulty was encountered, and the gas flow began surging through the unit, with the probable result that unburned hydrogen sulfide escaped into the atmosphere.

The available meteorological data indicate a pronounced low altitude temperature inversion, a high concentration of haze and nuclei, and a very slight wind movement prevailing.

The epidemiologic study indicated that the time of onset of the incident was about 4:50 A.M., and the acute phase was ended about 5:10 A.M. The victims were affected in a geographic area in direct proximity to the effluent stack of the Girbotol unit. During this short period the acute exposure to the atmospheric pollution caused the death of 22 people and the hospitalization of 320.

The onset, the symptoms and signs, and the pathological findings are consistent with hydrogen sulfide poisoning, and there were no findings that conflicted with this diagnosis. Therefore it is to be concluded that the hydrogen sulfide that caused this morbidity and mortality came from the effluent stack of the Girbotol unit.

London (4). During the period December 5–8, 1952, the fourth and worst air pollution disaster up to that time occurred. When these deaths were being recorded, the London area experienced periods of smog, culminating in one of much intensity. The onset was accompanied by meteorological factors of low temperature inversion and almost complete absence of wind or air movement. No radio or balloon-sonde ascents were made within the fog belt. Ascents at Cardington, 50 miles north of London, where visibility did not fall below 1000 yards during the period, show a continuous inversion from ground level up to heights between 500 and 1000 ft. Reports from aircraft landing at London Airport after the fog had cleared temporarily place the height of the inversion at 200 to 300 ft, with a haze layer extending to 2000 ft.

This is the first air pollution disaster for which air sampling was conducted before, during, and after the episode. The data collected showed that the particulate matter (obtained by sampling with filter paper and reporting the degree of color in milligrams of total weight per cubic meter of air) ranged from 0.30* to 0.84 mg/m³ of air, with an average of 0.50 for the 2-week period preceding the disaster. On the days when deaths were occurring, the values increased to a maximum of 4.46 mg/m³ meter of air, a ninefold increase over the previous average.

Sulfur dioxide ranged from 0.09 to 0.33 ppm, with an average of 0.15, for the 2-week period preceding the disaster. During the period when deaths were occurring, the SO_2 values increased to a maximum of 1.33 ppm, which was also a ninefold increase over the average of the preceding days.

During the weeks ending December 13 and December 20, 1952, at least 3000 more deaths occurred than would be expected. Although the increase was present in every age group, the greatest increase was in the age group of 45 years and over. More than 80 percent of these deaths occurred among individuals with known heart disease and respiratory disease. As in other disasters reported, the symptoms produced were associated with irritation of the respiratory tract.

The British Ministry of Health concluded: "The fog was, in fact, a precipitating agent operating on a susceptible group of patients whose life expectation, judging from their pre-existing diseases, must even in the absence of fog, have been short."

Bhopal, India. In the episodes described above, meteorological and topographical parameters contributed to the occurrences. In a more recent disaster, these factors were not significant. On December 3, 1984, 42 tons of liquid methyl isocyanate burst into the air at the Union Carbide plant in Bhopal, India, releasing a mixture of at least four deadly gases over a slum neighborhood, and creating a disaster whose effects are still being felt. The immediate official death toll was 2500, with deaths continuing to accumulate, and at present numbering 3415. According to estimates by medical and official sources, 60,000 victims were seriously affected. The gas seared lungs, burned eyes and skin, and led to ulcers, colitis, hysteria, neurosis, and memory loss. Pregnant women exposed to the gas often had children who were born abnormal or died in infancy. Subsequently 592,000 personal injury claims were filed, plus claims filed by families of 5100 dead. The Indian government, which had decreed it would litigate on the behalf of the victims, settled out of court with Union Carbide in February 1989, for $470 million; however, the amount of the settlement was greatly resented and the Indian Supreme Court is hearing appeals by victims demanding more (5).

In addition to these devastating incidents described, since 1952 there has been statistical evidence of increased deaths associated with abnormally high air pollution levels in various communities, such as in New York City (6) and in Pittsburgh in 1975, where there were 14 excess deaths (7) recorded during an inversion period.

A "natural" disaster occurred in 1980 on May 18th when Mount St. Helens, Washington State's most active volcano, erupted. Volcanic ash expelled during the eruption was deposited on much of eastern Washington and had a profound effect on local air quality. Although ash is relatively inert, analysis revealed a small but significant amount of free crystalline silica, the causative agent of silicosis. The fine

---

*To compare this figure with U.S. measurements, use a factor of 1.8 for approximate relationships.

particles of ash were of respirable size, and there was a remarkable increase in the volume of respiratory cases seen in emergency departments during the period of high airborne particulate levels (8).

Another natural disaster occurred more recently in 1986 in Cameroon, West Africa, when a massive release of gas occurred from Lake Nyos, a volcanic crater lake. The hospital saw or admitted 845 survivors; the clinical findings were compatible with exposure to an asphyxiant gas. Carbon dioxide was blamed for the deaths of about 1700 people. This disaster, and one at Lake Monoun two years previously, gave new information on the effects of carbon dioxide, although other toxic factors in these gas releases were not excluded (9).

Beginning in 1958 there was an acceleration of activity in the field of air pollution, especially in the United States, and the federal government has been extremely active since 1969 in passing legislation to control mobile and stationary sources of pollution, as well as establishing standards for emission, and for ambient air for six common pollutants: total particulates, carbon monoxide, sulfur dioxide, photochemical oxidants, nitrogen dioxide, and hydrocarbons. The Environmental Protection Agency (EPA) was established, having authority in all areas of pollution: air, water, and solid waste. (The Clean Air Act of 1990 requires EPA to establish standards for 189 pollutants.) All states have passed air pollution laws, as have most of the major cities. The most active state in its efforts to control air pollution has been California.

Since 1958 extensive studies have been conducted on health, vegetation, and economic damage. Control procedures have been refined for both stationary and mobile sources. Numerous studies have been devoted to finding a viable alternative to the internal combustion engine. The automotive industry has been successful recently in developing an electric car which shows promise, in speed and distance. Previous efforts had been disappointing. The petroleum industry recently began to market several new formulations of gasoline to reduce gaseous emissions. In another effort toward reducing vehicle emissions, pilot trucks are now in the process of utilizing natural gas for fuel, which will reduce exhaust emissions by 80 percent. Citizen groups, such as the Sierra Club and National Wildlife Federation, have been very active in pursuit of a cleaner environment. Many technical societies devote a good portion of their committee activities and annual programs to reporting on the technical aspects of evaluating and controlling air pollution. As a result of this extensive activity, there is widespread recognition and awareness of the undesirable effects of air pollution, from the youngest schoolchild in the nation to the president of the largest corporation. Billions of dollars have been spent on the evaluation and control of air pollution. In the early 1970s people believed that air pollution should be controlled at any cost. This was explicit in the language of the legislation enacted in 1969. Legal requirements were established, and the courts subsequently ruled that the financial inability of a company to pay for controls was *not* a legitimate excuse for not installing controls. In 1976 the Supreme Court of the United States ruled that federal approval of state clean air programs may not be overturned on the grounds that the plans cost too much or require new types of equipment. In a unanimous opinion, the court stated that Congress intended to have tough standards that would force the development of new equipment and methods to alleviate air pollution. The 1970 amendments to the Clean Air Act required the states to design clean air plans and submit them for approval by the EPA. The ruling upheld the U.S. Court of Appeals in Chicago, which had refused to consider a case filed by an electric utility in a metropolitan area of Missouri.

The policies of the EPA have been in keeping with the mandate from Congress. Leon Billings, a senior professional staff member of the subcommittee on Environmental Pollution of the Senate Public Works Committee, made a revealing statement in June, 1976:

> To understand the evolution of the controversies it is useful and necessary to review implementation of the 1970 Act. In 1970 the Congress set forth the basic goal of protection of public health from the adverse effects of air pollution, through achievement of fixed emission reductions by date certain. The goal was absolute. The timetables and tools for its achievement were not to be tempered by economic considerations. The purpose was to eliminate as quickly as humanly possible unhealthy and thus unreasonable levels of air pollution. The Clean Air Act of 1970 was in that respect a significant departure from previous pollution policy.

It is of interest to note that prior to 1970 air pollution was commonly associated with the outdoors, with emissions that emanated from sources such as factories, refineries, and automobiles. Since the early 1970s increased attention has been given to *indoor* air pollution (see Chapter 17, Volume 1A) from sources such as exposure to tobacco smoke, nitrogen dioxide from gas-fueled stoves, formaldehyde exposure, radon daughter exposure, sealed office buildings, off-gases from new furnishings, and solvents and chemicals from duplicating equipment.

3 LEGISLATION PRIOR TO 1977

3.1 Federal Level

In the past, federal environmental legislation reflected three major congressional viewpoints: (1) that there is an atmospheric concentration *below* which there would be no adverse effects on people or vegetation; (2) that atmospheric pollution is primarily created by human activity; and (3) that the cost of controls can be borne by the country's economy without adversely affecting its viability.

It is now clear that these three tenets have a shaky, if not a false foundation, and each is discussed further in this chapter.

During the earlier years of abatement effort, the federal government had a very limited involvement. The Public Health Service had conducted investigations relating to the Donora incident under the authority of the Public Health Service Act. The federal program in air pollution, under the aegis of the Department of Health, Education and Welfare, was established by Public Law 84-159 in July 1955. This legislation assigned primary responsibility for the control of air pollution to the states and to local governments, and gave the main objectives of the federal program as providing leadership and assistance to control programs throughout the country.

Before passage of this legislation, the automobile had emerged as a major source of pollution. Two subsequent congressional actions, the Schenck Act of July 1960 and the Air Pollution Control Act of 1962, called for the Surgeon General to study the exhaust emissions from motor vehicles. With the enactment of the Clean Air Act of 1963, federal policy underwent a significant evolution. Federal responsibility to state and local programs was reinforced. The act further singled out for special attention two of the major unsolved air pollution problems: motor vehicle exhaust, and sulfur oxide pollution from the burning of fossil fuels.

The Clean Air Act of 1963 was amended in 1965 and again in 1966. These amendments resulted in expansion of state and local control programs through federal grant support. Federal abatement activities were also strengthened. Motor vehicle exhaust emission standards applicable to the 1968 model year were promulgated (under the 1965 amendments).

The Air Quality Act of 1967 is distinguished from preceding legislation by its focus on regional activities. This act places emphasis on regional air pollution control programs implemented by state and local authorities, with the Department of Health, Education and Welfare assuming a leadership role. The act is interpreted as a clear invitation for government, industry, and private organizations to join together in constructive action to abate pollution.

The National Environmental Policy Act of 1969 was amended in January 1970 and on August 9, 1975. The purposes of this act were stated:

> To declare a national policy which will encourage productive and enjoyable harmony between man and his environment; to promote efforts which will prevent or eliminate damage to the environment and biosphere and stimulate the health and welfare of man; to enrich the understanding of the ecological systems and natural resources important to the Nation; and to establish a Council on Environmental Quality.

Executive Order 11514, dated March 5, 1970, spelled out the responsibilities of the federal agencies and of the Council on Environmental Quality.

On July 9, 1970, the President transmitted to Congress Reorganization Plan No. 3 of 1970, establishing the Environmental Protection Agency. Heading the new agency would be an administrator, appointed by the President with the advice and consent of the Senate. The responsibilities assigned to the new agency were the following:

1. All functions vested by law in the Secretary of the Interior and the Department of the Interior that are administered through the Federal Water Quality Administration, all functions that were transferred to the Secretary of the Interior by Reorganization Plan No. 2 of 1966 (80 Sta. 1608), and all the functions vested in the Secretary of the Interior or the Department of the Interior by the Federal Water Pollution Control Act or by provisions of law amendatory or supplementary thereof.

2. The functions vested in the Secretary of the Interior by the act of August 1, 1958 (72 Stat. 479, 16 U.S.C. 742d-1; an act relating to studies on the effects of

insecticides, herbicides, fungicides, and pesticides on the fish and wildlife resources of the United States), and the functions vested by law in the Secretary of the Interior and the Department of the Interior that are administered by the Gulf Breeze Biological Laboratory of the Bureau of Commercial Fisheries at Gulf Breeze, Florida.

3. The functions vested by law in the Secretary of Health, Education and Welfare or in the Department of Health, Education and Welfare that are administered through the Environmental Health Service, including the functions exercised by the following components thereof: (i) the National Air Pollution Control Administration, and (ii) the Environmental Control Administration: (a) Bureau of Solid Waste Management, (b) Bureau of Water Hygiene, and (c) Bureau of Radiological Health. Activities dealing with industrial health and related matters are excluded.

4. The functions vested in the Secretary of Health, Education and Welfare of establishing tolerances for pesticide chemicals under the Federal Food, Drug and Cosmetic Act, as amended (21 U.S.C. 346, 346a, 348), together with authority, in connection with the functions transferred

(a) to monitor compliance with the tolerances and the effectiveness of surveillance and enforcement.

(b) to provide technical assistance to the states and conduct research under the federal Food, Drug, and Cosmetic Act, as amended, and the Public Health Service Act, as amended.

In addition, many of the functions relating to the environment formerly vested in the Council on Environmental Quality and in the Atomic Energy Commission, as well as certain functions of the Secretary of Agriculture and the Department of Agriculture, were transferred to the EPA.

Periodically the laws relating to the environment are reviewed by Congress. In a series of hearings held during 1975 and 1976 the areas of greatest controversy were the timetable for automotive emissions and the nondegradation clause of the 1970 act. Congress adjourned without any action.

3.2 State and Local Programs

Each of the 50 states and many cities and county units have their own air pollution codes. California, especially the city of Los Angeles, has been in the forefront of enacting legislation, because of the severity of their problem, which is due to their unique topography and to meteorological factors, in addition to population density.

4 AMBIENT AIR QUALITY STANDARDS

Ambient air quality standards are the limits promulgated to protect the health of the public and to provide a quality of atmosphere conducive to well-being. Until the 1990 act, the approach in the United States had been to legislate such standards, establishing time periods for compliance to be achieved. In contrast, the World Health Organization and the Commission of European Communities established goals to which they aspire over an indefinite period of time.

4.1 United States

The Clean Air Act amendments of 1970 set mid-1975 as the deadline for achieving public health (primary) standards for ambient air quality. The quality of air in the United States is better at the time of this writing than it was in 1970, but the deadline has not been fully met, and air pollution levels at many locations still range above the primary standards.

Earlier in this chapter reference was made to the national ambient air quality standards designated for six major pollutants: total suspended particulates, sulfur dioxide, carbon monoxide, photochemical oxidants, nitrogen dioxide and hydrocarbons—the "criteria" pollutants. Table 29.1 presents pertinent data for these pollutants. As of mid-1975, the standard levels had been fully achieved in only 91 of the nation's 247 Air Quality Control Regions. At monitoring sites in nearly two-thirds of the regions, the pollution levels designated by the standards were still being exceeded for one or more pollutants. The standards are exceeded for suspended particulates in 118 of the 247 regions, for sulfur dioxide in 34 regions, for carbon monoxide in 69 regions, for oxidants in 79 regions, and for nitrogen dioxide in 16 regions, according to EPA estimates.

The highest levels of particulates and sulfur oxides, the major pollutants from stationary sources (smokestacks), occur mainly in the northeastern and north central states. California continues to have the most severe automotive pollution problems. Occurrences of poor air quality are not restricted to any region or state, however, and in such widely separated large cities as Los Angeles, Chicago, and Philadelphia, primary standards for all the criteria pollutants are still exceeded at times. There are also problems in many small cities, and even in some rural areas.

Efforts to reduce air pollution have encountered a number of obstacles. Scientific knowledge and available technological and legal tools are sometimes insufficient, and attempts to reduce the nation's economic and energy difficulties are not always compatible with environmental goals. Some facilities have been granted extensions or variances that permit the burning of coal or oil with a high sulfur content. The full effects of such variances on ambient air quality are still uncertain. However, between 1970 and 1990 the nation's efforts to improve air quality have accomplished a great deal.

The main factors governing ambient air pollution levels are as follows:

- Types and quantities of pollutants produced by human activities and natural sources
- Controls used to reduce emissions of pollutants
- Geographic location of pollution sources and of emissions over a given period of time

Table 29.1. Six Ambient Air Quality Standards Established by the 1970 Clean Air Act (11)

Pollutant[a]	Characteristics	Principal Sources	Principal Effects	Controls		National Ambient Standards[b] ($\mu g/m^3$)
Total suspended particulates (TSP)	Any solid or liquid particles dispersed in the atmosphere, such as dust, pollen, ash, soot, metals, and various chemicals; the particles are often classified according to size as settleable particles: larger than 50 μm; aerosols: smaller than 50 μm; and fine particulates: smaller than 3 μm	Natural events such as forest fires, wind erosion, volcanic eruptions; stationary combustion, especially of solid fuels; construction activities; industrial processes; atmospheric chemical reactions	*Health:* directly toxic effects or aggravation of the effects of gaseous pollutants; aggravation of asthma or other respiratory or cardiorespiratory symptoms; increased cough and chest discomfort; increased mortality *Other:* soiling and deterioration of building materials and other surfaces, impairment of visibility; cloud formation; interference with plant photosynthesis	Cleaning of flue gases with inertial separators, fabric filters, scrubbers, or electrostatic precipitators; alternative means for solid waste reduction; improved control procedures for construction and industrial processes	Primary Secondary Alert	Annual = 75 24-hr = 260 Annual = 60 24-hr = 150 24-hr = 375

206

Sulfur dioxide (SO$_2$)				Primary	Annual = 80
A colorless gas with a pungent odor; SO$_2$ can oxidize to form sulfur trioxide (SO$_3$), which forms sulfuric acid with water	Combustion of sulfur-containing fossil fuels, smelting or sulfur-bearing metal ores, industrial processes, natural events such as volcanic eruptions	*Health:* aggravation of respiratory diseases, including asthma, chronic bronchitis, and emphysema; reduced lung function; irritation of eyes and respiratory tract; increased mortality *Other:* corrosion of metals; deterioration of electrical contacts, paper, textiles, leather, finishes and coatings, and building stone; formation of acid rain; leaf injury and reduced growth in plants	Use of low sulfur fuels; removal of sulfur from fuels before use; scrubbing of flue gases with lime or catalytic conversion	Alert	24-hr = 365 24-hr = 800

Table 29.1 (continued)

Pollutant[a]	Characteristics	Principal Sources	Principal Effects	Controls	National Ambient Standards[b] (µg/m³)	
Carbon monoxide (CO)	A colorless, odorless gas with a strong chemical affinity for hemoglobin in blood	Incomplete combustion of fuels and other carbon-containing substances, such as in motor vehicle exhausts; natural events such as forest fires or decomposition of organic matter	*Health:* reduced tolerance for exercise, impairment of mental function, impairment of fetal development, aggravation of cardiovascular diseases *Other:* unknown	Automobile engine modifications (proper tuning, exhaust gas recirculation, redesign of combustion chamber); control of automobile exhaust gases (catalytic or thermal devices); improved design, operation, and maintenance of stationary furnaces (use of finely dispersed fuels, proper mixing with air, high combustion temperature)	Primary	8-hr = 10,000 1-hr = 40,000
					Alert	8-hr = 17,000
Photochemical oxidants (O_x)	Colorless, gaseous compounds that can comprise, photochemical smog [e.g.,	Atmospheric reactions of chemical precursors under the influence of sunlight	*Health:* aggravation of respiratory and cardiovascular illnesses, irritation of eyes	Reduced emissions of nitrogen oxides, hydrocarbons, possibly sulfur oxides	Primary	1-hr = 160
					Alert	1-hr = 200

Pollutant	Description	Sources	Effects	Control Methods	Standard	Standards
[continued]	peroxyacetyl nitrate (PAN), aldehydes, and other compounds]		tract, impairment of cardiopulmonary function *Other*: deterioration of rubber, textiles, and paints; impairment of visibility; leaf injury, reduced growth, and premature fruit and leaf drop in plants			
Nitrogen dioxide (NO₂)	A brownish-red gas with a pungent odor, often formed from oxidation of nitric oxide (NO)	Motor vehicle exhausts, high temperature stationary combustion, atmospheric reactions	*Health*: aggravation of respiratory and cardiovascular illnesses and chronic nephritis *Other*: fading of paints and dyes, impairment of visibility, reduced growth and premature leaf drop in plants	Catalytic control of automobile exhaust gases, modification of automobile engines to reduce combustion temperature, scrubbing flue gases with caustic substances or urea	Primary Alert	Annual = 100 24-hr = 282 1-hr = 1130
Hydrocarbons (HC)	Organic compounds in gaseous or particulate form (e.g., methane,	Incomplete combustion of fuels and other carbon-containing substances, such	*Health*: suspected contribution to cancer *Other*: major precursors in the	Automobile engine modifications (proper tuning, crankcase ventila-	Primary	3-hr = 160

209

Table 29.1 (continued)

Pollutant[a]	Characteristics	Principal Sources	Principal Effects	Controls	National Ambient Standards[b] ($\mu g/m^3$)
	ethylene, acetylene)	as in motor vehicle exhausts; processing, distribution, and use of petroleum compounds such as gasoline and organic solvents; natural events such as forest fires and plant metabolism; atmospheric reactions	formation of photochemical oxidants through atmospheric reactions	tion, exhaust gas recirculation, redesign of combustion chamber); control of automobile exhaust gases (catalytic or thermal devices); improved design, operation, and maintenance of stationary furnaces (use of finely dispersed fuels, proper mixing with air, high combustion temperature); improved control procedures in processing and handling petroleum compounds	

Source: Based on information compiled by Enviro Control, Inc.

[a]Pollutants for which national ambient air quality standards have been established.

[b]Primary standards are intended to protect against adverse effects on human health. Secondary standards are intended to protect against adverse effects on materials, vegetation, and other environmental values.

- Dispersal and movement of pollutants, which depend on meteorological conditions and topography
- Various chemical reactions of pollutants in the atmosphere often resulting in the formation of different types of pollutants
- Processes by which pollutants are removed from the atmosphere, such as gravity and precipitation

The first two factors determine the emissions that occur; the other four determine how the emissions affect air quality.

4.2 World Health Organization

The World Health Organization (WHO) Technical Report Series No. 506 (12) states that the degree of health protection to be selected above the minimum acceptable level is a matter for political decision. That is, the appropriate authorities must decide on the level of health protection desirable for their society. Increments of health protection above the minimum acceptable level are generally purchased at ever-increasing increments in cost of control. Furthermore, the costs of the control program are directly related to the deadline by which it is to be operational; for example, it is more expensive to achieve the desired goals in 3 years than in 10 years. The zone in which increased health protection (benefit) is obtained at increasing control costs (cross-hatched area in Figure 29.1) is also the region of social decision making. Of course the level of protection desired must reflect awareness of the existing air pollution effects, but other considerations are also important,

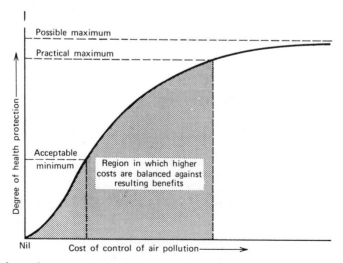

Figure 29.1. Schematic representation of degree of health protection as a function of cost of air pollution control.

Table 29.2. Long-Term Goals Recommended by WHO

Pollutant and Measurement Method		Limiting Level ($\mu g/m^3$)
Sulfur oxides[a]: British Standard Procedure[b]	Annual mean	60
	98 percent of observations[c] below	200
Suspended particulates[a]: British Standard Procedure[b]	Annual mean	40[b]
	98 percent of observations[b] below	120
Carbon monoxide: nondispersive infrared[b]	8-hr average	10
	1-hr maximum	40
Photochemical: oxidant as measured by neutral buffered K1 method expressed as ozone	8-hr average	60
	1-hr maximum	120

[a]Values for sulfur oxides and suspended particulates apply only in conjunction with one another.
[b]Methods are not those necessarily recommended but indicate those on which these units have been based. Where other methods are used an appropriate adjustment may be necessary. For example, to compare *annual* mean only to the U.S. procedure, multiply by 1.8 for appropriate comparison.
[c]The permissible 2 percent of observations over this limit may not fall on consecutive days.

including general social, cultural, and economic factors, as well as the magnitude of other health problems.

Another issue faces environmental administrators responsible for making recommendations for the control of sulfur oxides and suspended particulates: the ratio between these pollutants varies from country to country, and there is no information documenting equivalent effects for the various concentrations of the two pollutants.

In the WHO report, effects are described as related to pollutant concentrations measured over short periods and over longer periods, indicating that different exposure times may be associated with different effects. This raises a problem for the air pollution control agency, which must be sure that the air quality standards adopted will protect from the effects of *both* short-term and long-term exposures. To solve this problem one must know, for example, the relation between the annual mean 24-hr value and the daily 24-hr values. If the effects against which protection is sought are known to be produced by exposure for 24 hrs or less, any control measure stipulating an annual mean 24-hr value must take note of the variations expected, and must state the number of days per year on which the specified concentrations may be reached. Taking into consideration all the evidence available to it, the WHO committee agreed that in the light of present knowledge, the recommendations summarized in Table 29.2 could be offered as long-term goals intended to prevent undesirable effects from the air pollutants under discussion.

4.3 Commission of the European Communities

In January 1976 the Health Protection Directorate, Commission of the European Communities (CEC) (13) prepared air pollution criteria and levels of action for

Table 29.3. Health Protection Standards for Sulfur Dioxide and Suspended Particulates in Urban Atmospheres

Sulfur Dioxide		
Reference Period	Maximum Concentrations	Associated Concentrations of Suspended Particulates
Year	Median of daily means, 80 μg/m^3	Annual median of daily means >40 μg/m^3
Year	Median of daily means, 120 μg/m^3	Annual median of daily means <40 μg/m^3
Winter (October–March)	Median of daily means, 130 μg/m^3	Winter median of daily means >60 μg/m^3
Winter (October–March)	Median of daily means, 180 μg/m^3	Winter median of daily means <60 μg/m^3
24-hr	Arithmetic mean, 250 μg/m^3	Arithmetic mean of concentration over 24-hr >100 μg/m^3
24-hr	Arithmetic mean, 350 μg/m^3	Arithmetic mean of concentration over 24-hr <100 μg/m^3

Suspended Particulates[a]	
Reference Period	Maximum Concentrations
Year	Median of daily means,[a] 80 μg/m^3
Winter (October–March)	Median of daily means, 130 μg/m^3
24-hr	Arithmetic mean, 250 μg/m^3

[a]To compare annual mean only to the United States procedure, multiply by 1.8 for appropriate comparison.

the same pollutants studied by WHO. Essentially, their findings and philosophy were very similar to that of WHO. Because of the importance of sulfur dioxide and particulate matter as air pollutants, the data developed on these pollutants by CEC are presented in Table 29.3.

The criteria used to establish the relationships between given exposures and observable effects on humans for sulfur dioxide and suspended particulate matter were as follows:

1. When sulfur dioxide and suspended particulate matter (determined as "black smoke"*) exceed simultaneously a mean value of 500 μg/m^3 for several days, excess mortality and increase in the number of hospitalizations among aged persons having in particular severe cardiovascular symptoms are observed.

2. When sulfur dioxide and suspended particulate matter exceed simultaneously concentrations of 250 μg/m^3 for several days, a subjective exacerbation of symptoms

*The American high volume sampling method and the British method differ; although both express results in micrograms per cubic meter, they are not directly comparable.

is observed in patients having chronic bronchitis. This exacerbation is much less pronounced when only sulfur dioxide exceeds these levels.

3. For levels slightly lower than 250 $\mu g/m^3$ (daily concentrations) for sulfur dioxide and suspended particulate matter, there are indications that sensitive persons exhibit temporary changes in their pulmonary respiratory functions.

4. When sulfur dioxide and suspended particulate matter exceed simultaneously 100 $\mu g/m^3$ as long-term averages,* respiratory symptoms in the form of increased infection of the lower respiratory tract and decrease in the maximum expiratory flow rates are observed in children.

4.4 Examination of Ambient Air Quality Standards

The six ambient air quality standards established in the 1970 act remain as important in 1990 as they were when initially issued. Although the new act emphasizes maximum achievable control technology, the health effects of air pollution are still important. Tables 29.4 through 29.9 are presented as a summation of data through 1973 of pertinent literature on these contaminants.

4.4.1 Nitrogen Dioxide

Oxides of nitrogen are produced by the combustion of fossil fuels, and also by chemical and nitration industries. Most studies on the effects of oxides of nitrogen have focused on nitrogen dioxide (NO_2), because other oxides of nitrogen react in air to produce NO_2. However, the affinity of nitric oxide (NO) has not been well defined, and may be found to be as important as NO_2.

The number of community studies on human health effects of NO_2 are limited. A review of Table 29.4 brings out several effects of NO_2. The odor threshold for this gas is approximately 1 to 3 ppm, and it is capable of producing irritation to the eyes and nose at concentrations ranging from 10 to 15 ppm (15, 44). One study reported an increase in airway resistance in humans at concentrations of 1.6 to 2.0 ppm (28). However the only major epidemiologic studies involving oxides of nitrogen were carried out in Chattanooga, Tennessee, where a TNT factory was responsible for high concentrations of NO_2 in different areas of town. A survey of schoolchildren showed a significant decrease in the pulmonary function of youngsters exposed to high levels of NO_2 (0.06 to 0.10 ppm) when compared to individuals in a control area. Although SO_2 and "total particulates" were also present, these pollutants did not seem to account for the health effects observed (26). The same study showed that the incidence of respiratory illness in families was higher in the 0.06 to 0.1 ppm NO_2 area (25). There was an increased bronchitis rate in this area for children who had been exposed to 0.1 to 0.15 ppm of NO_2 for 2 to 3 years (29).

Animal studies have been carried out at different levels of NO_2 exposure, and the health effects generally produced by this contaminant were increased mortality from aerosol infection, histopathological and cellular changes in the lungs, and

---

*This amount is used in the British Standard Procedure.

respiratory distress. These effects were found at concentrations of NO_2 ranging from 2.0 to 40 ppm.

Because community studies of the effects of NO_2 on human health are limited, it is extremely difficult to make absolute conclusions about precise limitations. A WHO expert committee states: "While biological activity in animals and plants at low concentrations has been demonstrated, the Committee believes that there is insufficient information upon which to base specific air quality guides at this time" (12).

4.4.2 Photochemical Oxidants

Photochemical oxidants result from the chemical reactions of oxides of nitrogen, reactive hydrocarbons, carbon monoxide, and water vapor in the presence of sunlight and in relatively still conditions. The result of these chemical reactions are colorless, gaseous compounds: ozone, peroxyacetyl nitrates (PAN), aldehydes, and other oxidizing products.

Ozone is probably the most important constituent of photochemical smog and it is a frequent measure of the overall severity of this condition. Therefore most of the studies on the health effects of photochemical oxidants have been conducted using ozone. Table 29.5 reviews the health effects of photochemical oxidants. The symptoms observed are aggravation of respiratory illnesses, irritation of the eyes and respiratory tract, and impairment of cardiopulmonary function.

The perceptibility threshold and odor detection for ozone generally has been accepted at 0.02 to 0.05 ppm (44, 48). At this level most people regard the odor of ozone as irritating and unpleasant. In a study by Hammer et al. (64), student nurses in the Los Angeles area were asked to record symptoms using the diary system, to permit the calculation of dose–response thresholds for photochemical oxidants. The following thresholds were found: headache, 0.05 ppm; eye discomfort, 0.15 ppm; cough, 0.26 ppm; and chest discomfort, 0.29 ppm. The threshold for eye discomfort was similar to other findings, and the thresholds for cough and chest discomfort are in the range cited by other studies.

The acute effects of photochemical oxidants on mortality have been studied in the Los Angeles area, and there appears to be no relation between increased mortality and oxidant levels ranging from 0.50 to 0.90 ppm (128). A study by Schoettlin and Landau (49), however, revealed that there were a significantly greater number of attacks in asthmatic patients when oxidant levels were above 0.25 ppm. Various other acute effects have been reported within this range (i.e., 0.2 to 1.0 ppm). At ozone levels between 0.2 and 0.5 ppm, significant changes in such parameters as visual acuity and peripheral vision have been noted (44, 50). Young and Shaw (51) found pulmonary changes in the 0.2 to 1.0 ppm range. Their report cites significant differences in pulmonary alveolar diffusing capacity in 12 subjects following inhalation of ozone at 0.6 to 0.8 ppm for 2 hours.

Three important studies were conducted during the period 1969–1974 on the effect of ozone during short-term exposure. In two of the experiments, healthy volunteers were exposed to 0.37 or 0.75 ppm ozone for 2 hr. Both studies reported decreased pulmonary function, cough, dyspnea, and pharyngitis among the vol-

Table 29.4. Critical Literature Review of NO$_2$, 1961–1973.

Reference	Publication Date	Country	Type of Study		Exposure	Contaminants Measured
			Epidem.	Lab.		
15	1961	U.S.		×	Controlled exposure	NO$_2$
16	1962	U.S.		×	Controlled exposure	NO$_2$
17	1963			×	Controlled exposure	NO$_2$ and an aerosol of *K. pneumonia*
18	1964	U.S.		×	Controlled exposure	NO$_2$ and carbon particles
19	1965	U.S.		×	Controlled exposure	NO$_2$
20	1965			×	Controlled exposure	NO$_2$
21	1965			×	Controlled exposure	NO$_2$
22	1968 1969			×	Controlled exposure	NO$_2$
23	1968			×	Controlled exposure	NO$_2$
24	1969			×	Controlled exposure	NO$_2$

Concentrations[a]	Effects Studied	Effects Observed (Concentration)
1–50 ppm	NO_2 exposures on small number of volunteers	1–3 ppm Odor threshold 13 ppm Eye and nasal irritation 25 ppm For 5 min pulmonary discomfort 50 ppm For 1 min increased nasal irritation and pulmonary discomfort
0–104 ppm	Threshold concentrations for rats at various exposure times	Threshold concentrations based on borderline changes in lung to body weight ratios: 5 min, 104 ppm; 15 min, 65 ppm; 60 min, 28 ppm
0–3.5 ppm NO_2	Susceptibility of mice to *K. pneumonia* in presence of NO_2	2.5 ppm, no increased susceptibility; 3.5 ppm for 2 hr, mortality rate doubled
100–1000 ppm NO_2	Effect of NO_2 on lungs of mice and carbon as a carrier of NO_2	250 ppm or greater of NO_2 alone, respiratory distress during and following exposure
5 ppm	Effect of continuous exposure of NO_2 for 90 days on 50 rats and 100 mice	5 ppm for 90 days: rats, 18% mortality; mice, 13% mortality
5–25 ppm	Intermittent exposures for 18 months on laboratory animals	5–25 ppm: no increased mortality, no changes in body weight, hemoglobin values, or biochemical indices
5–15 ppm	Effect of NO_2 on the lung tissues of guinea pigs	Daily exposures of 5–15 ppm resulted in increasing titers of lung tissue serum
0.8–25 ppm	Effect of NO_2 on respiratory tract of rats	0.8–2 ppm, rats grew and survived normally; 4.0 ppm for 16 weeks, terminal bronchiolar epithelium was broadened and hypertrophied; 10–25 ppm, developed large air-filled heavy lungs and died of respiratory failure
26 ppm	Effect on pulmonary function in lungs of dogs	26 ppm for 6 months, marked histopathological changes in the lungs; however, 225 days after cessation of NO_2 exposure, no significant pulmonary function alterations
0–40 ppm	Effect on germ-free mouse lungs	40 ppm (6–8 weeks), proliferation of epithelial cells of bronchi to form multicellular membrane projecting into lumen; also increased epithelial cells in alveolar spaces

Table 29.4. (Continued)

Reference	Publication Date	Country	Type of Study Epidem.	Type of Study Lab.	Exposure	Contaminants Measured
25	1970	U.S.	×		Total exposure	NO_2; suspended nitrates
26	1970	U.S.	×		Total exposure	NO_2; suspended nitrates
44	1971	Japan	×		Total exposure	SO_2, CO, NO_2, O_3, TSP
45	1972	U.S.		×	Controlled exposure	NO_2, SO_2, CO, PbClBr
27	1972	U.K.	×		Total exposure	NO_2, NO
28	1973			×	Controlled exposure	NO_2
29	1973	U.S.	×		Total exposure	NO_2
30	1973		×		Total exposure	NO_2, CO, HC
31	1973			×	Controlled exposure	NO_2

[a]Primary national standard (United States): ~0.05 ppm/year.

unteers (58, 63). Bates (62) investigated the synergistic effects of ozone and sulfur dioxide and found a marked decrease in pulmonary function when healthy volunteers were exposed to 0.37 ppm ozone and 1000 $\mu g/m^3$ of sulfur dioxide while performing light exercise. It also was noted that healthy subjects doing light exercise and exposed to 0.37 or 0.5 ppm of ozone suffered measurable physiological and biochemical changes and felt physically ill (65).

The effects of oxidants on athletic performance have also been recorded. In one study the times of long-distance runners were compared to previous competition

Concentrations[a]	Effects Studied	Effects Observed (Concentration)
NO_2, 0.062–0.109 ppm; SN, 3.8–7.2 $\mu g/m^3$	Acute respiratory illness in families	0.062–0.109 ppm (24 weeks), 18.8% excess in acute respiratory illness in this range
NO_2, 0.062–0.109 ppm; SN, 3.8–7.2 $\mu g/m^3$	Pulmonary function in elementary school children	0.062–0.109 ppm (24 weeks), decreased pulmonary function in children exposed to high levels of NO_2
Varying levels of all contaminants	Effect of NO_2 exposure on humans is reviewed	1 ppm, perceptibility threshold; 10–15 ppm, irritation of eyes, nose, and upper airway; 25 ppm, safety threshold of short-term exposure
Varying levels of all contaminants	Effects on cynomolgus monkeys	6.78 ppm (NO_2 only), increased osmotic fragility of erythrocytes
88–167 ppm (NO_2 + NO)	Chronic lung disease in coal miners.	88–167 ppm, 84% of the men had emphysema based on residual volume of more than 150%
1–5 ppm (1880–9400 $\mu g/m^3$)	Pulmonary function of healthy and diseased adults	*Healthy adults*: 5 ppm (15 min), decrease in diffusion capacity *Diseased adults*: 4–5 ppm (60 min), decrease in arterial partial pressure of O_2 *Healthy and diseased adults*: 1–1.6 ppm (30 breaths); no effect; 1.6–2.0 ppm (30 breaths), increased airway resistance
0.10–0.15 ppm (188–282 $\mu g/m^3$)	Bronchitis rates of children exposed for 2–3 years	0.10–0.15 ppm, significantly higher bronchitis rates than children in low exposure areas
NO_2, 0.005–0.09 ppm; CO, 1–44 ppm; HC, 4.4–15.8 ppm	Chronic bronchitis in policemen	Increased chronic bronchitis, though not significantly, among policemen exposed for longer durations
2.9 ppm	Pulmonary effects on laboratory animals	2.9 ppm, significant pulmonary abnormalities

times, to determine the effect of the oxidant levels on the runners. It was found that if the oxidant level was above 0.1 ppm during the hour before the event, there was an increase in running times, (i.e., a decrease in athletic performance). However, no such association was found between the oxidant levels either 2 to 3 hr before the race or during the event (129). In another study, Smith (130) exposed 32 male college students to 0.3 ppm PAN while at rest and during a 5-min bicycle exercise. Oxygen uptake was increased over the control during the period of bicycling, but not while at rest.

Table 29.5. Critical Literature Review of Photochemical Oxidants, 1960–1974

Reference	Publication Date	Country	Epidem.	Lab.	Exposure	Contaminants Measured
48	1960			×	Controlled exposure	O_3
49	1961	U.S.	×		Total exposure	Total oxidant levels
50	1963			×	Controlled exposure	O_3
51	1964			×	Controlled exposure	O_3
52	1965			×	Controlled exposure	O_3
53	1967			×	Controlled exposure	Synthetic photochemical smog
54	1968			×	Controlled exposure	O_3
55	1969	Japan		×	Controlled exposure	O_3 (SO_2 and NO_2 were measured separately
56	1971			×	Controlled exposure	O_3
57	1971			×	Controlled exposure	O_3
44	1971	Japan	×		Total exposure	O_3, SO_2, NO_2, CO, TSP, HC
58	1972			×	Controlled exposure	O_3
59	1972			×	Controlled exposure	O_3
60	1973	Japan		×	Controlled exposure	O_3

Concentrations[a]	Effects Studied	Effects Observed (Concentration)
0.02–0.05 ppm	Odor detection in 10 volunteers	0.02–0.05 ppm, odor detection in 9 out of 10 subjects
25 ppm and over	Effect on asthmatics	0.25 ppm and over, significantly greater number of attacks in asthmatic patients
0.2–0.5 ppm	Effect on visual parameters	0.2–0.5 ppm, significant changes in visual parameters such as visual acuity
0.6–0.8 ppm	Exposure of 11 subjects	0.6–0.8 ppm, impaired diffusion capacity (DL_{CO})
1–3 ppm	Effect on 25 subjects	1–3 ppm, more profound effects and changes in pulmonary function than caused by smoking
Oxidant material, 0.50–0.75 ppm	Effect on the ultrastructure of mice	0.50–0.75 ppm, irreversible lesions of alveolar tissue
0.3–4.0 ppm	Effect on the pulmonary cells of rabbits	0.3–4.0 ppm, decreases in the percentage of alveolar macrophages and their ability to engulf streptococci
0.5 ppm	Effect on guinea pigs	0.5 ppm (2 hr), increase in air current resistance, increase in frequency of respiration, and decrease in tidal volume
1 ppm	Effect on mouse lungs	1 ppm, decrease in bacterial pulmonary deposition (67%), decrease in bactericidal activity
0.5 ppm	Effect on pulmonary alveoli in aging mice	0.5 ppm, lowered rates of DNA synthesis in alveolar cells
O_3, 0.1–1.0 ppm	Effect of O_3 exposure to humans reviewed	0.02 ppm, perceptibility threshold; 0.2–0.3 ppm, irritation of respiratory system; 0.1–1.0 ppm for 1 hr, resistance in the respiratory tract; 0.2–0.5 ppm for 3–6 hr, decrease in human sight
0.75 ppm	Effects on healthy volunteers, lightly exercising	0.75 ppm, substernal soreness, cough, some pharynaitis and dyspnea coupled with decreased pulmonary function
0.4 ppm	Effect of 10-month exposure on rabbits	0.4 ppm, emphysematous and vascular type lesions in the lung; small pulmonary arteries were thicker
1.0 ppm	Effect on erythrocytes of mice	1.0 ppm, increased resistance to erythrocyte hemolysis

Table 29.5. (Continued)

Reference	Publication Date	Country	Type of Study		Exposure	Contaminants Measured
			Epidem.	Lab.		
1.0 ppm			Effect on the lungs of rabbits		1.0 ppm for 3 hr, 1–3 days after exposure, the animals had reduced vital capacity; 7 day after, the vital capacity was reduced only slightly	
O_3, 0.37 ppm SO_2, 1000 $\mu g/m^3$			Effect on healthy subjects, lightly exercising		0.37 ppm, market decrease in pulmonary function among healthy subjects	
0.37–0.75 ppm			Effect on healthy volunteers, lightly exercising		0.37–0.75 ppm, decrease in pulmonary function	
—			Effect on student nurses, recorded in diaries		100 $\mu g/m^3$ (0.05 ppm), headache 300 $\mu g/m^3$ (0.15 ppm), eye discomfort 510 $\mu g/m^3$ (0.26 ppm), cough 580 $\mu g/m^3$ (0.29 ppm), chest discomfort	
0.37 or 0.50 ppm			Adult male volunteers with exercise		0.37 or 0.50 ppm (2 hr), measurable physiological and biochemical changes; subjects felt physically ill; sensitive subjects, 0.37 (2 hr), respiratory symptoms	

[a]Primary national standard (United States): 0.08 ppm, 1 hr.

A few animal studies are included in Table 29.5. Coffin et al. (54) reported that exposure of rabbits to 0.3 to 4.0 ppm ozone enhances the susceptibility of the animals to infective aerosols. In other laboratory studies cited, a range of ozone between 0.4 and 0.75 ppm appears to be responsible for lesions in the lungs, lower DNA synthesis in alveolar cells, decreased tidal volume, increased frequency of respiration, and decreased pulmonary function (53, 55, 57, 59).

4.4.3 Carbon Monoxide

Carbon monoxide (CO) is a colorless, odorless gas produced by incomplete combustion. The principal source is motor vehicle exhaust; however, industrial plants burning carbonaceous fuels contribute significant amounts as well.

The predominant characteristic of CO is its great affinity for hemoglobin—about 240 times that of oxygen. Therefore CO impairs the transport of oxygen at the tissue level and interferes with the release of oxygen from the hemoglobin molecule. There is a background level of carboxyhemoglobin in the blood of about 0.4 percent; however, smoking a pack of cigarettes a day increases this level to about 5 percent (138). It should be recognized that CO exposure does not necessarily raise the level of carboxyhemoglobin in the blood because of the equilibrium that is finally established between the blood and air. For example, 25 ppm CO results in 4 percent saturation regardless of the initial concentration in the blood (12). A person with

Concentrations[a]		Effects Studied		Effects Observed (Concentration)	
61	1973	Japan	×	Controlled exposure	O_3
62	1973		×	Controlled exposure	O_3 Simultaneous exposure SO_2
63	1973		×	Controlled exposure	O_3
64	1974	U.S. ×		Total exposure	Photochemical oxidants
65	1974		×	Controlled exposure	O_3

less than 4 percent carboxyhemoglobin will absorb the CO, whereas a person with a level greater than 4 percent will excrete it until he reaches an equilibrium at 4 percent. Table 29.6 indicates that the general health effects of CO are reduced tolerance for exercise, impairment of psychomotor function, impairment of fetal development, and aggravation of cardiovascular disease.

Carbon monoxide levels of 10 ppm produce carboxyhemoglobin saturation of about 2 percent; this level has been shown to decrease behavioral performance (35). At a concentration of 100 ppm, CO produces about 16 percent inactivation of hemoglobin to carboxyhemoglobin (33). This level has been reported not to have any effect on cranial or spinal nerve reflexes (Schulte, 32), yet carboxyhemoglobin levels of 5 percent and below were associated with the control of choice discrimination of psychomotor abilities, and with the degree of impairment related to increasing carboxyhemoglobin levels. The effects of varying levels of carboxyhemoglobin on performance and perception tests have also been evaluated. Values of 15 to 20 percent saturation were associated with headache and impairment of manual coordination (131). Bender et al. (47) exposed healthy volunteers to 100 ppm CO for 2.5 hr, then performed certain psychological tests. Significant decreases in visual perception, manual dexterity, and the ability to learn and perform certain intellectual tests were found. Studies by Hosko (132) have also shown effects on visual perception at levels above 20 percent saturation, although he found that

Table 29.6. Critical Literature Review of Carbon Monoxide, 1963–1974

Reference	Publication Date	Country	Type of Study Epidem.	Lab.	Exposure	Contaminants Measured
32	1963			×	Controlled exposure	CO
33	1963			×	Controlled exposure	CO
137	1966	U.S.		×	Controlled exposure	CO
34	1967			×	Controlled exposure	CO
35	1969	U.S.		×	Controlled exposure	CO
36	1970	U.K.		×	Controlled exposure	CO
37	1970			×	Controlled exposure	CO
131	1970			×	Controlled exposure	CO
132	1970			×	Controlled exposure	CO
133	1970			×	Controlled exposure	CO
46	1970	U.S.		×	Controlled exposure	CO
47	1971	Germany		×	Controlled exposure	CO

Concentrations[a]	Effects Studied	Effects Observed (Concentration)
100 ppm	Effect of CO on healthy adults' nervous systems	100 ppm, maximum of 20.4% carboxyhemoglobin with no changes in any spinal or cranial nerve reflexes
		5% Carboxyhemoglobin, definite impairment in cognitive discrimination and psychomotor abilities
100 ppm	Effect of CO on hemoglobin	100 ppm, 16% inactivation of hemoglobin to carboxyhemoglobin
	Effect of CO on healthy nonsmokers	4% Carboxyhemoglobin saturation, increase in oxygen debt with exercise
50 ppm	Effect on the impairment of sound perception on adults	50 ppm for 90 min, impairment of subjects' ability to discriminate between different durations of sound
10 ppm	Effect on the behavior and performance of humans	10 ppm, 2% carboxyhemoglobin level and degraded behavioral performance
500 ppm	Effect on the auditory flutter fusion threshold and critical flicker fusion threshold	500 ppm, no evidence of any depressant effect of CO; however auditory flutter fusion threshold increased a little
60 ppm	Effect on rat liver function	60 ppm, decreased ability of rat liver to metabolize 3-OH-benzo[a]pyrene
—	Effects of levels of carboxyhemoglobin on performance and perception tests	15–20% saturation, headache and impairment of manual coordination
—	Effects on vision and EEG activity	>20% carboxyhemoglobin level, modified visual evoked responses
		33% carboxyhemoglobin level, EEG activity affected
—	Effect of carboxyhemoglobin level on patients with coronary disease	5–10% carboxyhemoglobin level, coronary artery blood flow accelerated; >6% saturation, significant myocardial changes
50–100 ppm	Effect on dogs	50–100 ppm (chronic exposures) may produce functional disorders and morphologic changes in the heart and brain
100 ppm	Effect on healthy volunteers	100 ppm for 2½ hr, significant decrease in visual perception, manual dexterity, and ability to learn and perform certain intellectual tasks

Table 29.6. (Continued)

Reference	Publication Date	Country	Type of Study Epidem.	Lab.	Exposure	Contaminants Measured
38	1971			×	Controlled exposure	CO
39	1971			×	Controlled exposure	CO
40	1972			×	Controlled exposure	CO
41	1973	U.S.	×		Total exposure	Co, Pb, NO$_x$, TSP
42	1973	U.S.		×	Controlled exposure	CO
43	1973	U.S.		×	Controlled exposure	CO
134 135	1973			×	Controlled exposure	CO
136	1974			×		CO

[a]Primary national standard (United States): 9.0 ppm, 8 hr; 35.0 ppm, 1 hr.

Concentrations[a]	Effects Studied	Effects Observed (Concentration)
10,000 ppm	Survival of newly born chicks to CO	10,000 ppm, 50% of newly hatched chicks withstood exposure for 32 minutes; survival time to CO decreased with postnatal age; 1 day old, 10 min; 8–21 days old, 4 min
51, 96, 200 ppm	Effects of exposure on rats, guinea pigs, monkeys, and dogs	51, 96, and 200 ppm, not toxic effects were seen; only physiological change was increase in hemoglobin and hematocrit values for all species
90 ppm	Effect on pregnant rabbits	90 ppm (90 days), birth weights decreased 12%; exposure during first pregnancy resulted in carboxyhemoglobin 9–10%; neonatal mortality increased from 4.5 to 10%; mortality during following 21 days increased from 13 to 25%
CO, 63 ppm; Pb, 30.9 μg/m^3; NO$_x$, 1.38 ppm; TSP, 200 μg/m^3 (all values 30-day average)	Effect of pollution on bridge and tunnel workers	63 ppm, high percentage had symptoms suggestive of chronic bronchitis; airway resistance was elevated in 33% of the workers; almost all bridge and tunnel workers had an increase in closing volume, suggesting small airway disease
<2–500 ppm	Effect on time perception of healthy adult volunteers	<2, 50, 100, 200, and 500 ppm, carboxyhemoglobin saturation as great as 20% but has no detrimental effect on man's time sense
100 ppm	Effect of CO on monkeys with induced myocardial infarctions	100 ppm (24 weeks), significant and persistent characteristic elevations in the hematocrit, hemoglobin, and red blood counts after 3 weeks
—	Effect of CO on patients with atherosclerotic heart disease	2.5–3.0% carboxyhemoglobin level, patients experience chest pain earlier during exercise than at 1.0%
—	Effect of CO on men with peripheral atherosclerotic disease	3.0%, men with atherosclerotic disease develop leg pain after less walking

EEG activity was not affected until 33 percent saturation. The effect of 500 ppm CO on auditory flutter fusion and critical flicker fusion thresholds was investigated by Guest et al. (36), but the only reported effect was a slight increase in the auditory flutter fusion threshold. In another study, however, CO at 50 ppm for 90 minutes was found to impair the ability of subjects to discriminate between different durations of sound (34). The aggravation of cardiovascular disease can also be observed when carboxyhemoglobin levels become too high. Ayers et al. (133) studied the effects of carboxyhemoglobin levels on patients with coronary artery disease. When carboxyhemoglobin was raised to 5 to 10 percent saturation, coronary artery blood flow was accelerated, and at levels above 6 percent, myocardial changes were seen. Other studies have demonstrated that patients with atherosclerotic heart diseases experience chest pain earlier during carboxyhemoglobin levels of 2 to 3 percent when compared to 1 percent carboxyhemoglobin levels (134, 135). Patients with the same disease have also developed leg pain after less walking at 3 percent carboxyhemoglobin saturation (136). This indicates that people most sensitive to carbon monoxide exposure are those with cardiovascular or respiratory conditions.

Preziosi (46) chronically exposed dogs to 50 to 100 ppm CO and found that these levels may produce functional disorders and morphological changes in the heart and brain. Other work revealed that CO levels of 60 ppm decreased the ability of rat livers to metabolize 3-OH-benzo[a]pyrene (37). Pregnant rabbits exposed to 90 ppm CO during half the gestation period experienced carboxyhemoglobin levels of 9 to 10 percent. This resulted in decreased birth weights and increased neonatal and infant mortality (40). Several studies have also shown that CO exposures ranging from 51 to 200 ppm have increased hematocrit, hemoglobin, and red blood count values in rats, guinea pigs, dogs, and monkeys with induced myocardial infarctions (39, 43).

It has been generally accepted that carboxyhemoglobin levels of 4 percent and above are undesirable for humans. However, many people in our society already have carboxyhemoglobin levels exceeding 4 percent because of smoking. Although it may be readily inferred that it is the impaired individual with cardiovascular disease who is most affected by high CO levels, it is difficult to decide which segment of the population the standard should protect. As the WHO report states, "One is confronted by a similar dilemma when defining the fraction of the population that must receive absolute protection at all costs, for it is obvious that in any urban community, there will be some patients in extremis to whom any stress will prove ultimately intolerable" (12).

4.4.4 Hydrocarbons

Hydrocarbons are organic compounds in gaseous or particulate form, such as methane, formaldehyde, benzene, and acrolein. They are produced by the incomplete combustion of fuels and other carbon-containing substances; typical sources are motor vehicle exhaust and coal-burning heating plants.

Hydrocarbons promote the formation of photochemical smog, although there is little evidence relating to the direct health effects of hydrocarbons. Therefore the standards have been based almost entirely on the role of hydrocarbons as precursors of other compounds of photochemical smog.

Aliphatic hydrocarbons are basically inert and have virtually no demonstrable effect on health except at extremely high concentrations. The Criteria Document for Hydrocarbons (83) reports that exposure to 5000 ppm heptane and octane for 4 min can cause incoordination and vertigo, and 10,000 ppm octane or 15,000 ppm heptane will produce narcosis in 30 to 60 min. Ethylene at 5500 ppm for several hours, however, has been reported to have little or no effect.

Alicyclic hydrocarbons have also very little effect on health, although at high concentrations they can act as depressants or anesthetics (87). The Criteria Document states that cyclohexane vapor at 18,000 ppm for 5 min can produce slight muscle tremors in mice and rabbits, and the same concentration for 25 to 30 min will produce muscular incoordination and paralysis. Although chronic exposures of smaller concentrations of cyclohexane (3300 ppm for 6 hr/day in rabbits and 1240 ppm for 6 hr/day in monkeys) have no noticeable effects.

Aromatic hydrocarbons are much more irritating than aliphatic or alicyclic hydrocarbons. The review of vapor phase organic pollutants by the National Academy of Sciences (87) states that chronic exposures to some aromatic hydrocarbons have been associated with leukopenia and anemia, and that benzene, toluene, or xylene concentrations above 100 ppm may result in fatigue, weakness, confusion, skin paresthesias, and mucous membrane irritation; at concentrations above 2000 ppm prostration and unconsciousness could result. Absoy et al. (86) report six cases of Hodgkin's disease among workers chronically exposed to 150 to 210 ppm benzene for 1 to 28 years. The validity of this study is questionable, however, because there was no mention of how many other workers were exposed, or to what other pollutants they were exposed. In other studies it was reported that toluene at 50 to 100 ppm produces no effect; but fatigue, confusion, and paresthesia have resulted after acute exposure to 220 ppm for 8 hr (82).

Aldehydes are formed through photochemical reactions in the atmosphere and have very irritating properties. Of all the aldehydes present in the atmosphere, formaldehyde and acrolein are the two most important. Formaldehyde is recognized as being highly irritating to the mucous membranes of the eyes, nose, and throat. The odor threshold for formaldehyde has been set at 0.06 to 0.5 ppm (67, 68). Tuesday et al. (78) investigated the effect of formaldehyde on eye irritation, using a smog chamber. It was found that eye irritation occurred at levels as low as 0.15 ppm. Morrill (72) puts the level of irritant action of formaldehyde at 0.9 to 1.6 ppm, and it has been reported that acrolein irritates the eyes at exposures to concentrations of 0.25 ppm for 5 min (75). In addition, Ahmad and Whitson (85) reported rapid loss of consciousness in 10 new workers who had been exposed to formaldehyde concentrations ranging from 2 to 10 ppm.

Various experiments have been performed using animals to observe the effects of formaldehyde and acrolein (Table 29.7). The effect of aldehydes on the eyes have been observed, and in one study rabbits exposed for 30 days to 0.57 ppm acrolein experienced no effect, but at 1.9 to 2.6 ppm for 4 hr, enzyme alterations in eye tissues occurred (69, 70). However, Gusev et al. (80) have shown that

Table 29.7. Critical Literature Review of Hydrocarbons, 1957–1976

Reference	Publication Date	Country	Type of Study Epidem.	Type of Study Lab.	Exposure	Contaminants Measured
66	1957			×	Controlled exposure	Propionalde-hyde, butyr-aldehyde, iso-butyralde-hyde
67	1960					Formaldehyde
68	1963–1964					
69	1960			×	Controlled exposure	Aldehydes
70	1960					
71	1960			×	Controlled exposure	Formaldehyde
72	1961		×		Total ex-posure	Formaldehyde
73	1961		×		Total ex-posure	Aldehydes
74	1960			×	Controlled exposure	Formaldehyde
75	1962			×	Controlled exposure	Acrolein
76	1963			×	Controlled exposure	Acrolein
77	1964			×	Controlled exposure	Formaldehyde
78	1965			×	Controlled exposure	Hexane Oxides of nitro-gen; formal-dehyde
79	1966			×	Controlled exposure	Acrolein
80	1966			×	Controlled exposure	Acrolein
81	1967			×	Controlled exposure	Ethylene, pro-pylene, iso-butane, gas mixture and auto exhaust

Concentrations[a]	Effects Studied	Effects Observed (Concentration)
200 ppm, 134 ppm	Effects of hydrocarbons on humans	134 ppm (30 min) of propionaldehyde, mildly irritating to exposed mucosal surfaces; 200 ppm (30 min) butyraldehyde and isobutyraldehyde, almost nonirritating
0.1–1.0 ppm	Odor threshold	0.06–0.5 ppm, odor threshold
0.57–2.6 ppm	Effect on the eyes of rabbits	0.57 ppm, no apprent effect; 1.9–2.6 ppm (4 hr), enzyme alterations in eye tissues
1–3 ppm	Pulmonary effects on guinea pigs	1–3 ppm (1 hr), increased airflow resistance and respiratory work
—	Eye irritation of people working in paper processing	0.9–1.6 ppm, irritation of the eyes
0.035–0.35 ppm	Eye irritations in man exposed to smog in Los Angeles	0.035–0.35 ppm, direct relation to eye irritation
19 ppm	Toxicology of formaldehyde	19 ppm (10 hr), edema and hemorrhage in lungs and hyperemia of liver
0.25 ppm	Eye irritation of humans	0.25 ppm (5 min), irritation of the eyes
0.06 ppm	Physiological effects on guinea pigs	0.06 ppm, total pulmonary resistance; tidal volume increased and respiratory rate decreased
3.5 ppm	Biochemical effects on rats	3.5 ppm (18 hr), increased alkaline phosphatase activity in liver
0.3–2.6 ppm, 0.1–1.1 ppm, 0.15–0.3 ppm	Eye irritation in a smog chamber	High correlation of eye irritation with formaldehyde; 0.15 ppm, concentration of formaldehyde at which eye irritation first occurred
200 ppm	Lung damage in rats	200 ppm (10 min, once a week for 10 weeks), residual lung damage
0.57 ppm	Effect of chronic exposure to rats	0.57 ppm, loss of weight, decrease in whole blood cholinesterase activity, decrease in urinary coproporphyrin excretion and change in conditioned reflex activities
0.1–4.0 ppm	Eye irritation	0.1–4.0 ppm, high correlation of eye irritation with formaldehyde content

Table 29.7. (Continued)

Reference	Publication Date	Country	Type of Study Epidem.	Lab.	Exposure	Contaminants Measured
82	1968		×		Acute exposure	Toluene (aromatic HCs)
83	1970	U.S.				Aliphatic hydrocarbons
						Alicyclic hydrocarbons
84	1972	U.S.	—	—	—	Particulate polycyclic organic matter
85	1973		×		Industrial exposure (total)	Formaldehyde
86	1974		×		Industrial exposure (total)	Benzene
87	1976	U.S.	—	—	—	Volatile hydrocarbons

[a]Primary national standard (United States): 0.024 ppm, 3 hr.

concentrations of 0.57 ppm acrolein produce decreased weights, changes in conditioned reflexes, decreased urinary coproporphyrin excretion, and decreased cholinesterase activity in rats. At 0.06 ppm acrolein was seen to increase total pulmonary resistance and tidal volume and decrease the respiratory rate of guinea pigs (76). Catilina et al. (79) reported residual lung damage in rats after exposure to 200 ppm acrolein for 10 min once a week for 10 weeks. Formaldehyde has also been shown to affect the respiratory system, and increased airflow resistance and respiratory work in guinea pigs have been attributed (71) to concentrations of 1.0 to 3.0 ppm.

4.4.5 Sulfur Dioxide

Sulfur dioxide is a colorless gas with a pungent odor; it can be oxidized to form sulfur trioxide, sulfuric acid, and later, sulfate. The principal sources are the com-

Concentrations[a]	Effects Studied	Effects Observed (Concentration)
50–200 ppm	Effect of chronic exposure on man	50–100 ppm, no apparent effects; 200 ppm (8 hr), paresthesia, fatigue, confusion
	Criteria document	5000 ppm (4 min) heptane and octane, incoordination and vertigo; 10,000 ppm octane and 15,000 ppm heptane, produces narcosis in 30–60 min; 5500 ppm of ethylene, little or no effect; 350,000 ppm (5 min) acetylene, unconsciousness
	Criteria document	18,000 ppm (5 min) cyclohexane, slight muscle tremors in mice and rabbits; 3330 ppm (6 hr/day, 60 days) cyclohexane, no effect in rabbits; 1240 ppm cyclohexane, no effect in monkeys
—	Biological effects of polycyclic organic matter	Hypothesis that a reduction of 1 μg of benzo[a]pyrene per 1000 m³ of air will decrease the lung cancer death rate by 5%
2–10 ppm	Effects on 10 new female employees	2–10 ppm, rapid loss of consciousness
150–210 ppm	Effect of chronic benzene exposure on Hodgkin's disease	150–210 ppm (mean of 11 years of exposure), six cases of Hodgkin's disease
—	Biological effects of hydrocarbons	Discussion on effects of all hydrocarbons that are considered pollutants

bustion of sulfur-containing fossil fuels, the smelting of sulfur-bearing metal ores, and industrial processes. Literature on SO_2 (Table 29.8) shows the main health effects to be aggravation of respiratory diseases (including asthma, chronic bronchitis, and emphysema), reduced lung function, irritation of the eyes and respiratory tract, and increased mortality. In epidemiologic studies it is difficult to consider the health effects of SO_2 in isolation, because SO_2 and particulate matter tend to occur together in the same kinds of polluted atmosphere. The WHO states: "It follows, therefore, that unless the effect sought is highly specific, the use of epidemiologic techniques will seldom result in the attribution, with any degree of certainty, of an observed effect to a specific pollutant" (12).

The effect of SO_2 on animals and humans has been the subject of many toxicologic experiments. Although studies conducted in controlled atmospheres do not simulate urban air pollution accurately, they do shed light on the physiological

Table 29.8. Critical Literature Review of SO_2, 1964–1975

Reference	Publication Date	Country	Type of Study — Epidem.	Type of Study — Lab.	Exposure	Contaminants Measured
88	1964	U.K.	×		Total exposure	SO_2 TSP
89	1966	U.K.	×		Total exposure	SO_2 TSP
126	1966	U.S.		×	Controlled exposure	
90	1967	U.K.	×		Total exposure	SO_2
91	1967	U.K.	×		Total exposure	SO_2 Smoke
92	1967	Holland	×			
93	1968	U.K.	×		Total exposure	SO_2 Smoke
94	1968	U.S.				
95	1969	U.S.	×		Total exposure	SO_2
96	1970	U.S.	×		Total exposure	SO_2 TSP
97	1970	U.S.		×	Controlled exposure	SO_2

Concentrations[a]	Effects Studied	Effects Observed (Concentration)
75–115 μg/m³ (0.026–0.040 ppm) 80–160 μg/m³ (annual mean)	Effects on lung cancer and bronchitis in men and women in two communities	115 μg/m³ (0.040 ppm), increase in lung cancer mortality in men, and increase in bronchitis mortality in men and women when compared to SO_2 level of 75 μg/m³ (0.026 ppm)
130–148 μg/m³ (0.05–0.06 ppm) 91–138 μg/m³ (annual mean)	Effect on respiratory illness in children	130–148 μg/m³ (0.05–0.06 ppm), lower respiratory tract disease was increased
		Review by Hazelton Laboratories on the role and nature of SO_2, SO_3, H_2SO_4, and fly ash in air pollution
400–500 μg/m³ (0.15–0.19 ppm) (24 hr average)	Effects of SO_2, O_2, and soot on man	400–500 μg/m³ (0.15–0.19 ppm), increased mortality and disease
123–275 μg/m³ 97–301 μg/m³ (24-hr average)	Patterns of respiratory illness in school children	200 μg/m³, presence of both upper and lower respiratory infections significantly increased
	Review document on acceptable level of SO_2	500 μg/m³ (0.19 ppm), excess mortality; 300–400 μg/m³, increased respiratory illness as well as absenteeism
0.16 ppm–0.21 ppm (458–600 μg/m³) 300–400 μg/m³	Respiratory illness observations of 1000 men studied	0.16 ppm (458 μg/m³), illness attack rates increased (weekly average); 0.21 ppm (600 μg/m³), decrease in ventilatory lung function (daily)
—	Review article on the acute effects of SO_2	Author concludes that any effect air pollution might have on health "does not appear to involve SO_2 in its mechanism"
—	Effects on patients with bronchial pulmonary disease	0.25–0.30 ppm (710–858 μg/m³), illness rates increased significantly
119–500 μg/m³ >260 μg/m³ on about 15% of the days (24 hr average)	Effect on patients with chronic bronchitis	300–500 μg/m³ (0.11–0.19 ppm), increased hospital admissions with respiratory illness; 119–249 μg/m³ (0.05–0.09 ppm), substantial increase in illness of patients 55 years or older with more severe bronchitis
0.13, 1.01, 5.72 ppm	Long-term exposure of guinea pigs	0.13, 1.01, and 5.72 ppm (12 months), pulmonary function measurements indicated that no detrimental changes could be attributed to SO_2; 5.72 ppm increase in size of hepatocytes accompanied by cytoplasmic vacuolation in liver

Table 29.8. (Continued)

Reference	Publication Date	Country	Type of Study		Exposure	Contaminants Measured
			Epidem.	Lab.		
98	1970	U.S.		×	Controlled exposure	SO$_2$
102	1970	U.K.	×		Total exposure	SO$_2$ Smoke
99	1971	NATO/ CCMS	—	—	—	—
100	1971	U.S.		×	Controlled exposure	H$_2$SO$_4$ mist and particulate sulfates
101	1972	U.S.		×	Controlled exposure	SO$_2$
103	1974	U.S.	×		Total exposure	SO$_2$ TSP Suspended nitrates Suspended sulfates NO$_2$
104	1974	U.S.	—	—	—	—
105	1974	U.S.				

Concentrations[a]	Effects Studied	Effects Observed (Concentration)
650 ppm	Effect on Syrian hamsters with emphysema	650 ppm, mild bronchitic lesion with relatively minor changes in mechanical properties of lung
550 μg/m^3 250 μg/m^3 (24 hr average)	Effects on panels of bronchitis patients	500 μg/m^3, increased morbidity
—	Air Quality Criteria Document for SO_2	—
—		Literature review of H_2SO_4 mist and particulate sulfate toxicology studies
3 ppm (9000 μg/m^3)	Effect of SO_2 on humans	3 ppm (120 hr), increased small airway resistance and significant but minimal decrease in dynamic compliance of lung
40 μg/m^3 60 μg/m^3 2 μg/m^3	Effects on well adults and cardiopulmonary patients	40 μg/m^3, best judgment estimate of threshold for effect (24 hr average)
6 μg/m^3		
30 μg/m^3		
—	—	Toxicologic and epidemiologic review of the health effects of SO_x
	Health consequences of SO_x (CHESS studies)	*Best judgment effects for long-term exposure*: 95 μg/m^3 (100 μg/m^3 TSP, 15 μg/m^3 sulfates), increased prevalence of chronic bronchitis, 95 μg/m^3 (102 μg/m^3 TSP, 15 μg/m^3 sulfates), increased acute lower respiratory disease in children; 106 μg/m^3 (151 μg/m^3 TSP, 15 μg/m^3 sulfates), increased frequency of acute respiratory disease in families; 200 μg/m^3 (100 μg/m^3 TSP, 13 μg/m^3 sulfates), decreased lung function of children *Best judgment effects for short-term exposure*: >365 μg/m^3 (80–100 μg/m^3 TSP, 8–10 μg/m^3 sulfates), aggravation of pulmonary symptoms in elderly; 180–250 μg/m^3 (70 μg/m^3 TSP, 8–10 μg/m^3 sulfates), aggravation of asthma

Table 29.8. (Continued)

Reference	Publication Date	Country	Type of Study Epidem.	Lab.	Exposure	Contaminants Measured
106	1974	Denmark U.S.		×	Controlled exposure	SO_2
107	1975	U.S.	×		Total exposure	SO_2 TSP Suspended sulfates

[a]Primary national standard (United States): 0.03 ppm, 1 year; 0.14 ppm, 24 hr.

changes caused by known levels and durations of SO_2. This gas is a respiratory irritant that is absorbed in the nose and upper respiratory tract, but very high concentrations of it are needed to produce changes in the lungs. Exposures of Syrian hamsters to SO_2 levels of 650 ppm produced mild bronchitic lesions with relatively minor changes in the mechanical properties of the lungs (98). However, in one human study, after 120 hr of exposure to 3 ppm SO_2, increased small airway resistance and significant but minimal decrease in dynamic compliance of the lung were observed (101). In a study using guinea pigs, long-term exposure (12 months) of levels up to 5.72 ppm produced no detrimental changes on pulmonary function measurements, and the only effects observed were increased size of hepatocytes and cytoplasmic vacuolation in the liver (97). Amdur (100) and Hazelton Laboratories (126) have performed many toxicologic studies, and many of their results can be found in the review articles cited in Table 29.8. Animal data on the health effects of SO_2 can be very useful in helping to determine the mechanisms of action, but as one review on the health effects of sulfur oxides states, "The physiological responses found under controlled conditions give useful insights into mechanisms of action, but cannot be translated directly into adverse effects of exposures to contaminated urban air" (104).

Epidemiologic techniques have also been used to observe the health effects of SO_2. These techniques are useful in considering the effects that might produce chronic respiratory disease. Some of the epidemiologic studies conducted are reviewed below; one should recognize the many difficulties encountered when trying to establish dose-response effects, such as standardizing measurement techniques, allowing for smoking, sex, age, and socioeconomic factors, and considering meteorological variables.

Wicken and Buck (88) examined the mortality rates in two English communities with different kinds of pollution. They found an increase in lung cancer mortality in men and increases in bronchitis mortality in men and women who were exposed to long-term SO_2 levels of 0.04 ppm. Another study revealed a significant correlation between SO_2 pollution levels and deaths or disease with a 24-hr mean SO_2 level of 0.15 to 0.19 ppm, when there was a high soot content (90). Fletcher and his colleagues (93) observed 1000 men aged 30 to 59 for 5 years; they discovered

Concentrations[a]	Effects Studied	Effects Observed (Concentration)
1, 5, 25 ppm	Effect of 6-hr exposure on nasal mucus flow rate and airway resistance	5, 25 ppm, significant decrease in nasal mucus flow rate; 1, 5, 25 ppm, increased nasal airflow resistance
38–425 $\mu g/m^3$ 60–185 $\mu g/m^3$ 9–20 $\mu g/m^3$ (annual mean)	Acute respiratory effects on children	38–425 $\mu g/m^3$, excess acute lower respiratory disease morbidity in children

a significant relation between respiratory illness and SO_2 and smoke levels. Illness attack rates were increased when the weekly SO_2 levels exceeded 0.16 ppm with an accompanying smoke level of 400 $\mu g/m^3$. At daily SO_2 concentrations of 0.21 ppm and smoke concentrations above 300 $\mu g/m^3$, decreases in ventilatory lung function were observed. Major air pollution episodes were examined by Dutch scientists, and their review concluded that excess mortality resulted when 24-hr mean levels of SO_2 exceeded 500 $\mu g/m^3$ (0.19 ppm) for a few days. In addition, hospital admissions and absenteeism increased when SO_2 24-hr mean levels were 300 to 400 $\mu g/m^3$ (0.11 to 0.15 ppm) for 3 to 4 consecutive days (92).

Because SO_2 is known to aggravate respiratory symptoms, many of the epidemiologic studies have been carried out using bronchitic or asthmatic patients. Carnow et al. (95) performed a study of bronchitic patients in Chicago to determine what levels made their illness worse and found the critical level at which bronchitis illness rates increased to lie between 0.25 and 0.30 ppm. It should be noted, however, that although particulates were not measured, they were probably at high concentrations of about 148 $\mu g/m^3$ as determined by other studies. Carnow et al. (96) later reported that a group of patients in the age range of 55 years and older with chronic bronchitis had significant increases at daily levels of SO_2 between 0.05 and 0.09 ppm and in patients with more severe bronchitis. This relationship was not observed among patients under 55 years old, or with less severe bronchitis. Lawther et al. (102) studied a panel of bronchitic patients, who recorded their daily conditions of health in diaries. It was observed that the minimum pollution level leading to a significant response was about 500 $\mu g/m^3$ (0.19 ppm) of SO_2 with about 250 $\mu g/m^3$ of smoke.

A number of studies have been conducted using children, because they do not smoke cigarettes and might be more sensitive to pollution levels. Douglas and Waller (89), investigating the pollution effect on respiratory disease in London children, learned that lower respiratory disease was increased in areas with annual SO_2 levels of 0.05 to 0.06 ppm or higher, with accompanying levels of particulates. In another study, Lunn et al. (91) examined school children aged 5 years and 10 to 11 years, in four areas of Scheffield. Both upper and lower respiratory infections were found to be increased at annual SO_2 concentrations of 200 $\mu g/m^3$ (0.076 ppm) and smoke concentrations of 200 $\mu g/m^3$.

Epidemiologic studies of EPA were carried out under the CHESS program to develop dose–response information relating short-term and long-term exposures to adverse health effects. *Health Consequences of Sulfur Oxides: A Report from CHESS, 1970–1971* (105) reviews studies associated with sulfur dioxide exposure in New York City, the Salt Lake Basin area, five Rocky Mountain areas in Idaho and Montana, Chicago, and Cincinnati. The best judgment effects for long-term exposure were estimated at 95 $\mu g/m^3$ (0.036 ppm) for increased prevalence of chronic bronchitis and increased acute lower respiratory disease in children, 106 $\mu g/m^3$ (0.04 ppm) for increased acute respiratory disease in children and families, and 200 $\mu g/m^3$ (0.076 ppm) for decreased lung function of children. The best judgment effects for short-term exposure were cited at greater than 365 $\mu g/m^3$ (0.14 ppm) for aggravation of cardiopulmonary symptoms in the elderly, and 180 to 250 $\mu g/m^3$ (0.07 to 0.095 ppm) for aggravation of asthma. All these health effects, however, were accompanied with TSP levels greater than 70 $\mu g/m^3$ and sulfate levels greater than 8 to 10 $\mu g/m^3$. The validity of the CHESS studies has been questioned. The methods used, the techniques of sampling, and the internal inconsistency of the findings have been criticized. It should be noted that some of the recent evidence suggests that the epidemiological data may be more closely related to the presence and concentration of sulfates than to SO_2, from which sulfates are most commonly derived.

4.4.6   Particulates (TSP)

Particulate matter is any solid or liquid particle dispersed in the atmosphere, such as soot, dust, ash, and pollen. Table 29.9 gives the main health effects of TSP as the aggravation of asthma or other respiratory disease, increased cough and chest discomfort, and increased mortality. Therefore one recognizes the importance of deposition, retention, and disposition of particulates; for a detailed discussion of these factors the reader is referred to Chapter 11.

Because particulate matter and sulfur oxides tend to occur in the same kinds of polluted atmosphere, it is difficult to differentiate between the two pollutants in epidemiologic studies. Such factors as smoking, age, socioeconomic conditions, and sex are not always accounted for, and although particle size is important for the effectiveness of deposition, it is hardly ever considered in field studies. Therefore the results of many epidemiologic studies are of questionable value in establishing standards.

In several such studies, the effects particulates have on mortality were examined. Buck and Brown (108) studied 214 areas of the United Kingdom* to relate mortality ratios for 5 years to daily smoke and SO_2 concentrations greater than 200 $\mu g/m^3$. However, mortality from lung cancer was not found to be positively associated with these smoke and SO_2 levels. Martin (109) reported an increase in mortality from all causes when smoke levels rose above 1000 $\mu g/m^3$ with SO_2 levels above 715 $\mu g/m^3$, but because his pollutant measurements were made in only one part

*The American high volume sampling method and the British method differ; although both express results in micrograms per cubic meter, they are not directly comparable.

of central London, a variety of concentrations must be considered to have contributed to the effects observed. In 1964 Wicken and Buck (88) conducted a study of bronchitis and lung cancer mortality in six areas of northeast England. They compared deaths from bronchitis and lung cancer to nonrespiratory disease deaths for a period of 10 years, adjusting for age, sex, social class, and smoking. However, Eston was the only one of six areas in which SO_2 and TSP values were available. Eston was divided into two sections of pollution, and a positive association appeared between the incidence of bronchitis and lung cancer in the high pollution area of 160 $\mu g/m^3$ smoke and 115 $\mu g/m^3$ SO_2 when compared to smoke levels of 80 $\mu g/m^3$ and SO_2 levels of 74 $\mu g/m^3$.

Other epidemiologic studies have attempted to establish morbidity effects of particulates. In one conducted by the (British) Ministry of Pensions and National Insurance (110), a representative population was observed for sickness from bronchitis, influenza, arthritis, and rheumatism. There was a significant correlation between bronchitis incapacity and particulate and SO_2 levels. There was also more arthritis and rheumatism in areas with high smoke pollution. The lowest bronchitis inception rate observed in this study was set at smoke levels of 100 to 200 $\mu g/m^3$, with SO_2 levels of 150 to 250 $\mu g/m^3$. Holland et al. (111) studied outdoor telephone workers in London, rural England, and the east and west coasts of the United States to observe effects of SO_2 and particulates on health. They concluded that an increase in smoke concentrations from 120 to 200 $\mu g/m^3$ with an equivalent increase in SO_2 increases the risk to older workers of deteriorated pulmonary function and chronic respiratory disease.

Because particulate pollution has been known to cause or aggravate respiratory diseases, many epidemiologic studies have been conducted using bronchitis or asthmatic patients. For example, Lawther et al. (118) evaluated conditions of a group of bronchitic patients through the diary system. Worsening of patients' conditions was related to daily smoke and SO_2 levels. With daily levels of 250 $\mu g/m^3$ of smoke and 500 $\mu g/m^3$ of SO_2, a significant worsening of bronchitic symptoms occurred. Children are also considered to be more susceptible to particulate pollution, and their health effects have been studied. In the earlier work by Douglas and Waller (112), illness data on upper and lower respiratory symptoms were collected from the mothers and school doctors. Upper respiratory tract infections were found not to be related to the amount of pollution, but lower respiratory infections were related. Increased frequency and severity of lower respiratory diseases were observed when smoke levels were about 130 $\mu g/m^3$. However, in another study by Lunn et al. (91) levels of smoke of 100 $\mu g/m^3$ were associated with increased respiratory infections.

Subsequent evidence on the health effects of particulates has been obtained from the EPA and the CHESS programs. In many of the investigations children were used as the study populations. Shy et al. (121) examined the effects of particulates on the ventilatory function of school children. At TSP levels ranging from 96 to 133 $\mu g/m^3$ with accompanying SO_2 levels of 39 to 57 $\mu g/m^3$, and sulfate levels of 8.9 to 10.1 $\mu g/m^3$, they found a decrease in the pulmonary function of their subjects. At TSP levels of 103 to 109 $\mu g/m^3$ with SO_2 less than 25 $\mu g/m^3$, Hammer et al.

Table 29.9. Critical Literature Review of Total Particulates, 1964–1975

Reference	Publication Date	Country	Type of Study		Exposure	Contaminants Measured
			Epidem.	Lab.		
108	1964	U.K.	×		Total exposure	Smoke SO_2
109	1964	U.K.	×		Total exposure	Smoke SO_2
88	1964	U.K.	×		Total exposure	Smoke SO_2
110	1965	U.K.	×		Total exposure	Smoke SO_2
111	1965	U.K.	×		Total exposure	Smoke SO_2
112	1966	U.K.	×		Total exposure	Smoke SO_2
113	1967	U.S.	×		Total exposure	Smoke SO_2
91	1967	U.K.	×		Total exposure	Smoke SO_2
114 115 116	1967 1968 1969	U.S.	×		Total exposure	TSP
117	1969	U.S.	—	—	—	—
118	1970	U.K.	×		Total exposure	Smoke SO_2
119	1971	NATO/ CCMS	—	—	—	—

Concentrations[a]	Effects Studied	Effects Observed (Concentration)
>200 $\mu g/m^3$ >200 $\mu g/m^3$ (24 hr time)	Effects on mortality from lung cancer and bronchitis	>200 $\mu g/m^3$, excess bronchitis mortality
1000 $\mu g/m^3$ 715 $\mu g/m^3$	Effects on mortality and morbidity	1000 $\mu g/m^3$, increased mortality from all causes
80–160 $\mu g/m^3$ 74–115 $\mu g/m^3$ (annual mean)	Effects on lung cancer and bronchitis mortality	160 $\mu g/m^3$ (115 $\mu g/m^3$ SO_2), positive association between incidence of bronchitis and lung cancer and the level of TSP and SO_2
100–400 $\mu g/m^3$ 150–400 $\mu g/m^3$ (24 hr average)	Effects on absences due to bronchitis influenza, arthritis, and rheumatism	100–200 $\mu g/m^3$ (150–250 $\mu g/m^3$ SO_2), lowest bronchitis inception rates 400 $\mu g/m^3$ (400 $\mu g/m^3$ SO_2), highest bronchitis inception rates
120–200 $\mu g/m^3$ 30–300 $\mu g/m^3$ (annual mean)	Prevalence of chronic respiratory disease symptoms in outdoor telephone workmen	Increased smoke concentration from 120–200 $\mu g/m^3$, with an increase in SO_2, will increase the risk to older workers of poorer lung function and chronic respiratory disease
>130 $\mu g/m^3$ >130 $\mu g/m^3$ (annual mean)	Effect on respiratory infection of children	>130 $\mu g/m^3$, increased frequency and severity of lower respiratory diseases in school children
2.0–8.2 cohs 0.10–1.00 ppm	Mortality and morbidity during an episode of high pollution	2.0–8.2 cohs, immediate rise in daily deaths due to all causes and rise in exacerbation of bronchitis and asthma
97–301 $\mu g/m^3$ 123–275 $\mu g/m^3$ (24 hr average)	Patterns of respiratory illness in school children (5–6 years old)	100 $\mu g/m^3$ (>120 $\mu g/m^3$ SO_2), increased association of respiratory infections
80–>135 $\mu g/m^3$ (2 yr geometric mean)	Effect on total mortality and respiratory mortality in men	>135 $\mu g/m^3$, death rate twice as high as at <80 $\mu g/m^3$ >135 $\mu g/m^3$, death rate due to stomach cancer almost twice as high as at <80 $\mu g/m^3$
—	Air Quality Criteria based on health effects (API)	—
250 $\mu g/m^3$ 500 $\mu g/m^3$ (24 hr average)	Effect on exacerbation of bronchitis	250 $\mu g/m^3$, minimium pollution level leading to a significant worsening of bronchitis patients
—	Air Quality Criteria Document for particulates	

Table 29.9. (Continued)

Reference	Publication Date	Country	Type of Study Epidem.	Lab.	Exposure	Contaminants Measured
120	1973	U.S.	×		Total exposure	TSP
121	1973	U.S.	×		Total exposure	TSP Suspended sulfates SO_2
122	1973	U.S.	×		Total exposure	TSP SO_2
123	1973	U.S.	×		Total exposure	TSP SO_2 Suspended sulfates
124	1974	U.S.	×		Total exposure	TSP SO_2
125	1974	U.S.	×		Total exposure	TSP SO_2 Suspended nitrates Suspended sulfates
127	1974	U.S.	×		Total exposure	TSP SO_2 Suspended sulfates
103	1974	U.S.	×		Total exposure	TSP SO_2 Suspended sulfates NO_2
107	1975	U.S.	×		Total exposure	TSP SO_2 Suspended sulfates

Concentrations[a]	Effects Studied	Effects Observed (Concentration)
100–269 $\mu g/m^3$ (24 hr average)	Family surveys of symptoms during acute pollution episodes	100–269 $\mu g/m^3$, increased cough and chest discomfort; restricted activity
96–133 $\mu g/m^3$ 8.9–10.1 $\mu g/m^3$ (annual mean) 39–57 $\mu g/m^3$	Effect on ventilatory function of school children	96–133 $\mu g/m^3$, dereased pulmonary function in school children
100–150 $\mu g/m^3$ Steady decrease until <80 $\mu g/m^3$ (annual mean)	Effect on chronic respiratory disease of military inductees	100–150 $\mu g/m^3$, increased chronic respiratory disease symptom prevalence in adults
66–151 $\mu g/m^3$ 63–275 $\mu g/m^3$ 2.4–20.3 $\mu g/m^3$ (annual mean)	Effect on acute respiratory disease in adults and children	>100 $\mu g/m^3$, increased frequency and severity of acute lower respiratory disease in school children
103–168 $\mu g/m^3$ <25 $\mu g/m^3$ (annual mean)	Survey of acute lower respiratory disease of children	103–139 $\mu g/m^3$, authors' best judgment regarding level that produces lower respiratory disease in children
81–168 $\mu g/m^3$ 11–<25 $\mu g/m^3$ 2–3 $\mu g/m^3$ 10–13 $\mu g/m^3$	Survey of respiratory disease symptoms of adults	<150 $\mu g/m^3$, no strong relation to chronic respiratory disease symptoms
34–185 $\mu g/m^3$ 22–425 $\mu g/m^3$ 9–18 $\mu g/m^3$ (annual mean)	Survey of lower respiratory disease symptoms in school children	60–185 $\mu g/m^3$ (38–425 $\mu g/m^3$ SO_2, 9–20 $\mu g/m^3$ suspended sulfates), excess acute lower respiratory disease morbidity in children
<60–120 $\mu g/m^3$ <40–100 $\mu g/m^3$ <6–12 $\mu g/m^3$ <30–75 $\mu g/m^3$	Survey of a panel of well subjects and cardiopulmonary patients	70–120 $\mu g/m^3$ (significant SO_2), cardiopulmonary effects
34–185 $\mu g/m^3$ 22–425 $\mu g/m^3$ 9–18 $\mu g/m^3$ (annual mean)	Acute lower respiratory disease symptoms in children	85–110 $\mu g/m^3$ (175–250 $\mu g/m^3$ SO_2, 13–14 $\mu g/m^3$ suspended sulfates), "best judgment" estimate associated with excess childhood respiratory morbidity

(124) estimated increased production of lower respiratory disease in children. In another study of children, the best judgment estimate associated with excess childhood respiratory morbidity was levels of TSP of 85 to 110 $\mu g/m^3$, SO_2 of 175 to 250 $\mu g/m^3$, and sulfates of 13 to 14 $\mu g/m^3$ (107).

It is evident that there is no unanimity of results in the various studies. In one case TSP levels less than 150 $\mu g/m^3$ were found to have no strong relation to chronic respiratory disease symptoms (125), whereas in another TSP levels of 100 to 150 $\mu g/m^3$ were shown to increase chronic respiratory disease symptoms (122).

4.5 Commentary

A careful review of the literature in MEDLINE on sulfur dioxide, carbon monoxide, photochemical oxidants, nitrogen dioxide, hydrocarbons and total particulates revealed no substantial new data had been presented that would refute Tables 29.4 through 29.9; therefore nothing has been added to the tables since the literature review of 1977.

5   RECENT LEGISLATION

Early in 1990 the administration released figures showing that regulations governing large industrial sources have already resulted in emissions reductions of about 80 percent. The remaining problems are the "nonattainment areas," mostly major metropolitan areas that failed to meet EPA's standards for one or more of the six pollutants listed for cleanup. These areas, such as southern California, Denver, and New York, are in the process of revising plans to meet clean air standards. The new version of the Clean Air Act addresses three key issues: (1) bringing nonattainment areas into compliance; (2) controlling emissions of air toxics; and (3) dealing with the problem of acid rain.

During the 1980s acid rain became a matter of prime concern. The degree to which sulfur compounds contribute to the damages of lakes and forests has yet to be established. Such important data are crucial to developing a comprehensive program, which would allow, for instance, miners of the high-sulfur coal area of West Virginia to continue to work. It is conceivable that with the judicious use of atomic power plants and coal-fired plants using high-sulfur coal that the total emissions of sulfur to the atmosphere would be within acceptable limits.

After many months of conflict between the idealism of the goals of legislators, enthusiastically supported by environmentalists, and the practicalities of the economics of sustaining an expanding population along with control of air pollutants, both the House and Senate passed the 1990 Clean Air Act, and on November 15, 1990 President Bush signed this bill into law. In signing the bill he stated, "Every American expects and deserves to breathe clean air." Further, he called this the "most significant air pollution legislation in our nation's history."

This act is very significant in many facets. Under the expired air pollution act, ambient air standards were based upon health effects. The new act determines the standards based upon maximum achievable control technology (MACT). After the MACT reductions are in place, EPA must develop health-based standards that would limit the cancer risk to exposed individuals to no more than one case in one million. This is a significant departure from the past protocol.

A summary of the bill follows, taken from the *Environment Reporter*, dated November 2, 1990.

Title I: Urban Air Quality

Title I of the bill, on attainment of federal air quality standards, significantly changes the procedure for urban areas to come into compliance with ozone and carbon monoxide limits. The basic compliance process, however, remains unchanged. States will continue to submit plans to EPA that are designed to bring all non-attainment areas into compliance within specific timeframes.

The principal goal is to reduce ozone pollution. Carbon monoxide is expected to be a long-term problem in only a few cities; particulate pollution likewise is not expected to be a widespread problem. The ozone standard is 0.12 parts per million, measured as a one-hour average.

The bill divides areas into five classes by their ozone levels: marginal (0.121 ppm to 0.138 ppm), moderate (0.138 to 0.160), serious (0.160 to 0.180), severe (0.180 to 0.280), and extreme (0.280 and above). These areas, mostly cities or urban counties, will have different deadlines for compliance. Marginal areas have three years, moderate six years, serious nine years, severe 15 years, and extreme 20 years from enactment of the law. Severe areas with ozone levels above 0.190 ppm would have 17 years.

Los Angeles is the only city in the extreme category.

All but marginal areas must reduce emissions of volatile organic compounds by 15 percent within six years of enactment. After that point the areas must reduce VOC emissions by 3 percent annually.

The classifications also determine what control measures an area must adopt. One major new control measure that will be required is vapor recovery equipment on gasoline pump nozzles; all non-attainment areas except those classed as marginal ones must have them installed.

The other major control measure is automobile inspection and maintenance. Many areas that do not now have them must implement them, and severe and serious areas would need to enhance them. The Environmental Protection Agency is required to issue guidance on such programs.

Most ozone reduction strategies have focused on one major class of precursors, volatile organic compounds. Areas also may seek to control ozone by limiting the other major precursors, nitrogen oxides. State authorities must convince EPA, however, that the combination of VOC and NOx controls would yield equivalent or greater reductions in ozone than the required VOC reductions alone.

Another key aspect of the ozone control program is a definition of "major" sources. Typically they have been defined as those emitting more than 100 tons per year of an ozone precursor. Under the new Clean Air Act, that definition will remain for marginal and moderate areas, but in serious areas it would drop to 50 tons per year, in severe areas to 25 tons per year, and in extreme areas 10 tons per year.

The definition is important because major sources will be required to install reasonably available control technology, as defined by EPA.

Also important is the treatment of offset requirements. Under previous law, new sources in non-attainment areas would need to find offsets—reductions at other facilities—to make up for their contributions. The new law requires proportionally greater offsets based on the severity of the ozone problem.

In marginal areas new sources must offset emissions at a 1.1-to-1 ratio. In moderate areas the ratio is 1.15-to-1; in serious areas 1.2-to-1; in severe areas 1.3-to-1; and in extreme areas 1.5-to-1.

Carbon Monoxide, Particulates

The bill divides areas failing to meet the carbon monoxide standard into two categories, moderate and serious. Moderate areas, with eight-hour averages between 9.1 parts per million and 16.4 ppm, have five years to meet the standard. Serious areas, with levels above 16.5 ppm, have 10 years.

Carbon monoxide non-attainment areas would be required to adopt or enhance vehicle inspection and maintenance programs. Another requirement for most or all such areas is an oxygenated fuels program.

For both ozone and carbon monoxide, an area within a state may be classified by EPA as part of a multistate nonattainment area. For such areas, EPA may not approve state plans or revisions that fail to bring about compliance for the entire area. States would petition EPA, stating that their inability to meet a standard is due to another state's failure to implement required controls.

For particulates, all non-attainment areas would be classified as moderate, and compliance would be required within six years.

Title II: Mobile Sources

The new mobile source standards essentially take two tracks. One track is a tightening of emission standards. The second track is an ambitious attempt to force the fuel industry and automakers to produce alternative fuels and cars that run on them.

The tailpipe emission standards are divided into two tiers. Tier I standards for cars and light trucks are phased in between 1994 and 1998. The second tier of reductions would take effect in 2003 only if EPA decides it is necessary.

Under Tier I, the automobile hydrocarbon emission standard is 30 percent lower than the current standard, and the nitrogen oxides standard is 60 percent lower. Forty percent of manufacturers' automobiles must meet the standards in the 1994 model year. In 1995, 80 percent of the vehicles must comply, and in 1996 all vehicles must comply.

The Tier II standards would be roughly twice as stringent as those in Tier I.

Light trucks weighing less than 6,000 pounds are subject to a particulate standard of 0.08 gram per mile for a useful life of five years or 50,000 miles. A standard of 0.10 gram per mile would apply over 10 years or 100,000 miles.

Standards for trucks weighing more than 6,000 pounds similarly are tightened for hydrocarbons, carbon monoxide, nitrogen oxides, and particulates.

The bill also includes a requirement for automakers to control carbon monoxide emissions from automobiles during cold weather. The percentages of vehicles required to meet the standard are phased in from 1994 to 1996 as are the other automobile standards.

Within one year of enactment, EPA is required to issue regulations requiring installation in automobile fuel systems of equipment to trap evaporative emissions from refueling. EPA must first consult the Department of Transportation regarding the safety of such equipment.

The bill requires sale of reformulated gasoline in the nine cities with the worst ozone pollution. Gasoline marketed in these cities in 1995 must achieve at least a 15 percent reduction in emissions of volatile organic compounds. By 2000, the reduction must be at least 20 percent, and may be increased to 25 percent if EPA determines that it is technically feasible.

Gasoline sold in the winter in areas that are failing the federal carbon monoxide standard must contain at least 2.7 percent oxygen in 1992. This would typically be achieved by adding alcohol or methyl tertiary butyl ether. The requirement could be delayed for up to two years if EPA determines there is a supply problem.

In California, a pilot program will be launched that would require production of 150,000 clean-fuel vehicles per year in 1995 through 1998 and 300,000 such vehicles in 1999 and thereafter. California would need to ensure the availability of fuels for these vehicles.

Another major provision on clean-fuel vehicles is a requirement that EPA set standards for vehicle fleets. In 25 cities with air pollution problems, 70 percent of cars and light trucks and 50 percent of heavy trucks added to fleets must meet the California clean-fuel vehicle requirements.

California already has tighter motor vehicle standards than the rest of the country, and states would have authority to adopt those standards. States that so choose, however, would be restricted in the other types of controls they could impose.

Title III: Toxic Air Pollutants

The essence of the bill's toxic air pollution title is the list of 189 chemicals that EPA must regulate. The agency's task is to determine "maximum achievable control technology" for these substances.

EPA may add or delete chemicals from the list, and private parties may petition the agency on additions and deletions. The agency has 10 years to issue MACT standards for all sources of the 189 chemicals.

Sources that emit more than 10 tons per year of any substance or 25 tons per year of any combination of listed chemicals are covered by this title.

The MACT standards will differ for new and existing sources. MACT for new sources would be the most stringent level currently achieved with a similar source. MACT for existing sources would be the average control levels for the best 12 percent of similar sources. For any category with fewer than 30 sources, MACT would be the average of the five best-performing sources.

Any source making a 90 percent reduction of a listed chemical will have a six-year extension in the deadline for achieving the MACT standard.

Six years from enactment EPA must complete a study and report to Congress on the level of risk remaining after the MACT reductions are in place. If Congress does not act on these recommendations, EPA must develop health-based standards that would limit the cancer risk to exposed individuals to no more than 1 case in 1 million.

The bill also requires EPA to issue new source performance standards for municipal trash incinerators, and to conduct a study on toxic pollution in the Great Lakes, Chesapeake Bay, and other waters.

Title IV: Acid Rain Control

The basis of the acid rain control program is a system of sulfur dioxide emission allowances, which can be banked or sold by emitters. The bill establishes a system whereby EPA will issue allowances to existing sources. The allowances limit sulfur dioxide emissions to 8.9 million tons annually by 2000. An allowance is equivalent to 1 ton of sulfur dioxide emissions.

An initial phase of reductions will require the 111 highest-emitting plants to meet a standard of 2.5 pounds of sulfur dioxide per million British thermal units of heat input by 1995. Those sources are listed in the bill along with their Phase I allowances. Plants that use traditional scrubbers to meet the standard will have until 1997.

EPA is authorized to create a reserve of up to 3.5 million allowances, and these are intended to cover the utilities that postpone the compliance date by using scrubbers.

In Phase II, utilities must meet a standard of 1.2 pounds of sulfur dioxide per million Btu of heat input—equivalent to the current standard for new sources. The deadline for compliance is 2000. Total nationwide sulfur dioxide emissions would be capped at 8.9 million tons annually, although EPA will have a reserve of 530,000 extra allowances from 2000 to 2009.

EPA also is required, by 1993, to develop utility emission standards for nitrogen oxides. These would apply to plants that are subject to sulfur dioxide controls. The agency must issue standards for other types of coal-fired boilers by 1997.

Another provision authorizes EPA to give allowances to high-growth states—those that experienced population growth of 25 percent or more between 1980 and 1989. Up to 40,000 allowances may be provided under this clause.

A key part of the acid rain compromise is a program of assistance to displaced coal miners and other workers laid off as a result of the legislation. It would provide up to $250 million over five years for extended unemployment assistance, with a focus on retraining. To remain eligible, workers must enroll in a retraining program by the end of the 13th week of unemployment.

Title V: Permits

The bill establishes a new permit system in which sources will need permits that contain all applicable emission control requirements. EPA has one year to issue permit regulations, and the program is slated to go into effect four years from enactment.

Emitters defined as major sources, those subject to acid rain controls, those subject to new source performance requirements, and plants emitting hazardous air pollutants are required to obtain these permits.

States have three years to submit permit enforcement programs, which must meet minimum criteria in the legislation. EPA has authority to object to a state-issued permit if it conflicts with Air Act requirements. The permitting authority has 90 days to revise the permit; otherwise EPA must issue a final ruling on the permit.

The program includes a "permit shield" that is a presumption that a source is in compliance with the Air Act if it meets all of the requirements in its permit.

Another key provision adopted in conference sets up a technical assistance program for small businesses, to help them comply with the permit requirements. States also are authorized to reduce permit fees for small businesses.

Title VI: Stratospheric Ozone Depletion

The title requires a production phase-out of the five most destructive ozone-depleting chemicals by 2000, along with three halons and carbon tetrachloride. It also would ban methyl chloroform production by 2002.

For the less destructive hydrochlorofluorocarbons, production would be frozen in 2015, and new uses would be limited. Production would be phased out entirely by 2030.

EPA is required to issue regulations on safe use, recycling, and disposal, as well as rules establishing criteria for development of safe alternatives to ozone-depleting substances.

Title VII: Enforcement

The enforcement title generally increases EPA and state authority to impose civil and criminal penalties for violations of the Air Act. It allows EPA to take action more readily when sources violate state requirements of permits, particularly when states fail to act.

Agency officials have new authority to issue "field citations" that would not need to undergo formal review. The agency also could issue administrative penalties of $25,000 per day.

Criminal enforcement powers at EPA also are expanded, as the bill provides for criminal penalties for "criminal endangerment." This generally would apply to the knowing or negligent release of a hazardous or toxic air pollutant that endangers an individual.

The conferees adopted a clause specifying that an exception for non-senior management does not cover instances in which individuals knowingly commit violations.

Title VIII: Miscellaneous Provisions

A major transfer of authority contained in the bill make the EPA administrator responsible for setting emission standards for outer continental shelf oil and gas exploration and production. It previously was the responsibility of the interior secretary.

Various studies are authorized, including an analysis of renewable energy and energy conservation incentives. Studies also would be conducted on air quality in New Mexico and along the U.S.-Mexico border.

Provisions to require control of visibility problems in national parks were deleted from the bill, but Title VIII sets in motion various studies on visibility. This is an issue primarily in the West.

Title IX of the bill provides for various research activities related to clean air, including continuation of the National Acid Precipitation Assessment Program, which was slated to end in 1990.

6 SUMMARY

We recognize that air pollution is unpleasant and undesirable; however, we must also recognize that this earth's inhabitants have existed and multiplied through many centuries, in contaminated atmospheres, despite the sensitivity of certain individuals to pollens or unmeasurable trace quantities of some substances such as isocyanates. Billions of dollars have been spent and hundreds of person-years have been devoted to studying the levels that people and plant life can tolerate without adverse effects. Much of this work was summarized in developing the six national ambient air standards discussed earlier in this chapter. Implementation of the 1970 Clean Air Act has cost about $35 billion annually, costs borne by the consumers. The 1990 act will add an additional $20 billion to that figure.

That there is a dose–response relationship between air contaminants and human health has been established. In industrial hygiene, the standards established are to protect normal, healthy working people, and as a result, they are less stringent than are the standards for community air pollution, where the very young, the sick, those with allergies, and the aged must be considered. In order to protect the entire population, the dose–response value of any contaminant, or a combination of contaminants, would have to reach essentially *zero*.

It should be noted that the goal of the 1990 Clean Air Act is to determine "maximum achievable control technology" whereas previous acts have been based on health effects.

Finally, let us acknowledge that air pollution is but one factor in the total health and well-being of a community or the nation, and this factor, air pollution, is not necessarily the overriding, all-encompassing facet.

REFERENCES

1. *Meuse Valley Air Pollution*, Royal Academy of Medicine of Belgium, December 19, 1931, pp. 683–732.
2. H. H. Schrenk, H. Heimann, G. D. Clayton, and W. N. Gafafer, "Air Pollution in Donora, Pennsylvania," Public Health Bulletin No. 306, U.S. Government Printing Office, Washington, DC., 1949.
3. L. C. McCabe and G. D. Clayton, "Air Pollution by Hydrogen Sulfide in Poza Rica, Mexico," *Arch. Ind. Hyg. Occup. Med.*, **6**, 199–213 (1952).
4. Ministry of Health, "Mortality and Morbidity During the London Fog of December 1952," Reports on Public Health and Related Subjects, Her Majesty's Stationery Office, London, 1954.
5. Y. K. Tyagi and A. Rosencranz, "Some International Law Aspects of the Bhopal Disaster," *Soc. Sci. Med.*, **27**(10), 1105–1112 (1988).
6. L. Greenburg et al., "Report of an Air Pollution Incident in New York City, November 1953," *Public Health Rep.*, **77**, 7 (1962).
7. J. H. Stebbings, D. G. Fogleman, K. E. McClain, and M. C. Townsend, "Effect of the Pittsburgh Air Pollution Episode upon Pulmonary Function in Schoolchildren," *J. Air Pollut. Control Assoc.*, **26**, 6 (1976).
8. J. Nania, and T. E. Bruya, "In the Wake of Mount St. Helens," *Ann. Emerg. Med.*, April 1982, **11**(4), 184–191.
9. P. J. Baxter, M. Kapela, and D. Mfonfu, "Medical Effects of Large Scale Emission of Carbon Dioxide," *Brt. Med. J.*, May 27, 1989.
10. *Wall Street Journal*, Editorial, July 13, 1990.
11. Sixth annual report of the Council on Environmental Quality, U.S. Government Printing Office, Washington, DC, No. 040-000-00337-1, December 1975.
12. "Air Quality Criteria and Guides for Urban Air Pollutants," Report of a WHO Expert Committee, World Health Organization Technical Report Service, No. 506, Geneva, 1972.
13. "Sulphur Dioxide and Suspended Particulate Matter in Urban Environments," Health

Protection Directorate, Commission des Communautés Européennes, Brussels, Belgium, January 1976.

14. "The Environmental Protection Agency's Research Program with Primary Emphasis on the Community Health and Environmental Surveillance System (CHESS): An Investigative Report," prepared for the Committee on Science and Technology, U.S. House of Representatives, 94th Congress, November 1976, U.S. Government Printing Office, Washington, DC.

15. F. H. Meyers and C. H. Hine, "Some Experiences of NO_2 in Animals and Man," paper presented at the Fifth Air Pollution Medical Research Conference, Los Angeles, 1961.

16. T. R. Carson, M. S. Rosenholtz, F. T. Wilinski, and M. H. Weeks, *Am. Ind. Hyg. Assoc. J.*, **23**, 457 (1962).

17. M. R. Purvis and R. Ehrlich, *J. Infect. Dis.*, **113**, 72 (1963).

18. H. G. Boren, *Arch. Environ. Health*, **8**, 119 (1964).

19. K. C. Back, *Proc. 1st Ann. Conf. Atmospheric Contamination Confined Spaces*, Wright-Patterson Air Force Base, Ohio, 1965.

20. W. D. Wagner, B. R. Duncan, P. G. Wright, and H. E. Stokinger, *Arch. Environ. Health*, **10**, 455 (1965).

21. O. J. Balchum, R. D. Buckley, R. Sherwin, and M. Gardner, *Arch. Environ. Health*, **10**, 274 (1965).

22. G. Freeman et al., *Arch. Environ. Health*, **17**, 181 (1968); **18**, 609 (1969).

23. J. H. Riddick, Jr., K. I. Campbell, and D. L. Coffin, *Am. J. Clin. Pathol.*, **49**, 239 (1968).

24. R. D. Buckley and C. G. Loosli, *Arch. Environ. Health*, **18**, 588 (1969).

25. C. M. Shy et al., *J. Air Pollut. Control Assoc.*, **20**, 582 (1970).

26. C. M. Shy et al., *J. Air Pollut. Control Assoc.*, **20**, 539 (1970).

27. M. C. S. Kennedy, *Ann. Occup. Hyg.*, **15**, 285 (1972).

28. G. D. Von Nieding, H. Kreckler, R. Tuchs, H. M. Wagner, and K. Koppenhagen, *Int. Arch. Arbeitsmed.*, **31**, 61 (1973).

29. C. M. Shy, L. Niemeyer, L. Truppi, and T. English, "Re-evaluation of the Chattanooga School Children Studies and the Health Criteria for NO_2 Exposure," In-House Technical Report, National Environmental Research Center, Research Triangle Park, NC, March 1973.

30. F. E. Speizer and B. G. Ferris, Jr., *Arch. Environ. Health*, **26**, 313, 325 (1973).

31. E. C. Arner and R. A. Rhoades, *Arch. Environ. Health*, **26**, 156 (1973).

32. J. H. Schulte, *Arch. Environ. Health*, **7**, 524 (1963).

33. J. R. Goldsmith, J. Terzaghi, and J. D. Hackney, *Arch. Environ. Health*, **7**, 647 (1963).

34. R. R. Beard and G. A. Wertheim, *Am. J. Pub. Health*, **57**, 2012 (1967).

35. National Academy of Sciences and National Academy of Engineering, "Effects of Chronic Exposure to Low Levels of Carbon Monoxide on Human Health, Behavior and Performance," NAS-NAE, Washington, DC, 1969.

36. A. D. L. Guest, C. Duncan, and P. J. Lawther, *Ergonomics*, **13**, 587 (1970).

37. D. Rondia, *C. R. Acad. Sci. (D)*, **271**, 617 (1970).

38. J. J. McGrath and J. Jaeger, *Respir. Physiol.*, **12**, 46 (1971).

39. R. A. Jones, J. A. Strickland, J. A. Stunkard, and J. Siegel, *Toxicol. Appl. Pharmacol.*, **19**, 46 (1971).

40. Astrup et al., *Lancet*, **2**, 1220 (1972).

41. S. M. Ayres, R. Evans, D. Licht, J. Griesbach, F. Reimold, E. F. Ferrand, and A. Criscitiello, *Arch. Environ. Health*, **27**, 168 (1973).

42. R. D. Stewart, P. E. Newton, M. J. Hosko, and J. E. Peterson, *Arch. Environ. Health*, **27**, 155 (1973).

43. D. A. Debias, C. M. Banerjee, N. C. Birkhead, W. V. Harrer, and L. A. Kazal, *Arch. Environ. Health*, **27**, 161 (1973).

44. H. Hattori, *J. Sulphuric Acid Assoc.*, Tokyo, **24**, 13 (1971).

45. W. M. Busey, "Summary Report: Study of Synergistic Effects of Certain Airborne Systems in Cynomolgus," Hazelton Laboratories, Inc., Vienna, Va., Coordinating Research Project CAPM-6-68-5, June 1972.

46. T. J. Preziosi, "An Experimental Investigation in Animals of the Functional and Morphological Effects of Single and Repeated Exposures to High and Low Concentrations of CO," Preprint, New York Academy of Sciences, New York, 1970.

47. W. Bender, M. Goethert, G. Malorny, and P. Sebbesse, *Arch. Toxicol. (Berlin)*, **27**, 142 (1971).

48. D. Henschler, A. Stier, H. Beck, and W. Neumann, *Arch. Gewerbepathol. Gerwerbehyg.*, **17**, 547 (1960).

49. C. Schoettlin and E. Landau, *U.S. Public Health Rep.*, **76**, 545 (1961).

50. J. M. Lagerwerff, *Aerosp. Med.*, **34**, 479 (1963).

51. W. A. Young and D. B. Shaw, *J. Appl. Physiol.*, **19**, 765 (1964).

52. W. Y. Hallett, *Arch. Environ. Health*, **10**, 295 (1965).

53. R. F. Bils and J. C. Romanovsky, *Arch. Environ. Health*, **14**, 844 (1967).

54. D. L. Coffin et al., *Arch. Environ. Health*, **16**, 633 (1968).

55. Eiji Yokoyama, *Jap. J. Ind. Health*, **11**, 563 (1969).

56. E. Goldstein, W. Tyler, P. Hoeprich, and C. Eagle, *Arch. Intern. Med.*, **128**, 1099 (1971).

57. M. Evans, W. Mayr, T. Bils, and C. Loosli, *Arch. Environ. Health*, **22**, 450 (1971).

58. D. V. Bates, G. M. Bell, C. D. Burnham, M. Hazucha, J. Mantha, L. D. Pengelly, and F. Silverman, *J. Appl. Physiol.*, **32**, 176 (1972).

59. A. Pan, J. Beland, and Z. Jegier, *Arch. Environ. Health*, **24**, 229 (1972).

60. I. Mizoguchi, M. Osawa, Y. Sato, K. Makino, and H. Yagyu, *J. Japan. Soc. Air Pollut.*, **8**, 414 (1973).

61. E. Yokoyama, "The Effects of Low Concentration of Ozone on the Lung Pressure of Rabbits and Rats," Preprint, Japan Society of Industrial Hygiene 1973, pp. 134–135.

62. D. Bates, "Hydrocarbons and Oxidants," Clinical Studies, paper presented at Conference on Health Effects of Air Pollutants, National Academy of Sciences, Washington, DC, October 4, 1973.

63. M. Hazucha, F. Silverman, C. Parent, S. Fields, and D. V. Bates, *Arch. Environ. Health*, **27**, 183 (1973).

64. D. I. Hammer, V. Hasselblad, B. Portnoy, and P. F. Wehrle, *Arch. Environ. Health*, **28**, 255 (1974).

65. J. D. Hackney, "Physiological Effects of Air Pollutants in Humans Subjected to Secondary Stress," Final Report January 1, 1973–June 30, 1974, State of California Air Resources Board Contract No. ARB 2-372.

66. V. M. Sim and R. E. Pattle, *J. Am. Med. Assoc.*, **165**, 1908 (1957).

67. V. P. Melekhina, "Maximum Permissible Concentration of Formaldehyde in Atmospheric Air," in *Literature on Air Pollution and Related Occupational Diseases*, Vol. 3, B. S. Levine, Transl., Public Health Service, Washington, DC, No. TT 60 21475, pp. 135–140.

68. V. P. Melekhina, "Hygienic Evaluation of Formaldehyde as an Atmospheric Air Pollutant," in *Literature on Air Pollution and Related Occupational Diseases*, Vol. 9, B. S. Levine, Transl. (U.S.S.R.), Public Health Service, Washington, DC, 1963–1964, pp. 9–18.

69. C. H. Hine, M. J. Hogan, W. K. McEwen, F. H. Meyers, S. R. Mettier, and H. K. Boyer, *J. Air Pollut. Control Assoc.*, **10**, 17 (1960).

70. S. R. Mettier, H. K. Boyer, C. H. Hine, and W. K. McEwen, *AMA Arch. Ind. Health*, **21**, 1 (1960).

71. M. O. Amdur, *Int. J. Air Pollut.*, **3**, 20 (1960).

72. E. E. Morrill, Jr., *Air Cond., Heat., Vent.*, **58**, 94 (1961).

73. N. A. Renzetti and R. J. Bryan, *J. Air Pollut. Control Assoc.*, **11**, 421 (1961).

74. H. Salem and H. Cullumbine, *Toxicol. Appl. Pharmacol.*, **2**, 183 (1960).

75. C. W. Smith, Ed., *Acrolein*, Wiley, New York, 1962.

76. S. D. Murphy, D. A. Klingshirn, and C. E. Ulrich, *J. Pharmacol. Exp. Ther.*, **141**, 79 (1963).

77. S. D. Murphy, H. V. Davis, and V. L. Zaratzian, *Toxicol. Appl. Pharmacol.*, **6**, 520 (1964).

78. C. S. Tuesday, B. A. D'Alleva, J. M. Huess, and G. J. Nebel, "The General Motors Smog Chamber," Research Publication No. GMR-490, General Motors Corp., Warren, Mich., 1965.

79. P. Catilina, L. Thieholt, and J. Champeix, *Arch. Mal. Prof.*, **27**, 857 (1966).

80. M. Gusev, A. I. Svechnikova, I. S. Dronov, M. D. Grebenskova, and A. I. Golovina, *Hyg. Sanit.* **31**, 8 (1966).

81. J. C. Romanovsky, R. M. Ingels, and R. J. Gordon, *J. Air Pollut. Control Assoc.*, **17**, 454 (1967).

82. J. M. Peters, R. L. H. Murphy, L. D. Pagnotto, and W. F. Van Ganse, *Arch. Environ. Health*, **16**, 642 (1968).

83. Department of Health, Education and Welfare, Public Health Service, Environmental Health Service, National Air Pollution Control Administration. *Air Quality Criteria For Hydrocarbons*. NAPCA Publication No. AP-64, U.S. Government Printing Office, Washington, DC, 1970.

84. National Academy of Sciences, *Particulate Polycyclic Organic Matter*, in the series *Biologic Effects of Atmospheric Pollutants*, NAS, Washington, DC, 1972.

85. Ahmad and T. C. Whitson, *Ind. Med. Surg.*, September 1973.

86. M. Absoy, S. Erdem, K. Dincol, T. Hepyüksel, and G. Dincol, *Blut Band*, **8**, 293 (1974).

87. National Academy of Sciences, *Vapor-Phase Organic Pollutants: Volatile Hydrocarbons*

and Oxidation Products, in the series *Medical and Biologic Effects of Environmental Pollutants*, NAS, Washington, DC, 1976.

88. A. J. Wicken and S. F. Buck, "Report on a Study of Environmental Factors Associated with Lung Cancer and Bronchitis Mortality in Areas of North East England," Research Paper No. 8, Tobacco Research Council, London, 1964.

89. J. W. B. Douglas and R. E. Waller, *Br. J. Prev. Soc. Med.*, **20**, 1 (1966).

90. P. E. Joosting, *Ingenieur*, **79**, 50, A739 (1967).

91. J. E. Lunn, J. Knowelden, and A. J. Handyside, *Br. J. Prev. Soc. Med.*, **21**, 7 (1967).

92. L. J. Brasser, P. E. Joosting, and D. van Zuilen, "Sulfur Dioxide—To What Level Is it Acceptable?" Research Institute for Public Health Engineering, Report No. G-300, Delft, Netherlands, July 1967.

93. C. M. Fletcher, C. M. Tinker, I. D. Hill, and F. E. Speizer, "A Five-Year Prospective Field Study of Early Obstructive Airway Disease," Current Research in Chronic Respiratory Disease," *Proceedings of the Eleventh Aspen Conference*, Department of Health, Education and Welfare, Public Health Service, Washington, DC, 1968.

94. M. C. Battigelli, *J. Occup. Med.*, **10**:9, 500 (1968).

95. B. W. Carnow et al., *Arch. Environ. Health*, **16**, 768 (1969).

96. B. W. Carnow, R. M. Senior, R. Karsh, S. Wesler, and L. V. Avioli, *J. Am. Med. Assoc.*, **214**(5), 894 (1970).

97. Y. Alarie, C. E. Ulrich, W. M. Busey, H. E. Swann, and H. N. MacFarland, *Arch. Environ. Health*, **21**, 769 (1970).

98. I. P. Goldring, L. Greenburg, S. S. Park, and I. M. Ratner, *Arch. Environ. Health*, **21**, 32 (1970).

99. Committee on the Challenges of Modern Society NATO *Air Quality Criteria For Sulfur Oxides*, No. 7, November 1971.

100. M. O. Amdur, *Arch. Environ. Health*, **23**, 459 (1971).

101. F. W. Weir and P. A. Bromberg, "Further Investigation of the Effects of Sulfur Dioxide on Human Subjects," American Petroleum Institute, Project No. CAW C S-15, Washington, D.C., June 1972.

102. P. J. Lawther, R. E. Waller, and M. Henderson, *Thorax*, **25**, 525 (1970).

103. J. H. Stebbings and C. G. Hayes, "Frequency and Severity of Cardiopulmonary Symptoms in Adult Panels: 1971–1972 New York Studies," National Environmental Research Center, Environmental Protection Agency, Research Triangle Park, NC, August 1974.

104. D. P. Rall, *Environ. Health Perspect.*, **8**, 97 (1974).

105. *Health Consequences of Sulfur Oxides: A Report from CHESS, 1970–1971*. U.S. Environmental Protection Agency, Publication No. EPA-650/1-74-004 Research Triangle Park, NC, 1974.

106. I. Anderson, G. R. Lundquist, P. L. Jensen, and D. F. Proctor, *Arch. Environ. Health*, **28**, 31 (1974).

107. D. I. Hammer, F. J. Miller, A. G. Stead, and C. G. Hayes, "Acute Lower Respiratory Disease in Children in Relation to Sulfur Dioxide and Particulate Air Pollution: Retrospective Survey in New York City, 1972," Human Studies Laboratory, National Environmental Research Center, EPA, Research Triangle Park, NC, November 1975.

108. S. F. Buck and D. A. Brown, "Mortality from Lung Cancer and Bronchitis in Relation to Smoke and Sulfur Dioxide Concentration, Population Density, and Social Index," Research Paper No. 7, Tobacco Research Council, London, 1964.

109. A. E. Martin, *Proc. Roy. Soc. Med.*, **57**, 969 (1964).

110. "Report on an Enquiry into the Incidence of Incapacity for Work. Part II. Incidence of Incapacity for Work in Different Areas and Occupations," Ministry of Pensions and National Insurance, Her Majesty's Stationery Office, London, 1965.

111. W. W. Holland, D. D. Reid, R. Seltser, and R. W. Stone, *Arch. Environ. Health*, **10**, 338 (1965).

112. J. W. Douglas and R. E. Waller, *Br. J. Prevent. Soc. Med.*, **20**, 1 (1966).

113. M. Glasser, L. Greenburg, and F. Field, *Arch. Environ. Health*, **15**, 684 (1967).

114. W. Winkelstein, *Arch. Environ. Health*, **14**, 162 (1967).

115. W. Winkelstein, *Arch. Environ. Health*, **16**, 401 (1968).

116. W. Winkelstein, *Arch. Environ. Health*, **18**, 544 (1969).

117. M. C. Battigelli, Air Quality Monograph No. 69-2, American Petroleum Institute, New York, 1969.

118. P. J. Lawther, R. E. Waller, and M. Henderson, *Thorax*, **25**, 525 (1970).

119. Committee on the Challenges of Modern Society/NATO, "Air Quality Criteria for Particulate Matter," No. 8, November 1971.

120. C. J. Nelson, C. M. Shy, T. English, C. R. Sharp, R. Andleman, L. Truppi, and J. Van Bruggen, *J. Air Pollut. Control Assoc.*, **23**:2, 81 (1973).

121. C. M. Shy, V. Hasselblad, R. M. Burton, C. J. Nelson, and A. A. Cohen, *Arch. Environ. Health*, **27**, 124 (1973).

122. R. S. Chapman, C. M. Shy, J. F. Finklea, D. E. House, H. E. Goldberg, and C. G. Hayes, *Arch. Environ. Health*, **27**, 138 (1973).

123. J. G. French, G. Lowrimore, W. C. Nelson, J. F. Finklea, T. English, and M. Hertz, *Arch. Environ. Health*, **27**, 129 (1973).

124. D. I. Hammer, F. J. Miller, D. E. House, K. E. McClain, E. Tompkins, and C. G. Hayes, "Frequency of Acute Lower Respiratory Disease in Children; Retrospective Survey of Two S.E. Communities, 1968–1971, In-house technical report, U.S. Environmental Protection Agency, May 8, 1974.

125. W. Galke and D. House, "Prevalence of Chronic Respiratory Disease Symptoms in Adults: 1971–1972 Survey of Two S.E. United States Communities," In-house technical report, EPA, October 18, 1974.

126. W. O. Negherbon, "Sulfur Dioxide, Sulfur Trioxide, Sulfuric Acid and Fly Ash: Their Nature and Their Role in Air Pollution," Edison Electric Institute, EEI Publication No. 66-900, New York, 1966.

127. D. I. Hammer, F. J. Miller, A. G. Stead, and C. G. Hayes, "Air Pollution and Childhood Lower Respiratory Disease: Exposure to Sulfur Oxides and Particulate Matter in New York, 1972," paper presented at the American Medical Association Air Pollution Medical Research Conference, San Francisco, December 1974.

128. U.S. Department of Health, Education and Welfare, National Air Pollution Control Administration, Publication No. AP-63, 1970.

129. W. S. Wayne, P. F. Carroll, and R. E. Carroll, *J. Am. Med. Assoc.*, **199**, 901 (1967).

130. L. Smith, *Am. J. Publ. Health*, **55**, 1460 (1965).

131. R. D. Stewart, J. E. Peterson, E. D. Baretta, R. T. Bachand, M. J. Hosko, and A. Herrman, *Arch. Environ. Health*, **21**, 154 (1970).

132. M. J. Hosko, *Arch. Environ. Health*, **21**, 174 (1970).

133. S. M. Ayres, S. Giannelli, Jr., and H. Mueller, *Ann. N.Y. Acad. Sci.*, **174**, 268 (1970).

134. Aronow and Isbell, *Ann. Intern. Med.*, **79**, 392 (1973).

135. Anderson, Andelman, Strauch, Fortuin, and Knelson, *Ann. Intern. Med.*, **79**, 46 (1973).

136. Aronow, Stemmer, and Isbell, *Circulation*, **49**, 415 (1974).

137. R. B. Chevalier, R. A. Krumholz, and J. C. Ross, *J. Am. Med. Assoc.*, **198**, 1061 (1966).

138. R. L. Masters, "Air Pollution—Human Health Effects," in *Introduction to the Scientific Study of Atmospheric Pollution*, McCormac, Ed., D. Reidel, Dordrecht, Holland, 1971, pp. 97–130.

CHAPTER THIRTY

Hazardous Wastes

Lisa K. Simkins, P.E., C.I.H., Shri Nandan, P.E., and
Richard P. Thompson

1 INTRODUCTION

Industrial hygienists have traditionally concentrated their efforts on the protection of human health through the recognition, evaluation, and control of health hazards. A typical industrial hygiene problem once involved an employee exposure to a toxic chemical in an industrial environment. Increasingly, an industrial hygienist's role has expanded to include considerations that go beyond worker exposure. When hazardous materials are used in the workplace, the problem of waste is almost always an issue. The question of when a hazardous material becomes a hazardous waste is often difficult to answer, as is the question of when the industrial hygienist's role ends and the environmental engineer's role begins. Many professionals are now playing a dual role of industrial hygienist and environmental engineer. Thus the industrial hygiene professional must be knowledgeable of hazardous waste requirements and waste management issues.

This chapter provides an introduction and an overview of hazardous waste management. The topics include major regulatory requirements pertaining to hazardous waste, waste characterizations, management options, and health and safety considerations. Hazardous waste regulations proliferated during the 1980s, and are likely to continue to do so in the 1990s. Owing to the vast number of local, state, and federal regulations pertaining to this subject, the practicing professional must always consult the latest regulations before embarking on a hazardous waste man-

Patty's Industrial Hygiene and Toxicology, Fourth Edition, Volume 1, Part B, Edited by George D. Clayton and Florence E. Clayton.
ISBN 0-471-50196-4 © 1991 John Wiley & Sons, Inc.

agement program. This chapter can be viewed as a starting point that provides basic background information for the industrial hygienist or environmental professional involved in one or more facets of hazardous waste management.

1.1 Definition of Hazardous Waste

The standard definition of hazardous waste in the United States comes from the Resource Conservation and Recovery Act (RCRA), which describes it as "a solid waste, or combination of solid wastes, which because of its quantity, concentration, or physical, chemical, or infectious characteristics may (1) cause, or significantly contribute to, an increase in mortality or an increase in serious irreversible, or incapacitating reversible, illness, or (2) pose a substantial present or potential hazard to human health or the environment when improperly treated, stored, transported or disposed of, or otherwise managed" (1). Under the broad RCRA definition, solid waste could include semisolids, liquids, and contained gases.

A hazardous waste designation could result if the waste, or one of its components, is present on a federal or state list of hazardous wastes (2). If the waste exhibits certain characteristics, such as ignitability, corrosivity, toxicity, or reactivity, it may also earn a hazardous waste designation. These characterizations of hazardous waste are discussed further in Section 3.2.

Specifically excluded from RCRA's definition of a hazardous waste are (3):

- Domestic sewage or any mixture of domestic sewage and other wastes that pass through a sewer system to a publicly owned treatment works for treatment
- Irrigation return flows
- Industrial wastewater point source discharges subject to regulation under the Clean Water Act (CWA)
- Source, special nuclear, or by-product radioactive materials subject to the Atomic Energy Act (AEA) of 1954
- In situ mining wastes that are not removed from the ground as part of the extraction process
- Household wastes (e.g., garbage, trash, septic tank wastes)
- Municipal resource recovery wastes (e.g., refuse derived fuel)
- Agricultural residues
- Drilling fluids from the exploration, development, or production of crude oil or natural gas
- Samples of solid waste collected for testing to determine its characteristics or composition

These excluded wastes are not discussed in this chapter. Chapter 32, "Environmental Control in the Workplace," addresses management and handling of some of the wastes, such as domestic sewage, that are excluded from discussion here. RCRA regulations are discussed in Section 2.1.

1.2 Regulatory and Legal Liability

Increased regulation and concern for liability have changed industry's perspective and practice relating to hazardous waste over the past 20 years. To begin with, the financial liabilities caused by environmental problems can be significant. For example, the average cleanup cost of a Superfund site is about $25 million. Removing one ordinary underground storage tank can cost upwards of $10,000. Groundwater decontamination programs, involving recovery wells and air stripping towers, run into millions of dollars per year, and are ongoing projects until the groundwater is judged "clean."

Pollution liabilities are strict (meaning that liability is imposed regardless of fault), and also joint and several (meaning that liability can be imposed on any or all responsible parties independent of their relevant contribution to the pollution). There is also a growing trend among the Environmental Protection Agency (EPA) enforcement branch as well as among state and local prosecutors to file criminal charges against company officers for causing environmental pollution and workplace injury or death.

It is the responsibility of generators and handlers to be familiar with regulations governing their operation, and to institute practices necessary to remain in compliance. Industrial activities operated without knowledge of waste management regulations create an unnecessary risk of an environmental catastrophe.

2 REGULATORY REQUIREMENTS

Several environmental laws and regulations that emerged during the past 20 years resulted in new regulatory controls focusing on air and water pollution. More recently, the focus has expanded to include releases of toxic chemicals and disposal of hazardous waste. New and broader definitions of hazardous waste brought many industrial and commercial operations into the regulatory arena.

The cumulative result of these developments is that the need to address environmental issues has become a daily factor in industrial operations and business transactions. Businesses must consider an extraordinary web of regulatory requirements. These range from requirements for storing hazardous waste and associated bans on land disposal of certain wastes, tracked under RCRA, to the recognition of due diligence standards that limit potential liabilities under the Superfund Amendments and Reauthorization Act.

A better understanding of the requirements can be gained with an understanding of the framework of the federal rulemaking system. Environmental laws are developed by Congress and signed into law by the president. These laws mandate that the EPA develop regulations to implement them.

New regulations, as well as modifications to existing regulations, are printed in the *Federal Register*. The Code of Federal Regulations (CFR) is modified annually to take into account changes promulgated in the *Federal Register*. The CFR is divided into 50 titles covering the areas subject to federal regulation; Title 40 deals with protection of the environment.

Table 30.1. Resource Conservation and Recovery Act Section Titles and Descriptions (4)

PART NO.	TITLE	DESCRIPTION
260	Hazardous Waste Management System: General	Definitions, rulemaking, and delisting petitions
261	Identification and Listing of Hazardous Waste	Definitions of solid waste and hazardous waste; exclusions; requirements for conditionally exempt small quantity generators; characteristics of hazardous waste; lists of hazardous wastes
262	Standards Applicable to Generators of Hazardous Waste	Hazardous waste determination; EPA identification numbers; the manifest; pretransport requirements; 90-day accumulation requirements; requirements for 100-1,000 kg/month generators; recordkeeping and reporting
263	Standards Applicable to Transporters of Hazardous Waste	EPA identification numbers; compliance with the manifest system; recordkeeping, hazardous waste discharges during transportation
264	Standards for Owners and Operators of Hazardous Waste Treatment, Storage, and Disposal Facilities	General status (permit) standards that will be used to evaluate Part B permit applications and that will apply to facilities possessing a full RCRA permit
265	Interim Status Standards for Owners and Operators of Hazardous Waste Treatment, Storage, and Disposal Facilities	Operating requirements for existing facilities under interim status, until final disposition of their permit application

2.1 Resource Conservation and Recovery Act

The Resource Conservation and Recovery Act of 1976 (RCRA) marked the beginning of what has become the EPA's most far-reaching and complex regulatory program. RCRA tracks hazardous wastes from "cradle to grave." Generators of hazardous wastes are responsible for analyzing the material, maintaining appropriate records, submitting periodic reports of volumes, and most importantly, submitting reports on offsite shipments for disposal.

It took the EPA four years following the enactment of RCRA to promulgate the basic regulatory framework, which in turn has been amended many times since its initial publication in May 1980. The 1980 regulations established permit requirements for treatment, storage, and disposal (TSD) facilities handling hazardous wastes but granted "interim status" to allow those facilities to operate, if appropriate applications and notices were submitted. Groundwater monitoring and financial responsibility are two of numerous other requirements imposed on such facilities by these regulations.

As discussed in the beginning of this chapter, RCRA provided the standard definition of hazardous waste in the United States. The EPA then had to determine the scope of the regulated community, and what areas of hazardous waste man-

Table 30.1. (*Continued*)

PART NO.	TITLE	DESCRIPTION
266	Standards for the Management of Specific Hazardous Wastes and Specific Types of Hazardous Waste Management Facilities	Special (limited) requirements for certain types of waste being reclaimed or used as fuel. Materials so regulated include hazardous wastes and used oil burned for energy recovery, as well as precious metals and lead-acid batteries being reclaimed
268	Land Disposal Restrictions	Prohibitions on the direct land disposal of untreated hazardous wastes; treatment standards established to determine the acceptability of land disposal; provisions for extensions, petitions, and variances. A schedule is specified for the establishment of treatment standards for all listed and characteristic hazardous wastes
270	EPA-Administered Permit Programs: The Hazardous Waste Permit Program	Contents of the Part A and Part B permit applications; standard permit conditions, permit modifications, duration of permits, interim status
271	Requirements for Authorization of State Hazardous Waste Programs	Requirements for state programs to receive authorization to operate in lieu of EPA

agement were to be subject to these requirements. Table 30.1 briefly describes the RCRA sections found in Title 40 of the CFR.

RCRA's "cradle to grave" approach provides guidelines for generators, transporters, storage facilities, treatment operations, and disposal sites. Each of these waste management activities is subject to specific standards under these regulations.

2.1.1 Generators

A generator is any person, by site, whose act or process produces hazardous waste or whose act first causes a waste to be subject to regulation. Hazardous waste generators are regulated differently from treatment, storage, and disposal facilities, which makes it more difficult to estimate the number of generators that may be subject to provisions in RCRA.

At the most basic level, the requirements mandate that hazardous waste generators (except small-quantity generators):

1. Determine whether their wastes are hazardous.
2. Obtain an EPA identification number.
3. Initiate a manifest document that will follow the waste from cradle to grave.
4. Properly package and label the waste.
5. Maintain records of waste shipments.
6. Report to EPA if a copy of the completed manifest is not received from the designated disposal facility within 45 days of the date the waste was accepted by the initial transporter.

In addition, generators are required to submit biennial reports (every even-numbered year) covering the types and amounts of hazardous wastes shipped offsite. Other requirements include the training of personnel handling wastes, periodic documented inspections of storage areas, and the preparation of an emergency plan.

A generator is allowed to accumulate hazardous waste onsite for 90 days or less without a permit, provided that all of the above conditions are met. A generator who accumulates hazardous waste longer than 90 days is considered an operator of a storage facility, fully subject to requirements in 40 CFR Parts 264 or 265.

Special requirements apply to those who generate hazardous waste in small quantities of less than 100 kg (220 lb) per month. These generators are conditionally exempt small-quantity generators (SQGs) and, as such, qualify for reduced requirements for the management of their wastes. These requirements are found in 40 CFR Part 261.5.

Generators that produce more than 100 kg but less than 1000 kg of hazardous waste per month have more stringent requirements than the conditionally exempt SQGs, but less rigorous requirements than generators subject to the full set of regulations (4).

2.1.2 Transporters

All persons transporting hazardous waste off the site where it was generated must comply with the requirements in 40 CFR, Part 263, developed jointly by the EPA and the Department of Transportation (DOT). Both generator-owned and independent transporters must comply. Waste transporters are required to (5):

1. Obtain an EPA identification number.
2. Comply with the manifest signature requirements.
3. Deliver the waste in accordance with directions on the manifest.
4. Maintain records of the waste shipment.
5. Properly respond to any spills during transportation.

2.1.3 Waste Management Facilities

Any person who treats, stores, or disposes of hazardous waste is considered an owner or operator of a TSD facility and is subject to the requirements of Part 264 or 265. These requirements address general operation of facilities and the technical standards applicable to specific units, such as containers, tanks, surface impoundments, waste piles, land treatment, landfills, incinerators, and thermal treatment.

Other general standards applied to TSD facilities cover three major areas: (1) groundwater monitoring, (2) closure and post-closure processes, and (3) financial accountability.

The basic difference between Part 264 (Final Operating Standards) and Part 265 (Interim Status Standards) is that the interim status standards were written for facilities treating, storing, or disposing of RCRA hazardous waste when the RCRA regulations first went into effect on November 19, 1980. The EPA was required to establish interim standards that would allow facilities to continue to operate as though they had a permit. A facility received interim status by filing a RCRA Section 3010 notification (Notification of Hazardous Waste Activity), and a Part A application. An owner or operator that qualified for and obtained interim status remains subject to the Part 265 standards until the final administrative disposition of the facility's permit application is made. At that point Part 264 (Final Operating Standards) goes into effect.

The general facility standards are applicable to all RCRA hazardous waste management facilities unless specifically excluded. These compliance standards are essentially the same for both parts:

1. Every facility must have an EPA identification number.
2. The owner or operator of a facility receiving hazardous waste from an offsite source must inform the waste generator that the facility has obtained appropriate permits.
3. Before hazardous waste is treated, stored, or disposed of, the owner or operator must obtain a detailed chemical and physical analysis of a representative sample of the waste.
4. The owner or operator must prevent inadvertent and, to the extent possible, deliberate entry of people or livestock into the active portion of the facility.
5. The owner or operator must inspect his facility for malfunctions and deterioration, operator errors, and discharges, according to procedures found on a written inspection retained at the facility.
6. Facility personnel must complete a program of training in RCRA procedures and emergency response. A written description of the training program, attendees, and instructors must be kept at the facility.
7. All permitted facilities must have equipment adequate to deal with unexpected fires, explosions, or any unplanned sudden or nonsudden spill or other release of hazardous waste. Emergency response procedures must be detailed in a contingency plan and must describe the personnel in charge of emergency, emergency equipment, and evacuation procedures.

In addition to these general requirements, other substantive standards apply to specific permitted facilities and types of equipment including:

1. Containers
2. Tanks
3. Surface impoundments
4. Waste piles
5. Land treatment units
6. Landfills
7. Incinerators

Once the facility no longer accepts wastes, a period of closure begins. Operators are required to develop a closure plan that contains a schedule for closure, an estimate of the amount of wastes the facility will handle, a description of the closure process, and an outline of necessary decontamination procedures before receiving a permit to operate. Though the closure plan may be amended during the active life of the facility, once implemented, it must be strictly followed.

Following closure, a 30-year post-closure period commences, during which time the facility must continue groundwater monitoring and any maintenance activities necessary to preserve the integrity of the site. TSD facilities are required to prepare cost estimates and provide financial assurance of ability to implement the closure and post-closure plans (6).

2.2   Hazardous and Solid Waste Amendments

Unhappy with the EPA's progress with RCRA, Congress enacted the Hazardous and Solid Waste Act Amendments of 1984 (HSWA). The amendments imposed deadlines for the EPA to complete issuance of permits to existing TSD facilities, and for surface impoundments to meet stringent technological requirements to prevent leakage into groundwater. HSWA imposed tight requirements on landfills and other hazardous waste management units, and called for regulations to ban the land disposal of certain types of hazardous waste.

HSWA required the EPA to establish levels or methods of treatment which substantially reduce the likelihood of migration of hazardous constituents from the waste so that short-term and long-term threats to human health and the environment are minimized. Wastes that meet treatment standards may be land disposed.

There are a number of land disposal options, all subject to treatment standards and restrictions. These include landfill, surface impoundment, waste pile, land treatment facility, slat dome formation, underground mine or cave, and underground injection wells.

HSWA required the EPA to prepare a schedule for restricting the land disposal of all hazardous waste (excluding solvents, dioxins, and California List wastes covered under a separate schedule set by Congress). The schedule, based on the intrinsic hazard and volume ranking of the listed waste, established the following deadlines for hazardous waste land disposal restrictions:

1. At least one-third of all listed hazardous wastes by August 8, 1988 (First Third)
2. At least two-thirds of all listed hazardous wastes by June 8, 1989 (Second Third)
3. All remaining listed and characteristic hazardous wastes by May 8, 1990 (Third Third)

If the EPA failed to set a treatment standard by the statutory deadline for any hazardous waste in the first or second third of the schedule, the hazardous waste fell under the "soft hammer" provision. This provision allowed the land disposal of soft hammer hazardous waste as long as the generator could certify to the EPA that land disposal was the only practical alternative to treatment currently available, and that such land disposal took place in a landfill or surface impoundment that met minimum technology requirements (double liner and leachate collection system) specified in RCRA Section 3004 (0).

On May 8, 1990, any scheduled hazardous waste without a treatment standard became subject to the "hard hammer," which automatically prohibits land disposal. Exceptions are wastes for which "no migration" can be successfully demonstrated, those that have received a national capacity extension, or case-by-case extensions beyond the hard hammer date.

Treatment standards are based on the performance of best demonstrated available technology (BDAT) to treat the waste. The EPA may establish treatment standards either as (1) numerical standards (concentration limits) based on the performance of BDAT or (2) specific technologies (no defined concentration limits). The EPA is proposing a number of technologies as BDAT or required technology. These include, but are not limited to, the following:

- Thermal destruction
- Incineration
- Fuel substitution
- Recovery
- Carbon absorption
- Liquid–liquid extraction
- Steam stripping/carbon absorption
- Roasting or retorting
- Stabilization
- Deactivation
- Solubilization

- Encapsulation
- Thermal recovery
- Resmelting
- Recycling
- Wet air oxidation
- Chemical oxidation
- Alkaline chlorination
- Acid or water leaching
- Chemical precipitation
- Neutralization

The EPA estimates that the greatest volume of wastes covered in the proposed third third rule will be treated by precipitation or stabilization.

2.3 Comprehensive Environmental Response, Compensation, and Liability Act

The enactment of the Comprehensive Environmental Response, Compensation, and Liability Act of 1980, better known as CERCLA or Superfund, brought environmental concerns to the forefront for virtually every segment of American industry. By passing CERCLA, Congress authorized funds for the EPA to clean up abandoned dumps and other contaminated waste disposal sites.

The statute imposed strict, joint, and several liability for the EPA's cleanup costs on past and present site owners, past and present site operators, offsite generators who had arranged for disposal of hazardous wastes, and transporters who selected the site in question. In an effort to prevent future Superfund sites, Congress also imposed broad new reporting requirements for releases of hazardous substances into the environment.

Because the Superfund law is not fundamentally a regulatory statute, the primary obligation it imposes is liability for cleanup costs. The EPA's National Priority List (NPL) includes approximately 1200 sites and is growing rapidly. Most of these sites require a full Remedial Investigation/Feasibility Study (RI/FS), which is typically very time consuming and costly. Even higher costs are encountered in the Remedial Design/Remedial Action (RD/RA) phase of site cleanup, with cost estimates currently averaging $25 million per site.

2.4   Superfund Amendments and Reauthorization Act

A 3-year Superfund reauthorization process ended October 17, 1986, when former President Ronald Reagan signed the Superfund Amendments and Reauthorization Act of 1986 (SARA). It provided $8.5 billion over five years to the EPA and other federal agencies for the cleanup of abandoned and inoperative waste sites.

The amendments made major changes to the original Superfund law (CERCLA), including:

- Revisions that added strict cleanup standards strongly favoring permanent remedies at waste sites
- Stronger EPA control over the process of reaching settlement with parties responsible for waste sites
- A mandatory schedule for initiation of cleanup work and studies
- Individual assessments of the potential threat to human health posed by each waste site
- Increased state and public involvement in the cleanup decision-making process, including the right of citizens to file lawsuits for violation of the law

The amended law retained the concept of strict, joint, and several liability, which EPA has found to be its most powerful enforcement tool in inducing responsible parties to clean up waste sites.

During the reauthorization process, one of the key problems confronted was the lack of workable cleanup standards. SARA placed emphasis on remedial actions that permanently and significantly reduce the volume, toxicity, or mobility of the hazardous substances, pollutants, and contaminants. The discovery that sites where Superfund wastes had been dumped were also leaking led to language stating that offsite transport of wastes without treatment "should be the least favored alternative" where practical treatment technologies are available.

One of the major changes in the statute was the advent of schedules for accomplishments under the Superfund program. By 1991, the EPA was required to initiate cleanup activities at a minimum of 375 Superfund sites, a provision Congress included to increase what it saw as the slow pace of the program.

The revised law also contained two new separate sections not directly related to the task of cleaning up Superfund sites, but with important implications for the EPA's overall regulatory authority. A separate title within SARA requires industries that produce, use, or store hazardous chemicals or substances to report the presence of such substances to community authorities and to report routine and unauthorized releases of hazardous substances to the EPA. It also requires communities to improve emergency planning procedures for major chemical accidents.

A second section amended RCRA to require that owners of underground storage tanks take financial responsibility for cleaning up leaks and compensating third parties for property damage and bodily injury. A trust fund was established to pay for emergency cleanups where no responsible owner or operator of the tank could be found.

Finally, Superfund cleanup liability was extended to all owners of contaminated property, regardless of the circumstances of their ownership. CERCLA's liability standard had triggered several problems for property owners who took possession of land without knowledge that it was contaminated. Prompted by the perceived unfairness, Congress included a new defense in the amendments that was designed to protect innocent landowners.

Congress expanded the third-party exception by redefining the contractual relationship of property ownership. Innocent landowners who acquire property without having had reason to know that hazardous substances had ever been disposed of on the land are not liable as owners or operators. The landowner does not qualify as innocent unless he or she has completed a thorough examination of the land purchased.

2.5 Other Regulations

Another major federal regulation pertains to health and safety when handling hazardous wastes. This regulation, the Hazardous Waste Operations and Emergency Response Standard (29 CFR 1910.120), is discussed in Section 4.0.

In addition to the major federal regulations, there are numerous state and local regulations pertaining to hazardous wastes. State regulations vary in their application and are frequently updated. The federal regulations should therefore be considered the minimum standards that will apply and additional regulations should be consulted in each state in which a company operates a facility or disposes of waste.

3 WASTE MANAGEMENT

What began as waste disposal has evolved into waste management as an increasing number of regulations govern what the generator can and cannot do. As the preceding sections indicate, hazardous waste management has become a complex problem.

The first step in the development of a sound hazardous waste management program is a comprehensive inventory of all solid and hazardous wastes generated at a facility. Following an inventory, waste must be characterized. At that point the various management options available to the generator can be evaluated.

3.1 Inventory

A complete list of all waste streams and sources (processes that result in the generation of waste) needs to be created. To the extent known, constituents of individual waste streams must be identified at the outset. This will be helpful in the characterization process, which can be time consuming and expensive.

The discussion in the following sections focuses on the hazardous waste component of all solid wastes. However, *all* solid waste streams should be included in the inventory process. This should result in a clear distinction between hazardous waste regulated under RCRA, Subtitle C, or state laws, and solid wastes (non-hazardous) regulated under RCRA, Subtitle D.

In addition to identifying the chemical nature and source of all waste streams, a quantification must be made. The quantity may be computed on a daily basis, such as gallons per day, or an annual basis, such as tons per year.

3.2 Characterization

Other than the obvious municipal garbage and household debris, prudent strategy is to assume a waste stream is hazardous until tests prove otherwise. The characterization process is accomplished in one of several ways. A waste from a specific process may be designated hazardous by the EPA or a state environmental agency. These are known as listed wastes. A waste that is not listed by these agencies can be characterized through a series of tests.

If a generator produces a listed waste, but there is reason to believe it is not hazardous, the generator may wish to proceed with a process known as "delisting." If the tests show the waste to be nonhazardous, the listing agency is petitioned for an exemption. When considering this option, it's necessary for the generator to weigh the long-term benefit of disposing of the waste as nonhazardous against the cost of testing.

3.2.1 Listed Wastes

In developing the hazardous waste list, the EPA considers waste streams by source and segregates them into hazardous and nonhazardous categories. Those found to be hazardous are designated generically to be so. For instance, if selected drilling muds are assessed as hazardous, then drilling muds, in general, are listed.

The major advantages of this approach are easy implementation and enforcement. Once a designation is made, all wastes are categorized by source, and generators and regulators can quickly determine the status. The cost associated with characterization as hazardous or nonhazardous is relatively low, because the determination is made by referencing a list (7).

Federally listed wastes are found in Appendices VII and VIII of RCRA regulations. Since RCRA, several states have generated their own lists. These lists are revised periodically and it is important to consult the latest listing of hazardous waste in a given state. State lists are developed on the basis of either a "pure compound approach" or the waste-type approach used by the EPA. Table 30.2

Table 30.2. Sample List by Pure Compound Approach

Miscellaneous Inorganics		
Ammonium chromate	Sodium arsenite	Pentachlorophenol
Ammonium dichromate	Sodium bichromate	Picric acid
Antimony pentafluoride	Sodium chromate	Potassium dinitrobenzfuroxan
Antimony trifluoride	Sodium cyanide	(KDNBF)
Arsenic trifluoride	Sodium monofloroacetate	Silver acetylide
Arsenic trioxide	Tetraborane	Silver tetrazene
Cadmium (alloys)	Thallium compounds	Tear gas (CN) (chloroacetophenone)
Cadmium chloride	Zinc arsenate	Tear gas (CS) (2-chlorobenzylidene
Cadmium cyanide	Zinc arsenite	malononitrile
Cadmium nitrate	Zinc cyanide	Tetrazene
Cadmium oxide		VX (ethoxy-methyl phosphoryl-N,N-
Cadmium phosphate		dipropoxy-(2-2)-thiocholine)
Cadmium potassium cyanide	Halogens and Interhalogens	
Cadmium (powdered)		
Cadmium sulfate	Bromine pentafluoride	Organic Halogen Compounds
Calcium arsenate	Chlorine	
Calcium arsenite	Chlorine pentafluoride	Aldrin
Calcium cyanides	Chlorine trifluoride	Chlorinated aromatics
Chromic acid	Fluorine	Chlordane
Copper arsenate	Perchlory fluoride	Copper acetoarsenite
Copper cyanides		2,4-D (2,4-dichlorophenoxy-acetic
Cyanide (ion)		acid)
Decaborane	Miscellaneous Organics	DDD
Diborane		DDT
Hexaborane	Acrolein	Demeton
Hydrazine	Alkyl leads	Dieldrin
Hydrazine azide	Carcinogens (in general)	Endrin
Lead arsenate	Chloropicrin	Ethylene bromide
Lead arsenite	Copper acetylide	Fluorides (organic)
Lead azide	Copper chlorotetrazole	Guthion
Lead cyanide	Cyanuric triazide	Heptachlor
Magnesium arsenite	Diazodinitrophenol (DDNP)	Lindane
Manganese arsenate	Dimethyl sulfate	Methyl bromide
Mercuric chloride	Dinitrobenzene	Methyl chloride
Mercuric cyanide	Dinitro cresols	Methyl parathion
Mercuric diammonium chloride	Dinitrophenol	Parathion
Mercuric nitrate	Dinitrotoluene	Polychlorinated biphenyls (PCB)
Mercuric sulfate	Dipentaerythritol hexanitrate -	
Mercury	(DPEHN)	
Nickel carbonyl	GB (propoxy(2)-methylphosphoryl	
Nickel cyanide	fluoride)	
Pentaborane-9	Gelatinized nitrocellulose (PNC)	
Pentaborane-11	Glycol dinitrate	
Perchloric acid (to 72%)	Gold fulminate	
Phosgene (carbonyl chloride)	Lead 2,4-dinitroresorcinate (LDNR)	
Potassium arsenite	Lead styphnate	
Potassium chromate	Lewisite (2-chloroethenyl	
Potassium cyanide	dichloroarsine)	
Potassium dichromate	Mannitol hexanitrate	
Selenium	Nitroaniline	
Silver azide	Nitrocellulose	
Silver cyanide	Nitrogen mustards (2,2',2"-	
Sodium arsenate	trichlorotriethylamine)	
	Nitroglycerin	
	Organic mercury compounds	

provides a sample list of hazardous compounds. If any of these is present in a waste stream, the stream is designated hazardous.

There are weaknesses in relying solely on the pure compound approach for listing wastes. Assigning properties of a single constituent to a waste mixture fails to account for the interactions between constituents that may significantly alter the hazardous nature of the waste. Common interactions encountered are (7):

- *Additive effects* where constituents operate through similar mechanisms and thereby stress the receptors as if they were the same total quantity of just one of the constituents (e.g., a mixture of hydrocarbons of the same general molecular weight)
- *Synergistic effects* where the total effect of a combination of constituents is

greater than the sum of their individual effects (e.g., the mixture of chlorinated hydrocarbon pesticides and solvents, and cadmium and zinc in water)

- *Antagonistic effects*, the functional opposite of synergistic effects, where the total effect is less than the sum of effects of the constituents (e.g., arsenic and selenium)
- *Chemical interaction effects* where the presence of one constituent modifies the hazard potential of another through direct chemical reaction or modifies the availability of that constituent to the receptor [e.g., acid and alkali in the same solution (neutralization) and sulfate in solution with barium (immobilization)]

Similarly, a mixture of nonhazardous constituents may react to produce a hazardous product in the waste stream. These shortcomings can be resolved only by direct hazard testing of a waste stream.

Recognizing the inherent difficulties in relying solely on a listing approach to designation, most states have gone to alternative systems. One commonly used approach is the characteristics criteria method for designation of wastes.

3.2.2   Characteristics Criteria

This approach is a quantitative one that can be applied directly to wastes, or to chemicals if a pure compound approach is taken. Characteristics are measured by standard testing protocol. When the characteristic exceeds a set threshold value, the material is designated hazardous.

When the hazardous status of a waste is being determined, tests are conducted for the characteristics of ignitability, corrosivity, reactivity, and toxicity. A solid waste that exhibits any of these characteristics is a hazardous waste, whether it is listed or not. For example, a waste could be subjected to a test to determine its flash point (ignitability characteristic). If the waste was determined to have a flash point of 140°F or less, it would be designated hazardous.

The following description of these characteristics is found in 40 CFR, Parts 261.21 through 261.24 (1).

3.2.2.1 Ignitability. A solid waste exhibits the characteristics of ignitability if a representative sample of the waste is:

1. A liquid with a flash point less than 60°C (140°F)
2. A nonliquid capable under normal conditions of spontaneous and sustained combustion
3. Ignitable compressed gas
4. An oxidizer

Waste meeting this characteristic is designated by an EPA hazard code "I" and is identified by the Waste Code Number D001.

3.2.2.2 Corrosivity. A solid waste exhibits the characteristic of corrosivity if a representative sample of the waste is either:

1. A liquid with a pH less than 1 or greater than 12.5
2. A liquid that corrodes steel at a rate greater than $\frac{1}{4}$ in/year

Waste meeting this characteristic is designated by an EPA Hazard Code "C" and is identified by the Waste Code Number D002.

3.2.2.3 Reactivity. A solid waste exhibits the characteristic of reactivity if a representative sample of the waste has any of the following properties:

1. Normally is unstable—reacts violently
2. Reacts violently with water
3. Forms explosive mixtures with water
4. Generates toxic gases, vapors, or fumes when mixed with water
5. Contains cyanide or sulfide and generates toxic gases, vapors, or fumes between pH 2 and 12.5
6. Could detonate if heated under confinement or subjected to strong initiating source
7. Could detonate at standard temperature and pressure
8. Is listed by the Department of Transportation (DOT) as Class A or B explosive

Waste meeting this characteristic is designated by an EPA hazard code "R" and is identified by the Waste Code Number D003.

3.2.2.4 Toxicity. After several years of comments and changes, the EPA finalized revisions to the Toxicity Characteristic (TC) testing procedure used to identify wastes as hazardous. The new rule, published in the *Federal Register* on March 29, 1990, expanded the number of waste types considered hazardous under RCRA and added 25 new organic compounds to the list of toxic constituents that may render a waste hazardous. A greater number of waste generators will be regulated under the new rule. The 40 toxic constituents are listed in Table 30.3.

The new rule also replaces the extraction procedure (EP) toxicity test with the toxicity characteristic leaching procedure (TCLP). TCLP measures the leaching potential of toxic constituents, and is designed to determine the mobility of organic and inorganic contaminants in liquid, solid, and multiphasic wastes. If one or more of the constituents found in Table 30.3 can be leached from a waste in concentrations greater than the specified levels, the waste is regulated as hazardous. Waste meeting this characteristic is assigned a hazard code "E" and identified by Waste Code Numbers D004–D017.

Table 30.3. Toxicity Characteristic Constituent and Regulatory Levels

EPA HW No.[a]	Constituent	Regulatory Level (mg/l)
D004	Arsenic	5.0
D005	Barium	100.0
D018	Benzene	0.5
D006	Cadmium	1.0
D019	Carbon tetrachloride	0.5
D020	Chlordane	0.03
D021	Chlorobenzene	100.0
D022	Chloroform	6.0
D007	Chromium	5.0
D023	o-Cresol	200.0[b]
D024	m-Cresol	200.0[b]
D025	p-Cresol	200.0[b]
D016	2,4-D	10.0
D027	1,4-Dichlorobenzene	7.5
D028	1,2-Dichloroethane	0.5
D029	1,1-Dichloroethylene	0.7
D030	2,4-Dinitrotoluene	0.13[c]
D012	Endrin	0.02
D031	Heptachlor (and its hydroxide)	0.008
D032	Hexachlorobenzene	0.13[c]
D033	Hexachlor-1,3-butadiene	0.5
D034	Hexachloroethane	3.0
D008	Lead	5.0
D013	Lindane	0.4
D009	Mercury	0.2
D014	Methoxychlor	10.0
D035	Methyl ethyl ketone	200.0
D036	Nitrobenzene	2.0
D037	Pentachlorophenol	100.0
D038	Pyridine	5.0[c]
D010	Selenium	1.0
D011	Silver	5.0
D039	Tetrachloroethylene	0.7
D015	Toxaphene	0.5
D040	Trichloroethylene	0.5
D041	2,4,5-Trichlorophenol	400.0
D042	2,4,6-Trichlorophenol	2.0
D017	2,4,5-Trichlorophenoxypropionic Acid (Silvex)	1.0
D043	Vinyl chloride	0.2

[a]Hazardous waste number.

[b]If o-, m-, and p-Cresol concentrations cannot be differentiated, the total cresol (D026) concentration is used. The regulatory level for total cresol is 200 mg/l.

[c]Quantitation limit is greater than the calculated regulatory level. The quantitation limit therefore becomes the regulatory level.

3.3 Management Options

Once components of the waste streams have been defined, the generator is in a position to evaluate treatment and management options. This section addresses some of the more commonly used management methods, including waste reduction, incineration, chemical and biological treatment, and of course, land disposal.

Solid wastes, including hazardous wastes, have long been disposed of in landfills. Essentially, this amounts to long-term storage of the waste in a disposal site. Landfilling has been a favored method of disposal because of its relatively inexpensive cost. However, the available space in currently operating landfills is dwindling and approval for new sites is difficult to obtain, in part because of local opposition. Higher real estate costs and stricter federal and state regulations have also increased the costs of operations. In addition, under federal regulations, generators who contribute waste to a landfill can be held partially liable for any resulting contamination. RCRA goes so far as to ban some hazardous wastes from land disposal. Finally, the landfill option does not solve the problem of hazardous waste—it only stores and, we hope, contains it.

One way of reducing the hazardous waste that must be treated or stored is to reduce the amount of hazardous waste that is produced. Waste minimization techniques include inventory management, production process modification, volume reduction, and recovery of waste streams.

Waste reduction can lower the cost of materials and of waste disposal. It may also reduced a generator's liability for the hazardous waste produced and compliance costs for permits and monitoring. The method does not eliminate waste entirely though; sometimes it may not be feasible at all. Using materials more efficiently is also a more complex task than packaging waste and calling a hauler. Some generators may not be willing to expend the effort.

Incineration, or the burning of hazardous wastes at high temperatures in the presence of oxygen, is another waste treatment option. The EPA estimates that 60 percent of the hazardous substances produced in the United States can be effectively treated in this manner. The method is used for organic waste materials, for example, pesticides, solvents, and PCBs.

The heat produced by incineration of hazardous wastes has beneficial uses; for instance, it can be used to preheat combustion air. In addition to producing useful by-products, incineration detoxifies the hazardous waste by destroying the organic molecular structure, and reduces the volume of waste in the process. Another plus is that the incinerator operations do not require a large land area.

As with any method, incineration has its drawbacks. The process is one of the most expensive alternatives available; high capital and operating costs are associated with the method and the air pollution controls that must accompany it. The gaseous products and particulates are potential air pollutants that require expensive abatement equipment such as electrostatic precipitators, baghouse filters, wet air scrubbers, and activated carbon beds. Moreover, incineration is not popular with the general public—obtaining a permit is becoming difficult as communities protest local construction of facilities.

Certain hazardous wastes can be converted to a less hazardous form through chemical treatment, using reactions such as neutralization (for acidic or alkaline wastes), precipitation (for instance, of toxic metals), or oxidation and reduction. These processes make resource recovery possible, and can produce useful byproducts and environmentally acceptable residual effluents. These processes are also used to reduce the volume of hazardous waste.

Another volume reduction process is biological treatment, or the use of microorganisms that feed on hazardous wastes to decompose them. The process is useful for treatment of organic wastes such as sewage, wastes from a petroleum refinery, or paper mill sludges. Biological treatment can be used to lower the cost of downstream processes by reducing the organic load.

When using a biological treatment system, the proper conditions must be maintained. Considerations include the carbon and energy source (which may come from sunlight), nutrients such as nitrogen, phosphorus, and trace metals, a source of oxygen (or a substitute in some cases), controlled temperature and pH, and removal of toxic organisms. Experimental studies are required to design the unit and operation process.

These are just some of the more common treatment alternatives available to a generator faced with a disposal problem. The inventory and characterization processes help to better define disposal needs. Only after completing these tasks can the generator make an informed decision about the best treatment options.

4   HEALTH AND SAFETY

The Occupational Safety and Health Administration (OSHA) regulates health and safety during hazardous waste operations and emergency response under 29 CFR 1910.120. This OSHA standard applies to:

1. Cleanup operations required by a governmental body conducted at uncontrolled hazardous waste sites
2. Corrective actions involving cleanup at sites covered by RCRA as amended
3. Voluntary cleanup operations at sites recognized by federal, state, local, or other governmental bodies
4. Operations involving hazardous wastes that are conducted at treatment, storage, and disposal (TSD) facilities
5. Emergency response operations for releases of, or substantial threats of releases of, hazardous substances

This regulation includes numerous provisions that must be followed by employers involved in hazardous waste and emergency response operations. The following are the general categories of the requirements under this standard:

• Written safety and health program

- Site characterization and analysis
- Site control
- Training
- Medical surveillance
- Engineering controls, work practices, and personal protection for employee protection
- Monitoring
- Informational programs
- Drum and container handling
- Decontamination
- Emergency response
- Illumination
- Sanitation at temporary workplaces

Some of the requirements of the OSHA standard and good health and safety practices are discussed here. Because regulations are constantly being updated, the latest version of the actual regulation should be consulted for further detail.

4.1 Program Development

The type and degree of hazard vary greatly depending on the site conditions and on variety and quantity of chemicals. Health and safety programs also vary, depending on the type and magnitude of the hazards present, the size of the site, the number of employees, type of operations, type of business, and the overall philosophy toward safety and health. The following are key elements to most successful health and safety programs (8):

1. A policy that is explained and made available to all employees in order to insure a thorough understanding of program goals and individual responsibilities.
2. A clear definition of program objectives and a schedule for achievement.
3. An overall commitment which acknowledges management's responsibilities to the program.
4. A mechanism that will provide for mutual representation from all functional levels within the organization in the setting of priorities and the implementation of program objectives.
5. A clear definition of line and staff responsibilities and their reporting relationships. This is often best accomplished through the use of a functional organizational chart.
6. A means of periodically reviewing progress and accomplishments over the course of the program.

To meet the OSHA standard, 29 CFR 1910.120, the employer's safety and health program must include a written program designed to identify, evaluate, and control safety and health hazards, and provide for emergency response for hazardous waste operations. The written program must include:

- Organizational structure
- Comprehensive workplan
- Site-specific safety and health plan
- Safety and health training program
- Medical surveillance program
- Standard operating procedures for safety and health
- Interface between general program and site-specific activities

Major factors that should be considered in a health and safety program for dealing with hazardous wastes are discussed in the following sections.

4.1.1   Site Characterization

Site characterization and analysis of each individual site are necessary to identify specific site hazards and to determine appropriate safety and health control procedures. Based on the site characterization and hazard identification, personal protective equipment, engineering controls, and monitoring needs are determined, and employees are notified of the potential hazards and risks.

4.1.2   Air Monitoring

Airborne contaminants are one of the potential hazards that should be considered when developing a health and safety program. As with more typical industrial exposures, air monitoring to identify and quantify these contaminants is used to determine the need for personal protective equipment and medical surveillance. In the case of hazardous waste operations, the use of direct-reading instruments as well as laboratory analysis of air samples collected on sampling media are used.

Direct-reading instruments are widely used to evaluate airborne contaminants at hazardous waste sites owing to the need for immediate information at sites with multiple contaminants and changing conditions, such as an outdoor disposal site. The instruments are one of the primary tools used during initial site characterization. Often direct-reading instrument readings are used to determine when various levels of personal protective equipment are required. These requirements are typically delineated in a site health and safety plan.

However, when used in the field direct-reading instruments have a number of limitations that must be kept in mind:

- They are selectively responsive, detecting only specific classes of chemicals.
- They have lower detection limits that may be above the concentration of concern for very toxic chemicals.

- They are often subject to interferences from substances other than the substance of concern and to environmental conditions, such as humidity.
- They often have different responses to various chemicals.

The user must therefore be completely familiar with the instrument and its limitations in order to be able to interpret the readings. Interpretation should be conservative, particularly when multiple or unknown contaminants may be present. These direct-reading instruments must also be carefully calibrated frequently, at least before and after each use.

Direct-reading instruments are often not adequate to determine personal exposures to specific contaminants, especially those with low exposure limits. Therefore, full-shift personal air sampling is also applicable to hazardous waste operations. These samples may be collected by drawing air through sampling media with personal sampling pumps or by using passive dosimeters. In both cases, sample analysis is done in a laboratory. Personal monitoring of workers with the highest potential exposure is recommended. The frequency of monitoring depends on a number of site-specific issues, such as the number and types of contaminants, the number of different operations, and changes in site conditions.

Although the primary purpose of air monitoring is typically worker protection, environmental concerns and regulations may require that sampling be performed to check for migration of airborne contaminants offsite. Sampling at the perimeter of the site is conducted for this purpose. This sampling may be conducted using direct-reading and/or traditional sampling media analyzed in the laboratory. When choosing the location of perimeter samples, wind speed and direction must be considered for an outdoor site.

4.1.3 Personal Protective Equipment

When working around hazardous waste, personal protective clothing and equipment (PPE) is often needed to protect individuals from chemical, physical, or biological hazards. Proper selection, use, and care of PPE is essential to adequate protection of the respiratory system, skin, eyes, body, and hearing. PPE is selected based on the hazards and potential hazards identified during the site characterization and analysis.

The EPA has defined four levels of protection that are widely used for hazardous waste health and safety situations. Levels A, B, C, and D are summarized in Table 30.4. These levels are typically used as a starting point for choosing appropriate PPE; however, the PPE may need to be modified to meet the needs of the specific situation. In all cases, the limitations of PPE, such as respirator protection factors and permeability of protective suits by various chemicals, must be taken into account. The selection and limitations of specific respirators and other types of PPE are discussed further in Chapter 19.

On hazardous waste sites, the full extent of the hazard is often not defined at the beginning of a project. As more information becomes available, it may be necessary to modify PPE requirements by upgrading or downgrading the level of protection. Typical reasons for upgrading include (9):

Table 30.4. Levels of Protection (9)

Level of Protection	Equipment	Protection Provided	Should Be Used When	Limiting Criteria

A	**Recommended:**
	
	• Pressure–demand, full-facepiece SCBA or pressure–demand supplied–air respirator with escape SCBA
	
	• Fully-encapsulating, chemical resistant suit
	
	• Inner chemical resistant gloves
	
	• Chemical-resistant safety boots/shoes
	
	• Two–way radio communications
	
	Optional:
	
	• Cooling unit
	
	• Coveralls
	
	• Long cotton underwear
	
	• Hard hat
	
	• Disposable gloves and boot covers
		The highest available level of respiratory, skin, and eye protection	• The chemical substance has been identified and requires the highest level of protection for skin, eyes, and the respiratory system based on either:
			
			  – Measured (or potential for) high concentration of atmospheric vapors, gases, or particulates
			
			  or
			
			  – Site operations and work functions involving a high potential for splash, immersion, or exposure to unexpected vapors, gases, or particulates of materials that are harmful to skin or capable of being absorbed through the intact skin
			
			• Substances with a high degree of hazard to the skin are known or suspected to be present, and skin contact is possible
			
			• Operations must be conducted in confined, poorly ventilated areas until the absence of conditions requiring Level A protection is determined
				• Fully-encapsulated suit material must be compatible with the substances involved

- Known or suspected presence of dermal hazards
- Occurrence or likely occurrence of gas or vapor emission
- Change in work task that will increase contact or potential contact with hazardous materials
- Request of the individual performing the task

Typical reasons for downgrading include (9):

Table 30.4. (*Continued*)

Level of Protection	Equipment	Protection Provided	Should Be Used When	Limiting Criteria

B	**Recommended:**
	
	• Pressure–demand, full-facepiece SCBA or pressure–demand supplied–air respirator with escape SCBA
	
	• Chemical-resistant clothing (overalls and long-sleeved jacket; hooded, one-or two-piece chemical splash suit; disposable chemical-resistant one-piece suit)
	
	• Inner and outer chemical-resistant gloves
	
	• Chemical-resistant safety boots/shoes
	
	• Hard hat
	
	• Two-way radio communications
	
	Optional:
	
	• Coveralls
	
	• Disposable boot covers
	
	• Face shield
	
	• Long cotton underwear
		The same level of respiratory protection but less skin protection than Level A
		
		It is the minimum level recommended for initial site entries until the hazards have been further identified
			• The type and atmospheric concentration of substances have been identified and require a high level of respiratory protection, but less skin protection. This involves atmospheres:
			
			  – With IDLH concentrations of specific substances that do not represent a severe skin hazard
			
			    or
			
			  – That do not meet the criteria for use of air–purifying respirators
			
			• Atmosphere contains less than 19.5% oxygen
			
			• Presence of incompletely identified vapors or gases is indicated by direct–reading organic vapor detection instrument, but vapors and gases are not suspected of containing high levels of chemicals harmful to skin or capable of being absorbed through the intact skin
			
			• All criteria for the use of air–purifying respirators are met
				• Use only when the vapor or gases present are not suspected of containing high concentrations of chemicals that are harmful to skin or capable of being absorbed through the intact skin
				
				• Use only when it is highly unlikely that the work being done will generate either high concentrations of vapors, gases, or particulates or splashes of material that will affect exposed skin

- New information indicating that the situation is less hazardous than was originally thought
- Change in site conditions that decreases the hazard
- Change in work task that will reduce contact with hazardous materials

Training of the user is essential to the safety and effectiveness of PPE use. This is

Table 30.4. (*Continued*)

Level of Protection	Equipment	Protection Provided	Should Be Used When	Limiting Criteria
C	**Recommended:** • Full-facepiece, air-purifying, canister-equipped respirator • Chemical-resistant clothing (overalls and long-sleeved jacket; hooded, one- or two-piece chemical splash suit) • Inner and outer chemical-resistant gloves • Chemical-resistant safety boots/shoes • Hard hat • Two-way radio communications **Optional:** • Coveralls • Disposable boot covers • Face shield • Escape mask • Long cotton underwear	The same level of skin protection as Level B, but a lower level of respiratory protection	• The atmospheric contaminants, liquid splashes, or other direct contact will not adversely affect any exposed skin • The types of air contaminants have been identified, concentrations measured, and a canister is available that can remove the contaminant • All criteria for the use of air-purifying respirators are met	• Atmospheric concentration of chemicals must not exceed IDLH levels • The atmosphere must contain at least 19.5% oxygen
D	**Recommended:** • Coveralls • Safety boots/shoes • Safety glasses or chemical splash goggles • Hard hat **Optional:** • Gloves • Escape mask • Face shield	No respiratory protection. Minimal skin protection	• The atmosphere contains no known hazard • Work functions preclude splashes, immersion, or the potential for unexpected inhalation of or contact with hazardous levels of any chemicals	• This level should not be worn in the Exclusion Zone • The atmosphere must contain at least 19.5% oxygen

especially true at hazardous waste sites where the hazards can be high and the PPE ensembles complex.

OSHA regulations for hazardous wastes sites, 29 CFR 1910.120, require that a PPE program be prepared as part of the employer's written safety and health program. The written PPE program must address:

- PPE selection based upon site hazards
- PPE use and limitations of the equipment
- Work mission duration
- PPE decontamination
- PPE maintenance and storage
- PPE training and proper fitting
- PPE donning and doffing procedures
- PPE inspection procedures prior to, during, and after use
- Evaluation of the effectiveness of the PPE program
- Limitations during temperature extremes, heat stress, and other appropriate medical considerations

4.1.4   Training

All employees who engage in hazardous waste operations that could expose them to hazardous substances, health hazards, or safety hazards must receive health and safety training in accordance with requirements of 29 CFR 1910.120. Training requirements vary, depending on an employee's duties onsite. General site workers, such as equipment operators and general laborers, must receive 40 hr of instruction off the site and at least 3 days of actual field experience under the direct supervision of a trained, experienced supervisor. Workers who are on the site only occasionally for a specific limited task, such as groundwater monitoring, and who are unlikely to be exposed at levels greater than permissible exposure limits and published exposure limits, must receive at least 24 hr of instruction off the site and 1 day of actual field experience under the direct supervision of a trained, experienced supervisor.

In areas where a full characterization has indicated that (1) exposures are under permissible exposure limits and published exposure limits, (2) respirators are not necessary, and (3) there are no health hazards or the possibility of an emergency developing, workers regularly onsite must receive 24 hr of instruction offsite and 1 day of actual field experience under the direct supervision of a trained, experienced supervisor. Supervisors on hazardous waste sites must have the same training, plus an additional 8 hr of specialized training. Additionally, employees must receive 8 hr of refresher training each year.

Employees involved in emergency response to hazardous materials incidents require varying levels of training depending on their duties and functions. Most individuals that are actively involved in emergency response operations must receive at least 24 hr of training.

4.1.5   Engineering Controls and Work Practices

The use of engineering controls and work practices to reduce exposures to hazardous materials is more preferable than the use of PPE, and is required by OSHA for reducing exposures to materials with a permissible exposure limit (PEL), except to the extent that such controls and practices are not feasible. At hazardous waste sites, available engineering controls and work pratices may be limited.

Engineering controls that may be feasible for hazardous waste handling include the use of pressurized cabs or control booths on equipment, emergency alarm systems, forced air ventilation, and the use of remotely operated material handling equipment. Work practices that may be feasible at hazardous waste sites include controlling site access, minimizing the number of employees working in an area when a hazardous operation such as drum opening is performed, wetting down dusty operations, maintaining good housekeeping onsite, and locating employees upwind of potential sources of airborne hazards. Engineering controls and work practice design should take into account the potential offsite contamination/dispersion, as well as onsite employee protection.

4.1.6 Medical Program

A medical program is necessary to assess and monitor workers' health and fitness both prior to employment and during the course of work, and to provide emergency and other treatment as needed. The program also helps a company keep accurate records for future reference and meet OSHA requirements. OSHA requires a medical surveillance program to be instituted for (1) employees who may be exposed to hazardous substances at or above the permissible exposure limits (or published exposure limits if permissible exposure limits do not exist) for 30 days or more a year; (2) employees who wear a respirator for 30 days or more a year; (3) employees who are injured due to overexposure from an emergency incident involving hazardous substances or health hazards; or (4) members of hazardous materials response teams.

Typically, a medical program includes the following components:

- Surveillance, including:
  Pre-employment screening
  Periodic medical examinations
  Termination examination
- Treatment (emergency and nonemergency)
- Record keeping
- Program review

A medical program should be developed for each site based on the specific needs at the location and potential exposures at the site. The medical program should be designed and directed by a medical doctor who has experience in managing occupational health services.

4.1.7 Site Layout and Control

Proper site layout and control is necessary to minimize potential contamination of workers, protect individuals offsite, and prevent unauthorized entry to the site. An EPA training manual (10) suggests:

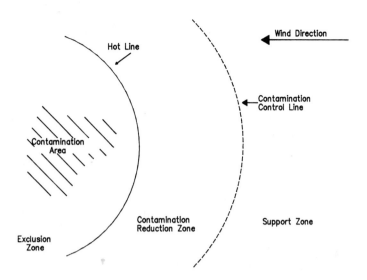

Figure 30.1. Site work zones.

- Setting up security and physical barriers to exclude unnecessary personnel from the general area
- Minimizing the number of personnel and equipment onsite consistent with effective operations
- Establishing work zones within the site
- Establishing control points to regulate access to work zones
- Conducting operations in a manner to reduce the exposure of personnel and equipment and to eliminate the potential for airborne dispersion
- Implementing appropriate decontamination procedures

To meet OSHA requirements, a site control program must include a site map; site work zones; the use of a "buddy system"; site communications, including alerting means for emergencies; standard operating procedures or safe work practices; and identification of the nearest medical assistance.

The work zones usually include an exclusion zone, a contamination reduction zone, and a support zone. This system provides an area for decontamination and reduces the likelihood of spreading hazardous substances from the contaminated area to clean areas.

Figure 30.1 shows a schematic representation of the three work zones. The exclusion zone is the contaminated area, where primary activities involving hazardous waste take place. The contamination reduction zone is the area where decontamination of personnel, clothing, and equipment takes place. The support zone is a clean area where administrative and other support functions are located.

Access between the work zones is through access control points, usually one for personnel and one for equipment.

5 SUMMARY

Landfill disposal used to be an easy "solution" to the problem of hazardous waste. Management of hazardous waste is now becoming more difficult owing to increasingly stringent regulation and restriction of disposal options. The consequences of improper management of hazardous waste affect everyone. Proper management is the responsibility of those who generate, transport, store, treat, and dispose of hazardous waste.

A logical approach for the generator faced with the task of developing and implementing a hazardous waste management program includes:

1. Identifying hazardous waste through an inventory and characterization process
2. Minimizing the generation of hazardous waste
3. Reviewing applicable federal, state, and local regulations
4. Reviewing treatment and disposal options
5. Tracking hazardous waste from "cradle to grave" with manifests
6. Transporting wastes to treatment or disposal sites with registered transporters, within 90 days of generation
7. Keeping adequate records of hazardous waste management practices
8. Following the health and safety requirements of RCRA and OSHA

The generation of hazardous waste is a reality that cannot be avoided. The management process of inventory, characterization, and treatment and the regulations governing these actions are pieces of a complex puzzle that, when complete, represent a solution to the problem of hazardous waste.

REFERENCES

1. Code of Federal Regulations, Title 40, Parts 261.20 to 261.24.
2. Code of Federal Regulations, Title 40, Parts 261.30 to 261.33.
3. Code of Federal Regulations, Title 40, Part 261.4.
4. G. F. Lindgren, *Managing Industrial Hazardous Waste*, Lewis, 1989.
5. T. Wagner, *The Complete Handbook of Hazardous Waste Regulation*, Perry-Wagner, Brunswick, ME, 1988.
6. C. A. Wentz, *Hazardous Waste Management*, McGraw-Hill, New York, 1989.
7. Dawson, Gaynor, Mercer, and Basil, *Hazardous Waste Management*, Wiley, New York, 1986.

8. S. P. Levine and W. I. Martin, *Protecting Personnel at Hazardous Waste Sites*, Butterworth, Stoneham, MA, 1985.

9. NIOSH, OSHA, USCG, EPA, *Occupational Safety and Health Guidance Manual for Hazardous Wastes Site Activities*, U.S. Department of Health and Human Services, October 1985.

10. Hazardous Response Support Division, National Training and Technology Center, *Personal Protection and Safety Training Manual*, U.S. Environmental Protection Agency, Cincinnati, OH, 1982.

CHAPTER THIRTY-ONE

Man-Made Mineral Fibers

Jaswant Singh, Ph.D., C.I.H., and Michael A. Coffman, C.I.H.

1 INTRODUCTION

Concerns about the adverse health effects of exposure to asbestos have prompted widespread removal of asbestos-containing materials, resulting in the increased use of substitutes composed of both naturally occurring and synthetic materials. Man-made mineral fiber asbestos substitutes are mineral fibrous materials such as fibrous glass, rock wool, slag wool, and refractory (ceramic) fibers.

Because of the similarity of the chemical composition and morphology of these substitute fibrous materials to those of asbestos, serious questions have been raised about their health implications. In particular, there is growing concern whether such substitutes pose a carcinogenic risk similar to that of asbestos. These health concerns are even more pronounced considering that man-made mineral fibers (MMMF) have found wide applications in commerce, in addition to their use as asbestos substitutes in the building industry.

1.1 Terminology

There is a variety of commercially produced and used fibers, both naturally occurring and man-made (Figure 31-1).

Artificial fibers are fibers in which the fiber-forming material is of vegetable or animal origin. These fibers include viscose rayon, cellulose ester, and protein fibers.

Organic fibers are fibers in which the fiber-forming material is derived from

Patty's Industrial Hygiene and Toxicology, Fourth Edition, Volume 1, Part B, Edited by George D. Clayton and Florence E. Clayton.
ISBN 0-471-50196-4   © 1991 John Wiley & Sons, Inc.

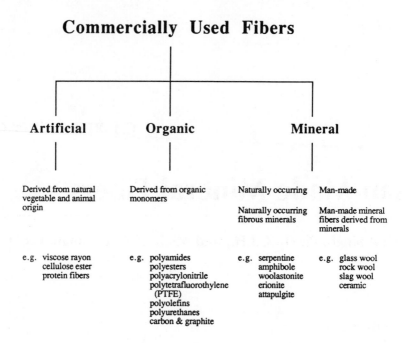

Figure 31.1. Commercially used fibers.

monomeric organic compounds. Examples of organic fibers are polyamides, polyesters, polyacrylonitrile, and polytetrafluoroethylene (Teflon®).

Carbon and graphite fibers are considered synthetic organic fibers because they are commercially produced by high-temperature processing of organic precursors such as rayon, pitch, or polyacrylonitrile. The terms carbon and graphite are often used interchangeably but there are important differences. Graphite fibers are generally stronger and stiffer because of their polycrystalline nature. Carbon/graphite fibers are used as reinforcing materials in composites for the aerospace and automobile industries. Although manufactured to nominal diameters of 5 to 8 μm, a considerable portion of these fibers may exist as respirable fibers (less than 3 μm in diameter).

Naturally occurring *mineral fibers* such as serpentine (chrysotile) and amphiboles (amosite and crocidolite) have been extensively used in commerce as thermal and acoustical insulation products. Other commonly known, naturally occurring mineral fibers are woolastonite, erionite, and attapulgite. Woolastonite ($CaSiO_3$) is widely used in the ceramic industry, in paints, plastics, and abrasives, and in metallurgy. Erionite is one of the two naturally occurring zeolites known to occur in the fibrous form. The main use of attapulgite is as an absorbent for oil and grease and pet wastes. Attapulgite is also used in drilling muds, fertilizers, cosmetics and pharmaceutical products.

MMMF, the main topic of this chapter, are further divided into three categories: fibrous glass, mineral wool, and refractory/ceramic fibers.

Table 31.1. Composition (% by Weight) of MMMF

Component	Glass Wool	Rock Wool	Slag Wool	Ceramic Fiber
SiO_2	34–73	45.5–52.9	40.6–41.0	0–53.9
Al_2O_3	2.0–14.5	6.5–13.4	11.8–12.5	0–95.0
MgO	3.0–5.5	—	—	0–0.5
CaO	5.5–22.0	10.8–30.3	37.5–40.0	0–0.7
FeO	—	1.0–5.8	0.9–1.0	—
B_2O_3	3.5–8.5	—	—	0–14
Na_2O	0.5–16.0	2.3–2.5	0.2–1.45	0–0.2
K_2O	0.5–3.5	1.0–1.6	0.3–0.4	0–0.1
TiO_2	0–8.0	0.5–2.0	0.4–0.44	0–1.6
Z_rO_2	0–4.0	—	—	0–92
PbO	0–59.0	—	—	—
Fe_2O_3	—	0.5–8.2	—	0–0.97
Y_2O_3	—	—	—	0–8

Note: Some of the mineral fibers shown also contain small percentages of other components including P_2O_5, CaS, S, F, MnO.
Source: IARC (2).

1. *Fibrous glass* includes glass wool, continuous filament, and special-purpose fibers. These materials are typically composed of oxides of silicon, calcium, sodium, potassium, aluminum, and boron. The main raw materials are silica sand, limestone, fluorspar, boron oxide, and glass fragments (cullet). The typical composition of glass wool and other MMMF is shown in Table 31.1.

2. *Mineral wool* includes both rock wool, derived from magma rock, and slag wool, made from molten slag produced in metallurgical processes such as the production of iron, steel, or copper. The main components of rock wool and slag wool are oxides of silicon, calcium, magnesium, aluminum, and iron (Table 31.1). Some researchers have also included glass wool under the term mineral wool.

3. *Refractory/ceramic fibers* are amorphous or partially crystalline materials made from kaolin clay or oxides of aluminum, silicon, or other metal oxides. Some ceramic fiber products developed for special applications contain little or no aluminum oxide or silicon dioxide. For example, zirconia fibers contain mainly oxides of zirconium and yttrium. Less commonly, refractory fibers are also made from non-oxide refractory materials such as silicon carbide (SiC), silicon nitride (Si_3N_4), or boron nitride (BN).

1.2 Historical Data

Glass fibers were reportedly used as early as 2000 B.C. (1). Early Egyptians are reported to have used coarse glass fibers for decorative purposes. The commercial use of MMMF was marked by the awarding of a patent to a Russian in 1840 for the preparation of fibrous glass. During this same time, rock wool/slag wool was

Table 31.2. Chronology of the Production of MMMF

Year	Fiber Type		
	Fibrous Glass	Mineral Wool	Ceramic Fibers
2000 B.C.	First reported production of glass fibers. Egyptians used coarse glass fibers for decorative purposes		
1840	Patent for method and apparatus for preparation of fibrous glass using a spinning process awarded to a Russian named Shamo	Rock/slag wool first produced in Wales	
1870		Slag wool first produced in Germany	
		First U.S. patent issued to produce slag wool by blowing molten slag produced in a blast furnace	
1885		First successful commercial production of slag wool	
1893	Edward Drummond Libbey exhibits a dress, lamp shades, and other articles woven of glass fibers at the Columbian Exposition in Chicago		
1895		German production of slag wool reaches 50 tons/year	
1906	Pollack, Pick, Pazsicky, and Bornkessel awarded patents for glass wool manufacture.		

	Patents for fibrous glass also issued in England and Germany during this time
1915	Allied blockade of Germany in World War I creates asbestos shortage resulting in full scale production of fibrous glass and slag wool
1925	Spun glass production starts in U.S. by drum-winding process
1932	Coarse glass fibers first used in an air filter. First fibrous glass thermal insulation used in U.S. Naval vessels
	First house installed with fibrous glass
1938	Owens Corning Fiberglas Company formed by Owens Illinos and Owens Corning
1941	Report on fibrous glass health hazard investigation by Walter J. Siebert with the aid of Owens Corning Fiberglass Corp. Evidence of pulmonary disease found
	Report by Leroy Gardner concludes that exposure to glass wool dust involved "no hazard to the lungs"

German production of mineral wool reaches 15,000 tons/year

Table 31.2. (continued)

| Year | Fiber Type | | |
	Fibrous Glass	Mineral Wool	Ceramic Fibers
1942	U.S. Patent Office issues 353 patents for glass wool products		
1942	Baer Sulzberger publishes "The Effects of Fiberglass on Animal and Human Skin in Industrial Medicine." Sulzberger notes skin irritation, but no permanent damage		
1932	First use of matted fibrous glass to reinforce plastic sheet		
1944		Germany has 23 cupolas producing 44,800 tons of slag and rock wool	
1951	Fibrous glass coveralls used by firefighters to withstand temperatures up to 2000°F		U.S. patent issued to Carborundum Corp. for its high melting Fiberfrax

1954	Babco and Wilson issued a patent for producing aluminoborosilicate fiber by fusing kaolin
1963	Johns–Manville assigned U.S. patent for manufacturing fibrous glass mats with a binder for insulation
Early 1970s	Commercial production of ceramic fibers begins
Mid-1980s	Production reaches 70–90 million
1987	International Agency for Research on cancer assigns glass wool, mineral wool and ceramic fibers to category 2B. Continuous filimentous glass is assigned to category 3

Source: IARC (2) and Peters and Peters (1).

first produced in Wales (2). Production of fibrous glass and slag wool was accelerated during World War I when the allied blockade of Germany created an asbestos shortage. In the United States in 1938, the formation of Owens Corning Fiberglas Company resulted in increased use and production of glass fibers. In the same year, German production of mineral wool reportedly reached 15,000 tons. In 1951, a U.S. patent was issued to Carborundum Corporation for its high-melting Fiberfrax®. The commercial production of ceramic fibers began in the early 1970s and has seen consistent growth, reaching a level of over $100 million U.S. sales by the mid-1980s.

The chronology of the production of MMMF is presented in Table 31.2.

2 PRODUCTION AND USES OF MMMF

2.1 Mineral Wool

In the production of mineral wool, the raw material (slag and/or rock) is loaded into a cupola furnace in alternating layers with coke and small amounts of other ingredients to impart special characteristics of ductility or size to the fibers. Approximately 70 percent of the mineral wool sold in the United States is produced from blast furnace slag (3). Most of the remainder is produced from copper, lead, and iron smelter slag. A small amount is produced with natural rock (rock wool). Combustion of the coke in the cupola furnace generates high temperatures (about 3000°F). The slag or rock is melted, then discharged from the furnace in a stream, and fiberized. In fiberizing mineral wool, the most commonly used techniques are (1) the steam jet process, (2) the Powell process, and (3) the Downey process. Each of these is described below in greater detail.

Steam Jet Process. Until the early 1940s, nearly all of the mineral wool was produced using a steam jet process. In this process, the molten stream of material discharged from the cupola furnace is fiberized by passing it in front of a high-pressure steam jet (Figure 31.2). The molten stream of slag or rock is broke up into many small droplets by the steam jets. Fiberization occurs as these droplets are swept out at high velocity in front of the jet.

Powell Process. The Powell process (Figure 31.3) uses a group of rotors spinning at high speed to collect and distribute the molten stream of slag in a thin film on the rotor surfaces and then fiberizing it by throwing it off the rotors with centrifugal force.

Downey Process. The Downey process (Figure 31.4) combines a spinning concave rotor with steam jets. The molten stream of slag or rock is distributed in a thin

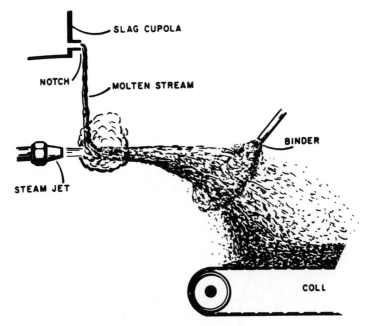

Figure 31.2. Steam-jet fiberization process.

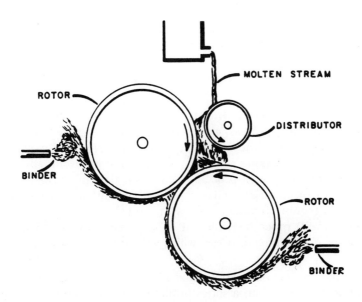

Figure 31.3. The Powell process.

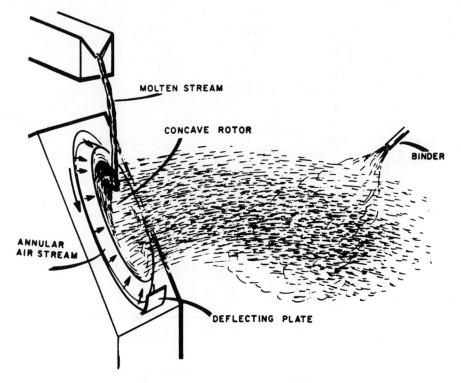

Figure 31.4.   The Downey process.

pool over the concave-surface of the rotor. The stream then flows up and over the edge of the rotor where it is swept out by the steam jets and is fiberized.

2.2   Fibrous Glass

Fibrous glass may be produced in two steps (marble melt) or in a single step (direct melt). In the marble melt process, glass is produced in a furnace that fuses the raw materials and homogenizes the melt. The glass melt exits the furnace through a forehearth to a marble making machine. The marbles are eventually remelted to make fibers. In the direct melt process, the molten glass proceeds from the forehearth directly to the fiberizer. Glass is commonly fiberized using one of four techniques: (1) extrusion (filamentous glass), (2) rotary, (3) flame attenuation, and (4) spinning (Powell) processes. Each of these processes is described in greater detail except for the spinning process, which was described previously.

Extrusion.   In the production of continuous filament textile glass, the molten glass flows from the forehearth into a series of platinum tanks fitted with hundreds of small-diameter orifices. The glass flows through these orifices, and the individual

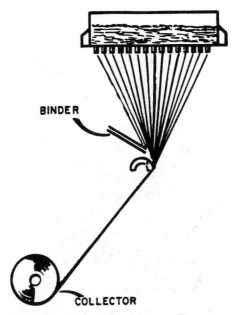

Figure 31.5.   Production of continuous filamentous glass.

filaments produced are collected together in a strand. A binder is applied and the strand is wound around a rapidly rotating drum (Figure 31.5).

Rotary Process. Similarly to the Downey process described above, glass wool is produced in the spinning process (Figure 31.6) by flowing the molten glass stream into the bowl of a concave rotor. The molten glass is distributed by centrifugal force to the sidewall of the rotor which contains many small holes. As the glass flows through these holes, it is further fiberized by high-velocity jets around the perimeter.

Flame Attenuation Process. In the flame attenuation process, the molten glass stream is passed through high-velocity gas jets to fiberize coarse primary filaments of glass, which are then fed in front of high-velocity jet flames (Figure 31.7).

2.3   Ceramic Fibers

Refractory or ceramic fibers are produced by (1) blowing and spinning, (2) colloidal evaporation, (3) continuous filamentation, and (4) vapor-phase deposition.

Blowing and Spinning. The blowing and spinning processes are essentially the same as those used for the production of glass and mineral wool fibers. Refractory or ceramic fibers are produced by fusing mixtures of natural minerals (such as kaolin clay), or synthetic blends of alumina, silica, or other metal oxides in an electric

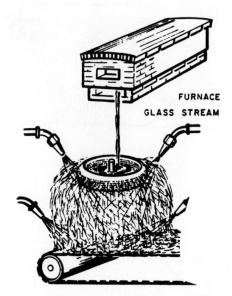

Figure 31.6. Rotary process for glass wool production.

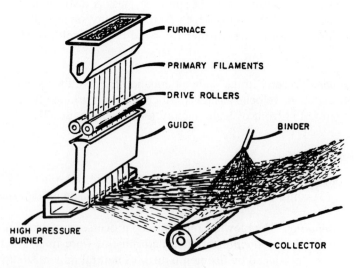

Figure 31.7. Flame attenuation process for producing glass wool.

furnace. Fibers are formed either by (1) passing the molten material through high pressure steam or other hot gas jets (blowing), or (2) forcing the molten stream onto rapidly rotating disks that fiberize the stream by throwing it off the disks with centrifugal force.

Colloidal Evaporation. Ceramic fibers of alumina, zirconia, silica, mixtures of zirconia and silica, and thoria have been produced by evaporating a colloidal suspension of these materials.

Vapor-Phase Deposition. Special ceramic fibers composed of non-oxide materials are produced by a vapor-phase deposition technique. In this technique, a volatile compound of the desired material is reduced or decomposed on a resistively heated substrate such as tungston wire. Monocrystalline ceramic materials known as whiskers are also produced by vapor-phase deposition techniques. These materials have high strength and micron-sized diameters. A large number of fibrous materials may be produced this way by using materials such as carbides, nitrides, halides, arsenates, vanadates, and silicates, to name a few.

2.4 Secondary Processes

The secondary or finishing processes are closely linked to specific final products. The secondary processes can present significant worker exposures. These processes include:

- Dust suppression by adding agents such as mineral oils, vegetable oils, or waxes to almost all insulation wools.
- Addition of binders such as natural resins, tars, or synthetic resins. More recently, the phenol–formaldehyde or urea–formaldehyde resins have meant the introduction of curing processes and posed exposure problems to volatile components such as formaldehyde.
- Other secondary operations include sawing, cutting, packing, and dust removal. In addition, operations such as painting or surface cutting may also impact the work environment.

2.5 Uses of MMMF

Products derived from MMMF have provided great benefits to society in terms of energy conservation, thermal comfort, factory thermal insulation, acoustic insulation, and fire protection, among others. Table 31.3 lists some major uses of MMMF. One researcher has identified more than 30,000 products based on MMMF (4).

Because of a variety of uses to which MMMF have been put, including their use as asbestos substitutes, production of MMMF has been steadily increasing. It is estimated that over 5 million tons of MMMF are being produced worldwide.

Table 31.3. Uses of MMMF

Factory thermal insulation
Building system thermal insulation
Acoustic insulation
Fireproofing
Aerospace insulation
Reinforcing material in plastics, cement, and textiles
Automotive components
Fiber optics
Air and liquid filtration
Refractory coatings
Gaskets and seals

Rapid increase in the production of MMMF is evident in the production figures (Table 31.4) for glass fibers in the United States between 1975 and 1986 (5).

3 HEALTH EFFECTS

Since their introduction in commerce almost 100 years ago, MMMF have been known to cause irritation of the skin and, under dusty conditions, irritation of the eyes and throat. Skin irritation is generally associated with fibers that have diameters greater than 4.5 to 5.0 μm (6), sizes commonly found in insulation wools and filamentous glass. Mechanical irritation can result in irritant dermatitis which in general is not severe and does not last long.

Table 31.4. U.S. Glass Fiber Production

Year	Quantity (million kg)
1975	247.88
1976	306.90
1977	357.30
1978	419.04
1979	460.36
1980	393.62
1981	472.61
1982	408.15
1983	530.27
1984	632.88

Source: Reference 5.

3.1 Nonmalignant Respiratory Diseases

Several epidemiologic studies have addressed the issue of nonmalignant respiratory disease (NMRD) among populations exposed to MMMF. Although some studies have suggested MMMF-related effects on the respiratory system, in general, no statistically significant increase in NMRD mortality has been observed in the studied populations in comparison with local rates (6). The two largest epidemiologic investigations (7, 8) conducted in Europe and the United States showed little evidence of excess mortality from NMRD.

Similarly, in the animal studies to date, there has been little or no evidence of fibrosis of the lungs. In all cases, the tissue reaction in animals exposed to MMMF was much less than the reaction in animals exposed to equal masses of chrysotile or crocidolite asbestos. Researchers have suggested, however, that the number of asbestos fibers reaching the lung may have been greater than MMMF.

3.2 Carcinogenic Effects

3.2.1 Experimental and In Vitro Studies

Numerous studies have been conducted to evaluate the carcinogenic potency of MMMF. These studies may be categorized according to the method by which the subject animals are exposed. In many of the studies, the test animals were exposed via implantation of injection techniques. These procedures involve the direct injection or implantation of MMMF into the body of the test animal, by one of three methods:

- Intrapleural—the material is deposited adjacent to the pleural lining of the lung cavity
- Intratracheal—the material is deposited into the windpipe (trachea)
- Intraperitoneal—the material is deposited adjacent to the lining of the abdominal cavity

Numerous inhalation studies have also been performed by examining.

The carcinogenic potential of MMMF has also been studied using mutagenicity screening or in vitro studies. The induction of cancer is thought to proceed in steps. The first of these steps, initiation, involves damage to DNA resulting in heritable alterations or in rearrangements of genetic information (mutations). It is safe to assume that all chemical carcinogens are mutagens. However, not all cellular mutations result in cancer; therefore, not all mutagens are carcinogens. The use of mutagenicity screening has, however, gained widespread acceptance as a rapid and inexpensive method of identifying potential carcinogens.

Fibrous Glass. Animal studies involving the intrapleural or intraperitoneal administration of glass fibers indicate some apparent carcinogenic potential (9, 10). In the in vitro studies reported in the literature, the investigators were able to detect morphological changes and chromosomal damage of mammalian cells (11–13). No mutagenicity was observed in tests using bacterial cultures (14).

Long-term inhalation studies, much better indicators of the harmful effects of occupational and environmental exposures to airborne fibers, have not shown the same positive results (15–21). In most of the inhalation studies cited, test animals were exposed to unusually high levels of the fibrous glass (as high as 400 mg/m^3). These levels are considerably higher than typical occupational exposures reported in the literature. The apparent discrepancy between the results of inhalation versus implantation studies may result from the well documented low durability of glass fiber in the lung.

Filamentous Glass. In animal studies involving the intraperitoneal administration of filamentous glass, no statistically significant carcinogenic response was observed (9).

Mineral Wool. In animal studies involving the intrapleural or intraperitoneal administration of mineral wool fibers, no statistically significant increases in the incidence of tumors of lung or pleura were found (16, 19, 22). Similar results were reported for inhalation studies involving mineral wool fibers (15, 16, 19).

Refractory Fibers. In animal studies involving the intrapleural or intraperitoneal administration of ceramic fibers, Smith et al. (19) and Pott et al. (9) reported increased incidence of tumors. No tumors were reported for animals exposed to ceramic fibers via intratracheal administration (19).

The results of animal studies involving the inhalation of ceramic fibers are conflicting. Davis (23) reported a statistically significant increase in lung tumors whereas in two other studies (19, 24), no increase in tumor incidence was reported.

3.2.2 Epidemiologic Studies

Although many studies have been conducted on the carcinogenic effects of MMMF on humans, much of the current state of knowledge about the carcinogenicity of MMMF is derived from three cohort studies conducted in the United States, Europe, and Canada. The three studies were based on a total of 7862 deaths among 41,185 workers in Europe and North America, and provided evidence of cancer mortality over a sufficiently long period of time. These studies indicated an excess of mortality (although not large in relative terms) caused by lung cancer among rock wool/slag wool workers but not a statistically significant increase among glass wool workers. In the United States and European studies, four cases of death from mesothelioma were reported. In view of the large cohort (involving about 800,000 person-years of observation), the incidence of mesothelioma was not considered excessive. Moreover, after further review, the researchers concluded that one of the mesothelioma cases could be attributed to that particular worker's exposure to asbestos in a naval shipyard. Table 31.5 summarizes the standardized mortality ratios (SMRs) for the three epidemiological studies (7, 8, 25). Their conclusions can be summed up as follows.

Table 31.5. Summary of Epidemiologic Studies for MMMF

Investigators	Number of Subjects	Estimated Exposure (fibers/cm³)	Lung Cancer Mortality Rate		SMR
			Observed	Expected	
Glass Wool					
Enterline et al. (8) (U.S.)	11,380	0.005 to 0.29	291	267	109
Simonato et al. (7) (Europe)	8,286	(Not reported)	93	91	103
Shannon et al. (25) (Canada)	2,557	<0.2	19	9.5	199
Mineral Wool					
Enterline et al. (8) (U.S.)	1,601	0.35	60	45	134
Simonato et al. (7) (Europe)	10,115	(Not reported)	81	124	65
Continuous Filamentous Glass					
Enterline et al. (8) (U.S.)	3,435	0.011	64	69	92
Simonato et al. (7) (Europe)	3,566	(Not reported)	15	97	15

Glass Wool. The Enterline (U.S.) study showed a slightly raised mortality rate from respiratory cancer compared to local rates. This increase, however, was not statistically significant. Moreover, mortality was neither related to the duration of exposure nor to exposure levels.

The European study also showed no overall excess mortality from lung cancer among glass wool workers as compared to local rates. There was an increase in mortality with time (since first exposure) but this was not statistically significant, and the trend was not related to the duration of exposure. The Canadian study of glass wool workers showed a statistically significant increase in lung cancer mortality but was not related to the duration of exposure or to the time since the first exposure occurred. The Canadian cohort was also much smaller compared to the U.S. and European studies and did not account for smokers in the study population.

Filamentous Glass. Both the U.S. and the European studies showed no increase in lung cancer mortality among workers exposed to filamentous glass. The observed standardized mortality ratios (SMRs) were lower than expected in both studies.

Mineral Wool. In the U.S. and European studies combined, there was a statistically significant excess of mortality from lung cancer among rock wool/slag wool workers. In the European study, a statistically significant lung cancer rate was found among workers who were first exposed to slag wool/rock wool in the early days of production when the dust levels were relatively high and before the introduction of dust suppressants and binders.

Refractory Fibers. There are no human data available on refractory fibers.

3.3   Factors Affecting Carcinogenic Potency

It is now generally agreed that it is the morphology (fiber shape and size) and not the chemical composition that drives cytogenic response. Fiber dimension dictates what is inhaled, what is deposited, and which fibers reach the target tissue.

Durability of fibers and their persistence in the body may also play a significant role in determining carcinogenic response. An important question arises on how long fibers have to stay in the bronchial wall or serossa tissue to cause tumors. To persist, the fibers must be chemically durable, although the opposite may not be true; that is, durable fibers may not always be persistent. The durability of fibers depends on their chemical composition and crystal structure. Therefore, even though chemical composition by itself is not considered a direct factor in inducing carcinogenic response, indirectly it may be an important factor because it determines durability.

The morphology and durability factors have led to the suggestion (26) that the carcinogenicity of fibers depends on three Ds:

1. Dimension
2. Dose
3. Durability

3.3.1   Dimension

In relation to technical products, a fiber is defined as a "long, thin filament of material, possessing the useful properties of high tensile strength and flexibility." Dimensional criteria of a fiber, as defined by the American Society for Testing and Materials (ASTM), is a length-to-width ratio of greater than 10 : 1. With respect to airborne concentrations, a fiber is defined as "a particle with an aspect ratio (length to width) of at least 3 : 1."

A respirable fiber according to the ILO definition (6) is "a particle with a diameter of less than 3 μm and with an aspect ratio of 3 : 1 or greater." In the United States, the National Institute for Occupational Safety and Health (NIOSH) recommends exposure limits for glass fibers that are less than 3.5 μm in diameter and greater than 10 μm long (27). Lippmann (28) concluded that mesothelioma is most closely associated with exposure to fibers longer than 5 μm, and lung cancer is most closely associated with fibers longer than 10 μm and with diameters greater than 0.15 μm.

Table 31.6. Nominal Diameters of Some MMMF

Fiber Type	Nominal Diameter (μm)
Refractory (ceramic) fibers (2)	1.2–6.0
Fiberfrax® HSA	1.2
Fibermax® bulk	2–3.5
Fiberfrax® bulk	2–3.0
Alumina bulk	3
Zirconia bulk	3–6
Glass wool (2)	
Thermal insulation	6–15
Molded pipe insulation	7–9
Lightweight aircraft insulation	1.0–1.5
Special purpose	0.05–3.0
Continuous filaments (6)	6–15
Electrical insulation	6–9.5
Nextel® 312	8–12
Mineral wool	6–9

For technical products, MMMF are manufactured to specific nominal diameters varying according to fiber type and use. Particle size is a major factor in imparting thermal properties to insulating wools. Table 31.6 shows the fiber size range for different fiber types.

Table 31.6 indicates that MMMF in general have much larger diameters compared to asbestos fibers. This fact is even more apparent in Table 31.7 (durability ranking of fibers). Some researchers believe that comparatively larger fiber diameters of MMMF may explain the observed differences in the carcinogenic potency between MMMF and asbestos in both human and animal inhalation studies. The larger MMMF fibers may be deposited in the respiratory tract and may not reach the lung tissue in the same quantities as finer asbestos fibers.

3.3.2 Durability and Persistence

There appears to be a growing consensus among researchers that fibers must be durable and persist for a certain time to induce a tumor. There is, however, no precise knowledge or agreement among researchers about the length of time that fibers must persist to alter cells of the bronchial walls or the pleural or peritoneal tissues.

Durability here means the relative resistance of the fiber to dissolution through attack by biological fluids. To persist in the body, a fiber must be chemically durable. On the other hand, a chemically durable fiber may not always persist. The persistence of a fiber may be affected by dissolution, disintegration, elimination, or simple migration in the body. The importance of fiber persistence in inducing tumors has perhaps been best stated by Pott (29):

Only a few seconds of exposure to ionizing radiation can cause damage to cells which

Table 31.7. Durability Ranking of Fibers

Rank	Material	Median Diameter (μm)	Dissolution Velocity (nm/day)	Fiber "Life" (years)
1	Glass wool TEL	3.5	3.45	0.4
2	Glass wool, Superfine	0.38	1.4	1.0
3	Diabase wool	4.0	1.14	1.2
4	Glass wool 475, JM, 104	0.41	0.90	1.7
5	Glass wool E, JM, 104	0.47	0.21	6.5
6	Slag wool	4.8	0.69	2.0
7	Refractory, Fiberfrax R	1.85	0.27	5.0
8	Refractory, Fiberfrax H	1.85	0.28	4.9
9	Refractory, Silica	0.77	1.1	1.2
10	Chrysotile	(0.074)	0.005	(~100)
11	Crocidolite	(0.17)	0.011	(~110)
12	Erionite	(0.005)	0.0002	(~170)

Source: Scholze and Conradt (34).

will lead to a tumor after years or even decades. With chemical carcinogens, too, a very short exposure time can cause the decisive biological alteration which will be followed by a tumor after a long latent period. A fibre has to be regarded as a physical carcinogen that works by its elongated shape. Clearly, a fibre which is both durable and persistent should have a stronger effect than a non-durable or non-persistent one.

In general, durability studies indicate that glass and mineral wool fibers are less durable than ceramic fibers and asbestos. Several researchers (30–33) have investigated the relative durability of man-made mineral fibers through in vivo studies. Other researchers (34–36) conducted in vitro experiments to determine the relative resistance of MMMF to attack by biological fluids. These studies are summarized in Table 31.8.

A durability ranking (Table 31.7) by Scholze and Conradt (34) is particularly interesting. Table 31.8 shows the durability of various fiber types expressed by the number of years it will take to "dissolve" a fiber of 1 μm diameter in a simulated extracellular fluid (pH 7.6 ± 0.2) derived from Gamble's solution. These data show striking differences in durability between MMMF and natural fibers (chrysotile, crocidolite). Although MMMF were estimated to dissolve in less than 10 years, natural fibers require approximately 100 years or longer to dissolve. Also noticeable in Table 31.7 are the differences among MMMF. In general, glass wool fibers were found to be the least durable, mineral wools somewhat more durable, and some ceramic fibers the most durable.

Table 31.8. Summary of Fiber Durability Studies

Investigator(s)	Type of Study	Results
Morgan and Holmes (30)	In vivo	Short fibers (≤ 10 µm) dissolved slowly and uniformly. Longer fibers (≥ 30 µm) dissolved much more rapidly, and less uniformly
Johnson et al. (31)	In vivo	Examination by electron microscopy of fibers removed from rat lung—surface etching observed
Le Bouffant et al. (32)	In vivo	Erosion of fiber surfaces observed using electron microscopy. Loss of sodium and calcium noted
Bellmann et al. (33)	In vivo	Residence times for different types of fibers: ceramic fiber > rock wool > glass wool > microfiber
		Fibers with a high calcium content dissolve most rapidly
Scholze and Conradt (34)	In vitro	Natural and refractory fibers more durable than glass fiber and mineral wool
Leineweber (35)	In vitro	High variability observed in the solubility of glass fiber. Chemical composition of the glass appears to be determinant
Klingholz and Steinkopf (36)	In vitro	Residence times for different types of fibers: slag wool dissolved more rapidly than glass wool, which dissolved more rapidly than rock wool

Another noticeable feature of Table 31.7 is the difference in the median diameter of MMMF as compared to natural fibers. Because median diameters of 1 µm (used by Scholze and Conradt for comparing the durability of fiber types) are unlikely for natural fibers, the researchers used a multiple of 5 of the average diameter for the natural fibers shown in Table 31.7.

3.3.3 Dose

It is a well-established principle in toxicology that physiological response is related to the dose. Dose is perhaps the most critical factor in explaining the observed differences in the carcinogenic potency of asbestos and MMMF. Exposures of workers in MMMF industries to airborne fibers are generally much lower than exposures in similar processes where asbestos is involved. This is mostly due to large fiber diameters of MMMF compared to asbestos. Unlike asbestos fibers, MMMF do not split longitudinally into fibers of small diameter but tend to break

Table 31.9. Worker Exposure to Fibrous Glass and Mineral Wool at Seventeen U.S. Plants

Process and Plant	Average Intensity of Exposure Fibers (<3 $\mu m/cm^3$)			
	Mean	Std. Error	Min.	Max.
All fibrous glass plants	0.039	<0.001	0.0	1.500
Plant 1	0.027	<0.001	0.0	0.032
Plant 2	0.021	<0.001	0.0	0.093
Plant 4	0.008	<0.001	0.0	0.032
Plant 5	0.003	<0.001	0.0	0.003
Plant 6	0.061	0.003	0.0	0.320
Plant 9	0.067	0.001	0.0	1.500
Plant 10	0.293	0.007	0.0	0.320
Plant 11	0.005	<0.001	0.0	0.032
Plant 14	0.023	<0.001	0.0	0.032
Plant 15	0.026	<0.001	0.0	0.032
Plant 16	0.005	<0.001	0.0	0.032
All mineral wool plants	0.353	0.006	0.0	1.413
Plant 3	0.427	0.011	0.0	1.413
Plant 7	0.195	0.011	0.0	0.344
Plant 8	0.367	0.016	0.0	1.342
Plant 12	0.215	0.009	0.0	1.074
Plant 13	0.238	0.011	0.0	0.888
Plant 17	0.391	0.013	0.0	1.355

Source: Enterline et al. (8).

transversely into shorter segments. It is postulated that inhaled coarser MMMF may not always reach the target issue to cause comparable (to asbestos) damage.

3.3.3.1 Industrial Workplace Exposures. In the most comprehensive epidemiologic study of MMMF workers conducted to date in the United States, Enterline et al. (8) estimated the average intensity of worker exposures to respirable (less than 3 μm) fibers at 17 glass wool and mineral plants shown in Table 31.9. The data in Table 31.9 show fibrous glass concentrations around 0.03 fibers per cubic centimeter of air (fibers/cm³). Exposures in the mineral wool plants were higher, but less than 0.5 fibers/cm³ in all cases.

In general, these results are consistent with those reported by Corn et al. (37) (Table 31.10) for exposure in a rock wool and a slag wool plant. Corn reported higher fiber concentrations in the rock wool plant compared with the slag wool plant. Fiber concentrations in glass wool production reported by European investigators (38, 39) (Table 31.11) also are in general agreement with exposures reported by Enterline. These exposures are much lower than the comparable historical exposures to asbestos. Such low worker exposures to glass wool and continuous

Table 31.10. Average Concentrations of Total Fibers in a Rock Wool and a Slag Wool Plant

Dust Zone	Number of Samples	Average Number of Total Fibers (fibers/cm³)
Rock wool		
Warehouse	3	1.4
Mixing-Fourdrinier ovens	3	0.14
Panel finishing	12	0.40
Fibre forming	10	0.20
Erection and repair	13	0.24
Tile finishings	22	0.31
All samples	63	0.34
Slag wool		
Maintenance	15	0.08
Block production	8	0.05
Blanket line	5	0.05
Boiler room	2	0.05
Yard	2	0.09
Ceramic block	7	0.42
Shipping	3	0.04
Main plant	11	0.01
Mold formation	19	0.03
All samples	72	0.10

Source: Corn et al. (37).

filaments led Enterline to conclude that "equivalent exposures (0.03 fibers/cc or less) to asbestos may not produce detectable respiratory cancer."

3.3.3.2 End-User Exposures. Exposures of workers installing acoustical ceiling ducts, attic insulation, and aircraft insulation products have been reported by Esmen et al. (40) (Table 31.12). The highest exposures were found for workers installing mineral wool insulation in attics. These exposures ranged from 0.04 to 14.8 fibers/cm³. The exposures for fibrous glass duct installers were found to be low, ranging from 0.005 to 0.2 fibers/cm³. Esmen found the majority of the fibers to be in the respirable range (less than 3 μm in diameter).

3.3.3.3 Nonoccupational Exposures. Several investigators have reported ambient air concentrations in buildings in which MMMF insulation products have been applied. In one study, the concentration of glass fibers in ambient air samples taken from several sites in California ranged from nondetectable to 0.009 fibers/cm³ with an arithmatic mean of 0.0026 (41) for fibers greater than 2.5 μm diameter as determined by phase-contrast microscopy.

Hohr (42) reported chrysotile, amphibole, and glass fiber concentrations in the air of several cities in the Federal Republic of Germany (Table 31.13). In all cases,

Table 31.11. Respirable Glass Fiber Concentrations

Exposure Process	At Four European Plants (38)			At Two Swedish Plants (39)		
	Number of Samples	Mean	Range	Number of Samples	Mean	Range
Preproduction	23	0.01	<0.01–0.03	—	—	—
Production	153	0.04	<0.01–0.62	49	0.22	0.056–0.65
Maintenance	63	0.04	<0.01–0.17	89	0.36	0.037–5.3
General	47	0.03	<0.01–0.06	34	0.19	0.034–0.53
Secondary process 1	131	0.03	<0.01–0.21	59	0.19	0.038–0.73
Secondary process 2	70	0.43	0.02–4.02	5	0.13	0.083–0.16
Cleaning	4	0.01	0.01–0.02	76	0.21	0.026–1.0

Table 31.12. Concentrations of Respirable Fibers Installation of MMMF Insulation

Product and Job Classification	No. of Samples	Fiber Concentration (fibers/cm³)		Average Respirable Fractions
		Average	Range	
Acoustical ceiling installer	12	0.003	0–0.006	0.55
Duct installation				
Pipe covering	31	0.06	0.007–0.38	0.82
Blanket insulation	8	0.05	0.025–0.14	0.71
Wrap around	11	0.06	0.03–0.15	0.77
Attic insulation				
Fibrous glass				
Roofer	6	0.31	0.07–0.93	0.91
Blower	16	1.8	0.67–4.8	0.44
Feeder	18	0.70	0.06–1.48	0.92
Mineral wool				
Helper	9	0.53	0.04–2.03	0.71
Blower	23	4.2	0.50–14.8	0.48
Feeder	9	1.4	0.26–4.4	0.74
Building insulation installer	31	0.13	0.013–0.41	0.91
Aircraft insulation				
Plant A				
Sewer	16	0.44	0.11–1.05	0.98
Cutter	8	0.25	0.05–0.58	0.98
Cementer	9	0.30	0.18–0.58	0.94
Isolated jobs	7	0.24	0.03–0.31	0.99
Plant B				
Sewer	8	0.18	0.05–0.26	0.96
Cutter	4	1.7	0.18–3.78	0.99
Cementer	1	0.12	—	0.93
Isolated jobs	3	0.05	0.012–0.076	0.94
Fibrous glass duct				
Duct fabricator	4	0.02	0.006–0.05	0.66
Sheet-metal worker	8	0.02	0.005–0.05	0.65
Duct installer	5	0.01	0.006–0.20	0.87

Source: Esmen et al. (40).

glass fiber concentrations were less than 0.002 fibers/cm³ and, in general, lower than the concentrations of chrysotile and amphibole fibers.

Measurements of MMMF in schools and office buildings have been reported by researchers in Denmark (43, 44) (Table 31.14). In all cases, the concentrations of respirable MMMF reported were 0.001 fibers/cm³ or lower. These researchers also

Table 31.13. Fiber Concentrations in Ambient Air In the Federal Republic of Germany

Measuring Site	No. of Samples	Concentration (fibers/cm³)			
		Total	Chrysotile	Amphiboles	Glass
Duisburg	17	0.041	0.0022	0.0019	0.00050
Dortmund	6	0.036	0.0026	0.0019	0.00170
Dusseldorf	21	0.027	0.0014	0.0013	0.00040
Krahm (rural area)	9	0.012	0.0005	0.0007	0.00004

Source: Hohr (42).

reported other respirable and nonrespirable fibers (including organics) in the buildings.

3.4 Carcinogenicity Evaluation

An important symposium on health implications of MMMF was held in Copenhagen October 28 to 29, 1986. This international symposium was organized by the Regional Office for Europe of the World Health Organization (WHO), IARC, and the Joint European Medical Research Board in association with the Thermal Insulation Manufacturers Association of America. Many researchers discussed their findings at this conference, including the three epidemiologic studies described earlier.

The central issue that the participants in the symposium were confronted with was the interpretation of the epidemiologic studies which showed a moderate increase in mortality rates among workers exposed to MMMF. Interpretation of these studies is made difficult not only because the excess mortality rates observed are only moderate, but because of the uncertainty of the fiber counts in the early days of the industry and the extent of other potentially carcinogenic materials present in the workplace. These extraneous exposures include arsenic in copper

Table 31.14. Mean Dust and Fiber Concentrations in Schools and Office Buildings in Denmark

No. of Buildings	Respirable MMMF (fibers/cm³)	Nonrespirable MMMF (fibers/cm³)	Other Respirable Fibers (Including Organics) (fibers/cm³)	Other Nonrespirable Fibers (Including Organics) (fibers/cm³)
10	0.0001	0.00002	0.18	0.013
6	0.0001	0.00004	0.15	0.011
8	0.00004	0–0.00008	0.17	0.012

Source: Schneider (43) and Rindel et al. (44).

slag, polycyclic aromatic hydrocarbons in cupola melting operations, silica (crystobalite) in refractory fiber production and use, and asbestos in some MMMF workplaces.

In spite of these complications, participants of the Copenhagen symposium reached some definitive conclusions summarized by the symposium chairman, Sir Richard Doll, as follows (45):

- There has been a risk of lung cancer in people employed in the early days of both the rock or slag and glass wool sectors of the MMMF industry. The risk has been approximately 25 percent above normal for 30 years after first employment. This risk is numerically substantial, however, because lung cancer is so common.
- The risk has been greater in the rock or slag wool sector than in the glass wool sector.
- No risk has been demonstrated in the glass filament sector.
- A variety of carcinogens has contributed to the hazard.
- Uncertainty about the fiber counts in the early days of the industry and the extent of the contribution of other carcinogens make it impossible to provide a precise quantitative estimate of the likely effect of exposure to current fiber levels.
- No specific hazard other than a hazard of lung cancer has been established.

IARC also took into consideration the available research, and presented its findings in the IARC monograph on man-made mineral fibers and radon, Volume 43.

In evaluating carcinogenicity, IARC takes into account the total body of evidence and describes the chemical according to the wording of one of the categories shown in Table 31.15. Assignment of IARC category is a matter of scientific judgment reflecting the strength of evidence based on animal and human studies and other relevant data.

IARC's evaluation of the available scientific evidence and the carcinogenicity groupings for the various MMMF is summarized in Table 31.16.

4   REGULATION OF MMMF

Despite major concerns about the potential health effects, particularly the carcinogenic effect of MMMF, only a few countries have adopted regulations to limit exposure to MMMF. It appears that most countries are awaiting the outcome of ongoing epidemiologic studies to determine to what extent they should regulate MMMF. Most countries regulate MMMF either as total dust or respirable dust, and in a few cases, as fibers (1, 6). In some countries, such as Italy and Japan, the silica content of the dust determines the permissible exposure limit. Typical exposure limits for MMMF are shown in Table 31.17.

Table 31.15. IARC Carcinogenicity Grouping

IARC Group	Evaluation	Evidence in Humans	Evidence in Animals
1	Carcinogenic to humans	Sufficient	—
2A	Probably carcinogenic to humans	Limited	Sufficient
2B	Possibly carcinogenic to humans	Limited	Absence of sufficient evidence
			or
		Inadequate	Sufficient
			or
		Inadequate	Limited
3	Not classifiable as to its carcinogenicity to humans	—	—
4	Probably not carcinogenic to humans	Evidence suggests lack of carcinogenicity	Evidence suggests lack of carcinogenicity

Source: IARC (2).

Table 31.16. Summary of IARC Findings on the Carcinogenicity of MMMF

Fiber Type	Evidence in Humans	Evidence in Animals	IARC Group	Overall IARC Evaluation
Rock wool	Limited	Limited	2B	Possibly carcinogenic to humans
Slag wool	Limited	Inadequate	2B	Possibly carcinogenic to humans
Glass wool	Inadequate	Sufficient	2B	Possibly carcinogenic to humans
Continuous filament	Inadequate	Inadequate	3	Not classifiable as to the carcinogenicity to humans
Refractory (ceramic)	No data	Sufficient	2B	Possibly carcinogenic to humans

Source: IARC (2).

Currently in the United States, there are no regulatory limits for MMMF. NIOSH has recommended an exposure limit of 3 fibers/cm^3 for fibers with a diameter of less than 3.5 μm and a length greater than 10 μm (27). Indirectly, however, the U.S. Occupational Safety and Health Administration (OSHA) hazard communication standard (46) specifically applies to MMMF. According to OSHA's standard, all materials with an IARC carcinogenicity rating of 2B must be labelled as suspect carcinogens on Material Safety Data Sheets.

5 SURVEYS AND INSPECTIONS

5.1 Identification of MMMF Products in Buildings

The first step in a survey for MMMF is to find the presence of such materials in construction products or pipe wrap. In general, visual observation is not sufficient to determine whether the suspect material is asbestos or MMMF. Bulk sampling of the material is therefore necessary to ascertain whether the material contains asbestos or MMMF.

The most difficult identification problem is encountered in the building where it is common to find both asbestos and MMMF in insulation, pipe wrap, and other building-related products. Because the health hazards of asbestos have become better known, MMMF have been increasingly used to replace asbestos-containing materials during renovation and maintenance and repair activities. The result is that very often several types of fibrous materials are encountered in the same facility. It is therefore important to identify these materials by taking an appropriate number of bulk samples and analyzing them by polarized light microscopy.

Table 31.17. Worldwide Exposure Limits for MMMF

Country	Fiber Type	Exposure Limit		Comments
		As Total Dust (mg/m³)	As Fiber Concentration (fibers/cm³)	
Bulgaria	All fibers with diameters >3 μm	3	—	
	Fibers with diameters <3 μm	2		
Czechoslovakia	Glass	8	—	
		4	0.2	Proposed limits
Denmark	All	5	—	Nonstationary workplaces
		—	2	Stationary workplaces
Federal Republic of Germany	All	6	—	Fibers with diameters <1 μm are listed as suspect carcinogens
German Democratic Republic	All	2	—	

Country	Material	Value		Notes
Italy	All		—	
Japan	All	$\dfrac{30}{\% \text{ quartz} + 3}$	—	
		$\dfrac{2.9}{(0.22)\,(\% \text{ quartz}) + 1}$	—	
Norway	All	5	1	
New Zealand	All	5	2	(diameter <3 μm)
Poland	All	4	1	(length <5 μm)
Sweden	All	—	2	
United Kingdom	All	5	1	Recommended for fibers with diameter <3 μm
United States				
OSHA	Fibrous glass	No limit established		
NIOSH		5	3	(diameter < 3.5 μm and length >10 μm)
USSR	All	2 mg/m^3	—	
Yugoslavia	All	12 mg/m^3	—	

Source: IARC (2) and ILO (6).

319

5.2 Air Sampling

The concentrations of MMMF in air have been determined on the basis of total dust or respirable dust in the air or on the number of fibers present per unit volume of air. Samples of air are drawn through a filter at a flow rate of 2.0 liters/min. The filters used are made of either mixed cellulose ester or polyvinyl chloride (PVC). For respirable mass, a cyclone sampler operated at a flow rate of 1.7 liters/min provides an adequate approximation of the respirable mass fraction. Filter faces should be directed downward to minimize the possibility of direct contamination of the filters by large nonrespirable particles ejected from a high speed process. To determine fiber concentrations, care must be taken not to overload the filters. This can be accomplished by taking several samples of short duration and time-weighting the average exposures of the samples for the entire shift.

5.3 Analysis of Air Samples

5.3.1 Gravimetric Analysis

Commonly employed in the past, the gravimetric method is easy to use, is efficient, and provides a reasonably good measure of the overall condition of the work environment.

Gravimetric determinations are used more extensively for MMMF as compared with asbestos. This is based on the observation that in many work environments containing MMMF, fibers constitute only a small fraction of the overall environmental burden.

In the past, several sampling strategies were used to characterize the work environment containing MMMF. Although most common techniques have been to measure either fiber concentrations or dust concentrations gravimetrically, some investigators have used "semi-specific analytical methods" for studies of MMMF exposure, particularly fibrous glass exposures. Johnson et al. (47) in 1987 performed chemical analysis of air samples for total silica content and estimated the amount of glass dust based on the silica content. This method has problems because silica content can differ considerably for the various types of glass. In another survey, Fowler (48) analyzed total dust samples for metals, such as cadmium (Cd), chromium (Cr), cobalt (Co), nickel (Ni), manganese (Mn), lead (Pb), and zinc (Zn), to determine the exposure of workers to mineral wool fibers.

The advisability of determining concentrations of MMMF on the basis of weight alone is questionable because it is the number and dimensions of the fibers and not the total weight that determines toxicologic response. Gravimetric determination is particularly unsuited as an indicator of exposure to fibers of smaller diameters. For fibers of the same length, fiber weight is a function of the square of the diameter. Therefore, a fiber with a diameter of 1 μm weighs 100 times as much as a 0.1 μm diameter fiber of the same length. The net result, therefore, is that the presence of just a few large-diameter fibers can increase the total weight appreciatively. The usefulness of total weight or even respirable weight determinations is particularly limited when respirable fibers are involved.

5.3.2 Microscopic Analysis

For fiber counting, the analytical methods employed are phase-contrast microscopy (PCM) or electron microscopy. The PCM method involves collecting samples with a membrane filter and counting the fibers by PCM at a magnification of 400 to $450\times$. The PCM method is more effective for analyzing MMMF than asbestos fibers because MMMF found in the workplace air are larger than asbestos fibers. Because most MMMF have relatively large diameters, they are easily resolved by optical microscopy. Using the PCM counting method, fiber levels as low as 0.01 fibers/cm^3 or lower can be determined. However, compared to asbestos, there has been much less standardization of MMMF techniques. In the United States, NIOSH has recommended a PCM method of fibrous glass determination. In the NIOSH-recommended method, the fibers with diameters less than 3.5 μm and length greater than 10 μm are counted. There has been, however, no evaluation of this technique through either intra-laboratory or inter-laboratory trials.

In Europe, the MMMF counting methods have been more extensively studied and evaluated. In the mid-1970s, the World Health Organization (WHO) in Copenhagen initiated a program known as the WHO/EURO Reference Scheme (49). The program's goals were to produce reference methods for sampling and evaluating MMMF and to minimize interlaboratory variations in the results obtained with these methods. The WHO/EURO reference method (50) is also based on the membrane filter technique used to determine asbestos fiber concentrations. In the WHO/EURO reference method, the number of respirable fibers in randomly selected areas of the filter is counted using a magnification of about $500\times$. The respirable fibers counted are fibers that are longer than 5 μm, have a diameter of less than 3 μm, and have an aspect ratio of 3 to 1 or greater. In contrast to the asbestos fiber counting rules, fibers in contact with particulate or other fibers are counted provided they meet the above criteria.

Although PCM is quite suitable to identify MMMF in most cases, there is a need to distinguish MMMF positively from asbestos fibers because, increasingly, both MMMF and asbestos fibers are present in the same environment. To identify and confirm the fiber types further, both scanning electron microscopy (SEM) and transmission electron microscopy (TEM) are used.

The WHO/EURO technical committee for monitoring and evaluating MMMF also established an SEM reference method in which samples are collected on a polycarbonate filter (Nucleopore®) or a PVC membrane filter. After preparation, the samples are observed with an SEM at a magnification of $5000\times$. The SEM method determines fiber concentration as well as fiber size. Most SEMs are also capable of performing energy dispersive X-ray diffraction analysis which enables the analyst to distinguish between fiber types by determining elemental composition of the fiber.

The specific identification of fiber type can be even more effectively made through TEM. Because most MMMF are amorphous fibers and are not crystalline in nature, it is possible to distinguish them from noncrystalline materials by electron diffraction of the individual fibers. The transmission electron microscope can provide very high resolution. For identification of fibers by morphology, the electron microscope is used at a magnification of 20,000×. A magnification of 20,000× is also used when performing selected area electron diffraction of individual fibers.

5.4   Other Environmental Measurements

In any work environment containing MMMF, industrial hygienists should also be aware of other occupational exposures. These exposures may be to excessive concentrations of dust, heavy metals such as lead, chromium, cadmium, cobalt, and nickel, polycyclic aromatic hydrocarbons, carbon monoxide, and other contaminants. These exposures should be given appropriate consideration while evaluating the work environment. Other industrial hygiene aspects such as ventilation, housekeeping, work practices, and personal protective measures must also be evaluated.

6   CONTROL OF MMMF

The specific types of controls necessary to prevent or reduce occupational exposures to man-made mineral fibers depend heavily on the type and application of the product containing the fibers. These control measures are not unique to MMMF, but rely on methods well established in the practice of industrial hygiene. These measures include engineering controls, work practices, and the use of personal protective equipment.

6.1   Engineering Controls

Fabrication operations, including cutting, shaping, and drilling of MMMF products can generate airborne fibers by mechanical disturbance. In production operations, such exposures are best controlled through the use of local exhaust ventilation systems. In maintenance operations and construction, use of power tools, such as drills and saws may also generate airborne fibers. Such tools may be ventilated using high-efficiency particulate air (HEPA) filter-equipped vacuum cleaners. Hoods designed to fit standard power tools are commercially available.

6.2   Work Practices

The use of work practices that minimize disturbance of fibers is critical. In many instances in both industry and construction, the use of engineering controls may not be feasible. This makes it critical to minimize the amount of airborne fibers generated while handling the MMMF-containing products and during any subsequent cleanup. The following work practices are recommended:

- Use of unventilated power tools should be minimized because the high operating speed can generate greater amounts of airborne dust. Hand tools are preferred because they generally produce less dust.

- Some products should be lightly sprayed with water, amended with a surfactant to enhance penetration prior to handling. This helps to minimize fiber release.
- Good housekeeping is an essential part of any safe construction or maintenance project. Materials that fall to the ground should be picked up as soon as possible because walking on scraps can break them into smaller pieces that can more easily become airborne or may be tracked further around the facility. Good housekeeping in the work area helps to minimize airborne dust. All external surfaces of equipment should be kept free of dust accumulation which, if dispersed, contributes to airborne concentrations. Such cleanup should be performed only using HEPA filter-equipped vacuum cleaners or wet methods, and never by dry sweeping.

6.3 Personal Protective Equipment

Persons working with loose insulation products should wear protective clothing to minimize skin contact and resulting irritation. Either disposable or launderable clothing is acceptable. Protective clothing should be laundered separately from other clothes and should never be worn or taken home by employees for laundering.

If engineering and work practice controls are not adequate to prevent exposure to airborne fibers, use of respiratory protection may be necessary. In selecting respiratory protection, guidelines established by OSHA for asbestos may be used.

7 CONCLUSIONS

Man-made mineral fibers are extremely important commercial products. Their use as commercial insulating materials and plastic reinforcement products have provided significant benefits to society. As asbestos substitutes, MMMF are finding wide application. The major epidemiologic studies conducted to date have shown only moderately increased mortality rates for a few specific diseases, in particular, lung cancer, for cohorts of industrial workers.

Although many questions still remain unanswered, evidence to date suggests that the carcinogenic risk from MMMF is much less when compared to asbestos. Perhaps the most definitive statement on this complex issue was made by Sir Richard Doll, Chairman of the 1986 Copenhagen symposium, in his concluding remarks (45):

> If, now, I abandon the firm basis of scientific fact and express a personal judgment, I do so because I know that in the absence of such a conclusion many people may think that the symposium has been a waste of time. Let me therefore add a seventh conclusion that, taking into account also the results of animal experiments, the experience of the asbestos industry and the experience of the glass filament sector of the MMMF industry, MMMF are not more carcinogenic than asbestos fibres and exposure to current mean levels in the manufacturing industry of 0.2 F_r mL^{-1}* or less is unlikely to produce a measurable risk after another 20 years have passed.

*0.2 F_rmL^{-1} = 0.2 respirable fibers per milliliter of air.

Because of the commercial importance of these materials, a number of studies are underway. Until we have a better understanding of the health effects of man-made mineral fibers, it is prudent for industrial hygienists to treat these materials with the same precautions as asbestos by instituting appropriate work practices and engineering controls to minimize worker and community exposures.

REFERENCES

1. G. A. Peters and B. J. Peters, *Sourcebook on Asbestos Diseases: Medical, Legal and Engineering Aspects*, Vol. 2, Garlanel Law Publishing, New Toni, NY, 1986, pp. 190–2102.

2. IARC Monographs on the Evaluation of Carcinogenic Risks to Humans, Vol. 43: *Man-made Mineral Fibers* (*Proceedings of a WHO/IARC Conference*), Vol. 2, World Health Organization, Copenhagen, 1984, pp. 234–252.

3. H. J. Smith, "History, Processes, and Operations in the Manufacturing and Uses of Fibrous Glass—One Company's Experience," in *Occupational Exposure in Fibrous Glass, A symposium*, National Institute for Occupational Safety and Health, Cincinnati, OH, 1976, pp. 19–26.

4. J. M. Dement, "Preliminary Results of the NIOSH Industry-wide Study of the Fibrous Glass Industry," DHEW(NIOSH) Publ. No. 1W3.35.3b; NTIS Pub. 40. PB-81-224693), National Institute for Occupational Safety and Health, Cincinnati, OH, 1973, pp. 1–5.

5. Anonymous, "Facts and Figures," *Chem. Engl. News*, **64**, 32–44 (1986).

6. International Labour Organization, "Working Document on Safety in the Use of Mineral and Synthetic Fibres of a Meeting of Experts," ILO, Geneva, 1989.

7. L. Simonato, A. C. Fletcher, J. Cherrie, A. Andersen, P. Bertazzi, N. Charnay, J. Claude, J. Dodgson, J. Esteve, R. Frentzel-Beyme, M. J. Gardner, O. Jensen, J. Olsen, L. Teppo, R. Winkelmann, P. Westerholm, P. D. Winter, C. Zocchetti and R. Saracci, "The International Agency for Research on Cancer Historical Cohort Study of MMMF Production Workers in Seven European Countries: Extension of the Follow-up," *Ann. Occup. Hyg.*, **31**, 603–623 (1987).

8. P. E. Enterline, G. M. Marsh, V. Henderson, and C. Callahan, "Mortality Update of Cohort of U.S. Man-made Mineral Fibre Workers," *Ann. Occup. Hyg.*, **31**, 625–656 (1987).

9. F. Pott, U. Ziem, and U. Mohr, "Lung Carcinomas and Mesotheliomas Following Intratracheal Installation of Glass and Asbestos," in W. I. Bergbau-Berufsgenossen, International Pneumoconiosis Conference, Bochum, 1984, pp. 746–757.

10. M. F. Stanton, M. Layard, A. Tegeris, E. Miller, M. May, and E. Kent, "Carcinogenicity of Fibrous Glass: Pleural Response in the Rat in Relation to Fiber Dimension," *J. Natl. Cancer Inst.*, **58**, 587–603 (1977).

11. A. M. Sincock, J. D. A. Delhanty, and G. Casey, "A Comparison of the Cytogenic Response to Asbestos and Glass Fibre in Chinese Hamster and Human Cell Lines," *Mutat. Res.*, **101**, 257–268 (1982).

12. T. W. Hesterberg, and J. C. Barrett, "Dependence of Asbestos and Mineral Dust Induced Transformation of Mammalian Cells in Culture on Fiber Dimension," *Cancer Res.*, **44**, 2170–2180 (1984).

13. M. Oshimura, T. W. Hesterberg, T. Tsutsui, and J. C. Barrett, "Correlation of Asbestos-induced Cytogenetic Efforts with Cell Transformation of Syrian Hamster Embryo Cells in Culture," *Cancer Res.*, **44**, 5017–5022 (1984).

14. M. Chamberlain, and E. M. Tarmy, "Asbestos and Glass Fibres in Bacterial Mutation Tests," *Mutat Res.*, **43**, 159–164 (1977).

15. L. LeBouffant, J-P. Henn, J. C. Martin, C. Normand, G. Tichoux, and R. Trolarel, "Distribution of Inhaled MMMF in the Rat Lung—Long Term Effects," in *Biological Effects of Man-Made Mineral Fibers* (Proceedings of a WHO/IARC Conference), Vol. 2, WHO, Copenhagen, 1984, pp. 143–168.

16. J. C. Wagner, G. B. Berry, R. J. Hill, D. E. Munday, and J. S. Skidmore, "Animal Experiments with MMM(V)F—Effects of Inhalation and Intrapleural Inoculation in Rats," in *Biological Effects of Man-made Mineral Fibres* (Proceedings of a WHO/IARC Conference), Vol. 2, WHO, Copenhagen, 1984, pp. 209–233.

17. H. Muhle, F. Pott, B. Bellmann, S. Takenaka, and U. Ziem, "Inhalation and Injection Experiments in Rats to Test the Carcinogenicity of MMMF," *Ann. Occup. Hyg.*, **31**, 755–764 (1987).

18. K. P. Lee, G. E. Barras, F. D. Griffith, and R. S. Warity, "Pulmonary Response to Glass Fibre by Inhalation Exposure," *Lab Invest.*, **40**, 123–133 (1979).

19. D. M. Smith, L. W. Ortiz, R. F. Archuleta, and N. F. Johnson, "Long-term Health Effects in Hamsters and Rats Exposed Chronically to Man-made Vitreous Fibers," *Ann. Occup. Hyg.*, **31**, 731–754 (1987).

20. B. Goldstein, I. Webster, and R. E. C. Rendall, "Changes Produced by the Inhalation of Glass Fibre in Non-human Primates," Proceedings of a WHO/IARC Conference in Association with JEMRB and TIMA, Copenhagen, April 20–22, 1982, World Health Organization, Regional Office for Europe, Copenhagen, 1984, pp. 273–286.

21. E. E. McConnell, J. C. Wagner, J. W. Skidmore, and J. A. Moore, "A Comparative Study of the Fibrogenic and Carcinogenic Effects of UICC Canadian Chrysotile B Asbestos and Glass Microfibre (JM 100)," in *Biological Effects of Man-made Mineral Fibres* (Proceedings of a WHO/IARC Conference), Vol. 2, World Health Organization, Copenhagen, 1984, pp. 234–252.

22. F. Pott, U. Ziem, F. J. Reiffer, F. Huth, H. Ernst, and U. Mohr, "Carcinogenicity Studies on Fibres, Metal Compounds and Some other Dusts in Rats," *Exp. Pathol.*, **32** 129–152 (1987).

23. J. M. G. Davis, J. Addison, R. E. Bolton, K. Donaldson, A. D. Jones, and A. Wright, "The Pathogenic Effects of Fibrous Ceramic Aluminum Silicate Glass Administered to Rats by Inhalation or Peritoneal Injection," in *Biological Effects of Man-made Mineral Fibres* (Proceedings of WHO/IARC Conference), Vol. 2, World Health Organization, Copenhagen, 1984, pp. 303–322.

24. G. H. Pigott, and J. Ishmael, "A Strategy for the Design and Evaluation of a 'Safe' Inorganic Fibre," *Ann. Occup. Hyg.*, **26**, 371–380 (1982).

25. H. S. Shannon, E. Jamieson, J. A. Julian, D. C. F. Muir, and C. Walsh, "Mortality Experience of Ontario Glass Fibre Workers—Extended Follow up," *Ann. Occup. Hyg.*, **31**, 657–662 (1987).

26. M. Corn, personal communication, 1988.

27. NIOSH, "Criteria for a Recommended Standard Occupational Exposure to Fibrous Glass," National Institute for Occupational Safety and Health, Publication no. 77-152, 1977.

28. Morton Lippmann, "Review of Asbestos Exposure Indices," *Environ. Res.*, **46**, (1) June 1988.

29. F. Pott, "Problems in Defining Carcinogenic Fibres," *Ann. Occup. Hyg.*, **31**, Hc. 4B, 799–802 (1987).

30. A. Morgan, and A. Holmes, "Solubility of Asbestos and Man-made Mineral Fibers in Vitro and in Vivo: Its Significance in Lung Disease," *Environ. Res.*, **39**, 475–484 (1986).

31. N. F. Johnson, D. M. Griffiths, and R. J. Hill, "Size Distribution Following Long-term Inhalation of MMMF," in *Biological Effects of Man-made Mineral Fibres* (Proceedings of a WHO/IARC Conference), Vol. 2, World Health Organization, Copenhagen, 1984, pp. 102–125.

32. L. Le Bouffant, H. Daniel, J. P. Henin, J. C. Martin, C. Normand, G. Tichoux, and F. Trolard, "Experimental Study on Long-term Effects of Inhaled MMMF on the Lung of Rats," *Ann. Occup. Hyg.*, **31**, 765–790 (1987).

33. B. Bellmann, H. Muhle, F. Pott, H. Konig, H. Kloppel, and K. Spurny, "Persistence of Man-made Mineral Fibres (MMMF) and Asbestos in Rat Lungs, 1987," *Ann. Occup. Hyg.*, **31**, 693–709 (1987).

34. J. Scholze, and R. Conradt, "An in Vitro Study of the Chemical Durability of Siliceous Fibres," *Ann. Occup. Hyg.*, **31**, 683–692 (1987).

35. J. P. Leineweber, "Solubility of Fibres in Vitro and in Vivo," in *Biological Effects of Man-made Mineral Fibres* (Proceedings of a WHO/IARC Conference), Vol. 2, World Health Organization, Copenhagen, 1984, pp. 87–101.

36. R. Klingholz, and B. Steinkopf, The Reactions of MMMF in a Physiological Model Fluid and in Water, in *Biological Effects of Man-made Mineral Fibres* (Proceedings of a WHO/IARC Conference) Vol. 2, World Health Organization, Copenhagen, 1986, pp. 60–86.

37. M. Corn, Y. Y. Hammad, D. Whittier, and N. Kotsko, "Employee Exposure to Airborne Fiber and Total Particle Matter in Two Mineral Wool Facilities," *Environ. Res.*, **12**, 59–74 (1976).

38. J. Cherrie, J. Dodgson, S. Groat, and W. MacLaren, Environmental Surveys in the European Man-made Mineral Fiber Production Industry, *Scand. J. Work Environ. Health.*, **12** (Suppl. I), 18–25 (1986).

39. Arbetarskyddsstyrelsen (National Swedish Board of Occupational Safety and Health), Measurement and Characterization of MMMF Dust (Partial Reports 3-9), Stockholm, 1981.

40. N. A. Esmen, M. J. Sheehan, M. Corn, M. Engel, and N. Kotsko, "Exposure of Employees to Man-made Vitreous Fibers: Installation of Insulation Materials," *Environ. Res.*, **28**, 386–398 (1982).

41. J. L. Balzer, "Environmental Data: Airborne Concentrations Found in Various Operations," in W. N. LeVee, and P. A. Schulte, Eds., *Occupational Exposure to Fibrous Glass* (DHEW Publ. No. (NIOSH) 76-151; NTIS Publ. No. PB-258869), National Institute for Occupational Safety and Health, Cincinnati, OH, 1976, pp. 83–89.

42. D. Hohr, "Investigations by Transmission Electron Microscopy of Fibrous Particles in Ambient Air" (Ger.), *Staub. Reinhalt. Luft*, **45**, 171–171 (1985).

43. T. Schneider, "Man-made Mineral Fibers and Other Fibers in the Air and in Settled Dust," *Deniron. Int.*, **12**, 61–65 (1986).

44. A. Rindel, E. Bach, N. O. Breum, C. Hugod, and T. Schneider, "Correlating Health

Effect with Indoor Air Quality in Kindergartens," *Int. Arch. Occup. Environ. Health*, **59**, 363–373 (1987).

45. R. Doll, "Symposium on MMMF—Overview & Conclusions," *Ann. Occup. Hyg.*, **31**, (4B), 805–819 (1987).

46. U.S. Dept. of Labor, Occupational Safety & Health Administration, Code of Federal Regulations, Title zq, 1910–1200.

47. D. L. Johnson, J. J. Healey, H. E. Ayer, and J. R. Lynch, "Exposure to Fibers in the Manufacture of Fibrous Glass," *Am. Ind. Hyg. Assoc. J.*, **30**, 545–550 (1969).

48. D. P. Fowler, "Industrial Hygiene Surveys of Occupational Exposures to Mineral Wool," National Institute for Occupational Safety & Health Publication No. 80-135, 1980.

49. WHO/EURO Technical Committee for Monitoring and Evaluating MMMF, "The WHO/EURO Man-made Mineral Fibre Reference Scheme," *Scand. J. Work Environ. Health*, **11**, 123–129 (1985).

50. WHO/EURO Technical Committee for Monitoring and Evaluating MMMF, "The Reference Methods for Measuring Airborne Man-made Minerals Fibres (MMMF)," WHO Regional Office for Europe, Environmental Health Report 4, 1985.

CHAPTER THIRTY-TWO

Environmental Control in the Workplace: Water, Food, Wastes, Rodents

Maurice A. Shapiro

1 INTRODUCTION

In the second and third, revised editions of this book, the chapter on industrial sanitation was introduced as follows:

> Industrial sanitation is essentially a specialized application of community environmental health services. Within the purview of industrial sanitation are the principles involved in controlling the spread of infection or other insults to the health of the employee not inherent in the manufacturing process per se.

The chapter heading change from "Industrial Sanitation" in the third edition to "Environmental Control In the Workplace" reflects temporal and attitudinal shifts manifested since 1978. Nevertheless, the definition is still valid; and because the objective of industrial hygiene is to safeguard the health of working people and improve their work environments, general environmental control in the workplace should be an intrinsic function of occupational safety and health. Exposures to pathogenic organisms and toxic substances can and do lead to illness among employees. For example, if an industrial establishment provides its employees with a food service but fails to supply sanitary food-handling facilities and practices af-

Patty's Industrial Hygiene and Toxicology, Fourth Edition, Volume 1, Part B, Edited by George D. Clayton and Florence E. Clayton.
ISBN 0-471-50196-4 © 1991 John Wiley & Sons, Inc.

fording maximum protection, it invites the disaster of widespread food-borne infection by harmful organisms and their toxins or by other poisonous materials. In the intervening years since the third edition, significant changes affecting the environment and environmental health regulation have taken place. Unchanged, the Williams–Steiger Occupational Safety and Health Act of 1970 includes sections relating to "General Environmental Controls," "Sanitation," "Temporary Labor Camps," and "Non-Water Carriage Disposal Systems." One major environmental regulatory change has been the passage of the Safe Drinking Water Act Amendments (SDWA) of 1986 (P.L. 99-339), which resulted from a growing concern about unregulated organic contaminant pollution of groundwater by industrial solvents and pesticides, as well as the parasite *Giardia lamblia*. The amendments are far-reaching with important new responsibilities placed on public and private entities purveying drinking water supplies. For example, mandated standards for 83 contaminants must be established and regulated by the Environmental Protection Agency (EPA) according to a three-year time table. By 1991, at least 25 more primary standards will be required. In addition, new criteria have been mandated for filtration of surface water supplies, and by 1990 new criteria must be specified for disinfection of surface and ground water.

The grandfather of environmental legislation, the Clean Water Act (amended by P.L. 95-217, P.L. 97-117, and P.L. 100-4), originally the Federal Water Pollution Act (P.L. 92-500), enacted in 1948, is the principal law governing water quality in the nation's streams, lakes, and estuaries. The act has two major sections: first, regulatory provisions that impose progressively more stringent requirements on industries and municipalities to treat their liquid wastes and meet the statutory goal of zero discharge of pollutants, and second, financial assistance for publicly owned wastewater treatment facilities. Because industrial establishments are significant contributors of wastewater to publicly owned sewerage and treatment systems, the Clean Water Act includes several provisions that require industrial wastewater sources to pretreat wastes that could disrupt the biological processes or could not be handled by treatment plants or would otherwise interfere with operation of the process. The National Pollutant Discharge Elimination System, as codified in 40 CFR Parts 122, 123, is a final rule making that requires that renewal permit applications include listed effluent data at the time of submission. The 1987 amendments to the Clean Water Act (P.L. 100-4) added a new Section 319 to the act, which requires states to develop and implement control of nonpoint sources of pollution, including construction and mining sites, and to focus on urban areas.

The Toxic Substances Control Act (TSCA) (P.L. 94-469) was enacted into law in 1976 to address the risks from hazardous chemicals. Continuing environmental contamination by toxic substances highlighted the fact that prior to the enactment of TSCA these damaging episodes were handled on an ad hoc basis through the Clean Air Act; the Clean Water Act; the Federal Insecticide, Fungicide and Rodenticide Act; the Consumer Product Safety Act; or the Occupational Safety and Health Act. TSCA legislation, originally proposed in 1971 was enacted in 1976, provides the EPA with the authority to:

- Induce testing of existing chemicals, those currently in widespread commercial production or use
- Prevent future chemical risks through premarket screening and regulatory tracking of new chemical products
- Control of unreasonable risks of chemicals already known or as they are discovered
- Gather and disseminate information about chemical production, use, and possible adverse effects on human health and to the environment

The Resource Conservation and Recovery Act (RCRA) of 1976 (P.L. 94-580), originally the Solid Waste Disposal Act enacted in 1965, established the federal program regulating solid and hazardous waste management. The act defines solid and hazardous waste. Subtitle C of the act regulates hazardous wastes and establishes "cradle to grave" management procedures of these wastes—from generation, through transport, to treatment and/or disposal stages. Solid waste, including solid waste from industry, is regulated under Subtitle D of RCRA. The regulation establishes criteria with minimum technical requirements for acceptable operation of municipal solid waste facilities. Current municipal solid waste management practices consist of (a) landfilling, (b) incineration, or (c) recycling. It is estimated that in 1976 there were 20,000 landfills operating in the United States; by 1986 the number was down to 7000. A further reduction to 3500 is anticipated by 1991. The issues that face legislators, communities, and industry are the declining availability of environmentally and economically viable landfilling disposal sites and who, under what auspices, should develop new solutions.

It has been estimated that some 240 million tons of wastes generated in the United States per year can be classified as hazardous according to the definition in Title C of RCRA. Four classes of definitions have been developed—irritability, reactivity, corrosivity, and toxicity. The materials subsumed under these classes are regulated because of the environmental damage they can cause if disposed in a municipal landfilling facility. In addition to the "cradle to grave" management of hazardous wastes detailed in the act, there is also the Comprehensive Environmental Response, Compensation and Liability Act (P.L. 96-150) (Superfund) which authorizes the federal government to remedy the conditions in the nation's worst hazardous waste sites and to respond to hazardous waste spills. The Superfund Amendments and Reauthorization Act of 1986 (SARA) (P.L. 99-499) set new cleanup standards and emphasized permanent solutions rather than merely transporting the wastes to a new location. The act also establishes a timetable for remediation, and under the Community Right to Know Program requires industrial concerns to provide information on the type of chemicals present at their facilities. A $500 million program was also authorized to clean up leaking underground fuel tanks. (See also Chapter 30.)

This synoptic review of laws related to environmental control in the workplace can only serve as an introduction to the subject. Moreover, new laws and amendments to existing laws are constantly being enacted. Also agencies and departments of government modify and update the regulations promulgated under legislative authority. A number of data systems are available to the reader that provide a variety of up to date legislative information. Some, such as Justice Retrieval and Inquiry System (JURIS), a computer-based storage and retrieval system, is meant primarily for legislative and litigation purposes. JURIS is a Department of Justice system. There are two commercial full text systems containing such information as the *United States Code*, the *Code of Federal Regulations*, and the *Federal Register*. LEXIS is available from Mead Data Central and WESTLAW from West Publishing Co.

The areas considered in this chapter are (*a*) the provision of a safe, potable, and adequate water supply, including such noningestion sources of waterborne disease as legionnaires' disease and inhalation and dermal absorption of chemical contaminants; (*b*) the collection and disposal of liquid and solid wastes; (*c*) the assurance of a safe food supply; (*d*) the control and elimination of insects and rodents, especially those known to be vectors of disease; (*e*) the provision of adequate sanitary facilities and other personal services; and (*f*) the maintenance of general cleanliness of the industrial establishment.

2 PROVISION OF A SAFE, POTABLE, AND ADEQUATE WATER SUPPLY

2.1 Source and Regulatory Control: The Safe Drinking Water Act

In the case of large industries, the facilities for providing a safe and adequate water supply rival the size and complexity of many a community system. Although the advent of the Safe Drinking Water Act of 1974 (P.L. 93-523) (SDWA) brought about major changes in water supply regulation, even more extensive and demanding requirements were enacted in the Safe Drinking Water Act Amendments of 1986 (P.L. 99-339).

New events and conditions such as outbreaks of waterborne giardiasis, previously a relatively unknown disease, brought pressure to assure that the protozoan, *Giardia lamblia*, is excluded from public water supplies. *Giardia lamblia* is more resistant than other pathogens to the most common disinfection process, chlorination, and therefore, if allowed to reach the final disinfection point, sufficient numbers of this pathogen to cause disease may survive in the distribution or plumbing systems. Although disinfection alone is not sufficient, when combined with coagulation, sedimentation, and filtration it is effective in removing and inactivating the organism. Besides such large communities as Seattle, Washington and New York, New York, which employ extensive water quality surveillance, there are many small unfiltered public water supplies in the United States that pose a danger of explosive giardiasis outbreaks; therefore new filtration criteria for surface water supplies leading to regulations have been mandated. Other developments, such as the growing awareness of the widespread contamination of drinking water sources such as shallow groundwater aquifers, by a variety of improperly disposed solvents, pesticides, and industrial chemicals, led to enactment of the extensive amendment embodied in the Safe Drinking Water Act of 1986 (P.L. 99-339). In addition, radon, a radionuclide naturally occurring in soil and groundwater, has been found in a number of localities. The public health significance of this contaminant, owing to the lung cancer that can result from long-term exposure to the gas, may be as great as any other. Similarly, past regulations did not control the use of lead in plumbing materials, and thus lead leached from piping as well as solder into potable water systems. Spurred by recent increased growth of knowledge of neurophysiological and neurochemical effects of low levels of lead, especially psychoneurological deficits and behavior disorders in children, Congress held hearings and in 1988 passed the Lead Contamination Control Act of 1988 (P.L. 100-57Z), which became Part F of the SDWA.

Of concern to industrial establishments is the SDWA definition of "public water system" [Section 1401 (4)]. As defined in the act, "public water system" means a system for the provision to the public of piped water for human consumption, if such a system has 15 or more service connections or regularly serves at least 25 individuals. The term includes (a) any collection, treatment, storage, and distribution facilities under control of the operator of such a system and used primarily in connection with such system, and (b) any collection or pretreatment storage facilities not under such control but that are used primarily in connection with such a system.

Privately owned as well as community systems are covered by the regulations. The EPA interprets service "to the public" to include "factories and private housing developments." Thus the individual, private industrial, or commercial supply previously designated as "private" or "semiprivate" and not specifically regulated by law must now meet the standards established for such systems under the SDWA (P.L. 99-339 June 19, 1986) and is subject to regulations promulgated by the EPA under the act. The regulations do not apply to or cover systems meeting all the following conditions:

1. System consists only of distribution and storage facilities and does not have any collection and treatment facilities.
2. System obtains all its water from, but is not owned or operated by, a public water system to which the regulations apply.
3. System does not sell water to any person.
4. System is not a carrier that conveys passengers in interstate commerce.

When a community water supply is not available to an industry at an economic cost, it must develop a source and treat the water in its private plant. To implement the act's provisions and regulations promulgated under the act, the EPA, the designated agency administering the SDWA, has transferred primary enforcement responsibility (primacy) to 48 states. In Wyoming and Indiana the EPA implements and enforces the act.

The drinking water regulations required in the SDWA are to: "protect health to the extent feasible, using treatment techniques, and other means, which the

[EPA] Administrator determines are generally available (taking costs into consideration) . . . ." To achieve this mandate the principal vehicle is "primary regulations," which specify maximum containment levels (MCLs) for contaminants that, in the judgment of the EPA administrator, may cause adverse health effects. The 1974 Act required that primary regulations be promulgated within 180 days of its enactment. Such regulations were issued on December 24, 1975 (40 CFR 59565), and July 6, 1976 (41 CFR 28402), which went into effect on June 24, 1977. The regulations established MCLs and monitoring requirements for three organic chemical groups (six pesticides and trihalomethanes), 10 inorganic chemicals, radionuclides, microbes, and turbidity. Other specifications included general operating and plant maintenance requirements.

The 1986 amendments to the SDWA mandate that the EPA develop new National Primary Drinking Water Regulations (NPDWRs). The EPA first is to propose maximum contaminant level goals (MCLGs), which are nonenforceable health goals, for each substance that, in the EPA administrator's judgment, may have any adverse health effect and that is anticipated or known to occur in public water supply systems. Each MCLG is to be set at a level at which no known or anticipated adverse effects on the health of persons occur and which allows an adequate margin of safety.

At the time an MCLG is set, the EPA is to propose an MCL, which is to be set as close to the MCLG as is "feasible" (which means utilizing best available technology or treatment, taking costs into consideration). Alternatively, EPA may require the use of specified treatment techniques instead of setting an MCLG if it is not technologically or economically feasible to establish such a level. However, based upon an adopted policy, EPA sets MCLGs at zero for all known or probable carcinogens.

Some other significant SDWA directives to EPA and its standard-setting procedures for drinking water contaminants are as follows (1):

- The statutory timetable to produce the MCLGs and NPDWRs/Monitoring requirements was as follows:
  - 9 by June 19, 1987
  - 40 by June 19, 1988
  - 34 by June 19, 1989
- MCLGs and NPDWRs/Monitoring must be set for other contaminants in drinking water that may pose a health risk.
  - The 1986 amendments required that EPA publish a Drinking Water Priority List (DWPL) of drinking water contaminants that may require regulation under the SDWA (Table 32.1).
  - The seven substituted contaminants must be included on the DWPL (Table 32.1).
  - The list must be published by January 1, 1988, and every three years thereafter.

Table 32.1. Original List of Contaminants Required to be Regulated under the SDWA Amendments of 1986

Volatile Organic Chemicals

Trichloroethylene	Benzene
Tetrachloroethylene	Chlorobenzene
Carbon tetrachloride	Dichlorobenzene
1,1,1-Trichloroethane	Trichlorobenzene
1,2,-Dichloroethane	1,1-Dichloroethylene
Vinyl chloride	*trans*-1,2-Dichloroethylene
Methylene chloride	*cis*-1,2,-Dichloroethylene

Microbiology and Turbidity

Total coliforms	Viruses
Turbidity	Standard plate count
Giardia lamblia	*Legionella*

Inorganics

Arsenic	Silver	Asbestos
Barium	Fluoride	Sulfate
Cadmium	Aluminum	Copper
Chromium	Antimony	Vanadium
Lead	Thallium	Sodium
Mercury	Beryllium	Nickel
Nitrate	Cyanide	Zinc
Selenium	Molybdenum	

Organics

Endrin	1,1,2-Trichloroethane
Lindane	Vydate
Methoxychlor	Simazine
Texaphene	PAHs
2,4-D	PCBs
2,4,5-TP	Atrazine
Phthalamates	Aldicarb
Chlordane	Acrylamide
Dalapon	Dibromochloropropane (DBCP)
Diquat	1,2-Dichloropropane
Endothall	Pentachlorophenol
Glyphosate	Pichloram
Carbofuran	Dinoseb
Alachlor	Ethylene dibromide (EDB)
Epichlorohydrin	Dibromomethane
Toluene	Xylene
Andipates	Hexachlorocyclopentadiene
2,3,7,8-TCDD (Dioxin)	

continued on next page

Table 32.1. (*Continued*)

Radionuclides

Radium 226	Beta particle and photo radioactivity
Radium 228	Uranium
Radon	Gross alpha particle activity

Removed from SDWA List of 83

Zinc	Sodium	Vanadium
Silver	Molybdenum	Dibromomethane
Aluminum		

Substituted into SDWA List of 83

Aldicarb sulfoxide	Heptachlor epoxide
Aldicarb sulfone	Styrene
Ethylbenzene	Nitrate
Heptachlor	

    – MCLGs and NPDWRs/Monitoring are to be set for at least 25 contaminants on the list by January 1, 1991.

    – MCLGs and NPDWRs/Monitoring are to be set for at least 25 contaminants every three years following January 1, 1991 (i.e., 1994, 1997, . . .), from subsequent triennial lists.

• Criteria must be established by which states must determine which surface water systems must install filtration. The SDWA deadline for promulgating these criteria was December 19, 1987. States with primacy enforcement responsibility must make determinations regarding filtration within 12 months of promulgation of these criteria and must adopt regulations to implement the filtration requirements within 18 months of promulgation.

• A treatment technique regulation must be promulgated that will require all public water systems to use disinfection.

    – Variances will be available. EPA will specify variance criteria (e.g., quality of source water, protection afforded by watershed management).

    – The disinfection treatment rule must be promulgated by June 19, 1989.

• The SDWA states that granular activated carbon (GAC) adsorption is feasible for the control of synthetic organic chemicals in drinking water. Any technology or other means found to be best available for control of synthetic organic chemicals must be at least as effective in controlling synthetic organic chemicals as granular activated carbon.

• The 1986 amendments banned the use of any pipe, solder, flux, or fittings that are not lead-free in a public water system or any building connected to a public water system. Flux and solder may not have more than 0.2 percent lead and pipe and fittings not more than 8 percent lead.

The final regulations provide that MCLs for organic chemicals and for inorganic chemicals other than nitrates are not applicable to "noncommunity" systems. To make clear which regulatory requirements apply to "community systems" and "noncommunity systems," the category covered is specifically indicated throughout the National Primary Drinking Water Regulations (NPDWRs).

The act provides that the states may be given primary enforcement responsibilities by the EPA administrator. If a system cannot reasonably meet the regulations, the state—or, in case a state has not assumed responsibility for regulation, the EPA—may grant variances and exemptions that do not impose unreasonable risk to the health of those served by the system. A schedule must be established for compliance, and a public hearing must be held before any public water system is granted such a schedule.

If a water supply system fails to comply with the primary regulations or fails to meet a compliance schedule established under a variance schedule granted by a state or the EPA, it must give notice to its customers or users of this failure. Furthermore, a supplier of water must report to the state or the EPA, within 48 hours, any such failure to comply with any drinking water regulations, including monitoring requirements. The regulations promulgated under the act are of two types, primary and secondary.

Primary regulations are those specifying contaminants that in the judgment of the EPA administrator "may have any adverse effect on the health of persons." The regulations apply to all "public water systems" as defined by the act, but they deal only with the basic legal requirements; an EPA guidance manual for use by public water systems and the states is forthcoming.

Secondary regulations specify the maximum contaminant levels that in the judgment of the EPA administrator will serve to protect the public welfare. Such contaminants are defined as those "(a) which may adversely affect the odor or appearance of such water and consequently may cause a substantial number of persons served by the public water system providing such water to discontinue its use, or (b) which may otherwise adversely affect the public welfare. Such regulations may vary according to geographic and other circumstances."

A major provision of the SDWA of 1986 important to industrial systems is "Part C, Protection of Underground Sources of Drinking Water," regulating the underground injection of fluids. The term "underground injection" refers to the subsurface emplacement of fluids by well injection. The act states that underground injection endangers drinking water sources (a) if it may result in the presence of contaminants in underground water supplies, (b) if it can reasonably be expected to do so, or (c) if the presence of such contaminant may result in a system failing to comply with any national primary drinking water regulation or may otherwise affect the health of persons. The regulations promulgated under the act are designed to allow states with different geological and other conditions to exercise judgment in their application to prevent underground injection practices from contaminating drinking water sources.

Section 1421(b)(2) of the SDWA states that regulations may not prescribe a requirement that interferes with or impedes underground injection related to oil and natural gas production or secondary or tertiary recovery of oil or natural gas, unless such a requirement is essential to ensure that underground sources of drinking water will not be endangered by such injection.

All current drinking water regulations that have been promulgated as of July 1 in any year may be found in the *Code of Federal Regulations* (CFR), Volume 40, Parts 141, 142 and 143. Regulations published between CFR editions may be found in the *Federal Register* (FR).

2.2   Drinking Water Supply

The SDWA amendments of 1986 and the regulations promulgated under the act apply not only to community supplies but to any factory or plant possessing its own drinking water supply and employing 25 or more persons. The National Primary Drinking Water Standards proposed in May 1989, set MCLs for eight inorganic chemicals, 30 synthetic organic chemicals, turbidity, and microbiological organisms. A separate set of regulations, developed for radioactivity, became effective June 24, 1977. It appears that the regulations relating to radioactivity apply only to "community" supplies. However, the hazard in industry should be of concern because of the danger of in-plant contamination with radioactive matter. Therefore a plant with its own drinking water supply should undertake an initial sampling to determine the level of radioactivity in its supply and, based on use, should undertake periodic reexaminations. The analytical methods utilized can be those detailed in the EPA handbook, "Interim Radiochemical Methodology for Drinking Water" (2) or any subsequent revision.

In addition to assuring safety, industrial plants should provide water for drinking and cooking that is acceptable to its employees. Drinking water should be supplied at a temperature within a range of 40 to 80°F (the optimal range is 45 to 50°F). When cooling is needed, mechanical refrigeration or ice can be used: the ice must be produced from water meeting drinking water standards and maintained to prevent post-production contamination. In non-food-purveying areas, ice that is used to cool drinking water should not be allowed to come in contact with the water. (In cafeterias and other food serving areas, the advent of ice-making machines makes it feasible to use ice to cool water and beverages.)

Water supplied in drinking fountains and food preparation centers of the plant must be safe, clean, potable, and cool. A sanitary drinking fountain of approved design is the most efficient method of providing drinking water for employees. The American National Standard Minimum Requirements for Sanitation in Places of Employment (3) state that: "Sanitary drinking fountains shall be of a type and construction approved by the health authorities having jurisdiction" (or meet local plumbing code requirements). "New installations shall be constructed in accordance with the requirements of the health authorities having jurisdiction, or, if there are not such requirements, in accordance with American National Standards Institute Standard for Drinking Fountains, ANSI/ARI 1010-78, or the latest revision approved by the American National Standards Institute" (4). To keep refrigeration needs to a minimum, individual disposable drinking cups may be supplied. Whenever it is not feasible to have a drinking fountain connected to the supply, an approved drinking water container with an approved fountain or individual disposable cups should be furnished. In general, location of fountains may be determined by an overall standard of one drinking fountain for each 50 employees. However, the distance the employee must travel to the nearest source may be a controlling factor in locating drinking water sources. (The ANSI/ARI Standard 1010-78 requires that this distance be no more than 200 ft.) Similarly, wherever the employees are subjected to above normal heat stress, this fact should be the controlling criterion for the location of a drinking water source.

The 1972 NIOSH criteria document on heat stress in the working environment (5) recommends a series of work practices to ensure that the employee body core temperature does not exceed 38.0°C (100.4°F). Among the seven work practices specified, the fifth is titled "Enhancing Tolerance to Heat."

> V. Regular breaks, consisting of a minimum of one every hour, shall be prescribed for employees to get water and replacement salt. The employer shall provide a minimum of 8 quarts of cool potable 0.1% salted drinking water or a minimum of 8 quarts of cool potable water and salt tablets per shift. The water supply shall be located as near as possible to the position where the employee is regularly engaged in work, but never more than 200 feet (except where a variance had been granted) therefrom.

In 1986 the National Institute for Occupational Safety and Health (NIOSH) developed revised criteria for a recommended standard for occupational exposure to hot environments (6). Two of the recommendations update prior recommendations for salinization of the drinking water supply in hot weather or hot environments. The 1986 revised criteria are:

> c. To ensure that water lost in the sweat and urine is replaced (at least hourly) during the work day, an adequate water supply and intake are essential for heat tolerance and prevention of heat induced illnesses.

> d. Electrolyte balance in the body fluids must be maintained to prevent some of the heat-induced illnesses. For heat-unacclimatized workers who may be on a restricted salt diet, additional salting of the food, with a physician's concurrence, during the first 2 days of heat exposure may be required to replace the salt lost in the sweat. The acclimatized worker loses relatively little salt in the sweat; therefore, salt supplementation of the normal U.S. diet is usually not required.

2.3   Water Uses in Industry

In addition to drinking water supply, industrial plants require water for cooling, processing, and cleaning, and different quality demands are associated with each. A pharmaceutical manufacturing plant, for example, may require deionized pyrogen-free water, whereas a steel mill uses very large amounts of cooling water with far different quality demands.

Manufacturing industries now withdraw about 100 billion gal/day (bgd) from freshwater sources such as lakes, rivers, underground aquifers, and estuarine salt water areas. Kollar and Brewer (7), reporting on water resources planning for industry, assert that although manufacturing industries in 1975 in the United States withdrew 50 bgd (The Water Resource Council reports 60.9 bgd were withdrawn), the actual gross need is more than 125 bgd. This need is met through a combination of one-through use and recycling of treated industrial effluents. Ranking industrial water uses by category is as follows:

1. Cooling
2. Steam generation
3. Solvent
4. Washing
5. Conveying medium
6. Air scrubbing

Cooling is the largest volume use, and electrical power generation is the major user of cooling water. Because cooling water typically is separated from process water (i.e., water utilized in the manufacturing process), it usually does not come in contact with the product. Except for the addition of heat, the other quality characteristics of cooling water may not be significantly changed. Additives designed to reduce fouling due to bacterial and algal growths, when recycling is practiced, do change its characteristics.

Water used for steam generation becomes contaminated with chemical additives or intermediate products that must be removed before reuse or discharge. In the petroleum refining industry, "sour water strippers" are used to remove sulfur and other polluting substances from condensed steam that has been employed to heat crude oil during the distillation process.

Water, the universal solvent, is extensively used in industry to dissolve compounds. To effect reuse, the renovation of such water demands careful evaluation of its contaminants. Wash water picks up and entrains a wide variety of contaminants that, if the water is to be reused, must be removed or at least greatly reduced in concentration. In addition, washing and degreasing compounds and solvents may be introduced into the wash water. Water is used extensively as a conveying medium and in the process is contaminated with solids removed from the material conveyed (e.g., soil from agricultural products and any other biological, organic, and inorganic matter that may be mechanically removed or dissolved).

With the advent of (8) more stringent requirements to control the emission of air pollutants, air-scrubbing devices that use water to entrap and entrain pollutants are being used in increasing numbers. These devices use large quantities of water and are a good example of the "cross-media" problem engendered. The material scrubbed out of the airstream is essentially a concentrated pollutant that must be dewatered and disposed of on land or otherwise utilized as a solid waste.

To be able to establish the most efficient waste-water treatment and renovation processes, industrial establishments at all times must be able to determine the following:

1. Where waste water streams originate
2. The contaminants present, their concentrations, and the variability
3. The diurnal, weekly, and so on, variability of the waste-water volume

McClure (9) reports that in the case of steel mill cooling water containing a variety of suspended solids and iron, removal of the offending material is accomplished by precipitation and subsequent cyanide destruction. This is followed by alkaline chlorination and phenol reduction by chlorination of the clarifier blowdown to enable utilization of a closed-cycle cooling system. Thus only the blowdown from the clarifier or thickener has to be treated, thereby minimizing makeup water needs.

In general, the distribution by category of water use in manufacturing industries in the United States is as follows:

Use	Percentage
Process water	28.3
Boiler feed water; sanitary	4.8
Heat exchange	3.2
Air conditioning	
Cooling	12.1
Steam, electric	
Other, condensing	51.6

Some of this water is not returnable because of evaporation, incorporation into products, leaks, and other losses.

2.4 Conservation and Reuse of Water in Industry

Another factor controlling the use of water in industry has been the growing necessity to conserve water resources, primarily but not exclusively in arid and semiarid regions. A new impetus of major importance was the advent of the Water Pollution Control Act Amendments of 1972 (P.L. 92-500). A most important element of the law, and the program of water pollution control it has spawned, is the system of effluent limitations and required permits under the National Pollutant Discharge Elimination System (NPDES) applicable to discharges of industrial and other wastes into the navigable waters of the United States. The EPA has been empowered to issue waste-water discharge permits to individual industrial establishments, power generating plants, refineries, municipal waste-water treatment plants, and similar facilities that are based on national effluent limitation guidelines. These guidelines designate the quantity and chemical, physical, and biological characteristics of effluent that industry may discharge into receiving bodies of water. Of crucial importance are the specific goals and objectives of Public Law 92-500 and the schedule of reaching water quality levels in the nation's waters which provide for the protection and propagation of fish, shellfish, and wildlife, and for recreation in and on the water. Second, the Water Pollution Control Act Amendments of 1972 mandated the elimination of the discharge of pollutants into navigable waters by 1985.

A major review of the act and its goals was authorized by Congress, which established the National Commission on Water Quality and charged it to ". . . make a full and complete investigation and study of all the technological aspects of achieving, and all aspects of the total economic, social and environmental effects of achieving or not achieving, the effluent limitations and goals set forth for 1983 in section 301(b)(2) of this Act" (10).

Studies conducted by the National Commission on Water Quality have indicated that in-plant changes, such as process modification and better internal control, can be joined to play an important role in meeting abatement goals. Flow reduction measures by means of increased water recycling and reuse are common to all the postulated strategies, as are better housekeeping procedures.

Since Congress mandated waste minimization as a policy in the 1984 Hazardous and Solid Waste Amendments to RCRA, the impetus to institutionalize such procedures and practices has increased. In addition, waste minimization can contribute to the reduction of a generator's liabilities under the provisions of the Comprehensive Response, Compensation and Liabilities Act (CERCLA or Superfund). The EPA working definition of waste minimization is that it consists of "source reduction and recycling." To assist in developing good hazardous waste management, EPA has published its *Waste Minimization Opportunity Assessment Manual* (11).

An example of reuse technology reported by Renn (12) suggests that a relatively low-level technology storage system for secondary treated waste water is feasible, particularly when water is in short supply. In this case the recycled water was utilized for air conditioning heat exchange, cooling towers, and a variety of machine tool cutting operations. Successful operation required modification of the industrial operations to accept water containing varying concentrations of organic matter, color, suspended matter, and particulates generated during storage or in transmission.

Storage and reuse of treated waste waters during drought periods is a practical method of extending the available water supply for some industries and communities. It also is a device for achieving zero effluent discharge during critical, low stream flow periods. Ultimately, however, the storage system must be discharged (13).

Conservation of water in industry refers to reduction in "net-intake" water requirements. This should be differentiated from "gross water applied" or "gross water use," terms that are directly related to production. Significant reduction of "gross water use" is desirable and may be achieved by new technology such as substitution of other fluids for heat transfer or process water. Conservation not only entails technical measures, such as leak detection and elimination, avoidance of spills, and reduction of evaporation losses; it also demands awareness and alertness on the part of individuals and groups of employees. In general, the techniques utilized in water conservation practice may be classified as follows: (*a*) using less water by avoiding waste; (b) recycling (using water over again in the same process); (c) multiple or successive use (using water from one process for one or more additional processes in the same establishment, or where possible using nonpotable waters); and (d) nonevaporative cooling techniques (using special methods to reduce the amount of water required for cooling).

Use of reclaimed water from sewage and industrial wastes has been practiced for many years in many parts of the world. In industry, its use is not yet as widespread or as extensive as in agriculture. Reclaimed waters have in the past been most suited for cooling purposes, because in this instance the quantity and temperature of the water are of greater importance than its quality. Industry can employ reclaimed water for plant processing water, boiler feed waters, certain sanitary uses, fire protection, air conditioning, and other miscellaneous cooling.

In addition to the general public health hazards involved in the utilization of a reclaimed effluent from a waste-water treatment plant, there are quality considerations that arise from the concentration effects of the treatment process. The concentration of total salts may be increased measurably, and the hardness of the water thus reclaimed can make its utilization impractical or costly.

However, recent developments in waste-water treatment and so-called advanced waste-water treatment auger well for the possibility of using renovated water in the growing of food crops, recharge of underground aquifers, and direct reuse in a wider variety of manufacturing plants. A study (14) reports that in 1975 reuse of treated municipal waste water in industry was continuous in 358 locations in the United States. Approximately 95 percent are located in the semiarid southwest. Arnold et al. (15) in their report on the reclamation of water from sewage and industrial wastes in Los Angeles County, drew the following, still valid conclusions:

(a) There are a number of important factors limiting the direct use of water reclaimed from wastewater in industry. In addition to public health hazards and total salt concentration, considerations of the hardness of the water may be of material importance.
(b) In general, the direct reuse in industry of acceptable water reclaimed from wastewater is tolerated and encouraged in manufacturing processes that do not involve contact between the reclaimed water and human beings or foods to be consumed by humans. (c) The most obvious direct reuse for reclaimed sewage water is for cooling and condensing operations.

With the advent of greatly increased effluent quality requirements, recycling of industrial waste water after treatment to meet future effluent guidelines should become more prevalent. Kollar and Brewer (7) reviewing the "20 best of file" plants with high recycling rates, concluded that treated wastewater is being recycled at a high rate by many of the major water-using industries.

The concept that highest quality water be reserved for drinking, culinary, and washing uses, with a lower-quality water derived from either nonpotable source or reclaimed waste water used to meet other needs, is not new. Sextus Frontinus, the water commissioner of Rome in *The Aqueducts of Rome* (16), describes the supply of lower-grade water delivered to the fountains and public baths of the city. Reclaimed water has long been used as a second supply to flush toilets in water-deficient areas such as the facilities at the north rim of the Grand Canyon. An early example of reclaimed water use in industry was the Bethlehem Steel Company plant at Sparrows Point, Maryland use of further-treated, City of Baltimore Back River treatment plant effluent for cooling and other nonpotable uses.

Until relatively recently such community and industrial use was highly restricted. However, since the 1970s in the United States and abroad, the number of such systems has been growing. An ambitious and carefully designed investigation of direct water reuse of renovated water is underway in Denver, Colorado. The objective of the study is to evaluate the water treatment unit process required when using waste-water treatment plant effluent as the water source. The goal is to be able to use such renovated water in the public water supply system. The unit process include flocculation, two-stage recarbonation, filtration, selective ion exchange, breakpoint chlorination, carbon adsorption, reverse osmosis demineralization, chemical oxidation, and disinfection (17).

Okun (18) has distinguished among the various types of reuse, a major source for the lower quality portion of a dual water supply:

- Indirect potable reuse—the abstraction of water for drinking and other purposes from a surface or underground source into which treated or untreated waste waters have been discharged
- Direct potable reuse—the piping of treated waste water into a water supply system that provides water for drinking
- Indirect nonpotable reuse—the abstraction of water for one or more nonpotable purposes from a surface or underground source into which treated or untreated wastewaters have been discharged
- Direct nonpotable reuse—the piping of treated waste waters directly into a water supply system that provides water for one or more nonpotable purposes

Okun further states that direct nonpotable reuse in a separate system is being adopted throughout the world. In the United States the practice is widespread in the southeast, southwest, and west. In contrast with a total of 358 reuse projects in the United States in 1975, in California alone by 1985 there were 380 projects in operation providing water for agricultural irrigation, cooling, industry, toilet flushing, construction, and other purposes not requiring drinking water quality.

As the practice inevitably grows, both public and private agencies must engage in study and research to provide a better foundation for increasing safety of such reuse. Among the problems to be addressed are quality standards for various nonpotable uses, treatment required to achieve these standards, monitoring requirements, plumbing codes for such supplies, distinguishing between the different supplies, for example, universal color codes, and managing the dual systems.

An American Water Works Association committee established to evaluate the potential for dual distribution water systems defined a dual distribution system, which although developed primarily with community conditions in mind, is similarly applicable for industrial establishments. The definitions are:

- Potable water—water of excellent quality intended for drinking, cooking and cleaning uses. This grade of water would conform to the quality requirements of state and federal regulatory agencies (SDWA).

- Nonpotable water—Water acceptable for uses other than potable. This water would be safe for occasional inadvertent human consumption and would generally conform to the water quality requirements of state and federal regulatory agencies for nonpotable water.

2.5   Cross-Connections and Other Means of Contamination

The great variety of water supplies and uses within a plant, as well as the growing need for reuse and recycling of industrial water, present actual and potential hazards to the cleanliness, safety, potability, and coolness of the drinking and culinary water supplies. Such danger occurs when there are any connections between a water supply of known potability (primary source) and a supply of unknown or lesser quality, and when there are plumbing defects that may allow waste water and toxic materials to enter the drinking water distribution system. There are many possibilities for contaminating a potable supply by a nonpotable supply when unauthorized or unsafe connections are made. Pressure variations in distribution systems are not uncommon, and when reduced pressure conditions occur in a potable system connected with another of unknown potability, having an even momentary higher pressure, the flow will be from the system of unknown potability into the potable system.

Spafford (19) reports that the 40 waterborne epidemics recorded from 1907 to 1953 in Illinois have resulted in 13,000 known cases of illness and 200 deaths. Eighteen of the outbreaks, almost half the total, were caused by some type of faulty piping or plumbing arrangement that permitted waste water, or other contaminated water, to enter the safe water system, either at the source or in the distribution facilities. This experience is in agreement with the general observation that cross-connections between safe and raw or unsafe water systems have been the greatest single cause of waterborne epidemics.

Contamination of the potable water supply by means of cross-connections or defective plumbing fixtures by "backflow" may take place (*a*) by the occurrence of a vacuum in the supply lines, causing back siphonage of contaminated material; (*b*) by the development in the fixture, appliance, or piping system to which the water supply is connected of pressure that is greater than that in the supply system itself; and (*c*) through the activities or actions of vermin, birds, or small animals in parts of the supply system not under pressure, or by dust reaching water-holding devices.

Backflow can occur under two conditions: (*a*) back pressure and (*b*) back siphonage. Back pressure in a supply system can take place whenever water temperature is elevated, or pressure or pumps are used in the system. Under such conditions it is possible that the pressure at the discharge point is higher than the supply pressure. Back siphonage occurs when atmospheric pressure exceeds supply pressure. The classic example of back siphonage is when the supply pressure is drastically reduced by the high-volume pumping of fire fighting equipment.

In general, cross-connection control requires, first, a thorough knowledge of the water supply systems in the plant and, second, the other liquid purveying and removal systems. Because changes to these systems are usually quite frequent, a record of such changes must be kept current and readily retrievable. Other requirements include the use of backflow prevention devices, hazardous activities isolation, air-gap separation and installation of atmospheric and pressure type vacuum breakers, and double check valve assemblies. Certain industries have special cross-connection conditions such as water-using equipment that deserve special attention.

A recent acute-nickel toxicity episode among electroplating workers who ingested a solution of nickel sulfate and nickel chloride was described by Sunderman et al. (20). The description of the exposure illustrates the perils of cross-connections very well.

The episode began on Saturday afternoon, June 13, 1987 when the main water supply to a metal-plating factory was temporarily shut down for repairs. The plant, which employs 338 workers, operates several automated electroplating lines used primarily to refinish automobile bumpers. Except for a few maintenance and repair personnel, the plant was unstaffed over the weekend until the evening shift reported for work at 1:00 p.m. on Sunday the 14th of June. After 3:00 p.m. several workers on a nickel-plating production line began to feel ill with symptoms that some attributed to bitter-tasting water from a drinking fountain. The supervisor on the night shift tested water samples from seven drinking fountains and found that the sample from the suspect fountain was green and contained nickel (approximately 2 g/liter) while the samples from the other fountains were colorless and uncontaminated with nickel. The water supply to the contaminated fountain, which was adjacent to a nickel-plating tank, was flushed and the fountain was disconnected at 3:00 a.m. on Monday, the 15th of June. Since the ambient temperature was warm, most workers on the evening and night shifts sweated profusely; the workers who developed symptoms evidently had ingested 0.5-1.5 liters of water from the contaminated fountain. The green tint of the water was inapparent under the factory lighting when viewed against the metallic gray background of the stainless-steel bowl.

The nickel contamination of the drinking water was caused by back-siphonage from a water recirculation system that cooled filtration pumps for the nickel-plating tank. Leakage of pump gaskets had allowed the nickel-plating solution (nickel sulfate, $NiSO_4 \cdot 6H_2O$, 310 g/liter; nickel chloride, $NiCl_2 \cdot 6H_2O$, 127 g/liter; boric acid, H_3BO_4, 36 g/liter) to seep into the water recirculation system. The recirculation system was connected via an open valve to a freshwater line. When the water main was shut down, nickel-contaminated water was sucked from the recirculation system into the freshwater line, from which the drinking fountain was directly tapped.

Periods when reduced pressure or partial vacuum situations may occur within a potable water system result from the following circumstances or by a combination of a number of them:

1. Interruption of the supply for maintenance of the municipal or service supply main
2. Excessive demand placed on the supply mains during fire or other emergencies
3. Complete failure of the supply due to breaks in the mains, earthquake damage, interruption of pumping by either mechanical failure or power shutoff, or deficient water supply
4. Freezing of mains or service lines in extremely cold weather
5. Excessive friction losses due to inadequate size of mains or service lines, or due to reduction of the effective diameter of pipes caused by deposits and encrustations
6. Occurrence of negative water hammer pressure waves
7. Operation of long pump suction lines
8. Condensation of steam within boilers, hot water systems, or units such as hospital sterilizers

2.6   Elimination of Cross-Connections

The safest method of making two water supplies available to a building is by the use of an unobstructed vertical fall through the free atmosphere between the lowest opening from any pipe supplying water to a reservoir and the highest possible level the water may reach in the reservoir. The minimum air gap thus provided should be twice the diameter of the effective opening and in no case less than 4 in. In addition, a float valve arrangement should be installed to cut off the supply when the water reaches the free overflow level. The overflow channel or pipe should allow free discharge to the atmosphere with no enclosed connection to the sewer. This is the recommended method of the Building Officials and Code Administrators International for safe water supply to tanks or cooling tower basins.

To prevent backflow from fixtures in which the outlet end may at times be submerged, such as a hose and spray, direct flushing valves, and other devices in which the surface of the water is exposed to atmospheric pressure, a vacuum breaker should be installed. There are two general types of vacuum breaker: moving part and nonmoving part. The type selected should meet test requirements of the American National Standard for Backflow Preventers in Plumbing Systems ANSI 40.6-1943 and the American Water Works Association (AWWA) Manual M-14, Recommended Practice for Backflow Prevention and Cross Connection Control (21), and the AWWA Standards for Backflow Prevention Devices (22).

All fixtures supplied by a faucet should have an air gap, which is measured vertically from the end of the faucet spout or supply pipe to the flood level rim of the fixture or vessel. The minimum air gap provided should be twice the diameter of the effective opening but no less than 1 in. When affected by a near wall, the minimum air gap should be $1\frac{1}{2}$ in. Other conditions should meet the requirements of the American National Standard for Air Gaps in Plumbing Systems, ASA40.4-1942 (revised 1973).

Connections to condensers, cooling jackets, expansion tanks, overflow pans, and other devices that waste clear water only should be discharged through a waste pipe connected to the drainage system with an air gap.

2.7 *Legionella pneumophila* and Legionnaires' Disease

The fact that the causative agent of legionnaire's disease is a common inhabitant of aquatic environments became known after *Legionella pneumophila* was identified. The organism is detected in high numbers in the plumbing systems of hospitals, dwellings, manufacturing plants, and service facilities. Two major sources of amplification of the organism are hot water heaters and cooling towers. Although *Legionella* has not always been associated with disease, the consensus is that legionnaires' disease is, in the majority of cases, water related. Therefore a short review of the disease is included in the water supply section of this chapter (23).

2.7.1 *Legionella*

The bacterium *Legionella* can cause two distinct diseases: legionnaires' disease and Pontiac fever. Legionnaires' disease is characterized by a pneumonia resulting in a 15 to 30 percent fatality rate (24). In contrast, Pontiac fever is a nonpneumonic, nonfatal, febrile disease with a very high attack rate. *Legionella pneumophila* is a Gram-negative, rod-shaped, fastidious, ubiquitous organism and can be recovered from various environmental sources, including soil, mud, showerheads, nebulizers, dehumidifiers, humidifiers, potable water, streams, lakes, evaporative condensers, and cooling towers. It has been recovered from industrial, commercial, hospital, and hotel plumbing systems and hot-water recirculating systems. Growth and multiplication of this fastidious organism in the environment is supported by metabolic products of protozoa, algae, and various bacterial species.

An outbreak of acute respiratory illness developed in 1976 among members of the Pennsylvania American Legion who had attended a state convention in Philadelphia. Approximately 7 percent of the attendees became ill and there was a 15 percent mortality rate among these cases. Laboratory investigation did not reveal the cause of the disease until six months after the beginning of the outbreak. In 1977, workers from the Centers for Disease Control isolated and characterized a bacterium that was later named *Legionella pneumophila*. The thoughtful preservation of sera and bacterial specimens from earlier explosive outbreaks of pneumonia disclosed that *Legionella* pneumonia was not a new syndrome but had occurred uncharacterized for some time.

As reported by States et al. (25) although both legionellosis syndromes appear to be caused by the same group of bacteria, the disease occurs in two distinct clinicoepidemiologic forms. The first is *Legionella* pneumonia (legionnaires' disease), a severe, acute pneumonia with multisystem involvement and an incubation period ranging from 2 to 10 days. Attack rates, which range between 0.1 and 4.0 percent of those exposed, are highest in immunocompromised individuals. Other risk factors include being male, being over 50 years old, being a smoker, heavy alcohol consumption, and underlying chronic disease. The other form of legionellosis, Pontiac fever, is a nonpneumonic, self-limited, nonfatal influenza-like syndrome of fever, myalgia, and headache. The incubation time is 5 to 66 hr. Pontiac fever is much less dependent on host factors, with an attack rate approaching 100 percent.

Although legionellosis is a reportable disease, it is difficult to ascertain its actual incidence. Based upon a retrospective study of pneumonia among enrollees of a prepaid medical group in Seattle, Washington, Foy et al. (26) estimated that in the United States there have occurred more than 25,000 cases of *Legionella pneumophila*. However, for the first 7 months of 1979, the Centers for Disease Control reported a total of 565 cases in the United States.

2.7.2 Pontiac Fever

The first known outbreak of Pontiac fever occurred in July of 1968 in Pontiac, Michigan, and was described by Glick et al. in 1978 (27). The early investigations of this outbreak failed to recognize an etiologic agent. Since these early investigations, a bacterium was isolated from guinea pigs exposed in the Pontiac Health Department building during the 1968 outbreak.

In July of 1973, 10 previously healthy 10- to 39-year-old males spent 9 hours cleaning a steam boiler condenser located on the James River in Virginia. All subsequently became sick for 2 days. No patient had pneumonia or objective evidence of other organ system involvement. The James River outbreak appears clinically and epidemiologically to resemble the Pontiac fever syndrome of legionellosis. The high attack rate, short incubation period, and absence of pneumonia are typical. The demonstration that five of the patients showed seroconversion and that three others had convalescent phase titers of 64 confirms the association with *Legionella*.

Another outbreak occurred among employees of an engine assembly plant in Windsor, Ontario, Canada. The outbreak occurred from August 15 to August 21, 1981. In the study of the outbreak, a case was defined as a worker who experienced and reported at least three of four symptoms: fever, chills, headache, and myalgia. A total of 695 employees (85 percent) completed a questionnaire. Of these, 317 (46 percent) met the case definition, 270 (39 percent) were not ill, and 108 (16 percent) were ill but did not meet the case definition. The illness had a mean maximum incubation period of 46 hr and was characterized by fever ranging from 99.5 to 104°F, severe myalgia, headache, and extreme fatigue. The illness was short (median duration of 3 days), but was severe enough to cause nearly 30 percent of the workers to miss work (median days of sick leave, 2). There were no fatalities, and only 4 of the workers reported similar illness in family members within 72 hr after the onset of the worker's illness.

A *Legionella*-like organism, designated WO44C, was isolated from a sample of coolant obtained on August 19 from system 17 in the piston department. No other legionellae or *Legionella*-like organisms were isolated from any of the other environmental samples. Like other *Legionella*, the organism did not grow on blood agar, required cysteine for growth, and produced catalase but did not produce urease, reduce nitrate to nitrite, or produce acid from carbohydrates. Unlike other previously described legionellae, it did not produce gelatinase, and WO44C and *L. pneumophila* are the only two species that hydrolyze hippurate (28). Pontiac fever is a relatively mild influenza-like, short-term disease with no fatalities. Because of these characteristics, it is believed that many outbreaks of Pontiac fever have not been recorded.

2.7.3 Legionella Pneumophila

Legionella spp. are widely distributed in nature and commonly found in many types of water systems; however, for reasons that are yet unexplained, infection in association with these systems appears to be the exception rather than the rule. The role of future research is to identify those sites that are likely to produce infection. It is not possible to treat routinely all potential sources in terms of practicability and cost, but if good engineering practices are followed in the operation and maintenance of water-cooled air conditioning and ventilation systems, this may reduce the possibility of *Legionella* sp. infections. Cleaning is indicated if organic growth is evident; particular attention should be paid to water storage tanks, which should normally be kept covered. Cooling water systems should be treated to prevent a buildup of slime, algae, and scale. Such treatment thus can not only minimize the loss of efficiency by the system but also inhibit rust, reducing the nutritional sources of the growth of *Legionella* sp. Although there is no current consensus as to which disinfection measure industries or other establishments should take, some institutions have resorted to the use of additional chlorination to reduce and even eliminate the presence of *Legionella* sp. in their plumbing and water distributions systems. Others, in particular hospitals where diagnosed cases of *Legionella* sp. have occurred and *Legionella pneuneophila* has been detected in the water distribution system, utilize a program of periodic heating of the hot water supply to 75°C and maintaining this temperature for 2 hr.

Fraser (29), summarizing the Second International Symposium on *Legionella*, postulated that six links in the chain of causation that lead from environmental sites to infection in humans are necessary: (1) there exists an environmental reservoir where legionellae live; (2) there are one or more amplifying factors that allow legionellae to grow from low concentration to high ones; (3) that there is some mechanism of dissemination of legionellae from the reservoir so as to expose people; (4) the strain of *Legionella* that is disseminated is virulent for humans; (5) the organism is inoculated at the appropriate site on the human host; and (6) the host is susceptible to *Legionella* infection.

The ubiquity of *L. pneumophila* makes the use of epidemiologic techniques all the more important in defining environmental sources for human infection. Air conditioning cooling towers or evaporative condensers have been clearly implicated as the source of outbreaks in the United States (Pontiac, Michigan; Burlington, Vermont; Memphis, Tennessee; and Eau Claire, Wisconsin). The risk the illness was directly related to was degree of exposure to the droplet-laden drift from the cooling tower or evaporative condenser, and the outbreak stopped after exposure ceased. In several other outbreaks, a cooling tower or evaporative condenser water was found to be contaminated with *L. pneumophila* and may plausibly have been the source of infection, but the epidemiologic proof was uncertain.

Monitoring of cooling systems is something that should be done in the interest of protecting the equipment. It will pay for itself in increased operation time and prolongation of equipment life; a secondary benefit is protecting the environment.

Cooling towers are considered to be man-made amplifiers of *Legionella* sp. Thus proper maintenance and choice of biocides are important. The only biocidal measure that has thus far been shown to be effective in field tests is the judicious use of chlorination. At the concentration recommended by the manufacturer of the biocide 1-bromo-3 chloro-5,5-dimethylhydantoin, neither the density nor the activity of *L. pneumophila* was affected. Fliermans and Harvey's (30) data indicate that concentrations of up to 2.0 ppm of this biocide were not effective in their cooling tower studies.

These same authors report that previous data indicate that the densities of *L. pneumophila* in groundwater obtained from deep wells are always less than those in surface waters, and that cooling towers receiving makeup water from underground sources have *L. pneumophila* densities substantially lower than towers receiving surface makeup water from lakes, rivers, or streams. It is recommended that in new buildings, cooling towers should not be located in such a position that the "drift" would readily enter a building through a ventilation system or by any other route.

As concluded by States et al. (25), research over the past decade has demonstrated that *Legionella* is a common inhabitant of aquatic environments and is capable of surviving and multiplying in water treated to meet drinking water standards. *Legionella* can proliferate within certain systems and bacterial amplifiers. Following amplification in cooling towers, for example, it is currently believed the organism is able to infect individuals, particularly those at high risk due to immunosuppression, underlying illness, advanced age, heavy smoking, or alcoholism.

Additional research designed to determine to what density the bacterium must be lowered in these heat-rejection devices is needed in order to prevent transmission of legionellosis in those who may inhale cooling tower aerosols.

3 LIQUID WASTES

Any individual process that includes water use produces water-carried wastes. These so-called industrial wastes, together with the sanitary wastes, must be treated before disposal. Industrial waste water may contain chemical, physical, or biological contaminants that pose dangers to plant personnel and to the population and plant and animal life in the surrounding community. Accidental discharges and spills of these possibly hazardous wastes are a constant environmental threat. Such events may result from pipeline breaks (sometimes hidden from view) or aboveground and underground storage tank rupture or leaks. The January 1988 collapse of a diesel fuel storage tank at Florette, Pennsylvania on the Monongahela River dumped nearly a million gallons of oil into the river and endangered the drinking water supply of hundreds of thousands of people. It is estimated that in the 1980s large manufacturing plants in the United States discharged between 275 and 300 billion gallons of industrial wastes per day. A portion of these wastes are conveyed to municipal waste-water treatment plants, another portion is treated on site and discharged into receiving bodies of water, and the remainder, mostly cooling water, is discharged directly.

The Federal Water Pollution Control Act and its 1987 amendments, the Clean Water Act (P.L. 100-4), are the principal law governing pollution abatement in the nation's streams, lakes, and estuaries, and the Marine Protection, Research and Sanctuaries Act (P.L. 92-532 and Amendments) regulates the dumping of all manner of matter into the nation's ocean waters. For ocean dumping the EPA has developed a permitting system, except for dredged material, which is the responsibility of the Corps of Engineers. The EPA criteria used in ocean dumping permit application review include (1) the need for the proposed dumping, (2) the effect of such dumping on health and welfare, including economic, esthetics, and recreational values, (3) the longevity of the effects of dumping, (4) the effect of dumping of particular quantities and concentrations of materials, (5) alternative methods of disposal, and (6) the effect of such ocean dumping upon scientific study, fishing, and other biological resource utilization.

In addition to the wastes of the more conventional industrial processes, the problem of safe disposal of wastes from the nuclear power industry (high level wastes) and other commercial, laboratory, medical, and manufacturing processes as well as nuclear power generation (low level wastes) is a continuing source of concern and regulation. Owing to the uniqueness of the radioactive waste disposal problem the federal government was, until the 1960s, responsible for all radioactive waste disposal. Many commercial disposal facilities developed in the 1960s failed to fulfill the objective of safe disposal. By 1981 the Nuclear Regulatory Commission promulgated new low-level waste disposal regulations (31).

The total low-level radioactive waste produced in the United States in 1985 had a volume of 75,907 m^3 and activity of 748,874 Ci. Fifty-seven percent of the volume was contributed by nuclear power generation and 36.8 percent by other industries. Of the total activity, 77.8 percent came from nuclear power generation and 21.1 percent from industry (32).

3.1 Industrial Waste Water

The original federal Water Pollution Act governing pollution in streams, lakes, and estuaries was passed by Congress in 1948. On October 18, 1972, Congress, over a presidential veto, passed the Federal Water Pollution Act Amendments of 1972 (P.L. 92-500). This act is substantially different from previous water pollution legislation. Besides increasing the level of federal grants for the construction of publicly owned waste-water treatment plants, it also supported planning for a new system and public participation in the planning process and created a new system of uniform effluent standards and a permit system (the National Pollutant Discharge Elimination System), for enforcement of effluent quality requirements on all point source dischargers.

The act had two primary goals:

- To reach, "wherever attainable," a water quality that "provides for the protection and propagation of fish, shellfish, and wildlife" and "for recreation in and on the water" by 1983
- To eliminate the discharge of pollutants into navigable waters by 1985

Eight policies were established.

1. To prohibit the discharge of toxic pollutants in toxic amounts
2. To provide federal financial assistance for construction of publicly owned treatment works
3. To develop and implement areawide waste treatment management planning
4. To mount a major research and demonstration effort in wastewater treatment technology
5. To recognize, preserve, and protect the primary responsibilities and roles of the states to prevent, reduce, and eliminate pollution
6. To ensure, where possible, that foreign nations act to prevent, reduce, and eliminate pollution in international waters
7. To provide for, encourage, and assist public participation in executing the act
8. To pursue procedures that drastically diminish paperwork and interagency decision procedures and prevent needless duplication and unnecessary delays at all levels of government

These goals and policies set the course and placed in motion development and construction activities leading to improved quality of waste-water discharges by industries and municipalities. Additional amendments were enacted by Congress in 1977 (P.L. 95-217) and 1981 (P.L. 97-117).

The act contained requirements and deadlines for achieving the goals and objectives in phases. Phase I, an extension of the program, embodied in many state laws and federal regulations based on prior legislation, required industry to install the "best practicable control technology currently available" (BPT) by July 1, 1977" [Section 3011](b)(1)(C)].

Phase II requirements were intended to be more rigorous and more innovative. Industries were to install the "best available technology economically achievable (BAT) . . . which will result in reasonable further progress toward the national goal of eliminating the discharge of all pollutants; including reclaiming and recycling of water, and confined disposal of pollutants" by July 1, 1983. The aim was to achieve the national goal of the elimination of the discharge of pollutants by 1985.

The act was intended to be more than a mandate for point source discharge control. It embodied an entirely new approach to the traditional way we have used—and abused—our water resources. Some of these mechanisms are found in Title I, the broad policy title; others can be perceived throughout the act in grants and planning, in standards and enforcement, and in permits.

One section of the act called for the development of comprehensive programs for preventing, reducing, and eliminating pollution, as well as research and development aimed at eliminating unnecessary water use. Section 208 directed the designation of areawide institutions (e.g., regional planning authorities) to plan, control, and maintain water quality and reduce pollution from all sources.

Construction grants for publicly owned treatment works are made available to encourage comprehensive waste treatment management, providing for the following:

1. The recycling of potential sewage pollutants through the production of agriculture, silviculture, and aquaculture products, or any combination thereof
2. The confined and contained disposal of pollutants not recycled
3. The reclamation of waste water
4. The ultimate disposal of sludge in a manner that will not result in environmental hazards

The grantees were encouraged to combine with other facilities and utilize each other's processes and wastes.

The amendments passed in February 1987 are now the Water Quality Act of 1987 (P.L. 100-4). These are the most comprehensive amendments to the act since 1972. Significantly, the act finds that despite great progress, the nation still faces many water quality problems. The EPA, the states, and industry are undertaking action to implement the act and meet the several deadlines in the legislation. The manner in which three problem areas are confronted will determine the effectiveness of implementation. These problems are:

1. The toxic pollutant control provisions
2. Nonpoint (urban, agricultural, mining, and forestry runoff) pollution management provisions
3. How effectively the "State Revolving Fund" provision, designed to transfer waste-water treatment funding responsibility to the states after 1994, functions

A major new provision is the establishment in the 1987 act (P.L. 100-4) of a program designed to alleviate the so-called "toxic hot spots," or toxic regions, in streams, lakes, and estuaries. It is now evident that toxic-containing discharges, in spite of previous laws and regulations and the construction and operation of control systems, continue to contribute to deterioration of water quality. The act requires each state to submit, for EPA approval, a list of these "toxic hot spots" within their borders. It also requires states to submit a strategy to bring affected waters into compliance with water quality standards within three years thereafter.

The act emphasizes point sources (individual industrial manufacturing facilities or municipal waste-water treatment plants) discharging toxic wastes. However, the act recognizes that to achieve water quality goals a combined approach to control of point and nonpoint sources of pollution is necessary.

As of June 1989 the EPA and the states have identified 879 point sources, including 240 waste-water treatment plants and 12 federal facilities that have been found to discharge toxic pollutants in quantities and nature that are causing water quality degradation.

As of July 1989, based on information supplied by the states, the EPA estimated that 10 percent of the nation's waters are degraded by a combination of toxic, sanitary, and nonconventional wastes derived from point to nonpoint sources.

The sanitary sewage portion of the liquid wastes of an industry, if kept separate from the industrial wastes, offers no special problem to the environmental engineer. The economics of any particular situation dictate whether the ultimate disposal facility will be the community's waste-water treatment plant or a private plant. If the industry is located within reasonable distance of a municipal system, connection to it for ultimate disposal through the community's treatment plant is preferable to construction and operation of disposal facilities by industry.

The technology is generally available to control the quality of industrial waste-water discharges to meet effluent limitations established by the EPA. Whenever industry discharges liquid waste (other than sanitary waste water) into a municipal system, the pretreatment standards promulgated under the 1987 act required the industry to reduce or eliminate "pollutants which are . . . not . . . susceptible to treatment by [publicly owned] treatment works" or any "pollutant interferes with, passes through, or otherwise is incompatible with such works."

The range of pollutants and contaminants in industrial wastewater is very wide. Of the wastewater produced in U.S. industry, about two-thirds is discharged cooling water. Electric power generation discharge is nearly all cooling water. In manufacturing industries the range is from less than 10 percent to nearly 90 percent cooling water discharge. Although cooling water with its elevated temperature can be "treated" by the use of various temperature reduction processes such as cooling towers, spray cooling, or ponds, it is the qualitative alternation by added chemical, physical, or biological matter that poses the greatest damage threat to receiving waters or the publicly owned waste-water treatment plant to which such industrial wastes are conveyed.

As great as the variations in quantity and quality of waste water produced by different industrial categories are, there also are significant differences between industries in the same category. Variations in design, age, and management practices all contribute to such differences. Therefore, when planning an industrial waste-water abatement program a waste characterization survey should be performed. The general objectives of such a study are to provide a definition of the waste-water sources, its qualitative aspects, flow and flow variation, and the state of the conveyance system. The EPA has published a sampling procedure manual and there are a number of handbooks devoted to such surveys.

One provision of the 1987 amendments to the Clean Water Act requires that

EPA periodically review and revise, as appropriate, existing regulations related to industrial categories. If deemed necessary, EPA is then empowered to initiate regulation of previously exempt industry categories that discharge toxic or non-conventional pollutants. Some facilities, such as hazardous waste treatment plants, may have escaped the clean water and other environmental regulations, whereas others may require review to enhance previously uncontrolled pollutants. At time of writing, the EPA was revising standards for three industry categories and studying 13 more.

The Clean Water Act (P.L. 100-8), in addition, requires states to review their water quality standards every three years. The 1987 amendments direct states to include specific quantitative standards when revisions are made. Where the numerical basis is not available through EPA guidance, states are directed to use biological assessment or monitoring procedures.

Aquatic bioassays are used to determine the toxicity of individual toxicants (contaminants) to aquatic life and/or provide information on toxicity of municipal and industrial waste-water treatment plant effluents to fish and aquatic invertebrates.

The types of tests currently used are (*a*) acute, lasting for 48 or 96 hr, (*b*) chronic, which expose the experimental animals throughout their life cycle, and (*c*) subchronic long-term tests that do not involve assessment of reproduction outcomes. The end-point measurement for the acute tests is mortality, but it is in the chronic test that growth, hatchability of offspring, growth in the larval stages, larval mortality, and number of broods per female is the information gathered. In abbreviated chronic tests, exposed fish eggs and their hatchability and larval growth are determined. Chronic exposure tests may last from 1 week (*Ceriodaphnia*) to several years in the case of trout. In the subchronic test change in growth rate is usually the measurement performed. The most commonly used test animals are *Daphnia Magna* (3 weeks), fathead minnows (*Pinephales promelas*), and several species of trout.

Biological monitoring usually involves collection of aquatic organisms in sections of the receiving body of water subject to a discharge, and sampling in an unpolluted control region. Most often, because of their sedentary nature, bottom invertebrates are the organisms collected. Fish are also collected. The assembled data are evaluated using species diversity and/or indicators of water pollution.

The rationale for the above requirement was based on the need for establishment of uniform standards for toxics. The record of state accomplishment in this sphere had been spotty. Such toxic standards are required to assure that states have the regulatory and enforcement tools to implement the toxic pollutant control requirements in the law.

Industrial establishments that convey their wastes to a publicly owned treatment works (POTW), rather than treating those wastes on site, are required to pretreat or otherwise remove toxic substances that would interfere with the operation of the receiving treatment plant. An additional problem can arise when toxic substances, although not detrimental to the waste-water treatment process, are concentrated in the sludge separated from the waste and thus interfere with the possibility of disposing of the sludge on land. To avoid duplication of treatment, the act allows industries to obtain credits for toxics adequately removed by the POTW and thus avoid pretreatment. These credits will not be available until EPA issues final sludge use and disposal rules. The process may take some time. At present, with no credits being issued, industries are compelled to install and operate treatment facilities. The General Accounting Office in its report "Improved Monitoring and Enforcement Needed for Toxic Pollutants Entering Sewers" (GAO/RCED-89-101), contends that a large number of municipalities fail to bring enforcement actions against industries in their jurisdictions that violate current pretreatment requirements (33).

The act also requires that all water quality permits (NPDES) for industrial and municipal waste-water treatment plants include a plan for sludge management. Regulations relating to sludge disposal proposed by the EPA in 1989 are being reviewed and analyzed. The final rule will be promulgated in 1990 or 1991. EPA is required to identify pollutants of concern in sewage sludge and to establish numerical limits for each identified pollutant and the management requirement to achieve them.

To contend appropriately with industrial wastes it is now recognized that prevention, minimization, and reduction of waste production is a desirable and cost-effective means of combating the hazardous and industrial waste problem. Waste minimization is made up of two major processes, source reduction and recovery/recycling. Source reduction is defined as any means that reduces or eliminates the generation of waste at the source. Usually that takes place in the unit process. Recovery/recycling is defined as the recovery and/or reuse (with or without treatment) of that which otherwise would be waste.

Waste minimization, a more general term, can be practiced at existing plants and is a requirement in the design, construction, and operation of new facilities. The EPA Hazardous Waste Engineering Research Laboratory has published a series of reports on waste minimization in several industry categories and developed audit procedures. An EPA *Waste Minimization Opportunity Assessments Manual* was published in 1988 (11).

Apart from the admonition of the old adage, there are a number of other good reasons for minimizing wastes:

- Economic viability, based upon reduction of waste treatment needs and disposal costs, and saving of raw material costs
- Enhanced ability to meet increasingly stringent regulations (comprising landfill disposal standards, permitting requirements for waste treatment, and reporting requirements)
- Reduced liability, including liability for environmental degradation and workplace safety and health
- Improving the plant's good neighbor reputation by a demonstration of environmental concern

The EPA has established industrial effluent guidelines and standards for 66 industrial categories. These are listed under *Code of Federal Regulations* 40CFR Parts 405 through 471. Industrial effluents containing toxic pollutants are to be found in the *Code of Federal Regulations*, Title 40, Part 401. For pesticides the information is in 40CFR 129.

Treatment of industrial waste waters to remove pollutants before discharge into a receiving body of water is now commonly referred to as "end of the pipe abatement." This means that waste flows are combined at a waste-water treatment facility to be treated before discharge. Cost and treatability and regulatory constraints considerations induce reduction of pollutant production by various means, ranging from better housekeeping, through reduction of spills and leaks, to education programs, and if necessary and possible, to modification of the process itself. For example, an evaluation of the design requirements in the food processing industry as described by Eckenfelder (34) indicates that significant reductions in waste load, with corresponding cost reductions, can be achieved in the brewing industry by collecting the fermentation sediment, reducing biochemical oxygen demand loads by modifying bottle and keg filling operations to minimize beer losses, collecting spilled grains, reusing rinse water, and establishing an educational program for the brewery workers.

The EPA has conducted plant-by-plant surveys of abatement measures. Reports of these surveys are available for a number of industries. This material serves as an excellent guide to current practice in these industrial segments.

3.2 Treatment

End of the pipe conventional treatment can be divided into two major categories, physical-chemical and biological. The two categories are employed individually or in combination. However, physical processes such as straining, sedimentation, and flotation are common to all methods and categories of waste-water treatment. The particles in wastewater exhibit a large range of sizes. Therefore a system devised to remove them can consist of (1) coarse screening with vertical or inclined "bar screens" (openings 0.75 to 6 in. wide) to remove accumulated rags, sticks, and stones: (2) "grit chambers" designed to remove more readily settled inorganic matter by reducing velocity of flow; (3) "flotation" in a sedimentation basin or "flotation unit" to remove organic materials that will float; (4) "sedimentation" to remove settleable material from incoming wastes or to remove materials rendered settleable by chemical or biological means ("sedimentation" or "settling" basins have two functions: (*a*) clarification—production of an effluent relatively free of settleable matter, and (*b*) thickening—production of an underflow that contains the solids removed in high concentration); and (5) "fine screening," which can also be utilized to remove settleable organics.

The manner in which particles settle out of suspension allows them to be categorized. When the particles (e.g., sand, grit, coal, and fly ash) tend to remain discrete and do not coalesce when in contact, the settling or clarification is termed class I. When the particles to be removed have been flocculated or have a tendency to form flocs, resulting in a change of size and settling rate, the clarification is termed class II. Design considerations are primarily (*a*) the rate at which clarification occurs and (*b*) the assurance that once a particle has settled, it will not be resuspended before being removed from the basin. The rate of sedimentation of particles in water is determined by their weight and the buoyant force and drag force opposing their subsidence.

The settling velocity of discrete particles, class I clarification, is influenced by fluid properties and the characteristics of the particle. From a consideration of the forces acting on the particle (fluid density and viscosity and gravity), the relationship determining settling velocity of discrete particles is

$$V_s = \sqrt{\frac{4}{3} \frac{g}{C_D} \frac{(\rho_s - \rho)}{\rho} d} \tag{1}$$

where g = gravitational constant
C_D = Newton's drag coefficient
ρ_s = density of the particles
ρ = density of the liquid
d = effective particle diameter

The coefficient of drag C_D at low (less than 1) Reynolds number, can be approximated from the equation

$$C_D = \frac{24}{N_{Re}} \left(\text{Reynolds number} = N_{Re} = \frac{Vd\rho}{\mu} \right) \tag{2}$$

and Equation 1 is transformed to become

$$V_s = \frac{g}{18} \frac{(\rho_s - \rho)}{\mu} d^2 \tag{3}$$

where μ = dynamic viscosity. Equation 3 is known as Stokes' law.

When particles are in high concentration in the liquid, streamlines about particles overlap so that their subsidence is retarded by the presence of neighboring particles. Under such conditions of hindered settling (class II settling), the subsidence rate of the suspension V_s becomes

$$V_s = V_p E^{4.65}$$

where E is the porosity (35).

When particles are agglomerated by flocculation, their settling characteristics are influenced by the nature of the floc. The larger the mass, the greater the settling rate. With high particle concentration, hindered settling is very pronounced. This causes the particles to settle in a fixed position relative to each other. The characteristic of this settling is the formation of a distinct liquid–solid interface, and it is termed zone settling.

If the concentration of a suspension is high enough that the particles are in contact, settling becomes dependent on deformation of particles or the destruction of the interparticle bonds. In this type of settling, compression takes place.

In biological treatment, the most commonly used processes employ microorganisms as the primary utilizers of the organic fraction. The process may operate under aerobic conditions, (e.g., trickling filtration or activated sludge) or anaerobic conditions (e.g., anaerobic sludge digestion). The basic process is the same, consisting of microbial growth and energy utilization.

3.2.1 The Trickling Filter Process

Trickling filters consists of shallow, usually circular tanks filled with crushed stone or, more recently, synthetic media. The liquid is applied by means of a distributor, continuously or intermittently, over the top surface. As the liquid percolates through the filter, it passes over the media and is collected at the bottom. The size of the voids permits the liquid to flow over the media and air to circulate. The designation of these units as "filters" is incorrect because the major removal is the result of an adsorption process that takes place on the surfaces of the biological growth encasing the filter media.

Because the organic substance is incorporated into the biological growth, it in turn releases to the liquid inorganic matter resulting from the oxidation of the adsorbed organics. Oxygen, supplied from the circulating air, allows aerobic oxidation in the surface layer of the biological growth. The composition of the biological growths or film on the media is very similar to that of the flocculent material in activated sludge. Besides bacteria, it supports large numbers of higher organisms that feed on the organic matter. These are insect larvae, spiders, aquatic earthworms, and so on. In spring, when warm weather arrives, the film is sloughed off, and large numbers of organisms are dislodged. Trickling filters may be subject to infestation of a mothlike fly of the genus *Psychoda*, which can become a distinct nuisance.

The biological mass on the media contains a wide variety of bacteria, the principal being heterotrophic bacteria. These in turn become the food for a number of saprobic organisms such as protozoans, rotifers, nematodes, and other invertebrates. Fungi also feed on chemical substances.

The trickling filter system includes a trickling filter and sedimentation, sludge handling and treating, and chlorination facilities.

3.2.2 Activated Sludge Processes

In activated sludge processes, in contrast with trickling filtration, the flocculated biological growths are in continuous circulation. The growths are always in contact with the waste water being treated and with oxygen. Oxygen is supplied by a variety of mechanical means such as agitators and turbines and by injecting air into the sludge mixture, causing turbulent mixing with small air bubbles. The aeration unit process is followed by liquid–solid separation, and a portion of the settled sludge is recirculated to the incoming waste water. In the aeration process, the segment of the organics amenable to biological degradation is converted to inorganics, and the remainder is converted to additional activated sludge.

Activated sludge consists of microorganisms, generally similar to those found in the slime covering the media in trickling filters, nonliving organic matter, and inorganic matter. The term "activated" is derived from the unique property of the sludge in which a series of steps occurs: (a) when in contact with activated sludge, suspended colloidal and soluble organic matter is rapidly adsorbed and flocculated; (b) there is a progressive oxidation and synthesis, proportional to the biological oxidation, of organic matter that is continuously removed from the solution. When the complete storage capacity of the sludge has been taken up, the sludge floc is no longer active and can no longer adsorb organic matter. The "activity" of the sludge can be restored by aeration, during which the stored material is utilized in oxidation and synthesis. Lawrence and McCarty (36) have demonstrated that although the overall reaction in biological waste treatment differs in anaerobic processes such as sludge digestion, which contrast with aerobic processes such as activated sludge, the processes of microbial growth and energy utilization are similar. On this foundation it is possible to develop general relationships applicable in a wide variety of processes mediated by microorganisms.

In anaerobic systems complex wastes are first hydrolyzed and fermented to organic acids. It has been demonstrated that the limiting step in anaerobic treatment of organic matter is the rate of fermentation of these acids to methane. McCarty (37) proposed that from a knowledge of the kinetics of fermentation of the key organic acids, it is possible to formulate the kinetics of the overall treatment process.

In practical terms, two clearly defined stages can be discerned as the mixture of incoming waste water and return activated sludge flows through an aeration tank. The first is a clarification stage, in which the major portions of the colloidal and suspended organics are adsorbed to the surface of the floc. The second is a stabilization stage, which takes place in the major portion of the aeration basin or tank. It is during this period that the organics that are stored during the clarification stage are utilized in growth and oxidation.

Trickling filtration or activated sludge, with separate anaerobic digestion of sludge, are waste-water treatment processes widely utilized in industrial waste-water treatment. Alternative biological systems are extended aerobic digestion of sludge and ponds.

The pond treatment system has been selected to receive untreated municipal sewage, industrial wastes and primary treatment plant (screening, sedimentation, and flotation) effluents, secondary treatment plant (biological treatment) effluents, excess activated sludge, and other unsettled wastes containing settleable solids. In general, the treatment process in a pond system depends on the bacteria that pervade the system to degrade nonrefractive or putrescible organic matter and the algae that provide the major portion of the oxygen utilized by bacterial respiration in the aerobic portion of nonaerated ponds. There seems to be new evidence that the larger organisms inhabiting ponds also play a role in degradation and elimination of organic matter. The similarity among lakes and streams and ponds is seen from the manner in which ponds develop facultative bacterial systems of varying types, similar to those found in lakes and streams. Algae, which have the capacity for autotrophic and heterotrophic growth, usually perform as facultative chemical organotrophs, utilizing sugars and organic acids as a source of energy and reduced carbon.

During daylight, the photosynthetic algae production of oxygen and utilization of carbon dioxide may be as much as 20 times the reverse reaction, which takes place in the absence of light. The factors that are important in regulating the growth rate of algae in ponds and streams are light intensity and duration, temperature, and available nutrients. In industry, the canning, meat processing, and pulp and paper industries have made extensive use of ponds to treat their waste. The canning industry has long used ponds and pioneered in the use of sodium nitrate for odor control.

In the meat packing industry, ponds are employed as anaerobic systems to provide pretreatment, complete treatment preceded by the recovery of grease and settleable solids, and complete treatment in a series of anaerobic–aerobic ponds. They are also used to effect advanced treatment following trickling filtration or activated sludge treatment.

In the pulp and paper industry, the use of ponds to reduce the biochemical oxygen demand of the effluent is probably the form of secondary treatment most frequently employed. The pond treatment process has demonstrated a high degree of flexibility, permitting shifts and alteration in treatment parameters to achieve a high degree of pollutant removal at the most economic cost.

3.2.3 Rotating Biological Contactor (RBC)

The rotating biological contactor is a fixed growth process. The growth takes place on rotating disks that are attached to a horizontal shaft. The disks are thus rotated through waste water flowing through a basin.

3.2.4 Anaerobic Treatment

The first modern use of anaerobic processes in waste-water treatment was to decompose the sludges produced in the primary and secondary settling units. However, such digestion was practiced earlier. The decomposition in pit privies and septic tanks is essentially anerobic. The anaerobic unit process is a biological process in which anaerobic microorganisms convert organic compounds into methane, carbon dioxide, cellular debris, and so on. The energy value of the combustible gas produced has approximately half the calorific content of natural gas.

Until relatively recently the major use of the anaerobic process in treating industrial wastes has been restricted to packinghouse waste. However, more recently the use of anaerobic treatment has been extended to other industrial wastes such as petrochemicals, distillery wastes, and a range of food processing wastes.

A measure of the growth of interest and use of the anaerobic process in the treatment of industrial wastes is the constantly increasing number of articles on the subject listed in the annual literature review (June 1989) issue of the research journal of the Water Pollution Control Federation (WPCF) (38). Other recent sources of information are the proceedings of a conference on granular anaerobic sludge held in Wageningen, the Netherlands (39), the annual proceedings of the Purdue International Waste Conference (40), and the proceedings of the biennial meeting of the International Association on Water Pollution Research and Control (41).

Apart from the increased recognition of the anaerobic process as a viable and economic biological treatment it is now also being developed for use in solid waste management. When utilized optimally the anaerobic process has the advantage of permitting energy recovery. This is feasible when treating municipal, agricultural, and decomposable industrial wastes. Hybrid advanced reactors under development will permit higher energy recovery levels. Study and research on anaerobic treatment have become·an interdisciplinary endeavor.

3.2.5 Physical-Chemical Processes

Physical-chemical processes, in addition to sedimentation, flotation, straining, filtration, coagulation, and flocculation, are increasingly employed in industrial wastewater treatment. Adsorption, the process in which a material is concentrated (i.e., becomes an adsorbate), takes place at the interface between two phases, (e.g., liquid–solid) and is employed to remove such pollutants as phenols from a liquid waste stream. In ion exchange, ions held by electrostatic forces or charged functional groups on the surface of a solid (a synthetic resin or natural solid) are exchanged for ions of similar charge in a solution in which the solid is immersed. Ion exchange is used extensively in water and waste-water treatment. In wastewater treatment it permits, apart from pollutant removal, recovery of reusable or otherwise valuable material. The problems to be overcome in the ion exchange treatment of industrial waste water are clogging of the resins, destruction, and fouling. The metal plating industry employs ion exchange to recover such metals as hexavalent chromium (Cr^{6+}) from waste streams, and treated water is reused (42). The system makes use of a cation exchanger to remove other metals such as copper, zinc, iron, nickel, and trivalent chromium. The hexavalent chromium is passed through the effluent as CrO_4^{2-}, subsequently to be removed in an anion exchanger. This results in a demineralized effluent capable of reuse. For recovery of the Cr^{6+} the anion exchanger is regenerated with NaOH, and Na_2CrO_4 is released. The Na_2CrO_4 solution is moved through a cation exchanger in which sodium is exchanged for hydrogen-releasing chromic acid (H_2CrO_4), which can then be recovered.

Separation processes such as reverse osmosis, electrodialysis, and ultrafiltration are three major membrane processes employed in waste-water treatment. As stated by Weber (42), reverse osmosis is a good technique for concentration, to reduce volume, to recover usable material, to be used in combination with other treatment processes, and to improve overall efficiency.

Chemical oxidation is employed in water and waste-water treatment to convert pollutant chemical species to species that are not harmful or otherwise undesirable. This implies that the process is not carried to completion. Therefore in water and waste-water treatment the process may be described as "a selective modification of objectionable and/or toxic substances" (42). In practice, only a few oxidizing agents meet treatability needs: economy, ease and safety of handling, and compatibility with the treatment train. The most widely used agents are air or oxygen, ozone, and chlorine. The finding that small amounts of haloforms are produced as a side reaction during chlorination of natural colored waters has led to investigations of other organic-containing waters. These studies have demonstrated the presence of an array of stable chlorinated organic compounds. The formation of halogenated organics in water supplies by chlorination was reported on by Morris (43) and Rook (44), whose original finding was that upon chlorination of natural colored water, as a side reaction, a small amount of haloform (e.g., $CHCl_3$) was formed. Haloform formation upon chlorination is complicated because inorganic bromide ions are oxidized by chlorine to a valence that renders bromine suitable for bromination of organic matter. Rook (45) has found mixed bromochlorohaloforms, which he believes indicate that one type of reaction is involved—that is, the haloform reaction, which occurs when compounds containing the acetyl group bond to H or RCH_2 or compounds such as ethanol that are oxidized to such substances when fluvic or yellow acids are combined with chlorine.

Basic and detailed information regarding the several processes employed and under investigation is provided in the general references listed at the end of the chapter. The annual review of literature published in the June issue of the research journal of the Water Pollution Control Federation (38) covers a wide spectrum of the world's literature on matters pertaining to waste water, water pollution, and treatment. Reports on studies and research conducted on contracted by government agencies (such as the EPA) are available through the National Technical Information Service.

Waste-water characteristics have a universal tendency to vary, often over wide ranges; it is therefore essential that studies be undertaken with representative samples to ascertain the costs associated with the available treatment possibilities. An example of such an undertaking was given by Zabban and Helwick (46), reporting on the defluoridization of waste water. Significant quantities of fluorides are discharged in effluents from glass manufacture, electroplating operations, aluminum and steel production and processing, and fertilizer manufacture. Fluoride is also contained in the discharge from certain organic chemical manufacturing processes. For example, boron trifluoride (BF_3) is a catalyst of ore containing feldspar; it is present in the form of a complex in the effluents from coke-oven plants and in the effluents from the manufacture of semiconductor and electronic components.

The beneficial effects of optimal fluoride intake in prevention of dental caries are well known. On the other hand, detrimental dental fluorosis can result from excessive fluoride intake, usually in drinking water. Therefore, the discharge of fluoride containing wastes into water bodies that are raw water sources is of concern. The "best practicable technology presently available" treatment process for defluoridization is the addition of excess of calcium salt, usually calcium hydroxide, with the attainment of a pH greater than 8 to precipitate the fluoride ion as insoluble calcium fluoride. Other processes utilize a mixture of calcium chloride and hydrated lime, or quicklime (CaO) and hydrated lime. These and other similar processes are capable of reducing the fluoride concentration to approximately 10 mg/liter. To reduce the concentration to approximately 1 mg/liter, other means have been employed, such as the use of apatite $Ca_9(PO_4)_6 \cdot CaCO_3$, tricalcium phosphate hydroxyapatite $Ca_5(PO_4)_3 \cdot OH$, ion exchange resins, magnesia, aluminum sulfate, calcined (activated) alumina, and bone char (tricalcium phosphate and carbon). At least three of these systems, aluminum sulfate (alum), activated alumina, and the bone char columns process, have been proved to be practicable. From this wide array of processes, the selection of the most effective and efficient method cannot be accomplished theoretically; process trials with representative samples must be made before a process is recommended.

The field of industrial waste-water collection and disposal engineering has reached a high state of development. It has become a specialty devoted to the complex problems of waste minimization, pretreatment, treatment, and renovation–reclamation. These professionals cooperate and collaborate with industrial and chemical engineers in process design, equipment development, and should be involved with industrial and chemical engineers in the design of processes and equipment that keep waste production to a minimum.

4 SOLID AND HAZARDOUS WASTES

Solid wastes are the discards of individuals, commerce, industry, and agriculture. The Resource Conservation and Recovery Act (RCRA) of 1976 (P.O. 94-580) inaugurated the federal program regulating solid and hazardous waste management and disposal. The act defines solid and hazardous waste; authorizes the EPA to establish standards for facilities that generate or handle waste; and establishes a permit system for hazardous waste treatment and storage and disposal facilities. The Hazardous and Solid Waste Amendments of 1984 (P.L. 98-616) are the last major change in the act (47). These amendments set deadlines for permit issuance, require double liners and leachate collection systems for most hazardous waste landfilling facilities, prohibit land disposal of many types of hazardous wastes, and establish a new program regulating underground storage tanks. Many problems have surfaced in the management of solid and hazardous wastes and it is anticipated that when RCRA is reauthorized the new act will include substantive changes (48).

Manufacturing, mining, oil and gas drilling, electrical utilities, and agriculture have been estimated to generate 11 billion tons of solid waste annually. Of these, mineral extraction and processing waste are the overwhelming contributors. Most of this waste is disposed at or near the site of generation. At one plant ash from coal-fired electrical generation and the flue gas desulfurization process are so voluminous as to fill a dammed valley to a depth of more than 100 ft.

The range of solid waste generation in other industries is from 10.6 lb/day/employee in the SIC 19-49 category, to 23.5 lb/day/employee in the electrical machinery SIC 36 category, and 25.5 lb/day/employee in the nonelectrical machinery SIC 35 category.

Households and commercial establishments annually generate an estimated 160 million tons of what is termed municipal solid waste. Currently, 83 percent is landfilled, 11 percent recycled, and 6 percent incinerated. With the advent of the Resources Conservation and Recovery Act of 1976 (P.L. 94-580) regulations have been promulgated that have made many former landfilling operations impractical. In the past 10 years the number of municipal solid waste landfills has declined by some 70 percent, from approximately 20,000 to 6000, and the decline is continuing. As existing landfills reach design capacity and more stringent regulations and standards are enforced, it is expected that nearly half of the remaining will be closed by 1991.

The major alternative methods to landfilling are recycling and incineration. Recycling reduces the waste volume by 20 percent. Although not a new concept, at the turn of the century a higher fraction of the solid waste volume was recycled than is currently; at present this alternative is being mandated legislatively and is still in the development stage. The technology for waste separation, either at the point of generation or at a dedicated facility is well developed, the difficulties facing the recycling process are (a) unstable markets and (b) remaining technical hurdles to the recycling of mixed materials.

Incineration can reduce the volume of combustible solid waste from 75 to 90 percent. Because incineration, well designed and operated, reduces the content of pesticides, putrescible material, and some other hazardous constituents, the residue can be landfilled. Of course heavy metals and other noncombustible matter are in the residue. Most recently constructed incinerators (operating at temperatures up to 2500°F) are designed to convert the heat value in the solid waste either to steam or generate electricity. Electricity generated by such facilities may be sold as is guaranteed by the Public Utilities Regulatory Act of 1978 (PURPA). Apart from concern with possible groundwater contamination by toxic substances in the ash, air emissions, in particular dioxin, are cited as potential environmental impacts of incineration. Experience with high-temperature incinerators will demonstrate whether they are environmentally benign or not.

Composting is a method of biodegradation of the organic constituents in solid waste. As in waste-water treatment the process takes advantage of the activity of microorganisms to decompose the organic constituents into a stable, humus like substance. In the decomposition process, usually an aerated process, heat is generated and pathogens are destroyed. Often sludge generated in waste-water treatment is co-composted with municipal solid waste. Because the temperature generated by the thermophilic bacteria in the composting piles are sufficient to destroy the bacteria, viruses, protozoans and helminths of concern, the primary problem is assurance that all parts of the pile will be subject to these temperatures. In recognition of the problem the EPA has promulgated criteria that refer to those processes that significantly reduce pathogens and processes designed further to reduce pathogens.

Heavy metals in waste-water sludge result from household and industrial discharges. The major concern is the content of these elements in composted material or waste-water sludge applied to the land. Some elements, under favorable soil chemistry, can accumulate in food crops grown on such soil. The elements of primary concern are cadmium, lead, arsenic, selenium, and mercury.

Hazardous wastes is a catchall term used to describe industrial wastes as well as certain waste materials from households, institutions, including universities, and industry that are deemed to pose unreasonable risks to health and safety. Property damage and environmental deterioration caused by hazardous wastes are also of concern.

Under RCRA Subtitle C and the Hazardous and Solid Waste Amendments of 1984 (HSWA) the control of hazardous wastes is undertaken by first identifying and listing those wastes deemed hazardous at the point of generation, assuring that these are transported safely, regulating their storage, treatment and/or disposal.

Hazardous wastes may be designated as such or so defined. Definition may be by one or all of four classes—ignitability, reactivity, corrosivity, and toxicity. So far, EPA has published four lists containing nearly 1000 chemical compounds. EPA has been authorized to implement the HSWA requirements according to a schedule through 1992. In 1989 EPA's hazardous waste program was required to implement land disposal bans for some of the most hazardous and refractory wastes. Under Subtitle C of RCRA land disposal of solvents and dioxin has been banned.

Hazardous waste management, that is, minimization, treatment, and disposal has become a major national problem. It is a problem that concerns industry, municipalities, institutions, and the average citizen. As of January 1989, EPA and the states had issued permits to 21 percent of the 4000 active hazardous waste management facilities. Over the period 1986–1989, 2000 such facilties were closed and 40 percent of these had no approved closure plan. The subject is of such significance that it is addressed in a separate chapter in this book; see Chapter 30.

5. ASSURANCE OF A SAFE FOOD SUPPLY

The objective of food sanitation is the prevention of food-borne disease. Technological and economic considerations of purveying food play an important role in the application of effective control measures. However, the individual food worker has a very significant and even crucial role in the safety of food that begins with its production, continues during processing and preparation, and ends when the food is served or otherwise made available to the employees. In addition, the aim of food sanitation programs is to prevent disease without impairing the nutritional value of food.

5.1 Food-Borne Disease

Diseases caused by ingestion of contaminated food may be divided into four major classes: (*a*) infections caused by the ingestion of living pathogenic organisms (bacterial or viral) contained in the food, (*b*) intoxication resulting from ingestion of food in which preformed microbial toxins have developed, (c) animal parasitism, and (d) poisoning by chemically contaminated food. In addition, poisonous plants and animals are implicated.

Foodborne disease reporting in the United States began some 50 years ago, and in 1966 a system of food-borne disease surveillance was established by the Centers for Disease Control (CDC). Reports of enteric disease outbreaks attributed to microbial or chemical contamination of food (and liquid vehicles) received by the CDC are incorporated into annual summaries. The Annual Summary of Food-Borne and Waterborne Disease Outbreaks (49) for 1982 lists 656 reported food-borne disease outbreaks involving 19,380 cases. Significantly, this is the largest number reported to the CDC Food-Borne Disease Surveillance Activity in a single year since publication of such outbreaks was initiated. An outbreak is defined as an incident in which (1) two or more persons experience a similar illness, usually gastrointestinal, after ingestion of a common food; and (2) epidemiologic analysis implicates the food as a source of the illness. An exception to this definition, for example, is that one case of botulism or chemical poisoning constitutes an outbreak.

Estimates of the total annual number of food-borne disease cases in the United States are quite varied. However a reasonable one cites the figure of 20 million/year.

In the 5-year period 1978–1982, the responsible pathogen was identified in 34 percent of food-borne disease outbreaks reported to CDC. As reported in the Annual Summary for 1982, many pathogens are not identified because of late or incomplete laboratory investigation. In yet other cases the responsible pathogen may have escaped detection despite a thorough laboratory investigation. In other cases the responsible pathogen cannot yet be identified by available laboratory techniques. There is a need for further clinical, epidemiologic, and laboratory investigations to permit the identification of these pathogens or toxic agents and the institution of suitable measures to control diseases caused by them. Pathogens suspected of being, but not yet determined to be, etiologic agents in food-borne disease include group D *Streptococcus, Yersinia enterocoliticus, Citrobacter, Enterobacter, Klebisiella, Pseudomonas*, and the presumed viral agents of acute infectious nonbacterial gastroenteritis.

5.1.1 Microbial Infections and Chemical Intoxication

The majority of reported and investigated foodborne disease outbreaks in the United States are caused by bacterial contamination. In 1982, 656 outbreaks (19,380 cases) of food-borne disease were reported to CDC and the etiologic agent was confirmed in only 34 percent of the reported outbreaks. Bacterial pathogens accounted for 151 (68.7 percent) of the total of 220 investigated outbreaks. *Salmonella* (25 percent) was the most frequently isolated organism, followed by *Staphylococcus aureus* (12.7 percent) and *Clostridium perfringens* (12 percent). In 1982 two outbreaks of a previously unrecognized *Escherichia coli* (0.9 percent) of the serotype 0157:H7, which causes a severe diarrhea characterized by grossly bloody stools, were reported (49). Other etiologic agents were *Clostridium botulinum* (2.1 percent) and *Bacillus cerus* (3.6 percent).

Chemicals, including heavy metals and toxins, accounted for 47 (21.4 percent) of the outbreaks. Parasitic infection was involved in one and viruses in 21 (9.4 percent) Hepatitis A was reported in 19 (8.5 percent) of the outbreaks. The relative incidence of food-borne disease of various etiologies is still unknown. Disorders characterized by short incubation periods, such as *Staphylococcus* infection, are more likely to be recognized as due to common source foodborne disease outbreaks than those with longer incubation periods. Viral disease, such as hepatitis A, with its typical incubation period of several weeks, is most likely to escape detection. Other detection problems are due to difficulties in the transportation and culturing of anaerobic specimens. Other problems relate to the fact that outbreaks caused by *Bacillus cereus* and *Escherichia coli* are probably less likely to be confirmed because these organisms are less often considered clinically, epidemiologically, and in the laboratory.

The Enterobacteriaceae of which *E. coli* is a member are facultative Gram-negative rods that ferment glucose (50, 51). Enterobacteriaceae are widely distributed in the environment, on plants, in soil, and in the intestines of humans and animals. Organisms in this family are associated with many types of human infections, including abscesses, pneumonia, meningitis, septicemia, and intestinal, urinary, and wound infections. These bacteria account for 60 to 70 percent of bacterial enteritis cases. As far as intestinal infections are concerned, although many Enterobacteriaceae have been implicated as a cause of diarrhea, only members of the genera *Escherichia, Salmonella, Shigella*, and *Yersimia* are clearly established as enteric pathogens. The nonpathogenic strain of *E. coli* inhabits the lower bowel of humans and animals, where it is often present in concentrations of 10^7 or more viable organisms per gram of fecal material. For many decades it has been used as an indicator organism for fecal contamination of drinking water supplies as well as in food.

However, *E. coli* is also associated with at least four types of human enteric disease, (1) enteropathogenic, (2) enterotoxigenic, (3) enteroinvasive, and (4) hemorrhagic. Of these, for example, hemorrhagic colitis was recently recognized as being acquired by consumption of partially cooked hamburger meat.

When contamination is heavy or when the particular bacteria are highly virulent, infection may follow ingestion of the food even though it has been stored under optimal conditions. More frequently, however, the bacteria originally contaminating the food multiply to dangerous numbers during prolonged storage at a temperature between 45 and 140°F (7 to 60°C), the so-called incubation or danger zone. Foods most commonly found to be vehicles of the bacterial infectious agents are milk and cream, ice cream, seafoods, meats, poultry, eggs, salads, mayonnaise and salad dressings, custards, cream-filled pies, and eclairs and other filled pastries. Unfortunately, food contaminated by disease agents is not necessarily decomposed or altered in taste, odor, or appearance.

Bacterial contamination of food may occur in different ways. Infected persons may transmit bacteria by droplets (sneeze or cough) from the respiratory tract, by discharges from skin infections of the hand, and by contamination of hands with feces, nasopharyngeal secretions, and discharges from open or draining sores elsewhere on the body (i.e., by failure to keep these adequately covered). In some instances, the infected person is without symptoms or signs of illness; therefore the infection is difficult to detect and control by, for example, assignment to other work. Another source of bacterial contamination of food is the multiplication of bacteria in improperly cleaned utensils, with transfer to foods subsequently prepared, stored, or served in them.

Mice, rats, and roaches may contaminate foods and utensils by bacteria carried mechanically on their feet and bodies from an infected to a noninfected area, or they may contaminate food with their urine and feces, which contain pathogenic organisms such as, in the case of rats, *Salmonella* and *Leptospira icterohemorrhagica* (cause of Weil's disease). The source of infection in meat and poultry may be infection before killing (or contamination may occur subsequent to killing, as with other foods).

The clinical symptoms of acute gastric enteritis due to bacterial infection usually have their onset 6 to 24 hours after ingestion of the infected food. Cramping, abdominal pain, diarrhea, nausea, and vomiting are the chief manifestations; often these symptoms are accompanied by headache, pyrexia, and general malaise.

5.1.2 Bacterial Food Intoxication

Bacterial food intoxication is caused by bacteria that, although usually harmless to humans, produce toxic substances when they grow in food. When ingested in sufficient amount these toxins give rise to serious or fatal disease, even though after ingestion there is no further multiplication of the organisms in the body. Bacteria of the *Staphylococcus* group, often found in skin infections (pimples, boils, carbuncles) and respiratory tract infections, may form a virulent enterotoxin when allowed to grow in food. The symptoms of staphylococcal intoxication are nausea, vomiting, cramping, abdominal pain, and diarrhea. In severe cases, blood and mucus may be found in the stool and vomit. The chief characteristic distinguishing food poisoning from toxins and enteric disease from that due to pathogenic bacteria carried in food is the time of onset of symptoms. In staphylococcal intoxication, this is often less than 3 hours after the ingestion of the contaminated food and rarely longer than 6 hours. On the other hand, in bacterial infections a delay of 6 to 24 hours between ingestion of food and onset of symptoms is more likely. The rapidity of onset and the uniform distribution of food poisoning symptoms in a group of people make it possible to determine the meal and particular foods in the meal responsible for poisoning.

5.1.3 Animal Parasitism

Some of the diseases caused by ingestion of food contaminated with animal parasites are amoebiasis, trichinosis, and tapeworm infestation. In the instance of infection with *Entamoeba histolytica* (causative organism of amoebic dysentery), the source and transmission of infection is similar to that of other enteric pathogens such as

Shigella and *Salmonella*. The problem is somewhat different for *Trichinella spiralis* and the tapeworms *Taenia saginata* (beef tapeworm) and *Taenia solium* (pork tapeworm), for in these cases the infectious organisms are present in cattle or swine before the time of killing. Prevention of disease caused by these parasites, as well as their control, depends on inspection and selection of food at the time of purchase as well as proper preparation, as the heat of prolonged cooking will destroy the organisms.

5.1.4   Toxic Chemicals

Accidental contamination of foods by toxic chemicals is an additional hazard. Acute poisonings are characterized by sudden onset, from a few minutes to 2 hours after ingestion of the chemical. The more common poisons reported as borne by foods are cadmium (from metal-plated utensils), sodium fluoride (from roach powder), and arsenic (from insecticides). Other substances frequently mentioned are anti-mony, zinc, lead, and copper.

Compounds such as aliphatic and aromatic amines used as boiler feedwater conditioners may present an additional hazard in the preparation of foods. Wher-ever steam may come in direct contact with food, the toxicity, type, and quantity of boiler feedwater conditioners should be ascertained. Complete physical sepa-ration of the food and steam from the boiler is recommended. The OSHA standard in Section 1910.141 (Sanitation) states:

(g) Consumption of food and beverage on premises:

1. *Application.* This paragraph shall apply only where employees are permitted to consume food or beverages or both, on the premises.

2. *Eating and Drinking Areas.* No employee shall be allowed to consume food or beverages in a toilet room nor in any area exposed to a toxic material.

3. *Waste Disposal Containers.* Receptacles constructed of smooth, corrosion-resistant, easily cleanable or disposable materials, shall be provided and used for the disposal of waste food. The number, size, and location of such receptacles shall encourage their use and not result in overfilling. They shall be emptied not less frequently than once each working day, unless unused, and shall be maintained in a clean and sanitary condition. Receptacles shall be provided with a solid tight-fitting cover unless sanitary conditions can be maintained without use of a cover.

4. *Sanitary Storage.* No food or beverages shall be stored in toilet rooms or in an area exposed to a toxic material.

5. *Food Handling.* All employee food service facilities and operations shall be carried out in accordance with sound hygienic principles. In all places of employment where all or part of the food service is provided, the food dispensed shall be wholesome, free from spoilage, and shall be processed, prepared, handled, and stored in such a manner as to be protected against contamination.

The revised American National Standards Institute Minimum Requirements for Sanitation in Places of Employment (3) states in Item 10.1.1 that:

In all places of employment where employees are permitted to lunch on the premises, an adequate space suitable for the purpose shall be provided for the maximum number of employees who may use such space at one time. Such space shall be separate from any location where there is exposure to toxic materials.

Furthermore, Item 10.1.3 states that "No employee shall be permitted to store or eat any part of his lunch or eat other food at any time where there is present any toxic material or other substance that may be injurious to health." Item 10.1.4 is as follows:

In every establishment where there is exposure to injurious dusts or other toxic materials, a separate lunch room shall be maintained unless it is convenient for the employees to lunch away from the premises. The following number of square feet per person, based on the maximum number of persons using the room at one time, shall be required:

Persons	Area (ft²)
25 and less	8
26–74	7
75–149	6
150–499	5
500 and more	4

5.2 Food Handling

Food handlers educated in the proper methods of handling foods, glassware, and utensils and in good personal hygiene are the cornerstone of a safe food supply. In food sanitation, the health of the food handler is of primary importance, and every effort should be made to exclude from food handling work any sick or injured employee who has a discharging wound or lesion. This statement does not constitute recommendation of a system of frequent medical and laboratory examinations of food handlers. On the contrary, these procedures have not been found to be cost effective. Education of the food handler to report promptly any illness is a more effective means of reducing food-borne disease. In many localities health departments are prepared to provide preemployment or in-service food handler training programs, and a variety of training manuals and visual aid materials are available.

In the handling of foods, particularly those previously mentioned as being sources of infection or intoxication, care must be taken to keep the materials at a temperature out of the incubation zone; that is, they must be refrigerated at a temperature below 45°F or kept hot at 140°F or above. The exception to this rule is that during preparation, serving, or transportation, food may be kept at intermediate temperature for a period not exceeding 2 hours. Indicating thermometers of proved accuracy should be installed in all refrigerators. Common deficiencies found in refrigerated food storage are the overloading of the refrigerator, which prevents the free circulation of cold air, and the storage of foods in containers so deep that the total mass of the food cannot be cooled to 70°F in 2 hours and 45°F or less within 4 additional hours. Further provisions for continual vigilance against possible chemical contamination must be exercised by prohibiting the use of cadmium-plated utensils, prohibiting the use of galvanized utensils for cooking, and banning the use in food preparation and serving areas of roach powders containing sodium floride, cyanide metal polish, and other hazardous substances (52).

In view of the widespread custom, especially in the United States, of providing cubed and crushed ice in drinking water and beverages, it is of interest to note two reports on the sanitary quality of such ice (53). Both studies report that such ice often fails to meet sanitary standards and is contaminated with *Escherichia coli*, clostridia, micrococci, and streptococci. A suggested control method is to provide chlorine disinfecting solutions (approximately 2 ppm) in which the ice is placed.

5.3   Kitchen and Kitchen Equipment Design

The ordinance and code prepared by the Public Health Service (54) and adopted by the American National Standards Institute as a minimum standard whenever local regulations do not apply to the industrial food handling establishment provides for certain materials of construction and gives minimum lighting and ventilation requirements. The design of kitchens represents a problem in materials handling, batch preparation, and small quantity distribution. The designer must provide for the greatest ease in performing a large variety of tasks, usually in crowded quarters.

It behooves the kitchen designer to allow sufficient space in this important working area to permit the sanitary performance of the tasks of preparing food and washing of dishes and utensils. Separation of activities—food preparation, baking, and dishwashing—is preferable to combining them in one room. The location of kitchen equipment in relation to ease of cleaning is of great importance. Too often stoves, steam kettles, peelers, and mixers are placed close to walls, allowing dirt to accumulate and providing excellent breeding and hiding places for roaches.

Tables and stands should have shelves no less than 6 in. off the floor. Fixed kitchen equipment should be sealed with impervious material when attached to the wall or, alternatively, kept far enough away from the wall to allow for easy cleaning. All work surfaces should be made of impervious, noncorrosive, and easily cleaned material. When wood is used, it should remain unpainted and should be scrubbed daily. Cutting, peeling, and other mechanical equipment should be designed to eliminate dirt-accumulating crevices and should be readily disassembled for cleaning. A voluntary organization, the National Sanitation Foundation, representing the public health profession, business, and industry, publishes a series of standards for soda fountains and luncheonette equipment, food service equipment, and spray-type dishwashing machines; others are in preparation (55). These standards are of value in identifying equipment that has been tested to meet exacting food sanitation requirements.

5.4 Food Vending Machines

Vending machines are available that serve hot beverages, cold carbonated and noncarbonated beverages, sandwiches, and pastries. A recommended ordinance and code has been published by the Department of Health and Human Services (56). The following considerations may serve as a guide for the selection of automatic food vending machines.

All surfaces that come in contact with the food should be constructed of smooth, noncorrosive, nontoxic materials. All flexible piping should be nonabsorbent, nontoxic, and easily cleanable; treatment with the more common disinfecting agents should not harm it.

Vending machine purveying perishable food should be refrigerated. Their design should include an automatic temperature-controlled shutoff that will prevent the machine from operating whenever the temperature in the storage compartment rises above 45°F. In general, the design of the machine should allow easy accessibility for cleaning and disinfecting all surfaces that come in direct contact with the beverage or food.

Whenever disposable cups are furnished they should be stored in protective devices, and it should be possible to reload these without touching the lips or interior surface of the cups.

The drip receptacles in beverage vending machines should be provided with a float switch that would prevent the machine from operating when the liquid in the receptacle reaches a certain level. The vending area of the machine should be protected from dust, dirt, vermin, and possible mishandling by patrons by means of self-closing gates or pans.

The cleaning of tanks, containers, and other demountable equipment should be done at a central location. Whenever cleaning is accomplished at the vending machine, a portable three-compartment sink should be provided, to permit proper washing, rinsing, and disinfecting of all demountable equipment. At such points water and sewer facilities should be provided.

The food and drink for the vending machines should be stored and handled as they would have to be to meet the requirements of food handling establishments. A good rule is to offer for sale only packaged foods that were prepared at a central commissary. All perishable food should be coded to show place of origin and date of preparation. Perishable food that remains unsold, even though it has been properly referigerated, should be disposed of at frequent intervals.

No vending machines should be located in areas where there is exposure to toxic materials.

6. CONTROL AND ELIMINATION OF INSECTS AND RODENTS

6.1 Insect Control

In general, insect control in an industrial setting requires that a high level of sanitation be practiced in and surrounding the plant. To control insect infestation, the primary prevention strategy is to eliminate breeding and harborage locations and to deprive the organisms of food. For example, depriving the *Aedes Aegypti* mosquito, the vector of yellow fever, of accumulated standing water assures its control. However, any small puddle, or rainwater accumulation in discarded cans, tires, or other containers, provides near-ideal breeding locations for this mosquito.

Whenever unrecognized infestation occurs it is possible to obtain idenfitication assistance and advice on control measures from the State or County Extension Service.

The insects of public health significance are flies, fleas, lice, mosquitoes, mites, ticks, and roaches. To be able to prevent or control infestation effectively one must possess an understanding of insect biology, morphology, breeding habits, and disease spreading mechanisms. Based upon such knowledge it is possible to mount the most effective chemical, physical, or biological control.

Except for the food and other industries dealing with products that may serve as nourishment for insects, the problem of controlling these pests depends largely on good housekeeping. The more general problem of control of arthropods and arthropod-borne disease cannot be accomplished on a restricted scale as in a single industrial establishment. To be effective, it must be a community-wide or regional undertaking of considerable magnitude. It is axiomatic that an industry located in an infested area should cooperate with the governmental agency responsible for effectuating control.

The number and prevalence of flies and roaches are good general indexes of the "sanitation" practiced in the establishment. Of direct interest is the proof obtained of the relation between fly prevalence and the infections, diseases, and deaths caused by the bacillary dysentery organism (57). Field studies showed that bacillary dysentery was materially reduced in a community when effective fly control measures were undertaken. The most direct method of dealing with an insect infestation problem is to eliminate breeding places. Wherever garbage, rubbish, or other organic matter is allowed to accumulate and become exposed to insects, it produces ideal conditions for their development. In food handling areas, nourishment is available, and cracks, crevices, voids, and other unused spaces harbor the pests.

The most common insect invaders are the roaches—German cockroach, *Blatella germanica*; American cockroach, *Periplaneta americana*; Australian cockroach, *Periplaneta australasiae*; Oriental cockroach, *Blatta orientalis*; brown-banded cockroach, *Supella supellectilium*—and flies, of which 90 to 95 percent are housefiles, *Musca domestica*, and blowflies, *Phaenicia* (57).

The use of insecticides, which should be undertaken only after the environmental conditions favoring breeding have been eliminated, poses health problems to the applicator and others. The problems and their prevention are extensively discussed in the literature. New insect control technology is being rapidly developed; for example, sex attractants and juvenile hormones are proposed as insect control materials. However, they too may have adverse health effects on mammalian species, including humans. On the other hand, biological insect control may have beneficial public health and ecologic consequences because it could prevent the positive feedback aspects inherent in the use of chemical insecticides.

Other biological control methods include the utilization of specific parasites, insect predators, and pathogens. The release of radiation-sterilized males, as practiced in the attempt to control Mediterranean fruit fly infestation, is an advanced biological control technique. Other means considered are use of genetic engineering and biologically produced compounds. Plant insecticides have been an extremely important group that include derris and its chief constituent rotenone, and pyrethrum, which contains pyrethrins. The latter are nontoxic to warm-blooded animals and have a very rapid "knock-down" effect on a large variety of insects. Because they are highly volatile and are broken down by light, their effect is transitory. They may be stabilized and their effect intensified by the addition of piperonyl butoxide and other compounds.

6.2 Rodent Control

Apart from the health hazards rodents produce as vectors of disease (bubonic plague, endemic or murine typhus fever, salmonellosis, and Weil's disease), their infestation of an industrial plant can be a tremendous economic drain. As a general principle, in the community the destruction of food, crops, merchandise, and property by rats is serious enough to justify suppressive measures, even if rats are not responsible for human disease. Because of their destructiveness and their pollution of food, rats have gained the unenvied fame of being the worst mammalian pest in civilization.

Sewer systems in cities and large industrial facilities provide a major habitat for rodents. In particular this is the case in older sewer systems in need of repair. With the low state of repair of the nation's infrastructure, including its sewer system, the sewer rodent infestation problem is of consequence. Sewers provide rodents with food, shelter, water, and an equable temperature year round. Sewers also provide easy egress and ingress. This is especially true in the case of combined sewers, which convey sanitary wastes and storm water in a single pipe. A separate sewer system, a requirement in all new construction, is less favorable to rodent infestation. Sewer rodents living in sewers are exposed to a variety of infectious diseases and thus afford, once in contact with people, an increased opportunity for spreading disease.

When old and deteriorated structures are to be demolished it is necessary first to conduct a rodent infestation survey and then to institute a program to rid it of rodents before they migrate.

In the industrial environment, other than the destruction of food, rodents pose a health problem with the possible transmission of Weil's disease or leptospirosis, a relatively rare disease caused by a spirochete bacterium. The organism is harbored in the rodent and excreted in the urine. The individual becomes infected by direct or indirect contact with infected rodent urine. The organism enters the body through the mucous membranes, cuts, or skin abrasions. There are about 100 cases of Weil's disease per year in the United States, with a few deaths.

Davis estimated about one rat per 35 people in New York City, and about one rat per 20 people in Baltimore (58). As pointed out by Davis, the regulatory factors in any problem of species management can be classed in three main groups: (*a*) environment, which includes availability of food and protective shelter; (*b*) predation, which includes animals that feed on the species, trapping, poisoning, and disease; and (*c*) competition, the fight for a limited supply of environmental necessities.

Control of the rodent population should be based primarily on a change of the environmental conditions so that the rodents are deprived of food and shelter. When this is accomplished, the rodent population is reduced by intraspecies competition. Because the average reproductive rate of Norway rats (they reproduce during all seasons of the year, with a maximum in spring and fall occurring in many areas) is 20 rats per year, it is evident that killing procedures will have little effect on reducing the population and maintaining it at low level if environmental conditions are favorable to rats (59).

6.3   Ratproofing (Building out the Rat)

"Ratproofing" is a term applied to procedures for controlling the portals of entry of rats into a structure. Being extremely cautious, nocturnal mammals, rats prefer to enter a building through small openings. Only when they are subjected to peril do they venture away from normal pathways. The points of ingress and egress of rats that must be guarded against in particular are any openings around pipes, unused stacks and flues, ventilators, and hatches.

All new buildings should be designed and constructed to be ratproof. Foundations must be continuous, extending not less than 18 in. below the ground level, and they should always be flush with the under surface of the floor above. Floor joists should be embedded in the wall and the space between the joists filled in and completely closed to the floor level. All construction materials should be as ratproof as possible (59). In industry, the most frequent shelters for rats are areas surrounding the cafeteria, lunchrooms, and other eating areas, although no place that affords attractive food and shelter is immune to rat infestation. The office in which employees are permitted to eat their lunch is often rat infested, the scraps left in the waste baskets providing the rats with a good food source. All other organic wastes, such as those found in sewers where they harbor, are a good additional source of food for rats.

6.3.1   Killing Rats

Trapping by means of snap and cage traps is a method of killing rats that requires great skill and ingenuity, for rats are rather wary mammals. Killing by use of rodenticides has advanced greatly in the past two decades. The most widely used and oldest poison is Red Squill, which has the advantage of having an emetic action that protects humans and all animals with the ability to vomit. Barium carbonate and phosphorus, utilized in the past, are not widely used now because of their unpredictable results. The newer rodenticides are Antu, a specific for Norway rats; sodium fluoroacetate (1080), an extremely powerful poison with all animals, which must be used with extreme caution and has been prohibited in many localities because of human deaths; and warfarin (a derivative of dicoumarol), an anticoagulant that causes internal hemorrhage in the rat. Although warfarin affects other animals, it has a large factor of safety because of the low dosage (0.025 percent) used in the baits.

Multiple dose anticoagulants are widely used (Warfarin, Pindone, Diphacinone, Chlorophacionone). It has been found that some bait formulations work better in one environment than others. Baits are usually in pellet form, in cereal or wax block. The single-dose anticoagulant Bromadiolone, is more toxic than warfarin, it is also more hazardous to nontarget animals, but owing to its effectiveness saves time and labor.

Some of the newer controls utilize sterilants or inhibitors of adenosine triphosphate, the chief energy-carrying chemical in the body. New rodenticides are being developed and it is prudent that if rodent control is undertaken by the industry itself those conducting the service should be qualified by training. Otherwise a licensed "pest control operator" should contracted.

In certain enclosed structures, primarily aboard ships, fumigants such as hydrocyanic acid gas or carbon monoxide are used for quick and nearly complete elimination of rats.

Control of rats and other rodents requires (a) environmental control to eliminate sources of food and harborage, (b) effective ratproofing, and (c) efficient rat killing programs. Controlling rat populations, not individual rats, is the key to a successful rodent control program.

7 GENERAL ENVIRONMENTAL CONTROL AND PROVISION OF ADEQUATE SANITARY FACILITIES AND OTHER PERSONAL SERVICES

The Occupational Safety and Health Standards promulgated under the Williams–Steiger Act of 1970 (84 Stat. 1593) were published in the *Federal Register* of June 27, 1974 (60). Subpart J (General Environmental Controls) contains Section 191D.141 (Sanitation), Section 1910.142 (Temporary Labor Camps), and Section 1910.143 (Non-Water-Carriage Disposal Systems).

To be applicable in a wide variety of circumstances, the regulations in these sections are, for the most part, very general. For example, insect and rodent control, which is described as "vermin control," requires that "enclosed workplaces shall be constructed, equipped and maintained, so far as reasonably practicable, as to prevent the entrance and harborage of rodents, insects and other vermin. Furthermore, a continuing and effective extermination program shall be instituted where their presence is detected." Such a statement provides little that is helpful to the plant manager and less to OSHA personnel inspecting the premises. On the other hand, the requirements for toilet and washing facilities (e.g., lavatories and showers) are enumerated and relatively detailed.

The personal hygiene practices of employees carry many health implications without reference to the individuals' particular employment. For instance, the disease-transmitting potential from failure to clean the hands after defecating, or the skin diseases resulting from poor body cleansing, may not be related to the person's job. There are, however, health problems relating to personal hygiene practices that occur in industry exclusively. Such problems are related to ingestion of chemical toxins or disease-producing organisms and to local skin, conjunctival, and mucous membrane inflammation, due to sensitivity or direct irritative action of industrial chemicals.

Schwartz et al. (61) have stated that uncleanliness "is probably the most important predisposing cause of occupational dermatitis." Lack of cleanliness in the working environment exposes the worker to large doses of external irritants. Personal uncleanliness not only does the same, but also permits external irritants to remain in prolonged contact with the skin. Workers wearing or carrying to their homes their dirty work clothes may even cause dermatitis in other members of the family who come in contact with soiled clothes, or among unsuspecting workers who clean them. Safeguards such as protective creams help, but personal hygiene is a necessity.

Where the employee is exposed to toxic materials or is a food handler, the need for the optimum in clean, well-lighted, and well-ventilated washing and locker facilities becomes imperative. From the point of view of protecting the health of the individual employee and minimizing the possibility of transmitting infections to others, toilet and washroom facilities should be easily cleanable, adequate in size and number, and accessible, or the personal hygiene habits of employees in industry will suffer. The American National Standards Institute Minimum Requirements for Sanitation (3) considers that "ready accessibility" has not been provided when an employee has to travel more than one floor-to-floor flight of stairs to or from a toilet facility. Some minimum standard on the basis of distance must be set. However, the advent of complex automatic and remote controls and unit operations may bring about modification of some distance standards. In Section 1910.141 (Sanitation) of the OSHA June 27, 1974, Standards item (c) Toilet Facilities (ii) has taken this into account and exempts establishments in which mobile crews work in normally unattended work locations "so long as employees working in these locations have transportation immediately available to nearby toilet facilities which meet the usual requirements."

The OSHA standards state that whenever employees are required by a particular standard to wear protective clothing because of the possibility of contamination with toxic materials, change rooms equipped with storage facilities for the protective clothing shall be provided. Although recommended a number of years ago, the following arrangement, which differs from the current OSHA requirement of locating work and street clothing lockers side by side, is good practice.

A good arrangement is a room or building divided into two sections—a street clothes section and a work clothes section, with bathing and toilet facilities between. The street clothes section has an outside entrance and lockers for street clothes, toilets and wash basins or wash fountains, and changing facilities for supervisors.

The work clothes section has room for work clothes, toilets and wash basins or wash fountains, showers, and laundry. Three connections between the street clothes section and the work clothes section are (a) through the supervisors' change room, (b) through the main shower room, and (c) a hall with "one-way traffic doors" from street to work clothes sections (62).

In the mineral, petroleum, and allied industries, the change house may be located at the entrance gate or near the parking lot to make it readily accessible. It may be a separate building connected to the working area by a covered walkway, and it should be constructed of fireproof or fire-resistant materials and be properly heated, lighted, and ventilated to meet the requirements of Section 4 of the American Standard Minimum Requirements for Sanitation (3). The interior arrangement of the equipment and facilities in change houses and in locker and toilet rooms is not standardized but is a matter of proper design to meet the space, cost, and other requirements of the individual plant.

The minimum facilities required for places of employment are set forth in the various state regulations of departments of labor and divisions of industrial hygiene. In addition, various industrial associations have established, through industry-wide sanitation committees, practice codes related to the particular industry. An example is the Code of Recommended Practices for Industrial Housekeeping and Sanitation of the American Foundrymen's Association (63). The Association of Food Industry Sanitarians, in cooperation with the National Canners Association, has published a comprehensive manual, Sanitation for the Food-Preservation Industries (64). For all places of employment, minimum requirements for housekeeping, light and ventilation, water supply, toilet facilities, washing facilities, change rooms and retiring rooms for women, and food handling requirements are set forth in the American Standard Minimum Requirements for Sanitation (3).

The OSHA requirements (OSHA Standards June 27, 1974, Section 1910.141, Sanitation) are somewhat less stringent for establishments with fewer than 150 employees, nor does the OSHA regulation require the provision of retiring rooms. In general, minimum standards for toilet facilities can be delineated quite specifically with respect to their construction and provision for ease of cleaning. Critics point out that in some instances, at least, the standard lavatories installed in washrooms fail to meet the need for ease of washing. The wash fountain with its foot pedal control supply, basin large enough to permit easy bathing of arms, face, and upper torso, free-flowing stream of thermostatically tempered water, and central dispenser of cleansing material offers a practical solution to the problem of providing good washing facilities. In addition, the wash fountain requires less floor space, initial installation is less complicated and costly, and it is maintained with greater ease. Experience with wash fountains in washrooms for female employees indicates that they are acceptable there.

8 MAINTENANCE OF GENERAL CLEANLINESS OF THE INDUSTRIAL ESTABLISHMENT

Housekeeping and plant cleanliness in their fullest sense imply not only the absence of clutter and debris in the working area and passageways, but also the maintenance of painted surfaces in a clean condition, the frequent removal of dust from lighting fixtures, and in general making the plant "a better place to work." The industrial plant that processes hazardous or potentially hazardous materials that cannot be prevented from escaping into the general atmosphere has a double responsibility to provide the utmost in housekeeping and plant cleanliness.

It is no longer sufficient to delegate responsibility for cleanliness of the plant and surroundings to a foreman or supervisor whose primary duties, interests, and training lie elsewhere. The complexity and cost of housekeeping machinery and supplies, the technical knowledge required for their proper and effective use, and the extensiveness of the work load make the establishment of a housekeeping department in large plants a virtual necessity. Working standards, schedules, and quality controls are also required in this routine plant operation. (See Chapter 15, Volume 1A.) Additional evidence of the economic importance of this operation is the fact that fire insurance premiums are in part determined by the status of plant cleanliness.

Section 1920.22 (General Requirements) of the OSHA regulations sets very general standards for housekeeping. For a continuing process, too often relegated to a low level of managerial supervision, it is possible to establish quantitative measures of housekeeping effectiveness by such means as dust counts, bacteria counts, color intensity, and light transmission measurements.

The food industry presents special problems of frequent or even continuous cleaning operations as well as at the end of each shift. Thus this industry, with its combination of wet and dry cleaning requirements, represents an extreme example for other industries. Perhaps the ultimate in "housekeeping" requirements are those for the "clean rooms" of the electronics and pharmaceutical industries, where continuous sanitation of the total environment is absolutely essential.

Housekeeping must be scheduled, and it must be the assigned responsibility of a trained group under proper supervision. A staggered work arrangement, for example, allows the use of a shift permitting the housekeeping crew to start work 1 to 2 hours after the production personnel. This permits some general cleaning during the morning before very much material has accumulated. It also permits scheduling cleaning during the regular lunch hour shutdown and completing the schedule after the day's operations have ended.

Housekeeping, a continuing indirect cost in industry, deserves uninterrupted scrutiny. The activity must be evaluated periodically with respect to effectiveness and cost.

9   THE ENVIRONMENTAL ENGINEER IN INDUSTRY

The environmental factors described in this chapter are inherent in varying degrees in every industrial establishment. The plants whose products are intended for human food may be placed in a special category, for they have a paramount responsibility to protect their product from contaminants that endanger the health of the consumer. Because of this responsibility, the roles of the environmental engineer and sanitarian in such industries should be obvious. There is great need to conserve the nation's water resources through effective minimization and treatment of industrial wastes, to provide production and management personnel with safe food and potable water at all times, and to maintain a sanitary working environment to protect the employee from the diseases that are not directly of occupational origin. Legislation emphasizes the need for such specialized professionals in industry.

The standards and regulations promulgated to meet environmental goals and objectives are in a state of flux. Timetables are under administrative and legislative review, and the technology to achieve them is evolving. In addition, it is the intent of Congress to have increasing public participation in the standard setting process. Most generally, the environmental engineer is employed by industry to apply his or her knowledge to the solution of problems of water supply and their corollary, liquid waste disposal, solid and hazardous waste management, and air pollution abatement. However, engineering and public health skills are additionally applicable in such areas as food sanitation, insect and rodent control, and plant maintenance. Although the environmental engineer may or may not be directly associated with the occupational health section of the industry, the roles filled by the engineer and by the industrial hygienist are directed toward the same end. The parallel may be further drawn by pointing out that just as the industrial hygienist aims to recognize hazardous working conditions in the design state, to be able to eliminate them completely or to make the application of control measures more effective, convenient, and economical, the environmental and environmental health engineer in industry applies the same principle to the control of environmental factors other than those related to the occupation per se.

REFERENCES

1. U.S. Environmental Protection Agency, Office of Drinking Water, Criteria and Standards Division, "Fact Sheet—Drinking Water Regulations Under the Safe Drinking Water Act," Washington, DC, June 1989.

2. "Interim Radiochemical Methodology for Drinking Water," Environmental Monitoring and Support Laboratory EPA- 600/4-75-008, Cincinnati, OH.

3. American National Standards Institute, "Sanitation in Places of Employment—Minimum Requirements," ANSI-Z4.1, 1986.

4. American National Standards Institute, "Drinking Water Fountains and Self Contained Mechanically Refrigerated Drinking Water Coolers," ANSI/ARI 1010-78, and ANSI/ARI 1020-84.

5. U.S. Department of Health and Human Services—Public Health Service—Centers for Disease Control—National Institute for Occupational Safety and Health, "Occupational Exposure to Hot Environments Criteria 1972."

6. U.S. Department of Health and Human Services–Public Health Service–Centers for Disease Control–National Institute for Occupational Safety and Health, "Occupational Exposure to Hot Environments-Revised Criteria 1986."

7. K. L. Kollar and R. Brewer, *J. Am. Water Works Assoc.*, **67**, 686 (1975).

8. M. P. Wanielista, Y. A. Yousef, J. S. Taylor, and C. D. Cooper, *Engineering and the Environment*, Brooks/Cole Wadsworth, Belmont, CA, 1984.

9. A. F. McClure, *J. Am. Water Works Assoc.*, **66**, 240 (1974).

10. National Commission on Water Quality, Report, Washington, DC, November 1975.

11. U.S. Environmental Protection Agency, *Waste Minimization Opportunity Assessment Manual*, EPA/625/7-88/003, Hazardous Waste Engineering Laboratory, Cincinnati, OH, 1988.

12. C. E. Renn, "Management of Recycled Waste—Process Water Ponds," Project WPD117 Environmental Technology Series, U.S. Environmental Protection Agency R2-73-223, 1973.

13. W. C. Ackerman, *J. Am. Water Works Assoc.*, **12**, 691 (1975).

14. C. J. Schmidt et al., *J. Water Pollut. Control Fed.*, **47**, 2229 (1975).

15. C. E. Arnold, H. E. Hedger, and A. M. Rawn, "Report Upon the Reclamation of Water from Sewage and Industrial Wastes in Los Angeles County, California," prepared for the County of Los Angeles, 1949.

16. Sextus J. Frontinus, "The Aqueducts of Rome (translated by Bennett), William Heineman, London, 1925.

17. Bogly W. J., Jr., "Experiments with Dual Water Distribution Systems," *Proceedings AWWA Seminar*, AWWA Publication 20189, June 23, 1985, p. 5.

18. D. A. Okun, "Overview of Dual Water Systems," *Proceedings AWWA Seminar*, AWWA Publication 20189, June 23, 1985, p. 1.

19. H. A. Spafford, *J. Am. Water Works Assoc.*, **46**, 993 (1954).

20. F. W. Sunderman Jr., B. Dingle, S. M. Hopfer, and T. Swift. "Acute Nickel Toxicity in Electroplating Workers Who Accidentally Ingested a Solution of Nickel Sulfate and Nickle Chloride," *Am. J. Ind. Med.*, **14**, 257 (1988).

21. American Water Works Association, *Manual M-14*, AWWA No. 30014, 1966, 22 pp.

22. American Water Works Association, "Standards for Backflow Prevention Devices— Reduced Pressure Principle and Double Check Valve Types," Standard c506-78/(R83), AWWA No. 43506, 1983, 20 pp.

23. G. Keleti and M. A. Shapiro, "Legionella and The Environment," *Environ. Controls*, **17**(2), 133 (1987).

24. J. E. McDade, C. C. Shepared, D. W. Fraser, T. F. Tsai, M. A. Redus, W. R. Dowdie, and the Laboratory Investigative Team, Legionnaires Disease, "Isolation of a Bacterium and Demonstration of its Role in Other Respiratory Disease," *New Engl. J. Med.*, **297**, 1197 (1977).

25. J. J. States, R. M. Wadowsky, J. M. Kuchta, R. S. Wolford, L. F. Conley, and R. B. Yee, "Pathogenic Organisms in Drinking Water: *Legionnella*," in *Advances in Drinking Water Microbiology Research*, G. A. McFeters, Ed., Science Tech Publishing, Madison, WI, 1989.

26. H. M. Foy, P. S. Hayes, M. K. Cooney, C. V. Broome, I. Allan, and R. Tobe, "Legionnaires Disease in a Prepaid Medical Care Group in Seattle, WA, 1963–75," *Lancet*, **1**, 767 (1979).

27. T. H. Glick, M. B. Gregg, G. W. Mallison, W. W. Rhodes, and B. Kassanoff, "Pontiac Fever: An Epidemic of Unknown Etiology in a Health Department. Clinical and Epidemiologic Aspects," *Am. J. Epidemiol.*, **107**, 149 (1978).

28. L. A. Herwaldt, G. W. Gorman, A. W. Hightower, B. B. Brake, H. Wilkinson, A. L. Reingold, P. A. Boxer, T. McGrath, D. J. Brenner, C. W. Moss, and V. V. Broome. "Pontiac Fever in an Automobile Assembly Plant, in Legionella," in *Proc. 2nd International Symposium*, C. Thornberry, A. Balows, J. C. Feeley, and W. Jakubowski, Eds., ASM, Washington, DC, 1984, p. 246.

29. D. W. Fraser, "Sources of Legionellosis," in *Proc. 2nd International Symposium*, C. Thornberry, A. Balows, J. C. Feeley, and W. Jakubowski, Eds., ASM, Washington, DC, 1984, p. 277.

30. C. B. Fliermans and R. S. Harvey, "Effectiveness of L. bromo-3-chloro-5,5-dimethyl-hydantoin Against *Legionella* Pneumophila in Cooling Towers," *Appl. Environ. Microbiol.*, **47**, 1307 (1984).

31. U.S. Nuclear Regulatory Commission, "Licensing Requirements for Land Disposal of Radioactive Waste," Code of Federal Regulations, Title 10, Part 61, 1981.

32. U.S. Department of Energy, National Low-Level Management Program, "The 1985 State by State Assessment of Low-Level Disposal Sites," DOE/LLW – 59T, Idaho Falls ID, 1986.

33. Water Pollution Control Federation, "Pretreatment of Industrial Wastes," *Manual of Operation FD-3*, 1981 Order No. MFD3P4.

34. W. W. Eckenfelder, Jr., *Water Works Eng.*, **13**, 83 (1976).

35. J. F. Richardson and W. N. Zaki, Transl., *Inst. Chem. Eng.*, **32**, 34 (1954).

36. A. W. Lawrence and P. L. McCarty, Transl., *Sanit. Eng. Div. ASCE*, **96**, SA3 (1970). Also *ASCE Procedure Paper 7364*, 1970, p. 757.

37. P. L. McCarty, "Anaerobic Treatment of Soluble Wastes," in *Advances in Water Quality Treatment*, E. F. Gloyna and W. W. Eckenfelder, Jr., Eds., University of Texas Press, Austin, TX, 1968, p. 336.

38. *Water Pollution Control Federation Research Journal*—Annual Literature Review Issue, June, 1989.

39. G. Lettinga et al., Eds., *Granular Anaerobic Sludge: Microbiology and Technology*, Pudoc, Wageningen, The Netherlands, 1988.

40. J. M. Bell, Ed., *Proceedings of the 43rd Industrial Waste Conference, Purdue University*, Lewis Publishers, Chelsea, MI, 1989.

41. *Proceedings Biennial IAWPRC*, London, 1988.

42. W. J. Weber, Jr., *Physicochemical Processes for Water Quality Control*, Wiley-Interscience, New York, 1972.

43. J. C. Morris, "Formation of Halogenated Organics by Chlorination of Water Supplies," U.S. Environmental Protection Agency Publication No. 600/1-75-002, 1975.

44. J. J. Rook, Jr., *Water Treat. Exam.*, **23**, 234 (1974).

45. J. J. Rook, Jr., *Am. Water Works Assoc.*, **68**, 168 (1974).

46. W. Zabban and R. Helwick, "Defluoridization of Wastewater," *Proc. 30th Annual Purdue Industrial Waste Conference*, West Lafayette, IN, 1975.

47. U.S. Congress Senate—Committee on Environment and Public Works, RCRA Oversight Hearings, 100th Congress, 1st Session, Part 2, U.S. Government Printing Office, Washington, DC, 1988, pp. 694.

48. U.S. Environmental Protection Agency—Office of Solid Waste, Report to Congress EPA/530-SW-89-019, 1989, pp. 70.

49. Centers for Disease Control, *Annual Summary of Food-Borne and Water-Borne Disease Outbreaks*, CDC, Atlanta, GA, 1982, p. 198.

50. M. T. Kelly, D. J. Brenner, and J. J. Farmer III, "Enterobacteriaceae," in *Manual of Electrical Microbiology, IV* E. H. Leunette, A. Balows, W. J. Housler, Jr., and H. J. Shadony, Eds., ASM, Washington, DC, 1985, pp. 263–277.

51. C. F. Clancy, in *Enterobacteriaceae Practical Handbook of Microbiology*, W. M. O'Leary, Ed., CRC Press, Boca Raton, FL, 1989, pp. 71–80.

52. U.S. Department of Health and Human Services, U.S. Public Health Services, "Sanitary Food Service Instructor's Guide," FDA 78-2081, 1976 Revision.

53. E. W. Moore, E. W. Brown, and E. M. Hall, "Sanitation of Crushed Ice for Iced Drinks," *J. Am. Public Health Assoc.*, **43**, 1265 (1953); V. D. Foltz, "Sanitary Quality of Crushed and Cubed Ice as Dispensed to the Consumer," *Public Health Rep.*, **68**, 949 (1953).

54. U.S. Public Health Service, "Ordinance and Code, Regulating Eating and Drinking Establishments" (Recommended DHHS, PHS), Washington, DC, 1976.

55. National Sanitation Foundation, "Soda Fountain and Luncheonette Equipment" Standard No. 1; "Food Service Equipment" Standard No. 2; "Spray-Type Dishwashing Machines," A Sanitation Ordinance and Code, Publication No. 546, 1965.

56. U.S. Public Health Service, "The Vending of Food and Beverages," A Sanitation Ordinance and Code, Publication No. 546, 1965.

57. D. E. Lindsay, W. H. Stewart, and J. Walt, "Effect of Fly Control of Diarrheal Disease in an Area of Moderate Morbidity," *Public Health Rep.*, **68**, 361 (1953).

58. D. E. Davis, "Control of Rates and Other Rodents," in *Preventive Medicine and Hygiene*, 8th ed., K. F. Maxcy, Ed., Appelton-Century-Crofts, New York, 1956, Chapter 8.

59. Centers for Disease Control, *Rat-Borne Disease; Prevention and Control*, CDC, Atlanta, GA, 1949.

60. Occupational Safety and Health Standards, U.S. Department of Labor, Occupational Safety and Health Administration, *Fed. Reg.*, **39**(125), 23502–23828 (June 24, 1974).

61. L. Schwartz, L. Tulipaw, and D. J. Birmingham, *Occupational Diseases of the Skin*, 3rd ed., Lea & Febiger, Philadelphia, 1957.

62. F. E. Cash, "Suggested Standards for Change Houses," paper presented at the annual meeting of the American Public Health Association, October 1951.

63. American Foundrymen's Association, "Code of Recommended Practices for Industrial Housekeeping and Sanitation," AFA, Chicago, IL, 1944.

64. Association of Food Industry Sanitarians and National Canners Association, *Sanitation for the Food Preservation Industries*, AFIS-NCA, Berkeley, CA.

GENERAL REFERENCES

American Water Works Association, *Water Quality and Treatment: A Handbook of Public Water Supplies*, 3rd ed., McGraw-Hill, New York, 1971.

American Public Health Association, American Water Works Association and Water Pollution Control Federation, *Standard Methods for the Examination of Water and Wastewater*, 17th ed., APHA, Washington, DC, 1989.

American Society of Civil Engineers, American Water Works Association, Water Pollution Control Federation, *Glossary: Water and Wastewater Control Engineering*, 3rd ed., New York, 1981.

American Water Works Association, *New Dimensions in Safe Drinking Water. 1986 Safe Drinking Water Act*, 2nd ed., No. 20235, Denver, CO, 1988.

American Water Works Association, *Water Conservation*, No. 20238, Denver, CO, 2nd ed., 1987.

American Water Works Association, *Recoverable Materials and Energy from Industrial Waste Streams*, No. 20234, Denver, CO, 1987.

American Water Works Association, *Cross-Connection* and *Backflow Prevention*, No. 20106, Denver, CO, 1974.

G. F. Craun, Ed., *Waterborne Disease in the United States*, CRC Press, Boca Raton, FL, 1986.

W. W. Eckenfelder, Jr., *Industrial Water Pollution Control*, 2nd ed., McGraw-Hill, New York, 1989.

H. Koren, *Handbook of Environmental Health* and *Safety*, Vols. I and II, Pergamon Press, Elmsford, NY, 1980.

Foundation National Restaurant Association, *Applied Food Sanitation. A Certification Course Book*, 3rd ed., Wiley, New York, 1985.

International Association of Milk, Food and Environmental Sanitarians, *Procedures to Investigate Foodborne Illness*, 4th ed., Ames, IA, 1987.

T. McLaughlin, *Industrial Housekeeping—The Cleaning, Hygiene* and *Maintenance Handbook*, Prentice Hall, Englewood Cliffs, NJ, 1973.

Water Pollution Control Federation, *The Clean Water Act of 1987*, Alexandria, VA, 1987.

Water Pollution Control Federation, *Environmental Audits—Internal Due Diligence*, Alexandria, VA, 1989.

Water Pollution Control Federation, *Safety and Health in Wastewater Systems*, Alexandria, VA, 1983.

U.S. Water Resource Council, *The Nation's Water Resources*, U.S. Government Printing Office, Washington, DC, 1979.

G. Tchobanoglous and E. D. Schroeder, *Water Quality: Characteristics, Modeling, Modification*, Addison-Wesley, Reading, PA, 1985.

J. E. Montgomery, Consulting Engineers Inc., *Water Treatment Principles and Design*, Wiley, New York, 1985.

P. A. Vesilind, P. J. Pierce, and R. F. Weiner, *Environmental Engineering*, 2nd ed., Butterworth, Stoneham, ME, 1988.

R. E. Corbitt, Ed., *Standard Handbook of Environmental Engineering*, McGraw-Hill, New York, 1989.

CHAPTER THIRTY-THREE

Asbestos Management in Buildings

Michael A. Coffman, C.I.H., and Jaswant Singh, Ph.D., C.I.H.

1 INTRODUCTION

In the past decade, asbestos, perhaps more so than any other environmental issue, has had a profound impact on our society. Litigation initiated as a consequence of asbestos exposure has affected thousands of individuals and hundreds of corporations and has shaken the insurance industry. For many years, the exposure of industrial workers to asbestos was the major issue. In recent years, however, there has been a growing concern over the potential for exposure of occupants of public and commercial buildings, including schools and office buildings.

Perhaps the greatest impact has been on U.S. school districts, where public concerns about the presence of asbestos in school buildings has led to stringent U.S. regulations. In October 1986, the U.S. Congress enacted the Asbestos Hazard Emergency Response Act (AHERA). AHERA required the U.S. Environmental Protection Agency (EPA) to establish a comprehensive regulatory framework of inspection, management planning, operations and maintenance activities, and response actions to control asbestos-containing materials in school buildings. It has been estimated that compliance with the AHERA requirements will cost U.S. school districts approximately $3 billion.

The problem of asbestos management in public and commercial buildings is much larger. In large commercial properties, asbestos removal programs may run into millions of dollars for a single building. In some commercial properties, leaving

Patty's Industrial Hygiene and Toxicology, Fourth Edition, Volume 1, Part B, Edited by George D. Clayton and Florence E. Clayton.
ISBN 0-471-50196-4 © 1991 John Wiley & Sons, Inc.

the asbestos in place may result in lost revenues as more and more companies make the decision to move to buildings that are "asbestos-free."

AHERA required that the EPA conduct a study to determine the extent of danger to human health posed by asbestos in public and commerical buildings. In its February 1988 report to Congress (1), EPA concluded that there were approximately 750,000 public and commercial buildings in the United States that contained asbestos. Of these, EPA estimated that 500,000 contained partially damaged asbestos and more than 300,000 buildings contained some asbestos that was significantly damaged. The EPA estimated the cost of abating asbestos at private and commercial properties to be $50 billion.

Management of asbestos in buildings requires the input of professionals in many disciplines, including industrial hygiene, toxicology, medicine, law, risk management, engineering, architecture, construction, and administration. The professional industrial hygienist, as a key member of this team, is uniquely qualified to coalesce all input into meaningful informed decisions.

In asbestos management, the industrial hygienist plays multiple roles:

- As a health professional, the industrial hygienist understands the health risks of asbestos and can relate that risk to human exposure.
- As an air monitoring specialist, the industrial hygienist has a firm grasp of the sampling and analytical techniques to ensure a clear and meaningful interpretation of sampling results.
- As an engineer, the industrial hygienist has an understanding of the practical issues involved in contamination control, work area isolation, work practices, and the use of personal protective equipment.
- As a teacher, the industrial hygienist is routinely called upon to communicate technical and health risk information to laypersons and to instruct workers in safe work practices.
- As a project manager, the industrial hygienist has experience in working with a team to achieve specific goals.

Industrial hygienists can also be effective facilitators in asbestos abatement. In its publication "Recommendations for Asbestos Abatement Projects," the American Industrial Hygiene Association (2) listed 13 critical points requiring industrial hygiene input (Table 33-1).

1.1 Mineralogical Classification of Asbestos

The term asbestos is a commercial name referring to a group of naturally occurring fibrous silicates. The asbestos minerals belong to two mineralogical groups, serpentines and amphiboles. The asbestiform varieties of serpentine and amphibole are shown in Table 33-2.

The most abundant type of asbestos is chrysotile, which is the only member of the serpentine group. Chrysotile is composed of curly fibers that can shear into smaller fibrils.

Table 33.1. Asbestos Management Activities
Requiring Industrial Hygiene Input

Building surveys
Pre-abatement sampling
Abatement recommendations
Cost estimates for remedial alternatives
Warning, notification, and record keeping
Asbestos management programs
Asbestos abatement specifications
Asbestos abatement contractor selection
Pre-abatement project meeting
Pre-abatement inspection
Monitoring compliance with specifications
Post-abatement inspections

Source: American Industrial Hygiene Association (2).

Unlike chrysotile's curled shape, amphiboles crystallize in straight double chains, resulting in needle-like structures. Amphiboles are typically more brittle than chrysotile and tend to cleave longitudinally. The amphiboles of greatest commercial importance are crocidolite and amosite, whose individual fibrils tend to be thicker than those of chrysotile. Other amphiboles, which are similar structurally but differ in chemical composition, include tremolite, anthophyllite, and actinolite.

1.2 Historical Perspective

To understand fully the magnitude of the asbestos management problem, it is useful to review the history of asbestos use and its eventual decline. The unique fire-resistant properties of asbestos, a naturally occurring mineral, have been known since ancient times. Roman historians wrote about asbestos. Plutarch described the asbestos lamp wicks that were used in temples. Herodotus mentioned the use of asbestos as a cremation cloth that made it possible to salvage the ashes of the honored dead from their funeral pyres. Pliny described asbestos fabric as "a rare

Table 33.2. Minerological Classification of
Asbestos

Mineral Group	Mineral	Chemical Formula
Serpentine	Chrysotile	$(Mg,Fe)_6(OH)_8Si_4O_{10}$
Amphibole	Actinolite	$Ca_2Fe_5(OH)_2Si_8O_{22}$
Amphibole	Tremolite	$Ca_2Mg_5(OH)_2Si_8O_{22}$
Amphibole	Amosite	$Mg_7(OH)_2Si_8O_{22}$
Amphibole	Anthophyllite	$(Mg, Fe)_7(OH)_2Si_8O_{22}$

and costly cloth, the funeral dress of kings." He also described the use of a transparent bladder skin as a respirator to avoid the inhalation of asbestos dust by slaves, perhaps the earliest, if oblique, inference of asbestos-related disease (3).

Later, even Charlemagne found a use for asbestos. Anticipating war with an Arabic prince, Harun-al-Raschid, Charlemagne had a tablecloth woven of asbestos fiber. He invited the prince and his coterie to a feast, after which Charlemagne entertained his guests by throwing the soiled asbestos tablecloth into a fire to "clean" it. Apparently, this feat so amazed the prince's followers that they advised their leader not to wage war against such a powerful magician as Charlemagne.

In modern times, the practical uses of asbestos were first demonstrated during the 1850s in Quebec by a young shopkeeper named Henry Ward Johns. His neighbors, who were farmers, complained of a curious fibrous rock which made plowing nearly impossible. The merchant, however, recognized the rock as asbestos and began tinkering to find a practical application. Johns mixed the asbestos with burlap, pitch, and manila paper, ran the mixture through his wife's clothes wringer, and developed the first asbestos roofing product. In an era of wooden buildings heated by wood or coal fires, Johns's fire-resistant roofing became an instant success. Next he turned his attention to gasket materials and eventually to pipe and boiler insulation. In 1898, before introducing his pipe and boiler insulation products, Johns died. It is ironic that Henry Ward Johns, the man who pioneered the modern use of asbestos fiber, apparently died of an asbestos-related disease. According to his death certificate and obituary, Johns died of "dust phthisis pneumonitis" (4) [the term asbestosis was first used to describe a fibrotic lung disease in 1927 by Dr. W. E. Cooke (5)]. Following his death, Johns's company was purchased by C. B. Manville to form the Johns–Manville Company.

Heralded as the "magic mineral" (6), asbestos grew in use throughout the first half of the twentieth century. The unique properties of asbestos, its high tensile strength, temperature resistance, and chemical resistance, plus its abundance in nature led to its use in a growing list of products.

In World War II, when the United States mobilized an unprecedented shipbuilding effort, asbestos was considered to be a "strategic material" and its production was controlled by the U.S. government. Not only was asbestos used to insulate the boiler and piping systems on these ships, but later in the war, an asbestos insulation product was sprayed onto the inside framework of the vessels. Following World War II, a similar type of sprayed-on insulation product was developed for fire protection on structural steel in multistory buildings. In 1970, sprayed "inorganic fiber" was reportedly used as a fireproofing agent in well over half of all large multistory buildings (7).

In the early 1970s, the use of asbestos began to decline following the publication of various health studies, particularly the work of Dr. Irving Selikoff and his associates at the Mt. Sinai School of Medicine in New York. Responding to Selikoff's and other studies, as well as growing public concern, the City of Philadelphia in 1970 enacted the first ban on asbestos fireproofing products (8). Other cities across the United States followed suit, and in 1973, the EPA enacted a nationwide ban on spray-applied fireproofing.

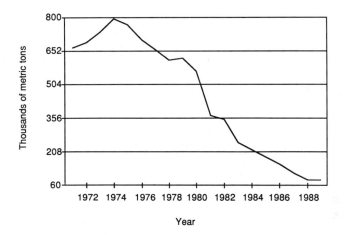

Figure 33.1. U.S. consumption of asbestos fiber 1970–1988. From Prevost (10) and Anderson et al. (9).

Since then, the use of asbestos in the United States has continued to decline (see Figure 33.1). From a peak in 1973 of almost 800,000 metric tons, U.S. consumption of asbestos plummeted to approximately 83,000 metric tons in 1988. Worldwide consumption (see Figure 33.2) also declined through 1983, after which

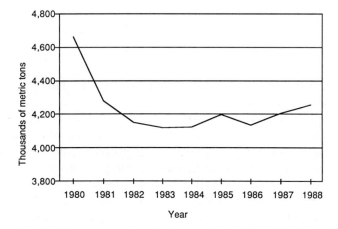

Figure 33.2. Worldwide consumption of asbestos fiber 1980–1988. From Prevost (10).

Table 33.3. Consumption of Asbestos Fiber in Nine Countries

	1985		1986		1987		1988	
Nation	Tons	%	Tons	%	Tons	%	Tons	%
Worldwide	4,194,987	100	4,135,637	100	4,204,267	100	4,253,645	100
USSR	1,963,010	47	1,970,663	47	2,081,549	49	2,080,825	49
Japan	265,358	6	259,569	6	277,117	7	324,000	8
Brazil	151,934	4	188,300	4	182,402	4	191,025	4
China	150,543	4	140,517	4	143,604	3	157,100	4
India	110,038	3	107,108	3	96,333	3	101,500	2
Italy	122,408	3	119,911	3	96,271	2	96,000	2
S. Korea	61,843	1	71,000	1	79,554	2	74,500	2
W. Germany	60,106	1	64,154	1	64,154	1	54,171	1
United States	154,232	4	112,872	4	84,279	2	83,000	2

Source: Michel Prevost (10).

it appears to have stabilized, even showing a small increase in 1987 and 1988. This recent increase in worldwide consumption of asbestos fiber is in contrast to the sharp decline in U.S. consumption. The increase in worldwide consumption can be attributed to increased use of asbestos in several countries including the Soviet Union, Japan, Brazil, and China (see Table 33.3).

The sharp decline in asbestos consumption in the United States parallels the rise in litigation against the manufacturers of asbestos products. In 1982, Manville (formerly Johns–Manville), the company started nearly a century earlier by Henry Ward Johns, filed for bankruptcy. Manville's financial problems were attributed to lawsuits brought by asbestos victims and property owners. By 1985, the claims against Manville totalled more than $80 billion (11).

1.3 Applications of Asbestos in Buildings

The EPA (12) estimated that at its peak use, asbestos was used in as many as 3000 products. Table 33.4 lists some of the most common applications in building products.

In building environments, the greater concern is for "friable" materials. Friable materials are defined by the EPA (13) as those that may "be crumbled, pulverized or reduced to powder by hand pressure." The friability of a product is generally thought to be related to its fiber release potential, with the more friable materials posing the greatest risk of fiber release. Of the products described in Table 33.4, four are generally considered to be friable, including fireproofing, acoustical plasters, pipe and boiler insulation, and ceiling tiles. In a 1984 survey conducted for EPA, 231 buildings in 10 cities were inspected to determine the prevalence of friable asbestos-containing materials in U.S. buildings. Based on the findings of that study, EPA and its contractors (14) estimated that friable, asbestos-containing

Table 33.4. Asbestos-containing Materials Commonly Used in Building Construction

Type of Material	EPA Category
Fireproofing Decorative acoustic plasters	Surfacing materials
Pipe and pipe fitting insulation Boiler and tank insulation HVAC duct insulation[a] Gaskets	Thermal system insulation
Floor tiles Roofing felts Roof coatings Asbestos/cement pipe Asbestos/cement sheet Fabrics	Miscellaneous materials

[a]HVAC = heating, ventilation, and air conditioning.

products are present in approximately 20% of all buildings in the United States. Further findings of this study by type of product are summarized in Table 33.5.

The use of asbestos products is not restricted to commercial or public buildings; single-family homes may also contain one or more asbestos-containing building materials. In a study of 62 selected homes in three U.S. cities, Rock (15) reported damaged asbestos-containing materials in 45 of the homes. The type of asbestos-containing material was found to vary according to geographic location with soft

Table 33.5. Prevalence of Asbestos-containing Materials

Type of Building	Number of Buildings Surveyed	Prevalence of Asbestos-containing Friable Material (%)			
		Sprayed- or Troweled-on	Ceiling Tile Insulation	Pipe/ Boiler	All Types
Federal govt.	66	16	2	25	39
Residential (10 or more rental units)	55	18	0.5	44	59
Private (nonresi- dential)	110	4	0	12	16
All buildings (combined)	231	5	0.1	16	20

Source: J. Strenio et al. (14).

acoustic plasters prevalent in homes in the San Francisco area and thermal system insulation found in homes in Cleveland and Philadelphia.

2   UNDERSTANDING THE HEALTH RISKS

2.1   Asbestos-related Diseases

The inhalation of asbestos fibers has been shown to increase the risks of lung disease, including asbestosis, lung cancer, and mesothelioma.

2.1.1   Asbestosis

It has been known since the early part of this century that workers who inhaled large amounts of asbestos dust sometimes developed a disabling or fatal fibrosis of the lungs (16,17). This condition is commonly called asbestosis. Asbestosis, a diffuse interstitial pulmonary fibrosis or fibrotic pneumoconiosis, occurs almost exclusively among workers who have chronic exposure to high concentrations of airborne asbestos fibers. The disease, which on an average requires 15 or more years to develop, is characterized by the formation of scars or collagen tissue in the lungs, and is associated with progressive deterioration of pulmonary function and work capacity.

2.1.2   Lung Cancer

Asbestos exposure was first linked to bronchogenic carcinoma (lung cancer) in 1935, but it was not until 1947 that epidemiologic evidence of a causal connection was established (18). Since then, numerous studies have confirmed an association between occupational exposure to asbestos and significantly elevated rates of bronchogenic cancer. This form of lung cancer generally occurs 20 to 30 years after exposure begins, and a strong relationship between bronchogenic cancer and asbestosis has been demonstrated. Smoking and asbestos exposure act synergistically to increase the risk of lung cancer.

2.1.3   Mesothelioma

The first link between asbestos exposure and pleural mesothelioma (a cancer of the lining of the lung cavity) was reported in 1943. A strong association was not established until 1960 when a study in a crocidolite mining area of South Africa found that 31 of 33 persons with pleural mesothelioma had experienced some asbestos exposure (19). Additional evidence of a causal relationship between asbestos exposure and pleural mesothelioma has since accumulated. This type of carcinoma is virually unknown in the general population; the incidence rate among asbestos workers has been estimated at approximately 5 to 10 percent. An estimated 1600 cases of mesothelioma were reported in the United States in 1980 (29). Mesothelioma is invariably fatal, usually within 2 years of diagnosis. The risk of contracting mesothelioma appears to be independent of smoking.

Epidemiologic evidence has not clearly demonstrated mesothelioma to be dose-related, although in general, exposure levels associated with this disease tend to be less than those that produce asbestosis. Mesothelioma has been reported in family members of asbestos workers and in persons living near asbestos mining areas (20).

2.1.4 Other Cancers

A relationship between asbestos exposure and other forms of cancer, in particular gastrointestinal cancer, has been reported in some studies. However, this relationship has not been clearly established.

2.1.5 Pleural Plaques

A relatively benign disorder related to asbestos exposure is the occurrence of pleural plaques. Pleural plaques refer to calcification of the pleura on the inner surface of the rib cage. They do not interfere with pulmonary function to a significant extent, nor do they necessarily indicate the presence of pulmonary fibrosis. The occurrence of pleural plaques has been strongly associated with past asbestos exposure (occupational or nonoccupational) (21) and is often considered a marker of such exposure.

2.2 Mechanisms of Fiber Toxicity

Estimating the health risks from low-level asbestos exposures (such as those typical in building environments) is complicated by several factors including the long latency period associated with asbestos-related diseases, the relationship of fiber size and type of asbestos with health risk, and concurrent exposures to other agents, most notably cigarette smoke.

2.2.1 Synergy with Tobacco Smoke

The relationship of asbestos and tobacco smoke with the incidence of lung cancer has been well documented. In studies of asbestos insulation workers, Selikoff et al. (22) reported that workers who smoked were 10 times more likely to die of lung cancer than nonsmoking asbestos workers (Table 33.6). In studies of factory workers exposed to amosite asbestos, Selikoff et al. (23) estimated that the combination of cigarette smoking and asbestos exposure increased the risk of lung 135cancer death about 80 times.

Several mechanisms have been proposed to explain this synergistic relationship. Sanchis et al. (24) and Cohen et al. (25) have suggested that toxic and irritant properties of cigarette smoke cause dysfunction and loss of ciliated and secretory cells lining the respiratory tract, preventing these cells from effectively capturing and clearing deposited particles. Gerde and Scholander (26) suggest that the asbestos fiber acts as a carrier for carcinogens in the smoke. They proposed that the adsorption of phospholipids from the lung onto the surface of asbestos fibers allows

Table 33.6. Asbestos and Smoking Age-Standardized
Lung Cancer Death Rates

Group	Asbestos Exposure	Smoker	Death Rate	Mortality Ratio
Control	no	no	11.3	1.00
Asbestos workers	yes	no	58.4	5.17
Control	no	yes	122.6	10.85
Asbestos workers	yes	yes	601.6	53.24

Source: Selikoff et al. (22).

lipophilic carcinogens in the tobacco smoke (e.g., polycyclic aromatic hydrocarbons) to diffuse more readily across the bronchial lining and through cell membranes of the bronchial epithelium.

Though the mechanism of this synergistic relationship is not completely understood, the management implications of this phenomenon are clear. Many contractors in the asbestos-abatement industry have chosen selectively to hire nonsmokers as a means of reducing their companies' long-term risk. Smoking abatement workshops have been used as an effective tool by many companies to assist employees occupationally exposed to asbestos to quit smoking, thereby reducing their overall lung-cancer risk.

2.2.2 Type of Asbestos

A controversial issue is the relative risk of different forms of asbestos, particularly chrysotile versus amphiboles. In many countries, including the United Kingdom and Canada, different exposure limits have been established (see Table 33.7) for the various forms of asbestos, reflecting a belief that amphiboles (particularly crocidolite) pose a greater health risk than chrysotile. In the United States, experts at the Occupational Safety and Health Administration (OSHA) have taken the view that there is insufficient evidence available to define clearly the relative risks of the different forms of asbestos and have therefore established a single limit (27).

Generally, studies on the relative risks of different forms of asbestos have focused on mesothelioma. Hughes and Weill (28) concluded that there is strong evidence that crocidolite and amosite pose a greater risk of mesothelioma than chrysotile. By contrast, the Committee on Nonoccupational Health Risks of Asbestiform Fibers of the National Research Council (29) concluded that the available epidemiologic literature "does not present a clear picture . . . on the relative ability of different fiber types to cause disease." The committee further concluded that experimental and animal studies have also failed to identify discernible differences in the carcinogenicity of different types of asbestos fibers.

2.2.3 Size of Asbestos Fibers

Another controversial issue is the size of asbestos fibers and the relative risk of large versus small fibers. This issue has become particularly important as trans-

Table 33.7. Worldwide Exposure Limits for Asbestos

	Form of Asbestos		
Country	Chrysotile	Amosite	Crocidolite
Canada (Ontario)	1 f/cc	0.5 f/cc	0.2 f/cc
Czechoslovakia			
<10% asbestos	4 mg/m³	4 mg/m³	4 mg/m³
>10% asbestos	2 mg/m³	2 mg/m³	2 mg/m³
Denmark	1 f/cc	1 f/cc	0.1 f/cc
Federal Republic of Germany	1 f/cc	1 f/cc	0.5 f/cc
Finland	2 f/cc	2 f/cc	2/f/cc
Hungary	200 particles/cc	200 particles/cc	200 particles/cc
India	2 f/cc	0.5 f/cc	0.2 f/cc
Japan	2 f/cc	2 f/cc	0.2 f/cc
Norway	1 f/cc	0.5 f/cc	0.2 f/cc
Republic of China	0.3 mg/m³	0.3 mg/m³	0.3 mg/m³
Switzerland	1 f/cc	1 f/cc	1 f/cc
United Kingdom	0.5 f/cc	0.2 f/cc	0.2 f/cc
United States	0.2 f/cc	0.2 f/cc	0.2 f/cc

Note: All values are for 8-hr time-weighted average exposures; f/cc = fibers per cubic centimeter, mg/m³ = milligrams per cubic meter, and particles/cc = particles per cubic centimeter.
Source: Cook (30).

mission electron microscopy (TEM) has become the preferred method for analyzing air samples in nonoccupational environments. Because TEM is able to detect the smallest asbestos fibers (fibrils), researchers are now able to measure concentrations in buildings of fibers that were previously undetected. Some experts disagree, however, on the significance of TEM sampling results as an indication of health risk.

It is well accepted in the practice of industrial hygiene and occupational medicine that particle size determines the extent to which particles are able to enter the respiratory tract and penetrate into the lung. The size of asbestos fibers may also determine how readily, once they reach the lung, they are engulfed by macrophages. Dupre et al. (31) reported that fibers shorter than 5 μm are readily engulfed by macrophages, whereas longer fibers are not.

In studies in which various sizes of fibers were implanted onto the pleura of rats, Stanton et al. (32, 33) found that the biological activity of such fibers was directly related to their dimension. Other investigators have reported similar findings (34) suggesting that the long, thin asbestos fibers are most apt to be pathogenic. In a comprehensive review of this issue, Lippman (35) concluded that asbestosis is most closely related to the surface area of retained fibers, mesothelioma is most closely associated with numbers of fibers longer than 5 μm and thinner than 0.1

Table 33.8. Summary of Historical Asbestos Exposure Levels In Industry

| Type of Exposure | Analysis Method | Mean Concentration (f/cc) | | | Ref. |
		Fibers >5 μm	Fibers <5 μm	Total	
Asbestos mining (worker exposures)	PCM	1.3	N.R.	—	Schutz et al. (36)
Asbestos milling (worker exposures)	PCM	10.0	N.R.	—	Schutz et al. (36)
Asbestos mining (worker exposures)	PCM	1.8	N.R.	—	Roberts (37)
Shipbuilding (worker exposures)	PCM	250	N.R.	—	Harries (38)
Textile manufacturing (worker exposures)	OM	3.4	N.R.	—	Daly et al. (39)

Note: PCM = phase-contrast microscopy; OM = optical microscopy; N.R. = not reported.

μm, and lung cancer is most closely associated with fibers longer than 10 μm and thicker than 0.15 μm.

2.3 Levels of Exposure

Occupational and general public asbestos exposure levels are summarized in Tables 33.8 through 33.10. These exposure data clearly show the large differences that exist between historical occupational exposures in shipyards, mining and milling operations, and commercial insulating and the levels being reported in building environments. The levels found in building environments are generally several orders of magnitude lower than levels in occupational environments. It is also interesting to note that the results of samples collected in occupied buildings invariably showed predominantly small fibers (shorter than 5 μm in length).

2.4 Extrapolation of Historical Data/Risk Models

All of the epidemiologic studies described previously involved occupationally exposed populations. These individuals were exposed to fiber concentrations far in excess of those which occur in building environments. Most estimates of risk at low levels of exposure, such as nonoccupational exposures in buildings, are mathematical extrapolations from data representing much higher levels of exposure. Some of the health risk (dose–response) models are summarized in Table 33.11. These models vary considerably in their estimate of health risks from asbestos exposures, particularly at the relatively low concentrations that may occur in public and commercial buildings.

In a February, 1988 report to Congress, EPA (1) estimated the number of asbestos-related deaths of noncustodial occupants of commercial and public build-

Table 33.9. Summary of Asbestos Exposure Levels In Construction-Related Occupations

Type of Exposure	Number of Samples	Mean Fiber Conc.	Range of Fiber Concentrations <5 μm	Ref.
Commercial insulating (worker exposure)		9		EPA (40)
Commercial insulating (worker exposure)		5–99		Reitze et al. (41)
Commercial insulating (worker exposure)		12		Nicholson et al. (42)
Building renovation				Paik et al. (43)
Carpenter		0.13		
Electrician		0.13		
Sheet metal worker		0.19		
Painter		0.08		
Building Renovation				
Carpenter	41	0.16	<0.01–0.69	Coffman et al. (44)
Electrician	39	0.11	<0.01–1.6	
Pipefitter	6	0.12	0.04–0.53	
HVAC worker	30	0.14	0.01–2.8	
Gen. laborer	21	0.14	0.02–0.87	
Installation of A/C sheet in cooling towers	102	0.10	<0.02–0.5	Coffman et al. (44)
Removal of insulation from pipes using glove bags	15	0.01	<0.01–0.02	Coffman et al. (44)
Wet removal of ACM				Coffman et al. (44)
Fireproofing	255	1.1	<0.01–170	
Pipes and boilers	57	0.46	<0.01–8.0	
Industrial ovens	23	0.22	0.03–3.0	
Refinery pipes	11	0.09	<0.01–0.57	
ACM removal (misc.)	479			
Wet methods	—	0.9	N.R.	Ewing and Simpson (45)
Dry removal	—	17.	N.R.	

Note: All sample results reported above are for phase-contrast microscopy. HVAC = heating, ventilation, and air conditioning; ACM = asbestos-containing material; N.R. = not reported.

ings. Between the years 1988 and 2118, a period of 130 years, EPA estimated that there would be 4280 deaths resulting from exposure to asbestos in buildings. This translates to an average of 33 deaths per year in the United States. Using its estimate of 35.6 million people exposed, EPA's estimated mortality rate may be expressed as 0.092 deaths for every 100,000 persons exposed per year.

Table 33.10. Summary of Asbestos Exposure Levels In Nonoccupational Environments

| Type of Exposure | Analysis Method | Mean Concentration (f/cc) | | | Ref. |
		Fibers >5 μm	Fibers <5 μm	Total	
Prevalent levels in 17 commercial and public buildings	TEM PCM	N.R. 0.007	N.R. —	0.001^a 0.007	Nicholson, et al. (46)
Prevalent levels in 25 school buildings	TEM	N.R.	N.R.	0.002^a	Constant et al. (47)
Prevalent levels in 19 Ontario buildings	TEM	<0.001	0.021	0.021	Pinchin (48)
Prevalent levels in 43 U.S. government buildings	TEM	N.R.	N.R.	0.00059 to 0.00073	Hatfield et al. (49)
Prevalent levels in 10 commercial buildings	TEM	N.D.	0.0006	0.0006	Coffman (50)

Note: TEM = transmission electron microscopy; PCM = phase-contrast microscopy; N.R. = not reported; N.D. = none detected.
^{a}Results were originally reported in units of nanograms per cubic meter. To allow comparison, these data were converted to fibers per cubic centimeter using a factor of 30 fibers/nanogram.

3 CONDUCTING BUILDING SURVEYS

To determine whether asbestos-containing materials were used in the original construction or subsequent renovations of a building, a detailed building survey, including the collection of bulk samples of suspect materials, is required.

3.1 Building Inspection

As the initial step of a building survey, the investigator should review building plans, remodeling records, and related documents to determine if asbestos-containing materials were specified. It should be emphasized that such information is not a substitute for a thorough building inspection. Information on the type and location of building materials that might potentially contain asbestos will help focus the inspection effort on asbestos-containing materials in the building. A building inspection might start with those areas where asbestos-containing materials are reported to be present and then expand to all parts of the building. If suspect materials are found, they should be sampled and analyzed as discussed below.

Materials that contain asbestos may be sprayed or troweled onto surfaces in either a relatively loose or cementitious form. Among the most common materials are asbestos-containing insulation products used on pipes and boilers. Pipes and boilers are not typically directly accessible to building occupants other than main-

Table 33.11. Mortality Estimates at Low Levels of Asbestos Exposure

Source	Estimated Exposures		Age at First Exposure	Mortality per Million Persons Exposed
	Concn. (f/cc)	Duration (years)		
RCA (51)	0.001	10	22	16
	0.001	10	7	31
CHAP (52)	0.01	10	20	15–150
OSHA (53)	0.1	20	25	2260
PH&B (54)	0.006	35	25	140
NAS (55)	0.0004	73	0	33
Doll and Peto (56)	0.0005	20	20	10
EPA (57)	0.002	30	N.R.	120
Enterline (58)	0.002	50	20	46

Note: N.R. = not reported.

tenance or custodial workers. It is not unusual, however, for these types of materials to be disturbed when repairing leaks, rebuilding valves, or during other routine maintenance and repair of mechanical and plumbing systems. Pipe and boiler insulation may look like chalk, bricks, cement, plaster, or corrugated cardboard. Often these insulation materials are enclosed in a fabric, metal, or plastic jacket. The fabric jacket may also be made of asbestos fiber. The boiler wrapping and pipe jackets should be carefully inspected for damage that could result in the release of asbestos fibers.

Nonfriable asbestos-containing materials include asbestos cement products, floor tile, and gaskets. These products normally release fibers only when cut, drilled, sanded, or removed during remodeling or demolition.

In performing a building survey, the investigator must visit and inspect every room in a building including the custodian's storage closets, steam tunnels, boiler room, attics, and so on. A person familiar with the building such as the building engineer or a custodian is an ideal guide in performing this survey.

3.2 Bulk Sampling and Analysis

As the building inspection is performed, bulk samples of suspect materials should be collected. At a minimum, three replicate samples should be collected of each encountered material. The investigator, however, should always consult appropriate regulations for sampling requirements.

It is important to collect bulk samples to the full depth of the suspect material to ensure that a complete and representative sample is collected. In many situations, additional layers of nonasbestos insulation may have been applied over the top of an older asbestos-containing material.

A convenient tool for bulk sampling is the cork borer. This sharpened metal tube is forced into the suspect material, and a plug is removed. Samples can also be collected using a pocket knife. Prior to sampling, the area should be gently wetted with a spray bottle to minimize fiber release. The cork borer must be cleaned between samples to prevent cross-contamination. Each sample is transferred from the cork borer to a labeled plastic vial and submitted to the laboratory for analysis.

In collecting bulk samples, precautions should be taken to minimize exposure of the investigator and other building occupants to asbestos fibers. Bulk samples should be collected at a time when the building is vacated.

The accepted method of analysis of building materials for the presence of asbestos is polarized light microscopy (PLM). The EPA (59, 60) has specified this technique in regulations developed for schools. The PLM procedure is based on the spectrographic method used by minerologists for many years to identify and classify different types of minerals. Briefly, the technique involves:

- Examination of the bulk material under a stereomicroscope (100X magnification) to determine the fibrous nature of the material, the color and morphology of the fibers, and the types of fibers present
- Mounting of each type of fiber on microscope slides in liquids of appropriate refractive indexes
- Examination of the mounted slides under a polarized light microscope at 100X magnification and with dispersion-staining objectives.

The fibers are identified based on (1) the morphology (the shape, color, and general appearance) of the fibers, (2) the optical characteristics of the fiber when viewed using polarizing filters, and (3) characteristic colors produced by various fibers under polarized light microscopy using the dispersion-staining objective. The relative abundance (percent composition) of each fiber type is estimated while examining the sample through the stereomicroscope.

3.3 Evaluating Exposure Risks

In addition to identifying asbestos-containing materials, a comprehensive building survey should include an evaluation of the potential exposure risk, for example, the likelihood that a particular asbestos-containing material will release asbestos fibers into the building air during day-to-day activities or during foreseeable maintenance or renovation activities. In evaluating this risk, several factors must be considered, including the condition of the asbestos material, its friability, and its accessibility to building occupants.

3.3.1 Condition

The condition of the asbestos-containing material may determine the likelihood of fibers being released into the air. An assessment of the condition depends upon a combination of the quality of the installation, adhesion of the material to the underlying substrate, deterioration, vandalism, and other damages including water damage.

3.3.2 Friability

The products containing asbestos vary widely in their relative degree of friability. The more friable the material, the greater the potential for asbestos fiber release.

3.3.3 Accessibility

If the asbestos-containing material can be easily reached, it is subject to accidental or intentional contact and damage. Material that is readily accessible (within reach) is most likely to be disturbed. Accessibility therefore is one of the most important indicators of exposure potential. The close proximity of the friable material to heating, ventilation, lighting, and plumbing systems requiring maintenance or repair indicates high accessibility. Existing damage is an obvious indicator of accessibility.

Table 33.12 describes one strategy for rating each of these factors that has been used by the authors in assessing potential exposure risks in commercial and public buildings.

EPA and other groups have developed algorithms by assigning numerical scores to the factors discussed above and to other factors. Algorithms proposed by EPA, the U.S. Navy, the Massachusetts Public Schools, and the Colorado Public Schools and the criteria used in each algorithm are summarized in Table 33.13. When results are compared between buildings, and for similar materials, such algorithms may provide a relative measure of the overall fiber release potential. Such algorithms, however, are based on individual subjective judgments. These algorithms should not be considered a substitute for professional judgment and for routine air monitoring as the means of establishing asbestos exposure risks.

3.4 Record Keeping

Careful documentation of field observations made by the investigator is another important aspect of a building survey. Survey records should include:

- Identification of all areas/rooms which were inspected whether or not suspect materials were identified
- The condition of the materials, their accessibility, and friability on an area-by-area basis
- Descriptions of all sampling locations
- Descriptions of each unique material observed and sampled

In describing areas of the building or mechanical systems, it is best to rely on descriptions provided in mechanical drawings or by facility personnel. In some cases, names given to building areas or systems may be unique to that building. By using these names, it is easier to identify and relocate specific materials, sampling

Table 33.12. Exposure Risk Factors for Asbestos-Containing Materials in Buildings

Factor	Description
Condition	
Good	Material is intact and shows no signs of deterioration
Fair	Visual inspection and physical contact indicate that the material is breaking up into layers or beginning to come loose from the substrate. There may be small areas (<10% of the total area) where the material is deteriorating. There may be signs of accidental or intentional damage
Poor	The material is noncohesive. Pieces are dislodged and debris in the area is evident. Parts of the material may be hanging loosely or may have fallen to the floor. The possibility of severe accidental or intentional damage should be investigated
Friability	
Low	Material that is difficult yet possible to damage by hand. This would include most troweled-on materials and manufactured items such as ceiling tiles
Moderate	Fairly easy to dislodge and crush or pulverize by hand. Material may be removed in small or large pieces
High	The material is fluffy, spongy, or flaking and may have pieces loose or hanging
Accessibility	
Staff only	The material is located above a suspended ceiling or is concealed by ducts or piping. The building occupants cannot touch the material. Only maintenance staff would normally have access to the asbestos-containing material
Moderate	The material is touched only during infrequent maintenance or repair activities. Building occupants rarely touch the material or throw objects against it
Easy	Material is touched frequently owing to routine maintenance, and/or the building occupants can touch the material during normal activity

points, and so on. Wherever possible, building drawings should be annotated to document location(s) of asbestos-containing materials. In commercial buildings, the location description should not rely on room numbers or names of office occupants or tenants, for such buildings undergo frequent renovation, often involving the reconfiguration of entire floors.

4 AIR SAMPLING AND ANALYSIS

The preceding sections outlined procedures for determining whether asbestos-containing materials are present in a building and for assessing the likelihood that these materials will release fibers into the air. To evaluate the extent to which

Table 33.13. Comparison of Evaluation Criteria for Several Algorithms

	Algorithm			
Criterion Used	EPA (61) (Sawyer)	Navy (62)	Massachusetts (63) (Ferris)	Colorado (64)
Condition	X	X	X	X
Accessibility	X	X	X	X
Friabity	X	X	X	X
% Asbestos content	X	X	X	X
Activity/movement	X	X	X	X
Air plenum/air movement	X		X	X
Water damage	X			
Exposed surface	X			
No. of occupants		X		
Occupancy duration		X		

Source: Adopted from Simkins and Hill (65).

building occupants may be exposed to asbestos fibers, it is necessary to collect air samples.

4.1 Sampling Strategies

Air sampling, as it is applied within the context of the building environment, may be divided into several basic categories.

4.1.1 Prevalent Level Sampling

Prevalent level or background air monitoring may be used to establish concentrations of airborne fibers in a building under "normal" building conditions. Such sampling is performed to establish background levels prior to asbestos abatement or renovation or may be a part of a long-term management plan. Prevalent level monitoring may be repeated annually to evaluate any significant change in fiber concentrations that may be indicative of deterioration of the asbestos-containing product. To obtain results that are representative of the entire building, at least one sample should be collected for every 10,000 ft^2. At a minimum, one sample should be collected per floor. Sampling media should be placed at "breathing zone" height (3 to 5 ft above floor level). Sampling should be performed during normal building occupancy; this is essential in obtaining representative results. Generally, it is necessary to use high-volume pumps (flow rate of 8 to 10 liters/min). When samples are to be analyzed by electron microscopy, ambient (outdoor) air samples should also be collected for comparison.

4.1.2 Personal Sampling

Personal sampling is used during maintenance, renovation, or abatement work to determine employee exposure (without consideration to the use of a respirator) to airborne fibers. The results of personal sampling are used to select proper respiratory protection and for documentation required under many state and federal regulations. Air sampling results can also be used as a means of evaluating those work practices that result in the lowest employee exposure.

4.1.3 Perimeter Sampling

During an abatement or renovation project, samples are collected from locations outside the work area (but inside the building) to determine how well asbestos fibers are being contained. These samples are especially important in situations where unprotected people occupy other areas of a building in which such work is being conducted.

The number of perimeter air samples collected varies with the size and nature of the project. Priority should be placed on sampling at potential leakage points outside the containment barriers. A typical daily sampling scheme might include two or three samples inside the building, one sample in the clean room of the decontamination unit and one sample outside the building. As discussed previously, the latter is necessary for comparison purposes, particularly when using electron microscopy for analysis.

4.1.4 Clearance Sampling

Prior to dismantling containment barriers for either abatement or renovation projects, "clearance" air samples are collected. Clearance air monitoring is performed in conjunction with visual inspection to ensure that the work area is free of asbestos contamination. Final clearance samplers are placed inside the work area with approximately one sampler for every 10,000 ft^2. A minimum of three samples, or at least one per room, should be collected.

As in prevalent level sampling, a large air volume is necessary to obtain the desired detection limit. Such air samples may be collected using passive or aggressive sampling techniques. Passive sampling implies monitoring an area as is, without creating any additional disturbance in the air. For clearance monitoring performed in school buildings, EPA requires that clearance air sampling be performed using "aggressive" methods. The objective of aggressive sampling is to resuspend fibers that may have settled onto the floor and other surfaces prior to sampling. Briefly, this technique uses a 1-horsepower leaf blower to blow off all walls, ceilings, floors, ledges, and other surfaces in the room. A 20-in. fan is then used to provide constant air movement (one fan for every 10,000 ft^2). Samples are collected while the fans are operating.

4.2 Analysis

Analysis of air samples is performed using one of three techniques: phase-contrast microscopy (PCM), scanning electron microscopy (SEM), or transmission electron

Table 33.14. Comparison of Analysis Procedures for Asbestos Fibers

Comparison Factor	Analysis Procedure		
	PCM	SEM	TEM
Standard methods	NIOSH P&CAM 239 method	No standard method	EPA provisional method and update
Quality assurance	Proficiency analytical testing program	No lab testing program	Proposed NVLAP accreditation program
Cost	$25–50	$100–300	$300–800
Availability	Readily available	Less available	Readily available
Time requirements	1 hr preparation and analysis, <6 hr turnaround	4 hr preparation and analysis, 6–24 hr turnaround	4–24 hr preparation and analysis, 2–7 days turnaround
Sensitivity (thinnest fiber visible)	0.15 μm at best; 0.25 μm typical	0.05 μm at best; 0.20 μm typical	0.0002 μm at best; 0.0025 μm typical
Specificity	Not specific for asbestos	Definitive for asbestos when used to its fullest capabilities	Definitive for asbestos when used to its fullest capabilities

Note: PCM = phase-contrast microscopy; SEM = scanning electron microscopy; TEM = transmission electron microscopy; NVLAP = National Voluntary Laboratory Accreditation Program; P&CAM = physical and chemical analysis methods.

microscopy (TEM). The relative capabilities of each of these three procedures are described in Table 33.14.

4.2.1 Phase-Contrast Microscopy

A phase-contrast microscope is a light microscope equipped with phase annuli in the objective and condensor lenses to provide enhanced contrast between the fibers and the background. Samples for PCM analysis are collected on cellulose ester membrane filters, which are then rendered opaque (cleared) with a chemical solution or solvent so that trapped particulate material can be viewed through the microscope at a magnification of approximately 400X. This method cannot distinguish between asbestos and other fiber types and can detect only those fibers longer than 5 μm and wider than about 0.25 μm. Because of these limitations, results of PCM analysis provide only a measure of the total airborne fiber concentration. As the proportion of the airborne fibers less than 0.25 μm in diameter increases, PCM becomes a less reliable analytical tool.

A principal advantage of PCM is the ability to perform analysis on-site. During asbestos removal or renovation projects, it is critical that air sampling results be available immediately. This is especially important in perimeter sampling where the results are used to determine the effectiveness of containment barriers. Onsite analysis allows the project manager to make this determination within a few hours, and if a problem is discovered, it can be quickly corrected.

PCM is the method of choice for evaluating occupational exposures to asbestos.

4.2.2 Scanning Electron Microscopy

SEM is a technique that directs an electron beam onto the sample surface and collects the reflected beams. A magnified image is produced on a viewing screen. Under the best conditions, SEM can provide a sharp, three-dimensional image of the filter surface. Air samples for SEM analysis may be collected on polycarbonate or cellulose ester filters.

When connected to an energy-dispersive X-ray analyser (EDS), SEM can distinguish between asbestos and nonasbestos fibers by performing an elemental analysis of individual fibers observed on the viewing screen. SEM can detect fibers with diameters as small as 0.05 μm.

4.2.3 Transmission Electron Microscopy

TEM is a technique that focuses an electron beam onto a thin sample. As the beam transmits through certain areas of the sample, an image, resulting from varying densities of the sample, is projected onto a fluorescent screen. Air samples are collected on polycarbonate or cellulose ester filters (depending on the specific protocol being followed).

TEM is the only method of these three techniques that can detect the smallest diameter asbestos fibers (fibrils). TEM does not, however, produce the same high-quality image as SEM. The images produced by TEM are essentially shadows of the particles being examined. TEM can be equipped with either or both EDS and Selected Area Electron Diffraction (SAED) for fiber identification. The latter procedure provides information on the crystalline structure of individual fibers viewed.

4.3 Direct-reading Instruments

Fibrous aerosol monitors (FAMs) are direct-reading devices designed for analyzing airborne fiber concentrations. Similar to other types of particulate monitors that rely on light scattering, FAMs have been modified to discriminate between fibrous and nonfibrous particulates. Results are reported as a fiber count. These devices operate by drawing air through a tube where two pairs of electrodes create electric fields perpendicular to each other and to the axis of the tube. The two fields are alternately activated, causing the fibers to align and oscillate as they travel through the tube. A laser beam shines along the axis of the flow tube and the fibers scatter this light in directions perpendicular to their long axis. A photomultiplier tube, positioned in the wall of the flow tube, gives a pulsed response as the scattered light scans over it. Fiber size is determined from the shape of these pulses.

As discussed previously, the conventional methods for measuring concentrations of airborne asbestos fibers involve collecting fibers on filters and analyzing them with either a light microscope or an electron microscope. Results of comparative testing using the fibrous aerosol monitors and conventional techniques do not always show good agreement (66). FAMs have nevertheless found considerable application in the asbestos abatement industry where, equipped with alarms, they are used frequently to warn of "barrier" failures.

Accurate, defensible results depend on careful calibration of direct-reading aerosol monitors. Unfortunately, calibration of these monitors is difficult. Aerosols cannot be obtained in commercial calibration mixtures nor are there aerosol-generation systems available in which the concentrations can be determined a priori. As a result, in routine use, calibration is performed by comparing field monitoring results with "reference" techniques involving filter collection and microscopic analysis of the filters.

5 CONTROL OF ASBESTOS IN BUILDINGS

5.1 Control Options

5.1.1 Encapsulation

The term "encapsulation" describes the spray application of a sealant to the asbestos-containing materials. The application of encapsulants helps bind the asbestos fibers together to reduce the fiber release potential.

Usually encapsulation involves the application of multiple coats using two types of sealants. A penetrating encapsulant is applied first. The penetrant is intended to infiltrate into the friable material to help bind the fibers together from within. The second coat is called a "bridging encapsulant." A bridging encapsulant forms an impervious skin at the exposed surface of the asbestos-containing material. Encapsulation is typically less expensive than removal. It is not, however, a permanent solution to the asbestos problem. Encapsulation does not prevent the release of fibers from direct mechanical disturbances of the friable material.

The performance of an encapsulant with any particular asbestos application is unpredictable. The most frequently encountered problem is a cohesive or adhesive failure of the original asbestos product. Many friable asbestos-containing products are not capable of supporting the additional weight of the encapsulant. Encapsulation should not be considered without first testing the cohesion/adhesion strength of the asbestos-containing product. A procedure for cohesion/adhesion testing has been established by the American Society for Testing and Materials (ASTM) (67).

5.1.2 Enclosure

Enclosure involves the construction of a rigid barrier between asbestos-containing products and building occupants. The purpose of enclosure is to prevent damage to the asbestos-containing material, thereby reducing the likelihood of fiber release. Enclosure can be an effective means of control where frequent entry into the enclosure is not necessary. Suspended ceilings may seem like an enclosure; however, in most cases, they do not provide effective control because:

- Maintenance workers frequently work above a suspended ceiling to gain access to electrical, ventilation, or lighting equipment
- The space above the suspended ceiling often functions as a return-air plenum for the building ventilation system.

Two types of enclosure systems are routinely employed: (*a*) fixed, mechanical barriers, and (*b*) spray-applied barriers. A fixed, mechanical barrier might consist of a plywood or drywall enclosure around the asbestos-containing material. Spray-applied barriers, often called encasements, differ from encapsulation in that the "cured" sealant forms a hard external surface. Also, commercially available encasement systems typically include some means of mechanically fastening the barrier to the substrate (e.g., pipe, structural steel) thus preventing delamination.

Enclosures should not be used when (*a*) there is a high probability of water damage, (*b*) where the asbestos-containing material is already damaged, or (*c*) in situations where it will frequently be necessary to work above or behind the enclosure.

5.1.3   Removal

Asbestos removal is the only permanent solution to the problem of asbestos in buildings. When competently performed, and with adequate protection for workers and building occupants, removal should minimize the risks of asbestos exposure in a building. Serious questions have been raised, however, about the effect of asbestos removal on subsequent asbestos fiber concentrations. In studies conducted in nine buildings before and after asbestos removal, Pinchin (68) reported equal or higher asbestos fiber concentrations after removal than were present before removal (Table 33.15).

Another concern with asbestos removal is the availability and relative safety of replacement products. In most cases, the asbestos product must be replaced with a suitable substitute. At a minimum, the substitute must provide the same acoustical or thermal insulation properties as the original material. Indeed, in some cases building owners are discovering that, as a result of local building codes changes since the original construction, they must replace the asbestos product with a product exhibiting superior insulation properties. The most widely used asbestos substitutes are fibrous glass, rock wool, and slag wool, commonly referred to as man-made mineral fibers. Since their introduction about 90 years ago, it has been recognized that these man-made mineral fibers can irritate the skin and eyes and, under dusty conditions, may also irritate the nose and throat. Recent concerns, however, have focused on the possible carcinogenicity of asbestos substitutes. Man-made mineral fibers are discussed in greater detail in Chapter 31.

Despite these limitations, asbestos removal and replacement may be the best option when the asbestos-containing material is badly deteriorated. In the United

Table 33.15. Concentrations of Asbestos Fibers Before and After Removal

Project Number	Before Asbestos Removal			After Asbestos Removal		
	PCM (f/cc)	TEM (f/cc)	TEM (fibers >5 μm/cc)	PCM (f/cc)	TEM (f/cc)	TEM (fibers >5 μm/cc)
1	0.022	0.027	<0.005	0.07	0.057	<0.004
2	<0.1	0.016	0.003	0.1	0.054	0.008
3	<0.1	0.015	0.001	<0.1	0.026	0.005
4	<0.1	0.023	0.003	<0.1	0.010	0.001
5	0.003	0.037	0.002	0.039	0.53	0.028
6	0.004	0.012	0.001	0.022	0.119	0.002
7	0.002	0.016	<0.001	0.003	0.182	0.012
8a	0.03	—	0.004	0.01	—	0.01
8b	<0.01	0.019	<0.004	0.03	0.256	0.012

Source: Adapted from Royal Commission on Asbestos Study Series No. 8 (69).

PLM = phase-contrast microscopy; TEM = transmission electron microscopy.

States, removal of friable asbestos is required by EPA prior to the demolition of a building (70).

5.1.4 Operations and Maintenance Program

The objective of an operations and maintenance (O&M) program is to minimize asbestos exposures of building occupants and workers by controlling access to the asbestos-containing material, maintaining the material in good condition, and using work practices that minimize fiber release. The O&M program is a set of specific procedures applied to routine activities such as cleaning, maintenance, and renovation to maintain the building as free of asbestos contamination as possible. Elements of an effective O&M program will vary with the circumstances of each building but generally will include the following items:

5.1.4.1 Periodic Surveillance.
Because the condition of asbestos-containing materials in a building may change with time owing to water damage, accidental disturbance, or aging, routine surveillance of these materials is an essential element of an asbestos management program. The incidental reporting of damage by maintenance or custodial workers or building occupants should be documented. Routine surveillance should be performed by an individual trained in the assessment protocol, and the results should be carefully documented.

5.1.4.2 Routine Air Monitoring.
Routine prevalent level monitoring is another type of periodic surveillance that should be included in a management program. Such monitoring is useful for assessing changes that may have occurred in the condition of the materials as well as for further documenting the airborne concentrations

during normal building occupancy. During routine air monitoring, maintenance and renovation activities underway should be documented.

5.1.4.3 Notification. Occupants of buildings, including maintenance and custodial personnel, should be informed of the location(s) of asbestos-containing materials, the potential health risks associated with asbestos exposure, and the air monitoring results. Further, occupants should be instructed to avoid contact with asbestos-containing materials and to advise building management of any observed damage.

5.1.4.4 Employee Protection. In most buildings it will sometimes be necessary for maintenance personnel to work in close proximity to or disturb asbestos-containing materials in order to effect a repair or adjust heating, ventilation, and air conditioning or lighting. With the use of proper equipment and good work practices, maintenance employees may perform such work with minimal exposure to asbestos.

5.1.4.5 Medical Surveillance Programs. Those employees whose jobs may necessitate occasional contact with asbestos should be included in a medical surveillance program including routine medical examination, chest X-rays, and pulmonary function testing.

5.1.4.6 Training. All employees who may potentially come into direct contact with asbestos-containing materials should be trained in following the correct work procedures, using personal protective equipment, and understanding the health risks of asbestos.

5.1.4.7 Special Work Practices for Maintenance Activities. Even relatively routine maintenance procedures may involve disturbance of asbestos-containing materials. This applies especially to work above suspended ceilings in buildings where the steel structural members are sprayed with asbestos fireproofing. Special equipment and work practices, such as those shown in Table 33.16, are necessary to prevent exposures of the maintenance employees as well as other building occupants.

5.1.4.8 Special Work Practices for Renovation. Plans for routine renovation in a building should be carefully evaluated to determine if such activities will disturb the asbestos-containing materials. If so, precautions such as those described in Table 33.17 must be taken. The cost impact of such controls may be minimized through careful planning and scheduling of the renovation project.

5.1.4.9 Episodic Releases. In the event of an unplanned disturbance of an asbestos-containing material, a quick response is essential to prevent exposures to occupants. A contingency plan, including making arrangements with appropriate outside resources (e.g., abatement contractors, consultants), should be developed for each building in which asbestos-containing materials are present.

Table 33.16. Asbestos-Control Procedures For Routine Maintenance Activities

Minimal occupancy	Whenever possible, maintenance activities involving possible disturbance of asbestos-containing materials should be performed when the affected area is unoccupied
Cleaning	Any asbestos-containing materials dislodged during such work should be cleaned up with a high-efficiency particulate air filter-equipped vacuum cleaner or by wet mopping and never by sweeping. The entire work area should be cleaned thoroughly using a combination of wet mopping and vacuuming at the completion of the maintenance work, but before the area is reoccupied
Personal protection	For such maintenance activities, employees should be provided with approved respirators and disposable coveralls with hoods or hats. The use of respiratory protection necessitates the development of a respirator program. Respirators should be individually fitted, and employees should be trained in their correct use and limitations
Disposal	Asbestos-contaminated wastes generated during such activities must be disposed of in accordance with applicable laws and regulations
Wet methods	Wet methods should be employed if the work involves direct disturbance of the asbestos-containing material

5.2 Selecting the Best Option

For the property owner and manager, as well as for the industrial hygienist, selecting the best asbestos control option can be a complex decision. The potential penalties for a poor decision may include regulatory fines, civil court actions, and potential health risks. Because each building is unique, asbestos management decisions must be made on a building-by-building basis. The risks posed by the presence of asbestos in buildings can vary considerably depending on:

- The condition, location, and friability of the asbestos-containing material(s)
- The way(s) in which the building is used (e.g., school, office, retail store)
- The type of ventilation systems in the building
- The amount and extent of maintenance and renovation work anticipated

The industrial hygienist must recognize and consider not only the actual health risks but also the risks perceived by the building occupants, parents of schoolchildren, and the general public.

From a practical point of view, there are only two alternatives: immediate removal or deferred removal. If the removal is deferred, the building owner must implement an ongoing asbestos management program, which may include controls such as enclosure or encapsulation.

Table 33.17. Asbestos-Control Procedures For Routine Renovation Activities

Isolation	The renovation area should be isolated from the remainder of the facility. Typically, this involves the construction of a temporary plastic barrier with an airlock for entry of workers. Ventilation in the work area should also be temporarily isolated and a portable exhaust fan [equipped with a high-efficiency particulate air (HEPA) filter] should be used to place the renovation area under a slight negative pressure with respect to the rest of the building. Wherever possible, this fan should discharge out-of-doors
Warning signs	Signs should be posted at all entrances to the renovation area to warn unprotected persons from entering
Cleaning	Any asbestos-containing materials dislodged during such work should be cleaned up with a HEPA filter-equipped vacuum cleaner or by wet mopping and never by sweeping. The entire work area should be cleaned thoroughly using a combination of wet mopping and vacuuming at the completion of the renovation work, but before the area is reoccupied
Personal protection	For such activities, renovation workers should be provided with approved respirators and disposable coveralls with hoods or hats. The use of respiratory protection necessitates the development of a respirator program. Respirators should be individually fitted, and employees should be trained in their correct use and limitations
Disposal	Asbestos-contaminated wastes generated during such activities must be disposed of in accordance with applicable laws and regulations
Wet methods	Wet methods should be employed if the work involves direct disturbance of the asbestos-containing material
Air monitoring	At regular intervals during the renovation work, air monitoring should be performed along the perimeter of work area to demonstrate the efficacy of the isolation and ventilation controls at preventing fibers from escaping. After the work area has been thoroughly cleaned, but before the barriers are removed, clearance air monitoring should be performed

Removal is generally the only viable alternative when asbestos-containing materials are badly deteriorated. For buildings in which the asbestos-containing materials are in good condition and do not pose an immediate health risk, the decision may be based on economic and practical considerations.

Deferring removal may postpone a large capital expense. The ongoing operating costs, however, of an asbestos management program may also be considerable. In addition to the costs of monitoring and controlling asbestos, future renovation/remodeling costs in an asbestos-containing structure will also be higher. In commercial properties, the loss or reduction of rental income may be a significant factor.

As with any critical business or technical decision, the key is to evaluate carefully all of the pertinent factors discussed above before making an "informed decision."

5.3 Planning an Abatement Action

No matter which control technique is chosen, it is important to employ specially trained and qualified professionals to perform this work and to require strict adherence to established operating procedures. All persons performing asbestos abatement work should be properly protected, trained, and supervised so that building tenants and other persons outside of the containment areas are not exposed to unacceptable concentrations of asbestos fibers as a result of abatement activities.

Before an asbestos abatement project is undertaken, the following steps are necessary:

- Contract specifications tailored to the individual project
- A thorough evaluation of contractors and their previous performance prior to awarding the contract
- On-site surveillance of contractor performance and air monitoring during abatement activities by an independent third-party expert working directly for the building owner or manager
- Clearance and other performance criteria to determine successful project completion

In addition to an industrial hygienist and the abatement contractor, planning for an asbestos abatement project requires input from several experts including:

- Architects
- Mechanical and electrical engineers
- Contract specialists
- Building managers
- Building maintenance people

Although often overlooked in the planning process, the building maintenance people may provide valuable insight owing to their intimate knowledge of the building and building systems.

6 RECOMMENDED READING

Topic	Reference
Health rules	National Research Council, Committee on Nonoccupational Health Risks of Asbestiform Fibers, *Asbestiform Fibers: Nonoccupational Health Risks*, National Academy Press, Washington, DC, 1984, p. 132

continued on next page

Topic	Reference
	Dupre, J. Stefan, J. F. Mustard, R. J. Uffen, D. N. Dewees, J. I. Laskin, and L. B. Kahn, "Report of the Royal Commission on Matters of Health and Safety Arising from the Use of Asbestos in Ontario," Vol. 1, 1984.
	M. Lippmann, Review: "Asbestos Exposure Indices," *Environ. Res.* **46**(1) (1988)
Control methods	"Commercial and Industrial Use of Asbestos Fibers; Advance Notice of Proposed Rulemaking," *Fed. Reg.*, **44** (202) (Oct. 17, 1989)
	A. R. Rock, "Preliminary Results from the Study of Asbestos in 45 Homes," in *Asbestos—Its Health Risks, Analysis, Regulation and Control*, American Pollution Control Association, 1987, Pittsburgh, PA.
	Environmental Protection Agency, "Guidance for Controlling Asbestos-Containing Materials in Buildings" ("Purple Book"), Publication No. EPA 560/5-85-024, EPA, Washington, DC, 1985
General information	L. M. Thomas, "EPA Study of Asbestos-Containing Materials in Public Buildings: A Report to Congress," U.S. Environmental Protection Agency, Washington, DC, February, 1988
	G. A. Peters and B. J. Peters, *Sourcebook on Asbestos Diseases: Medical, Legal and Engineering Aspects*, Garland STPM Press, New York & London, 1980
Sampling and analysis	*Asbestos Exposure Assessment in Buildings Inspection Manual*, U.S. Environmental Protection Agency, Washington, DC, October, 1982
	U.S. Environmental Protection Agency, "Measuring Airborne Asbestos Following An Abatement Action," EPA 600/4-85/049, EPA, Washington, DC, November 1985

REFERENCES

1. L. M. Thomas, "EPA Study of Asbestos-Containing Materials in Public Buildings: A Report to Congress," U.S. Environmental Protection Agency, Washington, D.C., February, 1988.
2. F. W. Boelter, and R. E. Sheriff, "American Industrial Hygiene Association Recommendations for Asbestos Abatement Projects," AIHA, Akron, Ohio, September, 1986.
3. D. Hunter, *The Diseases of Occupations*, 4th ed., Little, Brown, Boston, 1969, p. 1009.
4. P. Brodeur, "Annals of Law; The Asbestos Industry On Trial; *1. A Failure to Warn*," *The New Yorker*, June 10, 1985.

5. W. E. Cooke, "Pulmonary Asbestosis," *Br. Med. J.* **2**, 1024 (December 1927).

6. M. S. Badollet, "Asbestos, a Mineral of Unparalleled Properties," *Can* Mining *Metall. Bull.* (April 1951).

7. W. B. Reitze et al., "Application of Sprayed Inorganic Fiber Containing Asbestos: Occupational Health Hazards, *Am. Ind. Hyg. Assoc. J.*, **33**, 179–191 (1972).

8. "City Orders a Ban on Asbestos Sprays in Building Industry," *Philadelphia Inquirer*, December 3, 1970.

9. D. Anderson, C. Clark, B. McCrodden, M. McGivney, V. Ramachandiran, and J. Warren. "Asbestos Industry Profile, Trends and Outlook, and Substitute Products Analysis," Research Triangle Institute Report for U.S. EPA, Office of Pesticides and Toxic Substances, April, 1983.

10. Michel Prevost, The Asbestos Institute, Montreal, Quebec, personal communication.

11. "Manville Accord Would Settle Asbestos Claims," *Wall Street Journal*, October 16, 1985.

12. "Commercial and Industrial Use of Asbestos Fibers; Advance Notice of Proposed Rulemaking." *Fed. Reg.* **44**(202) (October 17, 1989).

13. D. Keyes and B. Price, "Guidance for Controlling Friable Asbestos-Containing Materials in Buildings," EPA 560/5-83-002, U.S. Environmental Protection Agency, Office of Pesticide and Toxic Substances, Washington, DC, March, 1983.

14. J. Strenio, P. C. Constant, Jr., M. Gabriel, D. R. Rose, and D. Lentzen, "Asbestos in Buildings: A National Survey of Asbestos-Containing Friable Materials," EPA360/5-84-006, United States Environmental Protection Agency, Office of Toxic Substances, Washington, DC, October, 1984.

15. A. R. Rock, "Preliminary Results from the Study of Asbestos in 45 Homes," in *Asbestos—Its Health Risks, Analysis, Regulation and Control*, American Pollution Control Association, 1987, Pittsburgh, PA.

16. R. R. Sayers, and W. C. Dreesen, "Asbestosis," *Am. J. Public Health* **29**(3), 467–479 (March 1939).

17. G. W. Wright, "Asbestos and Health in 1969," *Am. Rev. Respir. Dis.*, **100**(4), 467–479 (October 1969).

18. I. J. Selikoff, and D. H. K. Lee, *Asbestos and Disease*, Academic Press, New York, 1978.

19. J. C. Wagner, C. A. Sleggs, and P. Marchand, "Diffuse Pleural Mesothelioma and Asbestos Exposure in the North Western Cape Province," *Br. J. Ind. Med.*, **17**, 260–271 (1960).

20. I. J. Selikoff, H. A. Anderson, and H. Seidman, "Asbestos Disease Among Household Contacts of Asbestos Workers" in *Disability Compensation for Asbestos-associated Disease in the U.S.*, I. J. Selikoff, Ed., Environmental Sciences Laboratory, Mount Sinai School of Medicine of the City University of New York, 1982, pp. 73–76.

21. R. J. Levine, Cancer Control Monograph: *Asbestos*, SRI International, April, 1978, Palo Alto, California.

22. I. H. Selikoff, E. C. Hammond, and H. Seidman, "Asbestos Exposure, Cigarette Smoking and Death Rates," Ann. N.Y. *Acad. Sci.*, **330**, 473–490 (1979).

23. I. H. Selikoff, H. Seidman, and E. C. Hammond, "Mortality Effects of Cigarette Smoking Among Amosite Asbestos Factory Workers," *J. Natl. Cancer Inst.* **65**(3), 507–1513 (September 1980).

24. J. Sanchis, M. Dolovich, R. Chalmers, and M. T. Newhouse, "Regional Deposition and Lung Clearance Mechanisms in Smokers and Nonsmokers," in *Inhaled Particles III*, Vol. I, W. H. Walton, Ed., Unwin Brothers, Surrey, England, 1971, pp. 183–191.

25. D. Cohen, S. F. Arai, and J. D. Brain, "Smoking Impairs Long-Term Dust Clearance from the Lung," *Science* **204**, 514–517 (May, 1979).

26. P. Gerde and P. Scholander, "A Hypothesis Concerning Asbestos Carcinogenicity: The Migration of Lipophilic Carcinogens in Adsorbed Lipid Bilayers," *Ann. Occup. Hyg.*, **31**(3), 395–400 (1987).

27. U.S. Department of Labor, Occupational Safety and Health Administration Preamble to the OSHA Asbestos Standard, 29 CFR Parts 1910 and 1926, Occupational Exposure to Asbestos, Tremolite, Anthophyllite, and Actinolite; Final Rules, *Fed. Reg. Bol.* **51**(119) (June 20, 1986).

28. J. M. Hughes and H. Weill, "Asbestos Exposure-Quantitative Assessment of Risk," *Am. Rev. Respir. Dis.*, **135**, 5–13 (1986).

29. National Research Council, Committee on Nonoccupational Health Risks of Asbestiform Fibers, *Asbestiform Fibers: Nonoccupational Health Risks*, National Academy Press, Washington, DC, 1984, p. 132.

30. W. A. Cook, *Occupational Exposure Limits—Worldwide*, American Industrial Hygiene Association, 1987.

31. J. S. Dupre, J. F. Mustard, R. J. Uffen, D. N. Dewees, J. S. Dupre, J. J. Laskin, and L. B. Kahn, "Report of the Royal Commission on Matters of Health and Safety Arising from the Use of Asbestos in Ontario," Vol. One, 1984.

32. M. F. Stanton and C. Wrench, "Mechanisms of Mesothelioma Induction with Asbestos and Fibrous Glass," *J. Natl. Cancer Inst.* **48**(3) (March 1972).

33. M. F. Stanton and M. Layard, "The Carcinogenicity of Fibrous Minerals," in *Proceedings of the Workshop on Asbestos: Definitions and Measurement Methods, Gaithersburg, Maryland: 18–20 July, 1977*, NBS Special Publication 506, Washington, DC, 1978.

34. F. Pott, K. H. Friedrichs, and F. Huth, "Results of Animal Experiments Concerning the Carcinogenic Effect of Fibrous Dusts and Their Penetration with Regard to the Carcinogenesis in Humans," *Zbl. Bakt. Hyg.*, **162**, 467–505 (1976).

35. M. Lippmann, Review: "Asbestos Exposure Indices," *Environ. Res.* **46**(1) (June 1988).

36. L. A. Schultz, W. Bank, and G. Weems, "Airborne Asbestos Fiber Concentrations in Asbestos Mines and Mills in the United States, "Bureau of Mines Health and Safety Program, Technical Progress Report 72, Ps-222 611, U.S. Department of the Interior, Denver, Colorado, 1973.

37. D. R. Roberts, "Industrial Hygiene Report: Asbestos at Jaquays Mining Corp, PS81-241846, National Institute of Occupational Safety and Health, Cincinnati, Ohio, 1980.

38. P. G. Harries, "Experience with Asbestos Disease and its Control in Britain's Naval Dockyards," *Environ. Res.* **11**, 261–267, 1977.

39. A. R. Daly, A. J. Zupko, J. L. Hebb, and R. F. Weston, "Technological Feasibility and Economic Impact of OSHA Proposed Revision to the Asbestos Standard," Asbestos Information Association/North America, Washington, DC, 1976.

40. U.S. Environmental Protection Agency, Office of Toxic Substances, "Support Document for Final Rule on Friable Asbestos-containing Materials in School Buildings: Health Effects and Magnitude of Exposure, Washington, DC, 1982.

41. W. B. Reitze, W. J. Nicholson, D. A. Holaday, and I. J. Selikoff, "Application of

Sprayed Inorganic Fiber Containing Asbestos: Occupational Health Hazards," *Am. Ind. Hyg. J.*, **31**, 178–191 (1972).

42. W. D. Nicholson, "Dose–response Relationships for Asbestos and Inorganic Fibers," Environmental Sciences Laboratory, Mount Sinai School of Medicine of the City University of New York, 1981.

43. N. W. Paik, R. J. Walcott, and P. A. Brogan, "Worker Exposure to Asbestos during Removal of Sprayed Material and Renovation Activity in Buildings Containing Sprayed Material," *Am. Ind. Hyg. Assoc. J.*, **44**(6) (1983).

44. M. Coffman, and R. Walcott, "Asbestos Exposure in the Construction Industry," presented at the American Industrial Hygiene Conference, May, 1985, Las Vegas, Nevada.

45. W. M. Ewing and G. J. Simpson, "Air Sampling at 52 Asbestos Abatement Projects," Presented at the American Industrial Hygiene Conference, Philadelphia, Pa, May 24, 1983.

46. W. J. Nicholson, A. N. Rohl, and I. Weisman, "Asbestos Contamination of the Air in Public Buildings," EPA Publication No. EPA-450/3-76-004, October, 1975.

47. P. C. Constant, Jr., F. J. Bergman, G. R. Atkinson, D. R. Rose, D. L. Watts, E. E. Logue, T. D. Hartwell, B. P. Price, and J. S. Ogden, "Airborne Asbestos Levels in Schools," EPA Publication No. EPA 560/5-83-003, June, 1983.

48. D. J. Pinchin, "Asbestos in Buildings," Royal Commission on Asbestos Study Series No. 8, Royal Commission on Asbestos, Toronto, Table II, 1982, p. 6.7.

49. J. Hatfield, J. Stockrahm, and J. Chesson, Draft Report for Task 2-31: Preliminary Analysis of Asbestos Air Monitoring Data for the U.S. Environmental Protection Agency, Office of Toxic Substances, Washington, DC, 1987. Taken from L. M. Thomas, "EPA Study of Asbestos-Containing Materials in Public Buildings, a Report to Congress," EPA, Washington, DC, February, 1988.

50. M. Coffman, "Asbestos Management in a Changing Environment," *Proceedings of Symposium on Asbestos in Buildings: Management, Measurement & Risks*, Clayton Environmental Consultants, Inc., July 28, 1988, Southfield, Michigan.

51. Ontario Royal Commission on Asbestos, "Report of the Royal Commission on Matters of Health and Safety Arising from the Use of Asbestos in Ontario," Queen's Printer, Toronto, 1984.

52. Chronic Hazard Advisory Panel on Asbestos, "Report to the U.S. Consumer Product Safety Commission," U.S. Consumer Product Safety Commission, Washington, DC, 1983.

53. U.S. Occupational Safety and Health Administration, "Quantitative Risk Assessment for Asbestos-Related Cancers," Washington, DC, 1983.

54. Putnam, Hayes and Bartlett, Inc., "Cost and Effectiveness of Abatement of Asbestos in Schools," Draft report for U.S. Environmental Protection Agency, Office of Pesticides and Toxic Substances, 1984, Washington, DC.

55. National Academy of Sciences, Committee on Nonoccupational Health Risks of Asbestiform Fibers, National Research Council, *Asbestiform Fibers: Nonoccupational Health Risks*, National Academy Press, Washington, DC, 1984.

56. R. Doll and J. Peto, "Effects on Health of Exposure to Asbestos," U.K. Health and Safety Commission, Her Majesty's Stationery Office, London, 1985.

57. L. M. Thomas, "EPA Study of Asbestos-Containing Materials in Public Buildings: A

Report to Congress," U.S. Environmental Protection Agency, Washington, DC, February, 1988.

58. P. E. Enterline, "Extrapolation from Occupational Studies; a Substitute for Environmental Epidemiology," *Environ. Health Perspect.*, **42**, 39 (1981).

59. U.S. Environmental Protection Agency, *Fed. Reg.*, 40 CFR Part 763 (October 30, 1987).

60. D. E. Lentzen, E. P. Brantly, Jr., K. W. Gold, and L. E. Myers, "Interim Method for the Determination of Asbestos in Bulk Insulation Samples," EPA-600/M4-82-020, U.S. Environmental Protection Agency, Office of Pesticides and Toxic Substances, Washington, DC, December 1982; U.S. Environmental Protection Agency, *Fed. Reg.*, 40 CFR Part 763 (October 30, 1987).

61. "Asbestos Exposure Assessment in Buildings Inspection Manual," U.S. Environmental Protection Agency, October, 1982.

62. "U.S. Navy Procedure for Risk Evaluation: Asbestos Hazard Index," Chapter 5 of Report R883, *Management Procedure for Assessment of Friable Insulating Materials*, Civil Engineering Laboratory, Naval Construction Batallion Center, Port Hueneme, CA.

63. K. S. Irving, R. G. Alexander, and H. Bavley, "Asbestos Exposures in Massachusetts Public Schools," *AIHA* J. (April, 1980).

64. Baldwin, Beaulieu, Buchan, and Johnson, "Asbestos in Colorado Schools," *Public Health Rep.* (July–August, 1982).

65. L. Simkins and W. Hill, "Algorithms: Their Usefulness in Hazard Assessment," presented at the Annual Meeting of the National Asbestos Counsel, September, 1987, Oakland, California.

66. W. Phanprasit, V. E. Rose, and R. K. Oestenstad, "Comparison of the Fibrous Aerosol Monitor and the Optical Fiber Count Technique for Asbestos Measurement," *Appl. Ind. Hyg.*, **3**(1) (January 1988).

67. American Society for Testing and Materials, ASTM Method E736-86, Philadelphia, Pennsylvania.

68. D. J. Pinchin, "Asbestos in Buildings," Royal Commission on Asbestos Study Series, No. 8, Royal Commission on Asbestos, Toronto, 1982.

69. D. J. Pinchin, "Asbestos in Buildings," Royal Commission on Asbestos Study Series, No. 8, Royal Commission on Asbestos, Toronto, 1982, pp. 7–12, Table III.

70. National Emission Standard for Hazardous Air Pollutants (NESHAPS), 40 CFR part 265, subparts A & M, 1973.

CHAPTER THIRTY-FOUR

Safety and Health Effects of Visual Display Terminals

Raymond P. Briggs, Ph.D.

1 INTRODUCTION

This chapter describes safety and health problems related to the use of the visual display terminal (VDT). The first section defines and describes VDT safety problems. The second section summarizes what is known about VDT safety based on research findings. The third section indicates how one might evaluate VDT safety. The fourth section discusses the special issue of software safety, which may be the most intractable long-term problem related to VDT use. A comment on legislation concludes the chapter.

1.1 Definition of Problems

The increasingly specialized use of computers in employment has led to the creation of a new worker category that cuts across traditional job descriptions: the VDT operator. Such an operator might be a middle manager, accountant, or staff analyst—traditional "white collar" work. Alternatively, the VDT operator might be working in clerical support, dispatch operations, or records—traditional "blue collar" work. The VDT work might be critical, as in emergency dispatch work, or noncritical, as in data entry. It has been estimated that between 15 and 25 million Americans work with computer terminals in the office setting alone, with a projected increase to as many as 40 million by 1990 (1–3). In any case, VDT use

Patty's Industrial Hygiene and Toxicology, Fourth Edition, Volume 1, Part B, Edited by George D. Clayton and Florence E. Clayton.
ISBN 0-471-50196-4 © 1991 John Wiley & Sons, Inc.

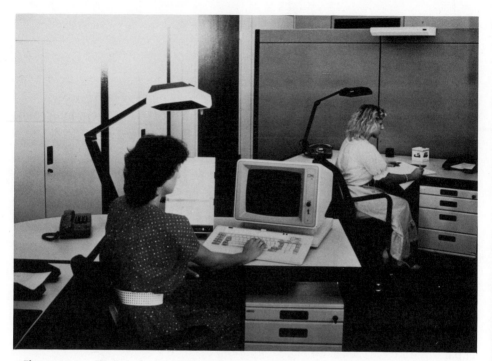

Figure 34.1. Well-designed VDT work stations (But note veiling glare for first user).

substantially changes the nature of work. Many of these changes promise gains in productivity and reduction in repetitive tasks, opening up work opportunities that are more rewarding to the ordinary worker. Unfortunately, other changes create new safety and health risks to the worker. These risks, left unattended, could undermine (or at least delay) many of the potential benefits of computer technology. A VDT (see Figure 34.1) is the typical interface between human and computer. The human can enter information by means of a keyboard. This information, or information generated interactively by the computer, is typically displayed on a cathode ray tube (CRT) television style screen. When a final information product must be permanently displayed, it is displayed on paper by means of a printer. Because all of these components (CRT display, keyboard, printer) were manufactured and used separately, before the advent of modern computer technology, neither the complaints nor possible safety risks were anticipated by manufacturers. The response to worldwide complaints and concerns has, perhaps predictably, been slow and skeptical. Nonetheless, certain concerns related to VDT safety have been recognized and addressed while others have emerged. In this chapter we review what is known about VDT safety in the broader context of computer technology safety.

Although potential VDT hazards would seem to be relevant to all VDT users, most of the complaints have come from workers engaged in "word processing," a task involving transfer of information to and from paper copy. Word processing tasks can place unusual demands on users, requiring many continuous hours of actively looking recursively between video displays and conventional paper displays. Word processing might represent the tip of the iceberg in VDT safety, involving problems that most users will ultimately face. Alternatively, VDT complaints might be based upon problems that are only tangentially related to VDT use significant to only a limited number of users. In either case, there *are* serious concerns related to VDT safety, and such concerns have surfaced consistently wherever extensive office automation using VDTs has been implemented in the United States and throughout the world. Manufacturers are increasingly coming to the realization that VDT safety must be part of work station design. For example, R. B. Dolan (4), editor of the *Hewlett-Packard Journal*, says that a VDT must meet user requirements: "It has to be ergonomically designed, because we've learned that people develop physical problems when they have to adapt to their work areas rather than the other way around." Many offices that have implemented VDT work stations have merely substituted a VDT for a typewriter, or placed a VDT on a writing desk—implicitly supposing that there is nothing significantly different about a VDT display. Many workers are given VDTs as replacements for their typewriters, without any corresponding changes in the job description and little or no specialized training in VDT use—implicitly supposing that there is nothing significantly different about a VDT job. Such assumptions about VDT displays/ apparatus and VDT jobs are clearly unsound (5).

Paper copy, whether in the form of individual typewritten pages or books, is a high contrast, stable, nonreflective, truly portable display. If there is glare, the display can be moved to any appropriate angle. If viewing distance is inappropriate, it can be changed. If viewing angle is uncomfortable, it can instantly be changed.

A video display is fixed, lower contrast, somewhat unstable, highly reflective, and limited to a fixed number of lines. If there is glare, it often must be tolerated. Likewise, if viewing distance and/or angle are uncomfortable, one must accommodate. Adjustments of "contrast" often produce apparent gains by smearing and reducing resolution, an unhappy tradeoff.

Non-VDT jobs are often linear: one inputs a report as a whole (with a typewriter), then makes corrections as a whole. VDT jobs are interactive: one must exchange information with a machine. Such people/machine interactions tend to create completely different tasks (as well as sources of frustration). The efficiency of the VDT operator depends both upon his or her skill and upon the quality of the software support package. Two workers with the same skills, using the same machine, might still differ greatly in apparent productivity if one has an appropriate software interface for the task whereas the other worker has an inappropriate software package. This test rarely is possible because very few users ever learn their software system. Carrol (6) (of IBM) says:

> But although text editors are potentially "easy" to learn to use, people are still having tremendous difficulty learning to use them. In most cases, they are learned incompletely and after significant confusion and error.

For the computer professional, research scientist, or upper level manager, the response is to complain—and complain bitterly. More often than not, for the word processor, the response is to feel inadequate.

The manager's frustration with training, software design, and unmet deadlines often leads him in the direction of computer evaluation of performance—key stroke monitoring. Rather than obtaining efficiency from computer skills, such monitoring can lead to efficiency through a speedup in the information processing assembly line, followed by loss in morale, stress, fatigue, and disability suits.

1.2   Software-related VDT Safety Problems

Most of the controversy in VDT safety involves possible "radiation" harm to the user. Most of the progress in VDT safety involves work station design. However, most of the truly intractable problems involve computer software used interactively between human and technology in critical tasks. Safety issues involve the reliability of the programming, whether the programming meets specifications (software safety), the decrease in human performance due to lack of practice when computer software controls performance on a task (skill obsolescence), the change in location and level of risk in a computerized system (transfer of risk), and the use of software to carry out or communicate critical user commands (instrumental software safety). Many of these concerns are embedded in complex systems/tasks such as flying airplanes, computer-assisted manufacturing, interacting with robots, monitoring nuclear power plants, or dispatching emergency vehicles. We discuss these emerging safety concerns in some detail.

2   STUDIES OF VDT USE

Most studies of VDT use are retrospective: VDT systems are designed and installed; worker complaints lead to investigation to determine the basis for complaints; possible safety hazards are identified; research is carried out to investigate the credibility of such hazards as well as ways to address worker concerns.

The literature has unfolded sufficiently to merit serious review, based upon surveys of complaints, laboratory demonstrations of specific problems, detailed analyses of aspects of computer design, and research and development projects for work station and job design. In some areas (work station design) broad consensus is leading to agreed upon standards. In other areas (radiation hazards) controversy reigns. Safety problems related to software designs are only beginning to be studied. In still other areas, employers and workers must be able to provide more specific information about the job or task before the safety of VDT technology can even be evaluated. Still, VDT-related safety and human factors research remains in its infancy and new discoveries are emerging. One example of the vitality of this field is an international scientific conference devoted to "Work with Display Units" (7) that was held in Stockholm, Sweden, in May of 1986 (and sponsored by major computer manufacturers such as IBM and Phillips). One very formal panel review of VDT use was carried out over a two-year period by the Committee on Vision of the National Research Council, on behalf of the National Institute for Occupational Safety and Health (NIOSH). Entitled *Video Displays, Work and Vision* (8), the study attempted to establish whether or not there was credible evidence that VDTs are responsible for eye damage. A less formal report, *VDTs in the Workplace* (9), was carried out by the Bureau of National Affairs, Inc. This report evaluated both research and litigation. Other reports and reviews abound, including those by Knave and Widebeck (10), Cakir et al. (11), Chapnik and Gross (12), Grandjean, (13), Nilsen (14), and Stammerjohn et al. (15).

2.1   Radiation Exposure Risks

There have been consistent and persistent attempts to relate worker harm to the discharge of energy. The initial concerns about X-ray radiation leakage has spread to ultraviolet, optical, and infrared emissions, radio frequencies, very low frequencies, extremely low frequencies, and electrostatic forces. The general finding of those who have investigated radiation is that biological effects of VDT radiation are inconsequential to health [Frankenhauser, 1986 (16); Suess, 1986 (17); Michaelson, 1986 (18); Guy, 1986 (19)]. Let us look at specific measurements.

1. **X-Ray Radiation.** Research by NIOSH has demonstrated that VDT X-ray emissions are measured at levels less than background levels (20, 21). In fact, VDTs failed to operate before they produced excessive levels of X-rays (22).

2. **Ultraviolet, Optical, and Infrared Emissions.** Many studies (21, 23, 24) indicate levels well below accepted levels of exposure. The most likely consequence of high levels of these emissions would be sunburn, leading to skin cancer. However, if anything, the levels of at least optical radiation may be too *low* (see Section 2.2). Measured radiation is on the level of 1/1000th of permissible levels.

3. **Microwave Radiation and Communication Frequencies.** There are no measurable levels of microwave radiation, and communication frequencies (UHF and VHF used for radio and television) are 1/10,000th the safety level (25).

4. **Very Low Frequencies and Extremely Low Frequencies.** There has been considerable interest in these frequencies, which produce pulsed (60 times per second) magnetic fields. Such fields are thought by some to be related to subtle biological problems and many contribute to stress. Because radiation standards are primarily thermal standards, and these radiation sources produce no thermal radiation, there are no standards for these radiation sources. However, the measured values of extremely low frequencies with VDTs are lower than with many home appliances (26), suggesting that if this pulsed radiation is a problem, it would affect far more than VDT use.

5. **Electrostatic Forces.** The surfaces of VDT screens sometimes are changed and release static electricity ions. These ions have been related to skin rashes in a few sensitive people (27, 28). Even if this were definitively linked to VDT use, the number of people involved are few and countermeasures can be taken (14).

2.2   Visual System Harm

Users of VDTs in very large numbers complain about visual discomfort or visual fatigue. In some studies over three-fourths of participants had vision complaints (29, 30); in most, over half had vision complaints (31–33). In most of these studies, if a comparison is made, there are more complaints from VDT users than nonusers (34). There is disagreement over the cause for visual complaints, the potential threat to health, and appropriate countermeasures to take.

One hypothesis is that complaints reflect actual changes in visual functioning. The usual method for investigating possible oculomotor changes is to subject users to presumed high visual overload for a period of time and then measure changes from a pretest on some visual functions test. The result of such studies is to find small changes in accommodation or vergence or dark focus, (35–37), or no changes for accommodation, vergence, pupillary response, eye movements, or saccade patterns (25, 38, 39). This suggests that visual complaints are not directly related to observable changes of the visual system, at least over relatively short periods of time.

Longer-term studies are equally negative. A study in Montreal in 1982, carried out by ophthalmologists, measured standard optometric functions (acuity, phoria, muscle balance, and vergence) on VDT users for five years and found no difference between VDT users and nonusers. Studies of pathology failed to show a relationship between VDT use and vision disease (40, 41).

When the study relates visual complaints to work station conditions, there are relationships between ergonomic conditions and VDT operator complaints (33). Some of the work station conditions related to visual fatigue include overuse of the eyes, poor screen images, fluctuating luminances, screen glare, improper overall illumination, and high contrast light sources in the work environment (32, 33, 42).

It is not clear whether visual discomfort or fatigue are serious health problems, and, if so, whether the health problem is in the visual, musculoskeletal, or some other system. Thus countermeasures other than good ergonomic design are questionable. Appropriate eye glasses are helpful for vision in general, but specialized glasses or additional testing of vision may not provide added benefits. Glare screens on the VDT may provide legibility benefits by increasing contrast between a target and background, or may reduce legibility by degrading contrast between target and background depending on a variety of environmental factors. Similar benefits may be obtained without a glare screen through sound ergonomic design.

2.3   Musculoskeletal Harm

Although not quite so prevalent as vision complaints, large percentages of users report musculoskeletal disturbances (43) and there are more complaints among users than nonusers (30). Unless care is taken in ergonomic design, the VDT causes poor posture, causes workers to spend more time in static seated postures, and leads workers to repetitive motions. These conditions lead to back discomfort, neck and shoulder aches, and palm and wrist disorders (carpal tunnel syndrome).

Although the evidence is often based upon case studies and anecdotal reports, there are converging studies that support concerns about chronic trauma to the back and wrists. Bammer (44) examined keyboard operators with shoulder–neck or wrist–hands pain. She found that the amount of keying activity was critical: 35 percent of those using VDTs and typewriters, 18 percent of those using VDTs alone, and 5 percent of those using typewriters alone reported problems. Sauter et al. (45) examined the arms and wrists of VDT operators reporting wrist trouble and observed one operator with inflammation and tenderness. In several workers' compensation cases, VDT users have successfully shown a relationship between VDT work and carpal tunnel syndrome (Bureau of National Affairs, 1984, (9).

Other "load studies" have confirmed that the long-term load of various back-related muscles is greater than recommended levels for seated long-term work (46, 47). In addition, circulation problems related to excessive blood pooling may occur in lower legs (48, 49).

2.4 Stress-Related Harm

Among the most common VDT user complaints are reports of perceived stress especially for workers in critical tasks. For example, Briggs (7) surveyed radio dispatch operators as well as data input workers. Both reported very high levels of perceived stress, but the radio dispatch operators indicated levels that were significantly greater. Ghiringhelli (50) found that 43 percent of VDT operators studied reported anxiety aggravation and 40 percent reported depressive disorders. Elias et al. (51) found a number of stress-related symptoms that varied with the type of activity. Smith et al. (29, 52) compared psychological disturbances and job stressors in professional VDT users, clerical VDT operators, and as a control, clerical workers not using VDTs. They found the highest levels of reported anxiety, depression, anger, confusion, and mental fatigue in the clerical VDT operators. Although stress may be higher in certain VDT tasks, especially critical, demanding tasks, reported stress is not unique to VDT work. Office workers in general, especially clerical workers, are at high risk for stress-related disease (53). It seems clear that it is not the computer hardware or work station that causes job stress, but rather the redesign of work, including the software interface. Segmenting the various demands on a person that are stressful into work, recreation, home life, and other categories is impossible, and associating health conditions with stress is also difficult. Nonetheless, it may be the mediating role of stress that is responsible for many worker complaints as well as various health consequences.

2.5 Miscarriages and Other Health Difficulties

There have been frequent reports of birth defect and spontaneous abortion "clusters" among VDT users. These clusters, in turn, have led many to assume a radiation cause emanating from the VDT. We have seen that there is no support whatsoever for this "radiation leak" hypothesis (Section 2.1). However, although studies of these clusters by government agencies have shown no relationship between VDT use and adverse reproductive outcomes, the VDT has not been ruled out as a potential contributor, according to NIOSH (54), Centers for Disease Control (55), and the U.S. Army Environmental Hygiene Agency (56).

McDonald (57) reports a study of reproductive hazards in the workplace that included the VDT as a potential adverse factor. More than 50,000 women were interviewed including about 10 percent that experienced a spontaneous abortion. VDT users had a slightly higher ratio of spontaneous abortions than nonusers. Kurppa et al. (58, 59) showed that VDT users are no more at risk for either defective births or spontaneous abortions. Because the Kurppa et al. and other studies that fail to show a difference have smaller samples, the various studies can be reconciled by concluding that if there is any effect of VDT use on reproduction, the effect is infinitesimally small. Such small effects are more consistent with worker stress than radiation harm.

2.6 VDT Risk and Remedy

When one considers that once-in-a-lifetime revolutionary changes in the performance of activities and tasks involving the purchase of VDT-controlled equipment are currently taking place, it seems unwise to wait a decade for possible conclusive studies (yet to be funded) on VDT safety. The implementation of VDT technology is not going to wait for the outcome of research. Safety need not wait, either. It would seem helpful to implement video safety practices as part of an effort to *increase* the productivity gains of VDTs while designing out unwanted possible risks or hazards. The outlines of such a strategy have already emerged and in many cases are ready for implementation. Table 34.1 presents an analysis of possible work station hazards, based on the currently available literature. There are three categories of possible harm: (1) direct physical harm: (2) work station design induced harm; and (3) job design induced harm. The preponderance of opinion currently suggests that all three categories of VDT-related safety problems (whether real or imagined) could be addressed based on existing knowledge with the expectation that overall cost effective *gains* in productivity would result. Such gains would more than offset any costs associated with VDT safety, (i.e., research, ergonomic improvements, training, software modifications). The second two categories of possible hazards (work station design and job design) are already the subject of considerable research, although even here the important older worker is often neglected in design considerations (60). Some recent studies suggest that work station redesign (to eliminate ergonomic problems) increases productivity 10 to 15 percent; whereas software redesign (to reduce stress and improve job design) increases productivity 30 to 35 percent (61–64). However, even the concerns about possible physical risks of radiation (perhaps unfounded) might be addressed indirectly—with no net loss to performance. For example, even if no forms of radiation exposure turn out to pose a threat, prudent work station design might suggest larger screens with high resolution, increasing overall efficiency. This would lead to greater screen/subject distances and less potential radiation risk. Alternatively, sound job design might suggest that pregnant women be exposed to less job stress during pregnancy. One consequence of such an approach would be to reduce the amount of daily VDT exposure, thus reducing any possible VDT related radiation risk. The two areas in which safety (in general) leads to productivity gains are in the areas of work station design and job design. There is broad agreement confirming work station design on the general issues of background lighting (diffuse and dim), furniture (movable and adjustable), and the physical VDT (adjustable angles for screen, detachable or extensively adjustable keyboard). Figure 34.2 illustrates an award-winning work station in which the movable chair has many adjustments for height, angle, and back support; the table can be raised or lowered by a micrometer and the height indicated; and the VDT screen and keyboard can be adjusted for angle and location.

However, there is no agreement on specifications of the display itself. VDT display screens are currently (at best) modified television screens, subject to apparent movement-related phenomena (due to phosphor decay, missing pixel elements, voltage irregularities, software inconsistencies, variable baud rates, screen distortion, limited resolution, and/or limited contrast, etc.). High-resolution displays with less distortion and less movement (as well as autocalibration) may soon be available. Non-CRT plasma screens, electroluminescent screens, and liquid crystal display (LCD) screens also might soon be used. Such screens could overcome most concerns about resolution as well as physical angle adjustment (because some non-CRT screens might be as little as 1 in. thick). Inflexible guidelines might hamper rather than aid the implementation of improved productivity and safety-related VDT display screens. Glare shields are also controversial because shields reduce glare by reducing contrast, and hence legibility. Careful arrangement of the work station (including illumination) can eliminate most glare.

The requirement for software that communicates between the human and computer technology also has some consensus of support. Recent software design reduces reliance on human memory for appropriate commands by means of a "menu" that displays alternatives, and a "mouse," which allows complex computer activities to be carried out analogically on a VDT display. However, such features may reduce efficiency in time contingent tasks because such "user friendly" features tend to be time consuming and repetitive. Eliminating the user friendly features may increase both speed and error frequency. In critical time contingent tasks software may lead to instrumental safety errors in which an incorrect command leads to a critical incident. Promoting safety under such conditions may require extensive software rewriting so that, for example, commands are not accepted unless they are safe as well as appropriate.

Job design and the related computer issue of software development offer potentially much more promise both for the reduction or elimination of VDT symptoms/complaints and for the improvement of efficiency.

However, any employer or designer who wishes to benefit from advances in software development must gather significant information about the job requirements, user needs, user qualifications, and administrative structure. By not gathering such information, it is impossible for the manufacturer to know what to design

Table 34.1. VDT Risk and Remedy Analysis

Possible Hazard	Complaints	Known Risk	Remedy	Implementation
Direct physical harm from machine				
Radiation exposure	Cataracts, miscarriages	Measured levels suggest no risk with modern design and maintenance	Shielding of VDT, greater screen distance, keep pregnant women away	Inspection/guidelines, guidelines, negotiations, research
Noise exposure from printer	Ear damage	Measure against ANSI standards	Shielding of printer to reduce noise	Enforce existing OSHA standards
Work station design				
Screen/lighting (glares, jitter, contrast)	Visual fatigue, disturbed OMS performance, physical (posture), eyes	Well documented Some documentation		Legislated guidelines, negotiations based on survey, users participate in purchase and plans, users' right to request modification of work-station, work site inspection by NIOSH, VDT-related on-site eye exam
Ergonomics	Fatigue Musculoskeletal performance, physical damage	Well documented Lawsuits won based on documented instances	Redesign workplace consistent with ANSI guidelines and *user* recommendations	

Job design				
Too many VDT hours per day or inadequate rest	Fatigue and stress	Lawsuits have been successful on some of these issues	Limit hours/day at VDT; mix VDT and non-VDT work, provide adequate rest	Negotiations based on surveys
Work pace too fast and rigid	Fatigue and stress	Very poorly studied	Pace should be set primarily by user, not machine	
High criticality of task	Fatigue and stress	Very poorly studied	Provide for backups for user	
Software unfriendly to user	Fatigue and stress	Very poorly studied	Guidelines; select software carefully, modify if necessary; better, more advanced training for user	
Software provides inadequate backup	Fatigue and stress	Very poorly studied	Guidelines; select software carefully, modify if necessary; better, more advanced training for user	
Software inappropriate to tasks			Guidelines; select software carefully, modify if necessary; better, more advanced training for user	
Monitoring of performance			Monitoring should be explicit and individualized; some privacy should be allowed to user	

Note: ANSI = American National Standards Institute.

Figure 34.2. Award-winning work station includes adjustable chair, table, VDT keyboard, and VDT screen.

and impossible for the worker to adapt his/her old job to the new technology. Subsequent failures and retrofitting lead to much higher human and financial costs.

3 EVALUATING VDT SAFETY

One can identify three components to any VDT work station: the technical equipment (hardware), the human/machine interface (software), and the human user. Any or all of these components might be involved in an incident. The safety-related consequences might include direct or indirect physical harm to the human user, instrumental harm to others, and/or system failure. One can carry out a safety evaluation either as part of design or after installation in an operating system. All these considerations complicate the evaluation process.

Because most VDT safety evaluations have resulted from complaints after installation, the focus of concern has been direct physical harm to the user from the technical equipment itself or from the layout of the equipment. As Table 34.1 (see Section 2) illustrates, the concern focuses on possible radiation risks and/or consequences of prolonged sitting in front of a VDT in a stereotyped posture. Much progress has been made on these issues. Although there is still concern about radiation exposure risk and epidemiologic studies continue to search for possible relationships between some source of energy and physical harm (e.g., miscarriages, birth defects, skin rashes), most of the recent progress in VDT technology relates to the proper design of a work station.

One way of evaluating a VDT work station is to construct a checklist of requirements based upon American National Standards Institute guidelines. One example of such an approach is illustrated in Table 34.2. The evaluation is derived by comparing a work station to a standard. Please note that this particular evaluation would be appropriate only for a "word processing" work station. There are other work stations that would require the user to walk around and/or use other equipment (multiple screens, telephones, multiple keyboards, etc.). Such work stations would need different checklists. Correction of work station design would consist of replacing or modifying equipment or the environment that fails to merit a "check." Obviously new work stations should be constructed to conform to the checklist.

The work station checklist evaluation should be carried out on a regular basis because the components of the work station can deteriorate. Comfortable chairs and well designed desks may be moved away. Especially vulnerable is the VDT screen. Visibility of the CRT screen can be reduced due to pixel decay (loss of individual viewing cells on the screen), jitter, distortion, overall loss of brightness, electrical interference, and other factors. Non-CRT screens have their own specialized problems. Screen legibility therefore needs to be maintained. Similarly, the keys on a keyboard may become variably resistant to touch, or the chair may lose its adjustment or support, the printer may lose its noise shield, the room lighting may change, or other changes may take place.

3.1 Job Design

If one purchases a VDT work station according to the checklist given in Table 34.2, and if there are no undetected leaks of energy or radiation, it is often supposed that the VDT is "safe." This assumption neglects both the uses of the VDT work station and the user. As Table 34.1 indicates, the possible harm to the user is closely related to the intensity, amount of time, and quality of the interaction between the user and the VDT. A user who must work long uninterrupted work shifts in isolation, day after day, using "unfriendly" software that requires constant attention and tolerates no errors, may still experience physical harm from even the best designed work station. Another user who interacts sporadically on a flexible basis with "friendly" software may not be harmed by very poor work station design. Any complex human/machine interaction should be evaluated. Work should be allocated so that both the computer and the human are carrying out suitable tasks. There has been a tendency to allocate as much work as possible to the computer, leaving the leftovers to the human user. It is a tribute to human versatility that humans so frequently are able to bail out the designer by carrying out such a hodgepodge of activities with few errors. However, there are formal functional analyses that can be carried out which make more optimal allocations that facilitate

Table 34.2. Work Station Evaluation Checklist

Chairs

Seat
() Adjustable, minimum range 15–20 in.
() Adjustable from seated position
() Contoured bucket, gently dished, not form fitting
() Seat pan forward tilt—7° forward tilt
() Minimum 17 in. wide, 16 in. long
() Seat front—waterfall design
() Firm padding
() Upholstery able to absorb perspiration

Base
() Five-leg caster design
() Hard Casters
() Swivel

Chair arms
() Rounded
() Short
() Removable or hinged
() Adjustable height and lateral movement

Back support
() High back support
() Lumbar support centered 9–10 in. above lowest point in seat
() Adjustable lumbar support
() Minimum 16 in. wide
() Back tilt angle between seat pan and seat back adjustable to 120°

Footrests

() Angled, wedge-shaped
() Non-slip surface
() Large enough to allow for variations in posture

Tables

() Adjustable between standard desk height, 30 in. and typing height, 27 in.
() Tabletop no thicker than 1 in.
() Space for knees and legs—area under table 27 in. wide by 27 in. deep (minimum)
() Minimum depth of table 28 in.—this allows minimum of 3 in. of space in front of keyboard
() Minimum 40 in. wide to provide sufficient room for documents and job aids—this will allow enough room alongside of CRT at screen height
() Table legs set back from front edge so knees are not hindered when moving in and out from table
() All adjustable equipment has measurement devices that enable quick and easy individual adjustment when one workplace is shared by several users

Document Holders

() Adjustable—height and angle
() Fitted with, or capable of being fitted with, task lighting

Table 34.2. (*Continued*)

Arm, Wrist Supports

() Adjustable
() Firm padding
() Upholstery should be able to absorb perspiration
() Able to be mounted on edge of table
() Minimum 3 in. wide
() Minimum 15 in. long

Noise

() Accoustical pads under keyboards
() Accoustical pads under printers
() Printers with accoustical covers
() Segregate printers from operators
() Substitute laser printers (quiet printers) for impact or dot matrix printers

Temperature Control

() Control room temperature levels—heat generated by equipment is equivalent to heat generated by a human body. This heat must be controlled by improvements in the air conditioning system
() Vents shielded to direct airflow away from operators to eliminate the "draft effect"

Electrical Outlets

() Placed for convenient equipment location
() Placed to eliminate electrical cord tripping hazards
() Placed to allow for movement of equipment without long cords

Lighting

Room light adequate to reduce eyestrain and glare
() Overhead lighting—30–50 footcandles (fc) measured at tabletop (NIOSH)
() Task lighting—50–70 fc measured at document (NIOSH)

Light Reflectances

() Ceilings Not less than 70%
() Walls Not less than 40%
() Other vertical surfaces Not less than 40%
() Floors Not less than 20%

Overhead Lights

() Parabolic louvers on overhead lights to diffuse lighting
() Illuminate entire ceiling uniformly with brightness ratio of 5:1 or less. When looking into mirrorlike CRT screen, areas of brightness in excess of 500 footlamberts (fl) or less than 100 fl should not be visible
() Avoid lighting that directs light downward in excess of 500 fl, at any spot on the ceiling or from any other angle

Bright Light Sources

() Mask bright light sources; cover windows, but allow adjustment so employees can control light
() Place VDTs at right angles to windows

continued on next page

Table 34.2. (*Continued*)

Glare

() Provide glare screens for CRTs
() Place terminals away from harsh light sources
() Mat finish casing on terminals
() Provide screen hoods
() Mat paint on walls—light colors on walls to give brightness

Task Lighting

() Horizontally adjustable positioning
() Limited swivel and vertical adjustment
() Able to be attached to document holder, or adjustable to be positioned at document holder
() Opaque shade, adjustable shade
() Adjustable light level (e.g., dimmer switch)
() Low power requirement, low heat output

Traffic Flow near Operators

() Reduce/eliminate traffic flow near operators; activity and movement can be distracting to operator's eyes, decreasing visual attention and concentration, causing fatigue

CRTs

Keyboards:

() Detachable
() Should have substantial touch
() Wedge Shape—angle 5–15°
() Concave key surfaces
() Mat finish key tops
() 1.5–2.5 in. from top of table to top surface of keyboard (or bottom row of keys)

CRT Screen

() Tiltable monitor—face of screen adjustable to 5° forward to 15° backward
() Rotatable monitor
() Antiglare screens
() Brightness/contrast adjustable
() Sharp (well-defined), nonflickering images
    () 60-Hz Refresh rate adequate for most applications
    () 70-Hz Refresh rate for short-persistence phosphors and reversed video (light background) displays
() Rapidly changing displays—short-persistence phosphors (e.g., P4 bluish-white)
() Static images—P38 orange or P39 green screen image
    Colors Listed in order of preference
    () White characters on cobalt blue field
    () Black characters on cream white field
    () Yellow orange characters on black field
() Sharp color contrast
() Flat panel displays—provide flickerless display

Table 34.2. (*Continued*)

CRT Casing

() Mat finish on casing
() Soft curves on casing
() Neutral colors on casing
() Soft surface texture
() Low heat output from CRT

Character Size

() Font size should maximize legibility (minimum $3/16$ in. high)
() Preferred dot matrix size is 7×9
   () Acceptable dot matrix size—5×7
() Matrices with square dots—preferred over oblique or round dots

Printers

() Accoustic pads under printers—integral with printer
() Accoustic cover on printer
() Noise generated by printer does not exceed 70 dBA, measured at distance of 5 ft from printer
() Provide equipment behind printer to collect printouts

CRTs

Screen height and viewing angle
() Top of screen no higher than eye level
() Top of screen 10° below horizontal at eye level

Screen Face Angle

() Face of screen tilted back 10–20° for easiest viewing

Viewing Distance

() For comfort, 18 in. minimum is recommended
() Distance adjustable from 14 to 32 in.

Platform

() Height adjustable
() Tiltable, rotatable platform if CRT is not tiltable and rotatable

Work Stations

() Optimum reach 14–18 in.
() Maximum reach 22–26 in.
() Optimum desk front length (for reach) 40 in.
() Maximum desk front length 60 in.
   Avoid: operator's back near the back of another machine, "clusters" where operators face each other in groups, (isolation of) worker, to avoid feelings of isolation and dehumanization

human/machine system performance. Flexible allocations between the human and machine are often possible because there is a software-controlled interface that can facilitate human/machine communication while allowing control of activities to rest with either the human or the machine. It is beyond the scope of this chapter to detail how to analyze CRT related jobs and tasks. An excellent introduction to this is given by Bailey (65). Job and task analyses are always helpful to reduce occupational stress on the human and improve system performance, but in certain activities they are critical. For example, a VDT interface may allow the user to monitor the safety of a nuclear power plant or chemical plant. If assigned inappropriate tasks, the user may make errors or fail to attend to the monitor. Such an error may lead to an accident or an incident. An investigation may blame the incident on human error and reduce the human role. This may lead to skill obsolescence on the part of the user and even less attentiveness to the task. When the human would be required to carry out a manual override of the computer or may be called to an emergency, the human may not notice the problem or know how to handle the emergency.

Although many of these incidents have been attributed to human error, they may be at least partly due to poor job design and function allocation between machine and human. Job design is a very important component of overall VDT safety, especially when the instrumental uses of the VDT are associated with life critical tasks.

4   SOFTWARE SAFETY

Perhaps the most neglected aspect of VDT safety is the role of software. In noncritical tasks, inadequate software contributes to human stress by causing the human to make errors, lose files, or even shut down the computer system altogether. The human is often forced to refer to manuals for commands or help that a software prompt or special "help" subroutine could provide. In critical tasks, software could cause or allow errors with catastrophic consequences.

Software is not only the cause of human/machine difficulties but is also the most likely source of a retrofit "fix" when difficulties are discovered. Software can prompt the user with alternatives, disallow inappropriate and dangerous responses, avoid presenting the user with unnecessary detail, and reallocate work between the computer and man. Despite its obvious importance, software has defied most analysis. No one has been able to evaluate either software reliability (at any level) or software safety.

The most basic analysis of software safety involves the issue of programming error. It is known, for example, that a program becomes more and more difficult to fix as it involves more and more commands embedded in increasingly complex architectures. At some point it is claimed that more errors will result in "fixing" a program than in leaving the error in the program (66). We thus know that large programs contain errors, but we do not know how many. Even if we knew how many, we would not be able to assess the significance of the errors because commands vary in frequency of use with different users.

One way to get around this difficulty partially is to identify requirements that software must perform and to evaluate the software against performance requirements. This sometimes works against software safety because performance requirements usually demand that frequent rather than critical command subroutines work. If one has a completely automated system, if the safety-related components are evaluated against specifications, and if the specifications themselves assure safety, then software safety can be at least partly evaluated, though usually not in terms like "probability of software failure." However, when one is using software to interface a VDT with a human, the evaluation of human/machine failure in terms of software is still more difficult. In a human/machine VDT system, the responsibility for monitoring an activity is given to the computer, but nonroutine safety decisions are given to the human. For example, an emergency dispatch operator receives a call asking for emergency help. The operator may have a VDT display that shows the location and availability of ambulances. A decision must be made to send or not send the ambulance, which ambulance to send, how immediate the priority is, and so on. In this situation, an error is almost always attributed to the human. However, truly adequate software would prompt the operator with questions to ask and would present alternatives for the operator to choose. If the operator inadvertently pushed a button that was in error, the software could either lock out the command or require a second response. Such problems create grave difficulties for software safety evaluations. There is no clear way to measure the likelihood of an incorrect computer or human response due to faulty software. Redesign to facilitate human performance may harm automated performance. The user friendly approach, involving user prompts, menus, and multiple confirmations described in the above example, all make the software more complex and error prone. They also make the system slower. In time critical tasks, this may trade one type of safety risk for another. Because the safety of the software is not measured or evaluated, incidents that could have been due to software failure or inadequate software design are often attributed to human failure.

In many human/machine systems, errors have been eliminated by eliminating the human component, often with the assumption that eliminating human responses would increase reliability and safety. Such approaches have led to airplanes that are primarily flown by computers, computer-integrated manufacturing or manufacturing with robots, and automated control of chemical or nuclear power plants. In these systems, the VDT operator is primarily a spectator, locked out of control of tasks, brought in only when problems arise. From a safety viewpoint, risk has been transferred from the VDT user to other parts of the system—for example, the humans who design the software that control the machine. When called in to take human control back from the computer, the user finds that lack of practice leads to "skill obsolescence." All of this leads to concerns about overall system safety, specifically in regard to catastrophic incidents.

Although complex software-controlled automated systems are often developed to increase system-wide reliability and are said to be extremely safe, there have been catastrophic incidents involving nuclear power plants, automated chemical plants, space shuttles, and robots in which human life has been lost. In many cases, analyses had suggested that a catastrophic incident would not happen. Given the difficulty of even evaluating software safety in the context of complex computers and the impossibility of evaluating the safety of a human/VDT controlled system, it is unclear how estimates of software-related risk are made at all. Clearly, the need to evaluate the safety of software in human/machine systems is a critical unsolved need.

4   LEGISLATION

There is considerable interest in passing legislation pertaining to VDT use at various levels of government. In December, 1989, the strictest VDT safety bill in the United States was passed when the New York City Council approved legislation to protect the health and safety of more than 12,000 city workers who use VDTs. The bill is the first law of its kind in a major city, according to the New York City Video Display Terminal Coalition of labor unions and occupational safety and health groups. The bill mandates that VDT operators at city agencies be given a 15-min break when they are required to work at a terminal for more than two consecutive hours. Under the law, VDT workers with certain medical conditions—including pregnancy—would be allowed to obtain transfers.

This legislation may be the precursor of similar actions by other governmental bodies as the use of computers within government and industries escalates.

REFERENCES

1. M. J. Smith, "Are VDT's a Pain in the Neck?" *Los Angeles Times* (July 24, 1988).
2. C. Gorman, *Time*, 51 (June 27, 1988).
3. D. Olsmos, "New Technology Opens Door for More Workers to Stay Home," *Los Angeles Times* (April 24, Business Section, 1989).
4. R. B. Dolan, *Hewlett-Packard J.*, 3 (1985).
5. B. DeMatteo, *Terminal shock*, NC Press Limited, Toronto, 1985.
6. Carrol, *Computer*, **15**(11), 50 (1982).
7. R. P. Briggs, "VDU Job Evaluations—California State Worker Survey," *Proceedings of International Scientific Conference: Work with Display Units*, National Board of Occupational Safety and Health, Stockholm, 1986, p. 325–328.
8. National Research Council, *Video Displays, Work, and Vision*, Washington, D.C., 1982.
9. Bureau of National Affairs, *VDT's in the Workplace*, 1982.
10. B. Knave and P. G. Wideback, *Work With Display Units*, Elsevier, Amsterdam, 1987.
11. A. Cakir, D. J. Hart, and T. F. M. Stewart, *The VDT Manual*, Ince-Fry Research Association, Darmstadt, FDR, 1979.
12. E. B. Chapnik and C. Gross, "Visual Display Terminals: Health Issues and Productivity," *Personnel*, 10–16 (1987).

13. E. Grandjean, *Ergonomics and Health in Modern Offices*, Taylor and Francis, London, 1984.

14. A. Nilsen, "Facial Rash in Visual Display Unit Operator," *Contact Dermatitis*, **8**(1), 25–28 (1982).

15. L. Stammerjohn, M. Smith, and B. Cohen, "Evaluation of Work Station Design Factors in VDT Operations," *Human Factors*, **26**(4), 347–356 (1981).

16. B. Frankenhauser, "Video Display Terminals: Electromagnetic Radiation and Health," *Proceedings of International Scientific Conference: Work with Display Units*, National Board of Occupational Safety and Health, Stockholm, 1986, pp. 57–59.

17. M. J. Suess, "Visual Display Terminals and Radiation," *Proceedings of International Scientific Conference: Work with Display Units*, National Board of Occupational Safety and Health, Stockholm, 1986, pp. 76–79.

18. S. M. Michaelson, "Health Implications of Exposure to Emissions from Video Display Terminal," *Proceedings of International Scientific Conference: Work with Display Units*, National Board of Occupational Safety and Health, Stockholm, 1986, pp. 49–52.

19. A. W. Guy, "Health Hazards Assessment of Radio Frequency Electromagnetic Fields Emitted by Video Display Terminals," *Proceedings of International Scientific Conference: Work with Display Units*, National Board of Occupational Safety and Health, Stockholm, 1986, pp. 40–44.

20. NIOSH, *A Report on Electromagnetic Radiation Surveys of Video Display Terminals*, Publication No. 78-129, National Institute for Occupational Safety and Health, Cincinnati, OH, 1978.

21. NIOSH, *Potential Health Hazards of Video Display Terminals*, National Institute for Occupational Safety and Health, Cincinnati, OH, 1981.

22. FDA 1981, cited in M. S. Smith, "Mental and Physical strain at VDT Workstations," Workshop presented at the Second International Conference on Human-Computer Interaction, Honolulu, Hawaii, 1987, p. 5.

23. W. E. Murray, "Video Display Terminals: Radiation Issues," *IEEE Computer Graphics Appl.* 41–44 (1984).

24. M. M. Weiss, "The Video Display Terminal—Is There a Radiation Hazard" *J. Occup. Med.*, **25**, 98–100, (1983).

25. S. E. Taylor and B. A. Rupp, "Display Image Characteristics and Visual Response," *Proceedings of International Scientific Conference: Work with Display Units*, National Board of Occupational Safety and Health, Stockholm, 1986, pp. 212–215.

26. M. A. Stuchly, D. W. Lecuycer, and R. P. Mann, "Extremely Low Frequency Electromagnetic Emissions from Video Display Terminals and Other Devices," *Health Phys.*, **45**, 713–722, (1983).

27. V. Linden, and S. Rolfsen. "Video Computer Terminals and Occupational Dermatitis," *Scand. J. Work, Environ. Health*, Research Department, Stockholm, 1981.

28. M. Rahimi, and K. Abedini, "Ergonomics and Safety in User—VDT Interaction," Industrial Safety Chronicle, 1987, pp. 164–172.

29. M. J. Smith, "An Investigation of Health Complaints and Job Stress in Video Display Operations," *Human Factors*, **23**(4), 389–400 (1981).

30. S. L. Sauter, *The Well-Being of Video Display Terminals Users*. National Institute for Occupational Safety and Health, Cincinnati, OH, 1983.

31. U. O. Bergqvist, "Video Display Terminals and Health: A Technical and Medical Appraisal of the State of the Art," *Scand. J. Work Environ. Health*, **10**, 1–87 (1984).

32. M. J. Dainoff, "Occupational Stress Factors in Visual Display Terminal Operation: A review of Empirical Research," *Behav. Inf. Technol.*, **1**, 141–176 (1982).

33. M. J. Smith, "Health Issues in VDT Work," *Visual Display Terminals, Usability Issues and Health Concerns*, Prentice-Hall, Englewood Cliffs, NJ, 1984, pp. 193–228.

34. S. J. Starr, C. R. Thompson, and S. J. Shute, "Effects of Video Display Terminals on Telephone Operators," *Human Factors*, **24**, 699–711 (1982).

35. E. Gunnarsson, and I. Soderberg, "Eye-strain Resulting from VDT Work at the Swedish Telecommunications Administration," National Board of Occupational Safety and Health, Stockholm, 1980.

36. O. Ostberg, "Visual Accommodation Before and After VDU Work With Split Screen Versus Traditional Screen/Paper Mode of Operation," *Proceedings of International Scientific Conference: Work with Display Units*, National Board of Occupational Safety and Health, Stockholm, 1986, pp. 367–370.

37. S. Yamamoto, K. Noro, S. Kurimoto, and T. Iwasaki, "VDT Operators Variation of the Accommodation of The Eyes during VDT Work," *Proceedings of International Scientific Conference: Work With Display Units*, National Board of Occupational Safety and Health, Stockholm, 1986, pp. 878–888.

38. L. Stark, "Visual Fatigue and the VDT Workplace," *Visual Display Terminals. Usability Issues and Health Concerns*, Prentice-Hall, Englewood Cliffs, NJ, 1984.

39. M. J. Dainoff, L. Frazier, and B. Taylor, 1984. "Workstation and Environmental Design Factors in Best and Worst Case Conditions and their Effects on VDT Operator Health and Performance," National Institute for Occupational Safety and Health, Cincinnati, OH, 1984.

40. A. B. Smith, S. Tanaka, W. Halperin, and R. D. Richards, "Cross-sectional Survey of VDT Users at the Baltimore Sun," National Institute for Occupational Safety and Health, Cincinnati, OH, 1982.

41. F. M. Grignolo, A. OiBari, B. Brogliatti, A. Palumbo, and G. Maina, "Ocular Tonometry in VDT Operators," *Proceedings of International Scientific Conference: Work with Display Units*. National Board of Occupational Safety and Health, Stockholm, 1986, pp. 578–581.

42. E. Grandjean, *Ergonomical and Medical Aspects of Cathode Ray Tube Displays*, Federal Institute of Technology, Zurich, 1979.

43. E. Gunnarsson, and O. Ostberg, "The Physical and Psychological Working Environment in a Terminal-Based Computer Storage and Retrieval System," Report 35, National Board of Occupational Safety and Health, Stockholm, 1977.

44. G. Bammer, "VDUs and Musculo-skeletal Problems at the Australian National University—A Case Study," *Proceedings of International Scientific Conference: Work with Display Units*, National Board of Occupational Safety and Health, Stockholm, 1986, pp. 243–246.

45. S. L. Sauter, L. J. Chapman, S. J. Knutson, and A. A. Anderson, "Wrist Trauma in VDT Keyboard Use: Evidence, Mechanisms and Implications for Keyboard and Wrist Rest Design," *Proceedings of International Scientific Conference: Work With Display Units*, National Board of Occupational Safety and Health, Stockholm, 1986, pp. 338–340.

46. B. Jonsson, "Measurement and Evaluation of Local Muscular Strain in the Shoulder during Constrained Work," *J. Human Ergology*, **11**, 73–88 (1982).

47. K. Schuldt, J. Ekholm, and K. Harms-Ringdahl, "Sitting Work Postures and Movements, and Muscular Activity in Neck and Shoulder Muscles," *Proceedings of International Scientific Conference: Work With Display Units*, National Board of Occupational Safety and Health, Stockholm, 1986, pp. 338–340.

48. F. J. Thompson, B. J. Yates, and D. G. Franzen, "Blood Pooling in Leg Skeletal Muscles Prevented by a 'New' Venopressor Reflex Mechanisms," *Proceedings of International Scientific Conference: Work with Display Units*, National Board of Occupational Safety and Health, Stockholm, 1986, pp. 212–215.

49. J. Winkel, "Macro—and Micro—Circulatory Changes during Prolonged Sedentary Work and the Need for *Lower* Limit Values for Leg Activity—A Review," *Work With Display Units*, Stockholm, 1986, pp. 497–500.

50. L. Ghiringhelli, "Collecting Subjective Opinions on the Use of VDUs," *Ergonomic Aspects of Visual Display Terminals*. Taylor & Francis, London, 1980, pp. 227–232.

51. R. Elias, F. Cail, M. Tisserand, and M. Christman, "Investigations in Operators Working with CRT Display Terminals: Relationship between Task Content and Psychophysiological Alterations," *Ergonomic Aspects of Visual Display Terminals*. Taylor & Francis, London, 1980, pp. 211–218.

52. M. J. Smith, "Video Display Operator Stress," *Ergonomic Aspects of Visual Display Terminals*, Taylor & Francis, London, 1980, pp. 201–210.

53. B. G. F. Cohen, *Proceedings of a Conference on Occupational Health Issues Affecting Clerical/Secretarial Personnel*. National Institute for Occupational Safety and Health, Cincinnati, OH, 1982.

54. NIOSH, *Health Hazard Evaluation of Southern Bell, Atlanta, Georgia*, HETA-83-329-1498, National Institute for Occupational Safety and Health, Cincinnati, OH, 1983.

55. Centers for Disease Control (1983), cited in M. S. Smith, "Mental and Physical Strain at VDT Workstations," Workshop presented at the Second International Conference on Human-Computer Interaction, Honolulu, Hawaii, 1987, p. 3.

56. AEHA, "Investigation of Adverse Pregnancy Outcomes," Occupational Health Special Study no. 66-32-1359-81, U.S. Army Environmental Hygiene Agency, Washington, DC, 1981.

57. A. D. McDonald, "Birth Defect, Spontaneous Abortion and Work with VDUs," *Proceedings of International Scientific Conference: Work with Display Units*, National Board of Occupational Safety and Health, Stockholm, 1986, pp. 669–670.

58. K. Kurppa, P. C. Holmberg, K. Rantala, T. Nurminen, and L. Saxen, "Birth Defects and Video Display Terminal Exposure during Pregnancy: A Finnish Case-Referent Study," *Scand. J. Work Environ. Health*, **11**, 353–356 (1985).

59. K. Kurppa, P. C. Holmberg, K. Rantala, T. Nurminen, J. Tikkanin, and S. Hernberg, "Birth Defects and Video Display Units: A Finnish Case-Referent Study," *Proceedings of International Scientific Conference: Work with Display Units*, National Board of Occupational Safety and Health, Stockholm, 1986, pp. 661-663.

60. R. P. Briggs, "Visual Changes Among Older Workers: Implications For Workstation Design," *Proceedings of Human Factors, 30th Meeting* 1986, p. 801.

61. K. Nussbaum, Ergonomics Workplace Improves Productivity. *Computerworld*, 102 (1985).

62. M. J. Paznik, "Ergonomics Does Pay," *Adm. Manage.*, 17–24 (August, 1986).

63. J. Soat, "Office Seating: A Productivity Enhancer," *Office Adm. Autom.*, 33–38, 84–85 (1985).

64. E. Frazelle, and C. L. Smith, Jr., "Worker Participation in Office Space Planning and Design Process Pays Large Dividends," *Ind. Ergonomics*, 34–37 (1984).

65. R. W. Bailey, *Human Performance Engineering*, Prentice-Hall, Englewood Cliffs, NJ, 1982.

66. Pagels, Heinz, Eds., "Computer Culture," *Ann. N.Y. Acad. Sci.*, **426** (1984).

CHAPTER THIRTY-FIVE

Integrated Loss Prevention and Control Management

Judith Stockman, R.N., C-A.N.P., C.O.H.N.

1 INTRODUCTION

In 1987, employee health care, which accounted for an amazing 50 percent of corporate pretax profits, had dramatically increased from just 9 percent in 1965 (1). What was once considered "just the cost of doing business" now threatens America's ability to compete in the marketplace. Because occupational health care and employee safety traditionally have been seen as fixed costs, little serious effort has been made to contain them.

Now that these expenses have started to have a significant impact on the corporate balance sheet, managers have begun to search for ways to reduce or at least slow the spiraling growth of these expenditures. It is possible to limit expenses related to employee health and safety; however, it requires a totally new approach to loss prevention and control.

An effective loss control program must utilize an integrated occupational health team approach involving occupational health clinicians, industrial hygienists, safety professionals, and workers' compensation professionals. The team should set goals, then create and, with senior management input, institute policies and programs designed to achieve those goals. In order to show the effectiveness of their programs, they must provide accurate statistical data relating to employee health and safety.

Patty's Industrial Hygiene and Toxicology, Fourth Edition, Volume 1, Part B, Edited by George D. Clayton and Florence E. Clayton.
ISBN 0-471-50196-4 © 1991 John Wiley & Sons, Inc.

2   COSTS OF HEALTH CARE AND RELATED COSTS TO EMPLOYERS

Members of the occupational health team can be viable contributors to today's economic climate. However, they must expand their roles dramatically from what was once perceived as a technical staff position to that of a cost-effective member of management. Historically, the perception of loss prevention and control has been limited to the prevention of industrial accidents and related costs, but that is not a viable position today.

Not only today, but also in the decades to come, skyrocketing costs of group medical benefits, expenses associated with tort litigation, impact of criminal litigation, and other expenses borne by industry add a completely new dimension to loss prevention and control. These costs place such a severe economic burden on industry that unless a proactive cost containment program is initiated, health-related costs and, ultimately, the ability of corporations to compete will be adversely affected.

2.1   Direct Workers' Compensation Costs

The most rapidly increasing cost associated with industrial accidents is workers' compensation. Employers paid $34.1 billion in 1986 for workers' compensation benefits; approximately 16 percent more than in 1985, and almost 36 percent higher than the $25.1 billion paid in 1984. Approximately 87 percent of all workers were covered by the system in 1986, with the benefit amounts varying from state to state (2).

Though workers' compensation continues to rise dramatically, it does not necessarily follow that workplace injuries are increasing. For example, although California employers pay more per capita in workers' compensation benefits than any other state in the Union, a comparison of the seven Pacific states surveyed by the California Division of Labor Statistics and Research, in cooperation with the U.S. Bureau of Labor and Statistics, demonstrated that California registered the lowest total Occupational Safety and Health Administration (OSHA) incidence rate in the private sector, 8.8 cases per 100 full-time employees (3, 4). This illustrates that the number of actual injuries may bear little relationship to the number of workers' compensation claims filed.

A significant reason for the dramatic rise in the cost of workers' compensation benefits is the prevalence of claims that are not in fact legitimate work-related injuries. In 1988 in California, the Application of Adjudication was the first notice of injury in 47 percent of all workers' compensation cases (5). In other words, almost half the claimants sought to seek the services of an attorney before notifying their employers of the injury. It seems highly unlikely that a truly injured worker would seek the services of an attorney before notifying the employer.

2.2   Indirect Workers' Compensation Expenditures

In addition to the direct (insured) cost of workers' compensation claims, another factor that escalates employers' injury expenditures is indirect costs. Available data indicate that for every dollar paid in workers' compensation, there is another dollar in indirect (uninsured) costs for which the employer is responsible. This figure is highly conservative though indirect costs vary dramatically from company to company. Published estimates range from 1:1 to 5:1, illustrating the necessity of developing in-house data (6).

2.2.1   Calculating Indirect Cost

The ratio of direct to indirect workers' compensation costs should reflect the specific experience of the individual company. Even a ratio developed from data within the same general industry serves little useful purpose. Each business must develop its own data relative to indirect costs based on the following:

1. Cost of productivity paid for time lost by workers witnessing an accident
2. Uninsured medical costs borne by the company
3. Costs of wages for time lost by the worker, other than workers' compensation benefits
4. Extra cost of overtime necessitated by the accident
5. Cost of wages paid to supervisors for time spent on activities relative to the accident
6. Cost associated with decreased productivity by an injured worker after returning to work
7. Cost inherent in training a replacement employee
8. Cost of damages to material and equipment
9. Cost of time spent by management and clerical personnel in processing forms, communicating with the employee, insurance company, health care providers, and so on
10. Any miscellaneous costs (6)

The cost of health care in general is rising at a rate that is equally as alarming as the increase in the cost of the workers' compensation system. As the cost rises, the employer is paying an ever increasing share of the bill. In 1986 it is estimated that U.S. firms paid $109 billion for its employees' health care. Additionally, $56.3 billion is estimated to have been paid in taxes that helped finance the nation's health care system (7). The occupational health team should work with upper management to reduce expenditures in this area, thereby increasing corporate profitability.

3   CAUSES FOR INCREASES IN HEALTH CARE COSTS AND OTHER RELATED LOSSES

The major reasons underlying the tremendous increases faced by employers relative to worker health care costs include the following:

1. Increased cost of the medical benefit system
2. An increasingly litigious society
3. The complexity of workplace hazards
4. Misuse of workers' compensation statistical data

3.1 The Medical Benefits System

Increasing costs of the health care system have far exceeded the inflation rate. A survey of over 2000 large and middle-sized businesses reported a rise of 7.9 percent in medical premiums between 1986 and 1987, about twice the rate of inflation for that period. In 1988, the overall increases for group medical plan expenditures ranged between 10 and 70 percent. The majority rose between 12 and 25 percent, at a time when the inflation rate was between 4 and 5 percent (7).

The aging worker population in America plays a significant role in the increased cost of medical benefits. As longevity increases and medical technology improves, the costs of complex modalities such as organ transplants and kidney dialysis place a tremendous burden on the medical benefit system, which is funded largely by business. The cost of an aging worker population is reflected in workers' compensation costs too. Primarily, these costs are associated with older employees who develop cumulative trauma diseases that can be aggravated and frequently caused by the aging process. Additionally, the cost of providing medical benefits to retirees will continue to escalate as the "baby boom" generation ages.

Conversely, a tremendous cost that is just beginning to have a financial impact on business primarily affects younger workers—acquired immune deficiency syndrome (AIDS). The treatment regime for this disease is expensive and protracted. Because there is no known cure and the number of victims who will ultimately succumb to this deadly disease can only be estimated, the financial impact cannot be projected. There are also complex legal issues regarding AIDS that have yet to be resolved and could affect the cost that this disease would represent to corporations (8).

Owing to the rapidly escalating and thus far unchecked rise in health care costs, the business community has attempted to contain costs utilizing a number of strategies, many of which have proven largely ineffective. The use of health maintenance organizations (HMOs) and preferred provider organizations has met with varying degrees of success. Unfortunately, the utilization of such plans has often resulted in "price shifting." One example is that many HMOs with a prepaid contract to provide medical benefits report medical conditions that are nonindustrial as work-related in order to be reimbursed via the workers' compensation system. In essence, they are being paid twice.

Other obstacles to containing medical costs are the use of inappropriate and expensive diagnostic tests, unnecessary surgical intervention, and oftentimes, treatment of dubious value. Additionally, health care practitioners frequently have financial agreements with, or outright ownership of, diagnostic treatment centers, hospitals, physical therapy centers, or other treatment modalities to which they refer. Obviously, such arrangements foster the potential for a conflict of interest.

3.2 Litigation Costs

The increasingly litigious nature of society in the United States has greatly contributed to the ever escalating cost of health care. In no area are increased litigation costs more evident than in workers' compensation. In California, as of 1988, when a workers' compensation claim becomes litigated the cost of the claim increases approximately 52 percent (2). Many of these costs are related to medical/legal examinations, the duplication of diagnostic examinations, and unnecessary and often inappropriate medical treatment (9). For example, in a case involving alleged injury to several body parts, that is, back, neck, and pulmonary, the costs of forensic medical examinations frequently exceed $10,000 if the case is litigated. This does not include any medical treatment and other benefits.

3.3 The Medical Cost of an Increasingly Diverse Workplace

Scientific advancement has contributed to the increased cost relative to occupational health in the workplace owing to the complexity of workplace hazards. Not only are the numbers of chemicals and other potentially toxic materials increasing, but also combinations of such chemicals in the workplace may pose dangers about which there is relatively little data available.

Occupational disease accounts for the largest of workers' compensation costs in the United States (10, 11). The industrial hygienist, by appropriate documentation, can often play the pivotal role in determining the work relatedness of many employee illnesses. Because of the tremendous cost involved, the potential for savings is equally tremendous.

3.4 Misuse of Statistical Data

An area that affects employer health care cost but has largely gone unrecognized is the misuse of workers' compensation data. Workers' compensation statistics are routinely used in research regarding the work relatedness of occupational injury and illness. Unfortunately, much of these data are greatly distorted by a number of factors. The statistics often include fraudulent claims, claims denied by the employer, or "companion cases" that have been included after the fact when a case is adjudicated.

Because occupational health professionals are the major contributors to research related to medical issues, there is a particular responsibility on their part to ensure that the data themselves are accurate in order that valid statistics are obtained. By ensuring the accuracy of statistics relating to workers' compensation, the occupational health researcher and author will be providing more accurate information. It is these data that are often key to the regulatory and legislative processes that formulate compliance standards as well as the determination of the work relatedness of employee injury and illness.

4 ORGANIZATIONAL STRUCTURE OF A COST-EFFECTIVE LOSS CONTROL TEAM

In order to design, implement, and audit a cost-effective, integrated, loss prevention and control program, it is imperative that all members of the occupational health team and the workers' compensation manager report to the same department head. Ideally, the manager should be a member of the occupational health profession, be it industrial hygienist, occupational health clinician, safety professional, or a professional in a related field.

Numerous benefits are derived from having an occupational health professional manage the loss prevention and control program. First, the territorialization of function that can occur between occupational health and workers' compensation is minimized. Second, by being part of a holistic approach to loss prevention and control management, the team learns more about one another's roles and develops more appreciation and respect, thereby encouraging communication rather than conflict. Third, and most important, in areas where there is a great deal of overlap, the roles can be delineated. It is only by utilizing such a team approach that a loss prevention program can be presented to corporate management as part of the overall business plan and is able to document cost effectiveness and promote employee health.

4.1 How a Program Works

Noise-induced hearing loss (NIHL) continues to be one of the 10 leading causes of occupational disease in the United States (11). It can serve as an example of how utilizing a comprehensive team approach to hearing conservation and presenting meaningful data to management should essentially eliminate workers' compensation claims for true noise-induced hearing loss.

Obviously, the first step in an effective hearing conservation program is the formation of written policy and procedure. The policy should be written with the input of the entire occupational health team and the workers' compensation manager. Health professionals often forget that the workers' compensation manager has little, if any, knowledge of the medical or safety issues related to hearing loss. Similarly, occupational health professionals are many times unaware of what constitutes a "compensable" hearing loss claim. Only by working together and integrating their skills can an effective policy be developed.

Certain components of the hearing conservation program will naturally be assumed by the health professional with the appropriate technical skills to perform them. For example, the industrial hygienist will perform the environmental monitoring and an audiometric technician will conduct the audiometric testing. However, other elements of a comprehensive program such as training and interpretation in the use of hearing protection, testing to ensure proper fit of hearing protection, and taking responsibility for compliance may be overlapping functions. The responsibility regarding such functions can be delineated and a system of accountability established by utilizing a team approach with an occupational health professional as the leader.

When a properly formulated policy with specific goals and a viable course for reaching those goals are presented to management along with statistical data showing potential cost savings, the probability of acceptance and therefore success is greatly enhanced.

The commitment of management is essential to the success of the hearing conservation program; it may be necessary for the loss prevention team to educate them as to their legal responsibilities and options. It is only with the commitment of management that the necessary compliance can be achieved. A successfully implemented program would essentially eliminate all occupational hearing loss and enable the company to defend itself against future claims.

5   THE ROLE OF THE INDUSTRIAL HYGIENIST

Traditionally, the efforts of the industrial hygienist have been aimed at the prevention of work-related diseases through the recognition, evaluation, and control of health hazards. The development of safety and health policies and procedures, employee training, environmental monitoring programs, and the maintenance of health hazard evaluation data by the industrial hygienist has been instrumental in improving the quality of the work environment. However, the economic picture has so dramatically changed during the past 10 years that these roles, by necessity, must change. It is incumbent upon the industrial hygienist to understand the financial impact of these programs and the impact industrial hygiene recommendations have on business. The industrial hygienist must learn to "sell" these programs to upper management based not only on their moral and compliance consideration, but also on their economic viability.

The successful industrial hygienist must possess a host of technical skills. Chemistry, physics, engineering, acoustics, ergonomics, biology, toxicology, and so on all come into play when an industrial hygiene program is implemented. The same skills provide the industrial hygienist with an opportunity to have a tremendous impact on the loss prevention and control program of business. These opportunities interface directly with marketing, product research and development, and legal departments.

The assumption of such expanded responsibilities makes the role of the industrial hygienist an integral part of the loss prevention and control program and translates into improved support for the overall industrial hygiene effort. A partial listing of areas in which the industrial hygienist must take a leadership role follows:

1. Support new product development by assessing the potential exposure hazards associated with use of the product in the customer's plants and identifying the required protective measures. In many ways this is a classic role for an industrial hygienist, but now they function as "money making" members of the marketing effort.

2. Assist in developing the premanufacturing notifications (PMN) required by the U.S. Environmental Protection Agency under the Toxic Substance Control Act. In the European economic community a similar premanufacturing (premarketing notification) process defining the potential hazards is also required. Often the very approval of new products requires an estimate of the exposures expected within the customer's workplace. The role of the industrial hygienist in this process is essential.

3. Assess the hazards and potential liabilities inherent in implementing new processes and installing new equipment. The industrial hygienist is generally best qualified to assess the toxic exposure potential, the potential environmental impact, and the compliance cost inherent in such purchases.

4. Provide a detailed industrial hygiene survey when a facility is being shut down or sold. The study serves to document exposures, the effectiveness of engineering controls, the use of personal protective equipment, employee training, as well as the operating procedures and their impact on employee exposure. Such information is invaluable when defending against future fraudulent workers' compensation and product liability claims.

5. Monitor facility-wide chemical usage and attempt to minimize the varieties of chemicals used within the organization. This can have a positive impact on procurement cost and significantly reduce hazardous waste disposal cost.

6. Join maintenance engineering in the development and installation of engineering control measures. Direct involvement by an industrial hygienist can greatly enhance the engineering effort and can eliminate costly mistakes. The installation of local exhaust ventilation systems is an area where such involvement pays large dividends.

7. Advise corporate management of pending legislation and its potential impact on the organization. This provides management the opportunity to comment, and greatly enhances the industrial hygienist's image as a member of the management team.

The areas discussed above generally go beyond the usual industrial hygiene functions. The point being made is simple; expand the role of the industrial hygienist while maintaining those functions that are currently being performed. Through maintenance of an active employee exposure monitoring program, documentation of general working conditions within the facility (i.e., good housekeeping), subjective assessment of air quality (lack of nuisance dust, odors, etc.), and procedures requiring the use of personal protective equipment and employee training, the industrial hygienist will have a significant impact on not only workers' compensation costs but also other related costs.

In all too many cases, the workers' compensation insurance carrier or administrative agency has no knowledge of the available industrial hygiene reports that could eliminate or greatly mitigate the potential for a workers' compensation claim. For example, at one California company, there were numerous complaints and resulting claims from alleged exposure to chlorine. However, when workplace monitoring was initiated by a certified industrial hygienist, the results revealed ambient levels of chlorine well below the permissible limits. The physician then had appropriate objective data on which to make a decision other than the employee's subjective history of exposure, and physical findings that may or may not have been work related. In fact, the exposures were so low that the level of chlorine present was below that contained in potable water, thus negating even a potential claim for allergy to chlorine. These claims were successfully denied based on data provided by the industrial hygienist to the treating physician.

It cannot be overemphasized that as industrial hygienists' roles expand, they must work even more closely with all other team members and upper management. In no area is this interface more critical than between industrial hygiene and safety.

6 THE ROLE OF THE SAFETY PROFESSIONAL

The role of the safety professional encompasses much more than the traditionally perceived function of preventing accidents. Such a limited role, although still very important, is not enough to ensure the economic viability of the function necessary in today's corporate climate. Safety professionals must broaden the scope of their function in order to become members of proactive, cost-effective loss control teams.

Prevention of accidents or illnesses is obviously the most effective way to limit health care cost. Unfortunately, all workplaces have some degree of hazard, and accidents can and will happen. When a mishap does occur, the safety professionals should employ their skills to limit both the human and economic severity of the incident (12).

6.1 Accident Investigation

Investigation of industrial accidents is one of the most important aspects of an effective safety professional's role. In the event of an accident, the safety professional must employ specialized training and analytical experience to compile a report that is comprehensive and objective. There are several methods that are generally used. The American National Standard Institute, V 16-2 Standard primarily focuses on acts and conditions. Other techniques involve investigation within the framework of defects in humans, machine, media, and the management (the four Ms), or education, enforcement, and engineering (the three Es). These are referred to as the statistical methods of analyses (13). A proper investigation of an accident may provide the company's workers' compensation administrator or insurance company with information that may help reduce the cost of the claim as well as providing management with a plan for preventing future incidents.

6.1.1 Investigating Suspect Injuries

One area that traditional accident investigation does not address is the legitimacy of an employee's claim that the injury sustained was work related. This should be one of the first issues addressed in any investigation. Many times the answer will be obvious, but there are some situations that should raise questions. These include:

1. Unwitnessed injuries
2. A report of injury more than 24 hr after occurrence
3. Injuries where the mechanism of injury does not correlate with medical fact
4. Injuries alleged by employees with personnel problems, for example, absenteeism, probation, and lack of productivity
5. Injuries reported immediately prior to a layoff
6. Injuries to employees who have previously litigated suspect claims

By investigating such claims and developing a detailed history of the injury, the safety professional can provide the health care clinician with the information necessary to make an objective determination regarding the legitimacy of the injury. Thorough investigation of all accidents also serves to deter other employees from filing fictitious claims.

6.2 Documentation of Safety Policies

Proper documentation is essential if a company is to defend itself successfully against specious workers' compensation claims. Because workers' compensation in many states is subject to "liberal construction" the employee's testimony is usually given more weight than that of the employer. Therefore it is necessary for employers not only to act responsibly with regard to their employees' health and safety, but also to be able to prove it in court.

Suppose an employee alleges a long history of exposure to toxic chemicals via inhalation and, as a result, has developed occupationally related cancer. What's more, the employee claims that he or she was never provided with, or trained in the use of, respirators. Regardless of whether instruction in the use of protective equipment was received, the case would most likely be decided against the employer, at great expense, unless proper documentation could be supplied. However, if the safety manager has maintained records that documented that the employee was provided the necessary respirator and properly trained in its use and care, the claim can be defended and the cost inherent either eliminated or greatly mitigated.

Documentation of all policies and procedures is an important aspect of the expanded role of the safety professional. Accurate data must be maintained regarding all written safety programs, safety committee meetings, and employee training, in addition to required OSHA record keeping. Such documentation is necessary not only for compliance, but also as a key factor in reducing workers' compensation and related costs.

6.3 Trend Analysis

The data regarding the health and safety of the worker population maintained by safety professionals also enable them to perform trend analysis. Trend analyses allow the safety professional to pinpoint high-risk areas and the areas of highest cost relative to work-related injuries. Using data generated from the Workers' Compensation Department, programs developed in-house, and OSHA records, an appropriate data base can be developed. This information allows the safety professional to determine which department has the highest incidence of injuries, the incurred cost of injuries, the type of injuries, the shifts on which injuries most often occur, and the supervisor responsible for that shift. Only when a problem has been identified can a problem solving approach be utilized and the cost associated with it controlled.

It is incumbent upon safety professionals to make the other members of the loss control team and management aware of the data and documentation they have compiled, thus enabling them to make use of it. By playing a more active role in loss control, the safety professional can alleviate the problem that has long plagued safety managers—lack of support.

Safety professionals have long been frustrated when management has failed to show the commitment necessary to achieve proposed goals. Development of a program that shows at least cost neutrality and often cost benefit will enhance the role of the safety professional and make the safety program substantially more efficient.

7 THE ROLE OF THE OCCUPATIONAL HEALTH CLINICIAN

The drastic rise in the cost of health care has brought about a transformation of the role of the occupational health care clinician; what was once a limited technical position should emerge as a coequal member of business management. Regardless of what degrees they hold, occupational health specialists must utilize their technical and managerial skills to ensure that the employee population receives optimal health care at the lowest cost to the employer.

The treatment of minor injuries, cursory examinations, employee counseling, and medical records maintenance was the traditional province of the occupational health professional. In addition to having an excellent medical/nursing background and a comprehensive knowledge of occupational health, today's occupational health clinician must have an understanding of the rapidly escalating cost of health care to employers and its financial effect on the company. The clinician must then be zealous in implementing cost containment measures having an impact on health care, as well as maintaining the occupational health of the employee population.

There are four programs known to be cost effective in industry:

1. Preplacement health screening
2. Appropriate management and treatment of injuries
3. Employee assistance programs
4. Absence surveillance/disability management (14)

7.1 Preplacement Health Screening

Many times injuries occur because a worker is physically unsuited to the task being performed. Additionally, employment of workers who are already injured or ill has a tremendous impact on the cost of health care. Preplacement health screening performed mindful of the OSHA mandate to "provide a safe and healthful workplace" and in accordance with Federal Employment Guidelines will serve to:

1. Place applicants in jobs for which they are physically qualified
2. Place workers in jobs that are both safe and healthful for them as well as for fellow workers
3. Provide base-line medical information that can be used as a standard in the event of future alleged work-related injury or illness
4. Identify individuals with previously unrecognized or inadequately managed health problems (15–17)

A preplacement health screening policy should be part of overall company occupational health policy. Standards utilizing job-related criteria should determine the fitness of the applicant. Employment criteria should be based on widely accepted standards that have been extensively researched or developed in-house, using data generated through a company's documented findings. Care should be taken to ensure that the standards are nondiscriminatory and that they are the same for all workers performing similar tasks.

Once a preplacement health screening policy has been written and the job-related criteria have been established, the occupational health clinician must utilize a comprehensive occupational health-medical history questionnaire and preplacement health screening examination format that documents negative as well as positive findings. Preplacement health screening exams preformed by outside providers may often be cursory and nondefinitive and, in the long term, of little value. To be effective, the preplacement health screening examination must be tailored to meet the needs of occupational health rather than the general population.

Basic diagnostic testing such as blood pressure, temperature, pulse, respiration, hemoglobin determination, urinalysis, and base-line audiogram should be performed on all employees in accordance with written policy. Specific diagnostic tests should be performed as needed, utilizing job related criteria (18, 19). For example, persons working in some solvents, heavy metals, and so on should have blood chemistry panels before a determination is made as to whether they are acceptable for the particular job. In many instances, it is neither prudent nor safe for employees to begin work in an area of potential exposure if they had preexisting abnormal blood chemistry findings that could be aggravated by potential workplace exposures.

Most injuries occur within the workers' first year of employment. Utilization of appropriate preplacement health screening can reduce injuries which are an "aggravation of a preexisting injury" or "preexisting medical conditions." The appropriate screening also reduces the number of cumulative trauma claims which arise over time.

However, for the above to be accomplished, the professional occupational health clinician, working with the Human Resources Department and any outside health care providers, must design, implement, and audit the preplacement health screening process.

7.2 Management and Treatment of Work Related Injuries

Inevitably, some workers do become injured. At this point, the technical skills of the occupational health professional come into play. Physicians and occupational health nurse specialists, working under standardized procedures, can provide quality primary health care in-house (20). This greatly reduces the number of work-related injuries that are referred to outside health care providers. Numbers of "return" doctors' visits and physical therapy sessions can also be greatly reduced by providing such care in-house. Lastly, the employees' progress can be more closely monitored by providing primary medical care in-house, thus preventing complications, promoting early recovery, and reducing time lost from work, all elements of health care cost containment (21).

A key element in the medical management of injuries is the history of injury. Oftentimes, failure to elicit appropriate information and a lack of documentation will have serious negative consequences including misdiagnosis, assumption of responsibility for an injury that is not work related, and inappropriate medical care.

When it is necessary to utilize outside clinics, or refer to specialists, diagnostic centers, emergency rooms, or hospitals, it is the responsibility of the clinician to obtain the best possible care for the injured employee at the lowest cost to the employer. It should be remembered that outside health care providers lack the incentive to take detailed histories or correlate medical findings, which may protect the employer. The occupational health specialist must work with other providers to make sure that the employer's needs, as well as those of the employee, are met.

The occupational health clinician plays a major role in the development of modified work programs. On a case by case basis, as part of a disability management program, it is advantageous to the injured employee and the employer to participate in a modified work program. In most instances, not only do workers who return to modified work have a speedier recovery, but also their progress can be monitored in-house, with basic physical therapy treatment provided, dressings changed, and minor complications treated before serious consequences result. This saves the employer not only the cost of temporary disability payments but also the cost of medical care that would otherwise be provided off-site (22).

At the time of an injured employee's discharge from medical treatment, the in-house occupational health professionals should provide their own assessment and documentation of case closure with specific notations regarding any positive physical findings, need for future treatment, and so on. The in-house clinician should also ensure that any outside medical professional provides appropriate documentation of case closure. Failure to provide such documentation can, and often will, lead to an employee's later claim that the injury never completely healed, or that there was an injury to another body part which was unreported at the time of the initial injury. Appropriate documentation will provide a defense against such fraudulent allegations. In the instance of true permanent disability, the same information provided to the workers' compensation carrier will speed the process of obtaining the injured worker's permanent disability benefit without need for legal counsel.

7.3 Employee Assistance Programs

Since the mid-1970s, the cost effectiveness of Employee Assistance Programs (EAPs) has been well documented by pioneers in the field, including the U.S. Navy and the United Parcel Service. The 1980s brought an increased awareness of the dramatic human cost, as well as tremendous financial burden, that has resulted from the use of drugs and alcohol in the workplace, making EAPs even more essential. The role of the occupational health professional in the development of EAPs is crucial because of the technical and confidential issues involved.

Even though substance abuse testing is a key element of the EAP, many types of testing are fraught with controversy. Preplacement testing for alcohol and drugs generally presents no legal problem for private industry as long as it is incorporated into an EAP policy, confidentiality is maintained, and the occupational health professional audits the collection mechanism, laboratory utilization, and documentation (23). Preplacement substance abuse testing is, in and of itself, cost effective. The range of positive tests varies from 12 to 25 percent.

Because it is estimated that 30 to 50 percent of all work-related injuries are the direct result of substance abuse in the workplace, it can be assumed that a substantial number of persons who are reported as "positive" on preplacement substance abuse testing would have been otherwise injured on the job (24).

Other areas in which substance abuse testing is used include "for cause" testing, "post injury" testing, and "random" testing. These have more potential for legal liability than preplacement testing; therefore the involvement of the occupational health professional is crucial in determining appropriateness of utilization (23, 25).

In the development of a successful employee assistance program, the occupational health clinician should (1) be involved in the selection of the employee assistance counselor, (2) provide education to management as to types of employee assistance programs, (3) play a key role in writing the company employee assistance policy, and (4) be responsible for the implementation and audit of the program (26).

7.4 Disability Management

Unnecessary surgeries, unexplained medical costs, prolonged hospitalizations, overuse of in-patient hospitalization, and overuse of expensive diagnostic tests have all contributed to the dramatic increase in medical costs (27). Three-fourths of all health care costs can be attributed to less than 10 health conditions, but the proper management of these cases will reduce their cost (28).

Case management is defined as the process of organizing and mobilizing health services and resources to offer quality cost-effective care for an individual's health condition (28). It is an all-encompassing approach for development of individual plans of care for patients. In one study, 120 insurance claims were reviewed. A cost savings of $430,000 resulted from 29 of these cases, an approximate savings of $3500 per case (29). Experts in the area of disability management suggest that case management works best when targeted at clients with very specific diagnoses and when cases are referred and managed soon after the onset of the condition. It is the role of the occupational health professional to develop appropriate statistical data to determine the areas of greatest cost and then present a plan to management for controlling such costs.

An increasing number of employers and insurance companies are utilizing occupational health nurses as case managers to contain health care costs. It is the responsibility of the in-house professional to determine what methodologies of disability management are the most effective for their company, and develop the program as well as an auditable method for determining quality assurance and cost effectiveness.

7.5 Group Medical Benefits

Because of the tremendous escalation in the costs of medical benefits, it is incumbent upon the occupational health professional to become knowledgable and involved in the company's purchase of medical insurance for employees and their families. Often decisions as to benefit structure, choice of insurance carriers, and utilization of preferred provider organizations are left to either insurance brokers or risk managers who have little if any knowledge of health care. Conversely, health care clinicians have little if any knowledge regarding insurance. In order to become viable contributors in the mainstream of employee health, occupational health clinicians must assume responsibility for becoming knowledgable and involved in the health care delivery system of the corporations in which they are employed.

7.6 Health Promotion

All too often, wellness programs are seen simply as an employee benefit or are purchased as part of a package from a private firm with little thought given to quality assurance and cost effectiveness. Even though wellness programs are shown to be beneficial and cost effective, such benefits are limited to only those employees actually utilizing the program. Unfortunately, most wellness programs have no incentives to ensure compliance or, as in the case of employee fitness centers, are utilized primarily by persons who are already in good physical condition.

It is well documented that health promotion programs that have the most likelihood of being cost effective have built-in incentives (27). Because wellness programs obviously involve numerous complex health issues, the occupational health clinician plays a key role in the design, implementation, and auditing of health promotion programs.

The role of the occupational health clinician has expanded tremendously. This has come about largely because of advances in technology and economic necessity. Unfortunately, role expansion has been largely underutilized by both clinicians and the corporate structures in which they are employed. As a key member of the loss control team, the occupational health clinician must use his or her technical skills to protect and promote worker health. Management skills must be employed to meet the challenges and the burden of rising health care cost to the employer. No longer can the health care professional be simply a reactive provider of aid to the injured (30).

8   MANAGEMENT OF AN INTEGRATED LOSS PREVENTION AND CONTROL PROGRAM

For a loss prevention and control program to be effective, a method of determining priorities, setting goals, and auditing program effectiveness must be developed. The loss prevention and control team must gain an understanding of the corporate philosophy, its financial situation, and its organizational structure. Such knowledge enables the occupational health and safety professionals to work within company policy and educate senior management as to the benefits of an integrated program. A loss analysis, utilizing statistical data from all members of the loss prevention and control team, can be used to obtain management commitment, which is essential for comprehensive program implementation and success.

8.1   The Loss Analysis

A loss analysis determines the areas of most serious concern and highest cost, thus enabling the loss prevention and control team, as well as corporate management, to establish priorities and set goals. The data used in the compilation of the loss analysis should be obtained from a minimum of three years' history prior to program development. Basic data must include the following:

1. Number of new occupational injuries
2. Number of medical-only workers' compensation claims
3. Number of indemnity workers' compensation claims
4. Number of litigated claims in relationship to nonlitigated indemnity cases
5. Total number of OSHA recordable injuries
6. Incidence rate (OSHA frequency)
7. Statewide incidence rate for industries of similar nature
8. Manual adjusted workers' compensation insurance premium
9. Total incurred costs for workers' compensation (includes paid claims plus reserves)

10. Total incurred cost subcategorized as: medical only, indemnity, and expenses related to litigation
11. Payroll
12. Person-hours worked
13. Loss ratio (total incurred cost in relationship to the manual insurance premium)

The figures derived from the loss analysis also serve as base-line data against which future statistics can be compared to audit program effectiveness. The positive economic impact further enhances the necessity of a team approach to loss prevention and control.

8.2 Case Studies

The following examples provide data exemplifying the effectiveness a properly designed and implemented integrated occupational health program has on the number of injuries and the attendant workers' compensation costs.

8.2.1 Company A—Price Pfister Plumbing, Inc.

Price Pfister Plumbing, Inc. is a California-based manufacturing company which employed approximately 1300 employees in 1980. There are collective bargaining units; the workers are largely non-English speaking and the plant operates in three shifts. The company also has a relatively high attrition rate. In 1980, in response to increasing occupational health and attendant workers' compensation costs, the senior management of Price Pfister Plumbing, Inc. decided to utilize a loss analysis as a tool to develop an integrated loss prevention and control program. The Occupational Health Department was reorganized, following the guidelines outlined in the preceding section. At the same time, a decision was made to implement a more proactive safety program. Safety committees were organized at all levels, frequent unannounced safety inspections were conducted by management, and regular employee safety meetings were held. The Workers' Compensation Department also became involved in the cost containment process for the first time, managing the self-insured workers' compensation program as part of the loss prevention and control team.

The results were readily apparent (see Figure 35.1). In the year prior to the implementation of the integrated program (1980), the company had 72 OSHA recordable injuries. OSHA recordable cases are more serious injuries that require more than first aid. Work may be restricted, or time lost from the job. During the first year of the program (1981), the number of OSHA recordable cases dropped to 26. The effect of the safety program was apparent. The more than two-thirds decrease in the OSHA recordable injuries in just one year provides evidence of the results that can be achieved by the institution of a comprehensive industrial safety program.

One method of auditing the effectiveness of an Occupational Health Department

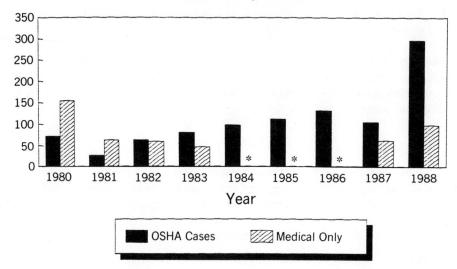

Figure 35.1. Company A: OSHA recordable injuries versus medical-only claims (Price-Pfister).

is to determine the number of less serious injuries (medical-only claims) being referred to outside health care providers. In 1980, Price Pfister Plumbing, Inc. sent 155 employees to outside providers with medical-only claims; in 1981, it referred 44 (see Figure 35.1). It is evident that the reorganization of the Occupational Health Department had an immediate impact on the costs of workers' compensation.

Because Price Pfister was self-insured for workers' compensation, the effect that an integrated program had on the attendant costs was highly visible immediately. In 1980, the company's workers' compensation cost was 75 percent of manual premium. Manual premium is the amount that it would cost to purchase insurance from an outside carrier at unadjusted (by experience modification) rates. The workers' compensation cost was reduced by one-third in the first year alone, totaling 50 percent of manual premium in 1981. This represented a cost savings to the company of several hundreds of thousands of dollars in one year. This trend continued until 1983, at which time workers' compensation costs constituted only 26 percent of manual premium (see Figure 35.2).

Price Pfister experienced a change in ownership in 1983. As a result, the company of necessity became insured for workers' compensation and costs once again began

Company "A"
Loss Ratio Data

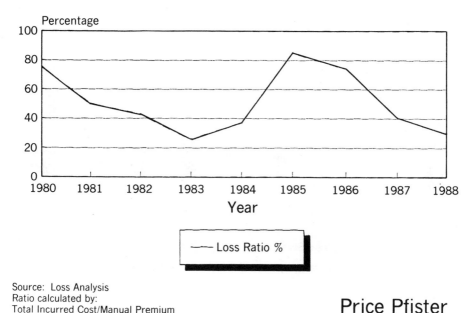

Source: Loss Analysis
Ratio calculated by:
Total Incurred Cost/Manual Premium

Price Pfister

Figure 35.2. Company A: Loss ratio data (Price-Pfister).

to rise. The cost of the workers' compensation insurance program reached a peak in 1985, totaling 85 percent of manual premium. In 1985, management utilized a consulting firm again to develop an integrated approach to contain workers' compensation costs.

In 1987, the new Price Pfister management made a decision to become self-insured. Working with their third party administrator, they were able to bring the cost of workers' compensation down to 41 percent of manual premium from 75 percent the year before. By 1988, workers' compensation costs were 30 percent of manual premium (see Figure 35.2).

The number of OSHA recordable injuries increased threefold in 1988 at Price Pfister, as opposed to 1987, even as the total costs of workers' compensation decreased. Similarly, the incidence rate increased from 7.3 to 19.8 during the same time frame. At first glance, this seems dichotomous. Such cost containment was possible only because of the effectiveness of the occupational health and workers' compensation management. The Occupational Health Department, staffed with highly trained occupational health nursing specialists working under protocol, that is, standardized procedures, and utilizing the major elements of the occupational programs known to be cost effective (Sections 7.1 to 7.3), were responsible for the

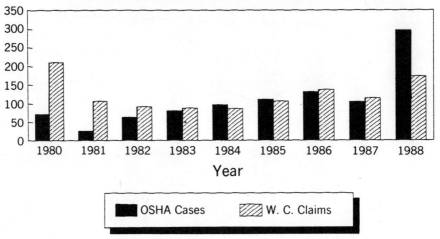

Figure 35.3. Company A: OSHA recordable injuries versus workers' compensation claims (Price-Pfister).

relatively few outside referrals of injured workers who needed more than "first aid" medical treatment (see Figure 35.3). For example, in 1988 there were 296 OSHA recordable cases with only 173 injured workers being referred to physicians for outside medical care. Treatment in-house not only reduced the amount spent for medical treatment, but also prevented both lost productivity due to time spent away from work and other indirect costs.

The dramatic rise in the number of work-related injuries at Price Pfister in 1988 prompted management to recommit itself to a vigilant safety program.

The case of Price Pfister Pluming, Inc. indicates what can be accomplished at any company if an integrated approach to loss prevention and control is undertaken. It also indicates how the loss analysis can be used to make decisions regarding workers' compensation funding and the implementation of specific programs.

8.8.2 Company B

Company B chose not to be identified but is also a California-based manufacturing operation. At the time of the study it employed approximately 300 workers who

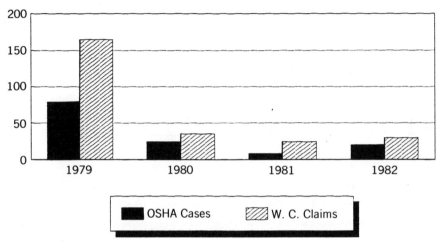

Company "B"
OSHA Recordable Cases vs.
Workers' Compensation Claims

Source:
1. OSHA 200 Summaries
2. Self Insurer's Annual Report

Figure 35.4. Company B: OSHA recordable cases versus workers' compensation claims.

were largely non-English speaking. The company had a collective bargaining unit. Production operated in two shifts with maintenance on a third shift. The company was self-insured, utilizing a third-party administrative agency for claims management.

In 1979, workers' compensation costs were at 96 percent of manual insurance premium. As a result, corporate management made a decision, after the data were analyzed, to construct an occupational health unit, employ an occupational health nursing specialist and a safety professional, and integrate workers' compensation management as part of a comprehensive program. The reductions in all areas were significant.

In 1979, there were only 79 OSHA recordable cases, whereas 165 workers were referred to physicians for medical treatment. Utilization of an occupational health nurse in-house decreased the total OSHA recordable cases to 24 and the total number of doctor's cases to 35 by 1980 (see Figure 35.4). Implementation of an effective safety program reduced the incidence rate from 22.8 to 7.44 percent, well below statewide incidence rates for like companies during the same time frame (Figure 35.5). The workers' compensation costs fell from $400,000 in 1979 to approximately $44,900 in 1980 (Figure 35.6). The total number of doctor's cases,

Company "B"
OSHA Incidence Rates

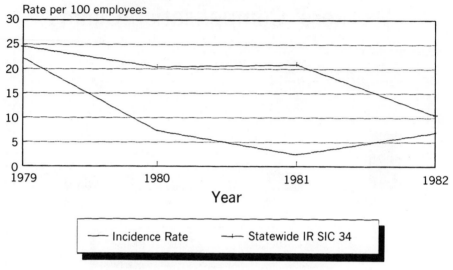

Source:
1. OSHA 200 Summaries
2. CA Dept. Industrial Relations

Figure 35.5. Company B: OSHA incidence rates.

as well as OSHA recordable injuries, continued to decrease through 1982, the last year for which data were available.

Company B demonstrates the dramatic savings that can be realized by even a small company if the policies and programs advocated are implemented. A large employee population is not necessary for such methods to be economically feasible. The savings of approximately $345,000 in workers' compensation costs and the reduction in fees paid to outside health care providers in the first year alone illustrate how small companies can recognize large savings.

8.2.3   Company C—Anaco, Inc.

ANACO, Inc. is based in Southern California and employs approximately 500 workers; the predominantly Spanish-speaking work force is not represented by a bargaining unit. Primarily a foundry operation, ANACO operates in three shifts. Multiple locations create additional challenges for management of occupational health and safety programs.

Utilizing information gathered from a loss analysis, the company decided in 1984 to employ the services of an occupational health nurse and build a separate Oc-

Company "B"
Manual Premium/Total Incurred Cost

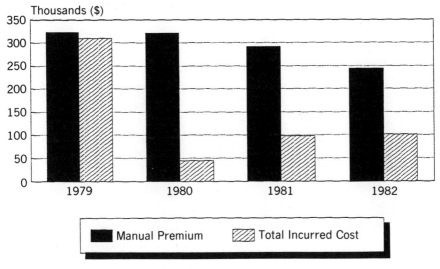

Source:
1. Loss Analysis
2. Self Insurer's Annual Report

Figure 35.6. Company B: Manual premium versus total incurred costs.

cupational Health Department. Major emphasis was placed on the control of injuries, and some basic safety practices were instituted by the occupational health nurse.

In the year prior to the implementation of an in-house occupational health program there were 101 "medical-only" claims. There were reduced to 55 in 1985. Moreover, the total number of OSHA recordable injuries decreased from 139 in 1984 to 86 in 1985 (see Figure 35.7). Though the company did not have an in-house safety professional, by utilizing the services of a safety consultant and with the continued involvement of the occupational health nurse, their incidence rate decreased in the first year of the program from 46.4 to 23.73, and it has escalated only slightly since (see Figure 35.8).

8.2.4 Company D

Company D is a large, self-insured company located in northern California. The company is seasonal in nature and operates in three shifts. There is a collective bargaining unit and a worker population that is largely non-English speaking. Because of the unusually high number of OSHA recordable injuries and the fact

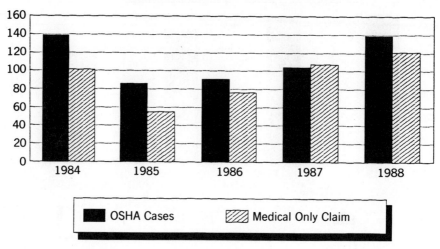

Figure 35.7. Company C: OSHA recordable cases versus medical-only claims (ANACO).

that approximately 20 percent of all workers' compensation costs were due to alleged chemical exposure claims, a limited industrial hygiene survey was undertaken in 1989.

The site selected was Plant 7. It is one of the company's largest facilities and has a large number of open chemical exposure claims. Otherwise it is not significantly different from the other operations. Industrial hygiene monitoring was conducted for chlorine and other potentially toxic substances. Special attention was directed to assessing the ambient concentrations of chlorine owing to the large number of workers' compensation claims that alleged excessive exposure. Ambient levels were consistently below 0.2 ppm, well below the threshold limit value of 1.0 ppm.

The data compiled by the industrial hygienist was provided to the physician specialist examining the employees alleging hazardous exposure to chlorine. Based on his findings, the decision was made to deny the claims. Subsequent to the industrial hygiene monitoring, all claims of chlorine exposure have been successfully defended. This incident points out one important facet of the role of the industrial hygienist in reducing workers' compensation costs.

As an integral part of the industrial hygiene monitoring program, employee

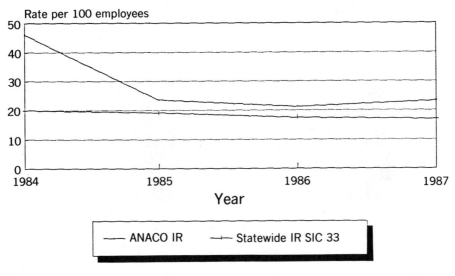

Company "C"
OSHA Incidence Rates

Figure 35.8. Company C: OSHA incidence rates (ANACO).

awareness meetings were conducted by the industrial hygienist during which the workers were shown actual monitoring results and encouraged to ask questions. The consultant also made specific recommendations regarding the safety program and, as a result, the total safety program received greater emphasis.

The results were soon apparent. In 1988, Plant 7 had 373 industrial injuries and all other facilities suffered a total of 412 (Figure 35.9). During 1989, the year of the industrial hygiene and safety survey, injuries at Plant 7 decreased 25 percent, numbering only 281, whereas the total number injuries in the other plants actually increased slightly. The person-hours worked increased in the other facilities by 5 percent in 1989 and the increase in the number of injuries was proportional. But at Plant 7 the number of person-hours worked increased 11 percent, further illustrating the effectiveness of the implementation of even a rudimentary safety and industrial hygiene program (Figure 35.10).

An integrated loss prevention and control policy should be part of the overall business plan. Management must be committed to the program, periodic audits must be conducted, and the progress of the program must be statistically evaluated to determine overall effect on the economic health of the company. The loss analysis enables the loss prevention and control team to provide objective evidence as to

Company "D"
OSHA Recordable Injuries following
Implementation Safety Program Plant 7

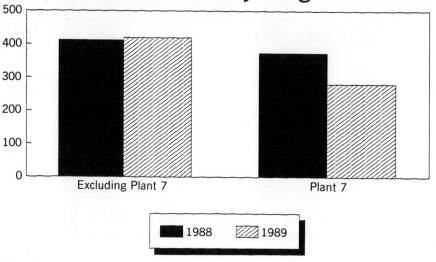

Source:
OSHA 200 Summaries

Figure 35.9. Company D: OSHA recordable injuries following implementation of safety program, Plant 7.

the cost effectiveness of the program. It is only when the occupational health team becomes responsible for developing such an approach to occupational health and safety that they will be viewed as significant contributors to the company's well-being. Obviously, such a perception is absolutely essential for promotion of employee health and corporate profitability in today's society.

9 SUMMARY

The ballooning costs of occupational health care generally, and workers' compensation specifically, challenge occupational health and safety professionals to accept responsibility for preventing and controlling such expenditures. This approach is essential in order to have a positive impact on the overall financial well being of business and promote optimum worker health. The areas posing the most serious current financial and human liability can be identified by the utilization of a loss analysis. Once problem areas are identified, policies and procedures designed to rectify them can be implemented and future problems can be averted.

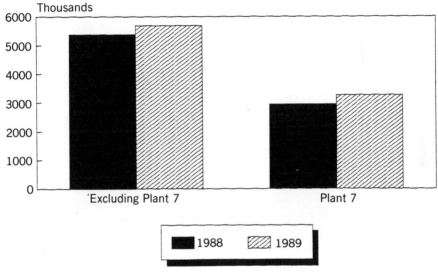

Company "D"
Manhours Data

Source:
Self Insurer's Annual Report

Figure 35.10. Company D: Man-hours data.

Policies developed by the loss prevention and control team should deal both with challenges that currently afflict the company, and those which provide proactive approach to health and safety. Policies should be designed that are broad in scope and aimed at controlling and preventing losses. Limited programs tend to be reactive and less effective. A comprehensive loss prevention policy requires a team approach. Integration of technical skills with a management perspective allows for an all-encompassing approach.

Industrial hygienists bring to the team a set of skills that enables them to monitor and assist in maintaining the quality of the environment of the workplace. Additionally, they can define and document potential hazards in order to prevent losses. Safety professionals must establish practices that are designed to prevent injuries and illness, and that should also provide information, training, and documentation that can reduce the cost of accidents both real and alleged. Occupational health clinicians possess a host of abilities that allow them to reduce the cost of employee health care while promoting employee health. Preplacement health screening, management of injuries, employee assistance programs, and disability management are areas where the knowledge that occupational health care professionals possess can affect the cost of health care dramatically.

The effectiveness of an integrated loss prevention and control program is determined primarily by the occupational health and safety team members. They must be willing to expand their knowledge, interface with other staff, and be accountable in order to meet the challenges presented by a new and complex set of problems relating to loss prevention and control. They must be able to demonstrate the positive financial impact of their roles and they must make their employers aware of the potential benefits they offer in expanded management positions.

REFERENCES

1. W. Nelson, "Workers' Compensation Coverage, Benefits and Costs, 1986," *Social Security Bull.*, **52**(3) (March 1989).
2. Staff, *California Workers' Compensation Institute Bulletin 89-6*, May 1, 1989.
3. Bureau of Labor Statistics, *Occupational Injury and Illness in the U.S. by Industry—1987*, Report No. 2259, 1989.
4. Staff, *California Workers' Compensation Institute Bulletin 89-16*, August 22, 1989.
5. National Safety Council, "Workers' Compensation," in *The Accident Prevention Manual for Industrial Operations: Administration and Programs*, 9th ed., Chicago, IL, 1988.
6. D. H. Chenoweth, *Health Care Cost Management: Strategies for Employers*, Benchmark Press, Indianapolis, IN, 1988.
7. J. Fielding, "Corporate Health Cost Management," *Occup. Med. State of the Art Rev.*, **4**(1), 121–143 (1989).
8. G. Shore, "Workers' Compensation and AIDS: Adaptation to New Occupational Diseases," *Occup. Med. State of the Art Rev.*, **4**(Special Issues 1989), 95–102 (1989).
9. S. Talo, N. Hendler, and J. Brodie. "Effects of Active and Completed Litigation on Treatment Results: Workers' Compensation Patients Compared with Other Litigation Patients," *J. Occup. Med.*, **31**(3), 265–269 (1989).
10. Division of Labor Statistics and Research, *Occupational Disease in California, 1987*, California Department of Industrial Relations, San Francisco, CA, 1989.
11. C. Lundeen, "Factors Affecting Workers' Compensation Claims Activity," *J. Occup. Med.*, **31**(8) 653–656 (1989).
12. J. V. Grimaldi, "Safety and Safety Management," in W. N. Rom, Ed., *Environmental and Occupational Medicine*. Little and Brown, Boston, 1983, pp. 957–969.
13. National Safety Council, "Accident Investigation, Analysis and Cost," in *The Accident Prevention Manual for Industrial Operations: Administration and Programs*, 9th ed., Chicago, IL, 1988.
14. P. Jacobs and A. Chovil, "Economic Evaluation of Corporate Medical Programs," *J. Occup. Med.*, **25**(4), 273–278 (1983).
15. Industrial Relations, *Title 8 California Code of Regulations*, Chapter 3, 1989.
16. Bureau of Business Practice, Inc., *Fair Employment Practices Guidelines* No. 198(1), 1981.
17. C. Freeman, "Importance of Pre-employment Physicals," *Occup. Health Nursing*, 35–37 (May 1983).

18. Canadian Public Health Association, *Norms for Grip Strength*, "Canada: Fitness and Amateur Sport," 1977.

19. Joint National Committee, National Heart, Lung, & Blood Institute, Bethesda, Md., 1984. *Detection, Evaluation and Treatment of High Blood Pressure*, 1984.

20. S. Boydstun, "The Policy and Procedure Manual: Essential Component of an Employee Health Unit," *Occup. Health Nursing*, 334–337 (July 1985).

21. H. A. Zal, "The OHNs Influence on Employee Attitude and Ability to Return to Work," *Occup. Health Nursing J.*, 600–602 (December 1985).

22. California Workers' Compensation Institute, "Vocational Rehabilitation Effectiveness in California," *CWCI Res. Notes*, (December 1989).

23. R. Engleking, "Employee Drug Screening," *AAOHN J.*, **34**(9), 417–420 (1986).

24. Bureau of Business Practice, Inc. "Drugs in the Work Place: Solutions for Business and Industry," *Employee Relations and Human Resources Bulletin* (Report No. 1540(3), March 21, 1983.

25. *EAP Digest*, **9**, 6 (Sept./Oct. 1989).

26. J. Nadolski and C. Sandonato, "Evaluation of an Employee Assistance Program," *J. Occup. Med.*, **29**(1), 32–37 (1987).

27. B. J. Spain, "Evaluation of an Occupational Health Cost Containment Program," *Occup. Health Nursing*, 328–333 (July 1985).

28. W. E. Hembree, "Getting Involved: Employers with Case Managers," *Business and Health* (February 8–14, 1985).

29. K. C. Brown, "Containing Health Care Costs: The Occupational Health Nurse as Case Manager," *AAOHN J.*, **37**(3), 141–142 (1989).

30. T. L. Giodotti and J. W. F. Cowell, "The Occupational Health Care System: An Overview," *Occup. Med.: State of the Art Rev.*, **4**(1), 153–169 (1989).

CHAPTER THIRTY-SIX

Lighting for Seeing and Health

Richard L. Vincent

1 INTRODUCTION

Nearly 2000 years ago it was stated, "The eye is the lamp to the body . . ." (1). Today a holistic approach to the understanding of the role of light continues. The objectives of this chapter are to relate the current status of our knowledge of how the absorption of light stimulates both seeing and health and thus influences human performance.

Most of life's activity is guided by the perception of visual cues that are being processed by the thousands and millions during the waking hours. Many of these cues are unconsciously handled and result in involuntary activity. For example, the unconscious detection by our peripheral vision of sudden movement may cause us to move out of the way of a bike or a bus while crossing the street. All of this can happen in a split second. A much smaller number of visual cues involve conscious processing in carrying out productive work. These are concerned with critical details that must be handled successfully for production and economic returns. There are a multitude of details in industrial processes that are critical for the production of good products; there are a tremendous number of symbols to be recognized and individually processed in the office. The illuminating engineer and lighting designer call each of these critical details the visual task. Each detail involves not only its own configuration but the background against which it is seen. The contrast of the detail with its background determines its ability to be seen.

Patty's Industrial Hygiene and Toxicology, Fourth Edition, Volume 1, Part B, Edited by George D. Clayton and Florence E. Clayton.
ISBN 0-471-50196-4 © 1991 John Wiley & Sons, Inc.

The total ability to see the details depends on (1) the characteristics of the visual task and (2) the characteristics of the observer. The characteristics of the task depend on the following:

1. The configuration of the detail
2. The narrowest size of the configurated detail or gap in the detail
3. The contrast of the detail with its background
4. The time of exposure of the detail to the pause of the eye or the glance of the observer
5. The luminance (or brightness) of the background to which the eyes become adapted
6. The specularity of the detail and its background

The characteristics of the observer depend on another set of factors:

1. The sensitivity of the individual visual system to
   a. Size
   b. Contrast
   c. Time
   d. Blur
2. Transient adaptation
3. Glare
   a. Disability
   b. Discomfort
4. Age
5. Motivational factors
6. Physiological factors

The ability to maintain a proper luminous input for biological health depends on understanding the interaction and role of light for:

1. Tissues (i.e., of the eye and skin)
2. Circadian rhythms (i.e., the daily regulation of body functions based on a light/dark cycle)
3. Hormone production and secretion

These biological responses are based on:

1. Irradiance
2. Wavelength
3. Duration
4. Time of exposure

2 THE LUMINOUS ENVIRONMENT

We have been provided with a luminous environment, which ranges from the sky, to the land to the sea—all of these natural elements send light to the eyes of the observer. Even on the darkest night, light is radiating from all elements of the field of view. Prior to the invention of electrically generated light nearly 100 years ago, most productive work ceased at sunset. Since then productive activity has steadily increased until today we live in a round-the-clock 24-hour society. Questions being raised for the design of man-made interiors today include: How should the designer proportion the light? For optimal seeing of critical details? For motivational values in the creative and energetic pursuit of tasks? For working longer periods without fatigue? For maintaining an optimum for health and safety?

Research is being carried out to find the answers in terms of the relationship of light to the characteristics of the task and the characteristics of the observers, to yield ultimately the optimal visual environment (2). Although much is known for us to begin providing an appropriate environment for quite accurate seeing, further work is needed to validate and predict the influence of lighting on overall task performance (3). In addition to the role of lighting for seeing, new frontiers are being forged to understand our need for light to regulate body rhythms, moods, and health. This is increasingly important for worker productivity and safety as we live and work in a competitive, global, 24-hour society (4).

3 LIGHTING TERMS AND UNITS

There are many terms and units used in illuminating engineering and lighting design. However, for this discussion it is necessary only to have a clear understanding of certain basic ones (3, 5).

Adaptation: The process that takes place as the visual system adjusts itself to the amount of color of light in the visual field. The term is also used, usually qualified, to denote the final state of the visual system when it has become adapted to a very low luminance.

Ambient Lighting: General lighting for an interior space.

Arousal: The state of alertness of the human organism.

Candela (cd): The System Internationale (SI) unit of luminous intensity, equal to 1 lumen/steradian.

Candlepower: A luminous intensity expressed in candelas. An ordinary candle about 1 in. in diameter produces about 1 candela in a horizontal direction.

Circadian Rhythms: 24-hour oscillations of human physiological function that are normally reset (entrained) daily by the natural light–dark period.

Color Rendering: A general expression for the appearance of surface colors when illuminated from a given source compared, consciously or unconsciously, with their appearance under light from some reference source.

Color Rendering Index: A measure of the degree of color shift objects undergo when illuminated by a light source as compared with the color of those same objects when illuminated by a reference source of comparable color temperature (See CIE Publication 13.2).

Contrast: A term used *subjectively* and *objectively*. *Subjectively*, it describes the difference in appearance of two parts of a visual field seen simultaneously or successively. The difference may be one of brightness or color or both. *Objectively*, the term expresses the luminance difference between an object and its immediate background. This luminance difference may be expressed in different ways depending on the relationship of the object to the background.

For periodic stimuli, such as gratings, contrast is defined as

$$\text{contrast} = \frac{L_{max} - L_{min}}{L_{max} + L_{min}}$$

where L_{max} = maximum luminance of the grating
L_{min} = minimum luminance of the grating

For small objects on large uniform backgrounds contrast is defined as

$$\text{contrast} = \left| \frac{L_t - L_b}{L_b} \right|$$

Sometimes for small luminance increments on a large background,

$$\text{contrast} = \frac{L_t}{L_b}$$

where L_t = luminance of the object
L_b = luminance of the background

Contrast Rendering Factor (CRF): The ratio of visual task contrast under a given lighting environment to its contrast under reference lighting conditions, that is, perfectly diffuse and unpolarized lighting by CIE standard illuminant A.

Critical Flicker Frequency (CFF): The frequency of intermittent stimulation of the eye at which flicker disappears.

Diffuse Reflection: The process by which incident flux is redirected over a range of angles.

Direct Lighting: Where all the light produced by a luminaire is focused downward.

Direct/Indirect Lighting: Where a proportion of the lighting produced by the luminaire is aimed upward toward the ceiling and the remaining proportion is aimed downward toward the floor.

Disability Glare: Glare that reduces visual performance and visibility and may be accompanied by discomfort.

Discomfort Glare: Glare that causes visual discomfort but does not necessarily interfere with visual performance or visibility.

Flicker: Visible oscillation in luminous flux.

Footcandle (fc): A non-System Internationale (SI) unit of illuminance having the same value as the lumen per square foot. One footcandle = 10.76 lumens/m^2.

Footlambert (fL): A non-System Internationale (SI) unit of luminance equal to 1/π candela/ft^2, or to the uniform luminance of a perfectly diffusing surface emitting or reflecting light at a rate of 1 lumen/ft^2. The brightness in footlamberts is equal to the illuminance on a surface multiplied by its reflectance. Note: This term, although widely used in the literature, is being deprecated in current use.

Glare: The sensation produced by luminances within the visual field that are sufficiently greater than the luminances to which the eyes are adapted to cause annoyance, discomfort, or loss in visual performance and visibility. Types of glare related to lighting include discomfort, disability, and reflected.

Illuminance: The luminous flux density at a surface, that is, the luminous flux incident per unit area.

Illumination: The act of illuminating or state of being illuminated.

Infrared Radiation: For practical purposes, any radiant energy within the wavelength range 770 to 10^6 nm. This radiation is arbitrarily divided as follow: near (short wavelength) infrared, 770 to 1400 nm; intermediate infrared, 1400 to 5000 nm; far (long wavelength) infrared, 5000 to 1,000,000 nm.

Irradiance: The radiant flux density at a surface, that is, the radiant flux incident per unit area of the surface.

Lamp: Any electric light-producing device, also interchangeably used with light source.

Light: Radiant energy that is capable of exciting the retina and producing a visual sensation. The visible portion of the electromagnetic spectrum extends from about 380 to 770 nm.

Lumen (lm): The System Internationale (SI) unit of luminous flux, used in describing a quantity of light emitted by a source or received by a surface. A small source which has a uniform luminous intensity of 1 candela emits a total of 4π lumens in all directions and emits 1 lumen within a unit solid angle.

Luminaire: A lighting *fixture* that controls the distribution of light given by a lamp or lamps and that includes all the components necessary for fixing and protecting the lamps and for connecting them to the supply circuit.

Luminance: The physical measure of the stimulus that produces the sensation of brightness; measured by the luminous intensity of the light emitted or reflected in a given direction from a surface element, divided by the area of the element in the same direction. The System Internationale (SI) unit of luminance is the candela/square meter.

Luminance Contrast: (see *objective* use of *contrast*).

Luminance Ratios: The ratio between the luminances (photometric brightnesses) of any two areas in the visual field.

Luminous Efficacy: The quotient of the total luminous flux emitted by a lamp and total lamp power input. It is expressed in lumens per watt, which is analogous to miles per gallon of gas for a car.

Luminous Flux: The light emitted by a source or received on a surface. The quantity is derived from radiant flux by evaluating the radiation in accordance with the spectral sensitivity of the standard eye as described by the CIE Standard Photometric Observer.

Luminous Intensity (Candlepower) Distribution: The distribution of the luminous intensity of a lamp or luminaire in all spatial directions. Luminous intensity distributions are usually shown in the form of a polar diagram or a table for a single vertical plane, in terms of candelas per thousand lumens of lamp luminous flux.

Lux (lx): The System Internationale (SI) unit of illuminance; 1 lux equals 1 lumen/m^2. One footcandle $\times$ 10.76 = 1 lux.

Mesopic Vision: Vision mediated by both rod and cone photoreceptors. It is generally associated with adaptation to luminances in the range 3.4 to 0.034 candelas/m^2.

Metamers: Two colors that are indistinguishable under a given light source but that have different spectral compositions.

Optical Radiation: That part of the electromagnetic spectrum from 100 nm to 1 mm.

Photopic Vision: Vision mediated essentially or exclusively by the cone photoreceptors. It is generally associated with adaptation to luminance of at least 3.4 candelas/m^2.

Photoreceptors: The sensors in the retina of the eye that convert light to electrical signals. There are two types, rods and cones, the classification being made on the basis of their shape, range of operation, and photochemical constituents.

Purity: A measure of the proportions of the amounts of the monochromatic and specified achromatic light stimuli that, when adequately mixed, match the color stimulus.

Radiance: At a point on a surface, the quotient of the radiant intensity emitted from an element of the surface in a given direction by the area of the element in the same direction.

Radiant Efficiency: The ratio of the radiant flux to the power consumed.

Radiant Exposure: At a point on a surface, the product of the irradiance and its duration.

Radiant Flux: The power emitted, transferred, or received as radiation.

Radiant Intensity: Of a source in a given direction; the quotient of the radiant flux emitted in a narrow cone containing the direction, by the solid angle of the cone.

Reflectance: The ratio of the luminous flux reflected from a surface to the luminous flux incident on it. Except for mat surfaces, reflectance depends on how the surface is illuminated, especially the direction of the incident light and its spectral distribution.

Reflected Glare: Bright luminous elements being reflected from a glossy surface toward the eyes, causing discomfort or disability (loss of visibility) for horizontal surfaces (i.e., a glossy desk top) or vertical surfaces (i.e., video display terminal screens).

Retinal Illumination: The luminous flux falling on the retina of the human eye. Retinal illumination is usually calculated from the luminance of the object and the pupil area of the viewer.

Scotopic Vision: Vision mediated essentially or exclusively by the rod photoreceptors. It is generally associated with adaptation to a luminance below about 0.034 candelas/m^2.

Solid Angle: A measure of that portion of space about a point bounded by a conic surface whose vertex is at the point. It can be measured by the ratio of intercepted surface area of the sphere centered on that point to the square of the sphere's radius. It is expressed in steradians.

Specular Reflection: Reflection without diffusion, in accordance with the laws of optical reflection, as from a mirror.

Steradian (sr): A solid angle subtending an area on the surface of sphere equal to the square of the sphere radius.

Stroboscopic Effect: An illusion caused by oscillation in luminous flux that makes a moving object appear as stationary or as moving in a manner different from that in which it is truly moving.

Task/Ambient Lighting: A nonuniform lighting strategy that lights the general working space with a lower level of light for circulation. The work station lighting is supplemented with a localized task light (either furniture integrated or movable). This provides the light level needed to perform the visual task (i.e., data entry from a paper task to a computer).

Task Performance: The performance of a complete task, expressed in measures of output such as speed and/or accuracy.

Threshold: The value of a variable of a physical stimulus that permits the stimulus to be seen a specific percentage of the time or at a specific accuracy level. In many psychophysical experiments, thresholds are presented in terms of accurate perception 50 percent of the time.

Ultraviolet Radiation: For practical purposes, any radiant energy within the wavelength range 100 to 380 nm. For convenience in describing biological effects of certain wavelength bands this range is further broken into:

UV-C	100–280 nm	Germicidal
UV-B	280–315 nm	Actinic (tanning)
UV-A	315–380 nm	Ocular effects (black light)

Veiling Reflections: Specular reflections superimposed on diffuse reflections from an object that partially or totally obscure the details to be seen by reducing the contrast. This effect sometimes is called "reflected glare."

Visibility: The quality or state of being perceivable by the eye. In many outdoor applications, visibility is defined in terms of the distance at which an object can be just perceived by the eye. In indoor applications it usually is defined in terms of the contrast or size of a standard test object, observed under standardized viewing conditions, having the same threshold as the given object.

Visual Acuity: A measure of the ability to discriminate fine details. Quantitatively it can be expressed as the reciprocal of the angular separation, in minutes of arc, between two lines or points that are just separable by the eye.

Visual Angle: The angle subtended by an object or detail at the point of observation. It usually is measured in minutes of arc.

Visual Field: The full extent in space of what can be seen when looking in a given direction.

Visual Performance: The quantitative assessment of the performance of a task, taking into consideration speed and accuracy.

Visual Task: Conventionally designates the details and objects that must be seen for the performance of a given activity, including the immediate background of the details or objects.

4   VISUAL PERFORMANCE

As a designer evaluates how to light a space, there are two goals that relate to the performance of work in the space: (1) the overall spatial order and architectural feeling to be reinforced and (2) the quantity and quality of light necessary to perform the visual work. For now we will look at the extent to which the quality and quantity of lighting can change performance of the task. Visual performance is the measurement of the observer response to the characteristics of the task. It may be the response to the visibility of the details of the task or the overall response to carrying out a given sequence of acts involving the visibility of the details.

4.1   General Visual Response To Different Task Parameters

It is helpful to know the general response of the visual system to each of the elemental characteristics of a task, its size, contrast, and time of seeing, and the luminance of the background of the detail. For more than a century, eye specialists have measured the response of human visual capability by having a person read the smallest line of letters on an "eye chart" bearing a graduated series of lines of different sized letters. A series of test objects that are symbolic of the details found in practical workplace tasks has evolved into the configurations shown in Figure 36.1. Each series gives a different curve of response, but the curves for each are largely parallel, showing the same general nature of the response (7). The ability to see decreasing size with changing illumination for black Landolt rings on white background (high contrast) is represented in Figure 36.2. The ability to see decreasing contrast with increasing illumination for a disk with visual angle of 4' subtense and 0.1-sec exposure is shown in Figure 36.3. The ability to see with

Figure 36.1. Commonly used test objects for determining size discrimination and visual acuity: Snellen E, Landolt ring, parallel bars, and disk (6).

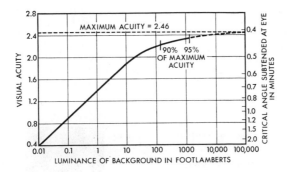

Figure 36.2. The variation in visual acuity and visual size with background luminance for a black object (Landolt ring) on a white background (8).

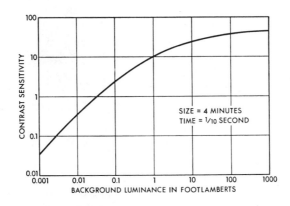

Figure 36.3. Variation of contrast sensitivity with background luminance (9).

decreasing amounts of time required as illumination increases can be understood from Figure 36.4. A further illustration of how these elemental characteristics influence the seeing ability with changing illumination is given by the eye charts in Figure 36.5, illustrating the effects of size, contrast, and reflectance of background. Of course reflectance of background makes a marked difference in seeing

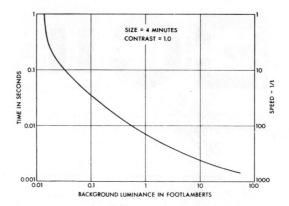

Figure 36.4. The effect of background luminance on the time required to see a high-contrast test object; right-hand scale represents speed of vision (9).

Figure 36.5. Series of eye charts showing effects of size, contrast, and reflectance. Black on white required 1 fc; gray on white, 120 fc; black or dark gray 520 fc (10).

the detail because the sensitivity of the observer depends on the adaptation luminance of the background (11), as further illustrated in Figure 36.6. Finally, Fry (12) has pointed out the need to recognize that the image received and processed by the retina of the eye is strongly influenced by blur. For the most part blur of the retinal image can be minimized by using optical aids where necessary, yet even when eyesight is optimal (with or without such aids) a certain amount of residual blur can occur. We need a better understanding of blur related to chromatic aberration, the ability of a person to focus, and the limits blur imposes on visibility. Further research is needed to understand the role of blur to quantify how the spectral distribution of light sources affects blur and visual performance (12).

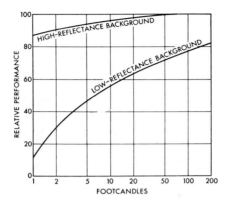

Figure 36.6. The effect of level of illumination on relative visual performance when the task involves high contrast (upper curve) and low contrast (lower curve) (11).

4.2 Effect of Surroundings

The ability to see size and contrast is influenced not only by the luminance of the background surrounding the detail but by the differing luminance and size of immediate surroundings around the background of the task, as illustrated in Figures 36.7 to 36.9. These effects appear to come from static viewing of the details involved, but it has been reasoned by some that the eyes could be stealing glances

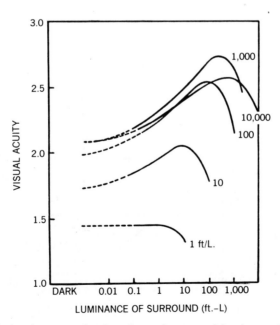

Figure 36.7. Relation between visual acuity and surround luminance for various central field luminances (120° surround to central field) (13).

to the surroundings, and the effect of sensitivity losses because of the differing surround luminance could be due to transient adaptation, discussed below.

4.3 Simulating Practical Tasks

Generally the productivity of practical tasks in commerce and industry is related to the speed and accuracy of carrying out those tasks. Correlations between productive performance of tasks at the work site have proved difficult to analyze because there are a number of factors that cannot be controlled. Thus laboratory tasks simulating practical tasks have been designed to measure the speed of response to the processing of test objects that would approximate the size and contrast of the range of details found in the field. Such a series, under the auspices of the Committee on Industrial Lighting of the National Research Council (1920s), appears in Figures 36.10 to 36.12. Unfortunately, the study did not cover values of illumination beyond 100 footcandles (fc).

A further practical simulation of a task vitally used in industry was the reading of a steel vernier rule to the nearest 1/1000 in. (13). In carrying out the simulation,

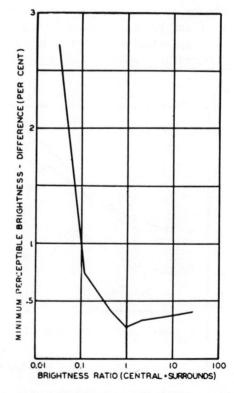

Figure 36.8. Relation between minimum perceptible brightness difference and the ratio of the brightness of the task to the brightness of the surrounding field (14).

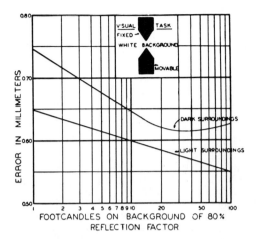

Figure 36.9. The effect of surroundings and illumination on the precision of performing a mechanical task guided by vision (15).

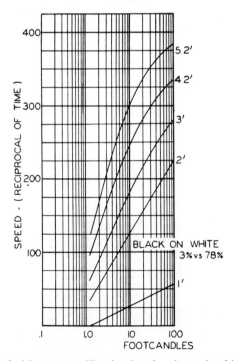

Figure 36.10. Speed of vision versus illumination for the task of identifying the position of a Landolt ring. Data from Ferree and Rand (16), under the auspices of the National Research Council.

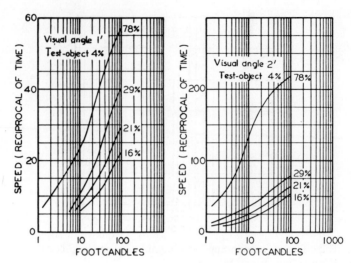

Figure 36.11.   Speed of vision versus illumination for task involving varying contrasts (16).

it was found that the speed of making the complete reading of the tenth-inch numbers, the quarter divisions within the tenth, and the exact position of the vernier divided into 25 parts varied with the luminance of the highlight of the background of the rule, against which the black divisions were seen in bold relief (Figure 36.13). The speed of reading the vernier with luminance is shown in Figure 36.14. Again the study did not carry values of luminance beyond 200 footlamberts (fL).

In the 1970s further work was provided on simulated industrial tasks. The threading of needles (Figure 36.15) represented fine detail work in industry; the reading of date digits on Lincoln pennies (Figure 36.16) represented metalworking details;

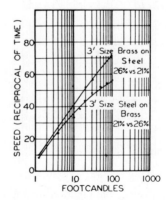

Figure 36.12.   Discrimination of a brass test object on a steel background and a steel test object on a brass background (17).

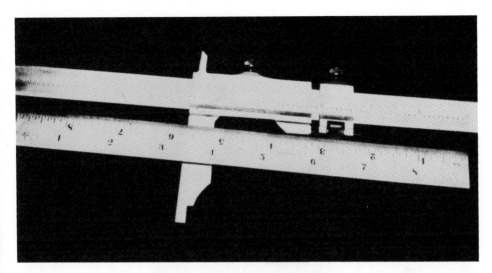

Figure 36.13. A steel vernier rule used in the practical simulation of a task down to the nearest 1/1000 in.

and the finding of symbols on a circuit board (Figure 36.17) illustrated intricate electronic assembly. Some of the results are given in Figures 36.18 through 36.20. When one considers labor costs, space ownership or rental, lighting costs, and electricity rates, it is found in many cases that monetary savings in productivity and decrease in errors are positive in relation to lighting levels (19) (Figure 36.21).

These productivity graphs are all in positive relation to the visibility of the detail. In other words, the better the visual cues are seen, the faster the speed of processing and the fewer the errors, but at a diminishing rate until a maximum is reached. Figure 36.22 illustrates this principle.

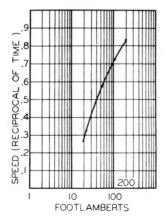

Figure 36.14. Reading a steel vernier rule: perception 1/1000 in. out of alignment.

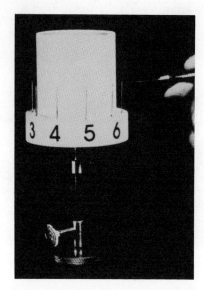

Figure 36.15. Simulated industrial task: the threading of needles represents fine detail work (18).

Figure 36.16. Simulated industrial task: reading date digits on Lincoln pennies represents metalworking details (18).

Figure 36.17. Simulated industrial task: a circuit board represents an intricate electrical assembly (18).

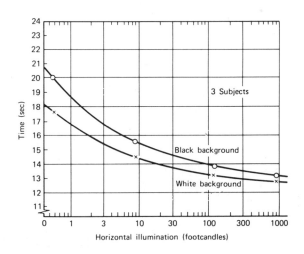

Figure 36.18. Relation between performance, time, and illumination to the needle probe task.

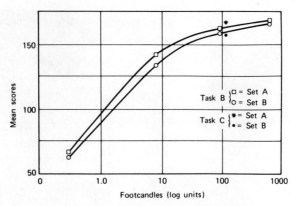

Figure 36.19. Results of response to the reading of date digits on Lincoln pennies.

There are field tasks that involve much mechanical activity or manipulation along with the visual piloting of the work. This activity may occupy so much of the time of doing the overall processing that the productivity curves (time and speed) reveal little or no change with illumination level or increasing visibility. The designer is then faced with a further consideration at this point. Either the emphasis is placed on providing good visual efficiency, regardless of the output curves, or

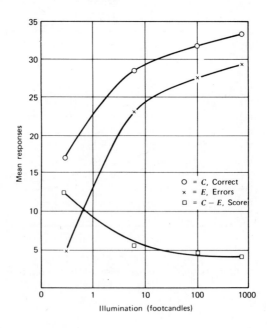

Figure 36.20. Results of response to the reading of a circuit board (finding of symbols versus illumination).

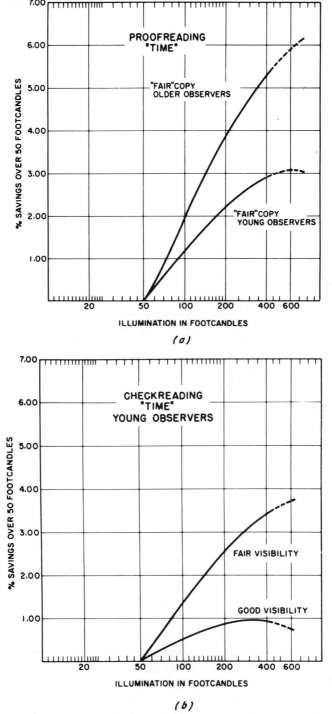

Figure 36.21. Percentage money savings from productivity response to lighting levels. (*a*) Proofreading time for older observers and young observers using "fair" copy. (*b*) Check reading time for young observers with fair visibility and good visibility.

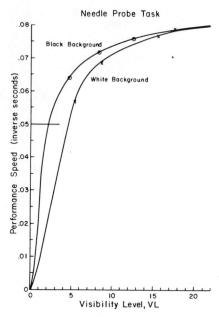

Figure 36.22. Relation between performance speed and visibility level (18).

visual efficiency is depreciated on the basis of the output curve, letting the economics force a lower level of illumination. This would result in a lower visibility than would be used if the visual component were the major work item. Unfortunately, the tests described are of short time duration, and data are not available on possible deterioration of performances over a long period, such as several hours or a day. Lowered visual efficiency induced by money-saving measures might result in lowered output over a period of time. Although this might be especially true of older workers whose visual capability is much less than that of college students, who were the subjects of the test, more research is needed to determine the long-term effects on performance.

Moreover, there is the issue of the workers' morale. The same visual details may be processed at different locations in a plant or office with differing components of visual versus mechanical. Shall the designer furnish much more illumination to workers where the visual component is the larger and much less to those having to do more mechanical work on the same visual details?

Finally, recent research to be described below indicates there may be a certain basic amount of lighting needed to stimulate the worker biologically to maintain alertness and prevent industrial accidents (4).

The same concepts apply to office lighting. In the middle 1970s tests similar to those described earlier for simulated tasks were conducted. These involved proofreading (19) of mimeographed material having good, fair, and poor quality copy (Figure 36.23), reading handwritten bank checks of varying quality (19) and comparing the dollar amounts against an adding machine tape listing that included

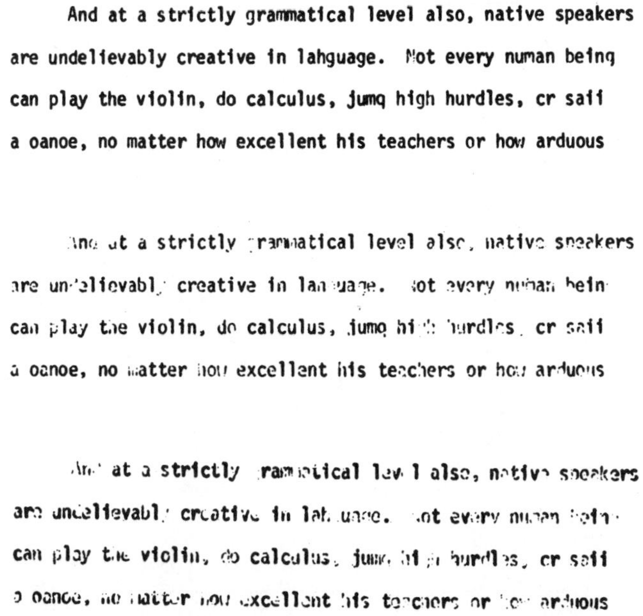

And at a strictly grammatical level also, native speakers
are undelievably creative in lahguage. Not every numan being
can play the violin, do calculus, jumq high hurdles, cr sail
a oanoe, no matter how excellent his teachers or how arduous

And at a strictly grammatical level also, native speakers
are undelievably creative in language. Not every numan bein
can play the violin, do calculus, jumq hi h hurdles, cr sail
a oanoe, no matter how excellent his teachers or how arduous

And at a strictly grammtical level also, native speakers
are undelievably creative in lat uage. ot every numan bein
can play the violin, do calculus, jumq hi i hurdles, cr sail
a oanoe, no matter you excellent his teachers or how arduous

Figure 36.23. Simulated office task showing good, fair, and poor quality copy.

systematic errors (Figure 36.24), reading and typing (19) from printed copy having 12-point and 6-point characters printed in black on white and light gray on white (Figure 36.25), and reading the Davis Reading Test in black and white and poor reproduction copy (Figure 36.26) for comprehension and speed of comprehension. Some of the results are presented in Figures 36.27 through 36.29.

The comprehension and speed of comprehension of the Davis Reading Test did not change with changes of illumination even for the poorer copy. Analysis of the results indicates that study and thinking occupied so much time that the gain in visual efficiency was masked. Here again the designer is faced with a choice. Because the overall performance remained the same under 1 fc as for 454, should the designer use 1 fc (used in street lighting); put in some amenity lighting, such as 10 to 30 fc, to make the task look lighted; or put in what visual efficiency would call for, 50–70–100 fc depending on the age of the observer? Furthermore, should the designer place 1 to 10 fc on the desk with the Davis Reading type of work but put, based on age, 50–70–100 fc on the desk with the "reading–typing" task that involves approximately the same visual details as the Davis Reading type of task? Compare Figures 36.25 and 36.26. In the 1930s Weston was faced with similar questions in England. He was director of research in occupational optics for the Medical Research Council. For many years he had studied the role of illumination in various types of industry, but he came to the conclusion that one could not isolate the gain in visual performance from the many contributing factors to industrial plant productivity. Under the aegis of his advisory committee, therefore, he set up a series of laboratory tests that would simulate industrial tasks, eliminating

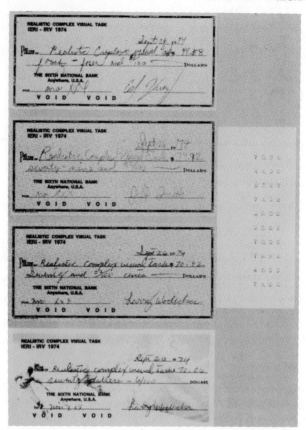

Figure 36.24. Simulated office task: variation in visibility of handwritten checks being compared with adding machine tape.

the many extraneous factors and determining the true function of "visual efficiency" in relation to illumination (20). He recognized that mechanical or "action" time, which varies greatly in field tasks, dilutes or masks the true visual efficiency; therefore he measured and subtracted this component to arrive at the net visual time and its relationship to illumination. This would be the true visual performance as related to illumination. Under the leadership of Crouch and Fry, since the early 1950s through the early 1980s in the United States, the Illuminating Engineering Research Institute used the Weston concept as it began its research program on required illumination for good visual performance. The concept of suprathreshold visibility for seeing was first introduced by Luckiesh and Moss as an attempt to quantify how well a stimulus can be seen. This was later used extensively by Blackwell, who, starting with the threshold of visibility for a base, extended to suprathreshold using visibility to unify his findings on size, contrast, time, and luminance (see Figure 36.30). Fry and Enoch (21), studying Weston's work, began relating it to patterns of eye movements. In conducting their visual performance

It was pet day at the fair. The children were waiting for the parade of animals to begin. They had trained their pets to do many different tricks. Among them was a tall boy whose goat made trouble for him. It kicked and tried hard to break away. When it heard the band, it became quiet. During the parade it danced so well that it won a prize.

It was pet day at the fair. The children were waiting for the parade of animals to begin. They had trained their pets to do many different tricks. Among them was a tall boy whose goat made trouble for him. It kicked and tried hard to break away. When it heard the band, it became quiet. During the parade it danced so well that it won a prize.

It was pet day at the fair. The children were waiting for the parade of animals to begin. They had trained their pets to do many different tricks. Among them was a tall boy whose goat made trouble for him. It kicked and tried hard to break away. When it heard the band, it became quiet. During the parade it danced so well that it won a prize.

It was pet day at the fair. The children were waiting for the parade of animals to begin. They had trained their pets to do many different tricks. Among them was a tall boy whose goat made trouble for him. It kicked and tried hard to break away. When it heard the band, it became quiet. During the parade it danced so well that it won a prize.

SAMPLES of two sizes and two contrasts of printed material used in the reading-typing studies.

Figure 36.25. Simulated office task showing variation in type size with contrast variability.

experiments using the Landolt C, a frequency of seeing curve similar to Figure 36.31 was derived. At the same time Blackwell was using the frequency of seeing concept in his studies. Fry noted that Blackwell's studies using frequency of seeing showed a good correlation between Weston's work. Ultimately this work culminated in an international model: CIE 19 and CIE 19/2 (22) relating lighting, visibility, and visual performance are described below. Today research evaluating the usefulness of the CIE 19/2 model and other models of vision and visibility is being sponsored by the IERI successor, the Lighting Research Institute.

4.4 Visual Performance Models

What is a model and why is it important? A model is a mathematical framework for understanding existing relationships. In its most useful form it not only conforms to existing data but also provides a rational basis for designers and engineers to

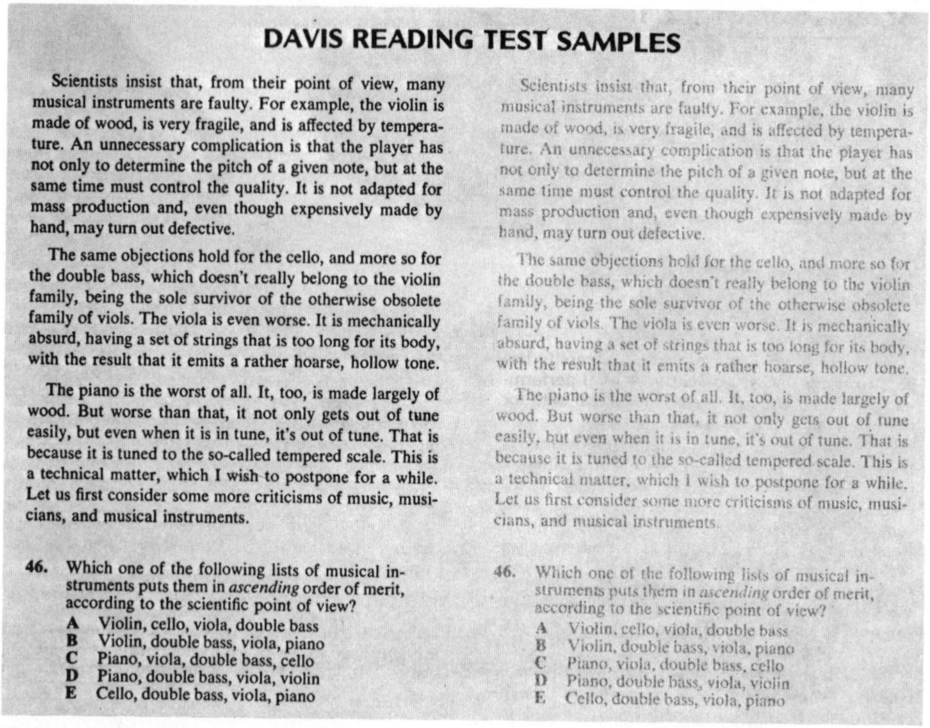

DAVIS READING TEST SAMPLES

Scientists insist that, from their point of view, many musical instruments are faulty. For example, the violin is made of wood, is very fragile, and is affected by temperature. An unnecessary complication is that the player has not only to determine the pitch of a given note, but at the same time must control the quality. It is not adapted for mass production and, even though expensively made by hand, may turn out defective.

The same objections hold for the cello, and more so for the double bass, which doesn't really belong to the violin family, being the sole survivor of the otherwise obsolete family of viols. The viola is even worse. It is mechanically absurd, having a set of strings that is too long for its body, with the result that it emits a rather hoarse, hollow tone.

The piano is the worst of all. It, too, is made largely of wood. But worse than that, it not only gets out of tune easily, but even when it is in tune, it's out of tune. That is because it is tuned to the so-called tempered scale. This is a technical matter, which I wish to postpone for a while. Let us first consider some more criticisms of music, musicians, and musical instruments.

46. Which one of the following lists of musical instruments puts them in *ascending* order of merit, according to the scientific point of view?
A Violin, cello, viola, double bass
B Violin, double bass, viola, piano
C Piano, viola, double bass, cello
D Piano, double bass, viola, violin
E Cello, double bass, viola, piano

Scientists insist that, from their point of view, many musical instruments are faulty. For example, the violin is made of wood, is very fragile, and is affected by temperature. An unnecessary complication is that the player has not only to determine the pitch of a given note, but at the same time must control the quality. It is not adapted for mass production and, even though expensively made by hand, may turn out defective.

The same objections hold for the cello, and more so for the double bass, which doesn't really belong to the violin family, being the sole survivor of the otherwise obsolete family of viols. The viola is even worse. It is mechanically absurd, having a set of strings that is too long for its body, with the result that it emits a rather hoarse, hollow tone.

The piano is the worst of all. It, too, is made largely of wood. But worse than that, it not only gets out of tune easily, but even when it is in tune, it's out of tune. That is because it is tuned to the so-called tempered scale. This is a technical matter, which I wish to postpone for a while. Let us first consider some more criticisms of music, musicians, and musical instruments.

46. Which one of the following lists of musical instruments puts them in *ascending* order of merit, according to the scientific point of view?
A Violin, cello, viola, double bass
B Violin, double bass, viola, piano
C Piano, viola, double bass, cello
D Piano, double bass, viola, violin
E Cello, double bass, viola, piano

Figure 36.26. The Davis Reading Test. Right, with poor reproduction; left: in original form.

predict what might happen in a space before it is built. Two major forms of models exist: (1) an empirical model that summarizes experimental data without offering a theoretical basis and (2) an analytical model that combines knowledge of the process being modeled as well as empirical data (23).

4.4.1 The CIE Visual Performance Model

The CIE 19/2 Visual Performance Model is an analytical model. It is based on the empirical studies of the measurement of speed and accuracy of real or simulated tasks versus illuminance or luminance. Some of these studies are described in Section 4.3 above. The analytical portion involves measurements of visual processes versus illuminance or luminance. First it uses a 4′ luminous disk as standardized target because its configuration is simple. Then to simulate more closely the visual performance of tasks, the Landolt C was chosen as the standardized test object. The CIE 19/2 model then mathematically combines the results of the practical performance of tasks with the present understanding of the visual response to light, contrast, size, and other parameters.

To provide a base line of understanding of visual performance, laboratory studies were first conducted under idealized conditions of uniform, glare-free light and

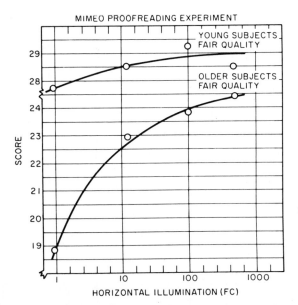

Figure 36.27. Results of mimeograph proofreading experiment of Figure 36.23. Score is a combination of speed and penalty for errors.

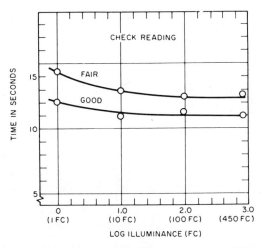

Figure 36.28. Results of check reading experiment of Figure 36.24.

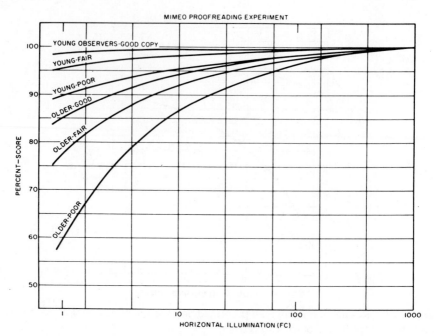

Figure 36.29. Relative results of "mimeograph" proofreading experiment from Figure 36.23 showing effect of quality of mimeograph material and age of observers.

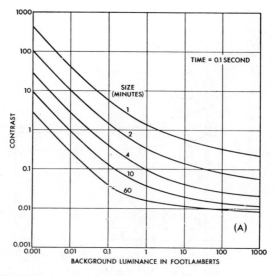

Figure 36.30. Relation between threshold contrast and background luminance for objects of various sizes (9).

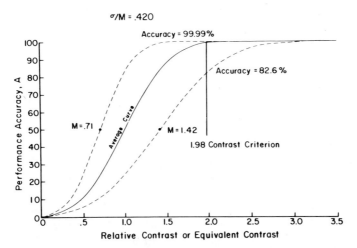

Figure 36.31. Variations in individual accuracy due to variation in *M*. Dashed curves represent limits of individual variability (i.e., limits of young population). *M* = midpoint (50 percent probability) of greater or less accurate performance.

using normally sighted 20 to 30 year old college students. Later real-life considerations such as the effects of glare, age, and nonuniform lighting were included. In summary form the model first describes the effect of light on the visibility of the visual task and then the effect of visibility on visual performance. Visibility then becomes the transforming link for the effect of light on visual performance. The following equations describe these relationships:

Equation 1: Visibility Level (VL)

$$VL = \frac{\tilde{C} \times RCS \times CRF \times DGF \times TAF}{0.923 \times m_1}$$

Equation 2: Relative Visual Performance (RVP)

$$RVP = w_1 P_1 + w_2 P_2 + w_3 P_3$$

This equation can be replaced by an equation in which its terms are related to VL:

$$RVP = \int \left[w_1 \left(\frac{\log VL - \bar{\alpha}_1}{0.145 + [0.278\,(\bar{\alpha}_1 - 0.050)]} \right) \right.$$
$$\left. + w_2 \left(\frac{\log VL - \bar{\alpha}_2}{0.180} \right) + w_3 \left(\frac{\log VL - \bar{\alpha}_3}{0.180} \right) \right]$$

Equation 3: Relative Task Performance (RTP)

$$RTP = w_{123}RVP + w_4P_4$$

A discussion of the mathematical variables that must be calculated for these formulas follows:

Visibility Level (VL): The key concept of the CIE model is visibility described in terms of visibility level (VL). At its most simple form, it is the contrast C of the visual detail divided by its contrast $\overline{C}$ brought down to threshold—the border between visibility and invisibility; VL = $C/\overline{C}$. This $\overline{C}$ value varies with individuals; therefore studies were conducted to find what kind of an average function an entire population of observers of the same age would produce using the concept of threshold. The first studies focused on sizes of details that are critical for seeing in commerce and industry. These were found to be in the range of 1′ to 10′ of arc subtended at the eye at the usual viewing distance (Figure 36.30). It was found over a 50-year span (10, 15–17, 20) that 3′ to 4′ size (Readers' Digest type or No. 2 pencil handwriting) represents the weighted average of the critical seeing tasks in working situations. The average eye pause to assimilate information in reading is 0.2 sec. Using this information a curve of threshold contrast versus luminance for a 4′ circular disk exposed for 0.2 sec was derived (Figure 36.32). Threshold defines the probability of seeing an object 50 percent of the time. Because no one would want to work under threshold conditions, suprathreshold levels of performance in the range of practical use were determined through empirical studies that produced curves of similar shape and for the most part parallel to the threshold

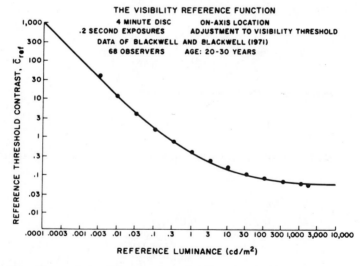

Figure 36.32. Curve shows 50 percent probability threshold for young normal sighted observers (25).

function. These formed the basis of the function for determining the illumination required for quite accurate seeing (9) for many years (Figure 36.33).

Equivalent Contrast of 4′ Disk. To link the laboratory data to the visibility of the visual detail being studied in the field, a measure of equivalency was derived: $\overline{C}$, equivalent contrast. It is the measure of the intrinsic difficulty of the detail being studied. At threshold all objects are equal in visibility. While the eye maintains a constant state of adaptation the contrast of the visual task in the field is reduced to threshold, usually using a visibility meter constructed with variable filters and optics described below. The degree of reduction of the task to its threshold point is the measure of how great its visibility is. The equivalent contrast then is a comparison of the visibility of a task measured in the field versus the visibility of the 4′ disk brought to its threshold under idealized reference conditions and at the same adaptation level of luminance. This is illustrated in Figure 36.33.

Relative Contrast Sensitivity (RCS): This variable is a measure of how sensitive the visual system is to the perception of contrast versus the illuminance or luminance provided (Figure 36.34). It represents the inverse of the threshold curve shown in Figure 36.33, which is put on relative basis as a matter of mathematical convenience. At first the RCS function represented only the average response of normally sighted 20 to 30 year old college students; however, further research resulted in a family of curves that can be derived based on age, whether the task is viewed on or off the line of sight, and the size of the detail being studied or combinations of these factors as shown in Figures 36.35 to 36.37.

Visibility Threshold Multiplier, m_1: The visibility threshold multiplier accounts for the change in contrast sensitivity, and hence the loss of visibility with increasing age (Figure 36.38).

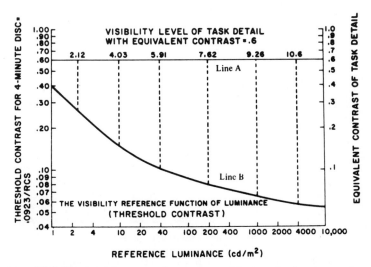

Figure 36.33. Visibility level increases for a given object as the luminance increases and threshold decreases (as in Figure 36.32) (25).

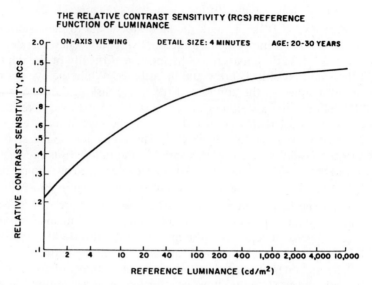

THE RELATIVE CONTRAST SENSITIVITY (RCS) REFERENCE FUNCTION OF LUMINANCE

Figure 36.34. Relative contrast sensitivity (RCS); the measure of the eyes' sensitivity to contrast. This is the inverse of the threshold curve shown in Figures 36.32 and 36.33 mathematically normalized at 100 cd/m² (25).

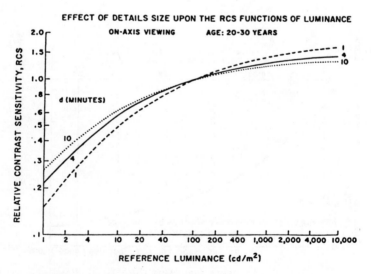

EFFECT OF DETAILS SIZE UPON THE RCS FUNCTIONS OF LUMINANCE

Figure 36.35. The RCS function changes based on the size of the object being viewed in terms of the angular size subtended at the eyes (here 1, 4 and 10 min) (25).

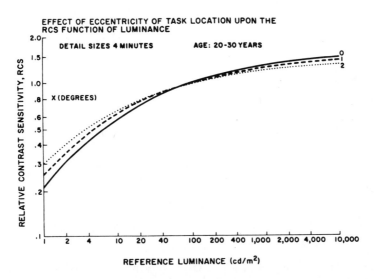

Figure 36.36. The RCS function changes based on the how far off the line of sight the object is being seen (here 0°, 1°, and 2°) (25).

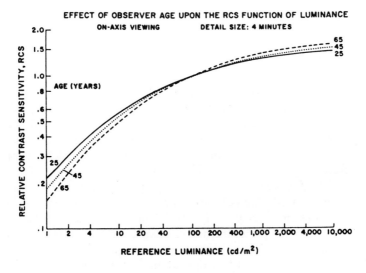

Figure 36.37. The effect of age of the observer on the RCS function (25).

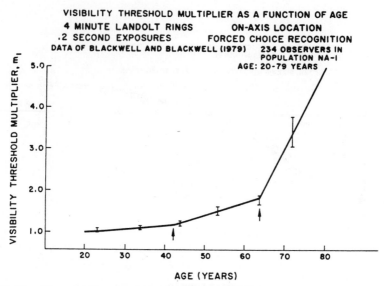

Figure 36.38. The visibility level varies inversely with the age of the observer and is expressed by the visibility threshold multiplier m_1 and the visibility level (VL) formula (25).

Visibility Threshold Constant for the 4′ Disk: 0.0923 is the visibility threshold for the 4′ standard disk at 100 cd/m², based on an average of the 20 to 30 year old population data.

Contrast Rendition Factor (CRF): This variable accounts for the degree of loss of visibility due to the reduction in contrast of the visual detail by light reflected from specular components of the task to the eyes of the observer, for example, from glossy photographs, magazines, or highly reflective metals. It is the ratio of the visibility of a task under an actual lighting system compared to the visibility using completely diffuse uniform reference lighting produced under laboratory conditions, usually an integrated sphere.

Disability Glare Factor (DGF): The disability glare factor accounts for the loss of visibility due to summation of the focused and scattered light in the field of view. The overall effect of disability glare is to produce an overlay of brightness on the image to be seen. This overlay of brightness is light being scattered in the interior of the eye from the glare sources in the field of view. DGF is different from earlier works of disability glare assessment in that it accounts for two different effects of ocular stray light produced by glare sources: (1) the reduction of the contrast of the image being viewed, and (2) the change in relative contrast sensitivity (RCS) due to the summation of focused and stray light. While the contrast of the image is being degraded by the glare, the higher adaptation luminance level produced by the glare increases the eye's sensitivity to the perception of contrast. Therefore the RCS factor modifies the total loss of visibility that would be determined by just considering the reduction of contrast produced by scattered light. For the interested reader, an extensive treatment and procedures for measurement of DGF are covered in the CIE 19/2 Report (22, 24). The significant influence of disability glare on age is discussed below.

Transitional Adaptation Factor (TAF): The eye for the most part operates in a dynamic mode in processing visual information. The transient adaptation factor (TAF) takes into account the dynamic eye movement and resulting changes in adaptation to the luminances encountered in these patterns of movement. As the eye moves from one point of reference to another in gathering information, it may encounter different luminances from its current adapted state. As transitional adaptation occurs it results in differences of visibility. This work was pioneered by Boynton and Boss and Rinalducci and Stearn and used by Blackwell and Tsou to develop the TAF. Ideally, one would like to be able to measure the continuous change in adaptation; to date, however, such a general model does not exist. Instead, only two dynamic conditions are covered: (1) glance transient adaptation factor, where an observer while looking at a visual task takes brief glimpses (totaling less than 2 sec) away to a brighter portion of the visual field; and (2) readaptive transient situation, where the observer looks away from the primary visual task for 10 or more sec, thus creating two visual tasks. This is an area under investigation today (24). It is especially needed to begin to assess luminance difference encountered between processing paper tasks as well as visual display terminals (VDT) tasks.

Relative Visual Performance (RVP): This second transfer function relates visibility to visual performance. In Blackwell's early studies the basic visual response was determined using the standardized 4' luminous disk, where the criterion used was the detection of the presence of the disk (the observer responded to whether it was present or not present). Later Blackwell chose the international test object used by Weston, the Landolt C, to conduct experiments that developed performance curves, where an observer scanned either a single Landolt C or a display of Landolt Cs to determine the orientation of the gap in the C under different conditions of size, contrast, luminance, and so on. This resulted in a family of S shape performance curves called ogives shown in Figure 36.39. The next step was to attempt to relate these laboratory studies to the performance of real field tasks or simulated field tasks. Fortunately a series of performance studies had been completed both in the United States as well as abroad. These data were transformed into ogive curves and compared with the laboratory performance curves. The results are shown in Figures 36.40 and 36.41. In most instances the comparison was quite similar between the practical performance curves and the laboratory performance curves. However, some of the ogive curves were of a slightly different shape, which challenged the thought there was only one process of seeing. Resulting studies determined that there are at least four processes; the first three are considered critical to gathering information and the fourth is a noncritical mechanical process. Relative visual performance depends on the following three processes and their weighting coefficient, w: (1) a sensory process, w_1P_1, which determines the detail to be seen; (2) fixation maintenance, w_2P_2, which guides and keeps the eye at a viewing point while gathering information; and (3) saccades, w_3P_3, the pattern of

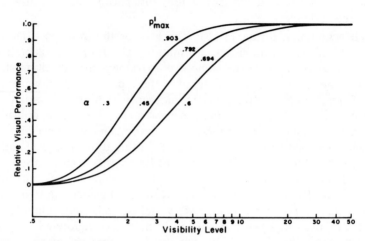

Figure 36.39. Family of ogives or "S" shaped curves developed from scanning studies of the Landolt rings. The flattest curve—that farthest to the right—represents the most difficult task, the steepest curve the easiest task (25). P' upper asymptote of original performance curves before mathematical normalization.

eye movements caused by the eye muscles responsible to move the eyes to turn to view a task and to sweep from one detail to the next. See Equation 2 above and Figure 36.42.

Relative Task Performance (RTP): Equation 3 above shows the fourth process, the noncritical motor response $w_4 P_4$ with its weighting coefficient. This process is composed of two subprocesses (Figure 36.43). As a hypothetical example of how these

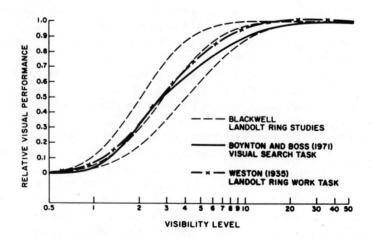

Figure 36.40. Meshing the ogives from the simulated tasks of two field researchers [Weston (20) and Boynton and Boss (22)] with Blackwell laboratory Landolt-ring ogives (25).

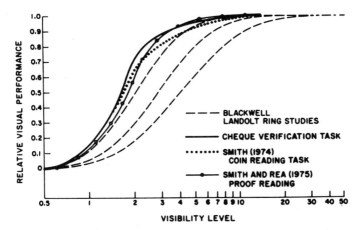

Figure 36.41. Meshing of ogives from industrial and task studies with Blackwell's laboratory Landolt-ring ogives. Coin reading simulates industrial task (25).

processes may work, one may represent the guidance of a pencil to accomplish a task while the other may indicate the cognitive act of accomplishing the task. This process is shown in relation to the others in Figure 36.44. The combination of all evaluates the total human performance. Although the International CIE committee that adopted this performance model placed a strong emphasis on relative visual performance, they also recognized that for any cost–benefit analysis, the RTP values would have to be considered.

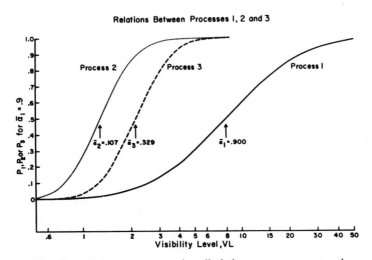

Figure 36.42. The first of three processes is called the sensory process, the seeing task. The next two are fixation maintenance and saccades; these are eye-directing adjustments. Ogives reflect a specific task (25).

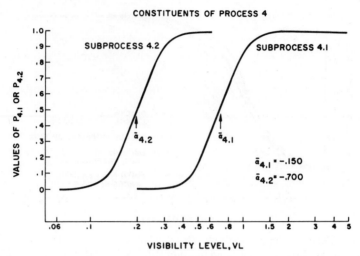

Figure 36.43. Subprocess 4.1 represents guidance to the point of work. Subprocess 4.2 represents execution of the work (25).

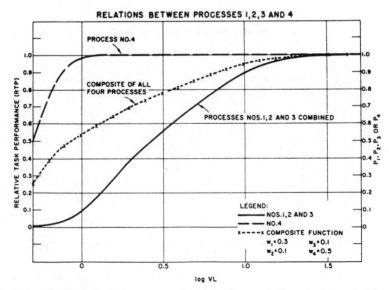

Figure 36.44. Combination of processes 1–4 based on specific task shown in Figure 36.43 (25).

4.4.2   Other Visual Performance Models

In addition to the CIE 19/2 model, in recent years several researchers (including Adrian, Indistsky, and Rea) have developed visual performance models to describe the influence of lighting parameters on visibility and hence performance (26). Of these, Adrian and Rea's models are the most applicable to interior lighting conditions. Adrian's model first report in 1969 has been extended in recent years and is quite complete and advanced. Although it is primarily being applied for roadway lighting purposes, it covers the range of luminances found in interior applications as well. Visibility as a characteristic of roadway lighting is being considered in the development of roadway lighting standards in North America. Of various models tested, Adrian's model most accurately predicts the visibility of small targets (20 cm × 20 cm) on a roadway as measured in the field using visibility (contrast reduction) meters among other techniques (27, 28). The elements considered in the Adrian model include background luminance, veiling luminance (disability glare), object size, observer age, and viewing time. Rea (26) has developed a performance surface model based on a numerical verification task that basically tests the performance of observers in detecting errors by comparing a standard column of numbers with another column of varying contrast and errors introduced. The results of his testings have been plotted three dimensionally as shown in Figure 36.45. This plot suggests that there is a wide range of physical conditions that can produce a high level of visual performance (as denoted by the flatter surface); however, once you move from this area, performance falls off rapidly. In comparing

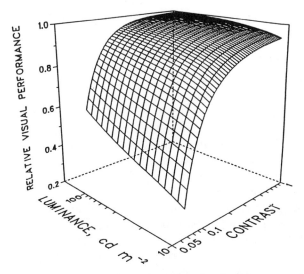

Figure 36.45.   Three-dimensional surface of relative visual performance derived from performance study on the numerical verification task. Courtesy of *Lighting Design and Application*, New York, May 1987, p. 10.

the model with empirical studies, Rea found that the work of McNelis could be predicted at the high levels of contrast but not at the lower levels. Further refinements of the model are being incorporated based on a series of reaction time studies in response to a square target of varying size, contrasts, polarity (dark letter on light background or vice versa), and luminances.

4.5 Limitations

Each of the models described above has particular limitations. Our discussion has represented the state of the art according to the best research knowledge available, yet the situation never remains constant because new knowledge is being obtained by continuing research. New improved techniques are being developed. The models described need refinement and validation so that they can be used subject to the knowledge of each's limitation. Probably the most universal criticism of each is its ability to predict actual performance accurately. We have basically discussed lighting in terms of enhancing the visibility of practical tasks, but it should be understood that some tasks have such poor inherent visibility no amount of ordinary lighting will give an adequate visibility level. In this case the contrast might be improved by special direction or color of light, to enhance the contrast. If there are indentations or raised characteristics to the detail, light striking these configurations obliquely will cause highlights and shadows that will increase the contrast. Magnification is another method of increasing visibility. Where all of these techniques fail, consideration should be given to altering the task if possible. Crouch and Vincent (29) have found that for some difficult lighting applications (strip mining and underground mining equipment, aircraft carriers, school stairwells, etc.) visibility of objects to be seen, for example, stairs, was improved by using highly reflective materials producing greater contrast and better visibility.

4.6 Field Application

A number of investigators (29, 30) have found that the visibility of unknown details in the field can be assessed by using a contrast threshold meter. Such a meter puts a variable veiling luminance over the detail of regard until its contrast has been reduced to threshold. The amount of veil needed to reduce the task to its threshold denotes its degree of visibility (above threshold), provided the luminance to which the eyes are adapted is maintained constant. This is done by coupling an absorbing filter for the external view with the mechanism for increasing the veil, so that viewed luminance through the instrument stays constant. Furthermore, because the eyes in scanning pause to take a picture at 200 msec, it is important to determine the threshold of visibility on the basis of exposure of the field for 200 msec. Once threshold has been reached for the field detail, then at the same luminance of the field object, the 4′ luminous disk is reduced to threshold. Because all objects at threshold are equal in visibility, there is now an equivalency between the 4′ disk with its resulting contrast and the field detail. Such instruments are now available, as illustrated in Figure 36.46. Attachments to these meters have been successfully

Figure 36.46. The Visual Task Evaluator designed to use 0.2-sec exposures that permit more realistic field measurements.

used to determine the disability glare factor (DGF) (31). More portable streamlined versions are being developed and used in field measurement. These refined visibility meters use the ambient light available to produce the veiling luminance and adaptation luminance. The result of the visibility meter measurements is to obtain an equivalent contrast of the standard 4' test object designated $\overline{C}$. When this is obtained, the value of luminance required can be obtained using calculational methods described in the CIE 19/2 or by referring to the current illuminance selection procedure of the Illuminating Engineering Society of North America (IES) (32).

4.7 Current Illumination Determination Method

While research continues to further the understanding of the various visual performance and vision models, by field testing and evaluation, professional standardizing bodies throughout the world have developed experience-driven consensus bases for recommending light levels for various tasks. In the United States, IES has developed an illuminance selection procedure where a range of illuminance required for a task is recommended (32). A designer selects what target level of illuminance is to be provided for the workplace based on the age of the occupants, an estimation of the difficulty of tasks involved, the importance of speed and accuracy, and so on. A combined weighting factor then shows whether the illuminance to be provided is at the low, middle, or high end of the range. Additionally, allowance is made for measurements with a visibility (contrast threshold) meter to determine the difficulty of specific tasks. A detailed description of the IES method is shown in Appendix A along with examples of target illuminance levels for various industrial tasks. It should be borne in mind that the recommended light level is for the task, *not* necessarily for the whole space. On the other hand, just lighting the overall space does not guarantee that the task will be lighted to the degree recommended either. A balance must be carefully determined by the designer in the overall lighting design program. There are a number of energy efficient as well as pleasing ways of providing the necessary light level without uniformly lighting the entire space. This is discussed below.

4.8 Industrial Tasks

Although many industrial tasks have remained the same, a revolution is taking place in the area of factory automation. The work force and production capabilities are being streamlined. As a result new tasks are being introduced such as computer-aided design and drafting (CADD) screens for easy reference to working drawings. These computerized tasks must be carefully treated from a lighting standpoint in order to eliminate or minimize distracting reflections in the screen (33). Figure 36.47 illustrates the problem of how luminous elements (whether the ceiling or luminaires) obscure the details to be seen by producing either a uniform luminance over the detail or an annoying pattern of reflections. Newly developed luminaires have sharp cutoffs to eliminate the majority, if not all, of the light in the offending zone (Figure 36.48). Careful guidelines for balancing the lighting needed for paper-based tasks being performed in conjunction with VDT work have been developed (34). A commonsense guideline for VDT use, such as next to windows, would suggest placing the VDT and desk perpendicular to the window. Adrian and Fleming (35) have conducted a current review of vision, VDTs, and lighting interaction. The review covers research on screen polarity and contrast of characters, polarity and adaptation, screen flicker, the effects of color, the effects of VDT use on

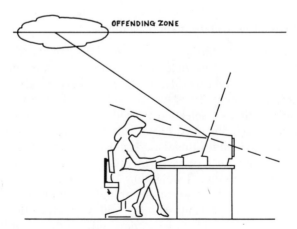

Figure 36.47. Indirect glare in VDT resulting from brightness in the "Offending Zone." Courtesy of *Lighting Design and Application*, New York, December 1988, p. 10.

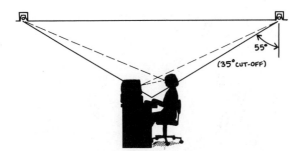

Figure 36.48. Illustration of luminaire cutoff distribution to eliminate or minimized indirect glare in VDT screen. Courtesy of *Lighting Design and Application*, New York, December 1988, p. 10.

accommodation and its impacts on fatigue, the effects of rest breaks on fatigue, and lighting systems for VDT use. From a lighting standpoint, further research is needed to determine better the luminance ratios to be used between the VDT task, surrounding tasks, other nearby surrounds, and the luminous elements related to the ceiling.

Beyond the specialized VDT task, there are almost innumerable industrial processes and products; it would be difficult to describe all the critical details from start to finished products. A number of details in various industries are described in lighting literature (36). The illumination required for various industries are given in the current edition of the IES Lighting Handbook (32).

4.8.1 Type of Tasks

The surface characteristics vary from mat to highly glossy finishes and from two-dimensional to three-dimensional shapes. Mat finishes do not present a lighting problem, although if there are three-dimensional objects to be seen, some unidirectional light beamed together with diffused overall lighting will bring out highlights and shadows that identify the three-dimensional form. Some mat surfaces have slight wrinkles or ridges in them that do not show up with general overhead lighting but do appear when a light beam is directed very obliquely across the surface (Figure 36.49). Some objects have a surface gloss that overlays the grain or textured base material. The glossy reflections of light room surfaces near the line of sight interfere with view of the texture beneath. In this case, light directed to the surface obliquely eliminates the reflections from the line of sight and penetrates the surface to illuminate the texture beneath. The metalworking industry largely involves glossy surfaces that act as mirrors. In this case, the use of large, low-brightness overhead luminaires serves to reflect low-brightness highlights in the surface, against which are seen any markings as interruptions of the highlight (Figure 36.50).

For the lighting solution of many industrial seeing tasks, one should refer to the American National Standard Practice for Industrial Lighting (36).

Figure 36.49.   Ridges showing up under grazing light.

5   VEILING REFLECTIONS

In offices and similar locations where paper work is done, it has long been recommended that mat paper be used to avoid glossy or mirrorlike reflections that
overlay the details to be seen and interfere with visibility. Research since the latter
1950s (37) has discovered the mat paper work continues to exhibit microscopic
mirrorlike reflections. Overhead lighting with a heavy downward component of
light rays (Figure 36.51) causes serious loss of visibility because of veiling reflections
in the tiny facets of the paper fibers and the inks or pencil marks used for the
printed or handwritten characters reducing the contrast. Such microscopic reflections cause loss of visibility in accordance with Figure 36.52. If one were to work
in the shade of a tree (Figure 36.53*a*, *b*), these glossy reflections would be eliminated
and light would come to the work from the sky beyond the tree. Reflections of
this light would go away from the eyes of the worker, and the details would be
seen ideally. A good compromise has been to reduce the downward components
and have the light come from the ceiling and side walls as though they represented
the hemisphere of the sky without sun (Figure 36.53*c*). To account for this loss of
visibility in practice, the concept of contrast rendition factor (CRF), described
above under the discussion of the CIE model, was developed. CRF compares the

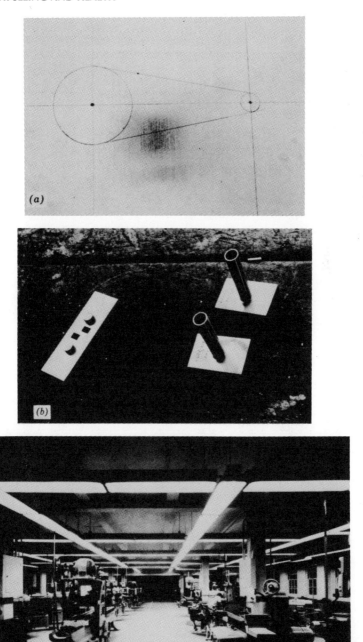

Figure 36.50. (*a*) Appearance of scribed lines on steel surface under large-area, low-brightness lighting equipment. (*b*) Highlight covering metal surfaces reveal contours and details. (*c*) Grid system of fluorescent lighting produces broad highlights on metal surfaces.

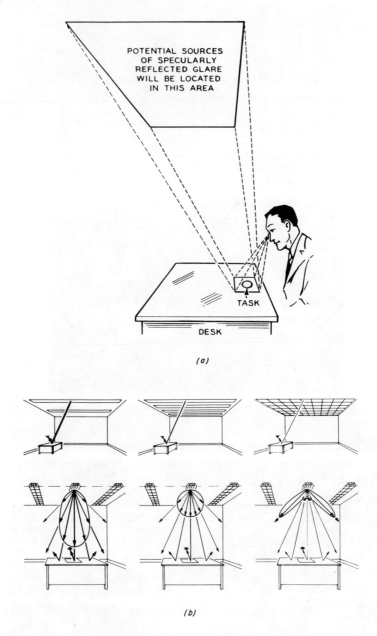

Figure 36.51. (*a*) Downward light from luminaires in this zone will cause veiling reflections and serious loss of visibility of office tasks. (*b*) Different luminaire downward components. The less the downward component in the reflection zone of the desk task, the greater the visibility of the task.

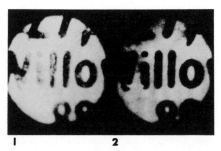
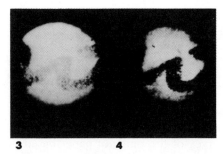

Figure 36.52. Veiling reflections reduce task contrast. (1) Magnified sample of type under a troffer causing veiling reflections. (2) Same sample with veiling reflection shielded out. (3) and (4) Similar situation with letters a, r, and t (parts only shown) handwritten in pencil.

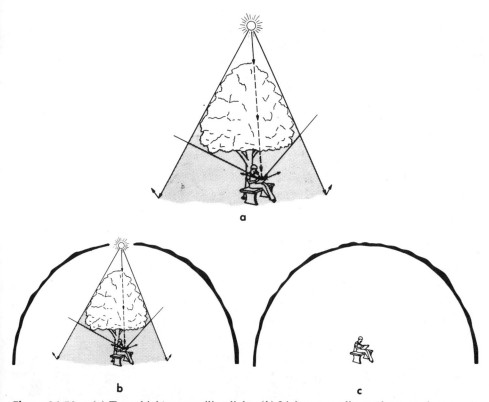

Figure 36.53. (a) Tree shields out veiling light. (b) Lights to reading task comes from vault of the sky outside veiling zone. (c) Vault of sky with no sun produces good results with low component from the veiling zone overcome by majority of light outside the veiling reflection zone.

effect on a task's contrast by the distribution of light of a particular lighting system to the contrast of the same task under reference (hemispherical) lighting conditions. The CRF formula is:

$$\text{CRF} = \frac{\text{contrast of the task under conditions of interest}}{\text{contrast of task under reference conditions}}$$

where reference conditions = complete diffuse, uniform illumination such as occurs in an integrating sphere. The need to consider veiling reflections was shown in a survey of the range of paper-based office tasks (38), which indicated that a large majority were of poor contrast.

A veiling reflection study (39) indicated that commonly used luminaire systems had only 15 to 50 percent effectiveness in revealing the details of the task. Therefore the lighting system should provide a minimum of downward light immediately over the desks, in comparison to the light coming from wider angles (40, 41), which does not cause glossy reflection toward the eye (Figure 36.54). Study of the angular reflections of a typical task (pencil handwriting) has resulted in a determination of the effectiveness of light coming toward the task from various angles (42). This effectiveness is represented in Figure 36.55, where the length of the radial lines indicates the relative effectiveness of light approaching the task in revealing the details of the task. This average effectiveness holds for light approaching from every direction of the compass (i.e., 360° in azimuth including body shadow). It

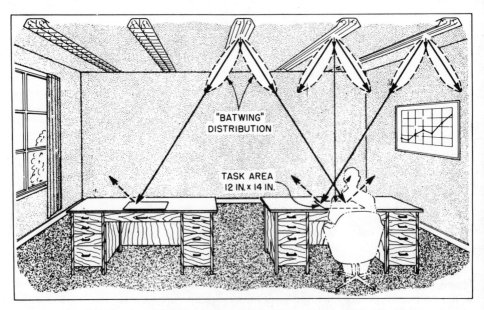

Figure 36.54. Batwing candlepower distribution of light from the luminaire minimizes light in veiling zone, distributes it from other luminaires outside the zone, and reflects it from the walls.

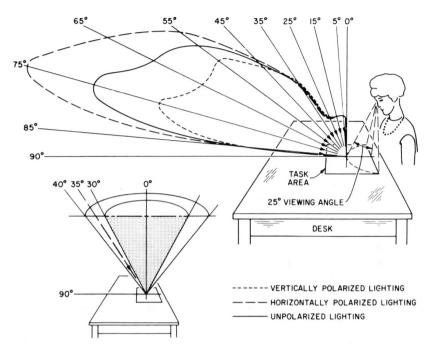

Figure 36.55. Average effectiveness of light approaching the task in different angular cones, of 360° azimuth.

will be noted that the maximum occurs at 75° with the vertical, indicating that the light from walls and windows is very important in rendering more effective visibility of the task. (The reflectance of the walls should not be greater than 60 percent to avoid glare.) Measurement techniques are commonly being employed by today's designers to evaluate a lighting system. CRF is being measured using such instruments as the Brüel and Kjaer meter or the Equivalent Sphere Illumination (ESI) meter. These have proved to be helpful evaluation tools. As with all evaluation techniques there are of course limitations to the CRF method of analysis. Depending on the task, veiling reflections can lead to low or high contrast and thus low or high CRF values. Researchers understand and can control veiling reflections for laboratory experiments; however, in practice, the magnitude of veiling reflections can vary greatly within a space because the incidence of veiling reflections depends on the geometry of the luminaire, the type of task, the observer, and where each of these is located within the space. Because the designer does not always know where people and tasks will be located within a given space, the consequences of veiling reflections on task performance can be readily estimated only for a specific task set at a fixed position and seen from a particular angle (23). Even with these limitations taken into consideration, CRF remains the best available measure for the assessment of veiling reflections.

The description of veiling reflections above primarily discusses the effects of lighting distributions on horizontal paper-based tasks. Existing lighting system technology is available that, when properly applied, can adequately handle these veiling reflections. Today however, with the introduction of VDTs into industry and offices, the issue of reducing annoying veiling reflections in the video screen must be addressed. Research is being conducted to determine how best to limit these reflections, as discussed under types of tasks above.

6   EFFECTS OF SURROUNDINGS: TRANSIENT ADAPTATION

The eyes are constantly scanning both the details of the task at hand and the immediate surroundings of the rest of the room. Time-lapse photography of office workers shows that when persons look up from their desk tasks, the peak of the frequency of their angles of viewing is at 90°, or a horizontal line of sight (43). Thus they look around the room and then back at their desk tasks with their immediate surroundings. This involves looking at various surfaces with their different luminances (brightnesses), as in Figure 36.56. These sudden changes from one luminance to another cause a serious loss of visual sensitivity or visibility, unless the luminances are kept within a suitable range of each other (44). The additional contrasts necessary to compensate for losses of sensitivity are plotted in Figure 36.57.

Figure 36.56.   Marked differences in luminance in the field of view cause serious momentary losses of visibility, thereby scanning sensitivity.

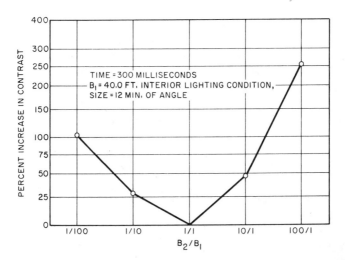

Figure 36.57.   Percentage increase in contrast necessary to compensate for loss of visibility due to change from 40 fL to 1/10 or 1/100 or to 10 to 100 times that of luminance.

The time-lapse studies show that the reaction between the luminance of the background of the details of the task and that of its immediate and more remote surroundings affects the ability to see the task after looking away from it (Figure 36.58). Furthermore, they demonstrate that the smaller the difference between the brightness of the visual surroundings and that of the visual task, the higher the efficiency of seeing. However, equal brightness in all areas of the room would result in a marked feeling of monotony. Investigation has shown that the immediate surroundings should be less bright than the task, but not less than one-third that of the task and never greater than five times that of the task. These ratios apply

Figure 36.58.   Serious loss of scanning sensitivity occurs when one encounters dark surrounding (or too bright surroundings).

to large areas of the surroundings and can be achieved by the use of ordinary lighting equipment and light-colored room surfaces. (Small areas with accent color do not interfere with the transient adaptation.) Every object reflects some portion of the light it receives, and the percentage reflected is known as *reflectance*. The reflectances of floor and furniture should be in the range of 30 to 50 percent, the walls 40 to 60 percent, and the ceiling 70 to 90 percent (Figure 36.59). Attempts to quantify the losses of visibility from transitional adaptation are described in the section on the CIE model above and under the discussion of transitional adaptation factor (TAF).

7 ELIMINATION OF DISCOMFORT GLARE FROM OVERLY BRIGHT LIGHTING SYSTEMS

For a number of years it had been hypothesized that the physiological basis for the feeling of discomfort is generated in the sphincter muscle controlling the opening of the iris (45). When one looks up from one's work and encounters an overly bright lighting system, the sphincter muscle closes down the iris for protection. If one continues to look near the horizontal, the sphincter muscle first closes down too much, then the system desires more light. It opens up too much, and the luminaire overbrightness causes it to close down again. This alternation continues and the phenomenon is called "hippus."

An age effect on the limiting luminance at the borderline of discomfort glare has been determined (47). It appears that on average, the limiting luminance varies inversely as the age of the observer. This may require greater protection for older workers than the current recommendations.

Figure 36.59. Reflectances of room surfaces to favor good scanning sensitivity.

Research on the sensitivity of people to peripheral brightness has developed a formula for discomfort glare that takes into account the luminances of the lighting units, the position of the lighting units in relation to the line of sight, the additive effect of a number of luminaires in the field of view, and the range of sensitivity of the population (48). The result of this research has brought about a standard known as "visual comfort probability" (VCP) for offices and classrooms. A VCP of 70 means that the visual comfort probability will be such that 70 percent of the occupants at the end or side of the room will be satisfied when the luminances of the luminaires are limited to a given value. From a layman's practical viewpoint, this means that when the lighting system is suddenly exposed to view 70 percent of the observers (Figure 36.60) will obtain no shock of overbrightness. At present VCP is applicable in interiors only for direct lighting systems; research is needed to consider new advances in fixture design.

Other applications where discomfort glare is being charted include roadway lighting. A North American approach for rating discomfort glare has recently been completed and is consistent with similar European work (46). Further recent research determined the sensitivity to discomfort glare for underground mining populations that showed an increased sensitivity to discomfort glare, having worked underground over a period of time. Formulas to take this into account were developed to assess the comfort of lighting systems to be developed for underground mining equipment (49).

a

Figure 36.60. Practical test for uncomfortable lighting units. (*a*) Stand or sit at one end of the room, look straight ahead. Shade eyes so that no units are seen and hold for a few moments for adaptation. (*b*) Suddenly remove hand; if there is a feeling of overbrightness, the lighting units are too bright for comfortable reading.

b

Figure 30.60. (*Continued*)

8 DISABILITY GLARE: LOSS OF VISIBILITY

Various researchers (50) have found that a high brightness source away from the line of sight has the same effect in reducing visibility as a uniform veil of brightness overlying the details to be seen on the line of sight. This equivalent veiling luminance is expressed as

$$L_v = 10\pi[E_1\theta_1^{-2} + E_2\theta_2^{-2} + \cdots + E_n\theta_n^{-2}]$$

where L_v = equivalent veiling luminance (footlamberts)

E = illuminance on a plane through the center of the pupil perpendicular to the line of sight contributed by each glare source (footcandles)

θ = angular displacement between the line of sight and each glare source

If L_v is put in the formula for the resulting contrast of the object to be seen, we arrive at the resulting contrast:

$$C' = \frac{L_b - L_o}{L_b - L_v}$$

where C' = contrast under glare effect
L_b = luminance of the background of detail
L_o = luminance of the detail

The original contrast without glare is expressed as follows:

$$C = \frac{L_b - L_o}{L_b}$$

The effect of surroundings around the immediate background of the details was presented in Figures 36.7 through 36.9. When the surroundings become brighter than the background of the task detail, the visibility decreases rather rapidly with this increased luminance. This is a disability glare effect. Even with uniform and equal surrounding field, there is a 7 percent loss, but this is better than a dark surrounding field. As an illustration of the disability glare effect even with comfortable surroundings, consider that one cannot see inside a mountain tunnel in bright daylight. As one approaches the mountain with its face illuminated in the sun, the tunnel looks like a dark hole with no details visible. If one could look through a tube at the tunnel opening, the details inside the tunnel could be seen. If one could look out at the sunlit sky from the bottom of a well, one could see the stars still shining. This form of apparently comfortable lighting can be a real danger in dark industrial processes and out on the roadway, where vital cues are missed because of overly bright surroundings.

It has been discovered that age has a significant effect on the veiling luminance, as much as a threefold difference between younger and older observers (51). This has resulted in a factor of age being put in the L_v formula (22, 24). Vos (1984) presented a formula to estimate the luminance of the scattered light for people of different ages over a wide range of deviations from the line of sight. The CIE model takes account of the disability glare changes on contrast thresholds and the resulting change in visibility. A thorough example of the modifications to present disability glare formulation is that by Adrian (27):

$$L_{seq} = k \sum_{i=1}^{n} \frac{E_{Gl_i}}{\theta_i^2} \text{ (cd/m}^2\text{)}$$

where E_{Gl_i} = the illuminance in lux at the eye from the glare source i
θ^2 = the glare angle between the center of the glare source and the target and valid over the range $1.5° < \theta < 30°$
k = the age dependent constant. For 20–30 years of age $k = 9.2$.

For age modifications:

$$k = (0.0752 \text{ Age} - 1.883)^2 + 9.2$$

Currently, a modified Fry glare lens that has been used with the Pritchard photometer is being developed for field assessment of disability glare. This can be used as another means to assess the disability glare beyond the factor DGF decribed in the CIE model and assessed using a modified Visual Task Evaluator (VTE).

9 AGE

Age is a significant factor in the ability to carry out visually guided activity. This is extremely important in safety and productivity. The age span is increased markedly, as Figure 36.61 indicates. As a result of a better quality of life in the Western nations, people are not only living longer but also are leading active lives both at work and leisure. In the United States alone it was estimated in 1988 that there were 25 million Americans 65 or older, and in the next 25 years this figure will double to 50 million. The effect of aging on requirements for light and vision is being addressed in a number of recent reviews (53).

Many physiological changes related to the aging eye are fairly well known. In the following discussion of the changes related to aging process, it will be helpful to refer to Figure 36.62. In considering the effects of age, two aspects of the function of the eye should be borne in mind, its optical function and its physiological makeup. First, the eye functions as an optical receiver, transmitter, and absorber of electromagnetic radiation (the visible portion known as light). As such, light that is processed by the eye obeys basic laws of optics until the optical energy that reaches the retina is transformed into electrochemical signals. These signals are then processed by the brain for vision and setting of biological rhythms. The second aspect to consider is the makeup of its physical structure. The eye is composed of various living tissues and a fluid-like substance (vitreous humor). These tissues and fluids respond in different ways over time. Some tissues continue to repair or replace themselves over time but others do not. For those that do not regenerate themselves, understanding cumulative effects of the absorption of optical energy is extremely important. Understanding the eyes' resilience over time is the key to setting safety standards for living with natural and man-made light sources. An excellent coverage of this topic has been completed by Sliney and Wolbarsht (54).

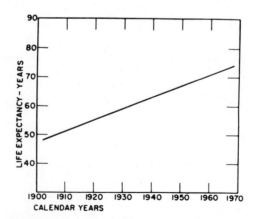

Figure 36.61.   Life expectancy since 1900 (52).

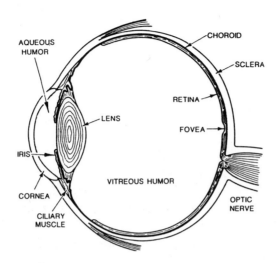

Figure 36.62. Horizontal cross-section of the human eye. Approximate length from cornea to cone layer of retina is 24 mm. Thickness of choroid is about 0.05 mm and the sclera 1.0 mm. Courtesy of *IES Lighting Handbook, Reference Volume*, 1984 (32).

Furthermore, industry standards on safety limits for optical sources have been adopted (55).

The quantity of light actually transmitted to the retina is greatly reduced with age. As optical radiation moves toward the eye, the first structure it may encounter is the eyelids. During the life of an individual these may become baggy or droopy, occluding some of the light striking the cornea. This outer shield of the eye initially is very transparent. Over its life the cornea gradually builds up fatty deposits that scatter available light before it enters the next structure, the pupil. It has been found that the eyes deteriorate considerably in their ability to adjust the pupil opening in proportion to the light available. Figure 36.63 shows that this eye gate becomes smaller and smaller, even in the dark. The mean difference in pupil size between ages 20 to 29 and 80 to 89 is 2.6 mm in the dark, and 1.7 mm at a level of 1 fL. The reduction in pupillary area with increasing age is equivalent to approximately a 0.3 log unit reduction in luminance. This decrease in the amount of light reaching the retina for image-forming purposes is one of the underlying causes of loss in visual sensitivity with age (56). Critical details to be seen in commerce and industry involve small size and, in many cases, poor contrast. Figure 36.64 demonstrates that the ability to see fine detail is seriously reduced with increasing age. Another closely related factor is the speed in the ability to see small objects, to be able to maintain safety and accurate production. Weston (58) found that this performance was radically reduced with age (Figure 36.65). One benefit of aging may be the better depth of field experienced from the smaller opening. The lens is the next structure in the path of light to the retina. At a very young age some portion of ultraviolet radiation can enter the eye. This early transmittance of ultraviolet is quickly diminished to a point of total absorption. Over its life the lens of the eye pays a price for guarding the interior of the eye by absorbing ultraviolet

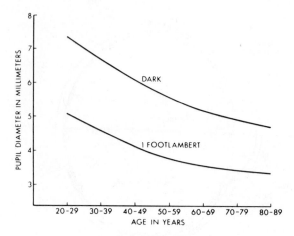

Figure 36.63. The change in pupil diameter with age as measured in the dark and at 1 fL for 222 observers (56).

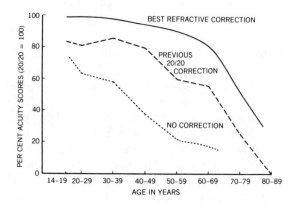

Figure 36.64. The reduction in visual acuity with age (57).

radiation. Gradually its transparency is reduced as it becomes more yellow. In some cases opaque cataracts are formed. At later stages it hardens and by age 40 starts to lose its ability to focus different distances flexibly. Finally the lens becomes a source of glare by scattering light in the interior of the eye. The yellowed lens acts as a filter absorbing more blue and violet. This changes color perception, with white looking more yellow and the fine discrimination of a difference between blues and greens diminished.

What other changes result from aging? A most vital factor in maintaining safety is the ability to see movement of objects representing a potential hazard out of the corner of the eye. Wolf (59) has found that the ability to see flicker (movement) 40 to 80° out from the line of sight is reduced as much as 60 percent (Figure 36.66).

A young person sees very keenly in the presence of very little light, but this sensitivity to low brightness is reduced to approximately one-half at 80 to 85 years

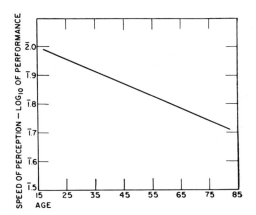

Figure 36.65. Speed of perception is reduced as the age of the eye increases.

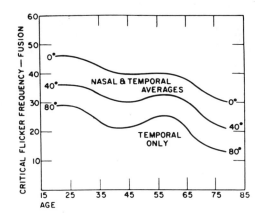

Figure 36.66. The ability to perceive movement at various degrees off center of view exhibits an undulating decline as age increases.

of age, as shown by McFarland et al. (Figure 36.67). Furthermore, the same investigators found (Figure 36.68) that it takes 30 min for the 85-year-old to adapt to lower outdoor night brightnesses. The shape of the curve shows that the older person would be partially blind for a while after having gone outdoors at night.

Most startling of all is the loss of ability to see in the presence of glare sources. Unshielded or poorly shielded light sources with high candlepower directed toward the eyes causes severe losses of visibility, especially to older people (Figure 36.69), yet inadequately shielded light sources are found in many industrial plants. This glare condition is found especially in night driving and in some sport arenas.

The increased need for more illumination for the older worker has been shown (Figures 36.27 and 36.29). Studies (62) have indicated that the working curve of contrast versus luminance (Figure 36.32) is displaced upward for every 10-year

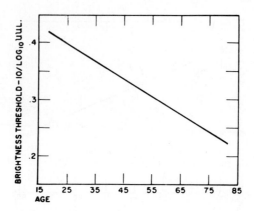

Figure 36.67. In moving from a bright to a dark area, the ability to see is sharply reduced with age (60).

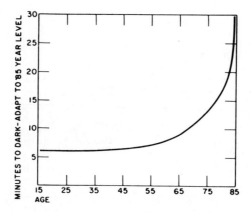

Figure 36.68. Eyes of "young normals" need time to adjust to brightness changes. Eyes aged 20 through 64 years gradually need more time; eyes older than 65 years need a sharp increase in time (60).

interval of age. The displacement is slight for the years 30 to 40, but it is increasingly larger with every succeeding decade of age (Figure 36.70). These displacements are in terms of static viewing (looking fixedly at detail to be seen), but studies are underway for determining the displacements for dynamic viewing such as the scanning of detail being carried out in commerce and industry. If the static viewing is taken as a possible criterion, the additional illumination needed would be as shown in Figure 36.71.

Although the eye decreases in its effectiveness with age there are a growing number of specialists and specific solutions that can help older people remain productive (53). In summary, when looking to provide a comfortable and productive environment for an aging population, consideration should be given to the following:

1. Light and glare can be a problem for older workers. Care must be taken to control light from windows and spotlights. A technique of backlighting of objects may help to reduce glare and focusing problems.

2. Older people have difficulty seeing at low, nighttime levels of light, reading small print, distinguishing similar colors, and coping with glare from VDTs. Consideration should be given to providing stronger, adjustable, task lighting. Other considerations include increasing contrast on stairs and circulation paths, repositioning desks to reduce glare on work and VDTs, and using light sources that would enhance color discrimination lost through the yellowed lens.

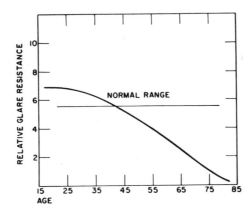

Figure 36.69. The older eye is much less able to resist glare that "washes out" the visual image (61).

10 MOTIVATIONAL FACTORS

Since ancient times, architects, designers, and artists have been using light as an expression of grace, beauty, and art. First, because of the lack of suitable man-made light sources, daylight, and particularly the sun, was used to create highlights and shadows and to focus attention. Now a whole artistic palette of electric light sources is available to give highlights, shadow, diffusion, and color in innumerable patterns depending on the skill and innovation of the designer. People are impressed with and artistically motivated by the beauty of the natural scenes about them, many scenes calling for activity, relaxation, or meditation. In the same way, the moods and behavior of people can be affected by the patterns of electric lighting combined with color and decor of reflecting surfaces. This concept opens up a whole vista of possibilities that seem to defy systematization short of the individual

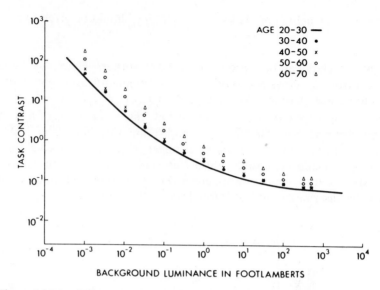

Figure 36.70.   Effect of age on required task contrast for task performance.

ingenuity of the designer. Some, however, have been inspired to make the attempt. Some of the basic elements of good lighting design (63) whether applied by the ancient designer or today's designer would include:

- Source—What source is best, daylight, an electric light (and which one), or a combination?
- All space has a clear hierarchy of importance and purpose. How should the lighting be planned to reinforce this hierarchy? The lighting design should
  - Reinforce spatial clarity
  - Promote productivity
  - Meet the client's budget
  - Fully utilize daylight if available
  - Be readily maintained
  - Be energy efficient
  - Have considered all state-of-the-art alternatives

10.1   Color Appearance And Color Preference

Color perception is influenced by direct and indirect aspects of light. Direct aspects include the physics of light as it is absorbed and reflected by surfaces and the physiology of the eyes' response to the reflected light rays. The physics of light in color perception is discussed in terms of hue, chroma, and value. The physiological aspects are discussed in terms of age, color constancy, and color response. Indirect

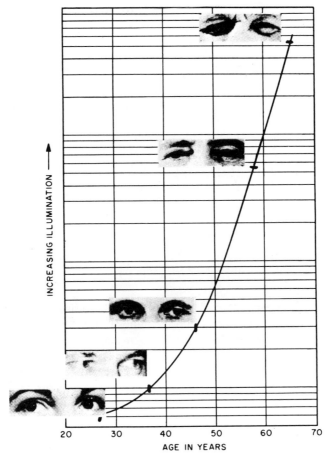

Figure 36.71. For fixed viewing, older eyes need much more light.

aspects of color perception are studied in terms of psychological interpretations of these light rays. These indirect aspects using subjective rating techniques study color memory, constancy, adaptation, and individual expectations of color.

Lighting plays two major roles in the perception of color. It provides the adaptation luminance needed to trigger the eye's photoreceptors and the spectral composition to trigger color vision. The range of luminances over which the eye operates is shown in Figure 36.72. The rod photoreceptors are active at luminances below about 0.034 cd/m² (scotopic vision), and no color vision is possible. The cone photoreceptors, which regulate color vision, are all fully operating at about 3 cd/m² and greater. The other factor affecting color perception is the spectral composition of light reaching the cone photoreceptors. Three types of cone photoreceptors are known today, each with different spectral sensitivities. Because different light sources emit different wavelengths, they produce different signals to the photoreceptors. When the output to the three photoreceptors is the same, colors are

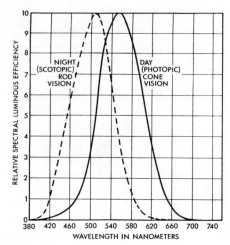

Figure 36.72. The relative spectral sensitivity for photopic (cone) and scotopic (rod) vision. Courtesy of *IES Lighting Handbook 1984 Reference Volume* (32).

indistinguishable from each other. Two colors that are indistinguishable from each other may have radically different spectral compositions. Any two colors that have different spectral compositions but are indistinguishable under a given light source are called metamers. Although metamers appear alike under one light source, they may be easily distinguishable under another. For example, in the purchase of pieces of clothing, under the particular store light source there appears to be a match of garments. Yet under daylight these same garments appear radically different. Other factors that affect color perception are size of the visual field occupied by the color and the context in which the colors are seen. Additionally, a significant minority of the population (8 to 9 percent) has defective color vision. Factors of aging that influence color perception were covered above.

Recent research by Berman et al. (64) has shown that the spectral composition of the light can affect performance of tasks that do not involve color judgment. They found the pupil contracts to different sizes when different light sources (high-pressure sodium and incandescent light) were seen over a large field and balanced for the same photopic luminance. When a large field of view is used, the spectral sensitivity of the mechanism controlling pupil size is dominated by rod photoreceptors. This finding might affect task performance because pupil size is related to visual acuity, retinal illumination, depth of field, and contrast sensitivity.

The resultant changes in task performance due to different light sources are somewhat mixed. Although Piper showed a slight reduction in performance of an achromatic reading task requiring changes in accommodation when lighted by either high-pressure sodium or fluorescent sources, Smith and Rea, on the other hand, found that performance of an achromatic task under light sources commonly used in practice has no affect on performance (65).

The Inter-Society Color Council has been a great aid in isolating and attacking color problems. Its leaders have been of great help to the lighting research community in planning and executing researches in the application of color to the lighting field. As a result, Helson and Judd made studies under average daylight and under various electric lighting sources (66). From this study and previous knowledge, an international system of indexes for rating color rendering of various light sources was developed (67, 68).

Helson and Lansford (69) studied the preferences of a sample of population for various color combinations, using a psychological scaling on a rating of 1 to 10 of viewing samples of color space (125 Munsell color cards) against large Munsell-rated backgrounds (25 widely different chromatic samples). The researchers were able to make 156,000 judgments under five different current color emission light sources (average daylight to incandescent to three fluorescent lights having different color temperatures). Ratings above 5 indicate increasingly preferred color combinations, and those below 5 increasing lack of preference (69). As a result of his studies and those of others, Judd developed a "flattery index" (70) for light sources in complimenting the appearance of human complexions and interior color schemes. This concept has been further developed by Jerome (71).

Today further work is being conducted to augment the early pioneer work of Helson and Judd. Modern researchers in lighting for color judgment work consider the type of light source and the level of illuminance to be the most important aspects. Because most industrial and commercial interiors fall within the illuminance range of 300 to 1000 lux, the choice of light source is the more important of the two.

Practical ramifications of lighting and color perception were considered by Collins (72). She has studied the ability of observers to determine safety colors under seven varying light sources. It was found that the perception of many standardized safety colors was distorted by the spectral composition of the light source used. Collins has suggested alternate safety colors that proved more stable in consistent recognition under a variety of sources, including high-intensity discharge (HID) sources. This emphasizes that choice of light source is critical for color identification.

10.2 Helpful Color Tools

A series of standardized nomenclature, color measurement techniques, color specification systems, color temperature specification, colorimetry of light sources, color rendering, and color application has been developed to promote the effective use of color data. These tools are described in the IES Lighting Handbook (73). An American Standard System of Specifying Color (D1535) designates the Munsell system as a standard of specifying color in terms of hue, value, and chroma. Collections of carefully standardized color chips in mat or glossy surface are available. A very handy tool for the field is the "Munsell Value Scales for Judging Reflectance," which contain a good sampling of color chips from the color space (various hues, value, and chroma), permitting rough judgments of the color specification of unknown color in the field and its reflectance.

A handy guide for the application of color in the lighting field, produced by the Color Committee of the Illuminating Engineering Society, is entitled "Color and the Use of Color by the Illuminating Engineer" (74).

10.3   Subjective Feeling of Interiors

Every interior "has a feeling about it," an impression, an impact that each occupant consciously or unconsciously senses. Naturally the senses record this impression, and the visual system shares a major responsibility. Architects and designers are constantly striving to create a stimulating and pleasurable experience, hoping for a reaction that will result in a behavioral pattern suitable to the function of the room. Psychological scaling has long been a method of rating people's feelings about objects or interiors, and a series of reactions is ranked on a numerical scale such as 1 to 10, or 1 to 100. This scaling procedure has now been developed into two approaches: semantic differential scaling (SDS) and multidimensional scaling (MDS). In semantic differential scaling a series of adjectives, both positive and negative, are used to describe the subjective characteristics of an interior. Such an approach employs a rating sheet with a series of positive adjectives on the left side and the opposing negative adjectives on the right side, with a scale of 6 to 10 divisions in between. The observer, after becoming acclimated to the interior, is to mark the scale to indicate his reaction to each pair of opposite adjectives. For instance, is the interior "neutral," "positive," or "negative"? Or is it negative? How negative? By marking the scales in answer to these questions, the observer has recorded his impression of the feeling of the room. These answers are then put in a computer program, and the important characteristics are brought out. These now appear to be impressions of (1) visual clarity, (2) spaciousness, (3) complexity (or liveliness), (4) personal prominence or anonymity (public vs. private), (5) relaxation or tension, and (6) pleasantness.

Multidimensional scaling involves trying to determine the psychological dimensions that underlie the judgment of differences or similarities between the impression of different lighted interiors. In this approach one takes a given lighted room or lighted arrangement as representing a base figure such as 10 to 100. Then all other rooms or other lighting arrangements in a given room are rated as bearing a higher or lower value, and a computer program is set up to bring out the two, three, or four psychological dimensions that governed the judgment of the higher or lower values.

Out of such studies (75) have come the following three dimensions:

1. The overhead–peripheral mode emphasizing vertical surfaces, unlike conventional overhead lighting, which emphasizes horizontal surfaces
2. The uniform–nonuniform mode (sometimes specular–nonspecular) affecting the articulation or modeling of forms or objects in the room
3. The bright–dim mode, affecting the perceived intensity of light on the horizontal activity plane

These are represented in a three-dimensional plot as shown in Figure 36.73a–c.

The color tone of light tends to warp recognition of these dimensions. Light color might also be a fourth multidimensional scaling element if properly manipulated during the experiment. Glare appears to be another modifier of subjective ratings of light settings, but the evidence is ambiguous, and investigation continues.

Both SDS and MDS provide reasonable consistency between spaces and between groups of subjects of similar backgrounds. Putting both methods together provides insight into the ways in which light settings affect judgment.

Beyond subjective ratings of impressions of spaces due to the lighting distribution, behavior appears to be an influence as well. Studies have shown that choice of passageway, selection, and interactions between people all have been influenced by lighting patterns (23).

11 PHYSIOLOGICAL FACTORS

"Physiological" is defined as "characteristic of or appropriate to an organism's healthy or normal functioning." Therefore fatigue and health effects on organic responses, not part of, but related to the visual process, come under this classification.

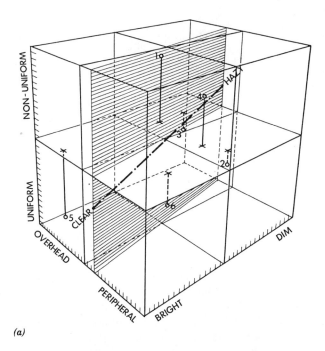

(a)

MULTIPLE REGRESSION COEFFICIENT

U UNIFORM	O OVERHEAD	B BRIGHT
NU NON-UNIFORM	P PERIPHERAL	D DIM

DIMENSION	CLEAR-HAZY SCALE
B/D	.950
B/D+O/P	.983
B/D+O/P+U/NU	.999

Figure 36.73.   Three-dimensional presentation to indicate lighting design decisions affecting (*a*) perceptual clarity, (*b*) pleasantness, and (*c*) relaxation (and tension) (IERI Project 92).

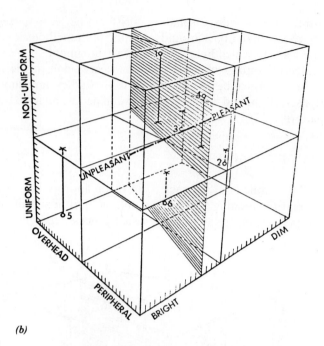

(b)

MULTIPLE REGRESSION COEFFICIENT

U	UNIFORM	O	OVERHEAD	B	BRIGHT
NU	NON-UNIFORM	P	PERIPHERAL	D	DIM

DIMENSION	EVALUATIVE FACTOR
O/P	.833
O/P+NU/U	.921
O/P+NU/U+B/D	.942

Figure 36.73. *(Continued)*

11.1 Fatigue and Eyestrain

Early literature (1900–1920) indicated studies of fatigue (76) that dealt with low levels (less than 5 fc) of illumination. Between 1913 and 1924 Ferree and Rand (77) made studies of the degree of fatigue from different distributions of light (indirect versus direct), different spectral emission light sources, and different colored paper backgrounds. During the same early period there was much discussion of eyestrain, which appeared to be related to the low levels and poor quality of electric lighting. Weston (78), conducting a series of studies in industrial plants of different types for the British Industrial Fatigue Research Board, and later for the Industrial Health Research Board, became convinced that fatigue and eyestrain were due to poor lighting, and upon review of Sir Duke Elder, famous British ophthalmologist, described strain from visual factors, strain from oculomotor factors, and "posteriol" eyestrain. Lancaster (79) describes eyestrain from faulty illumination as ocular discomfort and associated symptoms. Ocular discomfort shows

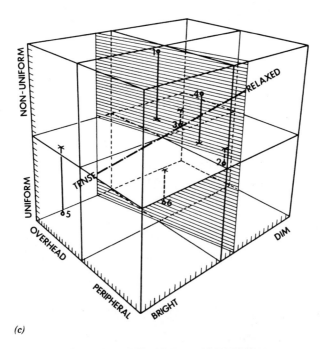

(c)

MULTIPLE REGRESSION COEFFICIENT

U	UNIFORM	O	OVERHEAD	B	BRIGHT
NU	NON-UNIFORM	P	PERIPHERAL	D	DIM

DIMENSION	RELAXED-TENSE SCALE
O/P	.770
O/P+U/NU	.978
O/P+U/NU+B/D	.987

Figure 36.73. *(Continued)*

up as "sandiness," tired, aching feelings of the eyeball, orbit, or head. Associated symptoms appear in terms of headache and feelings of fatigue. If eyestrain is bad enough, it may have indirect symptoms of vertigo and digestive and psychic reactions. However, both Weston and Lancaster were talking about levels comparatively lower than those being used in visual performance studies conducted by Blackwell, Smith, and Rea. As a result there appears to be little or no recent discussion of eyestrain. On the contrary, some ophthalmologists are saying that there is no such thing as eyestrain, which was medically called "asthenopia" by MacKenzie in 1843 and described in various ophthalmological textbooks throughout the years.

One of the forms of eyestrain mentioned by Weston is "eyestrain from oculomotor factors." Hebbard (80) had a similar theory that there might be more oculomotor adjustment activity under lower levels of illumination than under higher levels. He discovered that the amplitude of the microsaccades was significantly greater for 1 fL luminance of the background than for 306 fl (30 percent). Furthermore he found under sudden imposition of glare conditions (changing the surroundings from 0 to 306 fL) that the frequency of the microsaccades increased significantly (200 percent). This would indicate that the oculomotor adjustment system has to work much harder under adverse lighting conditions; therefore, over a working period of one-half to one day's time, there would be fatigue, or under prolonged conditions, eyestrain such as was described in the earlier literature.

11.2 Ease of Seeing

There has always been a dramatic contrast between the levels of illumination of daylight and electric lighting levels. Weston (81) pointed out that lace makers took their tasks involving fine detail and poor contrast out of doors. Campbell et al. (82) described the early weavers' houses in Coventry fitted with tall windows for daylighting. They stated, "The eye is accustomed to the natural illumination provided by sunshine and outdoor daylight, and it is natural for us to seek the best possible daylight for the performance of work which is difficult to see." Sir Duke Elder quoted in *Eyes in Industry* (83) wrote:

> In the care of the eyes the most important general principle is that they should have sufficient light whereby to work. It is to be remembered that the eyes of man were evolved to function in daylight; and although there is a popular superstition that artificial light is bad for them, the only fault lies in the fact that artificial light is rarely provided liberally or adequately distributed.

With the dramatic difference, it was natural for a feeling to arise that there was less fatigue (84) or a greater ease of seeing under daylighting and, by inference, under higher illumination levels. Luckiesh and Moss (10), by using a key for recording the completion of each page of reading (but really a pressure-sensitive key), were able to record the pressure exerted on the key during 308 half-hour sessions of 14 observers. The reading sessions were conducted under 1, 10, and 100 fc. The results showed a steady decrease of muscular tension as the level of illumination was increased (Figure 36.74). Glare was also introduced and made the nervous muscular tension greater.

The same authors, in another series of experiments, used the blink rate as an index of mental tension (Figure 36.75) and found apparent consistent trends related to illumination levels, glare, severe ocular scanning, and simulated refractive errors (85). Others could not reproduce the same effects. Much of the work depends on the careful design of the lighting conditions, such as veiling reflections, subtle differences in discomfort glare, and unrealized transient adaptation effects (86). Some of the nonconfirming researchers were not aware of these effects, but knowledge of them has been developed in later years (6). A recent summary of work on fatigue over the past decade concludes that there is sufficient evidence to support the indication that illuminance can affect the occurrence of fatigue (as measured in terms of decreased performance and increased variability), following the prolonged performance of some tasks. However, a widely applicable range of principles

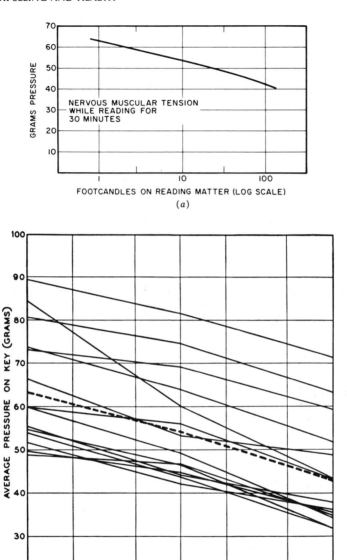

Figure 36.74. (*a*) Mean value of nervous muscular tension under three levels of illumination. (*b*) Relations between indicated nervous muscular tension and level of illumination for each of the 14 subjects; dashed line represents the geometrical mean of the results of the 14 subjects. The probable errors of the average values of nervous muscular tension are less than 1 percent.

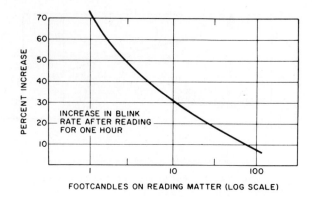

Figure 36.75.   Increase of blink rate with lower levels of illumination.

on the impact of lighting on fatigue can not be stated based on the evidence to date. Illustrative of the problem is a study in which a glare source was present during the paced performance of a task. Although the performance remained constant, variability increased with increased work duration. This seemed to indicate that gradually fatiguing subjects were compensating by increased exertion of effort. Therein lies the problem of how to measure both fatigue and possible compensation. This promises to be a fruitful area for further exploration (23).

11.3   Biological Effects

The ability to maintain a proper luminous input for human biological regulation depends on understanding the interaction and role of lighting for

1. Tissues (i.e., the eye and skin)
2. Circadian rhythms (i.e., the regulation of body functions based on a light/ dark cycle)

These biological responses (see Figure 36.76) are based on

1. Irradiance
2. Wavelength(s)
3. Duration
4. Time of exposure

Steinmetz (87) stated in 1916 that illuminating engineering embraced not only light for seeing but a knowledge of the effect of radiation from light sources involving the physiologist. Gage (88) in 1930 reviewed the work of physiologists who had discovered in the 1920s that the overcoming of rickets, the strengthening of bones, the reduction of colds, and the treatment of tuberculosis all are affected by sunshine

PHYSICAL PARAMETERS OF PHOTIC INPUT

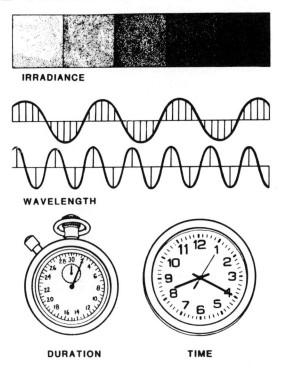

Figure 36.76.   Factors affecting biological responses in humans.

and simulated sources of sunshine such as mercury lamps. During the 1930s the lighting industry developed sunlamps S_1, S_2, RS, and later fluorescent sunlamps, which together with light control luminaires produced dual-purpose lighting—light for seeing and ultraviolet radiation in the zone found in sunlight from 280 nm to the visible region. Complaints regarding the use of the emerging fluorescent lamps for general lighting because of suspicions of harmful ultraviolet radiation (89) dampened the enthusiasm for dual-purpose lighting. Furthermore, some dermatologists became concerned with the carcinogenic effects of sunlight and urged that all sunlight be excluded. With the rise of skin cancer in the 1980s, dermatologists continue to urge use of protective lotions and avoidance of summer or tropical sun exposure from 11 a.m. to 2 p.m., and issue warnings concerning tanning booths. Some claims linked fluorescent lighting to skin cancer; however, further research did not establish this (90). Light therapy being used in overcoming respiratory diseases has served to renew interest in the use of sunlight and electrically simulated daylight (91, 92). Currently light sources such as white or blue fluorescent lamps are being successfully used to overcome jaundice in the newborn, a condition that occurs in 15 to 20 percent of all premature infants. The portion of the spectrum responsible for photodegradation of bilirubin was found to be in the blue region (Figure 36.77). Lucey (94), who first reported his work in 1968, led the United States in implementing the discovery of Cremer et al. (93).

Another study on calcium absorption to avoid osteoporosis was carried out by Neer et al. (95). It was found that the small ultraviolet component of currently available fluorescent lamps, when used at a level of 5000 lux, stimulated significantly more calcium absorption. This effect appears to be related to the role of vitamin D_3 in curing rickets (96). As people grow older, they go out less and less for daylight exposure, and stay behind ultraviolet-absorbing glass; this can result in less calcium being absorbed from their food, with a resulting weakness of the bones (osteomalacia). Mechanical shock may break their bones, and the recovery is very slow or indefinite. Wurtman (91) describes a study by Jean Aaron in England whose autopsies in winter reveal more cases of osteomalacia than those in the summer.

It has been established that light has other curative values for disease. Psoriasis is an affliction characterized by the "eruptions of large rough red areas that are very itchy and disfiguring" (92). Parrish et al. (97) have developed a successful treatment consisting of ingesting a photoactive drug called Psoralens, and after a 2-hr time delay, irradiating the area of the psoriasis with ultraviolet near the visible spectrum, 320 to 400 nm (UV-A). The photoactive chemical is found in nature in vegetables and fruits such as carrots, parsnips, celery, parsley, limes, lemons, and figs.

Diamond (98) has taken an avenue of approach for the treatment of cancer. He has used the property of the chemical hematoporphyrin, having been given in a dose, to stay in the malignant cells while clearing out of the normal cells. When hematoporphyrin is excited with a beam of red light, it generates an excited state of the oxygen molecule (oxygen singlet), which destroys the cell in which it is located.

With the increased success of cataract surgery, eye specialists have been studying safety limits for exposure to light from daylight and various man-made sources (54, 99). With the eye's natural protection removed, artificial means must be employed to remove effectively certain wavelengths that would damage the visual system.

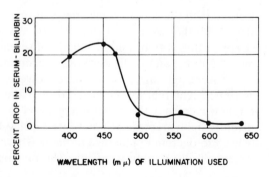

Figure 36.77. Wavelength effectiveness in reducing bilirubin in serum.

Further, eye specialists are looking at the cumulative effects of absorbed radiation to understand if cataracts and other eye afflictions can be retarded or prevented through screening certain wavelengths throughout life. These findings are being used to set safety standards (55).

In the 1970s the whole field of study of photobiology and biological rhythms, tracing the biological influence of light, began to catch researchers' and the public's attention. Wurtman (91) and others (100) described evidence that the duration and color of light affect the hormones, the bodily rhythms, and the sexual activity of animals and humans. Many anecdotal write-ups were given widespread coverage but when they were probed for a scientific basis, many of the studies could not be replicated because of lack of adequate controls (101). As a result of confusion in the field, the Lighting Research Institute began its program on the health and biological effects of light. Early basic animal research was conducted by Brainard et al. and Hoffmann et al. (102, 103) to establish a base line for further study of the human's response to light. It was found that if lighting parameters were not carefully controlled, results were inconclusive and unrepeatable. Light was shown clearly to have an impact on various animals' biological systems, activity level, growth, life-span, and reproductive responses (103). The impact of various light sources to produce biological effects were also studied. Three sources, fluorescent, high-pressure sodium, and fluorescent simulating 5500K natural daylight in the ultraviolet and visible regions were equalized on the basis of the visual sensitivity curve common to nocturnal animals. Although the animals responded to the intensity of light, there was no significant difference between the results based on the different spectral power distributions. This sets a further standard for future studies. Earlier published studies were flawed because they failed to distinguish between brightness and wavelength in nocturnal animals. Brainard has studied basic physiological mechanisms. He found that light entering through the visual system regulates the production of hormones and affects the metabolism of specific areas of the brain (102, 104). In humans, as in some other species, biological functions are regulated on a light–dark cycle. These approximate 24-hr oscillations of the physiology of the body are called circadian rhythms. These rhythms must be reset (entrained) every day by a natural or man-made light–dark cycle. Evidence is mounting that many in our 24-hour society are suffering from circadian disorders that range from jet lag, shift work accidents, and insomnia, to seasonal affected disorder (SAD) (105–108). By understanding circadian physiology new therapeutic approaches are being used to alleviate circadian disorders. Lewy et al. (105) have found that bright morning light (2000 to 2500 lux) is effective in treating SAD patients. This is four to five times current normal office levels. If research establishes that the general population is at risk because of isolation to lighting cues necessary each day to set the circadian clock, then some alteration of daily routine will be needed to ensure daily entrainment. Where light therapy is required professional guidance should be sought. A society of specialists has recently been organized to disseminate information on light therapy (109). Questions being studied today include: How can workers' alertness be maintained during shift work? What is the daily dose of light needed by the general population? Do current lighting standards meet this daily light requirement?

From all the studies mentioned above it would appear that light can very seriously affect the health and general welfare of humans. Much more knowledge is needed in the establishment, through design of the role of lighting for good health as well as good vision.

12   APPLICATION INFORMATION

12.1   Lighting Design

With the rapid advance of lighting technology and techniques, the increased concerns of environmental comfort for building occupants and the need to utilize our energy resources strategically, there is a growing need for lighting professionals who can integrate all these aspects. Traditionally architects and designers have turned to consultants such as electrical engineers to meet their lighting design requirements. Manufacturers and their representatives have also been heavily involved in lighting design. During the late 1960s, in addition to practicing illuminating engineers, a small group of independent lighting consultants evolved offering lighting design on a fee basis. A professional organization of lighting designers was then established, the International Association of Lighting Designers. Since then the role of lighting design as a profession has steadily grown (110). When consideration is being given to building a new structure or remodeling a space, a lighting professional should be involved as part of the design team of the architect, interior designer, and consulting engineer. The lighting professional will be able to establish a program and goals for lighting the space(s) to include integrating natural light if available, visualizing how the space will look, and then producing the desired results through the proper specification and installation of lighting systems to manipulate the color, intensity, direction, and distribution of light. There is a vast array of choices that make up a lighting system from light sources, luminaires (fixtures), and controls in possible combination with daylight. All of these aspects must address energy efficiency as discussed below.

12.2   Electric Light Sources

The designer has a vast repertoire of available "bulbs and tubes" in various shapes, sizes, wattages, and colors. These vary all the way from a single wavelength of light, such as low-pressure sodium with its yellow light, to the blue-green of mercury with its individual four-wavelength emissions, to the white multiwavelength emissions of metal halide and the golden high-pressure sodium. They vary in efficacy from 8 lumens of light per watt of power (small incandescent) to 180 lm/W (low-pressure sodium). There are two main categories of lamp, incandescent and gaseous discharge. Excellent references on these sources are available from the Illuminating Engineering Society of North America (111). A comparison chart of some aspects of the various sources commonly found in industrial applications is shown in Figure 36.78.

12.3 Incandescent Lamps

Hot tungsten wires that are high enough in temperature to glow with a pale yellowish white light are the basis of incandescent lights, which are available for general service in pear-shaped bulbs from 10 to 1500 W with the efficacy of 8 lm/W for the 10-W size to 24 lm/W for the 1000-W size. They are also available in parabolic and elliptical reflector bulbs. Different voltages include low, medium, and high voltages for special applications such as flashlights, bicycles, trains, locomotives, aircraft, and automotive vehicles. The life of the general service lamps varies from 750 to 1000 hr.

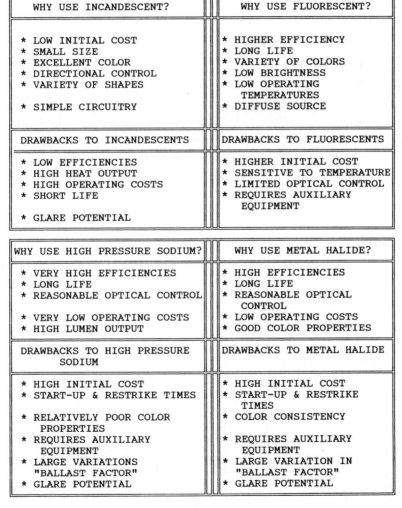

WHY USE INCANDESCENT?	WHY USE FLUORESCENT?
* LOW INITIAL COST * SMALL SIZE * EXCELLENT COLOR * DIRECTIONAL CONTROL * VARIETY OF SHAPES * SIMPLE CIRCUITRY	* HIGHER EFFICIENCY * LONG LIFE * VARIETY OF COLORS * LOW BRIGHTNESS * LOW OPERATING TEMPERATURES * DIFFUSE SOURCE

DRAWBACKS TO INCANDESCENTS	DRAWBACKS TO FLUORESCENTS
* LOW EFFICIENCIES * HIGH HEAT OUTPUT * HIGH OPERATING COSTS * SHORT LIFE * GLARE POTENTIAL	* HIGHER INITIAL COST * SENSITIVE TO TEMPERATURE * LIMITED OPTICAL CONTROL * REQUIRES AUXILIARY EQUIPMENT

WHY USE HIGH PRESSURE SODIUM?	WHY USE METAL HALIDE?
* VERY HIGH EFFICIENCIES * LONG LIFE * REASONABLE OPTICAL CONTROL * VERY LOW OPERATING COSTS * HIGH LUMEN OUTPUT	* HIGH EFFICIENCIES * LONG LIFE * REASONABLE OPTICAL CONTROL * LOW OPERATING COSTS * GOOD COLOR PROPERTIES

DRAWBACKS TO HIGH PRESSURE SODIUM	DRAWBACKS TO METAL HALIDE

* HIGH INITIAL COST
* START-UP & RESTRIKE TIMES
 	
* RELATIVELY POOR COLOR
  PROPERTIES
* REQUIRES AUXILIARY
  EQUIPMENT
* LARGE VARIATIONS
  "BALLAST FACTOR"
* GLARE POTENTIAL
	* HIGH INITIAL COST
	* START-UP & RESTRIKE
	  TIMES
	* COLOR CONSISTENCY
	
	* REQUIRES AUXILIARY
	  EQUIPMENT
	* LARGE VARIATION IN
	  "BALLAST FACTOR"
	* GLARE POTENTIAL

Figure 36.78. Criteria for selecting light sources: color, efficiency, life, size and shape, price.

A longer-life, concentrated incandescent lamp, called the tungsten halogen lamp, has been developed. In this design the filament is put in a small quartz tube with a negative-valence element such as iodine, bromine, chlorine, and fluorine. As this lamp burns, the tungsten gradually evaporates off the filament and normally would collect on the bulb wall as blackening; instead, it combines with the halogen that is in the quartz tube to form a gas. As this gas approaches the filament, it breaks down and deposits tungsten on the filament. These lamps maintain a good output over their life and last up to 3500 hr depending on a particular size and design. They are also available in reflector-type bulbs.

Architects and many designers like to use incandescent lamps because of the complimentary effects on human complexions and on warm colors. They are an extension of candlelight and the light of the fireplace. The emitted light emphasizes the yellows and reds but deemphasizes the blues and greens.

12.4 Fluorescent Lamps

Fluorescent lamps are essentially a mercury gaseous discharge in a long tube, coated inside with phosphor powder. The phosphor powder reacts to the invisible mercury emission spectral line 253.7 nm and converts the invisible energy into visible light. By using various chemical phosphors usually in a mixture, colors of visible light are emitted, and the emission becomes a continuous spectral output. Aside from particular spectral color emphasis such as blue, green, gold, pink, and red, the phosphor mix most commonly used is such that the emission tends to follow the CIE spectral luminous efficiency curve of the eye for photopic (or day) vision. By an appropriate shift in the mix, the emission emphasis moves toward the warm side (warm white), in having proportionately more yellow and red light, or cool white, with greater proportion of blue and green. For good portrayal of human complexion and all colors, one could use the deluxe cool white or deluxe warm white, where a proportionately greater red emission is present.

Fluorescent lamps require a ballast to limit the current and in many instances to transform the supply voltage to a suitable level. Lamp performance is influenced by the character of the ballast and luminaire, line voltage, ambient temperature, burning hours per start, and air movement.

The efficacy of fluorescent lamps (111) varies from approximately 30 to 80 lm/W for lamps of standard cool white color, depending on bulb size and shape [lamp length ranges from 6 to 96 in. (0.5 to 244 cm)]. This does not include losses in the ballast, typically on the order of 20 percent. Most fluorescent lamps are of conventional tubular design, but there are some special types such as circular, reflectorized, jacketed, and panel shaped. Recent developments in fluorescent technology include improved color rendition, longer life, compact sources, smaller diameters, and reduced wattages, as well as variety of wattages available. The compact fluorescents are becoming common replacements for incandescent sources. When considering retrofitting an incandescent source with a compact fluorescent source, careful consideration must be given to the selection in order to achieve the same light output and distribution.

12.5   High-Intensity Discharge Lamps

High-intensity discharge (HID) lamps (73) are also electric discharge sources. The basic difference between HID and fluorescent lamps is that the operating pressure of the arc is much higher in HID lamps. HID lamps include the groups commonly known as mercury, metal halide, and high-pressure sodium lamps. Spectral characteristics are different because the higher pressure arc emits a large portion of its visible light. HID lamps produce full light output when full operating pressure has been reached, which generally requires several minutes. Most HID lamps have both an inner and an outer bulb. The inner is made of quartz or polycrystalline aluminum, and the outer bulb is generally made of thermal shock-resistant glass. Light output is practically unaffected by surrounding temperatures. Like fluorescent lamps, HID lamps also require devices that limit the current. Recently Lewin and Stafford (112) have found different combinations of ballasts, lamps, and luminaires can result in up to a 40 percent reduction in actual light output versus what was predicted. These new equipment factors will allow more targeted lighting designs. Other recent advances in HID source technology have provided a wider range of available wattages, more compact sizes, improved re-strike times, and improved color rendition.

Mercury lamps for general lighting are available with either "clear" or phosphor-coated bulbs of various sizes and shapes with wattages of 40 to 1000 W. There is also a 3000-W clear mercury lamp. Typical efficacies range from 30 to 63 lm/W, not including a 6 to 20 percent power loss in the ballast. "Clear" mercury lamps produce light that is rich in yellow and green tones and almost lacking in red. Various types of phosphor-coated lamps are available to provide improved color, and they are more popular than clear lamps for industrial lighting. A number of special mercury lamps are available, including semireflector, reflectorized, and self-ballasted.

Metal halide lamps are very similar in construction to the mercury lamp, the major difference being that the arc tube contains various metal halides in addition to mercury. They are available with either clear or phosphor-coated bulbs of various sizes and shapes from 175 to 1500 W. Typical efficacies range from 70 to 100 lm/W not including power loss in the ballast. The metal halide additives improve the efficacy and color compared to a clear mercury lamp. Further color improvement is achieved with the addition of phosphor coatings to the lamp. Recent development in metal halide sources have also included better re-strike times, improvements in color, an extended range of wattages, and compact sizes.

In the high-pressure sodium (HPS) lamp, light is produced by electricity passing through sodium vapor. The lamps are available in wattages from 250 to 1000 W. Typical efficacies range from 100 to 130 lm/W, not including power loss in the ballast. This lamp produces a golden white light. HPS technology has benefited from the availability of compact sizes, a wider range of wattages, improvements in restrike times and some improvements in color.

In low-pressure sodium (LPS) discharge lamps, the arc is carried through vaporized sodium. The starting gas is neon with small additions of argon, xenon, or helium. To maintain the maximum efficacy of the conversion of the electrical input to the arch discharge into light, the vapor pressure must be at a given level and constant, and the proper operating temperature must be maintained. The arc tube must be enclosed in a vacuum flask or in an outer bulb at high vacuum, to ensure that the temperature is maintained. The light produced by low-pressure sodium arc is almost monochromatic, consisting of a double line in the yellow region of the spectrum at 589 and 589.6 nm.

Two types of arc tube construction are used in modern low-pressure sodium lamps—the hairpin or "U" tube and the linear. The efficacy is 180 lm/W without ballast.

13   DAYLIGHTING

Over the centuries humans have attempted to bring daylight to our indoor visual activities. Central courts open to the sky allowed the light to penetrate into the covered areas. In northern latitudes churches and guild halls had high windows to admit daylight as far as possible into the interior, permitting work to be carried on in spite of inclement weather. All this was necessary because of the weakness of man-made sources. Since World War II there has been decreasing dependence on daylighting because of availability of far more efficient electric light sources (fluorescent lights and the developing use of high intensity discharge sources). Architects and designers found they could reduce costs of buildings by lowering the ceilings (reducing the height of the windows). Although they continued to use large areas of glass, workers near the windows became overwhelmed by the high levels of daylight as well as the greater heat loads. Accordingly, low transmission glass (neutral tinted with 15 to 50 percent transmission) was developed to overcome these drawbacks. Furthermore, to reduce the amount of dirt accumulated, the windows were sealed and the interiors air conditioned. It was found that the daylighting, and particularly the sunload on the air conditioning, was appreciable. This has resulted in reflecting-type window glazing, where a large part of the heat is reflected back outside and a low transmission of light is maintained.

Our discussion of transient adaptation pointed out that encountering higher luminances than the task luminance greatly reduces the immediate scanning sensitivity of the visual system (Figure 36.57). The sky luminances as seen through clear glass windows may range between 10 and 100 times the task luminance; therefore the scanning sensitivity is severely reduced. This and a significant degree of discomfort glare probability accounted for the trend toward low transmission glass. In the older industrial plants with large expanses of glass in the walls or the roof, it is common to see the panes covered with paint to reduce the outdoor luminances, with the aim of achieving better balance with interior luminances.

Nevertheless window expanses with appropriate limited luminance do provide a significant illumination to the peripheral areas of a building. For peripheral offices the light that approaches desk tasks at wide angles with the vertical does a most effective job of overcoming veiling reflections and producing high visibility. Illumination from windows is not significant at a distance beyond twice the height of the windows above the floor.

Reflected ground light can make an important contribution by being directed onto a white ceiling as soft indirect illumination. A scheme of having low transmission vision strip opposite the eye level with clear glass above to admit ground light but with the view of the clear glass occluded by a horizontal shield or louvers is helpful (113).

Another system is to use silvered louvers that catch the sunlight (in high percentage sunlight areas) and direct the light out across the ceiling for indirect lighting (114).

Many structural forms such as unilateral, bilateral roof monitor, clerestory, sawtooth, and skylight have been studied as to the effective admission of daylight (115). A Recommended Practice of Daylighting (116), available from the Illuminating Engineering Society of North America, has guidelines for designing for daylight. Limitations of outdoor luminances with respect to transient adaptation and discomfort glare should be borne in mind.

During the mid-1970s architects in the United States rediscovered daylighting in the wake of the oil embargo. The federal government, through the Department of Energy, began studying how daylight could be used as a force to reduce the energy consumption of electric lighting. Research has been conducted on daylighting control, glare, chiller and heating, ventilation, and air-conditioning (HVAC) size, solar gain control, fading of interiors, and so on. Some benefits of a properly integrated daylighting–electric lighting and HVAC design include reduced electric lighting when daylight is available and reduced peak electrical demand for utilities. Sophisticated techniques are being developed in the laboratory including "smart windows" that can change their transmissiveness based on an electric current being applied to a specially impregnated window, special daylighting controls that dim the electric lights in response to daylight availability, occupant sensors to turn room lights on or off, and rating systems for various combination of windows and shading devices (117).

14 ENERGY MANAGEMENT

Beginning in 1944 through 1981 the Illuminating Engineering Research Institute (IERI), a trust affiliate of the Illuminating Engineering Society of North America (IESNA), was concerned with the efficient operation of the visual system for the most efficient use of human resources. This concern is being carried on by the IERI successor, the Lighting Research Institute. The pattern of these university-determined results has been reflected in the preceding pages of this chapter. The amount of light on the details of the visual task, the reduction or elimination of veiling reflections for optimal visibility, the proper brightness balance for optimal scanning sensitivity (transient adaptation), the avoidance of discomfort glare and disability glare, the motivation of good subjective feeling, and the provision of good biological health all contribute to optimal ability to perform efficiently.

Having provided the optimal use of human resources, one can begin to plan the strategy of the use of lighting energy (118). Research data do not indicate the need for illumination for a given visual performance over all the habitable space, but only on the task area, as long as the surrounding luminances are in balance with task luminance to avoid serious transient adaptation losses, disability glare, or discomfort glare. Therefore there can be nonuniform lighting with less light in the nontask areas (119, 120). These strategies must be skillfully employed; otherwise, as a recent post-occupancy evaluation study has shown, a large percentage of occupants might become dissatisfied (121). The IESNA has developed specific recommendations for nonuniform lighting, more efficacious light sources (fluorescent and HID), better and more efficient design of luminaires, combining daylight with electric lighting, providing controls for flexible use of lighting, using lighter finishes on room surfaces, and providing for better maintenance (122). To aid architects and designers, there is the "IES Recommended Lighting Power Budget Determination Procedure" (EMS-1), supplemented with "An Interim Report Relating the Lighting Design Procedure to Effective Energy Utilization" (EM2), and "Example of the Use of the IES Recommended Lighting Power Budget Determination Procedure" (EM3). Renewed concern for the environment and electrical generation shortages have focussed national attention on employing greater lighting efficiency (123). This must be done in context of the human requirements stated above. Education is the key to combining the need for energy efficiency with human productivity (124). New lighting courses and advanced technology guidelines are available and others are being developed that will meet the needs of nonexperts as well as professional practitioners.

Refer to Appendix A for currently recommended illuminance categories and values for industrial lighting design (target maintained levels).

REFERENCES

1. Matthew 6:22, *The Holy Bible, New International Version,* New York International Bible Society, Zondervan, MI, 1977.

2. R. L. Vincent, *LRI Research and Education Program: Status Report,* Lighting Research Institute, New York, 1989; "Research Priorities"; *1987 Annual Report,* Lighting Research Institute, New York, 1987; "LRI Research Agenda, *LD + A,* 23–26 (July 1984); "Illumination Roundtable III: Lighting Research and Education for the Eighties," *LD + A,* 23–26 (July 1984).

3. Commission Internationale de l'Eclairage, *Unified Framework of Methods for Evaluating Visual Performance Aspects of Lighting,* CIE Publication 19, CIE, Paris, 1972; Commission International de l'Eclairage, *An Analytical Model for Determining the Influence of Lighting Parameters on Visual Performance,* Vol. 1: *Technical Foundations,* Vol. 2: *Summary and Application Guidelines,* CIE Publication 19/2, CIE, Paris, 1981; P. R. Boyce, S. M. Berman, B. L. Collins, A. L. Lewis, and M. S. Rea, *Lighting and Human Performance: A Review,* Report to Lamp Division, National Electrical Manufacturers Association (NEMA) and the Lighting Research Institute, NEMA, Washington, DC, 1989; and W. Adrian, "A Model To Calculate The Visibility of Targets," *Lighting Research and Technology,* CIBSE, London, November 1989.

4. E. W. Bickford, *Human Circadian Rhythms: A Review*, Report to Lighting Research Institute, LRI Project 88 DR NEMA 2, Lighting Research Institute, New York, 1988.

5. C. L. Crouch, "Lighting For Seeing," Chapter 13, *Patty's Industrial Hygiene and Toxicology*: 3rd rev. ed., Vol. 1, *General Principles*, G. D. Clayton and F. E. Clayton, Eds., Wiley, New York, 1978.

6. "Light and Vision," in *IES Lighting Handbook*, 5th ed., Illuminating Engineering Society, New York, 1972, pp. 3–8.

7. S. K. Guth and J. F. McNelis, *Illum. Eng.*, **63**, 32 (1968); **64**, 99 (1969).

8. R. J. Lythgoe, "Measurement of Visual Acuity," Special Report No. 173, Medical Research Council, His Majesty's Stationary Office, London, 1932.

9. H. R. Blackwell, *Illum. Eng.*, **54**, 317 (1959).

10. M. Luckiesh, *Light, Vision and Seeing*, 1st ed., Van Nostrand, New York, 1944, pp. 158–165.

11. S. K. Guth, *J. Am. Ind. Hyg.*, **23**, 359 (1962).

12. G. A. Fry, "Blur of the Retinal Image of An Object Illuminated by Low Pressure and High Pressure Sodium Lamps," *J. IES*, 158–164 (April 1976).

13. W. R. Stevens and C. A. P. Foxell, *Light and Lighting* (London), **48**, 419 (1955).

14. P. W. Cobb and F. K. Moss, *J. Franklin Inst.*, **199**, 507 (1925).

15. P. W. Cobb, *J. Exp. Psychol.*, **1**, 540 (1916).

16. C. E. Ferree and G. Rand, *Trans. IES*, **22**, 79 (1927).

17. C. E. Ferree and G. Rand, *Trans. IES*, **23**, 507 (1928).

18. S. W. Smith, in: *1974 Annual Report*, Illuminating Engineering Research Institute, New York, pp. 10–22; *J. IES*, **5**, 235 (1976).

19. S. W. Smith and M. S. Rea, "Proofreading under Different Levels of Illumination," *J. IES*, **8**(1), 47 (1978); Annual Report of IERI, New York, 1975.

20. H. C. Weston, "The Relation Between Illumination and Industrial Efficiency: (1) The Effect of Size of Work," Joint Report of the Industrial Health Research Board (Medical Research Council) and the Illumination Research Committee (Department of Scientific and Industrial Research), Her Majesty's Stationary Office, London, 1935; "The Relation Between Illumination and Visual Efficiency—The Effect of Brightness Contrast," Joint Report of the Industrial Health Research Board (Medical Research Council) and the Illumination Research Committee (Department of Scientific and Industrial Research), Her Majesty's Stationary Office, London, 1944.

21. G. A. Fry, "Assessment of Visual Performance," *Illum. Eng.*, 426–437 (June 1962).

22. Commission Internationale de l'Eclairage, *Unified Framework of Methods for Evaluating Visual Performance Aspects of Lighting*, CIE Publication 19, CIE, Paris (1972); Commission International de l'Eclairage, *An Analytical Model for Determining the Influence of Lighting Parameters on Visual Performance*, Vol. 1: *Technical Foundations*, Vol. 2: *Summary and Application Guidelines*, CIE Publication 19/2, CIE, Paris, 1981.

23. P. R. Boyce, S. M. Berman, B. L. Collins, A. L. Lewis, and M. S. Rea, *Lighting and Human Performance: A Review*, Report to Lamp Division, National Electrical Manufacturers Association (NEMA) and the Lighting Research Institute, NEMA, Washington, DC, 1989.

24. O. M. Blackwell and H. R. Blackwell, "Individual Responses to Lighting Parameters for A Population of 235 Observers of Varying Age," *J. IES*, 205–232 (July 1980); O. M. Blackwell and H. R. Blackwell, "Recent Investigation of Glare as a Factor in Visibility at Night," *Night Vision Current Research and Future Directions, Symposium*

Proceedings, Working Group on Night Vision, Committee on Vision, National Academy Press, Washington, DC, 1987; J. J. Vos, "Disability Glare—A State of the Art Report," *CIE J.*, **3**(2), 39–53, 1984; R. E. Greule, "*Contrast Thresholds at Transient Adaptation*," *CIE J.*, **8**(1), 25–26 (1989).

25. 1979 Annual Report, Illuminating Engineering Research Institute, New York, 1979, pp. 5–22; H. R. Blackwell and O. M. Blackwell, *Final Report IERI Project 30B*, Report to the IERI, New York, 1984.

26. W. K. Adrian, "Die Unterschiedsempfindlichkeit des Augs und Die Moglichkeit Ihrer Berechnung," *Lichtechnik*, **21**, 118 (1969); B. Indistsky, H. W. Bodmann, and H. J. Fleck, Visual Performance—Contrast Metric—Visibility Lobes—Eye Movements," *Lighting Res. Technol.*, **14**, 218 (1982); M. S. Rea, "Toward a Model of Visual Performance: Foundation and Data," *J. IES*, **15**, 41 (1986); S. W. Smith and M. S. Rea, "Check Value Verification Under Different Levels of Illumination," *J. IES*, **16**(1), 21–38 (1987).

27. W. Adrian, "Visibility Levels Under Night-time Driving Conditions," *JIES*, 3–12 (Summer 1987); W. Adrian, "A Model to Calculate the Visibility of Targets," *Lighting Res. Technol.* **21**(4), 181–188 (November 1989).

28. M. E. Keck and R. E. Stark, "Evaluation of Visibility Models in the Roadway Situation," *Proc. CIE 21st Session, Venice*, 1987.

29. C. L. Crouch and R. L. Vincent, "Survey of Illumination Needs for Coal Processing Plant Areas both Interior and Exterior," Report to Bituminous Coal Research Institute by the IERI, New York, 1977; C. L. Crouch and R. L. Vincent, "Survey of Illumination Needs of Tasks for Safe Work Performance on Walkways and Work Areas of Mobile Surface Mining Machinery," Report to Mine Safety Appliances Company by the IERI, New York, 1980; C. L. Crouch and R. L. Vincent, "Survey of Illumination Needs for Safe and Productive Performance for Shipboard Tasks in: The Underway Replenishment Area, The Destroyer Refueling Area, The Jet Engine Test Cell Area, and the Hangar Bay Area," Report to the Naval Air Systems Command, Dept. of the U.S. Navy, by the IERI, New York, 1980.

30. L. A. Jones, *Phil. Mag.*, **39**, 96 (1920); C. Dunbar, *Trans. IES.* (London), **5**, 33 (1940); C. L. Cottrell, *Illum. Eng.*, **46**, 95 (1951); A. E. Simmons and D. M. Finch, *Illum. Eng.*, **48**, 517 (1953); A. A. Eastman, *Illum. Eng.*, **63**, 37 (1968); H. R. Blackwell, *Illum. Eng.*, **65**, 267 (1970).

31. C. L. Crouch and R. L. Vincent, "Disability and Discomfort Glare Field Studies Under Low Luminance Conditions," *Proc. CIE 20th Session, Amsterdam, 1983*. C. L. Crouch and R. L. Vincent, "Survey of Illumination Needs for Safe and Productive Performance of Flight Deck Tasks and Measurements of Disability Glare," Report to the Naval Air Systems Command, Dept. of the U.S. Navy, by the IERI, New York, 1981.

32. J. E. Kaufman, and J. F. Christensen, eds., *IES Lighting Handbook, Application Volume*, Illuminating Engineering Society of North America, New York, 1987; J. E. Kaufman and J. F. Christensen, eds., *IES Lighting Handbook, Reference Volume*, Illuminating Engineering Society of North America, New York, 1984.

33. M. Kohn, "Lighting Offices Containing VDTs," *LD+A*, 9–11 (December 1988).

34. IES Recommended Practice For Lighting Offices Containing Computer Visual Display Terminals, (RP-24) Illuminating Engineering Society of North America, New York, 1990.

35. W. K. Adrian and M. L. Fleming, *Review of the Literature on Vision and Visual Display Terminals (VDT) and Lighting Interactions*, University of Waterloo, Waterloo, Ontario, Canada, Report to the Lighting Research Institute, New York, 1989.

36. American National Standard Practice for Industrial Lighting (RP-7), Illuminating Engineering Society, New York, in press 1991.

37. J. F. Chorlton and H. F. Davidson, *Illum. Eng.*, **54**, 474 (1959); H. R. Blackwell, *Illum. Eng.*, **58**, 161 (1963); W. Allphin, *Illum. Eng.*, **58**, 244 (1963); C. L. Crouch and J. E. Kaufman, *Illum. Eng.*, **58**, 277 (1963).

38. R. L. Henderson, J. F. McNelis, and H. G. Williams, *J. IES*, **4**, 150 (1975).

39. F. K. Sampson, "Contrast Rendition in School Lighting," Technical Report No. 4, Educational Facilities Laboratories, New York, 1970.

40. I. Lewin and J. W. Griffith, *Illum. Eng.*, **65**, 594 (1970); N. S. Florence and S. B. Glickman, *Illum. Eng.*, **66**, 149 (1971).

41. I. Goodbar, L. J. Buttolph, E. J. Breneman and C. L. Crouch, *LD+A*, **4**, 45 (1974).

42. I. Goodbar, in 1973 Annual Report, Illuminating Engineering Research Institute, New York, pp. 22–29.

43. L. J. Buttolph, in 1973 Annual Report, Illuminating Engineering Research Institute, New York, p. 34.

44. R. M. Boynton, E. J. Rinalducci, and C. Sternheim, *Illum. Eng.*, **64**, 217 (1969).

45. G. A. Fry, and V. M. King, *J. IES*, **4**, 307 (1975).

46. Tungshang Liu, "Discomfort Glare: Luminaire Light Distribution, Vehicle Speed and Background In a Dynamic Roadway Lighting Simulation," Masters Thesis submitted to the Lighting Research Institute by Dr. S. Konz, Kansas State University, LRI Project 84 SP 7, 1989.

47. C. A. Bennett, "Discomfort Glare: Demographic Variables," Special Report No. 118, Kansas Engineering Experimental Station, Kansas State University, Manhattan, KS, 1976.

48. "Outline of a Standard Procedure for Computing Visual Comfort Ratings for Interior Lighting," *J. IES*, **2**, 328 (1973); S. K. Guth, *Illum. Eng.*, **58**, 351 (1963); "An Alternate Simplified Method of Determining Acceptability of a Luminaire from the VCP Standpoint for Use in Large Rooms," *J. IES*, **1**, 256 (1972).

49. C. L. Crouch and R. L. Vincent, "Disability and Discomfort Glare Field Studies Under Low Luminance Conditions," *Proc. CIE 20th Session, Amsterdam, 1983*.

50. L. L. Holladay, *J. Opt. Soc. Am.*, **12**, 279 (1926); W. S. Stiles, *Proc. CIE*, 1928, 220; G. A. Fry, *Illum. Eng.*, **50**, 31 (1955).

51. A. J. Fisher, and A. W. Christie, *Vision Res.*, **5**, 565 (1965).

52. "Public Lighting Needs," *Illum. Eng.*, **61**, 585 (1966); through The World Almanac, *New York World Telegram*, New York, 1964.

53. R. Cowen, *Eyes on the Workplace*, for the Committee on Vision, Commission on Behavioral and Social Sciences and Education, National Research Council, National Academy Press, Washington DC, 1988, and *Work, Aging, and Vision*, Working Group on Aging Workers and Visual Impairment, Committee on Vision, Commission on Behavioral and Social Sciences and Education, National Research Council, National Academy Press, Washington DC, 1987; P. R. Boyce, S. M. Berman, B. L. Collins,

A. L. Lewis and M. S. Rea, *Lighting and Human Performance: A Review*, Report to Lamp Division, National Electrical Manufacturers Association (NEMA) and the Lighting Research Institute, NEMA, Washington, DC, 1989; H. R. Blackwell and O. M. Blackwell, "Individual Responses to Lighting Parameters for A Population of 235 Observers of Varying Ages," *J. IES*, 205–232 (July 1980); R. A. Weale, "Do Years or Quanta Age the Retina," *Photochem. Photobiol.*, **50**(3), 429–438 (1989).

54. D. Sliney and M. Wolbarsht, *Safety with Lasers and Other Optical Sources*, Plenum Press, New York, 1980; D. Sliney, "Ocular Injury Due to Light Toxicity," *Int. Ophthalmol. Clinics*, **28**(3), (Fall 1988).

55. ANSI Standard Z311.1—Photobiological Safety for Lamps and Lighting Systems— General Requirements, Illuminating Engineering Society of North America, New York, Draft June 1990.

56. J. E. Birren, R. C. Casperson, and J. Botwinick, "Age Changes in Pupil Size," *J. Gerontol.*, **5**, 216 (1950).

57. L. B. Zerbe and H. W. Hofstetter, "Prevalence of 20/20 with Best Previous and No Lens Correction," *J. Am. Optical Assoc.*, **29**, 772 (1957–1958).

58. H. C. Weston, "The Effects of Age and Illumination upon Visual Performance with Close Sights," *Br. J. Ophthalmol.*, **32**, 645 (1948).

59. E. Wolf, "Effects of Age on Peripheral Vision," *Highway Res. Board Bull. No. 336*, 1962, p. 26.

60. R. A. McFarland, R. G. Domey, A. B. Warren, and D. C. Ward, "Dark Adaptation as a Function of Age and Tinted Windshield Glass," *Highway Res. Board Bull. No. 255*, 1960, p. 51.

61. E. D. Fletcher, "An Investigation of Glare Resistance and Its Relationship to Age," *Motorists' Vision*, **5**, 1 (1952).

62. O. M. Blackwell and H. R. Blackwell, "Visual Performance Data for 146 Normal Observers of Various Ages," *J. IES*, 1 (1971).

63. H. Brandston, "Lighting Design—A Practitioner's Problem," *Int. Lighting Rev.*, 67–68 (1983).

64. S. M. Berman, D. L. Jewett, L. R. Bingham, R. M. Nahass, F. Perry, and G. Fein, "Pupillary Size Difference Under Incandescent and High Pressure Sodium Lamps," *J. IES*, 3–20 (Winter 1987).

65. H. A. Piper, "The Effect of HPS Light on Performance of a Multiple Refocus Task," *LD+A*, **11**, 36 (1981); S. W. Smith and M. S. Rea, "Relationship Between Office Task Performance and Ratings of Feelings and Task Evaluation Under Different Light Sources and Level," *Proc. CIE 19th Session, Kyoto*, CIE, Paris, 1979.

66. H. Helson, D. B. Judd, and M. H. Warren, "Object-Color Changes from Daylight to Incandescent Filament Illumination," *Illum. Eng.*, **47**, 221–233 (1952).

67. D. Nickerson and C. W. Jerome, "Color Rendering of Light Sources: CIE Method of Specification and Its Application," *Illum. Eng.*, **50**, 262 (1965).

68. Commission Internationale de l'Eclairage, *Methods of Measuring and Specifying Color Rendering Properties of Light Sources*, CIE Publication 13.2, CIE, Paris, 1974.

69. H. Helson and T. Lansford, "The Role of Spectral Energy of Source and Background Color in the Pleasantness of Object Colors," *Appl. Opt.*, **9**, 1513 (1970).

70. D. B. Judd, " A Flattery Index for Artificial Illuminants," *Illum. Eng.*, **52**, 593 (1967).

71. C. W. Jerome, "The Flattery Index," *J. IES*, **2**, 351 (1973).

72. B. L. Collins, "Safety Color Appearance Under Different Illuminants," *J. IES*, **16**(1), 21–38 (1987).

73. J. E. Kaufman and J. F. Christensen, eds., "Color," in *IES Lighting Handbook, Reference Volume*, Illuminating Engineering Society of North America, New York, 1984.

74. "Color and the Use of Color by the Illuminating Engineer," *Illum. Eng.*, **57**, 764 (1962).

75. "The Psychological Potential of Illumination," Environmental Design Research Association Conference, Pennsylvania State University, University Park, PA, 1975; J. E. Flynn et al., "The Influence of Spatial Light on Human Judgment," *Proceedings of CIE 18th Session, London, 1975*, Publication CIE No. 36, 1976, pp. 39–46; J. E. Flynn, "A Study of Subjective Responses to Low-Energy and Non-uniform Lighting Systems," *LD + A*, **7**(2), 6–15 (February 1977); J. E. Flynn, et al., "A Guide to Methodology Procedures For Measuring Subjective Impression in Lighting," *J. IES*, (January 1979); J. E. Flynn and T. J. Spencer, "The Effects of Light Source Color On User Impressions and Satisfaction," *J. IES*, 167–179 (April 1977); J. E. Flynn, A. W. Segal, and G. R. Steffy, *Architectural Interior Systems: Lighting, Acoustic, Air Conditioning*, 2nd ed., Van Nostrand Reinhold, New York, 1988; R. J. Hawkes, D. L. Loe, and E. Rowlands, "A Note Towards the Understanding of Lighting Quality," *J. IES*, **8**, 111–120 (1979).

76. L. T. Troland, "An Analysis of the Literature Concerning the Dependency of Visual Functions upon Illumination Intensity," *Trans. IES*, **26**, 107 (1931).

77. C. E. Ferree and G. Rand, "Tests for the Efficiency of the Eye Under Different Systems of Illumination and a Preliminary Study of the Causes of Discomfort," *Trans. IES*, **8**, 40 (1913); J. R. Cravath, "Some Experiments with the Ferree Test for Eye Fatigue," *Trans. IES*, **9**, 1033 (1914); C. E. Ferree and G. Rand, "The Efficiency of the Eye Under Different Conditions of Lighting: The Effects of Varying the Distribution of Factors and Intensity," *Trans. IES*, **10**, 407 (1915); "Some Experiments of the Eye with Different Illuminants, Part I, "*Trans. IES*, **13**, 50 (1918); "Some Experiments of the Eye with Different Illuminants, Part I, "*Trans. IES*, **14**, 107 (1919); "The Effect of Variations in Intensity of Illumination on Functions of the Importance to the Working Eye," *Trans. IES*, **15**, 769 (1920); "The Effect of Variations of Visual Angle, Intensity and Composition of Light on Important Ocular Functions," *Trans. IES*, **17**, 69 (1922); "Further Studies on the Effect of Composition of Light on Important Ocular Functions," *Trans. IES*, **19**, 424 (1924).

78. H. C. Weston, *Sight, Light and Efficiency*, H. K. Lewis, London, 1949.

79. W. B. Lancaster, "Light and Lighting," in *The Eye and Its Diseases*, 2nd ed., C. Berens, Ed., Saunders, Philadelphia, 1950, p. 81.

80. F. W. Hebbard, "Micro Eye Movements: Effects of Target Illumination and Contrast," *Illum. Eng.*, **64**, 199 (1969).

81. H. C. Weston, *Sight, Light and Efficiency*, H. K. Lewis, London, 1949.

82. D. A. Campbell, W. J. B. Riddell, and A. S. MacNalty, *Eyes in Industry*, Longmans, Green, New York, 1951.

83. D. A. Campbell, W. J. B. Riddell and A. S. MacNalty, *Eyes in Industry*, Longmans, Green, New York, 1951, p. 2.

84. M. Poser, "Eye Fatigue in Industry," *Trans. IES*, **16**, 431 (1921).

85. M. Luckiesh and F. K. Moss, *The Science of Seeing*, Van Nostrand, New York, 1937.

86. C. L. Crouch, "Discussion," *Illum. Eng.*, **47**, 344 (1952).

87. C. P. Steinmetz, "The Scope of Illuminating Engineering," *Trans. IES*, **11,** 625 (1916).

88. H. P. Gage, "Hygienic Effects of Ultraviolet Radiation in Daylight," *Trans. IES*, **25,** 377 (1930).

89. C. Berens and C. L. Crouch, "Is Fluorescent Lighting Injurious to the Eyes?" *Am. Ophthalmol.*, **45,** 47 (1958).

90. "Fluorescent Lighting & Malignant Melanoma," Statement by International Non-Ionizing Radiation Committee of the International Protection Association, *Health Phys. J.*, **58,** 1,111–112 (January 1990); F. Urbach; P. D. Forbes; R. E. Davies, and C. P. Sambuco, "Artificial Lighting: Influence on the Production of Solar UVR-Induced Skin Tumors in Mice," Report to Lighting Research Institute, Project 84 DR NEMA 1, Lighting Research Institute, New York, April 1987.

91. R. J. Wurtman, "The Effects of Light on the Human Body," *Sci. Am.*, **233,** 68 (1975).

92. T. P. Vogl, "Photomedicine," *Opt. News*, 6 (Spring 1976).

93. R. J. Cremer, P. W. Perryman, and D. H. Richards, "Influence of Light on the Hyperbilirubinemia of Infants," *Lancet*, 1094 (May 1958).

94. J. R. Lucey, "The Effects of Light on the Newly Born Infant," *J. Perinat. Med.*, **1,** 1 (1973).

95. R. M. Neer, et al., "Stimulation by Artificial Lighting on Calcium Absorption in Elderly Human Subjects," *Nature*, 229 (1971).

96. W. F. Loomis, "Rickets," *Sci. Am.*, **233,** 77 (1970).

97. J. A. Parrish, T. B. Fitzpatrick, L. Tanenbaum, and M. A. Pathak, *New Engl. J. Med.*, **291,** 1207 (1974).

98. I. Diamond, A. F. McDonagh, C. B. Wilson, S. G. Granelli, S. Nielson, and R. Jaenicke, *Lancet*, **15,** 1175 (1972); S. G. Granelli, I. Diamond, A. F. McDonagh, C. B. Wilson, and S. Nielson, *Cancer Res.*, **35,** 2567 (1975); T. J. Dougherty, G. B. Grindey, R. Fiel, K. R. Weishaupt, and D. G. Boyle, *J. Nat. Cancer Inst.*, **55,** 115 (1975); J. F. Kelly, M. E. Snell and M. C. Berenbaum, *Br. J. Cancer*, **31,** 237 (1975).

99. D. G. Pitts, J. P. G. Bergtmanson, and L. W-F. Chu, "Rabbit Eye Exposure to Broad-Spectrum Fluorescent Lighting," *Acta Ophthalmol.*, Suppl. 159, Scriptor, Copenhagen, 1983; S. Zigman and R. J. Collier, "The Grey Squirrel Lens Protects the Retina from Near-UV Radiation Damage," in *Degenerative Retinal Disorders: Clinical and Laboratory Investigations*, J. G. Hollyfield, R. E. Anderson, and M. M. LaVail, Eds., Alan R. Liss, New York, 1987, pp. 571–586; S. Zigman, and R. J. Collier, "Comparison of Retinal Photochemical Lesion After Exposure to Near-UV or Short-wavelength Visible Radiation," in *Inherited and Environmentally Induced Retinal Degenerations*, M. M. LaVail, R. E. Anderson, and J. G. Hollyfield, Eds., Alan R. Liss, New York, 1989, pp. 569–575.

100. L. R. Ronchi, "An Annotated Bibliography On Some Basic Effects Of Optical Radiation on Humans", *CIE-J.*, **7**(2), 42–48 (September 1988).

101. D. L. Jewett, S. M. Berman, M. R. Greenberg, G. Fein, R. Nahass, "The Lack of Effect on Human Muscle Strength of Light Spectrum and Low Frequency Electromagnetic Radiation in Interior Lighting," *J. IES*, **15,** 19 (1986).

102. G. C. Brainard, P. L. Podolin, M. D. Rollag, and B. A. Northrup, "Circadian Rhythms of Melatonin in Primate (Cynomologus Macaque) Cerebrospinal Fluid," *Anat. Rec.* **221**(3), 25A, (March 1985); G. C. Brainard, P. L. Podolin, M. D. Rollag, and B. A. Northrup, C. Curtis, and F. M. Barker, "The Influence of Light Irradiance and Wavelength on Pineal Physiology of Mammals," *Pineal and Retinal Relationships*, Academic Press, 1986; G. C. Brainard, and F. M. Barker, "Regulation of the Pineal–Repro-

ductive Axis by Near Ultraviolet Radiation (UV-A) in three Rodent Species," in *Fundamentals and Clinics in Pineal Research*, G. P. Trentini, C. DeGaetani, and P. Pevet, eds., Raven Press, New York, 1987, pp 207–210.

103. R. A. Hoffman, and L. B. Johnson, R. J. Reiter, "Harderian Glands of Golden Hamsters: Temporal and Sexual Differences in Immunoreactive Melatonin," *J. Pineal Res.*, **2**, 161–168 (1985); R. A. Hoffman and L. B. Johnson, "Effects of Photic History and Illuminance Levels on Male Golden Hamster," *J. Pineal Res.*, **2**, 209–215 (1985); R. A. Hoffman, L. B. Johnson, and G. Brown, "Growth and Development of Golden Hamsters: Influence on the Pineal Gland, Melatonin and Photic Input," Chapter 42, *The Pineal Gland, Endocrine Aspects*, Proceedings of a Symposium held in conjunction with the Seventh International Endocrinology Congress, Canada, G. Brown and S. D. Winwright, Eds., Pergamon, April 1984; R. A. Hoffman, L. B. Johnson, and R. Corth. "The Effects of Spectral Power Distribution and Illuminance Levels on Key Parameters in the Male Golden Hamster and Rat with Preliminary Observations on the Effects of Pinealectomy", *J. Pineal Res.*, **2**, 217–233 (1985); R. A. Hoffman and L. B. Johnson "Interactions of Diet and Photoperiod on Growth and Reproduction in Male Golden Hamsters Growth," *Growth*, **49**, 380–399 (1985).

104. S. Shankman, "Light in the Laboratory," *LD + A*, **16**(5), 11–15 (May 1986).

105. A. J. Lewy, R. L. Sack, L. S. Miller, T. M. Hoban "Antidepressant and Circadian Phase-Shifting Effects of Light," *Science*, **16**(5), 11–15 (May 1986).

106. E. W. Bickford, *Human Circadian Rhythms: A Review*, Report to Lighting Research Institute, Project 88 DR NEMA 2, New York, December 1988.

107. A. L. Lewis, "Light Affects Health," *LD + A*, 32–36 (October 1987).

108. C. A. Czeisler, J. S. Allan, J. F. Duffy, M. E. Jewett, E. N. Brown, and J. M. Ronda, "Bright Light Induction of Strong (Type O) Resetting of the Human Circadian Pacemaker," *Science*, **244**, 1328–1333 (June 16, 1989).

109. Society for Lighting Treatment and Biological Rhythms, Inc., New York.

110. "Lighting Design," J. E. Kaufman and J. F. Christensen, eds., *IES Lighting Handbook, Application Volume*, Section 1, Illuminating Engineering Society of North America, New York, 1987; "Illuminating Engineers," Chronicle of Guidance Publication, Monrovia, NY, 1988; J. R. Benya, "The Lighting Design Professional, Making a Living at Lighting," *Architectural Lighting*, 40, 42 (February 1989); "The Profession of Lighting Design: A Roundtable Discussion," *Lightview International*, Vol. 2, No. 1, International Association of Lighting Designers, New York, 1989; W. J. Jankowski, "The Origins of Lighting Design," *Lightview International*, Vol. 2, No. 1, International Association of Lighting Designers, New York, 1989; "What to Consider When Looking for a Lighting Designer and Who to Chose?" *Lightview International*, Vol. 2, No. 1, International Association of Lighting Designers, New York, 1989.

111. Light Sources," in *IES Lighting Handbook, Application Volume*, J. E. Kaufman, and J. F. Christensen, eds., Illuminating Engineering Society of North America, New York, 1987, Section 8; IES Committee on Light Sources, *Choosing a Light Source For General Lighting*, IES CP-32, Illuminating Engineering Society of North America, New York, 1988.

112. I. Lewin and L. Stafford, "Photometric and Field Performance of High Pressure Sodium Luminaires," *J. IES*, 106–114 (Summer 1987); I. Lewin and L. Stafford, "Real World Use of Photometric Test Reports," *LD + A*, 18–24 (January 1988); and I. Lewin and L. Stafford, "Photometric and Field Performance of Metal Halide Luminaires," *J. IES*, 15–19 (Winter 1988).

113. "Daylighting," in *IES Lighting Handbook*, 5th ed., Illuminating Engineering Society, New York, 1972.

114. D. Rosenfeld, "Efficient Use of Energy in Buildings, Report on the 1975 Berkeley Summer Study," LBL 4411, Lawrence Berkeley Laboratory, University of California, Berkeley, 1976.

115. Bibliography, "Recommended Practice of Daylighting," *RP-5*, Illuminating Engineering Society of North America, New York, 1979.

116. "Recommended Practice of Daylighting," *RP-5*, Illuminating Engineering Society of North America, New York, 1979; IES Committee on Calculation Procedures, *IES Recommended Practice For The Lumen Method of Daylight Calculations, RP-23*, Illuminating Engineering Society of North America, New York.

117. S. Selkowitz, "Smart Windows," *Glass Mag.*, 86–91 (August 1986); S. Selkowitz and R. McClunney, *Commercial Building Fenestration Performance Indices Project; Phase I: Development of Methodology*, prepared by Windows and Lighting Program, Applied Science Division, Lawrence Berkeley Laboratory, and The Florida Solar Energy Center for the Lighting Research Institute, under contract to the Electric Power Research Institute, Palo Alto, CA, December 1988; K. Papamichael and F. Winkelmann, "Solar-Optical Properties of Multilayer Fenestration Systems," 1986 International Daylighting Conference Proceedings II, ASHRAE, 1988; S. Selkowitz, K. Papamichael, and J. Klems, "Determination and Application of Bidirectional Solar–Optical Properties of Fenestration Systems," *Proceedings, 13th National Passive Solar Conference*, June 19–24, 1988, Massachusetts Institute of Technology Cambridge, MA; S. Selkowitz, J. J. Kim, K. M. Papamichael, M. Spitzglas, and M. Modest, "Determining Daylight Illuminance in Rooms Having Complex Fenestration Systems," 1986 International Daylighting Conference Proceedings II, ASHRAE, 1988; R. Sullivan, D. Arasteh, K. Papamichael, and S. Selkowitz, "An Approach for Evaluating the Thermal Comfort Effects of Nonresidential Building Fenestration Systems," presented at the International Symposium on Advanced Comfort Systems for the Work Environment, Troy, NY, May 1–3, 1988, and to be published in the Proceedings, LBL-25060; R. Sullivan, D. Arasteh, K. Papamichael, J. J. Kim, R. Johnson, S. Selkowitz, and R. McCluney, "An Indices Approach for Evaluating the Performance of Fenestration Systems in Nonresidential Buildings," *ASHRAE Trans.*, **94**(Part 2) (1988); and K. M. Papamichael and S. Selkowitz, "Simulating the Luminous and Thermal Performance of Fenestration Systems," *LD+A*, 37–45 (October 1987); M. D. Parent and J. B. Murdoch, "The Expansion of the Zonal Cavity Method of Interior Lighting Design to Include Skylights," *J. IES*, 141–173 (Summer 1988); M. D. Parent and J. B. Murdoch, "Analysis and Comparison of Skylights Based on their Intensity Distribution Curves," *Lighting Res. Technol.*, **21**(3), 111 1989.

118. "Effective Seeing in an Era of Energy Conservation," Illuminating Engineering Research Institute, New York, 1973.

119. "Effective Seeing and Conservation Too," Illuminating Engineering Research Institute, New York, 1974.

120. J. E. Kaufman, "Optimizing The Use of Energy for Lighting," *LD+A*, **4**, 8 (1973).

121. B. L. Collins, W. S. Fisher, and R. W. Marans, "Second-Level Post-Occupancy Evaluation (POE) Analysis," J. IES, **19**, 2, 21–44 (Summer 1990).

122. IES Energy Management Committee: IES Recommended Procedure for Lighting Power Limit Determination, IES LEM-1-1982, Illuminating Engineering Society of North America, New York, 1983; IES Energy Management Committee: IES Recommended

Procedure for Lighting Power Limit Determination, IES LEM-1-1982, Illuminating Engineering Society of North America, New York, 1983; IES Energy Management Committee: IES Recommended Procedure for Lighting Energy Limit Determination, IES LEM-2-1984, Illuminating Engineering Society of North America, New York, 1985; IES Energy Management Committee: IES Design Considerations Effective Building Lighting Energy Utilization, IES LEM-3, Illuminating Engineering Society of North America, New York, 1987; and IES Energy Management Committee: IES Recommended Procedure for Energy Analysis of Building Designs and Installations, IES LEM-4-1984, Illuminating Engineering Society of North America, New York, 1985.

123. A. B. Gough, "Progress Report on Setting a National Agenda for Lighting Research and Education," Keynote speech IESNA Conference, Baltimore, MD (August 30, 1990), Lighting Research Institute, New York; *National Energy Strategy*: Interim Report (April 1990), U.S. Department of Energy, Washington, DC; and "Efficient Electrical Use: Estimates of Maximum Energy Savings," EPRI CUG 746 Project 2788, Electric Power Research Institute, Palo Alto, CA (March 1990).

124 *Introductory Lighting Education*, ED 100 Series, IES Education Committee, Illuminating Engineering Society of *Lighting Guidelines* (Pub. No. 400-90-014), California Energy Commission, 1516 9th St., Sacramento, CA (March 1990).

Appendix A
METHOD FOR PRESCRIBING ILLUMINATION*

Appendix A—Method for prescribing illumination

(This Appendix is not part of the "American National Standard Practice for Industrial Lighting," ANSI/IES RP-7-1983, but is presented as background material for the user of the Standard.)

A1 New illuminance selection procedure. (a) Since 1965 this Standard has included single-value illuminance recommendations based on a method established at that time.[1] In recent years it became apparent, through on-going research and design experience, that it was time to move away from the single-value recommendations to a range approach—illuminance ranges accompanied by a weighting-factor guidance system reflecting lighting-performance trends found in research. In 1979, such a new procedure was established.[2]

(b) It is intended that this new procedure will accommodate a need for flexibility in determining illuminance levels so that lighting designers can tailor lighting systems to specific needs, especially in an energy conscious era. Such flexibility requires that additional information be available to effectively use the new range approach—a *lighting task* must be considered to be composed of the following:

(1) The visual display (details to be seen).

(2) The age of the observers.

(3) The importance of speed and/or accuracy for visual performance.

(4) The reflectance of the task (background on which the details are seen).

(c) The visual display is the object being viewed—it will present some inherent visual difficulty. The age of the observer is a predictor of the condition of the observer's visual system. The importance of speed and/or accuracy distinguishes between casual, important and critical seeing requirements. The reflectance will determine the adaptation luminance produced by the illuminance. These characteristics, considered together, determine the appropriate amount of light for the *lighting task*. All four must be considered as comprising the lighting task.

(d) In applying the new procedure the first step is to determine a range of illuminances appropriate for the visual difficulty presented by the visual display, the first of the above characteristics (see **Table B1** in Appendix B) and then to determine a target value from that range on the basis of the remaining three characteristics.

(e) For a given visual display, a specific value of illuminance can be chosen from the recommended range only if the second, third, and fourth characteristics of the lighting task are known; i.e., observer's age, importance of speed and/or accuracy, and task reflectance. These should be determined at design time, by the designer in consultation with the user.

(f) A guide for using the second, third and fourth characteristics of the lighting task, to determine a specific target value of illuminance, takes the form of a table of weighting factors (see **Table 3**). The designer or user determines the weight of each characteristic. A combined weighting factor then indicates whether the lower, middle or upper value of illuminance in the range is appropriate (see **Table 1**).

(g) It can be seen that this procedure is an illuminance *selection* procedure, where consensus-determined recommended ranges combine with user supplied information and judgment. The result is the determination of a specific target value of illuminance appropriate for the lighting task under consideration.

A2 Limitations of the new selection procedure. This illuminance selection procedure is intended for use in interior environments where visual performance is an important consideration. It has been developed from a consideration of experience and research results from visual performance experiments. Its use is then limited to applications where this information can be applied directly. Thus, the illuminance selection procedure[2] is *not* used to determine the appropriate illuminances when:

(1) Merchandising is the principal activity in the space and the advantageous display of goods is the purpose of lighting.

*Excerpt from IES RP-7-1983. Used with Permission.

Appendix A—Method for prescribing illumination

(2) Advertising, sales promotion or attraction is the purpose of lighting.

(3) Lighting is for sensors other than the eye, as in film and television applications.

(4) The principal purpose of lighting is to achieve artistic effects.

(5) Luminance ratios have a greater importance than adaptation luminance, as when it is desired to achieve a particular psychological or emotional setting rather than provide for visual performance.

(6) Minimum illuminances are required for safety.

(7) Maximum illuminances are established to prevent nonvisual effects, such as bleaching or deterioration due to ultraviolet and infrared radiation in a museum.

(8) Illuminances are part of a test procedure for evaluating equipment, such as for surgical lighting systems.

A3 Example of Illuminance Selection. (a) A machine shop is to be relighted. The designer in consultation with the foreman has determined the following:

(1) The task is diemaking using metal of a reflectance of about 40 percent.

(2) The workers are young (under 40).

(3) The workers are under pressure to maintain close tolerances on die production. Speed and accuracy are considered to be important, but not critical to production.

(b) Using the above step-by-step procedure:

Step 1. The visual task is defined above.

Step 2. Referring to **Table B1,** an illuminance category of E is found under machine shops, "medium bench or machine work."

Step 3. Referring to **Table 1,** the illuminance range is found to be 500-750-1000 lux [50-75-100 footcandles].

Step 4. Referring to **Table 3** and the above information, the weighting factors selected are: −1 for workers' ages; 0 for speed and/or accuracy; and 0 for reflectance of task background. The algebraic sum is −1 +0 +0 = −1. Therefore, the illuminance to be selected is the mid-value, i.e., 750 lux [75 footcandles].

(c) If the task were considered to be fine bench work where the tolerances were critical, the metal used has a reflectance of 25 percent, and the average of the workers' ages was 50, the illuminance category would become G, the range 2000-3000-5000 lux [200-300-500 footcandles]; and the weighting factor for workers' ages would be 0, for speed and/or accuracy +1, and for reflectance +1. The new algebraic sum of the weighting factors is +0 +1 +2. Therefore, the illuminance to be selected is the high value in the new range, i.e., 5000 lux [500 footcandles].

(d) By referring to **Table A-1** after Step 2, the illuminance can be selected without referring to **Tables 1** or **3**.

References

1. IES Committee on Recommendations for Quality and Quantity of Illumination, "RQQ report no. 1—Recommendations for quality and quantity of illumination," *Illum. Eng.*, Vol. 58, No. 8, Aug. 1958, p. 422.

2. IES Committee on Recommendations for Quality and Quantity of Illumination of the IES; "RQQ report no. 6—Selection of illuminance values for interior lighting design," *J. Illum. Eng. Soc.*, Vol. 9, No. 3, Apr., 1980, p. 188.

3. *IES Lighting Handbook, 1981 Reference Volume,* New York: Illuminating Society of North America.

4. IES Lighting Design Practice Committee, "Zonal-cavity method of calculating and using coefficients of utilization," *Illum. Eng.*, Vol. 59, No. 5, May 1964, p. 309.

5. IES Design Practice Committee, "General procedure for calculating maintained illumination," *Illum. Eng.*, Vol. 65, No. 10, Oct. 1970, p. 603.

6. IES Lighting Design Practice Committee, "The determination of illumination at a point in interior spaces," *J. Illum. Eng. Soc.*, Vol. 3, No. 2, Jan. 1974, p. 171.

Table A1—Illuminance values, maintained, in lux, for a combination of illuminance categories and user, room and task characteristics

a. General Lighting Throughout Room

Weighting Factors		Illuminance Categories		
Average of Occupants Ages	Average Room Surface Reflectance (per cent)	A	B	C
Under 40	Over 70	20	50	100
	30–70	20	50	100
	Under 30	20	50	100
40–55	Over 70	20	50	100
	30–70	30	75	150
	Under 30	50	100	200
Over 55	Over 70	30	75	150
	30–70	50	100	200
	Under 30	50	100	200

b. Illuminance on Task

Weighting Factors			Illuminance Categories					
Average of Workers Ages	Demand for Speed and/or Accuracy*	Task Background Reflectance (per cent)	D	E	F	G**	H**	I**
Under 40	NI	Over 70	200	500	1000	2000	5000	10000
		30–70	200	500	1000	2000	5000	10000
		Under 30	300	750	1500	3000	7500	15000
	I	Over 70	200	500	1000	2000	5000	10000
		30–70	300	750	1500	3000	7500	15000
		Under 30	300	750	1500	3000	7500	15000
	C	Over 70	300	750	1500	3000	7500	15000
		30–70	300	750	1500	3000	7500	15000
		Under 30	300	750	1500	3000	7500	15000
40–55	NI	Over 70	200	500	1000	2000	5000	10000
		30–70	300	750	1500	3000	7500	15000
		Under 30	300	750	1500	3000	7500	15000
	I	Over 70	300	750	1500	3000	7500	15000
		30–70	300	750	1500	3000	7500	15000
		Under 30	300	750	1500	3000	7500	15000
	C	Over 70	300	750	1500	3000	7500	15000
		30–70	300	750	1500	3000	7500	15000
		Under 30	500	1000	2000	5000	10000	20000
Over 55	NI	Over 70	300	750	1500	3000	7500	15000
		30–70	300	750	1500	3000	7500	15000
		Under 30	300	750	1500	3000	7500	15000
	I	Over 70	300	750	1500	3000	7500	15000
		30–70	300	750	1500	3000	7500	15000
		Under 30	500	1000	2000	5000	10000	20000
	C	Over 70	300	750	1500	3000	7500	15000
		30–70	500	1000	2000	5000	10000	20000
		Under 30	500	1000	2000	5000	10000	20000

* NI = not important, I = important, and C = critical
** Obtained by a combination of general and supplementary lighting.

Table B1—Currently recommended Illuminance categories for Industrial interiors*

Area/Activity	Illuminance Category	Area/Activity	Illuminance Category
Inspection		**Paper manufacturing**	
Simple	D	Beaters, grinding, calendering	D
Moderately difficult	E	Finishing, cutting, trimming, papermaking machines	E
Difficult	F	Hand counting, wet end of paper machine	E
Very difficult	G	Paper machine reel, paper inspection, and laboratories	F
Exacting	H	Rewinder	F
Iron and steel manufacturing	(see Table B3)ᵃ		
Jewelry and watch manufacturing	G	**Petroleum and chemical plants**	(see Table B3)ᵃ
Laundries		**Plating**	D
Washing	D	**Polishing and burnishing (see Machine shops)**	
Flat work ironing, weighing, listing, marking	D		
Machine and press finishing, sorting	E	**Power plants (see Electric generating stations)**	
Fine hand ironing	E	**Poultry industry (see also Farm—dairy)**	
Leather manufacturing		Brooding, production, and laying houses	
Cleaning, tanning and stretching, vats	D	Feeding, inspection, cleaning	C
Cutting, fleshing and stuffing	D	Charts and records	D
Finishing and scarfing	E	Thermometers, thermostats, time clocks	D
Leather working		Hatcheries	
Pressing, winding, glazing	F	General area and loading platform	C
Grading, matching, cutting, scarfing, sewing	G	Inside incubators	D
Locker rooms	C	Dubbing station	F
Machine shops		Sexing	H
Rough bench or machine work	D	Egg handling, packing, and shipping	
Medium bench or machine work, ordinary automatic machines, rough grinding, medium buffing and polishing	E	General cleanliness	E
		Egg quality inspection	E
		Loading platform, egg storage area, etc.	C
Fine bench or machine work, fine automatic machines, medium grinding, fine buffing and polishing	G	Egg processing	
		General lighting	E
		Fowl processing plant	
Extra-fine bench or machine work, grinding, fine work	H	General (excluding killing and unloading area)	E
Materials handling		Government inspection station and grading stations	E
Wrapping, packing, labeling	D	Unloading and killing area	C
Picking stock, classifying	D	Feed storage	
Loading, inside truck bodies and freight cars	C	Grain, feed rations	C
Meat packing		Processing	C
Slaughtering	D	Charts and records	D
Cleaning, cutting, cooking, grinding, canning, packing	D	Machine storage area (garage and machine shed)	B
Nuclear power plants (see also Electric generating stations)		**Printing industries**	
		Type foundries	
Auxiliary building, uncontrolled access areas	C	Matrix making, dressing type	E
Controlled access areas		Font assembly—sorting	D
Count room	Eᶜ	Casting	E
Laboratory	E	Printing plants	
Health physics office	F	Color inspection and appraisal	F
Medical aid room	F	Machine composition	E
Hot laundry	D	Composing room	E
Storage room	C	Presses	E
Engineered safety features equipment	D	Imposing stones	F
Diesel generator building	D	Proofreading	F
Fuel handling building		Electrotyping	
Operating floor	D	Molding, routing, finishing, leveling molds, trimming	E
Below operating floor	C	Blocking, tinning	D
Off gas building	C	Electroplating, washing, backing	D
Radwaste building	D	Photoengraving	
Reactor building		Etching, staging, blocking	D
Operating floor	D	Routing, finishing, proofing	E
Below operating floor	C	Tint laying, masking	E

Appendix B
LIGHTING RESOURCE ORGANIZATIONS

CIE   Commission Internationale de l'Eclairage, A-1030 Vienna, Kegel-
         gasse 27, Austria

DOE   U.S. Department of Energy, 1000 Independence Avenue, SW,
         Washington, DC 20585

EPRI   Electric Power Research Institute, 3412 Hillview Avenue, Palo Alto,
         CA 94303

IALD   International Association of Lighting Designers, 18 East 16th Street,
         Suite 208, New York, NY 10003

IESNA   Illuminating Engineering Society of North America, 345 East 47th
         Street, New York, NY 10017

LRI   Lighting Research Institute, 345 East 47th Street, New York, NY
         10017

NLB   National Lighting Bureau, 2101 L Street, NW, Washington, DC
         20037

NIST   National Institute of Standards and Technology (formerly: National
         Bureau of Standards), Center for Building Technologies, Gaithers-
         burg, MD 20899

SLTBR   Society for Lighting Treatment and Biological Rhythms, Inc., P.O.
         Box 478, Wilsonville, OR 97070

CHAPTER THIRTY-SEVEN

Safety Interfaces, Profession and Practice

Ted Ferry, Ed.D.

1 INTERRELATIONSHIPS

The scope and depth of safety practice are enormous. They can be explored by looking at some interrelationships with other fields. Recent emphasis on the industrial hygienist and the safety professional working on each other's turf is now seen as working together for shared objectives. Increasingly we speak and write not of safety, industrial hygiene, or health, but of occupational safety and health (OSH). This view is reinforced by government regulations and marketplace demands.

The Occupational Safety and Health Act (OSHAct) itself provides for what amounts to a safety arm and a health-oriented arm. This is done though the Occupational Safety and Health Administration (OSHA) and the National Institute for Occupational Safety and Health (NIOSH) under the Department of Health and Human Services. Thus we see a merging of training, education, job preparation, and government funding for research and training to support the interface of safety and industrial hygiene. The expanded regulations on hazards communications and SARA Title III require an *interdisciplinary* approach—a safety and health approach (1).

Increased emphasis on the environment has led the safety professional deeply into industrial hygiene, fire science, wellness, health physics, and toxicology, to

Patty's Industrial Hygiene and Toxicology, Fourth Edition, Volume 1, Part B, Edited by George D. Clayton and Florence F. Clayton.
ISBN 0-471-50196-4 © 1991 John Wiley & Sons, Inc.

name only a few disciplines. Safety professionals are involved in environmentally regulated items and must work closely with many professionals to do their jobs.

Other professionals also find themselves solving their problems through adjunct disciplines. Hardly a ventilation, lighting, or disposal problem exists that does not also consider ergonomics, security, health physics, traffic, or production aspects. Safety-related problems with supervision, management, and engineering decisions can be solved more quickly than in other disciplines, but perhaps not as surely. This is quickly evident when a safety professional and an industrial hygienist, for example, evaluate a site together. Industrial hygienists often find that their tasks involve detailed measurement and cannot be completed swiftly. In three days, the safety person may uncover dozens of deficiencies plant-wide, while the industrial hygienist is taking measurements in one department. The common interests of safety and industrial hygiene reveal different methods of operation. The hygienist is a stakeholder in safety practice. This chapter, then, has one main objective: to provide an overview of the scope and depth of the safety profession for related professionals.

Industrial hygienists, health physicists, medical doctors, and others with toxicology backgrounds have headed major corporate safety and health departments. EG&G Idaho, a major Department of Energy contractor at the National Engineering Laboratory, favors a multi/interdisciplinary approach to safety and health. So *inter*related are their dealings with the most toxic substances that they use teams of industrial hygienists, health physicists, and safety engineers. People from any of the three disciplines may head their teams. They often learn much about each other's fields and change their own careers, such as from safety to health physics or from industrial hygiene to safety.

Major aerospace firms have led the way with multidisciplined approaches to safety and health. Frequently safety professionals lead combined departments. These multitalented persons recognize the value of other disciplines and must be knowledgable of them in the absence of reciprocated interest. In our growing litigious society, some corporations have found it prudent (not necessarily wise) for their safety and health function to report to an attorney. As often as not, it has been product and environmental concerns that drive this direction.

Government agencies are moving from physical hazard concerns to health-related emphasis. OSHA, which first devoted most of their effort to safety, now is over 80 percent health oriented. NIOSH does mostly safety research, and is occupational disease oriented. This stems partly from NIOSH's location under the Centers for Disease Control (CDC). Toxicology research is further diversified with a major arm under the Food and Drug Administration (FDA). The inherent research ability within the CDC is not available in OSHA. By default and by legislation, NIOSH has assumed safety-related research. This means concentrating on health issues and neglect of safety research. Other professions have similar problems. Toxicology research, for example, is diluted to a major arm in the FDA. Toxicology has its own identity crisis. In NIOSH's 10-item priority list for action in the mid-1980s, only two were clearly toxicology related. This does not mean that is how it is, only that it is the NIOSH perception of health-related risks. Only three of the 10 are plainly seen as workplace risks. Public perception of national problems changes regularly. To see this change, recall the major safety and health problems 10 years ago. Compare them with what are seen as the major problems today. Most earlier issues did not have an occupational disease aspect. Major problems perceived by the public, and in turn their elected representatives, relate to hazardous materials in the home, workplace, and recreational areas. A newspaper or public official gets more mileage out of "exposing" local toxic substances than almost any other subject. Landfillers trucking toxic materials on local highways or toxic releases from a local plant grab our attention. The public tends to look to a safety and health functionary to solve the matter. The concerned public seems to prefer safety and health regulation to prevention.

Toxic-based problems make strange bedfellows. We find fire departments in California (and in other states) drawn into Superfund activities because the first responders to hazardous materials spills are often fire departments. Thus they have become, in the past decade, a leading hazardous materials regulator. Fire departments that last year put out fires are busy this year, with augmented staffs, talking to safety and health groups.

Where a company is unionized, the union–safety interface cannot be understated. If an adversarial relation exists between the union and management, the union often makes safety issues the bone of contention and a point of dispute on contracts. The union acts as a worker representative and can force the company to act on safety and health issues. Where a nonadversarial relationship exists, the union may be a very strong lever to attain a safe and healthy workplace. The union may have safety and health resources unavailable to the company in terms of experts and data from outside the local union. The safety professional should use the leverage provided by union resources and objectives as long as it is truly in the worker interest.

Safety professionals are major stakeholders in many industrial environment-related professions. However, those professions seldom develop a reciprocal stake in safety practice. Their stake is not always recognized. An intent of this chapter is to inform the reader about the safety profession in the United States. Safety practice involves so many disciplines and professions, it is often difficult simply to call it a profession, in the singular. Where we fail to merge common interests it is often because we don't appreciate each other's values or it is to protect someone's "turf." It is *shared* turf—used by all.

2   THE SAFETY PROFESSION

Whether to refer to safety as a profession is a matter of definition. Although modern safety practice may meet the criteria for a profession, some see it as several disciplines. Every discipline, work area, or function interfaces with safety. The term "safety" itself is so imprecise that practitioners continually seek a better name for their work. We see terms such as risk management, loss control, loss prevention, safety engineering, system safety, and so on. The term "safety" remains the chief descriptor. Those favoring other titles for safety have cultivated their own activity centers, organizations, certifications, and publications, but all still fall under the "safety" umbrella.

2.1 Development of the Safety Movement

As with any profession establishing its roots, safety practitioners have often looked to the Bible and, in our case, also to ancient Babylon. So much for the safety roots. Formal safety started in Europe in the 1860s with the British Safety Act, and in Germany with national compensation legislation for injured workers. Although U.S. safety practitioners can refer to illustrious ancestors such as Benjamin Franklin, it was not until 1970 that the United States had a national safety and health act. When Heinrich's *Industrial Accident Prevention* appeared in 1931, only one other U.S. safety book had made a lasting impression. It was 1956 before the next landmark book appeared, Simonds and Grimaldi's *Safety Management*. Now in the fifth revision, it remains a leader. In the past 10 to 15 years, hundreds of safety books have appeared. With the increase has come more quality. Parallel development has taken place in Europe. As legislation and regulation have influenced safety, so has it involved industrial hygiene and other professions, often the same legislation and regulation.

Professionalism in safety has greatly increased in the last 15 years. The number of practitioners has grown slowly, but their qualifications, stature, and influence have increased dramatically. Now over 50 percent have graduate degrees. Ninety percent have bachelor's degrees. Safety research is a weak area that will improve as more qualified researchers enter the field.

2.2 Scope of the Field

The scope of safety, like that of health, communication, and management, is nearly unlimited. No matter what a person's interest or business, there are related safety positions. We find safety in any field. Disappointed pre-medical and pre-law students find safety a useful adjunct career for medicine and law interests. Book lovers find safety positions with libraries and publishers. Adventurers find that safety specialists deal with everything from space exploration and hang gliding to deep sea and underground mining. Fire or police careerists can use safety as a natural complement for an associated career. All aspects of construction safety have problems and provide many opportunities for safety expertise. Although material handling engineers have refined their own field, most workplaces lack in-house expertise. The safety person thus either augments the knowledge of the materials handling engineer or independently handles the safety-related problems. Safety specialists often liaise with the purchasing functionary to ensure that purchases meet safety and health standards. This helps keep safety and health problems from later arising in the workplace.

The scope of routine safety activity goes into every corner and includes machinery, pressure vessels, electric hazards, chemical exposures, noise, and vibration.

The practitioner commonly deals with atmospheric contaminants and thus toxic compounds, hazardous operations, ventilation, dust control, radiation, and so on. Several transportation areas provide careers such as with flight, marine, pipeline, rail, and vehicle safety. In vehicle safety there are several sub-areas such as driver safety education and fleet safety.

Safety relates closely to the human resources function, the safety officer often advising on personnel selection, training, and replacement. If a person cannot meet standards that ensure safe job performance it is a point of common interest. Human performance limits induced by design, drugs, alcohol, stress, or marital problems involve the safety professional, at least peripherally. This includes ergonomics, particularly in the safe design of tools and equipment. Without human factors specialists, these often surface as safety problems.

Safety professionals often work in product safety. This may come as early as the concept phase, particularly in complex systems. Unfortunately, this calls for well-developed conceptual and technical skills, often beyond the background of the small firm safety person, who is more likely to play a "catch-up" game after someone is hurt or damage is done through the product.

Increasingly the safety person acts as an environmental resource, background notwithstanding, and "grows" into the field as the firm tries to meet governmental mandates. Environmental engineers are scarce and most companies cannot afford the luxury of a specialist. Again, the safety person has assumed the duty by default and by being on the scene. Fortunately, environmental consultants are available as a company resource or to help the safety person.

Fire protection has long been in the safety domain. Until a company can justify a full-time fire resource it will likely remain that way. Related is the need for emergency and catastrophe planning. Because this infers injury and loss of resources, it, too, often falls to the safety person until the firm can support a skilled resource. In any event, the safety person usually plays a key role in contingency planning and assumes coordination roles with other companies and government entities.

The safety person seldom assumes medical staff duties. It follows, however, that if a doctor or occupational nurse is not on the staff, the safety person may assume some medical functions. When there is a medical staff the ties with safety are often close, sometimes as complementary roles in combined departments. When the company has an occupational nurse the nurse often assumes safety duties. Gradually, the nurse gains knowledge, and as the company grows, becomes the safety *and* health person.

Each federal agency, and to a lesser extent state and local governments, has staff safety personnel. This includes the armed forces, which have both military and civilian safety positions. In the local school district we also find a network of safety persons.

All safety fields named in this chapter have a common dominator in that safety specialists in these areas are also in the insurance industry. We estimate that over half (perhaps 150,000) of the safety practitioners work in the insurance business.

Non-workplace safety is common in fields connected with home and recreation safety. In short, then, safety covers all aspects of work, recreation, public, and home safety. It touches every occupation and deals with nearly every discipline. The variety of these interactions ensures opportunities for any interest or depth of study.

2.3 Cliches and Truisms

Safety, as an emerging field, carries a burden of cliches and truisms. Most have come into being over many years. Some should be put to rest immediately:

- "Safety First"—This old and worn slogan has done more to retard safety progress than any other single thing. Safety can never be first. It is profits, mission, product, return on investment, or service that comes first, and keeps the company alive. Anything the company does should contribute to its survival. Safety should have support and contribute to corporate goals of profits, service, or mission. This slogan sometimes appears as "Our First Concern is Safety."
- "Primary Cause"—This usually means the one thing that was responsible for a mishap, but several things always combine to cause a mishap. Picking one cause as primary tends to end the search for all causes leading to an event and thus leads to lost opportunity for prevention.
- "Single Cause"—There is always more than one cause for a mishap. If a person refers to "the cause," you know they are fond of the "primary cause" above, or have not looked at the mishap in enough detail to know the facts that brought it about.
- "Sequence of Events"—Reference to the sequence of events often shows a lack of knowledge about mishap causation. Things don't always happen in a sequence. Usually, they occur as several cause factors happening concurrently or at undefined times. The popularity of the term relates to ease of diagramming or explaining the mishap process.
- "The First-Line Supervisor Is the Key Man in Safety"—Granted, the supervisor is often in charge of a process or operation where a mishap occurs. He or she will likely be in charge of the person (operator) closest to the mishap. However, the supervisor responds to management and staff pressures. If they don't want safe performance over production, the first-line supervisor won't push it. Modern safety practice accepts the first-line supervisor's vital role in mishap prevention, but it also accepts that management and staff interact with an operation as well. To say that the supervisor is the "key" ignores that many others also contribute. The supervisor is important, but not the key.
- "Safety is Everyone's Responsibility"—This really means that people don't know their specific responsibilities. If it is everyone's responsibility it is no one in particular's responsibility. Specific safety responsibility should be assigned to a particular person.
- "By Definition It Was Not an Accident"—This is a "cop out" to keep from

reporting a mishap or dirtying the accident record. Because of the amount of dollar damage or time lost it may not count against the record. Therefore we believe we need not treat it as a mishap. Not so! The causes and contributions are the same and can lead to more mishaps with worse consequences. Investigating non-recordable events helps identify design errors and management problems that led to the event. Trying to evade recording an event keeps us from finding and acting on problem areas.

2.4 Cost of Mishaps versus Cost of Safety

Accidents have been studied for their true cost for over 60 years. Since the 1950s, the Simonds's studies (2) have been the standard on accident costs. Simonds did in-depth, landmark research on the indirect cost of accidents. His findings have been verified by further research each year. He found the ratio of indirect to direct costs is about 3:1. That is, for every dollar of direct cost there is an average of three dollars more for indirect costs. Direct costs are reimbursable costs such as medical care, workers' compensation, and insured property losses. Indirect costs, of which over 150 have been identified, are nonreimbursable. They include lost wages, lost production, lost business opportunities, influence of the event on the work of others, and so on.

There are also intangible costs that are not directly measurable but that can be sizable. Lowered morale as the result of worker concern over a mishap is reflected by lower production and lesser quality control. The public relations aspects of a mishap may shake confidence in a company by suppliers and buyers. This is particularly true if it raises the issues of mismanagement or liability. As public perception of a company is shaken by mishaps the results are similar to news that the company is facing bankruptcy. People do not want to have moneys owed, to furnish supplies on credit, or to purchase items or services that are not backed by an ongoing business. Although many mishaps cost far less, some researched indirect costs show 200 times the direct costs (some mining and construction mishap research).

With average direct mishap costs of $13,000 to $33,000, the indirect costs are $39,000 to $99,000 for each event. The total average costs then run from $52,000 to $132,000. This 3:1 ratio allows safety staffs to forecast savings. Most companies, serious about their safety programs, adopted an indirect cost ratio. Most federal agencies must do a cost–benefits analysis before imposing new regulations or procedures with a wide impact, such as a new reporting requirement. Safety and health benefits must outweigh the costs. Similarly, private business insists on proof before making nonmandated commitments for safety and health.

2.5 The Safety Language (What are They Talking About?)

Safety practitioners, like most professionals, tend to use terms unique to their function. This leaves others to wonder what was said or written. An intent of this chapter is to clarify safety language for the reader and take the mystery out of safety practice. Fortunately, much of this specialized safety language also pertains to other chapters in this book. Appendix A lists some acronyms that relate to safety practice.

2.6 Organization of the Safety Function

The safety function may be placed anywhere in an organization. Where safety's value is seen and the practitioner qualified, the latter may report to a vice-president, or in a few cases, to the chief executive. In the military the safety officer reports directly to the commander except in large organizations, where he or she reports to the vice or deputy commander.

The safety practitioner, alone or in a large safety and health organization, often reports to the person in charge of human services, perhaps a vice-president. The function may come under a corporate attorney or a manager of operations, whichever is seen to be most effective, or it might be found in the only place with time to oversee the function. The higher the perceived value of the safety function or incumbent, the higher it will be in the hierarchy. Section 6.2 has more information on this subject.

2.7 Mishap Investigation

Accident (mishap) investigation, is better described as recording or reporting. That is, based on government reporting requirements, industry must seldom do more than *record* what has happened on the proper forms (3). Sometimes a basic form of investigation calls for a few questions and a records review. This compares to newspaper reporting and thus is called *reporting*. Unfortunately, this recording or reporting level satisfies most requirements, but does little to prevent more mishaps. Without in-depth investigation we only record the facts and perhaps some causal factors. There is little information on why the event happened and what can be done to prevent it from happening again. The report may not even suggest corrective action. Mishap investigation should do more than record and report. It means a review of the facts around an event, finding out what really caused it to happen and what will keep it from happening again (3).

2.8 Workers' Compensation

Compensating workers for job-related injuries and illnesses is done in every state, territory, and the District of Columbia through workers' compensation acts. Each state act follows a national model so that they are somewhat standard. Their benefits may vary widely.

Workers' compensation (WC) was meant to advance workplace safety programs. It relates closely to safety because preventing injury or illness directly reduces the cost of workers' compensation payments. It was thought that this would motivate employers to develop safety programs. This it has done. The system is no-fault. That means that the worker only has to be injured or become ill related to the workplace to receive compensation. Only in the case of the most gross worker neglect or misbehavior is payment not given. Reducing WC costs is one of the few places that safety practice can show direct savings by mishap or illness prevention. WC may be administered in a company by the insurance department, the medical section, the personnel function, a separate workers' compensation section, or the safety office. Regardless of who does it, the safety function has a direct stake in the costs and the causes of an injury or illness bringing the compensation.

WC is funded through state-sponsored agencies, insurance companies, or a company being self-insured. A firm pays a set amount for each $100 of payroll. The exact amount depends on past and expected WC payouts. The states permit different arrangements, sometimes allowing only the state-sponsored agency to operate. In some states all three plans coexist. Financial plans for self-insurers are strictly regulated to ensure that enough funds are on hand for payouts.

2.9 License and Certifications

There are several certificates or licenses that help one measure the experience or knowledge level of a safety person.

The major safety license is Registered Professional Engineer (PE). Every state has a registration board for professionals. Registration requires proof of experience and passing tough written tests. The license must be renewed at set periods and the state monitors licensee performance. About 15 states have special PE *safety* registration.

The best known certificate is the Certified Safety Professional (CSP). This reflects the equivalent of 15 years of full-time professional experience. For some experience and college degrees the requirement can be reduced, sometimes to a very few years. At least two thorough, monitored exams must be taken. Once designated a CSP, holders enter a continuance program. The program requires building credits through academic activities, further examinations, or professional pursuits. This ensures currency in the field. There are other certificates but they do not have the credibility of the CSP.

The Board of Certified Safety Professionals and the American Board of Industrial Hygiene have joined to administer a certificate known as the Occupational Safety and Health Technologist. This illustrates the close relationship between safety and industrial hygiene practice at the technologist/technician level.

2.10 Professional Organizations

Professional organizations seek to promote the competence of members. Many safety groups border on being a "professional" group, but can better be described as social, fraternal, or stakeholder-based. These groups include those meant to promote vested interests rather than individual professionalism. Four safety-related groups stand out as professional and safety oriented.

2.10.1   American Society of Safety Engineers

The American Society of Safety Engineers (ASSE) dominates professional safety groups in the United States. It has 25,000 members and 125 chapters with members in 50 countries. This is the oldest "professional" group and serves its members many ways:

- National and regional professional development conferences
- Safety training seminars and programs sponsored by the national and local chapters
- A monthly journal, *Professional Safety*
- Extensive sale of publications at reduced prices published by ASSE and other publishers
- Full service to local chapters
- Several divisions such as the management, construction, and consultant's divisions that serve a membership segment
- Active development of ASSE student chapter sections

The ASSE is wholly member oriented, backed by a full-time professional staff, and very responsive to member's needs.

2.10.2   National Safety Management Society

The National Safety Management Society (NSMS) has perhaps 600 active members devoted to management aspects of safety. It was started in the late 1960s, when little formal focus was given to safety management aspects. Membership has dropped in recent years as the ASSE has become active in the management side of safety.

2.10.3   System Safety Society (and Techniques)

As the title suggests, this group specializes in the system safety aspects. The society is roughly the same size as the NSMS and has local chapters. It is known for the quality of its journal, *Hazard Prevention*, and an excellent annual conference. True system safety practice is exacting. Most members have engineering degrees.

System safety analysis looks at every situation that could arise in each part of a system during its lifetime, starting with the system concept. It has grown to include effects after disposal of a system. A specialty field of system safety analysts, engineers, and technicians has emerged. Common techniques are fault tree analysis (FTA) and failure modes and effect analysis (FMEA). System safety specialists have also adapted other techniques for their use and can list over 30 germane techniques. That includes THERP (technique for human error rate prediction, PERT (program evaluation review technique), and CPM (critical path method).

System safety practice gained reknown after being used to solve early missile problems that threatened our national defense. It is firmly embedded in the U.S. defense contract structure. Most critical defense systems require the use of system safety analysis techniques in development. The cost is a part of contracts. Unfortunately, nondefense contracts seldom provide for using system safety techniques to locate or solve problems. Thus parties to recent disasters such as Bhopal, highway and bridge collapses, or high-rise fires seldom benefit from system safety techniques. In other spectacular failures, such as space-related mishaps, system safety techniques were used, but did not consider the entire system. System safety use in industrial hygiene practice is nearly nonexistent.

2.10.4 National Safety Council

The National Safety Council (NSC), strictly speaking, is not a professional society. However, as long as there has been an organized safety movement, the council has filled the "professional needs" of many practitioners. The NSC is the largest single safety organization in the world (The British Safety Council is second), with over 10,000 corporate members representing 30 million employees. Membership dues depend on the number of workers in a member company. All workers of corporate members become members of the NSC.

The NSC is chartered by Congress and is a nonprofit group. The NSC has a network of 77 local safety councils in the United States and Canada that are run by local boards of directors. They sponsor safety seminars and courses and sell nationally sanctioned promotional items.

The NSC is organized into regions that often sponsor regional versions of the National Safety Congress. Organized largely to support industrial safety, it has 27 sections, backed by a staff of 350 full-time employees and 2000 volunteers to serve specialized needs. These sections range from aerospace to trades and services sections. Member benefits include discounts on educational and promotional goods and attendance at training seminars. This includes the annual National Safety Congress, the world's largest safety meeting (4, 5).

2.10.5 Other

Many professional groups relate to safety as seen by Appendices B and C. Few of those listed are strictly for safety, but they typify groups with which safety professionals interact. Where, for example, a safety person works with heavy industry and welding they might often contact or be a member of the Iron and Steel Institute or the American Welding Society. Many of these groups have their own safety directors.

3 TRAINING AND EDUCATION

Safety training in the United States is handled by government agencies and by public and private groups. Among government agencies the Department of Labor, through its OSHA Training Institute, handles the most students. Their courses are open to the private sector. NIOSH, through grants and contracts, conducts education programs to staff its needs and presents programs on safety and health equipment. Other agencies have in-house safety training, but limit public access. At least one government safety contractor conducts safety training for the private sector. The dominant private U.S. firm giving safety training is The International Loss Control Institute. Many smaller firms and consultants also offer safety training for both the public and private sector.

Safety education by U.S. colleges and universities have some 30 graduate and undergraduate safety and health programs. Perhaps 200 more offer safety courses and certificates to meet local needs. The ASSE seeks to be the accreditation group for safety degree programs. At this writing, they have accredited four schools. Most schools that offer safety degree programs have a healthy input of industrial hygiene and health-related courses, if not degree programs.

Safety education and training outside of the United States is common. Nearly every European country has active programs, some of high quality and reputation. At least two provide safety education at the doctoral level. European safety seminars and exhibitions are widely attended by people from countries other than the host country. European concepts are advanced, but often fragmented by country. As the European Community (EC) advances we can expect these ideas and thoughts to combine in outstanding programs. Pacific Rim and Middle East countries, with less safety training and education, take much schooling in other countries. Notable are the British Safety Council courses, whose student lists often resemble a United Nations roster. Increasingly common is the participation of Eastern Bloc countries. As their activities become known an unexpected degree of sophistication is sometimes seen. The International Labor Organization (ILO) does an excellent job of safety development with third-level countries. Their success with hands-on, practical safety is notable. Anyone dealing with emerging nations should build a knowledge of ILO ventures.

4 INFORMATION BASES

Information bases that relate directly to safety are often only record-keeping devices to track injuries, mishaps, protective clothing and equipment, and mandated safety training. In a few cases they track related legislation. The extent to which safety interfaces with health, engineering, hazardous materials, and environmental matters is a safety bonanza because major data bases exist in these areas. Thus safety practice has extensive data base access to help do the job. The ASSE alone, for example, has printed three volumes (6–8) of computer-related safety resources. Other countries, particularly in the EC, are also active with safety data bases.

5 PUBLICATIONS

Many publications relate to safety, but the interaction with other professions means that few U.S. periodicals deal with safety alone. For example, the three most widely read magazines relate to health either in their name (*Occupational Health and*

Safety and *Safety and Health*) or by inference (*Occupational Hazards*). Safety-related U.S. research journals are rare. The best come from Europe. Books on safety do tend to specialize, but it is hard to write a safety book without frequent reference to health and environmental matters.

5.1 Journals and Periodicals

Most professional organizations have a "journal," but the quality of published research and other material varies. Some safety-related journals such as ASSE's *Professional Safety* do fill a professional need. However, they do not have the quality of a journal from the American Medical Association, the American Industrial Hygiene Association, or the Human Factors Society. Nevertheless, *Professional Safety* could not be replaced by a scholarly research journal. It would not properly serve the membership.

Appendix D lists safety-related periodicals. There is no attempt to assess quality, only to provide a listing. Information on publishers' addresses and subscriptions may be found in most libraries (9).

5.2 Books

Only in the past 15 years have many quality books been published to direct the safety practitioner toward professionalism. Perhaps 100 closely related safety books published within the past five to seven years are worth considering. Another 75 or so older books are worth having for a large safety library. A few have value as timeless references. Outside of original research and some reference works, rapid changes in safety practice make most books over 10 years old outdated. See Appendix E for some books printed in the past decade.

6 DEPTH OF PRACTICE

On any day a staff safety person may work with the medical director, purchasing agent, chief engineer, comptroller, first-line supervisor or middle manager, fire department personnel, security chief, or other contacts. Safety problems tend to be short range and involve complex interactions between several plant resources. Whereas a hygienist may take three days to measure and solve a problem, the safety person may deal with several problems and as many people and functions during the same time. It is this interaction between internal and external functions that give depth to safety practice. Many disciplines and resources may be used to find or solve safety problems. The effective safety professional needs many management and communications skills.

The depth of the safety profession is not finalized. A firm with resources to hire only one person and needing a safety person, a hygienist, and a health physicist, for example, will most often hire the safety person. This is partly because of OSHA's high visibility for fines and inspections. Because the safety person tends to work with hygiene-related fields he or she is a natural staff choice to examine environmental problems also. Lack of expertise is augmented by working in or near a field, contact with experts on a timely basis, recent OSHA emphasis on health aspects of the regulations, and schooling in related aspects of safety practice.

6.1   Hands-On Practitioner

The hands-on practitioner works at the base level of safety practice. Dealing first hand with safety problems is at the base of the safety practice hierarchy. It is the basic knowledge of a safety practitioner. This basic practice tends to be the domain of the first-line or intermediate supervisor who uses the safety person as an on-call resource.

6.2   Staff Resource

The ideal safety practitioner role is as a staff resource with access to the entire workplace. This resource should be within reach of all other workplace functions. The safety function can be found nearly anywhere in the organization. This ranges from reporting to the corporate attorney to working for the production supervisor. This depends, however, on the stress given safety by the chief executive. It may not be practical, because of the flatness of an organization, to have many staff functions report to the top person. Only in the military does the safety officer report directly to the head person (commander) or a deputy. In practice, the safety function most often comes under the Director of Human Services or a similar title.

The safety position advises staff and is a resource for field units. With decentralized field safety functions, the corporate staff still carries the "safety ball" to provide high level backing. The higher-level safety functionary will probably be active in field evaluations and critical reviews.

The main point is that the safety practitioner should be free to serve the entire organization and not be handicapped by functional borders. The safety practitioner should be placed so as to evaluate line and staff operations freely and critically and to solve problems. If safety is a staff function, then final responsibility for safety rests with the line and ultimately the chief executive.

It may appear that the staff safety person does not have authority to see a job through and cannot be effective. To the contrary, the safety person can have good authority even if not in a "line" position. This stems from (1) authority to warn and be heard, (2) position as a staff person, and (3) reputation as a trusted resource to help solve problems. Much authority comes from being able to move freely throughout the company and coordinate solutions to safety- and health-related problems. Few other people have such free staff and management access to seek out and fix safety and health problems. This makes the safety person a valuable resource and ally for the line and for the staff.

7 TECHNIQUES OF PRACTICE

Every employer covered by OSHAct must provide each employee with a place to work that is free from known hazards likely to cause death or serious physical harm. Safety practice is largely driven by the need to comply with this duty, to avoid litigation, and to achieve corporate objectives. Whatever the drive, it is increasingly seen that sound safety programming *beyond compliance* offers an excellent return on investment.

Most safety programs appear to have a two-way approach, safety engineering and safety management (2). This does not begin to cover modern safety practice. A step forward is the "three E's" of safety. Briefly, this means that we should first seek *engineering* solutions to prevent or remove health or safety hazards. The next step is to *educate* or train workers to work safely. If they should fail to do so, *enforcement* (probably punishment) is the last step under the three E's. These three E's of safety have long been a basis for safety programs, but modern safety practice can be far more sophisticated and complete. More information on more sophisticated safety programs can be found in most of the references and Appendices D and E.

7.1 Standards

The Secretary of Labor develops and promulgates safety and health standards. The Department of Health and Human Resources, through NIOSH, routinely publishes documents that recommend permissible exposure levels for toxic substances and propose safety standards. The documents then go to OSHA for consideration for promulgation as standards. These interim documents are advisory only. Very few, if any of them, are actually proposed as standards by OSHA. The law requires NIOSH to produce criteria documents. Where federal standards do not exist stakeholder groups such as the American National Standards Institute (ANSI) shape guidance for their members. These are sometimes used by the federal government, under the general duty clause of the OSHAct, perhaps with little change. The federal government may publish emergency standards without regard to rule-making processes, but they should be followed within six months by permanent standards. These must go through a complete rule-making process and modify or replace interim standards.

7.2 Regulations

In the United States our concern with federal regulations is how they affect safety practice. Many states, and the military services, also have their own regulations covering the same areas. The best example concerns federal OSHA; about half the states at any one time have their own OSHA programs as well. We then see, for example, CALOSHA or NEVOSHA for California or Nevada. Businesses in those states comply with the state OSHA, but know it is at least as strict as the federal requirements. Between the federal and state plans each employer in a business affecting commerce in every state and several adjunct parts of the United States is covered, at least once. Despite the mandate to conform, there is much truth in the adage that "you cannot legislate safety." Regulating hazards does not ensure their control. Regulation has come to its present state largely because the public concern has willed it.

7.3 Company Plans

In Section 6 we discussed the three E approach to safety programs. Whether a company uses a technical or management approach to safety, or both, is a matter of experience or guidance. All techniques have their backers and successes. One approach to safety management, the human approach, emphasizes working with the individual as subject to the usual motivations and responses (10). Most approaches use certain common elements. They start with a policy statement, usually signed by the chief operating executive. Although compliance with regulations is necessary, the policy must also bring procedures to the performance level. The operating unit adopts government, corporate, or local guidelines to follow in the workplace. See Section 7.3.3.

7.3.1 For Compliance

OSHAct allows the Department of Labor to inspect the place(s) of business of any employer covered by the act. This is to ensure compliance with the act. The act also gives NIOSH the right of entry to conduct health hazard evaluations, but NIOSH itself has no enforcement authority. Seldom acknowledged is that full compliance with federal and state safety and health regulations and standards is a minimum safety strategy. It is only what must be done by law. This may do a good job but it is the least a company must do.

Thus a company safety plan must at least provide for full compliance with governmental requirements. It includes those imposed by the community, chiefly in environmental affairs. A major part of this compliance concerns industrial hygiene.

7.3.2 For a Higher Level of Safety

For a level of safety beyond more compliance, a company must go farther. A program must also consider the special needs of the company, the community, and the work force. Companies with a good safety record exceed the minimum requirements of the law with their own added provisions. Sometimes a company has such an extensive program that compliance with OSHA, EPA, and NIOSH requirements is a small part of the overall plan. This extra development is typical of the world's leading safety and health programs. Some programs are so successful in reducing unwanted losses that they become marketable items.

7.3.3 Typical Plan Provisions

7.3.3.1 Policy. Corporate and local safety policy is the broadest guidance. Usually developed with the help of corporate safety resources it dovetails with and supports other policies. This policy also gives direction for subordinate units. Failure to provide proper policy guidance leads to weak procedures and performance at lower levels. Corporate policy reflects senior management posture. Someone with overall, top-level responsibility for safe workplace activities signs the policy. This is not likely to be anyone lower than a vice-president or general manager at a separate location. Policy for other plants in a multi-plant company should fit into a total corporate plan.

7.3.3.2 Procedures. Details of executing safety policy are given in procedures. These provide guidance on the type of program to use and provide the resources (people, training, equipment, expertise, etc.) for the task. This is the function level for most safety persons. If the procedures level of staff and management doesn't provide the proper procedures and resources, then the performance level cannot safely do its tasks.

7.3.3.3 Performance. This is the operating level of production and service. It is where mishaps and injuries occur. Safe activity comes at this level only if proper emphasis is given to safety and if the right procedures and resources are present. The immediate supervisors are given the task of safe operation but they can only do it with training and resources provided by higher management and staff. Failure to provide this is a management and staff failure.

7.3.4 Safety Committees

Most committees have a shaky reputation for action and efficiency. When properly conceived and managed, safety committees can do wonders and serve as a valuable communication and action tool. A company safety plan using the policy, procedures, and performance approach has a natural ladder framework for safety committees. The Performance level committee finds problems and makes recommendations on workplace safety and health hazards. If that committee cannot resolve problems it refers them to the next higher committee at the procedures level of staff and supervisory management. A problem seldom exists that cannot be solved by expertise or authority found at this level. In those rare cases needing higher action the top-level, senior management safety committee can review the matter and take action. Committees function in this three-tiered structure by sending upward reports of their meetings, including actions taken and needed. Their committee reports also inform management of problems and activities at the performance level. It is common for this top level to have weekly or monthly safety committee meetings to assess operations. Such a committee structure requires the next higher level to respond, act, or, as needed, commend those who submit reports.

8 CHANGING SAFETY WORLD

The major safety concerns of 1991 were not major safety concerns in 1981. Today's major problems are seldom those of a decade ago. We can be certain that the major safety problems of the year 2000 will not be today's major problems. As this is written there is heavy emphasis on workplace health and community environmental problems. That direction did not exist for the safety profession 10 years ago.

The educational level of safety practitioners has steadily increased over the past two decades. Corporations now routinely require a bachelor's degree for placement, with a graduate degree being a plus feature. Government entities are nearly the only ones who do not require a degree for new safety hires. Even there, however, a degree is a positive feature and is required for some jobs. A degree need not be safety related as long as it applies somewhat to the field. A specialty area such as environmental safety, business administration, industrial hygiene, or an engineering branch lends marketability to the jobseeker. Salaries for degreed applicants compete with other professions and often exceed them. We can expect even more emphasis on formal education as the scope of the safety profession is recognized.

Qualified minority applicants are in demand. Second language capability is a plus. Women are so successful in safety practice that they seem to have a natural affinity, if such a thing is possible, for the field. Every sign leads us to expect the percentage of minorities in safety to increase.

Rapid changes in our social, political, and regulatory outlooks ensure that a high degree of professionalism will dominate safety practice. Safety requires constant upgrading of professional skills, as do those of a physician, accountant, or attorney. Without this, safety practice has a "half-life" of about 5 years. This means that 10 years from now, without upgrading, our present safety knowledge will be about 25 percent effective. Some leading futurists see this half-life continually shrinking and well below the figure given here. If the safety professional has not continually upgraded during the past 10 years, his or her present professional ability is deeply degraded and out-of-date.

Still to be widely grasped is the effect of the European Community (EC) on U.S. process, production, and safety. New safety and health standards will be common to all EC members. To compete or even exchange information with Europe means that U.S. firms must meet and deal with their standards, sometimes more stringent than our own. A united EC presents a far different perspective, market, and competition than we have faced (11).

9 SUMMARY

This chapter presents an overview of the interactions between safety and some related disciplines. The most prominent bonds are the need to protect the worker and the need for communication between disciplines. These needs can be met only if the scope and depth of the present and coming safety practice are seen.

It is not so important what keeps workers from doing their job properly, whether it be a health, injury, or property damage problem. It is of little concern whether the event is recordable as an accident, injury, or illness. The larger problem is a professional concern for underlying causes and working together. Of the work-related injuries and illnesses, about 3 percent (and growing) are illnesses. The rest are logged as injuries, although many of them probably relate to illness. It can be seen that those concerned with the safety and health aspects have a common stake in preventing industrial diseases, injuries, and mishaps. They become mutual stakeholders who must work together to solve their problems, problems that invariably involve each other's disciplines, professions, and turf. Finally, the common ground of safety and health interfaces will be greatly influenced by international aspects such as the EC and Asian partnerships.

ACKNOWLEDGMENT

The help of Dr. James O. Pierce of the University of Southern California in preparing parts of this chapter is gratefully acknowledged.

REFERENCES

1. T. S. Ferry, *Safety and Health Management Planning*, Van Nostrand Reinhold, New York, 1990.
2. J. V. Grimaldi and R. H. Simonds, *Safety Management*, 5th ed., Richard D. Irwin, New York, 1989.
3. T. S. Ferry, *Modern Accident Investigation*, 2nd ed., Wiley, New York, 1988.
4. *Accident Prevention Manual for Industrial Operations: Administration and Programs*, 9th ed., National Safety Council, Chicago, IL, 1988.
5. *Accident Prevention Manual for Industrial Operations: Engineering and Technology*, 9th ed., National Safety Council, Chicago, IL, 1988.
6. *Directory of Computer Related Safety Resources*, Vol. 1, American Society of Safety Engineers, Des Plaines, IL, 1987.
7. *Directory of Computer Related Safety Resources*, Vol. 2, American Society of Safety Engineers, Des Plaines, IL, 1987.
8. *Directory of Computer Related Safety Resources*, Vol. 3, American Society of Safety Engineers, Des Plaines, IL, 1988.
9. T. S. Ferry, *Safety Program Administration for Engineers and Managers*, Charles C Thomas, Springfield, IL, 1984.
10. D. Petersen, *Safety Management: A Human Approach*, Aloray, Goshen, NY, 1988.
11. Seminar "Product-Life: The Overall Design," subtitled "Life Cycle Engineering: The Key to Risk Management, Safer Products and Industrial Environmental Strategies," January 16 and 17, 1989, Swiss Federal Institute of Technology, Zürich, Switzerland. T. Ferry, "Riskomanagement von der Konstruktion bis zur Entsorgung," *gdimpuls*, Gottlieb Duttweiler Instituts, Zurich, Switzerland, 1989. "Bridging Engineering Gaps and Vacuums from Design to Disposal," *Journal of Occupational Accidents*, **13**, 17–31 (1990).

Appendix A

ACRONYMS—SAFETY AND HEALTH RELATED ORGANIZATIONS

AAEE	American Academy of Environmental Engineers
AAIH	American Academy of Industrial Hygiene
AAOHN	American Association of Occupational Health Nurses
ACGIH	American Conference of Government Industrial Hygienists
ACS	American Chemical Society
ADTSEA	American Driver and Traffic Safety Education Association
AGC	Associated General Contractors
AICH	American Institute of Chemical Engineers
AIHA	American Industrial Hygiene Association
AIIE	American Institute of Industrial Engineers
AIMPE	American Institute of Mining, Metallurgical and Petroleum Engineers
ANS	American Nuclear Society
ANSI	American National Standards Institute
APCA	Air Pollution Control Association
API	American Petroleum Institute
ASA	Acoustical Society of America
ASAE	American Society of Agricultural Engineers
ASCE	American Society of Civil Engineers
ASIS	American Society for Industrial Security
ASME	American Society of Mechanical Engineers
ASPA	American Society for Personnel Administration
ASQC	American Society for Quality Control
ASSE	American Society of Safety Engineers
ASTD	American Society for Training and Development
ASTM	American Society for Training and Materials
CEC	Consulting Engineers Council
CSA	Campus Safety Association of National Safety Council
CSAA	Construction Safety Association of America
CSSE	Canadian Society of Safety Engineering
HFS	Human Factors Society
HPS	Health Physics Society
IEEE	Institute of Electrical and Electronics Engineers
IIA	Insurance Institute of America
IIRSM	International Institute of Risk and Safety Management
IMMS	International Material Management Society
IOSH	Institution of Occupational Safety and Health
ISASI	International Society of Air Safety Investigators
ISEA	Industrial Safety Equipment Association
NFPA	National Fire Protection Association
NIFS	National Institute for Farm Safety

NSC	National Safety Council
NSMS	National Safety Management Society
NSPE	National Society of Professional Engineers
PEPP	Professional Engineers in Private Practice
RIMS	Risk and Insurance Management Society
RoSPA	Royal Society for the Prevention of Accidents
SAE	Society of Automotive Engineers
SAME	Society of American Military Engineers
SEDA	Safe Equipment Distributors Association
SES	Standards Engineering Society
SFPE	Society of Fire Protection Engineers
SME	Society of Manufacturing Engineers
SSS	System Safety Society
VOS	Veterans of Safety

Appendix B

ORGANIZATIONS CLOSELY RELATED TO SAFETY

Academy of Hazard Control Management

Academy of Product Safety Management

Academy of Safety Educators

Acoustical Society of America

Air Line Pilots Association (Air Safety Division)

Air Pollution Control Association

American Academy of Industrial Hygiene

American Association of Occupational Health Nurses

American Board of Industrial Hygiene

American Chemical Society

American College of Toxicology

American Conference of Governmental Industrial Hygienists

American Industrial Hygiene Association

American Nuclear Society

American Nurses Association, Inc.

American Occupational Medical Association

American Public Health Association

American School and Community Safety Association

American Society for Industrial Security

American Society of Mechanical Engineers (Safety Division)

American Society for Quality Control

American Society of Safety Engineers

American Society for Training and Development

American Welding Society

Aviation Safety Institute

Board of Certified Hazard Control Management

Board of Certified Safety Professionals

British Fire Protection Systems Association

British Safety Council

Campus Safety Association

Canadian Center for Occupational Health and Safety

Center for Auto Safety

Chemical Industry Safety, Health and Environmental Association, Ltd.

Construction Industry Research and Information Association

Fire Protection Association
Flight Safety Foundation
Health and Safety Executive
Health Physics Society
Human Factors Society
Industrial Accident Prevention Association of Ontario
Insurance Loss Control Association
Institute of Electrical and Electronics Engineers
Institute of Environmental Health Officers
Institute of Environmental Sciences
Institute of Occupational Medicine
Institute of Transportation Engineers
Institution of Occupational Safety and Health
Inter-American Safety Council
International Hazard Control Manager Certification Board
International Healthcare Certification Board
International Institute of Risk and Safety Management
International Labor Organization
International Society of Air Safety Investigators
Insurance Institute for Highway Safety

Laser Institute of America
Medical Commission on Accident Prevention
National Environmental Health Association
National Fire Protection Association
National Safety Council
National Safety Management Society
Occupational Medicine Association
Pan American Health Organization
Risk and Insurance Management Society
Robens Institute of Industrial and Environmental Health and Safety
Royal Society for the Prevention of Accidents
Safe Association
Society of Automotive Engineers
Society of Fire Protection Engineers
Society of Manufacturing Engineers
Society of Occupational Medicine
Society for Risk Analysis
St. John Ambulance Service
System Safety Society
Veterans of Safety
World Health Organization
World Safety Organization

Appendix C

TRADE ASSOCIATIONS CLOSELY RELATED TO SAFETY

Where can you go for help when your personal contacts don't work out? Try a trade association associated with your problem. Your local library can assist you with the addresses through directories of associations.

American Foundrymen's Association
American Gas Association
American Iron and Steel Institute
American Institute of Chemical Engineers
American Mining Congress
American Paper Institute, Inc.

American Petroleum Institute
American Pulpwood Association
American Road and Transportation Builders
American School and Community Safety Association
American Trucking Association, Inc.
American Waterways Operators, Inc.
American Water Works Association
American Welding Society
Associated General Contractors of America, Inc.
Association of American Railroads
Bituminous Coal Operator's Association
Chemical Manufacturers Association, Inc.
Compressed Gas Association, Inc.
Edison Electric Institute
Graphic Arts Technical Foundation
Industrial Safety Equipment Association, Inc.
Institute of Makers of Explosives
International Association of Drilling Contractors
International Association of Refrigerated Warehouses
Iron Castings Society
National Association of Manufacturers
National Coal Association
National Constructors Association
National Health Council, Inc.
National LP-Gas Association
National Petroleum Refiners Association
National Restaurant Association
National Rural Electric Cooperative Association
National Sanitation Foundation
National Soft Drink Association
New York Shipping Association, Inc.
Portland Cement Association
Printing Industries of America, Inc.
Robot Institute of America
Scaffolding, Shoring and Forming Institute
Society of Automotive Engineers
Society of Mechanical Engineers
Steel Plate Fabricators Association, Inc.

Appendix D

SELECTED SAFETY AND HEALTH PERIODICALS

AAOHN Journal
Accident Analysis and Prevention
Accident Facts
Accident Prevention
Air Force Safety Journal
Air Pollution Control Association
  Journal
American Industrial Hygiene Associ-
  ation Journal
American Journal of Epidemiology
American Journal of Nursing
American Journal of Public Health
Applied Ergonomics
Approach (U.S. Navy)
Army Aviation Digest
Australian Safety News
Bests Review
British Journal of Industrial Medicine
Business and Health
Business Insurance
Canadian Occupational Safety
CIS Abstracts
Computers in Safety and Health
Concern
Constructor
Consumer Product Safety Guide
Corporate Fitness
Corporate Health
Driver
Emergency Medical Service
Employee Services Management
Environmental Health Journal
Environmental Management News
Environmental Science and Health
Environmental Science and Technol-
  ogy Environment Reporter
Ergonomics
Family Safety and Health
Fathom
Fire Command
Fire Journal

Fire News
Fire Service Today
Fire Technology
Fire House
Fire International
Fire Surveyor
Fire Prevention
Fire Safety Journal
Fitness in Business
Flight Comment Propos de vol
Flight Safety Facts and Reports
Flying Safety
Forum
Hazardous Materials Intelligence Report
Hazardous Materials Journal
Hazardous Waste News
Hazard Prevention
Health and Safety Bulletin
Health and Safety Newsletter
Health Physics Journal
Healthy Companies
Human Factors
ILO Information
Industrial Engineering
Industrial Hygiene Digest
Industrial Hygiene News
Industrial Safety and Hygiene News
Industrial Safety and Loss Control
Industrial Safety Loss Control
Industrial Security
Insurance Review
International Journal of Radiation
Institute of Transportation Engineers
Insurance Journal
International Journal of Aviation Safety
  (England)
Journal of the Acoustical Society of
  America
Journal of American Insurance
Journal of the Air Pollution Control
  Association

Journal of the American College of Toxicology

Journal of the American Medical Association

Journal of Emergency Medical Services

Journal of Environmental Health

Journal of Environmental Pathology and Toxicology

Journal of Environmental Science and Health

Journal of Fire Safety

Journal of The Human Factors Society

Journal of Occupational Accidents

Journal of Occupational Medicine

Journal of Products Liability

Journal of Safety Research

Journal of Security Administration

Journal of Sound and Vibration

Journal of Traffic Safety and Education

Laser Focus

Loss Control Survey

Mac Flyer (Air Force)

Machine Design

Manufacturing Engineering

Mech, Naval Aviation Safety Review

Mine Safety and Health

Modern Metals Handling Magazine

Noise Control Report

NTSB Reports

National Traffic Safety Newsletter

Nuclear Safety

Occupational Accidents Journal

Occupational Hazards

Occupational Health and Safety

Occupational Health and Safety (Canada)

Occupational Health and Safety News Digest

Occupational Health and Safety Practice

Occupational Health Nursing

Occupational Medicine

Occupational Safety and Health

OSHA Reference Manual

OSHA Report

Prevention

Prevention Express

Prima

Proceedings of the Marine Safety Council

Product Liability Reporter

Product Liability Reports

Product Safety & Liability Reporter

Product Safety News

Product Safety Newsletter

Product Safety Up-to-Date

Products Liability International

Products Liability Journal

Professional Safety

Professional Safety (South Africa)

Protection of Assets

Protection Officer

Quality Progress

Record

Risk Analysis, an Internal Journal

Risk Control Review

Risk Management

Risk Management Manual

Safe Journal

Safety and Health

Safety Management (England)

Safety Manager's Newsletter (England)

Safety Advocate (Philippines)

Safety Canada

Safety in Australia

Safety Journal

Safety Management

Safety Management Planning

Safety Sciences Abstract

Safety Surveyor (England)

Security

Sentinel

Sound and Vibration

System Safety Newsletter

The Safety Practitioner (England)

Toxic Material News

Toxic Materials Transport

Toxicology and Environmental Health
Traffic Safety
Transportation Engineering Journal
Trial Magazine
TSCA Chemicals in Progress Bulletin
Water Pollution Control Federation
  Journal

Welding Journal
Your Health and Fitness
Your Health and Safety
Your Health and Safety Report

Many of the periodicals listed above are from outside the United States. Whims of the publishing business preclude this list from being current, even on the day of publication.

Appendix E

101 SELECTED RECENT SAFETY PUBLICATIONS

1. *Accident Prevention*, International Labour Office, Geneva, 1983.
2. *Accident Prevention Manual for Industrial Operations: Administration and Programs*, 9th ed., National Safety Council, Chicago, 1988.
3. *Accident Prevention Manual for Industrial Operations: Engineering and Technology*, 9th ed., National Safety Council, Chicago, 1988.
4. T. H. Allegri, Sr., *Handling and Management of Hazardous Materials*, Chapman and Hall, New York, 1986.
5. W. W. Allison, *Profitable Risk Control*, American Society of Safety Engineers, Park Ridge, IL, 1986.
6. J. A. Allocca and H. E. Levenson, *Electrical and Electronic Safety*, Reston Publishing Company, Reston, VA, 1982.
7. R. C. Anderson, *Promoting Employee Health: A Guide for Worksite Wellness*, American Society of Safety Engineers, Park Ridge, IL, 1985.
8. G. A. Peters and B. J. Peters, Eds., *Automotive Engineering and Litigation*, 2 vols., Garland Law Publishing, New York, 1988.
9. L. Bass, *Products Liability*, Shepards/McGraw Hill, Colorado Springs, CO, 1986.
10. P. B. Beaumont, *Safety at Work and the Unions*, Croom Helm and Biblio, London, 1983.
11. *Best's Safety Directory*, A. M. Best Company, Oldwick, NJ, annual.
12. D. L. Bever, *Safety: A Personal Factor*, Times Mirror/Mosby College Publishing, Santa Clara, CA, 1984.
13. F. E. Bird, Jr. and G. L. Germain, *Practical Loss Control Leadership*, International Loss Control Institute, Loganville, GA, 1985.
14. H. W. Blakeslee and T. M. Grabowski, *A Practical Guide to Plant Environmental Audits*, Van Nostrand Reinhold, New York, 1985.
15. M. Bryant, *Success With Occupational Safety Programmes*, International Labour Office, Geneva, 1984.
16. J. R. Cashman, *Hazardous Materials Emergencies: Response and Control*, Technomic, Lancaster, PA, 1983.

17. P. N. Cheremisinoff, *Management of Hazardous Occupational Environments*, Technomic, Lancaster, PA, 1984.

18. F. Church, Jr., *Avoiding Surprises*, Boston Risk Management Corporation, Boston, 1983.

19. H. D. Crone, *Chemicals in Society*, Cambridge University Press, New York, 1986, reprinted 1987.

20. K. Denton, *Safety Management: Improving Performance*, McGraw-Hill, New York, 1982.

21. B. S. Dhillon, *Human Reliability With Human Factors*, Pergamon, New York, 1986.

22. *Directory of Safety Related Computer Resources*, Vol. I, *Software and References*, American Society of Safety Engineers, Park Ridge, IL, 1987.

23. *Directory of Safety Related Computer Resources*, Vol. II, *Databases*, American Society of Safety Engineers, Park Ridge, IL, 1987.

24. *Directory of Safety Related Computer Resources*, Vol. III, *Systems and Hardware*, American Society of Safety Engineers, Park Ridge, IL, 1988.

25. D. A. Dodge, *Safety Manual for Municipalities*, American Society of Safety Engineers, Des Plaines, IL, 1986.

26. G. W. A. Dummer and R. C. Winton, *An Elementary Guide to Reliability*, 4th ed. Pergamon, New York, 1986.

27. G. Eads and P. Reuter, *Designing Safer Products: Corporate Responses to Product Liability Law and Regulation*, The Institute for Civil Justice, Rand Corporation, Santa Monica, CA, 1983.

28. L. Perggiani, Ed., *Encyclopedia of Occupational Health and Safety*, 3rd rev., International Labour Organization, Geneva, 1983.

29. G. S. Everly, Jr. and R. H. L. Feldman, *Occupational Health Promotion*, Wiley, New York, 1985.

30. T. S. Ferry, *Accident Investigation for Supervisors*, American Society of Safety Engineers, Des Plaines, IL, 1988.

31. T. S. Ferry, *Modern Accident Investigation*, 2nd ed., Wiley, New York, 1988.

32. T. S. Ferry, *New Directions in Safety*, American Society of Safety Engineers, Park Ridge, IL, 1985.

33. T. S. Ferry, *Readings in Accident Investigations*, Charles C Thomas, Springfield, IL, 1984.

34. T. S. Ferry, *Safety Program Management for Engineers and Managers*, Charles C Thomas, Springfield, IL, Publisher, 1984.

35. T. S. Ferry, *Safety and Health Management Planning*, Van Nostrand Reinhold, New York, 1990.

36. Paul L. Tung, Ed., *Fracture and Failure: Analysis, Mechanisms and Application*, American Society for Metals, Metals Park, OH, 1980.

37. *Fundamentals of Industrial Hygiene*, 4th ed., National Safety Council, Chicago, 1988.

38. T. Gassert, *Health Hazards in Electronics, A Handbook*, Asia Monitor Resource Center, Hong Kong, 1985.

39. D. S. Gloss and M. Gayle, *Introduction to Safety Engineering*, Wiley, New York, 1984.

40. D. Goldsmith, *Safety Management in Construction and Industry*, McGraw-Hill, New York, 1987.

41. J. V. Grimaldi and R. H. Simonds, *Safety Management*, 5th ed., Richard D. Irwin, Homewood, IL, 1989.

42. V. L. Grose, *Managing Risk*, Prentice-Hall, Englewood Cliffs, NJ, 1987.

43. *Guidelines for Controlling Hazardous Energy During Maintenance and Servicing*, DHHS (NIOSH) Pub. No. 83-125, Washington, DC, 1983.

44. W. Hammer, *Occupational Safety Management and Engineering*, 3rd ed., Prentice-Hall, Englewood Cliffs, NJ, 1985.

45. Lawrence Slote, Ed., *Handbook of Occupational Safety and Health*, Wiley, New York, 1987.

46. K. Hendrick and L. Benner, Jr., *Investigating Accidents With Step*, Dekker, New York, 1987.

47. D. Hetzler, *Industrial Fire Protection*, Fire Protection Publications, Stillwater, OK, 1982.

48. A. E. Green, Ed., *High Risk Safety Technology*, Wiley, New York, 1982.

49. D. Himmelfarb, *A Guide to Product Failures and Accidents*, Technomic, Lancaster, PA, 1985.

50. E. Eggleston, Ed., Human Factors Group, Eastman Kodak Company *Ergonomic Design for People at Work*, Vol. 1, Lifetime Learning Publications, Belmont, CA, 1983.

51. H. H. Hurt, *Motorcycle Accident Cause Factors and Identification of Countermeasures*, NHTSA, Washington, DC, 1981.

52. P. J. Imperato and G. Mitchell, *Acceptable Risks*, Viking Penguin, New York, 1985.

53. *Introduction to Occupational Health and Safety*, National Safety Council, Chicago, IL, 1986.

54. W. G. Johnson, *MORT Safety Assurance Systems*, Dekker, New York, 1980.

55. B. M. Kantowitz and R. S. Sorkin, *Human Factors: Understanding People–System Relationships*, Wiley, New York, 1983.

56. P. L. Kirk, *Kirk's Fire Investigation*, Wiley, New York, 1983.

57. R. L. Kuhlman, *How to Develop Your Safety Manual*, Institute Press, Loganville, GA, 1984.

58. R. E. Levitt and N. M. Samelson, *Construction Safety Management*, McGraw-Hill, New York, 1987.

59. J. G. Liebler, et al., *Management Principles for Health Professionals*, Aspen Systems, Rockville, MD, 1984.

60. G. F. Lindgren, *Guide to Managing Hazardous Waste*, Butterworth, Woburn, MA, 1983.

61. J. B. Mackie and R. L. Kuhlman, *Safety and Health in Purchasing, Procurement and Materials Management*, Institute Press, Loganville, GA, 1981.

62. G. Marshall, *Safety Engineering*, Brooks-Cole Engineering Division, Wadsworth, Monterey, CA, 1982.

63. W. M. Mazer, *Electrical Accident Investigation Handbook*, Electrodata, Glenn Echo, MD, 1982.

64. E. J. McCormick and D. R. Llgen, *Industrial Psychology*, 7th ed., Prentice-Hall, Englewood Cliffs, NJ, 1982.

65. J. M. Mendelhoff, *The Dilemma of Toxic Substance Regulation*, The MIT Press, Cambridge, MA, 1988.

66. F. Morgenstern, *Deterrence and Compensation*, International Labor Organization, Geneva, 1983.

67. *Motor Fleet Safety Manual*, 3rd ed., National Safety Council, Chicago, IL, 1988.

68. V. Kjellen, Ed., *Occupational Accident Research Proceedings of the International Seminar on Occupational Accident Research, Saltsjobaen, Sweden, 5–9 Sept. 1983*, Elsevier, Stockholm, 1984.

69. J. LaDou, Ed., *Occupational Safety and Health*, 6th ed., National Safety Council, Chicago, IL, 1986.

70. A. M. Ottoboni, *The Dose Makes the Poison: A Plain Language Guide to Toxicology*, Vincinte Books, Berkley, CA, 1984.

71. C. Perrow, *Normal Accidents: Living with High-Risk Technologies*, Basic Books, New York, 1984.

72. R. W. Perry and A. H. Mushkatel, *Disaster Management*, Quorum Books, Westport, CT, 1984.

73. G. A. Peters and B. J. Peters, *Automotive Engineering and Litigation*, Vols. 1 and 2, Garland Law, New York, Vol. 1 1984, Vol. 2 1988.

74. D. Petersen, *Human Error Reduction and Safety Management*, Garland STPM Press, New York, 1982.

75. D. Petersen, *Safe Behavior Reinforcement*, Aloray, Goshen, NY, 1989.

76. D. Petersen, *Safety Management, A Human Approach*, 2nd ed., Aloray, Goshen, NY, 1988.

77. H. Petroski, *To Engineer is Human*, St. Martins Press, New York, 1985.

78. P. L. Polakoff, *Work & Health—It's Your Life*, Press Associates, Washington, DC, 1984.

79. *Preventing Illness and Injury in the Workplace*, Office of Technology Assessment, U.S. Government Printing Office, Washington, DC, 1985.

80. *Protecting Workers' Lives: A Safety and Health Guide for Unions*, National Safety Council, Chicago, 1983.

81. *"Readings in" Series*, eight edited books from *Professional Safety*, American Society of Safety Engineers, Park Ridge, IL, 1984–1985.

82. *Refresher Guide for The Safety Professional*, American Society of Safety Engineers, Park Ridge, IL, 1984.

83. J. L. Reid, *What to Do When the Sky Starts Falling, A Guide to Emergency Planning for the Construction Industry*, Secant, Northglenn, CO, 1987.

84. H. E. Roland and B. Moriarty, *System Safety*, Wiley, New York, 1983.

85. C. W. Ross, *Computer Systems for Occupational Safety and Health Management*, Dekker, New York, 1984.

86. *Safety and Health Practices of Multinational Enterprises*, ILO, Geneva, 1984.

87. J. D. Worrall, Ed., *Safety and the Work Force*, Ithaca, NY, ILR Press, 1983.

88. J. Ridley, Ed., *Safety at Work*, 2nd ed., Butterworths, London, 1986.

89. G. Peters, Ed., *Safety Law: A Legal Reference for the Safety Professional*, American Society of Safety Engineers, Park Ridge, IL, 1983.

90. M. Seiden, *Product Safety Engineering for Managers*, Prentice-Hall, Englewood Cliffs, NJ, 1985.

91. D. L. Stoner, et al., *Engineering a Safe Hospital Environment*, Wiley-Interscience, New York, 1982.

92. *Successful Accident Prevention*, Institute of Occupational Health, Helsinki, Finland, 1987.

93. *Supervisors Safety Manual: 6th Edition*, National Safety Council, Chicago, IL, 1985.

94. E. J. Terrien, *Hazardous Materials and Natural Disaster Emergencies*, Technomic, Lancaster, PA, 1984.

95. P. R. Timm and C. G. Jones, *Business Communication*, 2nd ed., Prentice-Hall, Englewood Cliffs, NJ, 1987.

96. C. Trost, *Elements of Risk: The Chemical Industry and Its Threat to America*, Times Books, New York, 1984.

97. G. Well, *Safety in Process Plant Design*, Wiley, New York, ?.

98. G. K. Wilson, *The Politics of Safety and Health*, Oxford, England, 1985.

99. W. E. Woodson, *Human Factors Design Handbook: Information and Guidelines for the Design of Systems, Facilities and Equipment*, McGraw-Hill, New York, 1981.

100. P. M. Strubhar, Ed., *Working Safely With Industrial Robots*, Robotics International of SME, Dearborn, MI, 1986.

101. *Workstation Design for Current Office Environments*, American Society of Safety Engineers, Park Ridge, IL, 1985.

Note: Some recent books have not been selected or reviewed. Books published before 1980 are not included in the interest of keeping the list manageable.

CHAPTER THIRTY-EIGHT

Ionizing Radiation

George M. Wilkening, C.I.H.

1 INTRODUCTION

The mystique of ionizing radiation is undoubtedly derived from the fact that dangerous exposures may be experienced without perception or adequate warning by the human sensory systems. This tragic lesson was learned by the early workers with radioisotopes and X-rays near the turn of the twentieth century. Because of the widespread concern about potential detrimental genetic and somatic effects of ionizing radiation, there has been a steady flow of laws and regulations governing the manufacture, transport, use, and disposal of radioactive material and radiation-producing devices.

In recent years, emphasis has been placed on a more positive approach to radiation protection, an approach that goes beyond the setting of exposure limits and the issuance of laws and regulations for the enforcement of control measures. The promotion of the philosophy of ALARA, which is to keep all exposures "as low as reasonably achievable," has not only resulted in significant reductions in individual doses, it has actually modified the role of exposure limits in achieving adequate protection against ionizing radiation. An accompanying development just beginning to emerge among the lay public is that ionizing radiation must be treated with respect, not fear, and that if used properly it will remain a minor portion of the plethora of risks we all face in everyday life. The overall objective in modern radiation protection efforts is to provide an appropriate standard of protection for all persons without unduly limiting the benefits of ionizing radiation.

The principal governmental agency responsible for the development of infor-

Patty's Industrial Hygiene and Toxicology, Fourth Edition, Volume 1, Part B, Edited by George D. Clayton and Florence E. Clayton.
ISBN 0-471-50196-4 © 1991 John Wiley & Sons, Inc.

mation on the biological effects of ionizing radiation and regulations for control was the Atomic Energy Commission (AEC). However, in 1975, the research functions of the AEC were transferred to the Energy Research and Development Administration (ERDA), and the regulatory functions were assumed by the Nuclear Regulatory Commission (NRC). The Environmental Protection Agency (EPA) has some responsibility for the regulation of nuclear power plant discharges, and the Food and Drug Administration (FDA) is active in the development of standards on radiation limits for medical devices and electronic products. The Occupational Safety and Health Administration (OSHA) of the Department of Labor also can inspect radiation facilities within the industrial workplace under the provisions of the Occupational Safety and Health Act of 1970.

The general standards for protection against ionizing radiation are contained in Title 10, *Code of Federal Regulations*, Part 20 (10 CFR 20). Regulations covering by-product nuclear materials are found in 10 CFR 30 through 36, those for source materials (e.g., uranium and thorium) in 10 CFR 40, and those for special nuclear material (e.g., plutonium, ^{233}U or ^{235}U) in 10 CFR 70. The reader is urged to obtain copies of the appropriate *Code of Federal Regulations* from the U.S. Government Printing Office in Washington, DC.

In making practical distinctions between radiations that might produce ionization in tissue and those that will not, a value of 10 eV is often used (see Chapter 39). That is, radiant energies above 10 eV would be expected to produce ionization (10 eV = 1.6×10^{-12} erg). This should be kept in mind when reviewing the interaction of radiation with matter in Section 2.

The following terms are used in this chapter.

Absorbed Dose, *D*: The quotient of $d\overline{E}$ by $d\overline{m}$ where $d\overline{E}$ is the mean energy imparted by ionizing radiation to the matter in a volume element and $d\overline{m}$ is the mass of the matter in that volume element. $D = d\overline{E}/d\overline{m}$. The special name for the SI unit of absorbed dose (J/kg) is the gray (Gy). The earlier unit of absorbed dose was the rad, which is equal to 0.01 Gy.

Quality factor (*Q*): A factor used for radiation protection purposes that accounts for differences in biological effectiveness between different radiations. It is the ratio of the slope of the curve of risk versus dose for a given radiation to that of a reference radiation in the range of doses where the curves are assumed to be linear.

Rad: A special unit for absorbed dose, kerma, and specific energy imparted. One rad is 0.01 J absorbed per kilogram of any material. (Also defined as 100 ergs per gram.) Now being replaced by the gray. One rad equals 0.01 gray.

Gray (Gy): The special name for the SI unit of absorbed dose, kerma, and specific energy imparted. 1 Gy = 1 J/kg = 100 rad.

Dose Equivalent (*H*): A quantity used for radiation protection purposes that expresses, on a common scale for all radiations, the irradiation incurred by exposed persons. It is defined as the product of the absorbed dose (*D*) and the quality factor (*Q*). The name for the SI unit of dose equivalent (J/kg) is the sievert (Sv). Still in use temporarily is the rem; 1 rem = 0.01 Sv.

Effective Dose Equivalent (H_E): The sum over specified tissues of the products of the dose equivalent in a tissue (T) and the weighting factor for that tissue (w_T), that is, $H_E = \Sigma\, w_T H_T = H_{wb}$.

Sievert (Sv): The special name for the SI unit of dose equivalent. Sv = 1 J/kg = 100 rem. 1 rem = 0.01 Sv.

2   INTERACTION OF RADIATION WITH MATTER

When photons interact with matter, they may be scattered, reflected, or absorbed. These processes reduce or attenuate the number of photons in a beam in an exponential fashion as expressed by the equation

$$I = I_0 e^{-\mu x}$$

where I is the fluence at a certain depth x, I_0 is the fluence rate incident at the surface, and μ is the attenuation coefficient. The attenuation coefficient may be expressed in terms of thickness as a linear attenuation coefficient (cm^{-1}) or as a mass attenuation coefficient (cm^2/g), the latter obtained by dividing the linear coefficient μ by the density p of the absorbing material. Hence the equation may read

$$I = I_0 e^{-(\mu/p)x}$$

where μ/p is the mass attenuation coefficient. Mass attenuation coefficients for some common materials such as aluminum, iron, lead, water, and concrete are given in Table 38.1 (1).

2.1   Absorption of Radiation

The absorption of photons occurs primarily through the photoelectric effect, the Compton effect, and pair production.

2.1.1   The Photoelectric Effect

At some energy threshold, incident photons are absorbed by matter through the transfer of energy to orbital electrons, causing these to be ejected, thus producing ionization in the medium. This is the so-called photoelectric effect. The ejected electrons possess kinetic energies equal to the difference between the incident photon energy and the binding energy of the shell from which the electron has been ejected. Subsequent to this transfer of energy, X-rays are emitted as the shell vacancies are corrected. The portion of photons interacting by the photoelectric process increases with increasing atomic number and decreasing photon energy. The photoelectric effect predominates when the photon energy is less than 0.5 MeV.

Table 38.1. Mass Attenuation Coefficients (1)

Photon Energy (MeV)	Mass Attenuation Coefficients (cm²/g)				
	Aluminum	Iron	Lead	Water	Concrete
0.01	26.3	173	133	5.18	26.9
0.02	3.41	25.5	85.7	0.775	3.59
0.05	0.369	1.94	7.81	0.227	0.392
0.1	0.171	0.370	5.40	0.171	0.179
0.5	0.0844	0.0840	0.161	0.0968	0.087
1.0	0.0613	0.0599	0.0708	0.0707	0.0637
5.0	0.0284	0.0314	0.0424	0.0303	0.0290
10.0	0.0231	0.0298	0.0484	0.0222	0.0231

2.1.2 The Compton Effect

As the photon energy is increased above 0.5 MeV, the dominant photon interaction is incoherent scattering from electrons, causing the electrons to recoil. A fraction of the incident photon energy is imparted to the electron; the remaining energy is retained by the scattered photon. The energy of the scattered photon is expressed as follows:

$$hv' = \frac{m_0 c^2}{m_0 c^2 / hv + 1 - \cos \theta}$$

whereas the energy of the recoil electron is

$$\frac{(hv)^2 (1 - \cos \theta)}{m_0 c^2 + hv(1 - \cos \theta)}$$

where θ is the scattering angle for the photon. In all cases, h is Planck's constant, v is the photon frequency, and c is the speed of light. The essential effect is that energy is imparted to a free or loosely bound electron, and at the same time there is a loss in frequency and energy in the scattered photon. As the energy of the incident photon decreases, the fraction of its energy imparted to the electron also decreases. The angular relationships between recoil electrons, scattered photons, and incident photons largely depends on the original incident photon energy. As the incident photon energies decrease to values approximately 0.5 MeV, the angles at which the electrons are scattered become greater.

2.1.3 Pair Production

Pair production occurs by the interaction of photons with the electric field surrounding a charged particle. When the incident photon energy is above 1.02 MeV, or twice the rest mass energy of an electron, the photon may be annihilated in the high electric field of the nucleus and converted into an electron–positron pair in accordance with the following relationship:

$$hv = 2m_0c^2 + E_{e-} + E_{e+}$$

where hv is the incident photon energy and m_0c^2 is the rest mass energy of the electron or positron, E_{e-} is the kinetic energy of the electron, and E_{e+} is the kinetic energy of the positron. The positron produced in this reaction will rapidly interact with electrons, causing electron–positron annihilation, which in turn produces two gamma photons, each with an energy equal to 0.51 MeV. The proportion of incident photons that interact with the nucleus by pair production increases with increasing atomic number.

3   PARTICULATE MATTER

A very large percentage of available radionuclides emit beta radiation. Beta radiation, in turn, is frequently accompanied by X- or gamma radiation. The decay of the nuclide by way of beta particle emission is characterized by a continuous spectrum of beta energies, with the maximum energy being unique to the nuclide. The range of beta particle energies is quite wide, varying from a few kiloelectron volts to slightly over 4 MeV. Beta particles have a wide range of penetration depending on their initial energy and the density of the material through which they traverse. Figure 38.1 shows the range of beta particles in air as a function of their energy. By dividing the range of a beta particle by the density of the material being traversed, the thickness of material that will completely stop that particular beta radiation can be obtained. As beta particles traverse matter, they are decelerated; the lost energy is emitted as photons or electromagnetic radiation, sometimes called bremsstrahlung. Bremsstrahlung increases with increasing energy of the beta particles and atomic number of the absorber. The beta particles or electrons emit photons having a range of energy from zero up to the maximum energy of the beta particles being stopped. Only modest thicknesses of commonly available materials are sufficient to stop beta radiation completely.

Alpha particles are physically identical to helium nuclei in that they contain two neutrons and two protons. They are emitted spontaneously during the radioactive decay of certain radionuclides, primarily those of high molecular weight. Because of their comparatively large size and double positive charge, alpha particles do not penetrate matter readily. Even the most energetic alpha radiation is completely stopped by the skin. Once alpha particles are inhaled, ingested, or otherwise absorbed into the body, however, they produce a high degree of ionization in tissue. Figure 38.2 illustrates the range of alpha particles in air as a function of energy. Although not revealed by Figure 38.2, the energy range of alpha particles is predominantly between 4 and 8 MeV.

Although not an ionized particle, the neutron readily interacts with matter to generate secondary sources of ionization, usually by inducing radioactivity in or

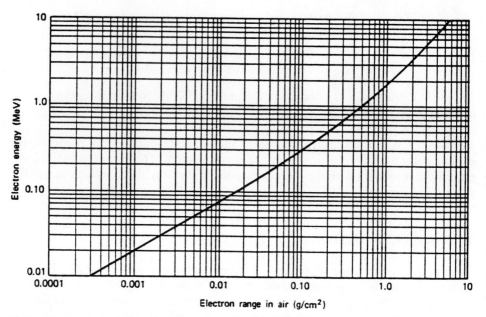

Figure 38.1. Range of beta particles in air as a function of energy. From "Physical Aspects of Irradiations," NBS Handbook No. 85.

ejecting particles from materials in its path. Neutron energies are often classified into three groups: slow, intermediate, and fast. A reasonable classification for the range of energies for each group would be: less than 1 eV, 1 eV to 0.1 MeV, and greater than 0.1 MeV, respectively. Thermal neutrons are so named because they are in thermal equilibrium with the medium in which they are found. The mean value of thermal neutron energy is 0.025 eV at 20°C. Ordinarily neutrons are not emitted by radionuclides, except for a few that undergo spontaneous fission. The more common neutron sources depend on the interaction of alpha or gamma radiation with the nuclei of certain target materials. These sources emit neutrons with an energy spectrum characteristic of the radionuclide and target material. The attenuation of the neutron usually occurs as a result of elastic and inelastic scattering, capture, and induced nuclear reactions. Moderation or slowing down of the neutrons by elastic collisions progressively changes the energy spectrum. The probability of such interactions taking place is specified in terms of cross sections, expressed in units of barns (1 barn = 10^{-24} cm²). Because the various neutron interactions result in the production of secondary radiations, particularly gamma rays, the person who is evaluating exposures in a neutron facility must not lose sight of the significance of secondary radiations.

At energies beginning at 5 to 13 MeV, except for ^{2}H and ^{9}Be, photons have enough energy to eject a neutron from the nucleus and produce a radioactive nucleus. The liberated neutron, in turn, interacts with surrounding matter to produce secondary ionization. Lower energy electrons can also produce ionization,

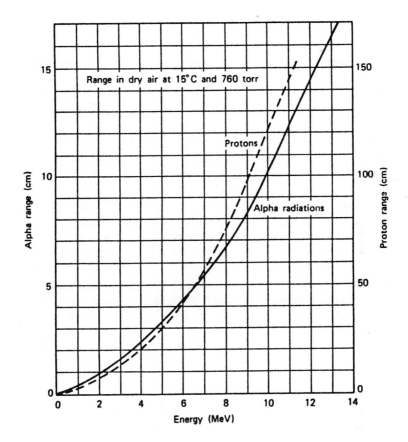

Figure 38.2. Range of alpha particles and protons as a function of energy. From E. F. Gloyna and J. O. Ledbetter, *Principles of Radiological Health*, Dekker, New York.

especially in gases. Protons and heavier charged ions ionize the medium through which they pass; high-energy protons interact with nuclei to induce radioactivity in target material and expel secondary particles that become sources of secondary ionization. The secondary ionization then continues until the energies of the secondaries are reduced below the ionization threshold.

4   X- AND GAMMA-RADIATION

X-Rays originate in the extranuclear part of the atom, whereas gamma rays are emitted from the nucleus during nuclear transitions or particle annihilation. X-Rays are commonly produced in an evacuated tube by accelerating electrons from a heated element to a metal target with voltages in excess of 16 kV. The electrons interact with orbital electrons of atoms in the target, causing energy level changes that result in the emission of characteristic X-rays, or the electrons may interact with the nucleus of the atom to produce electromagnetic radiation having a continuous spectrum (bremsstrahlung). The production of bremsstrahlung increases at higher electron accelerating voltages. The energy spectrum or quality of the X-ray beam may be expressed either in terms of an "effective energy" or in terms of its half-value layer. The accelerating voltage may be constant or continuous or pulsed, thus producing different photon energy distributions. X-Ray tubes and their housing are arranged to provide shielding in all directions except for a window, where the useful beam is emitted. The solid angle and shape of the useful beam is determined by the size of the window and by collimating devices such as diaphragms and cones. Some low-energy X-rays are absorbed in the target, others are removed from the useful beam by the material in the tube window, and still others are removed by filters and preferentially absorb the less penetrating radiation. This, in effect, hardens X-ray beams used for medical diagnostic purposes, preserving the more useful penetrating radiation.

Gamma rays are emitted by the nucleus of certain radionuclides during their decay scheme. Each of the nuclides emits one or more gamma rays having specified energy. Gamma rays may also be produced by neutron interactions with nuclei. Typical gamma ray energies vary from 8×10^3 to 10^7 eV; the corresponding frequencies are 2×10^{18} to 2.5×10^{21} Hz.

5   SOURCES AND USES OF IONIZING RADIATION

In gaining some broad insight into the overall pattern of radiation insult to humans, one should have some appreciation for the radiation received from the natural background. Most of this background is derived from cosmic radiations that vary in intensity as a function of altitude and latitude, and from terrestrial gamma radiations that vary in intensity with the amount of natural radioactive material present in the earth. Cosmic radiation, for example, increases by a factor of 3 in going from sea level to 10,000 ft and by 10 to 20 percent in going from 0° to 50° geomagnetic latitude (3). Internal exposures arise from body deposition of naturally produced radionuclides that have been inhaled or ingested; the major contributors are radon and its daughter products and ^{40}K, respectively. Other naturally occurring radiation sources include uranium, actinium, and thorium, and their decay products, ^{14}C, and tritium, the latter through neutron activation (neutrons supplied by cosmic radiation).

Nuclear reactors are prime examples of man-made radiation sources. In the operation of reactors, materials such as uranium, thorium, and plutonium break up into two or more fragments when bombarded with neutrons, thereby emitting more than one neutron, and the neutrons further sustain chain reactions. High intensities of neutron, beta, and gamma radiation exist within the reactor compartment. Fission weapons could be considered as separate and distinct sources of radiation; however, they are actually special cases of uncontrolled fast burst reactors. Fission weapons and all other fission processes convert nuclear fuel into elements of higher atomic number in a series of exothermic reactions. Nonfissionable materials generally absorb neutrons and become radioactive isotopes of the original material, thereby producing additional sources of radiation. Other well-known sources of ionizing radiation are to be found in the uranium industry (i.e., mines, mills, fabricating plants, and fuel reprocessing plants). With the increased attention to energy sufficiency and the need to develop multiple sources of power, it is possible for the level of activity in uranium mining and fabrication to be stepped up, despite some widespread reservations about nuclear power.

Another prominent man-made source of radiation is the high-energy, charged-particle accelerator. Such accelerators can present a variety of potential radiation hazards depending on the particle being accelerated, the interaction with the target, and the mode of acceleration. In addition to the hazards of an electron beam, X-rays may be produced. Also, acceleration of heavy particles such as protons and deuterons, tritium, or heavy positively charged particles may produce secondary electrons that are accelerated in a direction opposite to that of the main beam toward the high-voltage region of the device where X-rays can be produced. It has been noted that positive ion accelerators can produce X-rays in the vicinity of the target if the target is a good insulator (4). Normally, heavy ions produce negligible X-radiation as they are absorbed in the target. When the target is an insulator, however, the ions build up a charge on the target material; the resultant field accelerates electrons in its vicinity to produce significant localized radiation levels. Some accelerators produce neutrons upon interaction of the beam with the target material, causing activation of the target and adjacent components. Activation can also be produced directly by higher-energy particles. In these instances, the operators of these devices encounter not only the machine-made transient radiation hazards but also material with long-term radioactivity.

Devices such as medical and dental X-ray tubes, X-ray diffractometers, and X-ray radiographic and fluorescence equipment are well-known examples of ionizing radiation sources. X-Ray generators usually accelerate electrons onto a target material in a vacuum. As the electron is slowed in the target, it radiates energy in the form of X-rays. The maximum kinetic energy attained by the electrons before impinging on the target is equal to $1/2 m_0 v^2 = V_e = h\nu_{max} = hc/\lambda_{min}$, where m_0 is the electron mass, v is the maximum electron velocity, V is the anode to cathode potential difference, e is the charge on the electron, ν_{max} is the maximum X-ray frequency, λ_{min} is the minimum X-ray wavelength, h is Planck's constant, and c is the speed of light.

Any device that employs accelerating voltages above approximately 20 kV may generate X-rays of sufficient energy to penetrate the device envelope and present a source of potentially hazardous ionizing radiation. Magnetrons, klystrons, thyratrons, cathode ray tubes, and electron guns are examples. Many devices (e.g., alphatron gauges, electron tubes, fire alarms) have some radioactive materials sealed within their envelopes to improve their ionization characteristics. Gas and aerosol detectors such as those used in fire (smoke) detection and alarm systems employ the use of an alpha emitter to ionize the gas between two electrodes and cause the steady flow of a weak current. When smoke or thermal degradation products enter the air gap between the electrodes, the resulting change in current is relayed to a monitoring and alarm system. Most of the systems marketed for use in commercial buildings use ^{241}Am as the alpha source, in amounts ranging up to approximately 100 μCi, with an average of 50 μCi in each unit.

High-voltage rectifiers, electron microscopes, radioluminous devices, and static eliminators represent additional sources of ionizing radiation. Radioactive static eliminators may use ceramic microspheres that contain ^{210}Po, an alpha emitter. The size of the spherical particles is large enough to preclude inhalation and deposition in the alveolar spaces if the microspheres happen to get loose from the encapsulation. Reportedly some loss of the microspheres actually occurs during use (5), but the biological implications appear to be negligible.

Both the manufacture and use of gas mantles may involve exposure to beryllium oxide and the daughters of radioactive thorium (6). Some mantles are still manufactured as novelty items.

The following is a list of typical examples of some current uses of ionizing radiation.

1. In radiography: X-rays and gamma rays; beta emitters for thin film and low density materials; neutrons
2. In gauging levels in tanks, thickness of sheet materials and films, and moisture content
3. In tracing the flow of materials in pipes
4. In studying wear rates on metal parts
5. As tracers in chemical and biological research
6. In sterilization of food and medical supplies
7. As cross-linking agents to improve the properties of plastics
8. As energy sources for navigational beacons
9. In implanted cardiac pacemakers
10. In nuclear power reactors
11. In chemical analyses (e.g., neutron activation)
12. In research

6 BIOLOGICAL ASPECTS OF IONIZING RADIATION

The process of ionization in tissue may alter the structure of atoms and molecules in living cells to such an extent that irreparable damage is produced. Damage to the DNA of living cells is the event of critical importance and consequence; it may prevent the survival or reproduction of the cell or it may result in a viable but modified cell through repair mechanisms. Damage in cells is also possible through exposure to carcinogens or mutagens produced by sources other than ionizing radiation.

The effects of ionizing radiation on exposed individuals can be divided into the categories of somatic and genetic, even though such effects cannot always be distinguished from each other, especially when irradiation of body tissues and germ plasm occurs simultaneously. If a sufficient number of cells are killed or prevented from reproducing, a loss of organ function occurs, an effect that is usually called a nonstochastic process, that is, a nonrandom process. Some radiation experts are beginning to use the more descriptive term "deterministic" rather than "nonstochastic." On the other hand, a modified somatic cell may still retain its reproductive capacity, but it may, as a result of its modification, produce a malignant tumor. Or a modified germ cell may transmit faulty genetic information to the descendants of an exposed individual, thus potentially causing considerable harm to subsequent generations. The somatic and genetic effects produced from a single modified cell, as described above, are collectively known as stochastic (random) processes.

6.1 Dosimetric Quantities

The basic dosimetric quantity in radiation protection is the *absorbed dose, D,* defined as the energy absorbed per unit mass in units of joule per kilogram, which is given the name gray (Gy). With very few exceptions, absorbed dose, as used in the radiation protection discipline, means the average dose to a tissue or volume of tissue, including a body organ. The probability of producing a stochastic biological effect is usually estimated from a linear dose–response relationship plotted over a limited range of dose, whereas nonstochastic (deterministic) effects are estimated from the absorbed dose only when the dose is uniform over the tissue or organ.

Various types of radiation, such as alpha, beta, gamma, X, or neutron radiation, produce different responses in tissue even if the individual doses are the same. This difference in response depends not only on the type of radiation but also on the way in which the energy is distributed within the tissue. This difference in the way energy is distributed or is transferred is known as the linear energy transfer (L or LET). The "relative biological effectiveness" (RBE), a term that is receiving less attention and usage, denotes the ratio between the doses of high L and of low L radiations that produce equivalent effects. The RBE value is reflected indirectly in the term quality factor Q described below. Usually the RBE reference value of 1.0 is assigned to the biological effectiveness of 250-kV X-rays or ^{60}Co gamma rays. Examples of high L radiations are alpha particles, protons, and fast neutrons; examples of low L radiations are electrons and X- and gamma rays. Table 38.2 lists the average L values for various nuclear particles.

The biological risk per rad of low L radiations decreases to a greater extent with a decrease in dose equivalent and dose rate than does the risk with high L radiations. Data obtained from sources other than those from Hiroshima and Nagasaki (e.g., uranium miners, radium-treated patients, or radium dial painters) indicate that for cancer induction, alpha particles that are delivered at relatively low dose rates may be some 10 times more effective per rad average tissue dose than X- or gamma rays delivered at high dose rates.

Table 38.2. Average L (LET) Values[a]

Particle	Mass (amu)	Average L (keV/μm)	Tissue Penetration (μm)
Electron	0.00055	12.3	0.01
		2.3	1
		1.42	180
		1.25	5,000
Proton	1	90	3
		16	80
		8	350
		4	1,400
Deuteron	2	6	700
		1	190,000
Alpha	1	260	1
		95	365
		5	20,000

[a]The average amount of energy lost per unit of particle spur-track length: low L, radiation characteristic of electrons, X-rays, and gamma rays; high L, radiation characteristic of protons or fast neutrons. Average L is specified to even out the effect of particle that is slowing down near the end of its path and to allow for the fact that secondary particles from photon or fast neutron beams are not all of the same energy.

For purposes of radiation protection, it is important to develop a method for unifying all the complicating factors that determine a given biological effect in tissue. An indication of such an effect on an organ or tissue is given by the term "dose equivalent," which is basically an expression of absorbed dose that has been weighted or modified by certain factors. The dose equivalent H is equal to the product of D, Q, and N at a given point in tissue, where D is the absorbed dose, Q is the quality factor, and N is the product of any other modifying factors (10). A value of 1 is usually given to N for external sources. Estimating the probability of stochastic effects also requires knowledge of the type and energy of the radiation producing the dose. The *quality factor*, Q, is a dimensionless unit used to modify D by embodying a crude measure of the relative biological effectiveness of various radiation types and associated energies. It is defined as a function of the linear energy transfer L (also designated as LET) produced by the particular type of radiation being considered. An average value of Q, namely, $\overline{Q}$ is used to designate the average of a range of linear energy transfer values. For a range of radiation qualities the expression becomes $H = \overline{Q}DN$, where Q is the mean value. Because dose equivalent (H) is a weighted absorbed dose, its unit is the joule per kilogram, but it now has been named the sievert (Sv) to distinguish it from the gray (Gy) for absorbed dose. The value of Q accounts for the influence of radiation quality on biological effect. It is usually thought to express the microscopic distribution of

Table 38.3. Relations Between Linear Energy Transfer L and Quality Factor Q (11)

L (keV/μm)	Q
≤ 3.5	1
7	2
23	5
53	10
175	20

absorbed radiation energy, but it is also a function of the collision-stopping power of water. The collision-stopping power of water is synonymous with linear energy transfer, or L. To determine H, one needs to know the spectrum of absorbed dose in L for all values of L. When this spectrum is known for the specific mass of tissue of interest, an average value of Q ($\overline{Q}$) can be calculated for that mass of tissue.

$$\overline{Q} = \frac{1}{D} \int_0^\infty D_L dL$$

If one does not know the spectrum of absorbed dose, certain estimates are acceptable; for example, when the neutron energy spectrum in unknown, a $\overline{Q}$ value of 10 may be used to convert the measured absorbed dose into dose equivalent. The relationship between L and Q is shown in Table 38.3.

When the dose equivalent has been determined for a particular exposure and one wishes to compare the result with primary protection limits for either internal or external radiation, it is important to remember that the protection limits differ for different organs and tissues in the body and for various combinations of these organs, that the organs and tissues are at different depths within the body, and that the dose equivalent DQN is a function of position in the body (12). The numerical values of the absorbed dose and the dose equivalent are equal if the linear energy transfer of the charged particles that deliver the dose is less than 3.5 keV/μm. This is almost always the case for electrons. It is also largely true for X- and gamma radiations with energies of up to 10 MeV. Above this value, nuclear reactions may contribute particles of high collision-stopping power.

A review of the sources of information on which the present radiation protection guidelines are based elicits the striking fact that the basic data are taken from cases of victims of the atomic bombs at Hiroshima and Nagasaki, the underground miners exposed to radon gas and radioactive decay products, populations such as radium dial painters, with high body burdens of alpha-emitting radionuclides, and persons with ankylosing spondylitis who had received long-term treatment with X-rays. A point of interest is that most of the tumors observed in uranium miners have been located in the bronchial epithelium. The data from these populations indicate that the excess mortality from all forms of cancer correspond to roughly 50 to 165 deaths per million persons per rem (0.01 Sv) during the first 25 to 27 years after irradiation

(13). One of the most controversial subjects is the extrapolation of available mortality data from the high-exposure level groups just mentioned to an estimate of probable mortality resulting from a continuing low-level irradiation (0.10 to 0.20 rem) of the general population. Estimates of this type are bound to be made, despite complexities or lack of appropriate data. One such estimate is that the most likely number of deaths per 0.10 rem (1 mSv) per year in the U.S. population would be expected to be of the order of 1350 to 3300 (13).

6.2 Somatic Effects

6.2.1 Chronic Exposure

The estimation of biological risk at relatively low radiation dose levels is made by extrapolation from data obtained at higher dose levels. Certain assumptions are then made about dose–response relationships, the probable mechanisms involved, and the susceptibility of the population at risk. Because of the uncertainties about the dose–response relationships, extrapolation is usually made of the linear dose–incidence function at high exposure levels down to the origin at zero dose, on the assumption that the incidence at zero dose is also a point on the extrapolated line (7). The most important chronic effect of radiation on human populations is carcinogenesis, including leukemogenesis; yet cancers induced by radiation are indistinguishable from those occurring naturally. Hence the existence of cancer can be inferred only in terms of an excess over what is regarded as the natural incidence. The natural incidence of cancer varies over several orders of magnitude depending on the type and the site of the neoplasm, age, sex, and other factors. The period of time that elapses before the appearance of a clinically detectable neoplasm is characteristically long—years, or even decades. Such a long induction time complicates the prospective followup observations of an irradiated population. It also complicates the retrospective evaluation of patients for possible history of relevant radiation exposure. Many of the existing human and animal data on radiation-induced tumors come from populations exposed to internally deposited radio-nuclides where the dose–incidence relation is obscured by nonuniform distribution of dose to body tissues. In certain instances, data have been derived from studies of therapeutically irradiated patients in whom the effects of radiation may have been complicated by effects of the underlying disease and medication (8).

Earlier information derived from the Hiroshima and Nagasaki bombings led to the conclusion that susceptibility to the induction of leukemia was several times higher in persons irradiated *in utero*, during childhood, or late in adult life than it was in individuals of intermediate ages (9). However, the latest report of the BEIR V committee shows that risks are initially higher for those exposed at under 20 years of age, but decrease more rapidly with time after exposure than for those exposed at older ages. There was no clear indication that the risks for those under 10 years of age were significantly greater than those for persons 10 to 20 years old at the time of the bombing. The dose–response curve for the total number of excess cases of leukemia appears to increase in slope (with increasing mean dose

Table 38.4. Representative Dose–Effect Relationships in Humans for Whole Body Irradiation

Nature of Effect	Representative Absorbed Dose of Whole Body X- or Gamma Radiation	
	(rads)	(mSv)
Minimal dose detectable by chromosome analysis or other specialized analysis but not by hemogram	5–25	50–250
Minimal acute dose readily detectable in a specific individual (e.g., one who presents himself as a possible exposure case)	50–75	500–750
Minimal acute dose likely to produce vomiting in about 10% of people so exposed	75–125	750–1250
Acute dose likely to produce transient disability and clear hematological changes in a majority of people so exposed	150–200	1500–2000
Median lethal dose for single short exposure	300	3000

Source: Reprinted with permission of the National Council on Radiation Protection and Measurements from NCRP Report No. 39, Washington, D.C., 1971.

to the marrow) through a maximum in the dose range of 3 to 4 Gy, and to decrease with further increase in the dose.

6.2.2 Acute Exposure

From considerations of low-level radiation exposure levels and their long-term chronic effects, we now turn to acute somatic effects and radiation exposures that produce injury. Tables 38.4 and 38.5 present information on acute effects in humans: Table 38.4 lists the dose–effect relationships in humans for acute whole body irradiation, and Table 38.5 presents symptoms and probable mortality rates as a function of acute whole body absorbed dose.

The dose entries in Table 38.4 should be taken as representative compromises only of a surprisingly variable range of values that could be offered by well qualified observers asked to complete the right-hand column. This comes about in part because whole body irradiation is not a uniquely definable entity. Midline absorbed doses are used. The data are a mixed derivative of experience from radiation therapy (often associated with "free air" exposure dosimetry) and a few nuclear industry accident cases (often with more up-to-date dosimetry). Also, the interpretation of qualitative terms such as "readily detectable" is a function of the conservatism of the reporter.

In connection with Table 38.5, it should be noted that little information is available on long-term protracted exposure of humans. Animal experiments such as the continuous gamma irradiation of beagles at Argonne National Laboratories show that with daily exposures in the 10 to 40 R range, anemia is the major cause of death, and with daily exposures of 50 R or greater, infection is the major cause

Table 38.5. Somatic Injury Chart[a]

Exposure-Acute Whole Body (R)	Type of Injury	Medical Care Required	Able to Work	Probable Acute Mortality Rate During Emergency
10–50	Asymptomatic	No	Yes	0
50–200	Acute radiation sickness; see level I	No	No	Less than 5%
200–450	Acute radiation sickness; see level II	Yes	No	Less than 50%
>450 (450–900)	Acute radiation sickness; see level III	Yes	No	Greater than 50%
>600 (600–1000)	Acute radiation sickness; see level IV	Yes	No	100%
>1000–3000	Acute radiation sickness; see level V	Yes	No	100%

Level I symptoms	Less than half this group vomit within 24 hr of onset of exposure. Minor subsequent symptoms. Fewer than 5% require medical care. All others can perform their customary tasks.
Level II symptoms	More than half this group vomit soon after onset of exposure and are ill for a few days. This is followed by a period of 1–3 weeks with few symptoms. After the latent period, epilation is seen in more than half, followed by a moderately severe illness due to loss of white blood cells and infection. Most of the people in this group require medical care; more than half survive. However, essentially, all are unable to work.
Level III symptoms	A more serious version of the sickness described in level II: the initial period of gastric distress is more severe and prolonged; the main episode of illness is characterized by extensive hemorrhages and complicating infections. People in this group need medical care and hospitalization. Fewer than half survive, even with excellent medical care.
Level IV symptoms	An accelerated version of the sickness in level III. All in this group begin to vomit soon after the onset of exposure, and this continues for several days or until death. Damage to the gastrointestinal tract predominates, manifested by uncontrollable diarrhea which becomes bloody. Changes in the blood count occur early. Death occurs before the end of the second week and usually before the appearance of hemorrhages or epilation. Clinical problems resulting from the low exposure rate are related to the failure of the bone marrow. All in this group need care and few, if any, survive.
Level V symptoms	An extremely severe injury in which hypotensive shock secondary to vascular damage predominates. Symptoms and signs or rapidly progressing shock appear almost as soon as the dose has been received. Death occurs within a few days.

Reprinted with permission of National Council on Radiation Protection and Measurements; adapted from NCRP Report No. 42, Washington, DC.

[a]The results reported above are for whole body irradiation; the effect is less for partial body exposures. Delayed effects may sometimes occur after the early effects have been ameliorated, the extent depending on the dose received. In addition to the possible loss of hair, one possible effect that is of general concern is sterility.

of death. With protracted exposure below 10 R per day, myeloproliferative diseases, potentially leukemic, are causes of death.

Spermatogonia are among the most radiosensitive cells in the body. A dose of 50 rems (0.5 Sv) delivered in a single brief exposure may result in cessation of sperm formation (14). Because acute whole body irradiation has not been observed to cause permanent sterility, however, the sterilizing dose for men as well as for other male mammals is thought to exceed the lethal dose if applied to the whole body in a single brief exposure (15). In contrast to the testes, the ovary possesses its entire supply of germ cells early in life and lacks the ability to replace them as they are subsequently lost. Therefore ionizing radiation may cause a lasting reduction in the reproductive potential of the affected ovary. Data obtained from Japanese survivors of the atomic bomb and from investigations of women exposed to fallout in the Marshall Islands suggest that a minimum of 300 to 400 rems (3 to 4 Sv) must be given in a single exposure to cause permanent sterility and that a larger dose, of the order of 1000 to 2000 rems (10 to 20 Sv), is required for sterilization if administered to young women in fractionated exposures over a period of 10 to 14 days (16).

The erythemal dose to skin is generally believed to be of the order of at least several hundred rads of X-rays; the erythema appears in a matter of hours postirradiation. This threshold varies depending on the energy of the radiation, the dose rate, size of the area of skin irradiated, and the region of the body exposed. In order of increasing severity, the skin would be expected to react in the following sequence: erythema, dry desquamation, vesiculation, sloughing of the skin layers, and chronic ulceration.

Approximately 20 R of X- or gamma irradiation in a single dose is required to produce opacities in the human lens. The value of 200 R is probably close to the maximum single dose that produces opacities that do not interfere with vision. For multiple doses of low L, i.e., LET radiation, approximately 400 to 550 R is required to produce detectable opacities, depending on the fractionation of exposure time (10).

6.2.3 Genetics Effects

The two most notable genetic effects of ionizing radiation are the production of mutant genes and aberrations in chromosomes. Both profoundly affect the future of human progeny. On the other hand, it is well to remember that mutations and chromosome aberrations also occur spontaneously. Therefore a major question for all persons, but particularly those in the radiation protection profession, is the extent to which man-made ionizing radiation should be permitted to add to the load of defective genes already present in the population. Should humans be exposed to man-made radiation that is above the background level of radiation? If so, to what extent?

Because of these profound considerations, the National Academy of Sciences Committee on the Biological Effects of Atomic Radiation (BEAR Committee) introduced the concept of regulating the dose to the population (17). Specifically, the committee recommended that man-made radiation be kept at such a level that the average individual exposure is less than 10 R before the mean age of reproduction (i.e., 30 years). In 1956 the genetically significant medical dose—that is, the prereproductive gonad exposure—was estimated to be about half of the recommended 10 R limit. However, the Federal Radiation Council (18) did not include medical radiation in its calculations and recommended a 5 R average as the 30-year limit for the population. Using these values provides a limiting dose rate to the population of 0.17 R per year, or, 170 mR per year. There is still no limitation on population exposures resulting from medical practice. In reaching its recommended level of exposure for the population, the Federal Radiation Council made certain assumptions, including the following:

1. Mutations at any given dose rate increase in direct linear proportion to the genetically significant dose.
2. Mutations are irreparable and are almost always harmful.
3. The mutations caused by man-made radiation are similar to those which occur naturally.
4. There is no known threshold dose below which an effect will not occur.

The differences between genetic and somatic effects should be noted with care. There is unequivocal evidence that high radiation doses produce cancer and leukemia. That such effects are produced by low radiation doses is less clear. On the other hand, there is mostly indirect evidence that human genetic effects can be produced even at high radiation doses. The main basis for claiming genetic effects is the rather convincing studies of radiation effects in animals. As a result of these studies, there is general agreement that humans must be affected in the same way. The principal human concern about radiation-induced somatic effects is the possible induction of malignant diseases, whereas the concern about radiation-induced genetic effects is that all conceivable bizarre anatomical and physiological alterations may be passed on to future generations.

Considering all of the studies performed by the original BEAR Committee and subsequent BEIR Committees most of the conclusions about genetic effects were based on information about spontaneous abortions, congenital malformations, altered conformation of proteins, chromosome abnormalities, and premature death in experimental animals, mostly mice, supplemented with information obtained from the atomic bomb survivors. Ionizing radiation has been found to be mutagenic in all organisms studied thus far and it is assumed that the human organism is no exception. A consistent finding is that the majority of new mutations are harmful.

The BEIR V Committee (1990 Report) believes that the values in Table 38.6 give the best estimates of risk based on their conclusion that the doubling rate in humans is probably not smaller than 1 Sv (100 rem) based on studies in mice. The doubling rate of 1 Sv approximates the lower 95 percent confidence limit for the Hiroshima/Nagasaki bomb survivors and is in essential agreement with the previous studies of UNSCEAR conducted in 1972, 1977, 1982, and 1986. It is possible that

Table 38.6. Estimated Genetic Effects of 1 rem Per Generation[a]

Type of Disorder	Current Incidence per Million Liveborn Offspring	Additional Cases/10⁶ Liveborn Offspring/Rem/Generation	
		First Generation	Equilibrium
Autosomal Dominant			
Clinically severe[b]	2500[c]	5–20[d]	25[e]
Clinically mild[f]	7500[g]	1–15[d]	75[e]
X-Linked	400	<1	<5
Recessive	2500	<1	Very slow increase
Chromosomal			
Unbalanced translocations	600[h]	<5	Very slow increase
Trisomies	800[i]	<1	<1
Congenital abnormalities	20,000–30,000	10[j]	10–100[k]
Other orders of complex etiology[l]			
Heart disease[m]	600,000		
Cancer	300,000	Not estimated	Not estimated
Selected others	300,000		

Source: Health Effects of Exposure to Low Levels of Ionizing Radiation, BEIR V Committee on the Biological Effects of Ionizing Radiations, Board on Radiation Effects Research, Commission on Life Sciences, National Research Council, Washington, DC, 1990.

[a]Risks pertain to average population exposure of 1 rem per generation to a population with the spontaneous genetic burden of humans and a doubling dose for chronic exposure of 100 rem (1 Sv).
[b]Assumes that survival and reproduction are reduced by 20–80% relative to normal (s = 0.2 to 0.8).
[c]Approximates incidence of severe dominant traits.
[d]Calculated using s = 0.2 − 0.8 for clinically severe and s = 0.01 − 0.2 for clinically mild.
[e]Calculated with mutational component = 1.
[f]Assumes that survival and reproduction are reproduced by 1–20 percent relative to normal (s = 0.01 − 0.2).
[g]Obtained by subtracting an estimated 2500 clinically severe dominant traits from an estimated total incidence of dominant traits of 10,000.
[h]Estimated frequency from UNSCEAR (UN82, UN86).
[i]Most frequent result of chromosomal nondisjunction among liveborn children. Estimated frequency from UNSCEAR (UN82, UN86).
[j]Based on worst-case assumption that mutational component results from dominant genes with an average s of 0.1; hence excess cases $<30{,}000 \times 0.35 \times 100^{-1} \times 0.1 = 10$.
[k]Calculated with the mutational component 5–35%.
[l]Lifetime prevalence estimates may vary according to diagnostic criteria and other factors. The values given for heart disease and cancer are round number approximations for all varieties of the diseases.
[m]No implication is made that any form of heart disease is caused by radiation among exposed individuals. The effect, if any, results from mutations that may be induced by radiation and expressed in later generations, which contribute along with other genes to the genetic component of susceptibility. This is analogous to environmental risk factors that contribute to the environmental component of susceptibility. The magnitude of the genetic component in susceptibility to heart disease and other disorders with complex etiologies is unknown. Most genes affecting the traits are thought to have small effects, and new mutations would each contribute a virtually insignificant amount to the total susceptibility of the individuals who carry them. However, a slight increase in genetic susceptibility among many individuals in the population may produce, in the aggregate, a significant effect overall. Because of great uncertainties in the mutational component of these traits and other complexities, the committee has not made quantitative risk estimates. The risks may be negligibly small, or they may be as large or larger than the risks for all other traits combined.

the estimates of risk are too high, but for the purposes of setting standards it was considered to be an appropriate position to take until more definitive data are available.

7   GENERAL RADIATION PROTECTION CONSIDERATIONS FOR INTERNAL AND EXTERNAL SOURCES

Some appreciation of the genetically significant radiation resulting from natural as well as man-made sources is prerequisite to a full appreciation of radiation protection requirements. The dose equivalent of the highly penetrating and uniformly distributed cosmic radiation has an average value of 28 mrem (0.28 mSv) per year. The population weighted absorbed dose rate in air in the United States from external terrestrial radionuclides is estimated to be 40 mrad (0.4 mSv) per year; however, the absorbed doses must be corrected by a housing factor of 0.8 and a body screening factor of 0.8 to obtain a dose equivalent rate of 26 mrem (0.26 mSv) per year. This dose is largely due to gamma and X-rays. External terrestrial radiation dose is largely determined by the concentrations of ^{40}K and the uranium and thorium series in the soil. To illustrate the variability in external terrestrial whole body dose equivalent rates in terms of geographic area, it is worth noting that the values for the Atlantic and Gulf Coastal plains, the noncoastal plains excluding Denver, and the Colorado plateau, are 0.15, 0.30, and 0.55 mSv/year, respectively. Adding the average dose equivalent rates from internal and cosmic radiation to the gonads (approximately 0.50 mSv/year), the total respective dose equivalents for each of the three geographical areas mentioned are found to be 0.65, 0.80, and 1.05 mSv/year. The dose equivalent rate for persons living at about the altitude of Denver, Colorado, would be increased by approximately 0.20 mSv/ year from cosmic radiation. Therefore, the highest whole body dose received in the United States from natural radiation would approximate 1.25 mSv/year (19).

If we consider the additional whole body radiation received by the U.S. population from man-made sources, we find that medical diagnostic radiology accounts for at least 90 percent of the total. One estimate of the average dose rate to the population attributable to medical and dental exposures is 0.73 mSv/year (20). By comparison, the average dose rate attributable to occupational exposures is 0.008 mSv/year, and that due to nuclear power is well below 0.01 mSv/year. This means that the average total whole body annual dose equivalent received by each person in the country is approximately 200 mrem (2.0 mSv) and of that, at least one-third is due to medical and dental X-rays. Although the medical and dental component of radiation received by the human population is not figured in the calculation of radiation protection guides (RPG), major educational efforts have been conducted for the benefit of physicians and dentists to reduce the patient dose to the minimum, consistent with diagnostic and therapeutic requirements.

The methods used for evaluating exposures to external radiation sources and those involving sources that have been deposited internally in the body are quite different. In both cases, significant variation exists in the characteristic properties of the radiation source, the absorbed energy, the quality factor, the duration of exposure, and the identity of critical organs. The assumption is usually made that any radiant energy from external sources is uniformly distributed over or throughout the part of the body that is irradiated. In the case of inhaled or ingested radioactive material, the usual result is a rather complex distribution pattern throughout the body in accordance with biochemical rather than nuclear or radioactive properties. Questions of residence time in the body, concentration, and proximity of the radioactive material to organs and tissues become major considerations. Absorption through the skin may be significant for some compounds, such as tritium. If airborne radioactive particulate matter is inhaled, one would expect its deposition in the respiratory tract to follow well-established patterns, which depend primarily on the particle size distribution of the inhaled particulate and the construction of the respiratory system. Particles with diameters of less than a few micrometers are more important physiologically because they penetrate deeply into the lung to reach the alveolar spaces. Particles larger than this are preferentially deposited in the bronchioles, bronchi, and upper respiratory tract, including the nasal passages. Soluble particles that reach the alveolar spaces are usually circulated throughout the body by the bloodstream; however, their subsequent deposition, retention, and clearance from the body depend on the passage of xenobiotic material through multiple media compartments in the body as well as the chemical and toxicological properties of the contaminants. The importance of the chemical properties of the inhaled or ingested material is illustrated by the fact that the biological half-life for both stable and radioactive isotopes is the same.

When exact knowledge of the internal distribution of the radioisotope is lacking, the International Commission on Radiological Protection (ICRP) has recommended the assumption that 25 percent of the inhaled aerosol or particulate matter has been exhaled, 50 percent has been deposited in the upper respiratory passages, and 25 percent has been deposited in the lungs. The ICRP recommends the further assumption that all the material in the upper respiratory passages has been swallowed, and that half of the amount assumed to be in the lungs has also been swallowed. In other words, 37.5 percent of the original is assumed to be in the gastrointestinal tract (21). Needless to say, these assumptions should be replaced with quantitative microdosimetry wherever possible in order to obtain precise measures of the distribution within the body.

The dose equivalent received from radioactive materials that have been embedded in the skin surface is based on the assumptions that the irradiation is confined to a local body area and that biological transport through the body by way of the bloodstream is very slow. External alpha sources are of little significance in terms of exposure because of their limited range. Beta particles are usually localized at the skin surface or within the outer skin layers. The depth of penetration is proportional to the energy of the beta particle.

Although the dose equivalent received from external radioactive sources can be estimated by means of field measurements, the significance of exposure to radioactive materials that have been inhaled or ingested must be evaluated by other means. In the latter case, one should refer to the maximum permissible exposure data and compare the actual inhalation or ingestion conditions to these limiting concentrations. Sometimes special consideration of residence time in the body and metabolic factors is required.

Many situations feature a combination exposure to external radiation and internal emitters. Because the various types of radiation affect organ systems in different ways, it is essential that the dose equivalent to the most sensitive organs be given primary consideration. In the case of whole body irradiation, the blood-forming organs such as the bone marrow, the lens of the eye, and the gonads are more susceptible to detrimental effects. Hence these organs are considered to be "critical organs" for external exposure. For internal sources, where the radiation distribution is nonuniform, the critical or limiting organs are generally considered to be the lung, the gastrointestinal tract, bone, muscle, fatty tissue, thyroid, kidney, spleen, pancreas, and prostate. The radioactivity level in the critical organ is the limiting factor, but the maximum permissible dose (MPD) for the critical organ indicates the total activity that should be permitted in the entire body.

7.1 Plutonium

The use of plutonium serves to illustrate some of the problems associated with internal emitters in the work environment. More is known about this alpha-emitting element than others, and plutonium may well play a role of significance in satisfying the future energy requirements of the United States. The accidental release of plutonium from nuclear power reactor sites appears to be an extremely unlikely possibility.

If plutonium particles become lodged in body tissue, bone sarcomas or liver tumors might eventually be produced. The maximum permissible body burden of plutonium has been deduced from the level established for ^{226}Ra, namely, 0.1 μg. A value of 5 μg was first established for ^{239}Pu, then successively reduced to 0.04 μCi of ^{239}Pu. Estimations of the amounts of plutonium in living human organisms are complicated by the fact that the X- and gamma emissions during decay are of very low energy, and hence are almost completely absorbed in tissue. Photon measuring devices placed at or very near the surface of the body are severely handicapped in their ability to detect radiation. Urinary and fecal assay methods therefore have been useful in estimating body burdens, but continuous collection of all excrement over a period of time is necessary to detect very low levels.

In an analysis of occupational exposed groups, as well as those of the general population, Nelson et al. (23) found that concentrations of plutonium in all groups were highest in the tracheobronchial lymph nodes when compared with other organs and that liver concentrations in the occupationally exposed groups were higher than those found in the skeleton. There is evidence that the particle size of inhaled aerosol plays a role in determining the level of deposition not only in the respiratory systems but also in the systemic distribution. For example, larger particles tend to be trapped in the tracheobronchial lymph nodes, whereas smaller particles, which as a rule are more readily soluble, find their way to the bloodstream and to other organs such as the liver. Once the plutonium is in the bloodstream, the chemical and physical properties of the material are of paramount importance. Very small (<0.01 μm) particles tend to concentrate on bone surfaces; those with diameters in the 0.01 to 7 μm range tend to be taken up by the liver. Of the total amount of plutonium that is present in the bloodstream, 45 percent is deposited in the skeleton and 45 percent in the liver; 10 percent is deposited in other tissues or is excreted (24). The rate of clearance of plutonium from the human body is notoriously slow when compared with ^{226}Ra. For example, the plasma clearance is 0.1 to 0.2 liter/day, whereas approximately 100 liters of plasma per day is cleared with ^{226}Ra (25).

Detailed clinical studies (26, 27) made of a group of men who worked with plutonium during World War II revealed body burdens ranging from 0.005 to 0.42 μCi, but without apparent detrimental effect on health. The assumption is often made that the critical body organ for plutonium is the skeleton, yet there is still considerable uncertainty about this. Furthermore, although lung cancer has been produced experimentally in animals, no such cases have been demonstrated in humans, nor can any firm conclusions be drawn about long-term cumulative effects. Of those who have been exposed to plutonium, the major mode of entry to the body has been through inhalation. Because the latent period for the induction of physiological changes may be of the order of 50 years or more, it is difficult to predict whether injuries or detrimental physiological effects will be realized. A relatively small number of persons have lived for a considerable time with plutonium body burdens that are well above maximum permissible levels yet no detrimental effects have been observed.

A commonly employed technique for estimating the body burden of plutonium is the measurement of radioactivity in the urine (28). This can be done by X-ray spectrometry, proportional counting, or the counting of neutron-induced tracks in plastic. A rough assessment of the absorbed dose in tissue may be made by measuring the low-energy X- and gamma rays from plutonium by means of instruments placed on the surface of the body. Also, alpha ray spectrometry of urine samples has found wide application as a plutonium bioassay technique.

Although it is generally conceded that the urine assay is the method of choice in assessing plutonium body burden, it is also true that the correlation between urine assays taken on living human subjects and analyses of tissue from the same subjects after death has not been particularly reassuring. Table 38.7, for example, illustrates one set of urine assays that predicted much higher body burdens than were actually found when quantitative analyses were made of tissue obtained at autopsy. The same study (29) showed a general correlation between the amount of plutonium in the body, exclusive of the respiratory tract, and the amount of plutonium estimated by urine assay, especially at higher levels of body plutonium. The urine assay does not correlate well with plutonium burden in the respiratory tract.

The situation is considerably better in the case of tritium. Where environmental levels of tritium are low, urine analysis is sufficient for personal monitoring; changes at exposure levels well below regulatory limits can be detected. Proportional counters are capable of measuring ^{239}Pu in amounts as low as approximately 4 nCi to values

Table 38.7. Plutonium Body Burden
Calculations (29)

Case Number	From Urine Assay (nCi)	From Tissue Analysis (nCi)	
		Systemic	Total
1-039	24.6	18.3	21.0
2-030	10.0	4.2	8.8
2-100	2.8	0.3	0.3
2-126	2.4	0.0	0.5
2-130	3.2	1.0	1.2
3-016	3.4	0.7	0.9
5-150	2.7	0.0	0.01
7-066	30	0.6	0.8
7-084	0.9	0.1	0.4

of the order of 40 nCi, depending on counting gas, chamber characteristics, and counting time.

7.2 Cardiac Pacemakers

Most of the present-day pacemakers are powered by batteries, but some may be powered by a radioactive source such as plutonium. According to one source (30), the effects of distributing 10,000 plutonium-powered pacemakers to the public would be as follows:

1. Exposure to family members living with the pacemaker wearer would not be expected to produce a dose equivalent greater than 7.5 mrem (0.075 mSv) per year and a dose to the entire U.S. population of 128 man-rems (1.28 man-Sv) per year. These values are compared with the average natural background dose equivalent of 102 mrem (1.02 mSv) per year to individuals and the total natural background dose to the U.S. population of approximately 20×10^6 man-rems per year.

2. Doses to critical organs and the whole body received by pacemaker patients are well below the occupational exposure limit of 5 rems (50 mSv) per year.

3. The surgical implantation and removal of pacemakers results in a very small radiation dose to medical personnel. (Surface dose rate from the pacemaker is on the order of 5 to 15 mrem/hr, that is, 0.05 to 0.15 mSv/hr.)

Assuming that the plutonium-powered pacemakers perform safely and efficiently, the advantages, such as long-term, trouble-free maintenance, avoidance of surgical replacement operations, and the like, far outweigh any disadvantages.

The standards for permissible levels of plutonium or permissible body burden, assuming that bone is the critical organ, are not based on calculation of the maximum permissible dose rate to bone; rather they are based on a comparison of the relative toxicity of plutonium and radium in animals and on correlations with human radium exposure data. There is some concern among professionals in radiation protection that the quantification of dose and the criteria on which maximum permissible intake levels are based are all too heavily founded on the radium experience. Even newer exposure criteria seem to be mere modifications of the radium-based information and data rather than the result of tests made on specific radionuclides.

8 RADIATION MEASUREMENT

From the radiation protection standpoint, the primary purpose of radiation measurement is to make quantitative assessments of potential environmental exposure conditions, to permit the installation of proper controls, particularly if the exposure conditions appear to exceed established standards. Such procedures clearly indicate that the primary role of instrumentation is to prevent possible detriment to the health and safety of individuals working with radiation-emitting sources. Unfortunately, preventive measures are not always completely effective; hence measurements may be required to assess environmental conditions or the probable absorbed dose after individuals have been irradiated.

Many techniques are employed to measure energy transfer, absorption, or ionization phenomena caused by incident radiation. Some of the instruments and devices used to measure such phenomena include ionization chambers, proportional counters, scintillation counters, Geiger–Müller detectors, luminescent detectors, photographic emulsions, chemical reaction detectors, and fissionable material and induced-radiation detectors. The choice of detector depends on the radioactive source and its emitted radiation, its temporal and spatial characteristics, and the quality of radiation. Because radiation-measuring instruments are often designed to measure a specific type of radiation under very specific conditions, the following broad characterization of parameters that are important for proper measurements is in order.

8.1 General Considerations

When the radiation source consists of a mixture of unknown radionuclides, it is possible to differentiate the particles by means of their individual energy spectra. If one of the interfering isotopes is an isotope of the same element being measured, chemical methods will not distinguish one from the other. However, all the different elements in the source can usually be distinguished from one another by chemical means. Gamma rays are often distinguished by spectrometry, using scintillation or semiconductor detectors, and the ability of gamma rays to ionize gaseous molecules within ionization chambers is often employed to measure exposure rate. Beta particles have continuous energy spectra that permit only limited discrimination among different radionuclides. For certain beta decay schemes, however, coincidence techniques may provide additional selectivity. Alpha particles are identified by gas ionization according to their energies (alpha spectrum), and by scintillation and semiconductor detection techniques. To avoid contributions to background from low L (LET) radiation, one should avoid using detector elements such as anthracene, whose response to absorbed energy shows a pronounced inverse dependence on L. Alpha particles are emitted by many naturally occurring radioisotopes (most common construction materials show alpha particle emission at freshly cut surfaces), but man-made isotopes are essentially limited to the transuranic elements, and of these, only plutonium, americium, californium, and curium are potentially found in the work environment.

8.2 Detection and Measurement

8.2.1 Thermoluminescent Detectors

The use of thermoluminescent dosimeters (TLD) for personnel monitoring is a widespread practice. The TLD consists of a small crystalline detector, usually lithium fluoride, lithium borate, calcium fluoride, or calcium sulfate. Traces of other metal ions serve as activators. When the crystalline structure is exposed to radiation, energy is quantitatively absorbed in crystalline traps. When the material is subsequently heated, the stored energy is quantitatively released in the form of light, thus permitting an estimate of radiation exposure. Lithium fluoride is most frequently used because its response is relatively independent of the type and energy of incident radiation, and it has a useful range of about 0.003 to 10,000 rems (0.03 to 1000000 mSv) (31). A schematic general response and a readout cycle for a typical TLD chip are given in Figure 38.3. The sensitivity and accuracy of the TLD (Table 38.8) are adequate for personnel monitoring, and the reproducibility is excellent. A reproducible rate of heating must be used in measuring the light output from the TLD. By making simultaneous measurements using fluorides of ^{6}Li, which is very sensitive to neutrons, and ^{7}Li, which is less sensitive, it is possible to evaluate the neutron contribution to the overall dosage. A major advantage of the TLD over film is its energy-independent response. On the other hand, the energy-dependent response of film can be used to obtain a rough estimate of the energy spectrum of the radiation. In most routine applications, however, the measurement of the energy spectrum is unnecessary. One word of caution is that the lithium fluoride disks exhibit energy-dependent directional sensitivity, which may be further complicated by badge-filter orientation when the disks are placed in personal monitoring badges (32).

Both TLDs and radiophotoluminescent (RPL) detectors are replacing film badges in many applications. A comparison study between film and thermoluminescent dosimetry for routine personnel monitoring revealed that film is less reliable than the thermoluminescent dosimetry for monitoring radiation exposures above 10 R (33).

Glass dosimeters using RPL have been developed. One system based on thermally stimulated exoelectron emission (TSEE) is similar to TLD except that instead

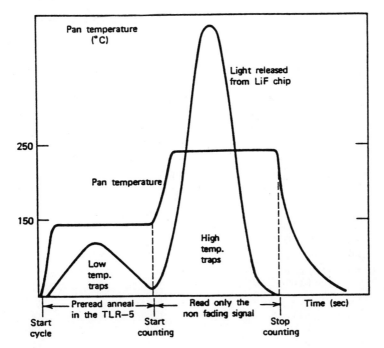

Figure 38.3. Typical "glow curves" for TLD reader (Eberline TLR-5). Data on typical measurement precision and accuracy of TLD badge appear in Table 38.8. Courtesy of Eberline Instrument Corporation.

of measuring light that is emitted when the material is heated, electrons are measured with a Geiger–Müller tube. The emission of electrons may be stimulated by light or heat; also, ultraviolet radiation may be emitted following optical stimulation (34).

8.2.2 Neutron Dosimetry

Personal neutron dosimetry has traditionally been accomplished by counting proton recoil tracks in exposed special film. However, one can also use dielectric materials such as plastic or glass for detecting and measuring neutron flux. The plastic or glass material may be covered with a layer of foil of fissionable material such as uranium. Neutrons cause fission action in the foil, and the highly energetic fission fragments cause damage in the dielectric material. When the material is chemically etched, the damage becomes apparent as visible tracks, or etch pits, which can be counted under the microscope. The proper measurement of neutrons is highly complex, especially if the energy-dependent spectral shape is not known.

One method for measuring neutron absorbed dose over the range of 500 to 5000 mSv has been developed (35) using fission fragment track etching. It was discovered that by using combinations of ^{232}Th or ^{237}Np as fissile materials and organic and

Table 38.8. Typical Measurement Precision and Accuracy of TLD Badge

Measurement Precision						
Exposure (mR)	5	10	50	300	3000	30,000
Standard deviation (%)	20	10	4	4	4	4

Accuracy—Personnel Badge		
	100 mrem (1 mSv) Dose Reported as	
Type Radiation	Whole Body	Skin
Gamma, ^{226}Ra	94	114
Gamma, ^{137}Cs	100	101
Gamma, ^{60}CO	90	110
X-ray, 175 keV	104	111
X-ray, 145 keV	102	113
X-ray, 100 keV	107	113
X-ray, 82 keV	111	119
X-ray, 58 keV	114	123
X-ray, 43 keV	129	143
X-ray, 23 keV	80	130
X-ray, 16 keV	21	116
Beta, ^{90}Sr, ^{90}Y	30	70
Beta, uranium slab	20	60

Courtesy Eberline Instrument Corporation.

inorganic track detectors, both automatic spark counting and visual track counting methods can be developed to cover the dose range mentioned. Similarly, neutron fluence distributions from ^{252}Cf sources have been measured in a phantom, using the track etching method. Thin polycarbonate sheets are first subjected to fragments from neutron-induced fission in foils of ^{235}U, ^{238}U, ^{232}Th, then counted under the microscope after etching (36). Fast neutron fission track dosimeters have appreciable directional dependence. In one study (37) the recorded ratio of sensitivities for 14 MeV neutrons incident normal and parallel to the plane of the detector was approximately 1.6. For fission neutrons and thorium radiators, the value was 1.4.

8.2.3 Pocket Dosimeters and Films

The pocket dosimeter (ion chamber) is normally used in conjunction with film badge for measuring X- and gamma radiation. The walls of the dosimeter, as well as the internal components and ionizable gas, are usually air equivalent or tissue equivalent. The indirect-reading gamma dosimeter is similar to the direct-reading dosimeter except that it must be read out on a charger–reader. The direct-reading instrument merely has to be held up to the light to make a reading. The reading is directly proportional to the discharge of current from a precharged value, and the discharge is proportional to the integrated radiation dose. The range of most pocket dosimeters is from 0 to 200 mR; some are available in the 0 to 200 R range. The energy dependence is usually less than 10 percent from 50 keV to 1.33 MeV (^{60}Co). The minimum measurable energy is approximately 10 to 15 keV. Although the models of dosimeters have vastly improved over the past two decades, they are still subject to accidental discharge.

Films are photographic emulsions that may be used to detect various types of radiation. One type of film blackens if exposed to beta, gamma, or X-radiation, whereas another type is used to record tracks that are produced by charged particles such as protons from fast or thermal neutron interactions. The blackening of the film by beta, gamma, and X-rays is not proportional to air or tissue absorbed dose at different energies. Therefore various types of moderating metal and plastic or loaded absorbing material are usually placed adjacent to the film to minimize the nonproportionality when the film is used in a radiation monitoring badge. Shielding material placed over films may also permit the discrimination between gamma rays and X-rays and beta radiation.

The minimum energy measured by films is of the order of 20 keV for X-rays and 200 keV for beta radiation. Some of the main shortcomings of film for radiation protection purposes are:

1. False readings may be produced if film is exposed to heat, pressure, or chemicals.
2. Variations in film quality occur from batch to batch.
3. Energy dependence for low-energy X-rays is strong.

Chemical reaction detectors are systems in which radiation produces a chemical change in a material in such a manner that a chemical analysis or direct-reading indicator will measure the amount of change. Such detectors are not used extensively because of their low sensitivity.

Induced radiation detectors are materials in which radiation interacts to form radionuclides whose radiation is subsequently measured. They are particularly useful for detecting or measuring neutron radiation.

8.2.4   Geiger–Müller Survey Meters

The detector tube in the Geiger–Müller (GM) survey meter is usually encased in a protective outer metal shield. Some tubes have openings at the extremities; others have openings along their length. These openings are usually covered with a thin sheet of mica or Mylar whose density does not exceed 2 mg/cm^2. By making provision for shielding these openings with metal, discrimination between beta and gamma radiation is possible. The output of the GM counters is not proportional to the exposure or absorbed dose rate unless the device is carefully calibrated for the specific radiation being measured. Significant errors can occur if the meter readings have not been specifically calibrated. GM tubes are usually halogen quenched to prevent undue avalanching of signals. The efficiency of beta measurement ranges from approximately 30 to 45 percent in GM tubes equipped with beta discrimination shields; gamma sensitivity with the shields closed is approximately 5000 counts per minute per milliroentgen. The minimum energy reliably measured is approximately 20 keV for X-rays and 150 keV for beta radiation. The main advantages of GM equipment for surveying purposes are reasonably high sensitivity and rapid response. Some of the disadvantages include saturation at high counting rates, strong energy dependence, and possible interference by ultraviolet and microwave radiation.

Typical full-scale readings of GM equipment range from 0.2 to 20 mR/hr; response time is from 2 to 10 sec.

8.2.5 Ionization Chambers

Ionization chambers have been specifically designed to measure dose or dose rate from beta, gamma, and X-radiation. The principle of operation of the chamber is that ions are produced when radiation impinges upon a preselected gas contained within the chamber. The ions move in a field supplied by a continuously applied voltage to produce small measurable currents whenever ionization events occur because of incident radiation.

The voltage required to produce a saturation current in any chamber is proportional to the rate of ionization. At saturation, the ionization current is related to the product of the number of ion pairs per unit time and the electronic charge. Because the chamber is essentially an integrating device for a large number of ionization events, the time constant of the electrical current readout device is long, to make possible the averaging out of wide fluctuations. Another variation in detection schemes is to measure the rate of voltage change on a capacitor whose charge depends on the rate of ionization within the chamber. The number of ion pairs formed per centimeter of path length is proportional to the density of gas in the chamber.

Most gas-filled ion chambers operate in the proportional or gas multiplication region, which yields a nominal ± 10 percent photon energy response from about 50 keV to 1.3 MeV. The minimum accurately measurable energy is usually of the order of 20 keV (for X-rays). Advantages include low energy dependence and the capability of measuring a wide range of air doses (e.g., 3 mR/hr to more than 500 R/hr). Disadvantages include slow response and relatively low sensitivity. In selecting an ion chamber for use in radiation surveys, certain desirable features should be kept in mind: drift-free response, solid state circuitry to eliminate warm-up time, lightweight construction, compactness, capability of installing an audible alarm that is triggered at a level selected by the user, capability of accurately measuring X-ray dose rate independently of the spectral energy distribution, remote operation, and interchangeable ion chambers. For example, it would be desirable to be able to interchange a rate meter that measures 0 to 1000 mR/hr using a high-sensitivity chamber with a low-sensitivity chamber that measures in the 0 to 1000 R/hr range.

Ion chambers must be compensated for use in measuring tritium concentrations in high gamma fields because as little as 1 mR/hr gamma dose rate may produce a reading that is 10 times the MPC for tritium in air (38, 39).

8.2.6 Scintillation Detectors

Scintillation detectors operate on the basis of producing light from the interaction of ionizing radiation (X, beta, and gamma) with constituents in a scintillation crystal. Scintillation counters should have good pulse height resolution; for example, a 3 × 3 in. unit for ^{137}Cs should deliver at least a 7.5 percent resolution. Detector performance is closely related to crystal purity. Minimum measurable energy from X-rays is of the order of 20 keV; full-scale range of commercially available detectors is typically from 0.02 to 20 mR/hr. Special attention must be paid to obtaining a good light shield around the photomultiplier tube, as well as to the need for using antimagnetic mu metal in the housing to minimize changes in gain due to magnetic fields. A typical scintillation crystal consists of sodium iodide doped with tellurium ions. Light pulses from the phosphor are detected by means of a photomultiplier tube whose output voltage is measured on a voltmeter. The magnitude of the light signal is proportional to the energy absorbed in the scintillation crystal. The operating voltage of the detector is variable, depending on the type and intensity of radiation. In the case of alpha scintillation probes, the window thicknesses are usually of the order of 0.5 mg/cm^2 of aluminized Mylar. The maximum efficiency using this particular window thickness is of the order of 30 percent. If a fine wire mesh screen is placed over the Mylar for protection, however, the efficiency is reduced to approximately 20 percent. In general, scintillation counters are known for rapid response and high sensitivity, but they are also relatively fragile and expensive.

8.2.7 Proportional Counters

Proportional counters usually consist of a gas cylinder containing a central wire to which a potential is applied. The potential may be adjusted to render the output voltage proportional to the energy released by the radiation-induced ionization events in the chamber. Because of this feature, the type of radiation to be measured can be selected. For example, the proportionality between alpha and beta particles can be determined because of the considerably greater ionization produced by the former. Proportional counters may be used for general surveys or for performing sophisticated spectrometric analyses, but in the majority of cases they are employed in the measurement of neutron and alpha radiation. In the case of gas flow proportional probes to measure alpha radiation, the window thickness is usually of the order of 0.5 to 0.9 mg/cm^2 of aluminized Mylar. The efficiency of these devices, expressed as a percentage of radiation received from a 2π surface, is of the order of 35 to 50 percent. A well-known tritium probe consists of a windowless gas flow proportional chamber for measuring the very low energy beta particles emitted by tritium. This device may be modified to measure alpha or high energy beta particles, but it can also be used for the assessment of alpha and beta swipe samples.

8.3 Calibration

Standard sources, such as those available through the National Institute for Standards and Technology (NIST), should be used for purposes of instrument calibration. In the absence of a NIST source, a secondary standard that is directly traceable or relatable to NIST material may be used. In many cases, sources can be sent to NIST for calibration. Another approach to instrument calibration is the development of a standard for use within a specified instrument facility, irrespective of whether the standard is relatable to a NIST primary standard.

When obtaining a calibration source for photon measuring instruments, the effective energy (usually in kilo- or megaelectron volts), half-life, and the specific exposure rate of that source must be ascertained. In general, the source should be much smaller than the intended detector, and the distance between the source and detector should be such that the detector measures essentially the radiation from a point source. It is common practice to make the source-to-detector distance at least seven times that of the largest dimension of either the source or the detector and to keep the source and detector far removed from scattering surfaces. The calibration source should have an energy spectral distribution identical to, or at least similar to, that of the radiation to be measured. It should also have a half-life long enough to permit a reasonable number of calibrations to be performed before replacement. The response of the instrument to be calibrated to various radiation energies will establish its energy dependence characteristics. If X-rays are used as a calibration source, the excitation voltage, the spectral energy characteristics, and patterns of radiated energy must be known.

The instruments most often used for the standardization of photon fields include cavity ion chambers, free air ion chambers, and calorimeters. Detectors must be operated within specified pressure, temperature, and humidity limitations. Whenever possible, a determination should be made of the extent to which radiations other than those of primary interest might interfere with the sensitivity and accuracy of the device. Specific examples of interferences include radio-frequency and microwave radiations generated by a wide variety of sources. Some of these sources may escape the attention of the unsuspecting user because of their common everyday presence (e.g., electrical appliances, coils, electrical discharges, solar radiation, and microwave ovens).

When NIST standards cannot be used and for reasons of convenience it is preferred to employ a secondary standard, every effort should be made to keep the accuracy of the secondary source calibration to within ±2 percent of the primary or NIST standard.

The following factors should be considered in the design of any calibration facility:

1. Background radiation levels must be low and quantifiable during the actual calibration procedure.
2. Neutron and photon calibrations should use free space geometry.

3. Neutron sources must be described in terms of effective energy and flux density.

4. Photon-emitting sources should be described in terms of exposure rates at specified distances from the source.

5. Exposure of persons conducting the calibration should be kept as low as practicable but at least within permissible limits.

6. The calibration radiation field should closely approximate that found in the test condition, if possible.

7. Proper records must be maintained on all calibrations.

When specifying the activity of a primary calibration source, especially when the source consists of an equilibrium mixture, the total activity of the radionuclides in equilibrium must be determined. For example, in the case of the two beta emitters ^{90}Sr and its daughter ^{90}Y, the two in equilibrium have double the activity of freshly separated ^{90}Sr. Within a period of five half-lives, the ^{90}Y activity is 95 percent of the ^{90}Sr activity.

8.3.1 Beta Calibration

Calibrations are often made by placing the detector window of the instrument in close proximity to a properly calibrated beta source such as ^{90}Sr. If an attempt is made to use a free space geometry calibration, one must be aware of the significant absorption of beta particles in air, that is, their limited range in air. Typical beta calibration sources include ^{85}Kr, ^{204}Tl, ^{90}Sr-^{90}Y, ^{14}C, and ^{42}K.

8.3.2 Alpha Calibration

Alpha calibration sources can be purchased commercially. Typical alpha sources include ^{241}Am, ^{210}Po, ^{252}Cf, ^{148}Gd, ^{230}Th, and ^{239}Pu. Because the half-life of ^{210}Po is 138.4 days, the usefulness of this isotope is limited.

8.3.3 Neutron Calibration

Three types of radionuclide neutron sources exist. First there are (α,n) emitters such as ^{210}Po, ^{238}Pu, ^{239}Pu, or ^{241}Am, which emit alpha particles in close contact with low Z elements, such as lithium, beryllium, bismuth, and fluorine. The energy involved with the alpha-emitting types ranges from 0 to tens of megaelectron volts, neutron yield is of the order of 10^7 to 10^8 neutrons per second (n/sec), and the sources are dangerous to handle unless sealed. Wipe tests are definitely required before handling. Second, there are (γ,n) sources consisting of a high-energy gamma emitter such as ^{124}Sb or ^{226}Ra placed in close proximity to low Z elements such as beryllium or deuterium. Monoenergetic neutrons are produced along with intense photon fields. Although ^{124}Sb has a relatively short half-life (160 days) and yields low neutron energies (30 keV), ^{226}Ra sources produce neutron yields in the 10^6 to 10^7 n/sec range. Third, there are spontaneous fission neutrons, such as those produced by ^{252}Cf; these are produced in yields of 10^8 to 10^9 n/sec, but the associated photon emission is of low intensity. Because ^{252}Cf can be fabricated in relatively small size, it acts as a point source; also NIST is in a position to calibrate ^{252}Cf sources, thereby rendering them derived standards.

Particle accelerators also produce intense neutron fields by accelerating charged particles such as deuterons, protons, or tritons onto low Z materials such as deuterium, lithium, and tritium. High neutron fluxes (10^{12} n/sec) are obtainable, but these are highly variable depending on the accelerator and target characteristics. Therefore the output of the accelerator must be monitored constantly if it is to be used for calibration purposes.

Neutron fields from reactors can vary from an unmoderated fission spectrum to heavily filtered slow neutrons from a thermal column. The flux densities [10^{10} $n/(cm^2)(sec)$] are high enough for calibration purposes, but they are highly variable and require constant monitoring.

Alpha-, beta-, and photon-emitting calibration sources may be obtained commercially and from such agencies as the National Institute for Standards and Technology, Gaithersburg, Maryland and the International Atomic Energy Agency, Vienna, Austria.

Calibrating sources should have high activity per unit mass, high chemical and radiochemical purity, and sufficient half-life to suit the purposes of a good instrumentation program. Radionuclides that are suitable as calibration sources include the following:

1. Alpha sources: ^{148}Gd, ^{210}Po, ^{230}Th, ^{238}Pu, ^{239}Pu, ^{241}Am, ^{244}Cm, and ^{252}Cf
2. Beta sources: ^{3}H, ^{14}C, ^{32}P, ^{35}S, ^{45}Ca, ^{85}Kr, ^{90}Sr-^{90}Y, ^{111}Ag, ^{185}W, ^{204}Tl, ^{210}Bi, and ^{238}U
3. Photon sources: ^{24}Na, ^{51}Cr, ^{57}Co, ^{60}Co, ^{137}Cs, ^{241}Am, and ^{226}Ra

The following general caveats are made with respect to the measurement of ionizing radiation for personal protection purposes:

1. Several methods of surveying and monitoring should be used rather than one. For example, one should not rely exclusively on film badges; a combination of film badges and direct-reading dosimeters is preferred. A dangerous situation might be missed while waiting for the film badge readout if no instant readout device is employed. Survey-type instruments should be supplemented with fixed station monitoring equipment whenever possible.
2. The function of all equipment should be checked periodically with a calibrating source.
3. The energy response of the instruments should be checked frequently.
4. Special care should be taken in the choice of instruments where pulsed sources are to be monitored. Instruments should have a time constant that permits accurate measurement of the pulse duration. For more detailed techniques and procedures, the reader is referred to the *Handbook of Radioactivity Measurements Procedures*, Report No. 58 of the National Council on Radiation Protection and Measurements (NCRP), Washington, DC.

9 EXPOSURE EVALUATION

The contribution of occupational exposures to the total ionizing radiation dose to the general population is far less than 1 percent (42). Except for cases involving the natural radioactive materials radon and radium, industrial operations have not caused any ill effects in humans from internal deposition of radioactive materials. This is a tribute to the excellent preventive programs established in industry. Also, the health of people employed in industry is generally better than that of comparable age groups in the population and the employees are better trained than the average citizen in the safe use of radioactive sources. Because the general population is made up of individuals from the very young to the extremely old, including the sick or disabled who may collectively be exposed for 24 hr a day, 365 days a year, it stands to reason that the permissible levels of radiation exposure for the general population must be set lower than those for radiation workers. The permissible exposure levels for the two groups (i.e., general population vs. radiation workers) usually differ by a factor of 10 or greater.

A complete tabulation of maximum permissible concentrations of radionuclides in air and water for occupational exposures can be found in NCRP Report No. 22 (43). As has been the case with past NCRP recommendations, individual radiation exposures resulting from necessary medical and dental procedures have not been included in the dose limiting recommendations, and such medical and dental procedures are presumed to have no effect on the radiation status of the individual.

Certain basic assumptions have been made in the derivation of all radiation protection guides: (1) that the effects of all types of radiation on any given individual can only be approximated; (2) that there is still uncertainty as to whether there exists a radiation level below which biological damage will not occur; and (3) that one should assume a linear relationship between biological effect and radiation dose, especially at low levels of continuous exposure (stochastic processes).

One method for estimating the retention of radionuclides that have been ingested or inhaled is to use an instrument that measures the gamma radiation being emitted from the whole body or a critical organ. The dosage to the thyroid, for example, can be estimated reasonably accurately by means of a crystal detector placed next to the thyroid. For practical purposes, the use of such instrumental techniques is limited to gamma or high-energy beta emitters. Alpha radiation, which is the most serious type of radiation once inside the body, cannot be detected on the external surfaces of the body. Because in some cases beta emitters (e.g., plutonium) radiate small amounts of gamma and X-rays, devices can detect the presence of particular radionuclides. A reasonable estimation may be made of absorbed dose equivalent in the case of internal contamination through the use of a whole body counter. This device consists of a battery of very sensitive detectors that have been well shielded to minimize the effect of background radiation. Under these conditions, a spectrometric analysis may be made of radionuclides that are present in the body. In the case of external irradiation of the body tissues, contamination may be deposited on the skin or in the clothing. For such situations, suitable instruments equipped with probes that can monitor body surfaces and clothing may be used.

The clothing is sometimes removed and the body washed, to help differentiate internal from external radiation.

10 STANDARDS AND REGULATIONS

Many of the basic standards on radiation protection have been formulated by the National Council on Radiation Protection and Measurements and the International Commission on Radiological Protection. The principal set of legally binding standards, however, consists of those issued and enforced by the Nuclear Regulatory Commission. Other regulatory agencies that have responsibilities for radiation protection include the Environmental Protection Agency, the Department of Health and Human Services, the Occupational Safety and Health Administration, and state government agencies.

A summary of exposure limits recommended by the National Council on Radiation Protection and Measurements (NCRP) (55) is given in Table 38.9. The recommended mean values of the quality factor for various types of radiation are given in Table 38.10.

In using Table 38.9 and 38.10, the following facts should be kept in mind:

A. For occupational exposures: (1) the age proration formula $5(N - 18)$ rem, where N is age in years, is discontinued, (2) the annual limit of 50 mSv (5 rem) is continued, (3) cumulative exposures should not exceed the age of the individual in years X 10 mSv (years X 1 rem), (4) for pregnant women under occupational conditions, the limit for the fetus [5 mSv (0.5 rem)] should not be received at a rate greater than 0.5 mSv (0.05 rem) per month, and (5) all limits apply to the sum of external and internal exposures.
B. For public exposures to man-made sources other than medical and natural background, the annual limits of 1 mSv (0.1 rem) for continuous exposure and 5 mSv (0.5 rem) for infrequent exposures are reaffirmed. Also reaffirmed are remedial action levels for the public of 5 mSv (0.5 rem) annual average for external exposure (from all sources except medical) and an annual average of 0.007 Jh/m^3 (2 WLM) for total exposure to radon and its decay products.
C. An annual negligible individual risk level is specified as 10^{-7}, corresponding to a dose equivalent of 0.01 mSv (0.001 rem).
D. For Planned Special Occupational Exposures
   In order to allow certain essential tasks to be performed, it may be necessary, on rare occasions, to permit a few workers to receive an annual effective dose equivalent in excess of 50 mSv (5 rem). The NCRP recommends that if exposure under these conditions is essential, these guidelines be followed:
   1. No worker should receive an effective dose equivalent[10] of more than 100 mSv (10 rem) in a single planned event.
   2. Nor should the effective dose equivalent received in such special planned exposures exceed 100 mSv (10 rem) over the working lifetime.

Table 38.9. Summary of Recommendations[a,b]

A. Occupational exposures (annual)[c]		
1. Effective dose equivalent limit (stochastic effects)	50 mSv	(5 rem)
2. Dose equivalent limits for tissues and organs (nonstochastic effects)		
a. Lens of eye	150 mSv	(15 rem)
b. All others (e.g., red bone marrow, breast, lung, gonads, skin, and extremities)	500 mSv	(50 rem)
3. Guidance: cumulative exposure	10 mSv × age in years)	(1 rem × ag
B. Planned special occupational exposure, effective dose equivalent limit[c]	see text	
C. Guidance for emergency occupational exposure[c]	see text	
D. Public exposures (annual)		
1. Effective dose equivalent limit, continuous or frequent exposure[c]	1 mSv	(0.1 rem)
2. Effective dose equivalent limit, infrequent exposure[c]	5 mSv	(0.5 rem)
3. Remedial action recommended when:		
a. Effective dose equivalent[d]	>5 mSv	(>0.5 rem)
b. Exposure to radon and its decay products	>0.007 Jhm^{-3}	(>2 WLM)
4. Dose equivalent limits for lens of eye, skin, and extremities[c]	50 mSv	(5 rem)
E. Education and training exposures (annual)[c]		
1. Effective dose equivalent limit	1 mSv	(0.1 rem)
2. Dose equivalent limit for lens of eye, skin, and extremities	0.5 mSv	(0.05 rem)
F. Embryo–fetus exposures[c]		
1. Total dose equivalent limit	5 mSv	(0.5 rem)
2. Dose equivalent limit in a month	0.5 mSv	(0.05 rem)
G. Negligible individual risk level (annual)[c] Effective dose equivalent per source or practice	0.01 mSv	(0.001 rem)

[a]Excluding medical exposures.
[b]See Table 38.10 for recommendations on Q.
[c]Sum of external and internal exposures.
[d]Including background but excluding internal exposures.

   3. Such planned special exposures should be authorized in writing by senior management prior to the exposure.
   4. Older workers with low lifetime effective dose equivalents should be selected whenever possible.
   5. Exposures resulting from planned special exposures shall be included in the lifetime record of exposure for each worker but separately identified.

Table 38.10. Recommended Values for Q for Various Types of Radiation

Type of Radiation	Approximate Value of Q
X rays, γ rays, β particles, and electrons	1
Thermal neutrons	5
Neutrons (other than thermal), protons,[a] alpha particles and multiple-charged particles of unknown energy[b]	20

[a]In circumstances where the human body is irradiated directly by high-energy protons, the RBE is likely to be similar to that of low-LET radiations, and therefore a Q of about unity would be appropriate for that case. Only low-energy protons (e.g., like those generated in tissue by fast neutrons) can be expected to require the high value of Q listed here.

[b]If energies are known, Q values can be obtained for neutrons by the procedures and values given in NCRP Report No. 38 (NCRP, 1971b) but multiplied by 2. For other particles of known energy, see ICRP Publication 21 (ICRP, 1973), but again the values would have to be multiplied by 2.

      The effective dose equivalent specified here must be the sum of external and internal effective dose equivalent, if both exist.

E. Guidance for Emergency Occupational Exposures

      Only actions involving lifesaving justify acute exposures in excess of 100 mSv (10 rem). The use of volunteers for exposures during emergency actions is desirable. Older workers with low lifetime accumulated effective dose equivalents should be chosen from among the volunteers whenever possible. Exposures during emergency actions that do not involve lifesaving should be controlled by the occupational exposure limits, including the planned special exposure recommendations.

      When the exposure may approach or exceed 1 Gy (100 rad) of low L (LET) radiation (or an equivalent high LET exposure) to a large portion of the body in a short time, the worker not only needs to understand the potential for acute effects but also should have an appreciation of the substantial increase in his or her lifetime risk of cancer. If the possibility of internal exposures also exist, these must be taken into account (55).

11   ELEMENTS OF A RADIATION PROTECTION PROGRAM

The elements of radiation protection programs vary according to the size, number, and type of radiation sources and their specific application(s), the facility in which the radioactive sources are housed, training level of personnel, complexity of the operations, and many other factors; therefore an attempt is made to describe general principles and elements that are common to most situations. The following items are covered:

1. Administration
2. Orientation and training

3. Control measures
4. Surveys and monitoring
5. Emergency procedures
6. Medical surveillance
7. Records
8. Notification of incidents

11.1 Administration

The most important requirement for an effective radiation protection program is an informed and highly supportive executive officer, manager, or owner who has responsibility and accountability for the basic safety of the operation. This is essential for a sustained level of attention to the many requirements in the program. The elements of the radiation protection program must be recorded in writing as an official part of organizational policy.

11.2 Orientation and Training

Of all the ingredients in a radiation protection program, the one that is most indispensable is establishing a fail–safe mechanism for informing the employee of potential hazards. An employee who knows about radiation hazards, their potential biological effects, and necessary control is in a position to cooperate to meet common objectives. Under these circumstances, the employee is sufficiently well motivated to use ingenuity to keep exposure as low as practicable, even when environmental controls are less than ideal. The specifics of the training program should include nature of radioactivity, background radiation, interaction of radiation with matter, biological effects (acute and chronic, somatic and genetic, internal and external), exposure criteria (permissible doses), standards and regulations, monitoring procedures, control measures (time, distance, shielding, engineering design, cautionary procedures, protective equipment, warning signs and labels, waste disposal practices), emergency procedures, and medical surveillance. Most important is a highly visible declaration of strong support of the program by those persons who are morally and legally responsible and accountable for the safety of operations, usually the owners or top management personnel.

The person or persons conducting the orientation and training should be qualified professionals in radiation protection. If the radiation facility warrants a full-time professional, such a person should be designated as the radiation safety officer (RSO) and given total responsibility for administering the program with major emphasis on the training functions of the program. Part-time personnel or consultants in radiation protection must be professionally qualified.

11.3 Control Measures

Primary emphasis and reliance should be placed on engineering control measures whenever possible. For example, if radiation levels can be controlled at the source, many administrative or procedural practices may be eliminated.

11.3.1 Ventilation and Facility Layout

The layout and design of a facility should be such that there is minimal risk of contamination. For example, high radiation areas should be separated from lower level operations wherever possible. Ventilation systems can be designed to place the high radiation area hoods at the end of the line, nearest the blower; then "once-through" (no recirculation air) ventilation systems can entrain radioactive contaminants and pass them through high-efficiency filters and/or scrubbers before discharge to the external environment. Ducts on the discharge (positive pressure) side of the blower should not be housed within buildings, because a leak in the discharge duct could release contaminants within the building.

For a more detailed treatment of ventilation system design, the reader is referred to the *Industrial Ventilation Manual*, published by ACGIH (46), and other references (47, 48).

11.3.2 Shielding

The thickness of a specified substance that when introduced into the path of a given beam of radiation reduces the value of a specified radiation quantity by one-half is referred to as the "half-value layer" (HVL). The HVL is often used to characterize the effectiveness of shielding materials. Table 38.11 presents HVL and TVL (tenth-value layer) data on lead, concrete, and steel for attenuation of X-rays produced at various peak voltages. Also included are some HVL and TVL values for ^{137}Cs, ^{60}Co, and ^{226}Ra. Similar data for the gamma radiation from a number of radionuclides are presented in Table 38.12. The gamma ray constants in the right-hand column may be used to determine the dose rates at various distances from the source. In using HVL data, it should be remembered that a shield thickness of 2 HVL reduces the exposure rate by a factor of 4, a thickness of 3 HVL by a factor of 8, and so on.

The shielding designs should take into account such factors as human occupancy and use, but the overriding consideration in the design should be to reduce all possible exposures to the lowest practicable value, and in every case, to well below the maximum permissible dose.

The choice of shielding materials can be crucial to the proper attenuation of radiation. In the case of photon emissions, such as X-rays, the selection of materials usually becomes a choice between concrete or lead, or combinations of both, or combinations with other materials (e.g., lead-lined lath and wall boards, lead-lined concrete blocks, lead glass, lead–steel combinations, loaded concrete, loaded concrete–lead). The attenuation of photons in concrete depends more on the uniformity of the density of the material than on a special chemical composition; in

Table 38.11. Half-Value (HVL) and Tenth-Value (TVL) Layers[a]

| Peak Voltage (kV) | Attenuating Material | | | | | |
| | Lead (mm) | | Concrete (in.) | | Steel (in.) | |
	HVL	TVL	HVL	TVL	HVL	TVL
50	0.05	0.16	0.17	0.06		
70	0.15	0.5	0.33	1.1		
100	0.24	0.8	0.6	2.0		
125	0.27	0.9	0.8	2.6		
150	0.29	0.95	0.88	2.9		
200	0.48	1.6	1.0	3.3		
250	0.9	3.0	1.1	3.7		
300	1.4	4.6	1.23	4.1		
400	2.2	7.3	1.3	4.3		
500	3.6	11.9	1.4	4.6		
1,000	7.9	26	1.75	5.8		
2,000	12.7	42	2.5	8.3		
3,000	14.7	48.5	2.9	9.5		
4,000	16.5	54.8	3.6	12.0	1.08	3.6
6,000	17.0	56.6	4.1	13.7	1.2	4.0
10,000	16.5	55.0	4.6	15.3		
^{137}Cs	6.5	21.6	1.9	6.2	0.64	2.1
^{60}Co	12	40	2.45	8.1	0.82	2.7
^{226}Ra	16.6	55	2.7	9.2	0.88	2.9

Source: Reprinted with permission of the National Council on Radiation Protection and Measurements, Washington, DC.

[a]Approximate values obtained at high attenuation for the indicated peak voltage values under broad beam conditions; with low attenuation these values are significantly less.

the case of loaded concretes, however, the addition of magnetite, steel, lead ferrophosphorus, and the like, produces a considerably improved attenuation, particularly when these additives are uniformly distributed in the mix.

Variations in the chemical compositions of concrete can be critically important in the case of neutron shielding. Such variations in composition have a more pronounced effect on the gamma radiation produced secondarily to neutron interaction than on the transmission of neutrons through the shield. Also, when dealing with photoneutron sources, gamma shielding may prove to be more important from a protection standpoint than neutron shielding.

In any case, before any shielding material is installed, special attention must be given to such details as overlapping joints, eliminating voids or nonhomogeneities in the material, need for structural support for non-load-bearing material such as lead, need to ensure proper attenuation through notoriously leaky areas in the shield (e.g., glass windows, joints, seams, pipes, conduits, service boxes, and doors), need to be certain that the correct shielding materials are being used for the type of radiation in question, and need for continuous maintenance of the shielding structure, to prevent deterioration.

Table 38.12. Selected Gamma Ray Sources

| Radioisotope | Atomic Number | Half-Life | Gamma Energy (MeV) | Half-Value Layer[a] | | | Tenth-Value Layer[a] | | | Specific Gamma Ray Constant[b] |
				Concrete (in.)	Steel (in.)	Lead (cm)	Concrete (in.)	Steel (in.)	Lead (cm)	[(R/cm²)/(mCi-hr)]
^{137}Cs	55	27 years	0.66	1.9	0.64	0.65	6.2	2.1	2.1	3.2
^{60}Co	27	5.24 years	1.17, 1.33	2.6	0.82	1.20	8.2	2.7	4.0	13.0
^{198}Au	79	2.7 days	0.41	1.6	—	0.33	5.3	—	1.1	2.32
^{192}Ir	77	74 days	0.13–1.06	1.7	0.50	0.60	5.8	1.7	2.0	5.0[c]
^{226}Ra	88	1622 years	0.047–2.4	2.7	0.88	1.66	9.2	2.9	5.5	8.25[d]

Source: Reprinted with permission of National Council on Radiation Protection and Measurements Washington, DC.

[a]Approximate values obtained with large attenuation.

[b]These values assume that gamma absorption in the source is negligible. Value in roentgens per millicurie-hour at 1 cm can be converted to roentgens per curie-hour at 1 m by multiplying the number in this column by 0.10.

[c]This value is uncertain.

[d]This value assumes that the source is sealed within a 0.5-mm thick platinum capsule, with units of roentgens per milligram-hour at 1 cm.

For a more detailed treatment of shielding design, the reader is referred to References 49 to 52.

11.3.3 Protective Equipment

Every effort should be made to control the potential radiation exposure by engineering means rather than through the use of personal protective equipment. Protective equipment does have a place in the program, however. If respirators must be used, for example (usually for nonroutine, intermittent operations), it is important to ensure that such respirators meet the approval of the National institute for Occupational Safety and Health. Special training is necessary for the proper use of protective equipment (e.g., the proper fit and maintenance, removal procedures that avoid recontamination, laundering, and disposal). Therefore such training responsibilities should be placed in the hands of a qualified radiation safety officer.

11.3.4 Radioactive Wastes

Radioactive wastes have half-lives ranging from minutes to thousands of years, complicating the procedure of delay and decay in achieving reduced radioactivity levels. Storage of radioactive solidified waste can be accomplished in excavated salt formations underground where heat can be generated and dissipated without producing seismic activity (53). For gaseous waste, the usual procedure is to delay (store) the material to permit decay. Filters have been used to collect radioactive particles that have been formed when a gaseous parent nuclide decays to a particulate radioactive daughter or becomes attached to other particles. Low-temperature adsorption may be useful in providing a delay-decay mechanism for short-lived noble gases (54). Special attention must be paid to leaks around filters, particularly with materials such as ^{234}Pu; more than one filter bank may be necessary. A major objective of plant design and operation is to process and recycle all waste streams in a manner serving to minimize both volume and level of activity. Continuous monitoring of the effluent is necessary to be certain that permissible levels are not exceeded. In any case, before any procedures are adopted with respect to the handling and disposal of radioactive wastes contact must be made with all appropriate governmental authorities to ascertain the nature and extent of all federal, state, and local requirements for the specific waste materials.

Incineration may not be used as a means for disposing of radioactive materials unless the case is specifically approved by the NRC, and in many cases, approved by local government authorities.

11.4 Surveys and Monitoring

11.4.1 Surveys

Periodic surveys should be conducted to evaluate any potential hazards associated with the production, use, release, disposal, or presence of radioactive materials and radiation-producing devices. All radioactive sources and devices producing ionizing radiation should be periodically surveyed to determine their whereabouts, condition, how they are being handled, and the possible levels of radiation associated with their use. Areas should then be designated in accordance with NRC requirements.

Radiation areas should be posted with the standard radiation symbol and the words "CAUTION, Radiation Area" or "CAUTION, High Radiation Area," as appropriate. High radiation areas require the establishment of special cautionary operating procedures, including the use of interlocks and alarms.

Wherever licensed radioactive material (exclusive of natural uranium and thorium) is used or stored in an amount exceeding the quantity specified by the NRC, the area must be posted with a sign bearing the radiation caution symbol and the words "CAUTION, Radioactive Material(s)." The use or storage of natural uranium or thorium in amounts exceeding NRC requirements requires the identical posting format. The outside surfaces of containers of radioactive materials must be similarly labeled.

The periodic survey of work environments should include an assessment of gross contamination levels. For this purpose, contamination working counts should be established. One example of such limits is the following:

Contamination	Limit per 100 cm^2 Area
Removable	
Alpha emitters other than uranium and thorium	100 c/m
Uranium and thorium	1000 c/m
Beta emitters other than ^{90}Sf and ^{129}I	1000 c/m
^{90}Sr and ^{129}I	250 c/m
Nonremovable	
Alpha emitters other than uranium and thorium	1 c/m per cm^2
Uranium and thorium	10 c/m per cm^2
Beta emitters other than ^{90}Sr and ^{129}I	0.2 mR/hr
^{90}Sr and ^{129}I	0.5 mR/hr

11.4.2 Monitoring

Various forms of monitoring and surveillance are required to ensure the continuing effectiveness of the radiation protection program. Some general principles and practices include the following:

1. Monitoring instruments should have sufficient sensitivity, precision, accuracy, response time, and dynamic range to accommodate the type of radiation being generated and its operational characteristics.
2. Radiation protection personnel must establish mechanisms for anticipating changes in the radiation environment. This means being made aware of the introduction of new radiation sources or changes in the operating modes of

existing radiation-producing devices. One cannot rely solely on the readout from monitoring devices to signal the advent of a changed environmental condition.

3. All monitors should be designed for a fail–safe response. This means that when instrument components fail, they fail in a mode that alerts people to an unsatisfactory condition (e.g., a visual and audible signal or alarm is used to signify a component failure).

4. The monitoring level on the instruments should be set high enough to avoid spurious signals but low enough to ensure the safety of personnel. Shielding or adjustments to the operating characteristics of monitoring instruments should enable the user of the instruments to eliminate interferences from external sources (e.g., other sources of ionizing, radio-frequency, microwave, and ultraviolet radiation).

5. Periodic calibration and maintenance of all monitoring instruments are absolutely essential to the program.

6. Periodic judgments by a qualified radiation protection specialist will be needed on such matters as the following: (*a*) the circumstances under which personnel should wear personal monitoring devices such as film badges, personal dosimeters, nuclear accident dosimeters, and thermoluminescent dosimeters; (*b*) frequency of environmental surveys; (*c*) frequency of monitoring of internally deposited radionuclides; (*d*) specific analytical measures to be taken (e.g., bioassay), and (*e*) periodic determination of whether the area, sampling points, or conditions being monitored are the appropriate ones to monitor, or whether program changes are necessary.

11.5 Emergency Procedures

A stepwise emergency procedure should be established and posted in each radiation area.

Because conditions change over time, the emergency procedure must be reviewed and updated periodically. The essential elements of an emergency procedure include simple, direct, readily understood stepwise instructions on the course of action to be taken. For example:

1. "In the event of fire or a radioactive spill, leave the room immediately. Close the door behind you."
2. "Call extension ____on nearest phone and report conditions.
3. "Call Radiation Safety Officer on extension ____."

Prior arrangements should be established with local police and fire departments, hospitals, in-house and outside emergency squads, and the medical department if one is readily available. Evacuation routes and assembly points must be designated. Periodic mock drills should help ensure the continuing effectiveness of emergency procedures.

The preparation of a formal written Emergency Response Plan and tangible provisions for its implementation is now standard practice in industry.

11.6 Medical Surveillance

A preplacement examination for radiation workers is recommended, with emphasis on medical history, complete blood analysis, previous radiation exposure history, and eye examination, particularly if the employee has had previous exposure to neutrons or plans to work with neutrons in the future. The maintenance of complete medical records on each radiation worker is essential.

11.7 Records

The Nuclear Regulatory Commission requires each licensee to keep records on the radiation exposure of all individuals who are required to be monitored and on the disposal of radioactive wastes.

The following records must be kept until the NRC authorizes disposition:

1. Surveys used to determine compliance with exposure limits
2. Exposure of all individuals, including visitors
3. Bioassay data
4. Surveys used to evaluate the release of radioactive effluents to the environment (airborne and waterborne)
5. Surface contamination and removal
6. Reviews of new designs and processes

Appropriate records must be kept to control the use of radiation sources and to facilitate the investigation of accidents and incidents. Because the medical condition that is closely related to exposure to radiation may make its appearance a long time after significant exposures, it is generally a legal requirement to keep records well past the date of the employee's termination of employment.

Records on material receipts, incidents and emergencies, training of personnel, contamination levels, and instruments and their calibration are also required as part of an effective radiation protection program.

11.8 Notification of Incidents

Immediate notification of the NRC Regional Office is required in the event of the occurrence of any of the following situations:

1. Any individual receives a whole body radiation exposure of 25 rems (250 mSv or more), and exposure to the skin of the whole body of 150 rems (1500 mSv) or more, or an exposure to the feet, ankles, hands, or forearms of 375 rems (3750 mSv).

2. The release of radioactive materials in concentrations that when averaged over a 24-hr period would exceed 500 times NRC limits.
3. A loss of workdays has occurred in the affected facility.
4. Damage to property in excess of the NRC dollar limit has occurred.

Procedures must be established to ensure shipment or transportation of all radioactive materials in accordance with the rules and regulations of the U.S. Department of Transportation, the Coast Guard, the Federal Aviation Agency, the Nuclear Regulatory Commission, and the International Atomic Energy Agency.

Appendix A

NCRP PUBLICATIONS

NCRP publications are distributed by the NCRP Publications' office. Information on prices and how to order may be obtained by directing an inquiry to:

NCRP Publications

7910 Woodmont Ave., Suite 800

Bethesda, MD 20814

The currently available publications are listed below:

Proceedings of the Annual Meeting

No.	Title

1 *Perceptions of Risk*, Proceedings of the Fifteenth Annual Meeting, held on March 14–15, 1979 (including Taylor Lecture No. 3) (1080)
2 *Quantitative Risk in Standards Setting*, Proceedings of the Sixteenth Annual Meeting, held on April 2–3, 1980 (including Taylor Lecture No. 4) (1981)
3 *Critical Issues in Setting Radiation Dose Limits*, Proceedings of the Seventeenth Annual Meeting, held on April 8–9, 1981 (including Taylor Lecture No. 5) (1982)
4 *Radiation Protection and New Medical Diagnostic Procedures*, Proceedings of the Eighteenth Annual Meeting, held on April 6–7, 1982 (including Taylor Lecture No. 6) (1983)
5 *Environmental Radioactivity*, Proceedings of the Nineteenth Annual Meeting, held on April 6–7, 1983 (including Taylor Lecture No. 7) (1984)
6 *Some Issues Important in Developing Basic Radiation Protection Recommendations*, Proceedings of the Twentieth Annual Meeting, held on April 4–5, 1984 (including Taylor Lecture No. 8) (1985)

7 *Radioactive Waste*, Proceedings of the Twenty-first Annual Meeting, held on April 3–4, 1985 (Including Taylor Lecture No. 9) (1986)
8 *Nonionizing Electromagnetic Radiation and Ultrasound*, Proceedings of the Twenty-second Annual Meeting, held on April 2–3, 1986 (including Taylor Lecture No. 10) (1988)
9 *New Dosimetry at Hiroshima and Nagasaki and Its Implications for Risk Estimates*, Proceedings of the Twenty-Third Annual Meeting, held on April 5–6, 1987 (including Taylor Lecture No. 11) (1988)
10 *Radon*, Proceedings of the Twenty-fourth Annual Meeting, held on March 30–31, 1988 (including Taylor Lecture No. 12) (1989)
11 *Radiation Protection Today—The NCRP at Sixty Years*, Proceedings of the Twenty-fifth Annual Meeting, held on April 5–6, 1989 (including Lecture No. 13) (1989)

Symposium Proceedings

The Control of Exposure of the Public to Ionizing Radiation in the Event of Accident or Attack, Proceedings of a Symposium held April 27–29, 1981 (1982)

Lauriston S. Taylor Lectures

No.	Title and Author

1 *The Squares of the Natural Numbers in Radiation Protection* by Herbert M. Parker (1977)
2 *Why be Quantitative About Radiation Risk Estimates?* by Sir Edward Pochin (1978)
3 *Radiation Protection—Concepts and Trade Offs* by Hymer L. Friedell (1979) [available also in *Perceptions of Risk*, see above]
4 From "*Quantity of Radiation*" and "*Dose*" to "*Exposure*" and "*Absorbed Dose*"— *An Historical Review* by Harold O. Wyckoff (1980) [available also in *Quantitative Risks in Standards Setting*, see above]
5 *How Well Can We Assess Genetic Risk? Not Very* by James F. Crow (1981) [available also in *Critical Issues in Setting Radiation Dose Limits*, see above]
6 *Ethics, Trade-offs and Medical Radiation* by Eugene L. Saenger (1982) [available also in *Radiation Protection and New Medical Diagnostic Approaches*, see above]
7 *The Human Environment—Past, Present and Future* by Merril Eisenbud (1983) [available also in *Environmental Radioactivity*, see above]
8 *Limitation and Assessment in Radiation Protection* by Harald H. Rossi (1984) [available also in *Some Issues Important in Developing Basic Radiation Protection Recommendations*, see above]
9 *Truth (and Beauty) in Radiation Measurement* by John H. Harley (1985) [available also in *Radioactive Waste*, see above]
10 *Nonionizing Radiation Bioeffects: Cellular Properties and Interactions* by Her-

man P. Schwan (1986) [available also in *Nonionizing Electromagnetic Radiations and Ultrasound*, see above]

11 *How to be Quantitative about Radiation Risk Estimates* by Seymour Jablon (1987) [available also in *Dosimetry at Hiroshima and Nagasaki and its Implications for Risk Estimates*, see above]

12 *How Safe is Safe Enough?* by Bo Lindell (1988) [available also in *Radon*, see above]

13 *Radiobiology and Radiation Protection: The Past Century and Prospects for the Future* by Arthur C. Upton (1989) [available also in *Radiation Protection Today—The NCRP at Sixty Years*, see above]

NCRP Commentaries

No.	Title

1 *Krypton-85 in the Atmosphere—With Specific Reference to the Public Health Significance of the Proposed Controlled Release at Three Mile Island* (1980)

2 *Preliminary Evaluation of Criteria for the Disposal of Transuranic Contaminated Wate* (1982)

3 *Screening Techniques for Determining Compliance with Environmental Standards* (1986, rev. 1989)

4 *Guidelines for the Release of Waste Water from Nuclear Facilities with Special Reference to the Public Health Significance of the Proposed Release of Treated Waste Waters at Three Mile Island* (1987)

5 *Living Without Landfills* (1989)

NCRP Reports

No.	Title

8 *Control and Removal of Radioactive Contamination in Laboratories* (1951)

22 *Maximum Permissible Body Burdens and Maximum Permissible Concentrations of Radionuclides in Air and Water for Occupational Exposure* (1959) [includes Addendum 1 issued in August 1963]

23 *Measurement of Neutron Flux and Spectra for Physical and Biological Applications* (1960)

25 *Measurement of Absorbed Dose of Neutrons and Mixtures of Neutrons and Gamma Rays* (1961)

27 *Stopping Powers for Use with Cavity Chambers* (1961)

30 *Safe Handling of Radioactive Materials* (1964)

32 *Radiation Protection in Educational Institutions* (1966)

35 *Dental X-Ray Protection* (1970)

36 *Radiation Protection in Veterinary Medicine* (1970)

73 *Protection in Nuclear Medicine and Ultrasound Diagnostic Procedures in Children* (1983)
74 *Biological Effects of Ultrasound: Mechanisms and Clinical Implications* (1093)
75 *Iodine-129: Evaluation of Releases from Nuclear Power Generation* (1983)
76 *Radiological Assessment: Predicting the Transport, Bioaccumulation, and Uptake by Man of Radionuclides Released to the Environment* (1984)
77 *Exposures from the Uranium Series with Emphasis on Radon and its Daughters* (1984)
78 *Evaluation of Occupational and Environmental Exposures to Radon and Radon Daughters in the United States* (1984)
79 *Neutron Contamination from Medical Electron Accelerators* (1984)
80 *Induction of Thyroid Cancer by Ionizing Radiation* (1985)
81 *Carbon-14 in the Environment* (1985)
82 *SI Units in Radiation Protection and Measurements* (1985)
83 *The Experimental Basis for Absorbed-Dose Calculations in Medical uses of Radionuclides* (1985)
84 *General Concepts for the Dosimetry of Internally Deposited Radionuclides* (1985)
85 *Mammography—A User's Guide* (1986)
86 *Biological Effects and Exposure Criteria for Radiofrequency Electromagnetic Fields* (1986)
87 *Use of Bioassay Procedures for Assessment of Internal Radionuclide Deposition* (1987)
88 *Radiation Alarms and Access Control Systems* (1987)
89 *Genetic Effects of Internally Deposited Radionuclides* (1987)
90 *Neptunium: Radiation Protection Guidelines* (1987)
91 *Recommendations on Limits for Exposure to Ionizing Radiation* (1987)
92 *Public Radiation Exposure from Nuclear Power Generation in the United States* (1987)
93 *Ionizing Radiation Exposure of the Population of the United States* (1987)
94 *Exposure of the Population in the United States and Canada from Natural Background Radiation* (1987)
95 *Radiation Exposure of the U.S. Population from Consumer Products and Miscellaneous Sources* (1987)
96 *Comparative Carcinogenesis of Ionizing Radiation and Chemicals* (1989)
97 *Measurement of Radon and Radon Daughters in Air* (1988)
98 *Guidance on Radiation Received in Space Activities* (1989)
99 *Quality Assurance for Diagnostic Imaging Equipment* (1988)
100 *Exposure of the U.S. Population From Diagnostic Medical Radiation* (1989)
101 *Exposure of the U.S. Population From Occupational Radiation* (1989)
102 *Medical X-Ray, Electron Beam and Gamma-Ray Protection For Energies Up To 50 MeV (Equipment Design, Performance and Use)* (1989)
103 *Control of Radon in Houses* (1989)
105 *Radiation Protection for Medical and Allied Health Personnel* (1989)
106 *Limits of Exposure to "Hot Particles" on the Skin* (1989)

Binders for NCRP Reports are available. Two sizes make it possible to collect into small binders the "old series" of reports (NCRP Reports Nos. 8-30) and into large binders the more recent publications (NCRP Reports Nos. 32-106). Each binder accommodates from five to seven reports. The binders carry the identification "NCRP Reports" and come with label holders that permit the user to attach labels showing the reports contained in each binder.

The following bound sets of NCRP Reports are also available:

Volume I. NCRP Reports Nos. 8, 22
Volume II. NCRP Reports Nos. 23, 25, 27, 30
Volume III. NCRP Reports Nos. 32, 35, 36, 37
Volume IV. NCRP Reports Nos. 38, 40, 41
Volume V. NCRP Reports Nos. 42, 44, 46
Volume VI. NCRP Reports Nos. 47, 29, 50, 51
Volume VII. NCRP Reports Nos. 52, 53, 54, 55, 57
Volume VIII. NCRP Reports No. 58
Volume IX. NCRP Reports Nos. 59, 60, 61, 62, 63
Volume X. NCRP Reports Nos. 64, 65, 66, 67
Volume XI. NCRP Reports Nos. 68, 69, 70, 71, 72
Volume XII. NCRP Reports Nos. 73, 74, 75, 76
Volume XIII. NCRP Reports Nos. 77, 78, 79, 80
Volume XIV. NCRP Reports Nos. 81, 82, 83, 84, 85
Volume XV. NCRP Reports Nos. 86, 87, 88, 89
Volume XVI. NCRP Reports Nos. 90, 91, 92, 93
Volume XVII. NCRP Reports Nos. 94, 95, 96, 97

(Titles of the individual reports contained in each volume are given above). The following NCRP Reports are now superseded and/or out of print:

No.	Title

1 *X-Ray Protection* (1931) [superseded by NCRP Report No. 3]
2 *Radium Protection* (1934) [superseded by NCRP Report No. 4]
3 *X-Ray Protection* (1936) [superseded by NCRP Report No. 6]
4 *Radium Protection* (1938) [superseded by NCRP Report No. 13]
5 *Safe Handling of Radioactive Luminous Compounds* (1941) [out of print]
6 *Medical X-Ray Protection Up to Two Million Volts* (1949)
7 *Safe Handling of Radioactive Isotopes* (1949). [superseded by NCRP Report No. 30]
9 *Recommendations for Waste Disposal of Phosphorus-32 and Iodine-131 for Medical Users* (1951) [out of print]

56 *Radiation Exposure from Consumer Products and Miscellaneous Sources* (1977) [superseded by NCRP Report No. 95]
58 *A Handbook on Radioactivity Measurement Procedures* [superseded by NCRP Report No. 58, 2nd ed.]

Other Documents

The following documents of the NCRP were published outside of the NCRP Reports and Commentaries series:

"Blood Counts, Statement of the National Committee on Radiation Protection," *Radiology*, **63**, 428 (1954)
"Statements on Maximum Permissible Dose from Television Receivers and Maximum Permissible Dose to the Skin of the Whole Body," *Am. J. Roentgenol., Radium Ther. Nucl. Med.*, **84**, 152 (1960) and *Radiology*, **75**, 122 (1960)
Dose Effect Modifying Factors in Radiation Protection, Report of Subcommittee M-4 (Relative Biological Effectiveness) of the National Council on Radiation Protection and Measurements, Report BNL 50073 (T-471) (1967) Brookhaven National Laboratory (National Technical Information Service, Springfield, Virginia)
X-Ray Protection Standards for Home Television Receivers, Interim Statement of the National Council on Radiation Protection and Measurements (National Council on Radiation Protection and Measurements, Washington, 1968)
Specification of Units of Natural Uranium and Natural Thorium (National Council on Radiation Protection and Measurements, Washington, 1973)
NCRP Statement on Dose Limit for Neutrons (National Council on Radiation Protection and Measurements, Washington, 1980)
Control of Air Emissions of Radionuclides (National Council on Radiation Protection and Measurements, Bethesda, Maryland, 1984)

Copies of the statements published in journals may be consulted in libraries. A limited number of copies of the remaining documents listed above are available for distribution by NCRP Publications.

REFERENCES

1. *Radiological Health Handbook*, rev. ed., Public Health Service, Washington, DC, January 1970.
2. E. C. Barnes, "Ionizing Radiation," in *The Industrial Environment—Its Evaluation and Control*, U.S. Department of Health Education and Welfare, 1973.
3. "The Effects on Populations of Exposure to Low Levels of Ionizing Radiation," National Academy of Sciences-National Research Council, Washington, DC, November 1972, p. 12.
4. M. M. Weiss, "X-Radiation from Positive Ion Beams Incident on Insulator Targets," *Health Phys.*, **15**, 4, 372 (October 1968).

5. M. K. Robertson and M. W. Randle, "Hazards from the Industrial Use of Radioactive Static Eliminators," *Health Phys.*, **26**, 245 (1974).

6. K. Griggs, "Toxic Metal Fumes from Mantle Type Camp Lanterns," *Science* **181**, 842 (1973).

7. K. Griggs, in Reference 3, p. 86.

8. "Plutonium—Health Implications for Man," *Proceedings of the Second Los Alamos Life Sciences Symposium*, Los Alamos, N.M., May 22–24, 1974; *Health Phys.*, **29**(4) October 1975).

9. See Reference 3, p. 87.

10. W. D. Norwood, *Health Protection of Radiation Workers*, Thomas, Springfield, IL, 1975, p. 72.

11. "Radiation Quantities and Units," Report No. 19, International Commission on Radiological Units and Measurements, Washington, DC, 1971.

12. "Conceptual Basis for the Determination of Dose Equivalent," Report No. 25, International Commission on Radiation, Units and Measurements, Washington, DC.

13. See Reference 3, p. 89.

14. W. H. Langham, Ed., "Radiobiological Factors in Manned Space Flight," National Academy of Sciences–National Research Council Publication No. 1487, National Academy of Sciences, Washington, DC, 1967.

15. United Nations Scientific Committee on the Effects of Atomic Radiation, Report General Assembly, Official Records, 17th Session, Supplement No. 16 (A5216), United Nations, New York, 1962.

16. T. G. Baker, "Radiosensitivity of Mammalian Oocytes with Particular Reference to the Human Female," *Am. J. Obstet. Gynecol.*, **110**, 746–761 (1971).

17. "Biological Effects on Atomic Radiation," National Academy of Sciences–National Research Council, National Academy of Sciences, Washington, DC, 1956.

18. Federal Radiation Council, "Background Material for the Development of Radiation Protection Standards," staff reports of the Federal Radiation Council, 1960, 1962.

19. "Natural Background Radiation in the United States," Recommendations of the National Council on Radiation Protection and Measurements, Washington, DC, issued November 15, 1975.

20. See Reference 3, p. 50.

21. M. Eisenbud, *Environmental Radioactivity*, McGraw-Hill, New York, 1963, p. 36.

22. W. H. Langham, J. N. P. Lawrence, J. McClelland, and L. H. Hempelmann, "The Los Alamos Scientific Laboratory's Experience with Plutonium in Man," *Health Phys.*, **8**, 753 (1962).

23. I. C. Nelson, K. R. Heid, P. A. Fugua, and T. D. Mahony, "Plutonium in Autopsy Tissue Samples," *Health Phys.*, **22**, 925 (1972).

24. "The Metabolism of Compounds of Plutonium and Other Actinides," International Commission on Radiological Protection, Pergamon Press, Oxford, England, 1972.

25. W. H. Langham, S. H. Bassett, P. S. Harris, and R. E. Carter, "Distribution and Excretion of Plutonium Administered to Man," Los Alamos Report No. 1151, Los Alamos Scientific Laboratory, Los Alamos, NM, 1950.

26. L. H. Hempelmann, W. H. Langham, C. R. Richmond, and G. L. Voelz, "Manhattan Project Plutonium Workers: A Twenty-Seven Year Follow-Up Study of Selected Cases," *Health Phys.*, **25**, 461 (1973).

27. L. H. Hempelmann, C. R. Richmond, and G. L. Voelz, "A Twenty-Seven Year Study of Selected Los Alamos Plutonium Workers," Report No. LA-5148-MS, Los Alamos Scientific Laboratory, Los Alamos, NM, 1973.

28. S. A. Beach, G. W. Dolphin, K. P. Dolphin, K. P. Duncan, and H. J. Dunster, "A Basis for Routine Urine Sampling of Workers Exposed to Plutonium-239" *Health Phys.*, **12**, 1671 (1966).

29. H. F. Schulte, "Plutonium: Assessment of the Occupational Environment," *Health Phys.*, **29**, 613–618 (1975).

30. R. L. Shoup, T. W. Robinson, and F. R. O'Donnell, "Generic Environmental Statement on the Routine Use of Plutonium Powered Cardiac Pacemakers," Annual progress report, period ending June 30, 1976, No. ORNL-5171, Health Physics Division, Oak Ridge National Laboratory, Oak Ridge, TN, 1976, p. 41.

31. Data obtained from Eberline Instrument Corporation, P.O. Box 2108, Santa Fe, NM.

32. E. H. Dolecek and R. A. Wynveen, "Evaluation of Employing Lithium Fluoride-Teflon Dosimeters in a Personnel Monitoring Program," paper presented at 20th Annual Meeting of the Health Physics Society; *Health Phys.*, **29**(6), 906 (December 1975).

33. W. L. Beck, R. J. Cloutier, and E. E. Watson, "Personnel Monitoring with Film and Thermoluminescent Dosimeters for High Exposures," *Health Phys.*, **25**, 425 (1973).

34. "Radiation Protection Instrumentation and Its Application," Report No. 20, International Commission on Radiological Protection, Washington, DC, 1971.

35. K. Becker and J. S. Jun, "Transfer Dosimeters for Fast Neutron Sources," *Health Phys.*, **29**(6), 915 (December 1975).

36. R. A. Oswald, L. H. Lanzl, and M. Rozenfeld, "Use of Fission Track Detectors in a Tumor-Mouse Phantom Irradiated with ^{252}Cf Neutrons," *Health Phys.*, **29**(6), 916 (December 1975).

37. W. G. Cross and H. Ing, "Directional Dependence of Fast Neutron Fission Track Dosimeters," *Health Phys.*, **29**(6), 907 (December 1975).

38. W. R. Busch, "Assessing and Controlling the Hazard from Tritiated Water," AECL-4150, 1972.

39. R. V. Osborne and G. Cowper, "The Detection of Tritium in Air with Ionization Chambers," AECL-2604, 1966.

40. "Measurement of Low Level Radioactivity," Report No. 22, International Commission on Radiological Units and Measurements, Washington, DC, 1972.

41. American National Standards Institute, "Radiation Protection Instrumentation Test and Calibration," Standard N 323, (N13/N42), ANSI, New York, 1976.

42. See Reference 3, p. 18.

43. Maximum Permissible Body Burdens and Maximum Permissible Concentrations of Radionuclides in Air and in Water for Occupational Exposure, 1959 (includes Addendum issued in August 1963).

44. "Standards for Protection Against Radiation," Part 20, *Fed. Reg.* **25**, 10914 (November 17, 1960). Nomenclature changes appear *Fed. Reg.* **40**, 8783 (March 3, 1975).

45. *Instruction Concerning Prenatal Radiation Exposure*, USNRC Regulatory Guide 8.13, Nuclear Regulatory Commission, Washington, DC, March, 1975.

46. Committee on Industrial Ventilation, American Conference of Governmental Industrial Hygienists, *Industrial Ventilation*, 13th ed. ACGH, Lansing, MI, 1974.

47. A. D. Brandt, *Industrial Health Engineering*, Wiley, New York, 1947.

48. W. C. L. Hemeon, *Plant and Process Ventilation*, Industrial Press, New York, 1954.

49. T. D. Jones and F. F. Haywood, "Transmission of Photons Through Common Shielding Media," *Health Phys.*, **28**, 630 (1975).

50. E. D. Trout, J. P. Kelley, and G. L. Herbert, "X-Ray Attenuation in Steel—50 to 300 kVp," *Health Phys.*, **29**, 163 (1975).

51. *Concrete Radiation Shields for Nuclear Power Plants*, USNRC Regulatory Guide 1.69, Nuclear Regulatory Commission, Washington, DC, 1975.

52. O. Bozyap and L. R. Day, "Attenuation of 14 MeV Neutrons in Shields of Concrete and Paraffin Wax," *Health Phys.*, **28**, 101 (1975).

53. R. L. Bradshaw, F. M. Empson, W. C. McClain, and B. L. House, "Results of a Demonstration and Other Studies on the Disposal of High Level Solidified Radioactive Wastes in a Salt Mine," *Health Phys.*, **18**, 63 (1970).

54. "Nuclear Power and the Environment," International Atomic Energy Agency, Vienna, 1972, p. 22.

55. "Recommendations on Limits of Exposure to Ionizing Radiation," Report No. 91, National Council on Radiation Protection and Measurements, Washington, DC, June 1, 1987.

CHAPTER THIRTY-NINE

Nonionizing Radiation

George M. Wilkening, C.I.H.

1 INTRODUCTION

1.1 Current Interest in Nonionizing Radiations

The development and proliferation of electronic devices that either intentionally or inadvertently emit nonionizing radiation have brought about immense interest in the subject. The promise of an increase in the number and use of such devices has concerned many persons who believe that the radiation hazards have not been sufficiently studied. Those who express concern about an inadequate understanding of the biological effects of nonionizing radiations point out that many electronic devices have already found their way into common use (e.g., microwave ovens, radar for pleasure boats, scanning lasers in supermarket checkout counters, near-ultraviolet radiation in fluorescent lighting fixtures, and a variety of high-intensity light sources). Other concerns include the many infrared, ultraviolet, microwave, and laser devices that might produce excessive occupational exposures. Because of the heightened public interest in electromagnetic radiation hazards, Congress enacted the Radiation Control for Health and Safety Act (1). The declared purpose of the Act is to establish a national electronic product radiation control program, including the development and administration of performance standards to control the emission of electronic product radiation. The Act covers both ionizing and nonionizing electromagnetic radiations emitted from any electronic product. This includes X-rays and gamma rays, and particulate, ultraviolet, visible, infrared, millimeter wave, microwave, radio-frequency, and interestingly enough, sonic,

Patty's Industrial Hygiene and Toxicology, Fourth Edition, Volume 1, Part B, Edited by George D. Clayton and Florence E. Clayton.
ISBN 0-471-50196-4 © 1991 John Wiley & Sons, Inc.

infrasonic, and ultrasonic radiation. Since the inception of the Act, the federal government has conducted or funded research into the biological effects of radiation, with special emphasis on low-level effects. Standards have been developed and promulgated for TV set receivers, medical X-rays (amendments to existing standard), cathode ray tubes, microwave ovens, and lasers. Calibration, measurement, and product testing laboratories have been established to ensure proper evaluation of accessible radiation from electronic products, and a compliance program has been developed to obtain manufacturers' adherence to established standards. During the course of standards development efforts, FDA has been required to consult with the Technical Electronic Product Radiation Safety Standards Committee (TEPRSSC), an advisory body established under the Act. Unlike some federal advisory committees, TEPRSSC has the authority to develop and recommend its own standards directly to the Secretary of Health and Human Services.

Other federal agencies that are actively concerned with nonionizing radiation hazards include the Occupational Safety and Health Administration (OSHA), the Environmental Protection Agency (EPA), the American Conference of Governmental Industrial Hygienists (ACGIH), and the National Institute of Environmental Health Sciences (NIEHS). The National Council on Radiation Protection and Measurements (NCRP) has enlarged its scope of interest to include nonionizing radiation; its first effort was Scientific Committee 39 on Microwaves. The product of SC39 was its NCRP Report 67 "Radiofrequency Electromagnetic Fields-Properties, Quantities and Units, Biophysical Interaction and Measurements" published in 1981. Other NCRP reports include Report 74, "Biological Effects of Ultrasound: Mechanisms and Clinical Implications (1983), and Report 86, "Biological Effects and Exposure Criteria for Radiofrequency Electromagnetic Fields" (1986). Attention is now being given to the biological effects of extremely low frequency magnetic fields, such as those associated with power transmission lines. See Section. 7.2.

1.2 Nature of Electromagnetic Radiation

One important aspect of electromagnetic energy is that it may be considered to be wavelike, having characteristics such as frequency, wavelength, and velocity of propagation; or it may be considered to consist of discrete quanta or energy packets (photons) having characteristics similar to those of particles (energy, momentum, etc.). This wave–particle duality is necessary to explain such physical phenomena as the photoelectric effect or diffraction. The photoelectric effect is most readily understood by considering the incident radiation to consist of a flux of discrete photons, each with a characteristic energy E. Diffraction of electromagnetic energy, on the other hand, can be approached by assuming the incident radiation to be wavelike. However, what at first appears to be a contradiction is resolved by Planck's law, which states that the photon energy E is directly proportional to the frequency of radiation v; that is, $E = hv$, where E is expressed in joules, h is Planck's constant equal to 6.625×10^{-34} J, and v is the frequency, expressed in reciprocal seconds (sec^{-1}).

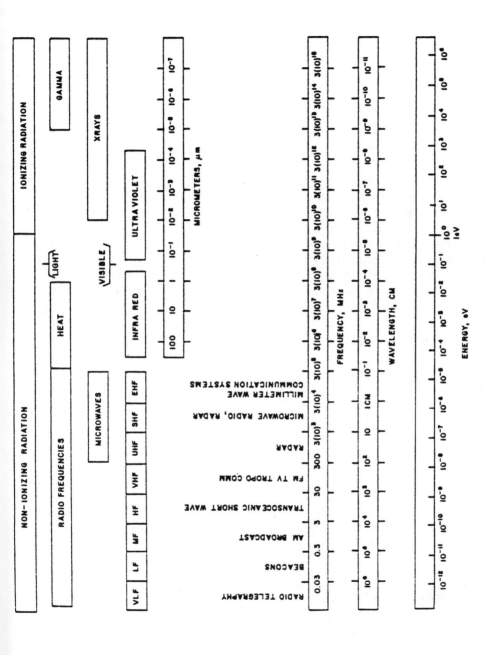

Figure 39.1. The electromagnetic spectrum. After Mumford (95).

In a more commonly used nomenclature, the photon energy is expressed in electron volts (eV), where 1 eV equals 1.602×10^{-19} J.

As just indicated, the photon energies of electromagnetic radiations are proportional to the frequency of radiation and hence inversely proportional to wavelength. The electromagnetic spectrum extends over a broad range of wavelengths, from less than 10^{-12} cm to greater than 10^{10} cm. The shortest wavelengths, hence the highest energies ($>10^2$ eV) are associated with X- and gamma rays, whereas the lower energies ($<10^{-5}$ eV) are associated with radio-frequency and microwave radiation. Ultraviolet, visible, and infrared radiations occupy an intermediate position. For example, radio-frequency wavelengths may range from 3×10^{10} to 3×10^2 μm, infrared from 3×10^2 to about 0.7 μm, visible from approximately 0.7 to 0.4 μm, ultraviolet from approximately 0.4 to 0.1 μm, and gamma- and X-radiation below 0.1 μm. Figure 39.1 is a graphic representation of the entire electromagnetic spectrum.

The nuclear binding energy of protons may be of the order of 10^6 eV or greater; hence these energies are much greater than the binding energy of chemical bonds, which is of the order of 1 to 15 eV. The thermal energy associated with molecules at room temperature is even lower, that is, approximately 0.03 eV. Because the photon energy necessary to ionize atomic oxygen and hydrogen is of the order of 10 to 12 eV, it seems appropriate to adopt a value of approximately 10 eV as a lower limit at which ionization is produced in biological material. The electromagnetic radiations that do not produce ionization in biological systems may be presumed to have photon energies less than 10 to 12 eV and are therefore termed "nonionizing." Nonionizing radiation may be absorbed, causing changes in the vibrational and rotational energies of tissue molecules, thus leading to possible dissociation of the molecules; or more often, the energy is dissipated in the form of fluorescence or heat (2, 3). Exposure to very high levels of nonionizing radiation may inflict considerable thermal injury.

This chapter is arbitrarily divided into sections on ultraviolet (UV), infrared (IR), visible, laser, radio-frequency (RF), microwave, and extremely low frequency (ELF) radiation. Laser radiation is included as a separate section because of the widespread special interest in lasers, even though lasers operate at wavelengths covered in the other sections. For this reason, the reader should consider the laser section to be an extension of material contained in the ultraviolet, visible, and infrared sections. In particular, the exposure limits contained in the laser section are generally applicable to UV, visible, and IR noncoherent sources.

2 ULTRAVIOLET RADIATION

2.1 Physical Characteristics

The ultraviolet wavelength range of interest extends from the vacuum UV (0.16 μm) to the near UV (0.4 μm) as follows:

UV Region	Wavelength Range (μm)	Photon Energy (eV)
Vacuum	0.16	7.7
Far	0.16–0.28	7.7–4.4
Middle	0.28–0.32	4.4–3.9
Near	0.32–0.4	3.9–3.1

The photon energy range for wavelengths between 0.1 and 0.4 μm lies between 12.4 and 3.1 eV.

A different but commonly used UV classification scheme, oriented toward biological effects, is as follows:

UV Region	Wavelength Range (μm)	Biological Implications
UV-A	0.4–0.32	Black light region Solar UV that reaches earth's surface UV-induced fluorescence
UV-B	0.32–0.28	Erythemal (sunburn) region
UV-C	<0.28	Germicidal region

In the wavelength region below 0.16 μm, UV radiation is completely absorbed by air and materials such as quartz; radiation can exist only in a vacuum. At somewhat longer wavelengths (0.16–0.2 μm), UV is poorly transmitted through air or quartz. At wavelengths in the range of 0.2 to 0.32 μm, UV is absorbed by ordinary window glass and by the epithelial layers of the skin and cornea; however, transmission through air and water can occur. The ozone layer above the earth's surface absorbs solar radiation at wavelengths below 0.29 μm. At wavelengths in the range of 0.3 to 0.4 μm, UV is transmitted through air but is transmitted only partially through ordinary glass, quartz, and water.

2.2 Sources of UV Radiation

The sun represents the major source of UV energy at the earth's surface, even though the atmospheric ozone layer filters out all UV wavelengths below 0.29 μm. Significant man-made sources include high- and low-pressure mercury discharge lamps, plasma torches, and welding arcs. More than 85 percent of the radiation emitted by low-pressure mercury vapor discharge lamps is at a wavelength of 0.2537 μm. At lamp pressures that are fractions of an atmosphere, the characteristic mercury lines predominate, whereas at high pressures (up to 100 atm), the lines broaden to produce a radiation continuum. In typical quartz lamps the amount of energy at wavelengths below 0.38 μm may be much greater than the radiated visible energy, depending on the mercury pressure. Other man-made sources include xenon discharge lamps, lasers, and certain fluorescent tubes that emit UV radiation at wavelengths above 0.315 μm, reportedly at an irradiance less than that measured outdoors on a sunny day (3).

The UV radiation in welding operations emits photons with sufficient energy (5 to 9.5 eV) to produce ozone and oxides of nitrogen from the dissociation of oxygen and nitrogen molecules (4).

2.3 Biological Interaction with UV Radiation

The penetration of UV radiation into human tissue is very limited. The deepest penetration may reach just below the epidermis if the wavelength is above 0.3 μm and the subject is essentially nonpigmented. The decrease in transmission through the skin as one proceeds from visible wavelengths to approximately 0.3 μm appears to be linear, indicating the lack of specific absorption bands in this wavelength range (5). Wavelengths between 0.28 and 0.32 μm penetrate appreciably into the corium or dermis; those between 0.32 and 0.38 μm are absorbed primarily in the epidermis, and those below 0.28 μm appear to be absorbed almost completely in the stratum corneum of the epidermis. The lens and tissues in the anterior segments of the eye may be exposed to UV if incident wavelengths are above 0.3 μm; the cornea, on the other hand, absorbs throughout the entire wavelength range of 0.1 to 0.4 μm, although much reduced between 0.32 and 0.4 μm. Because the cornea absorbs most of the energy up to 0.32 μm, radiation in this range generally produces corneal damage before lenticular effects are observed (6, 7).

The relative response of a biological system to irradiation at different wavelengths is called the "biological action spectrum" of that system. In some cases the action spectrum may be the same as the absorption spectrum; this is rarely true with complex biological systems, however. The maximum erythermal response at 0.29 μm in the biological action spectrum corresponds to absorption in tyrosine and other aromatic amino acids; the peak at 0.265 μm corresponds to absorption in pyrimidines and is thought to indicate changes in nucleic acids. The response at wavelengths longer than 0.3 μm shows no maxima or minima.

The advent of supersonic transport has raised speculation about detrimental effects to the atmospheric ozone layer caused by flight at exceptionally high altitudes. The same concern has been expressed about the interaction of halogenated chemicals (propellants) with the ozone layer. One estimate of damage is that a halving of the effective volume of the ozone layer would result in a two- to tenfold increase in the photobiological effects from UV (8).

2.4 Effects on the Skin

The immediate effects of UV radiation on the skin can be considered in terms of erythema, increased pigmentation (i.e., migration of melanin granules and the production of new melanin granules), darkening of pigment, and changes in cellular growth. Depending on the total UV dose, the latent period for the production of erythema may range from 2 to several hours. Usually an immediate erythema appears, followed by a second, more intense erythema after a period of some 2 to 10 hr. The intensity of the erythema and the complicating sequelae (e.g., edema and blistering) is proportional to the UV radiation dose. Irradiation of the skin at wavelengths above 0.3 μm still produces an erythema, but the efficiency is low. Despite this relatively low efficiency, the amount of solar UV at wavelengths above 0.295 μm is sufficient to cause significant erythema in the population.

Photosensitizing agents may have biological action spectra in the UV range. The combined effect of skin contact with these agents and exposure to UV radiation may result in severe irritation and blistering. For example, it is common knowledge that workers who routinely expose themselves to coal tar products while working outdoors experience photosensitization of the skin.

The biological action spectrum for pigmentation changes is very similar to that of erythema. The first effect of UV irradiation appears to be a spreading of existing pigment granules, followed by the production of new pigment granules. Migration of melanin granules from the basal cells to the Malpighian cell layers of the epidermis may also cause a thickening of the horny layers of the skin. The process of increased pigmentation occurs immediately after irradiation. The mechanics of action are thought to be either oxidation of premelanin granules or free radical scavenger action of melanin. The period of increased pigmentation is usually followed by another period in which the skin is more vulnerable to an acute erythema induced at the same wavelength (9).

There appears to be no direct evidence that melanomas are produced by UV radiation; furthermore, melanomas are induced in experimental animals only with great difficulty. On the other hand, irradiation of persons suffering from xeroderma pigmentosum, an inherited disease caused by an insufficiency of DNA repair enzymes, will result in the early production of skin tumors (10).

Ultraviolet phototherapy has been used with infants to correct problems of icterus; however, improper use has produced cases of erythema (11). There is widespread concern about the therapeutic use of UV radiation for herpes, because this virus is already considered to be carcinogenic (12, 13).

In addition to erythema and increased pigmentation, slowing or cessation of cell mitoses in the basal and superficial layers of the epidermis often appears immediately after UV irradiation. An increase in mitoses is delayed for some 24 hr postirradiation (14), after which superfluous cellular material is sloughed off.

2.5 Effects on the Eye

The most common clinical sign of overexposure to UV at wavelengths shorter than 0.3 μm is photokeratitis, appearing some 2 to 24 hr after irradiation, along with acute hyperemia, photophobia, and blepharospasm if exposure is severe. The duration of such difficulties is usually of the order of one to five days. However, there are usually no residual lesions.

2.6 Cancer

Although the premise that skin cancer may be induced by UV radiation is generally accepted in scientific circles, the evidence is largely indirect (i.e., epidemiologic and statistical). The mechanism of UV-induced carcinogenesis is unknown. In spite of these uncertainties, however, certain epidemiologic studies have revealed a strong correlation between skin cancer and terrestrial solar UV-B levels found at given altitudes and ground elevations (15).

The types of skin cancer that are possibly related to UV radiation include basal cell carcinoma, spindle cell carcinoma, and melanoma, the latter generally considered to be a remote possibility. The induction of tumors is believed to occur as a result of irradiation at wavelengths below 0.3 μm (16), with the maximally effective wavelengths being in the 0.26 to 0.27 μm range, corresponding to specific interaction with nucleic acids.

The alleged high incidence of skin cancer in outdoor workers who come into contact with chemicals such as coal tar derivatives, benzopyrene, methylcholanthrene, and other anthracene compounds raises questions about the role played by UV radiation in these cases.

There is no direct evidence that the UV radiation will produce tumors in the cornea or anterior chamber of the eye; however, it is of some interest that melanoma of the eye is much more common in blue-eyed persons. Melanoma of the iris, for example, was found exclusively in blue-eyed persons in one study (17).

Abiotic effects from exposure to ultraviolet radiation occur in the spectral range of 0.24 to 0.31 μm, where most of the incident energy is absorbed by the corneal epithelium. Although the lens is capable of absorbing 99 percent of the energy below a wavelength of 0.35 μm, only a small portion of the radiation reaches the anterior lenticular surface. Photon energies of about 3.5 eV (0.36 μm) may excite the lens of the eye or cause the aqueous or vitreous humor to fluoresce, thus producing a diffuse haziness inside the eye that can interfere with visual acuity or produce eye fatigue. The phenomenon of fluorescence in the ocular media is not of concern from a bioeffects standpoint; the condition is strictly temporary and without detrimental effect.

An important conclusion about the effects of UV radiation on biological systems is that tissue damage appears to depend on the total energy absorbed (dose) rather than on the rate of energy absorption; hence the dose must be controlled to afford protection.

2.7 Threshold

Fairly good agreement exists on the amount of radiant exposure necessary to produce minimal photokeratitis. The threshold has been found to be 4 mJ/cm² for primates and humans when exposed to wavelengths between 0.220 and 0.310 μm. The minimal radiant exposure of 4 mJ/cm² was obtained at the most sensitive wavelength of 0.27 μm (19).

The energy necessary to elicit a barely perceptible reddening* of the skin is approximately 30 mJ/cm². Depending on skin type, the erythema thresholds may vary from 8 mJ/cm² for untanned skin to 50 mJ/cm² for tanned Caucasian skin. For white Caucasian skin with an average degree of tanning, the value for the minimal perceptible erythema is more on the order of 8 to 10 mJ/cm² when irradiation takes place at wavelengths between 0.24 and 0.29 μm. Any threshold number that is used should be identified with its wavelength, for some wavelengths are more effective than others in producing a given biological effect (action spectrum).

2.8 Exposure Criteria

The exposure criteria adopted by the Council on Physical Medicine of the American Medical Association based on erythemal thresholds for 0.2537 μm radiation are as follows: 0.5×10^{-6} W/cm² for exposure periods up to 7 hr (12.6 mJ/cm²); 0.1×10^{-6} W/cm² for exposure periods up to and exceeding 24 hr (8.6 mJ/cm²). Although these criteria are thought to be stringent, they have nevertheless been successfully used for many decades. Several important qualifications must be kept in mind when the AMA guidelines are used: (1) since the guidelines apply only to radiation at 0.2537 μm (germicidal), corrections must be made for the biological effectiveness at wavelengths other than germicidal; and (2) a tenfold safety factor has been applied to the germicidal photokeratitis threshold in deriving the recommended numerical values.

The ACGIH (19) recommended threshold limit values (TLV) for UV irradiation of unprotected skin and eyes for actinic wavelengths between 0.2 and 0.315 μm (200 and 315 nm) are given in Table 39.1. It should be recognized that these TLVs have been derived on the basis of irradiating skin that has not received previous UV radiation. Note in this respect that the minimum TLV is somewhat below the photokeratitis threshold previously mentioned (i.e., 3 mJ/cm² rather than 4 mJ/cm²). Conditioned or tanned individuals can tolerate skin exposure in excess of the TLV without erythemal effect. However, such conditioning may not protect persons against possible skin cancer.

Exposure to wavelengths between 0.32 and 0.4 μm should not exceed the following criterion:

Wavelength, λ	TLV	Exposure Duration
0.32–0.4 μm	1 mW/cm²	$>10^3$ sec
	1 J/cm²	$<10^3$ sec

*The minimal erythema dose (MED) is the smallest radiant exposure that produces a barely perceptible reddening of the skin that disappears after 24 hours. The MED depends on skin color, degree of pigmentation, age, and the irradiated body site.

Table 39.1. Threshold Limit Values for
Exposure to Ultraviolet Radiation (0.2–
0.315 μm)

Wavelength (nm)	TLV (mJ/cm²)
200	100
210	40
220	25
230	16
240	10
250	7.0
254	6.0
260	4.6
270	3.0
280	3.4
290	4.7
300	10
305	50
310	200
315	1000

To determine the effective irradiance of a broad-band source weighted against the peak of the spectral effectiveness curve (270 nm), the following weighting formula should be used:

$$E_{eff} = E_\lambda S_\lambda \Delta_\lambda$$

where E_{eff} = effective irradiance relative to a monochromatic source at 270 nm (0.270 μm)

E_λ = spectral irradiance [W/(cm²)(nm)]

S_λ = relative spectral effectiveness (unitless)

Δ_λ = bandwidth (nm)

Table 39.2 shows relative spectral effectiveness as a function of wavelength.

Permissible exposure time in seconds for exposure to actinic ultraviolet radiation incident upon the unprotected skin or eye may be computed by dividing 0.003 J/cm² by E_{eff}, expressed in watts per square centimeter. The exposure time may also be determined using Table 39.3, which provides exposure times corresponding to effective irradiances in microwatts per square centimeter (W/cm² × 10⁻⁶).

The TLVs do not apply to lasers or to an exposure duration less than 0.1 sec. Also no corrections have been made in these criteria for photosensitization to chemicals, drugs, and cosmetics (19).

Table 39.2. Relative Spectral Effectiveness by Wavelength

Wavelength (nm)	Relative Spectral Effectiveness, S_λ
200	0.03
210	0.075
220	0.12
230	0.19
240	0.30
250	0.43
254	0.5
260	0.65
270	1.0
280	0.88
290	0.64
300	0.30
305	0.06
310	0.015
315	0.003

Table 39.3. Permissible Ultraviolet Exposures

Duration of Exposure per Day	Effective Irradiance, E_{eff} ($\mu W/cm^2$)
8 hr	0.1
4 hr	0.2
2 hr	0.4
1 hr	0.8
30 min	1.7
15 min	3.3
10 min	5
5 min	10
1 min	50
30 sec	100
10 sec	300
1 sec	3,000
0.5 sec	6,000
0.1 sec	30,000

Table 39.4. Sensitivity, Impedance, and Response Time of Junction Detectors

Type (circular)	Sensitivity [$\mu V/(\mu W)cm^2)$]	Impedance (Ω)	Response Time, 1/e (sec)
1 Junction: constantan–Mn	0.005	2	0.1
4 Junction: Cu–constantan	0.025	5	0.5
4 Junction: Bi–Ag	0.05	5	0.5
8 Junction: Bi–Ag	0.10	10	1.0
16 Junction: Bi–Ag	0.20	25	2.0
(linear)	(0.05)	(10)	(0.5)
12 Junction: Bi–Ag	0.05	10	0.5

2.9 Measurement of UV Radiation

One device that is particularly useful for measuring UV radiation is the thermopile. The coatings on the thermopile receiver elements are usually lampblack or gold black, to approximate the properties of a black-body radiator. A special window material transparent to the UV is necessary to surround the thermopile element to minimize the detrimental effects of air currents. Commonly used materials for windows are quartz crystals, calcium and lithium fluoride, sodium chloride, and potassium bromide.

Other UV detection devices include (1) photodiodes, (e.g., silver–gallium arsenide, silver–zinc sulfide, and gold–zinc sulfide: peak sensitivity of these diodes is at wavelengths below 0.36 μm; the peak efficiency or responsivity is of the order of 50 to 70 percent); (2) thermocouples (e.g., Chromel–Alumel); (3) Golay cells; (4) superconducting bolometers, and (5) certain photomultiplier and vacuum photodiodes (20).

Care must be taken to use detection devices having the proper rise time characteristics (some devices respond much too slowly to permit the obtaining of meaningful measurement). Also, when measurements are being made, special attention should be given to the possibility of UV absorption by many materials in the environment (e.g., ozone or mercury vapor), thus adversely affecting the readings. The possibility of photochemical reactions between ultraviolet radiation and a variety of chemicals also exists in the industrial environment.

Table 39.4 lists the sensitivity, impedance, and response time for certain junction detectors. It is possible to obtain a low radiation intensity calibration by exposing these and other detectors to a secondary standard furnished by the National Institute for Standards and Technology.

Other devices that have been used to measure UV radiation include photovoltaic cells, photochemical detectors, photoelectric cells, and photoconductive cells. Photovoltaic and photoconductive cells are usually the same device operated in different modes (in series with an ammeter or across a load and voltmeter). It is common practice to employ selective filters, such as a monochrometer grating in front of the detecting device, to isolate the portion of the UV spectrum that is of interest to the investigator. Certain semiconductors such as copper or selenium oxide deposited on a selected metal develop a potential barrier between the layer and the metal. Light falling on the surface of the cell causes the flow of electrons from the semiconductors to the metal. A sensitive meter placed in such a circuit will record a signal proportional to the intensity of radiation falling on the cell. Some photocells used for UV measurement take advantage of the property of certain metals to exhibit quantitative photoelectric responses to specific bands in the UV spectrum. Thus a photocell may be equipped with metal cathode surfaces that are sensitive to the UV wavelengths of interest.

One of the limitations of photocells is "solarization" or deterioration of the envelope, especially with long usage or following the measurement of high-intensity radiation. Frequent calibration of the cell is necessary under these conditions. The readings obtained with these instruments are valid only when measuring monochromatic radiation or when the relationship between the response of the instrument and the spectral distribution of the source is known. A measurement technique for which Bassi et al. (21) claim some promise involves the use of beryllium oxide and lithium fluoride thermoluminescent crystals that have been preirradiated with gamma rays. The response of the detector, after several calibrated gamma predoses, appears to be a function of the UV exposure received from germicidal lamps.

For environmental measurements that are intended to correlate with a given biological effect, the ideal instrument would be designed in such a way that its spectral response closely approximates that of the biological action spectrum under consideration. Such an instrument is unavailable at this time. Because available photocells and filter combinations do not closely approximate the UV biological action spectrum, it is necessary to calibrate each photocell and meter at a number of wavelengths. Such calibrations are generally made at a great enough distance to ensure that the measuring device is in the "far field" of the source. In performing these calibrations, special care must be taken to control the temperature of the mercury lamps as secondary standards, for the spectral distribution of the radiation from the lamps depends on the pressure of the vaporized mercury.

2.10 Control of Exposure

Because UV radiations are so easily absorbed by a wide variety of materials, appropriate attenuation is accomplished in a straightforward manner. The information given in Section 2.8 (Exposure Criteria) should be used for the specification of shielding requirements. The data of Pitts (18) may be used for ultraviolet lasers because of the narrow band UV source employed in his experiments to determine the thresholds of injury to rabbit eyes. It is important to remember that photosensitization may be induced in certain persons at levels below the suggested exposure criteria.

Personal protection against radiation at wavelengths below 0.32 μm can be accomplished through the use of eyeglasses, goggles, plastic face shields, protective clothing, or sunscreen creams or lotions. The wearing of tinted glasses or goggles is seldom recommended for protection against the UV intensities normally encountered in industry; however, such devices may be used in a supplementary manner to reduce the high visible brightness that often accompanies UV emissions.

3 INFRARED RADIATION

Infrared (IR) wavelengths extend from 0.75 to approximately 10^3 μm. This range can be arbitrarily divided into three subareas known as the near infrared (0.75 to 3 μm), the middle infrared (3 to 30 μm), and the far infrared (30 to 10^3 μm).

Infrared radiation is emitted by a large variety of sources including the sun, heated metals, home electrical appliances, incandescent bulbs, furnaces, welding arcs, lasers, and plasma torches. The energy and wavelength characteristics emitted from IR sources largely depend on the source temperature. For example, at a temperature of 1000°C about 5 percent of the total energy is emitted at wavelengths below 1.5 μm, whereas at 1500 and 2000°C, the respective values are 20 and 40 percent (22).

Infrared radiation exchange between two heated bodies proceeds in accordance with the fourth power of the absolute temperature of each body (Stefan–Boltzmann law). When persons are placed in the vicinity of large IR sources, the transfer of the heat load to the human body can represent a severe environmental stress. The problem of heat stress on the human body resulting from convective and radiative (as exemplified by large IR sources) heat loads is covered in detail in Chapter 21.

Infrared energy is not sufficiently energetic (<1.5 eV) to cause the removal of electrons from orbital shells; therefore IR seldom enters directly into chemical reactions in biological systems and does not usually cause ionization. However, the possibility exists that IR energy at certain wavelengths and intensities may affect the vibrational or rotational properties of certain tissue molecules in such a way as to cause detrimental effects. Even in these cases the consensus is that the effects produced are largely thermal. The body organs at greatest risk from IR radiation are the eye and skin.

Most biological systems are considered to be opaque to wavelengths greater than 1.5 μm; transmission of IR through the ocular media at wavelengths from 1.3 to 1.5 μm is also poor. For skin, the wavelength region of high transmission is between 0.75 and 1.3 μm, with a maximum of 1.1 μm. At a wavelength of 1.1 μm, 20 percent of the energy incident upon the stratum corneum will reach a depth of 5 mm.

3.1 Eye

Thermal damage to the cornea usually results from the energy absorbed in the epithelium rather than in the deep stroma. The iris is especially susceptible to radiations below 1.3 μm. Because the iris can dissipate its heat load only to the surrounding ocular media, it is often regarded as a heat sink serving to mitigate the amount of radiation reaching the lens. It is thought that a radiant exposure of 4.2 J/cm² within a wavelength range of 0.8 to 1.1 μm will produce a minimally perceptible lesion on the iris. To obtain a radiant exposure of 4.2 J/cm^2 at the iris, it would be necessary to irradiate the cornea at a level of 10.8 J/cm^2 (23).

The transmission characteristics of the lens apparently vary with age and nuclear sclerosis. Selective IR absorption bands in the lens exist at wavelengths from 1.4 to 1.6 μm and from 1.8 to 2.0 μm (24).

Although the principal effect of infrared radiation on the eye and skin appears to be acute thermal injury, a controversy has raged throughout most of the twentieth century over the etiology of "glass blower's cataract," specifically, whether IR radiation is a cataractogenic agent. Although Dunn (25, 26) in 1950 was unable to uncover cases of cataract in employees who had been exposed to IR for more than 20 years, the evidence now seems to favor IR as the etiologic agent (27–31).

The formation of a cataract depends on initial heating of the anterior portions of the eye, especially the cornea and iris, followed by heat transfer from the iris to the lens epithelium. The elevation of temperature at the anterior portion of the lens is the primary etiologic factor in glass blower's cataract, according to Goldmann and co-workers (29, 30). In some recent studies of various glass and steel plants, a significant number of employees have shown lenticular changes after some 10 to 15 years of exposure to infrared sources (22). The reported range of irradiance to which employees were exposed was 0.08 to 0.4 W/cm^2; however, no attempt was made to reconstruct exposure time-irradiance patterns for each employee. In practical cases where the eye is exposed to high temperature sources, there is often a significant near-infrared component, as well as some visible wavelength radiation. Special care is necessary to make distinctions between visible and IR emissions when industrial hygiene or epidemiologic studies are made of infrared radiation and its possible relationship to cataract formation.

The retina is at risk from IR radiation only when near-infrared wavelengths are being generated by the source. The dissipation of energy within the retina is accomplished through conduction of heat to adjacent structures, notably the choroid. When the size of the retinal image and the rate of energy deposition are such that heat cannot be conducted away quickly enough to keep the tissue temperature below approximately 45°C, protein denaturation or destruction of tissue (microsteam generation and burn) may occur. More near-infrared energy than visible wavelength energy is required to produce a minimal lesion on the retina (threshold of damage as observed through an ophthalmoscope). For a discussion of maximum permissible exposure levels for infrared radiation, the reader is referred to Section 5 (Lasers).

3.2 Skin

Effects on the skin include vasodilatations of the capillary beds and increased pigmentation. The skin is normally able to dissipate a heat load imposed by IR radiation because of capillary bed dilatation, increased blood circulation, the production of sweat, and ambient air movement.

The perception of warmth by the skin is related to the rate at which the skin temperature is raised. For example, for skin temperatures of 32 to 37°C, the threshold of warmth perception is reached when the rate of skin temperature increase is of the order of 0.001 to 0.002°C per second. Such a perception threshold depends on the size of the skin area irradiated by IR as well as the density of thermoreceptors in that area. The thresholds for perceiving warmth decrease with increasing area of irradiated skin. On the other hand, precooled skin can be rapidly heated without eliciting the sensation of warmth (2).

Unlike the thresholds of warmth sensation, the thresholds for pain appear to depend directly on skin temperature, despite wide individual variation in their values (16). Wertheimer and Ward (32) have recorded a range of skin temperature pain thresholds of 44.1 to 44.9°C, depending on body site. The experiments of Hardy (33) indicate that the threshold of pain is reached when the skin temperature is raised to 45°C (44.5 ± 1.3°C). The same value also appears to be critical for producing a skin burn. Pain thresholds therefore appear to be related to skin temperature, whereas the extent of tissue damage depends on skin temperature and duration of exposure.

3.3 Measurement of Infrared Radiation

An almost unlimited variety of IR instrumentation is available: spectrophotometers, radiation pyrometers, stationary or scanning radiometers, and other equipment. The various detectors usually fall into one of two categories:

1. *Thermal detectors*, where absorbed energy heats the detector, which in turn causes a change in the properties of the detector material. Response is to all IR wavelengths, but the response time is slow. Typical examples are thermopiles, thermocouples, liquid crystal, bolometers, and pyroelectric crystals.

2. *Quantum detectors*, where all incident photons are sufficiently energetic to free a bound electron. Response is to a limited range of IR wavelengths, but response time is fast. Examples of this type of detector are photodiodes, photomultiplier tubes, and semiconductor devices such as gallium arsenide diodes.

4 VISIBLE RADIATION

From the industrial hygiene standpoint the use of visible wavelength radiations generally involves two considerations: (1) proper lighting practices to achieve a pleasant visual environment and the efficient performance of tasks that require optimal visual acuity, and (2) the control of radiation to prevent damage to the human body, usually the eyes and skin. The first subject is covered in considerable detail in Chapter 36; the second is considered in this section.

The visible portion of the electromagnetic spectrum consists of a very narrow band of wavelengths between the ultraviolet and the infrared. The photon energies associated with the visible wavelength range of 0.38 to 0.75 μm lie between 3.1 and 1.65 eV. These energy levels are rather low and innocuous in terms of producing direct biological effects on the skin surfaces of the human body; however, radiation in the 0.38 to 0.75 μm range acts on a body organ that is uniquely and exquisitely designed to transduce the incident photons into intelligible vision by means of complex and incompletely understood photochemical reactions. This exquisite organ is the retina of the eye. The photoreceptors of incident photons are the rods and cones. The differences in function of these two receptor systems are worthy of note. Cones respond to higher levels of ambient illumination than do the rods; hence their level of activity is normally greater during daylight hours. The cones also come into play for the discrimination of fine visual detail. Because the rods are in greater numerical abundance in the peripheral paramacular area of the retina, they provide a wide area peripheral view of the visual field. The rods and cones respond differently to different wavelengths: the spectral luminous efficiency for rods is at a wavelength of 0.510 μm, whereas the peak efficiency of the cones is at 0.555 μm (34). The retina is able to adapt to the wide range of light intensities, but the adaptation time varies with the time-luminous flux. Light adaptation occurs within a short time duration, usually of the order of seconds, whereas dark adaptation may require more than an hour.

Visible light is not hazardous to the eyes under ordinary circumstances, even when high-intensity light sources are encountered. For example, the adaptation of the eye to high-intensity sunlight by such mechanisms as restricted pupil size, light adaptation, partial closing of the eyelids (squinting), blinking, and shading of the eyes by the eyebrows and periorbital socket is usually sufficient to prevent excessive radiation from reaching the retina. On the other hand, certain conditions involving exposure to light may produce significant hazards for humans. For example, if light is pulsed or gated at a frequency near the alpha rhythm (brain function), certain light-sensitive individuals may experience an epileptiform seizure. The effect may be produced when such persons observe a pulsating light pattern such as flickering sunlight coming through a stand of trees as the light is observed from a moving vehicle, or when watching a pulsed light pattern from a television set. In some cases it is the pattern of light generation rather than its brightness that is critical to the production of seizures. Flash blindness is another condition that may result from exposure to high-intensity light sources. This condition is caused by the bleaching of visual pigments that produce an "afterimage" or temporary blind area (scotoma) in the field of vision. The greater the light intensity and time of exposure, the more persistent will be the afterimage. If the original stimulation by a high-intensity light source is sufficient to obliterate focused perception completely, some protracted period of time may be required before normal visual function is restored. The sun may damage the retina when viewed without proper protection; eclipse blindness is a familiar example, where retinal damage is caused by exposing the eye to a combination of excessive visible and IR radiations.

On a lesser scale, ordinary glare caused by improper placement of light sources (fixtures) may produce severe visual discomfort and may represent a safety hazard because of interference with visual performance. A more serious condition is one in which a veiling glare may be temporarily superimposed on the retinal image, obscuring one's ability to discriminate details in the object being viewed. Such a condition exists on a more permanent basis in persons with cataracts.

High-intensity light sources such as lasers, flashbulbs, spotlights, welding arcs, and carbon arcs can produce retinal burns under appropriate viewing conditions. Because of its coherent properties, a laser beam may be focused to an extremely small diameter on the retina. This results in the creation of highly intense beams of light, an obvious hazard to personnel.

Of the factors that have entered into the development of acceptable exposure levels for high-intensity light sources, three seem to be of special importance:

1. The magnification factor of the eye as visible light passes through and is focused by the cornea and lens is of the order of 10^5. This means that the acceptable irradiance (W/cm^2) and the radiant exposure (J/cm^2) at the retina must have much greater numerical values than those present in front of the cornea where the industrial hygienist makes field measurements. Permissible exposure levels are those that enter or impinge upon the eye at the cornea.

2. Acceptable exposure levels for wavelengths in the near infrared and infrared are somewhat less stringent than those for the visible wavelengths, whereas exposure levels for wavelengths in the blue and ultraviolet are more stringent, probably because of the photochemical mechanisms coming into play at the shorter wavelengths.

3. When exposed to a highly intense visible light source, most persons exhibit an "aversion response," that is, a tendency to blink and turn the eyes and head. Presumably the elapsed time before the aversion response occurs gives some indication of the exposure time to be considered in establishing an exposure limit. In the case of the ANSI standard (35), the aversion time has been assigned a value of 0.25 sec. The use of the aversion response concept in establishing safety guidelines for exposure to visible radiations has drawn criticism from some quarters because not all persons exhibit such a response. On the other hand, to establish safety standards strictly on the basis of experiments that irradiate the eyes of sedated animals whose eyelids are immobilized in an open position, to receive radiation pulses in excess of 0.25 sec duration, means that practical conditions and normal physiological response are ignored.

5 LASERS

Lasers are widely used for a variety of purposes such as alignment, welding, trimming, spectrophotometry, interferometry, flash photolysis, fiber optics communications systems, nuclear fusion experimentation, and surgical removal or repair procedures. A review of medical and biological applications may be found in Wolbarsht and in many applied science journals (36).

The word "laser" is the acronym for "light amplification by stimulated emission of radiation." When an atom has been stimulated into an excited state, it may lose its excess energy through spontaneous decay, with the random emission of electromagnetic energy at a specific wavelength. When the atoms are raised to an excited state through an applied electromagnetic field, however, the radiated wave has the same phase, direction, and polarization as the stimulating (pumping) source. This process is known as stimulated emission. Because the population of atoms at a lower or ground state is normally greater than the number existing at a higher energy (excited) state, the energy of most radiations is usually absorbed by the atoms, thus tending to reduce or attenuate the field. However, if the population of atoms can be reversed by stimulating a greater number to energy levels higher than those existing at ground state, the energy radiated by the stimulated emission will exceed the energy absorbed by the atoms; hence a phase coherent amplification of the incident radiation will occur. The net power difference (gain) will be proportional to the number of atoms that undergo a radiative decay to a lower energy level.

The energy stimulation and decay scheme just described is representative of a three-level (ground state, broad energy absorption band, and upper laser level) system, such as the ruby laser. Simply stated, atoms in a three-level system are pumped from the ground state to a broad energy absorption band, whereupon they rapidly decay by a radiation-less transition into an upper laser level. Here they remain for a relatively long residence time (long compared with the time spent within the broad energy absorption band) before they return to the ground level, with an attendant emission of radiation at a specific frequency. The frequency of the radiation is equal to the difference in energy between the metastable upper laser level and the ground state divided by Planck's constant.

Many gas lasers and rare earth lasers (e.g., neodymium–yttrium–aluminum–garnet) operate in a four-level system, where a relatively small number of atoms (compared with a three-level system) are pumped from the ground state to achieve lasing.

Figure 39.2 depicts the energy diagram of a four-step laser system. Atoms are raised from ground state to a broad energy level (absorption band) through absorption of electromagnetic radiation, as in the ruby three-level system, or through electron impact as in the case of most gas lasers systems. Fast nonradiative relaxation occurs before the atoms are brought to level 2, the metastable level, where relaxation time is long and an accumulation of atoms occurs. Level 1, the lower laser level, has fast relaxation time; in fact, the dumping of atoms keeps the atom population low, to ensure population inversion. In passing from level 2 to level 1, radiation is emitted at the characteristic wavelength of the laser. The relative magnitude of the relaxation time at each level and the relative populations of atoms at each level are important parameters for achieving laser action.

Stimulated emission in gaseous systems was first reported in a helium–neon mixture (37). Since the early 1960s, lasing action has been reported at hundreds of wavelengths from the ultraviolet to the far infrared. Helium–neon lasers are typical of gas systems in which stable, single-frequency operation is important. The very popular helium–neon system can be operated in a pulsed or continuous wave mode at wavelengths of 0.6328, 1.15, or 3.39 μm, depending on resonator design. Typical power output for the helium–neon systems is of the order of 1 to 500 mW.

The carbon dioxide (CO_2) laser belongs to a class known as molecular lasers. The energy for the system is derived from transitions between vibrational and

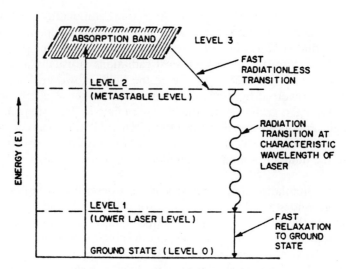

Figure 39.2. Four-level laser system.

rotational energy levels of CO_2 molecules. Such laser systems most frequently operate at wavelengths of 10.6 μm in either the continuous wave, pulsed, or Q-switched mode. A Q-switch is a device for enhancing the storage and quick discharge of energies to produce extremely high power pulses. The peak power output of a CO_2–N_2 system may range from several watts to greater than 10 kW. The CO_2 laser is attractive for terrestrial and extraterrestrial communication systems because of the low absorption window in the atmosphere between 8 and 14 μm. High-powered DF–CO_2 (deuterium fluoride–carbon dioxide) transfer lasers are based on the combustion of carbon monoxide and oxygen, which subsequently thermally dissociates the fluoride before supersonic injection into a laser cavity. Both the DF–CO_2 laser and high powered CO lasers (design goal is to obtain more than 10 kJ output in subnanosecond pulses at an efficiency greater than 1 percent) are being used in laser fusion coupling experiments. The use of heavy ion beams, coupled with high-energy accelerators of the storage ring type, seems to give a competitive edge over laser activation in experiments on inertial confinement fusion. However, major activity continues with the development of high power transfer lasers for use in laser fusion research.

Of major significance from the personal hazard standpoint is the capacity of far infrared lasers to radiate enormous power at wavelengths that are invisible to the human eye.

The argon ion gas systems operate in the blue wavelengths, typically 0.4578, 0.488, and 0.5145 μm, in either the continuous wave or pulsed mode. Power generation is usually less than 25 W.

Of the many ions in which laser action has been produced in solid-state crystalline materials, perhaps neodymium (Nd^{3+}) in garnet or glass and chromium (Cr^{3+}) in aluminum oxide (ruby) are most noteworthy (Table 39.5). Neodymium–glass lasers are already operational in the terawatt peak power range. Garnet or YAG (yttrium–

Table 39.5. Certain Ions That Have Exhibited Lasing Action

Active Ion	Wavelength (μm)
Nd^{3+}	0.9–1.4
Ho^{3+}	2.05
Er^{3+}	1.61
Cr^{3+}	0.69
Tm^{3+}	1.92
U^{3+}	2.5
Pr^{3+}	1.05
Dy^{2+}	2.36
Sm^{2+}	0.70
Tm^{2+}	1.12

aluminum–garnet) is an attractive host for the trivalent neodymium ion because the 1.06 μm laser transition line is sharper than in other host crystals. Frequency doubling to 0.53 μm, using lithium niobate crystals, may produce power approaching that available in the fundamental mode at 1.06 μm. Also, through the use of electrooptic materials such as potassium dihydrogen phosphate (KDP), barium–sodium niobate, or lithium tantalate, "tuning" or frequency scanning over wide ranges may be accomplished (38). Perhaps the most versatile device for obtaining a wide band of frequencies is the dye laser.

Semiconductor lasers are usually moderately low power devices (milliwatts to several watts) having relatively broad beam divergence, a factor that tends to reduce the potential ocular hazard. On the other hand, certain semiconductor lasers may be pumped with multikilovolt electron beams, thus introducing a potential ionizing radiation hazard (39). As a matter of fact, the X-rays from certain laser plasmas are already being considered for use in cancer detection.

Examples of certain ions that have exhibited lasing action appear in Table 39.5. The wavelength range of tunable infrared lasers is given in Table 39.6.

Through the use of carefully selected dyes, it is possible to tune through broad wavelength ranges. For example, if one is interested in producing lasing wavelengths in the vacuum ultraviolet (VUV), the beams from two dye lasers (one fixed at a double-quantum resonance wavelength, the other tuned over its tunable wavelength range) can be combined to obtain the VUV source. Through the use of four dye solutions, Sorokin et al. (40) were able to cover the range from 0.1578 to 0.1957 μm. Table 39.7 lists some additional lasers operating in the ultraviolet, and Table 39.8 gives the operating wavelengths and associated photon energies of certain lasers in common use.

5.1 Biological Effects

Despite the dramatic benefits of technology, it is axiomatic to say that the improper use or design of any apparatus can produce undesirable effects. The laser is no

Table 39.6 Tunable IR Lasers

Laser Type	Wavelength (μm)
Diode	<1–34
Spin flip Raman	5–6.5
	9–14.6
Nonlinear device, parametric oscillators	<1–11
Difference frequency generator	2–6
Two-photon mixer	9–11
Four-photon mixer	2–25
Gas lasers	
Zeeman tuned	3–9
High pressure CO_2	9–11

exception. Nevertheless, although technical achievements often move in advance of any understanding of hazards, it has been encouraging to witness from the inception the serious attention given to the evaluation of laser radiation effects on biological systems (41).

The primary hazard from laser radiation is exposure of the eye and, to a lesser extent, the skin. If the radiation levels are kept below those that damage the eye, there will be no harm to other body tissues. Therefore the material to follow emphasizes the effects of laser radiation on ocular tissue, the radiation exposure conditions required to produce detrimental effects (usually a threshold of injury), and some practical means of preventing undue exposure. In describing the laser radiation that is incident upon tissue, it has become customary to use certain terms. For example, the output of pulsed lasers is described in terms of energy (joules), and that from continuous wave (CW) lasers in terms of power (watts). The corresponding beam density parameter of a pulsed laser is known as the "radiant

Table 39.7. Recent Lasers at Wavelengths 360 nm and Below (40)

Method	Medium	Wavelength (nm)	E/Pulse	Efficiency[a] (%)
Nonlinear	$KB_5O_8 4H_2O$	217.3–234.5	0.4 μJ	2×10^{-3} (c)
	ADP (cooled)	208.0–212.4	0.5 mJ	5×10^{-3} (c)
				50–60 (p)
	Rb-Xe	364.7	30 mJ	5–10 (p)
Molecular	H_2	109.8–161.3	5 kW/cm^2	
	Xe_2	172.0	0.76 J	1 (c)
	Xe_2[b]	172.0	30 μJ	
	Ar-N_2	357.7	45 kW/cm^2	2.3 (c)

[a]Abbreviations = c, conversion; p, photon.
[b]Continuous wave.

Table 39.8. Lasers in Common Use

Laser	Wavelength (μm)	Photon Energy (eV)
Ruby (Cr^{3+})	0.69	1.79
Nd-YAG	1.06	1.17
Nd–glass	1.06	1.17
GaAs	0.90	1.47
Dye	0.36–0.65	3.5–1.9
He–Ne	0.633–3.39	1.96 ($\lambda = 0.633$)
He–Cd	0.325–0.4416	3.81 ($\lambda = 0.325$)
CO$_2$	10.6	0.117
Ar	0.4519–0.5145	2.54 ($\lambda = 0.488$)
CO	5.5	0.225
H$_2$	0.160	7.74
H$_2$O	7.0–220	0.0104 ($\lambda = 118.6$)
HCN	773	0.0016

exposure," expressed in joules per square centimeter (J/cm^2); that for a CW laser is known as "irradiance" and is expressed in watts per square centimeter (W/cm^2). Other parameters such as pulse duration, wavelength, spot size on the retina, and pulse repetition rate (or pulse repetition frequency) are commonly used.

Effects of laser radiation on the eye can range from an annoying glare and mild bleaching of the photoreceptors to massive damage to the foveal region of the retina. The extent of injury depends on exposure conditions and the characteristics of the laser radiation incident upon the eye. In practical situations it is the short-term or intermittent type of exposure that is to be expected, rather than long-term, chronic exposure. Although the effects of long-term, chronic exposure have not been systematically studied, there is some indication that irreversible retinal effects may be produced when the eye is continuously bathed in a luminous flux, without interruption, for extended periods of time (41).

For the most part, work on the biological effects of laser radiation has been aimed at the elucidation of photobiological phenomena and the determination of tissue damage thresholds. "Damage" has usually been described in terms of detrimental changes that are grossly apparent or, in the case of the eye, observable with optical instruments such as microscopes and ophthalmoscopes.

The cornea, lens, and ocular media are largely transparent in the visible region (0.380 to 0.750 μm). Of the visible energy transmitted through the ocular media to reach the retina, most is absorbed in the neuroectodermal coat in the pigment epithelium. The greater part of visible radiation is absorbed in the melanin granules in the retinal pigment epithelium and choroid, which underlie the rods and cones. Figures 39.3 and 39.4 show the percentage transmission of various wavelengths of radiation through the ocular media and the percentage absorption in the retinal pigment epithelium and choroid, respectively. It is at the pigment epithelium that the greatest absorption of energy occurs; hence this thinly pigmented (10 μm) layer

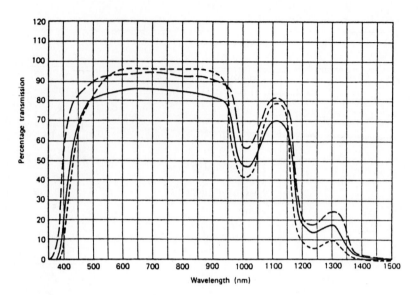

Figure 39.3. Percentage transmission for light of equal intensity through the ocular media of human (short dashes), rhesus monkey (solid curve), and rabbit (dashes). Curves are mean values.

is the most susceptible to damage. Indeed, optical radiation damage may be produced in the pigment epithelium while the photoreceptor layer remains essentially unaffected. The graphs illustrate why the retina is the organ at risk with visible wavelength radiation, whereas the cornea and skin surfaces are at risk with IR and UV radiation.

Laser radiation damage may be attributed to thermal, thermoacoustic, or photochemical phenomena. For pulse durations of the order of 10^{-9} sec and shorter, there is evidence that nonlinear mechanisms, Raman and Brillouin scattering, ultrasonic resonance, and acoustic shock waves may be brought into play (42). The wavelength shifts caused by scattering of electromagnetic radiation from acoustic modes (Brillouin) are small compared with scattering due to molecular vibrations and electron plasma oscillations (Raman). The precise mechanisms for producing damage under these short pulse conditions are not yet fully understood.

The millisecond pulses from a solid-state, visible wavelength laser may cause destruction of tissue by thermal means, whereas ocular and skin tissue may be damaged by photochemical mechanisms when exposed to UV and blue laser light. The consensus seems to be that several damage mechanisms exist. The mere stimulation of the photoreceptors caused by an irradiance that is only slightly higher than that presented to the eye under normal ambient conditions may be sufficient to increase cellular activity to the point of failure, especially if the irradiance is presented to the eye for an extended period of time (43).

When speaking of the thermal effects caused by laser radiation, one means the

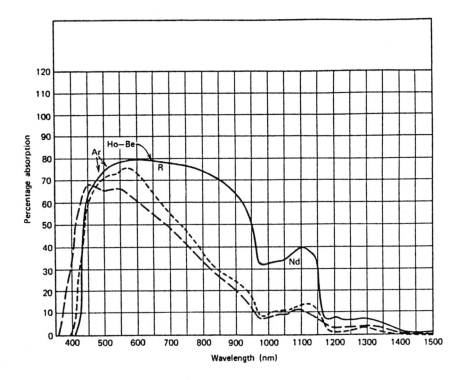

Figure 39.4. Percent absorption of light of equal intensity, incident on the cornea in the retinal pigment epithelium and choroid for Dutch chinchilla rabbits (dashes), rhesus monkey (solid curve), and humans (short dashes). Curves are mean values.

denaturation of protein through the absorption of energy and the consequent conversion to heat. Thermal injury is generally considered to be a rate process; in all probability, therefore, there is no single critical temperature at which injury takes place independent of exposure time. Also, because the molecules of the melanin granules in the pigment epithelium of the retina are relatively large, a broad spectral absorption would be expected to occur. Therefore, the monochromaticity of laser radiation would not be expected to produce biological effects significantly different from those produced by radiation exposure to more conventional light sources. In fact, the coherence of the laser beam is not considered to be a significant factor in producing chorioretinal injury.

The biological response to UV laser radiation is expected to be similar to that produced by noncoherent UV sources. Photophobia, tearing, conjunctival discharge, surface exfoliation, and stromal haze accompanied by complaints of "sand in the eyes," are the expected consequences of excessive exposure. Damage to the corneal epithelium probably results from photochemical denaturation of proteins. Exposure to UV-C radiation (100 to 280 nm) and UV-B (290 to 315 nm) may produce photokeratitis. The latency period for photokeratitis varies from 30 min to as long as 24 hr, depending on the severity of the exposure. Chronic high level exposure to UV-A radiation (315 to 380 nm) may produce cataracts or, for that matter, retinal changes (7).

There are three wavelength regions in the infrared that have significance from a biological standpoint: IR-A (700 to 1.4 nm), IR-B (1.4 to 3 μm), and IR-C (3 μm to 1 mm). A transition zone occurs between the far end of the visible region where retinal effects are observed and the infrared region where corneal effects are produced. The biological effects in the IR-B region are primarily lenticular and corneal damage; in the IR-C region the damage is primarily corneal, for absorption by water (a primary constituent of the cornea) is very strong at these wavelengths. For wavelengths beyond the IR-C region, in the far infrared, absorption occurs in the peripheral layers of tissue. It is in this region and in the ultraviolet A and B regions that the threshold for damage to the cornea is comparable to that of the skin.

To translate these data into the potential hazards associated with specific laser systems, we can say that lasers operating in the visible wavelengths, such as ruby (0.6943 μm) or helium–neon (0.6328 μm), produce damage primarily in the pigment epithelium of the retina. Those that operate in the ultraviolet, such as helium–cadmium (0.3250 μm) and nitrogen (0.3371 μm), produce damage primarily in the cornea, although there is some evidence that effects may be produced in the lens and retina if the radiation levels are sufficiently high (7). Lasers that emit radiation in the near infrared, such as the semiconductor type, may produce injury to the lens, although the output power is usually too low to cause damage. Far infrared lasers such as the carbon dioxide (10.6 μm) have the potential for producing damage to the cornea. Unlike the visible wavelengths, there is no magnification of UV or far infrared irradiation levels.

A retinal injury that occurs in the macula is more serious than one that occurs in the paramacula, for visual processes are more highly developed in the former. Injury to the paramacular region may have only a very minor effect on vision. Somewhat surprisingly, in some cases where direct injury to the macula has occurred, near-normal visual acuity has apparently been restored in a matter of months. Speculation is that optical signals reaching the retina may be refocused to various retinal sites "around the damage" site thus permitting satisfactory acuity.

There is evidence (44) to indicate that short wavelength light (0.4416 μm) produces retinal damage in primates by means of photochemical rather than thermal processes. Also, the thresholds of retinal damage were found to be considerably lower for irradiation at 0.4416 μm than for irradiation at 1.064 μm. For example, an irradiance of 24 W/cm^2 at a retinal temperature of 23°C above the retinal ambient produced a threshold lesion in 1000 sec using a wavelength of 1.06 μm (Nd–YAG laser). Using a wavelength of 0.4416 μm (He–Cd laser), however, required only 30 mW/cm^2, with a negligible temperature rise, to produce a lesion in a 1000-sec exposure. The appearance of the lesion induced by the 0.4416 μm blue light was that of a light yellowish patch on the fundus, whereas the lesion induced by the 1.064 μm radiation had a typical center core "burn" characteristic. Table 39.9

Table 39.9. Retinal Irradiance for Threshold Lesion as a Function of Wavelength and Exposure Duration[a] (44)

Laser	Wavelength (nm)	Irradiance (W/cm²) ± SD[b]				Transmittance (T)
		1 sec	16 sec	100 sec	1000 sec	
Nd–YAG	1064	56.1 ± 5.3 / 55°	37.5 ± 4.2 / 37°	32.5 ± 3.1 / 32°	24.0 ± 3.0 / 23°	0.76
He–Ne	632.8	29.9 ± 1.4 / 49°	15.2 ± 3.0 / 25°	8.4 ± 2.8 / 14°	5.4 ± 2.3 / 9°	0.93
Ar-pumped dye	610	22.0 ± 1.9 / 36°	12.3 ± 1.1 / 20°	8.1 ± 1.1 / 13°	5.8 ± 0.3 / 9.5°	0.92
laser	580	26.1 ± 4.5 / 43°	11.5 ± 2.4 / 19°	7.6 ± 0.8 / 12.5°	4.0 ± 1.7 / 6.6°	0.91
Ar	514.5	14.5 ± 3.3 / 25°	10.3 ± 2.3 / 18°	2.2 ± 0.5 / 4°	0.32 ± 0.1 / 1°	0.87
Ar	488	9.4 ± 1.2 / 17°	6.1 ± 0.8 / 11°	0.77 ± 0.2 / 1°	0.15 ± 0.08 / 0.25°	0.83
Ar	457.8	5.1 ± 0.8 / 10°	3.2 ± 0.6 / 6°	0.52 ± 0.1 / 1°	0.06 ± 0.02 / 0.1°	0.69
He–Cd	441.6	0.91 ± 0.03 / 2°	0.41 ± 0.02 / 1°	0.20 ± 0.02 / 0.4°	0.03 ± 0.01 / 0.05°	0.45

[a]Maximum retinal temperature (°C) above ambient estimated from mathematical model of Clarke; transmittance (T) through ocular media. Beam diameter on retina 500 μm to $1/e^2$ points. Entire laser beam, TEM_{00} mode, enters eye of anesthetized animal with pupil dilated to 8 mm. A beam splitter and fundus camera are used to view and direct beam on the retina. Criterion for a threshold lesion is the appearance 24 hr after exposure of a funduscopically visible lesion.
[b]SD = standard deviation.

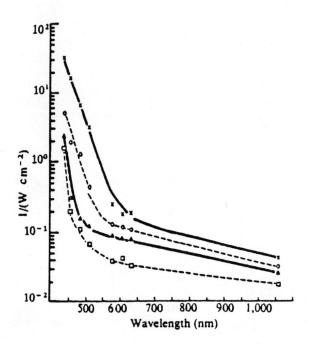

Figure 39.5. Action spectrum of retinal sensitivity to threshold damage in the rhesus monkey. Reciprocal of retinal irradiance $1/(W/cm^2)$ is plotted semilogarithmically against wavelength for four different exposure durations. Beam diameter on the retina was 500 μm to the $1/e^2$ points of the Gaussian distribution. Exposure durations (seconds): squares, 1; triangles, 16; circles, 100; crosses, 1000. From Ham et al. (42).

supplies threshold lesion data for the wavelength range of 0.4416 to 1.064 μm (441.6 to 1064 nm).

Until recently the thresholds for producing retinal lesions as a result of irradiation at all visible wavelengths were considered to be of the same order of magnitude (i.e., 5 to 10 W/cm^2). However, the data of Ham et al. (44) indicate a rather steep slope in the action spectrum at approximately 0.460 μm and below, particularly at exposure durations of 16 sec or less. Figure 39.5 illustrates the effect of two factors (decreasing wavelength and increasing exposure duration) on an increasing retinal sensitivity to laser radiation in the visible wavelengths. It is of considerable interest to note that although retinal sensitivity seems to increase as one shifts from the visible through the blue portions of the electromagnetic spectrum, there are fewer photons reaching the retina because of absorption in the lens and ocular media. Also, as expected, the site of absorption and damage changes from the retina to the lens.

Zuclich and Connolly (7) have reported the UV induction of cataracts at energy doses that are lower than those required to cause corneal damage. The mechanism for inducing corneal damage or cataracts is believed to be photochemical, and that for the production of lenticular cataracts is thought to be thermal. The same investigator reports his finding that the cornea must be irradiated at a level tenfold its injury threshold to produce reversible clouding of the lens. On the other hand, immediate cataracts are produced from irradiation by nitrogen lasers operating at 0.3371 μm and 1 J/cm^2. This level of irradiation is reportedly an order of magnitude lower than the corneal injury threshold. Also, retinal lesions are produced using 0.3507 and 0.356 μm radiation reportedly at levels an order of magnitude below the corneal damage threshold. Therefore it would appear that the hazard to all eye structures is increased as one proceeds into the blue and UV wavelengths. Furthermore, the damage threshold appears to depend on the presence of oxygen in that a much higher incident energy is required to produce lesions when the cornea is flushed with nitrogen.

Exposure of rhesus monkeys to ultrashort single pulses of 25 to 35 psec from an unfocused mode—locked Nd–YAG laser at a wavelength of 1.064 μm—resulted in a threshold injury to the retina at a mean energy of 13 $\pm$ 13 μJ (45). This compares with a threshold value of 68 $\pm$ 13 μJ determined by the same authors for a Q-switched Nd–YAG laser operating with pulse durations in the 9 to 30 nsec range. Therefore it appears that shortening the pulse durations from the nanosecond to the picosecond range—or three orders of magnitude—results in a lowering of retinal injury threshold by a factor of approximately 5. The mechanism of injury appears to be the destruction of the retinal pigment epithelium through the combined effects of acoustic and shock wave pressure and the extremely high temperature rise in the melanin granules of the pigment epithelium.

Another study (46) using ultrashort pulse trains from Nd–glass gave much lower thresholds of injury than the Ham study: 1.7×10^{-8}J versus 1.3×10^{-5}J, using similar pulse durations. Because the threshold of 1.3×10^{-5}J (13 μJ) was obtained with single pulses rather than pulse trains, the study (46) will be repeated with single pulses. These results have complicated the setting of safety standards in that the present maximum permissible exposure (MPE) levels are largely based on pulse durations no shorter than nanoseconds.

Work has also been completed on the effect of ultrashort visible radiation pulses on rhesus monkeys (47). These results indicate that the melanin granules are much more severely damaged by the visible radiation than by 1.064 μm radiation; however, the threshold of injury was higher for the visible wavelengths.

Studies of the biological effects of laser radiation have largely abated in the past 10 years. Yet we still have a need to develop techniques that detect functional as well as histologic changes in the eye structure. The development of such techniques could have an important bearing on the viability of the ophthalmoscope as a primary means for determining ocular change. Some investigators have already observed irreversible decrements in visual performance at exposure levels that are 10 percent below the damage threshold determined by observations through an ophthalmoscope. McNeer et al. (48) found that at 50 percent of the ophthalmoscopically determined threshold, the ERG B wave amplitude was irreversibly reduced. Davis and Mautner (49) reported severe changes in the visually evoked cortical potential at 25 percent of the ophthalmoscopically determined threshold. In addition to the problem of determining the true acute ocular injury threshold, there is a general deficiency of knowledge concerning the effect of long-term, chronic exposure to laser radiation.

5.2 Skin

The potential damage caused by irradiating the skin with lasers is considered to be less than that caused by exposure of the eye, for skin damage is often repairable or reversible. On the other hand, exposure of the skin to high-intensity radiation can cause depigmentation, severe burns, and possible damage to underlying organs. To keep the relative eye and skin hazard potential in perspective, one must not overlook possible photosensitization of the skin caused by use of drugs or cosmetic materials. In such cases the permissible exposure limits can be considerably below the currently recommended values.

Ultraviolet radiation in the wavelength region of 0.25 to 0.32 μm appears to be particularly injurious to the skin. Exposure to shorter (0.20 to 0.25 μm) and longer (0.32 to 0.40 μm) wavelengths is considered less harmful. Because of the structural inhomogeneities of the skin, any incident radiation will undergo multiple internal reflections; however, most radiation is absorbed in the first 3.5 mm of tissue. The shorter wavelengths are absorbed in the outer dead layer of the epidermis (stratum corneum), whereas exposure to the longer wavelengths has a pigment-darkening effect. Chronic exposure to ultraviolet radiation accelerates skin aging and probably the development of skin cancer. Please refer to Section 2 on Ultraviolet Radiation.

5.3 Exposure Limits

5.3.1 Eye

The exposure limits developed by the American National Standards Institute (ANSI) (35) seem to have been generally accepted (with modification in certain cases) by international standards-setting organizations such as the World Health Organization (50) and the International Electrotechnical Commission. The ANSI limits are presented in Tables 39.10, 39.11, and 39.12. Table 39.10 illustrates the exposure limits for direct ocular exposure (intrabeam viewing) to a laser beam; Table 39.11 shows the exposure limits for viewing a diffuse reflection of a laser beam or an extended source laser, and Table 39.12 gives the exposure limits for skin. The ANSI limits are used here because most laser safety criteria have been derived from those developed by the ANSI Z136 committee activities.

In the ANSI standard, exposure limits are expressed in terms of the radiant exposure (J/cm^2) or irradiance (W/cm^2) required to produce a minimal lesion, usually a retinal lesion, which is visible through an ophthalmoscope 24 hr after exposure. All the ANSI criteria are based on acute exposure conditions, for very little information is available on the long-term, chronic effects of laser radiation.

Table 39.10. MPE for Direct Ocular Exposure Intrabeam Viewing, to a Laser Beam

Wavelength, λ (μm)	Exposure Duration, t (sec)	Maximum Permissible Exposure (MPE) (J/cm^2)	Notes for Calculation and Measurement
Ultraviolet			
0.200–0.302	10^{-9} to 3×10^4	3×10^{-3}	Or $0.56t^{1/4}$ J/cm^2,
0.303	10^{-9} to 3×10^4	4×10^{-3}	whichever is lower.
0.304	10^{-9} to 3×10^4	6×10^{-3}	1-mm limiting ap-
0.305	10^{-9} to 3×10^4	1.0×10^{-2}	erture
0.306	10^{-9} to 3×10^4	1.6×10^{-2}	
0.307	10^{-9} to 3×10^4	2.5×10^{-2}	
0.308	10^{-9} to 3×10^4	4.0×10^{-2}	
0.309	10^{-9} to 3×10^4	6.3×10^{-2}	
0.310	10^{-9} to 3×10^4	1.0×10^{-1}	
0.311	10^{-9} to 3×10^4	1.6×10^{-1}	
0.312	10^{-9} to 3×10^4	2.5×10^{-1}	
0.313	10^{-9} to 3×10^4	4.0×10^{-1}	
0.314	10^{-9} to 3×10^4	6.3×10^{-1}	
0.315–0.400	10^{-9} to 10	$0.56t^{1/4}$	
0.315–0.400	10^3 to 3×10^4	1	
Visible and near infrared			
0.400–0.700	10^{-9} to 1.8×10^{-5}	5×10^{-7}	7-mm limiting aper-
0.400–0.700	1.8×10^{-5} to 10	$1.8t^{3/4} \times 10^{-3}$	ture
0.400–0.550	10 to 10^4	10×10^{-3}	
0.550–0.700	10 to T_1	$1.8t^{3/4} \times 10^{-3}$	
0.550–0.700	T_1 to 10^4	$10\,C_B \times 10^{-3}$	
0.400–0.700	10^4 to 3×10^4	$C_B \times 10^{-6}$ W/cm^2	
0.700–1.050	10^{-9} to 1.8×10^{-5}	$5\,C_A \times 10^{-7}$	
0.700–1.050	1.8×10^{-5} to 10^3	$1.8\,C_A t^{3/4} \times 10^{-3}$	
1.051–1.400	10^{-9} to 5×10^{-5}	5×10^{-6}	
1.051–1.400	5×10^{-5} to 10^3	$9\,t^{3/4} \times 10^{-3}$	
0.700–1.400	10^3 to 3×10^4	$320\,C_A \times 10^{-6}$	
1.4–10^3	10^{-9} to 10^{-7}	10^{-2}	See Table 39.17 for
	10^{-7} to 10	$0.56t^{1/4}$	apertures. 1-mm
	>10	0.1 W/cm^2	limiting aperture for 1.4–100 μm; 11-mm limiting aperture for 0.1–1 mm
1.54 only	10^{-9} to 10^{-6}	1.0	

Notes: $C_A = 1$ for $\lambda = 0.400$–0.700 μm.

$C_A = 10^{2.0(\lambda-0.700)}$ for $\lambda = 0.700$–1.050 μm.

$C_A = 5$ for $\lambda = 1.050$–1.400 μm.

$C_B = 1$ for $\lambda = 0.400$–0.550 μm.

$C_B = 10^{15(\lambda-0.550)}$ for $\lambda = 0.550$ μm.

$T_1 = 10 \times 1020^{20(\lambda-0.550)}$ for $\lambda = 0.550$–0.700 μm.

Table 39.11. MPE for Viewing a Diffuse Reflection of a Laser Beam or an Extended-Source Laser

Wavelength, λ (μm)	Exposure Duration, t (sec)	Maximum Permissible Exposure (MPE)	Notes for Calculation and Measurement
Ultraviolet			
0.200–0.302	10^{-9} to 3×10^4	3×10^{-3} J/cm^2	Or $0.56t^{1/4}$ J/cm^2,
0.303	10^{-9} to 3×10^4	4×10^{-3} J/cm^2	whichever is lower.
0.304	10^{-9} to 3×10^4	6×10^{-3} J/cm^2	1-mm limiting aper-
0.305	10^{-9} to 3×10^4	10×10^{-2} J/cm^2	ture
0.306	10^{-9} to 3×10^4	16×10^{-2} J/cm^2	
0.307	10^{-9} to 3×10^4	25×10^{-2} J/cm^2	
0.308	10^{-9} to 3×10^4	40×10^{-2} J/cm^2	
0.309	10^{-9} to 3×10^4	63×10^{-2} J/cm^2	
0.310	10^{-9} to 3×10^4	10×10^{-1} J/cm^2	
0.311	10^{-9} to 3×10^4	16×10^{-1} J/cm^2	
0.312	10^{-9} to 3×10^4	25×10^{-1} J/cm^2	
0.313	10^{-9} to 3×10^4	40×10^{-1} J/cm^2	
0.314	10^{-9} to 3×10^4	63×10^{-1} J/cm^2	
0.315–0.400	10^{-9} to 10	$0.56t^{1/4}$ J/cm^2	
0.315–0.400	10^3 to 3×10^4	1 J/cm^2	
Visible			
0.400–0.700	10^{-9} to 10	$10t^{1/3}$ J/(cm^2)(sr)	1-mm limiting aperture
0.400–0.550	10 to 10^4	21 J/(cm^2)(sr)	or α_{min}, whichever is
0.550–0.700	10 to T_1	$3\ 83t^{3/4}$ J/(cm^2)(sr)	greater
0.550–0.700	T_1 to 10^4	21 C_B J/(cm^2)(sr)	
0.400–0.700	10^3 to 3×10^4	21 $C_B 10^{-3}$ W/(cm^2)(sr)	
Near Infrared			
0.700–1.400	10^{-9} to 10	10 $C_A t^{1/3}$ J/(cm^2)(sr)	
0.700–1.400	10 to 10^3	$3\ 83\ C_A t^{3/4}$ J/(cm^2)(sr)	
0.700–1.400	10^3 to 3×10^4	$0\ 64\ C_A$ W/(cm^2)(sr)	
1.4–10^3	10^{-9} to 10^{-7}	10^{-2} J/cm^2	See Table 39.17 for
	10^{-7} to 10	$0.56t^{1/4}$ J/cm^2	apertures
	> 10	$0\ 1$ W/cm^2	
1.54 only	10^{-9} to 10^{-6}	10 J/cm^2	

Notes: $C_A = 1$ for $\lambda = 0.400$–0.700 μm.
  $C_A = 10^{2.0(-0.700)}$ for $\lambda = 0.700$–1.050 μm.
  $C_A = 5$ for $\lambda = 1.050$–1.400 μm.
  $C_B = 1$ for $\lambda = 0.400$–0.550 μm.
  $C_B = 10^{15(0.550)}$ for $\lambda = 0.550$ μm.
  $T_1 = 10 \times 1020^{20(0.550)}$ for $\lambda = 0.550$–0.700 μm.

Table 39.12. MPE for Skin Exposure to a Laser Beam

Wavelength, λ (μm)	Exposure Duration t (s)	Maximum Permissible Exposure (MPE) (J/cm²)	Notes for Calculation and Measurement
Ultraviolet			
0.200–0.302	10^{-9} to 3×10^{4}	3×10^{-3}	Or $0.56t^{1/4}$ J/cm²,
0.303	10^{-9} to 3×10^{4}	4×10^{-3}	whichever is lower.
0.304	10^{-9} to 3×10^{4}	6×10^{-3}	1-mm limiting ap-
0.305	10^{-9} to 3×10^{4}	1.0×10^{-2}	erture
0.306	10^{-9} to 3×10^{4}	1.6×10^{-2}	
0.307	10^{-9} to 3×10^{4}	2.5×10^{-2}	
0.308	10^{-9} to 3×10^{4}	4.0×10^{-2}	
0.309	10^{-9} to 3×10^{4}	6.3×10^{-2}	
0.310	10^{-9} to 3×10^{4}	1.0×10^{-1}	
0.311	10^{-9} to 3×10^{4}	1.6×10^{-1}	
0.312	10^{-9} to 3×10^{4}	2.5×10^{-1}	
0.313	10^{-9} to 3×10^{4}	4.0×10^{-1}	
0.314	10^{-9} to 3×10^{4}	6.3×10^{-1}	
0.315–0.400	10^{-9} to 10	$0.56t^{1/4}$	
0.315–0.400	10 to 10^{3}	1	
0.315–0.400	10^{3} to 3×10^{4}	1×10^{-3} W/cm²	
Visible and near infrared			
0.400–1.400	10^{-9} to 10^{-7}	$2 C_A \times 10^{-2}$	1-mm limiting aper-
	10^{-7} to 10	$1.1 C_A t^{1/4}$	ture for C_A see
	10 to 3×10^{4}	$0.2 C_A$ W/cm²	notes in Tables 39.10 and 39.11
Far infrared			
1.4–10^{3}	10^{-9} to 10^{-7}	10^{-2}	1-mm limiting aper-
	10^{-7} to 10	$0.56t^{1/4}$	ture for 1.4–100
	>10	0.1 W/cm²	μm. 11-mm limit-ing aperture for 0.1–1 mm
1.54 only	10^{-9} to 10^{-6}	1.0	

For this reason caution must be exercised in applying such criteria to conditions when long-term viewing of optical sources is required. On the other hand, the use of the ANSI criteria should prove to be satisfactory for the overwhelming majority of practical situations in which accidental exposure to laser radiation occurs. Some questions have been posed regarding whether the so-called natural aversion response (the strong tendency to blink and look away from a dazzling or extremely bright light source) is truly protective under all circumstances. It is speculated that certain persons might be able to override the aversion response if sufficiently well motivated; that is, some might be able to force themselves to stare at the extremely bright laser beam. The aversion response is generally thought to occur in a period of 0.1 to 0.25 sec. In the ANSI Z136 standard, a value of 0.25 sec has been used in calculating the permissible exposure limits for lasers operating at visible wavelengths.

The Federal Laser Product Performance Standard establishes accessible radiation emission limits for different classes of lasers, but these are identical to the ANSI standard for all practical purposes.

The most recent revisions of the ANSI Z136.1 standard accessible emission limits for continuous wave and single pulsed lasers are given in Tables 39.13 and 39.14.

5.4 Protection Guidelines

The objective of these guidelines is to provide reasonable and adequate guidance for the safe use of lasers and laser systems. A practical means for accomplishing this is first to classify lasers and laser systems according to their relative hazards and then to specify appropriate controls for each classification.

The basic consideration of the hazard classification scheme is the ability of the primary laser beam or reflected primary laser beam to cause biological damage to the eye or skin during intended use. A Class 1 laser is one that is considered to be incapable of producing damaging radiation levels and is therefore exempt from any control measures or other forms of surveillance. Class 2 lasers (low-power) are divided into two subclasses, 2 and 2a. A Class 2 laser emits in the visible portion of the spectrum (0.4 to 0.7 μm) and eye protection is normally afforded by the aversion response including the blink reflex. Class 3 lasers (medium-power) are divided into two subclasses, 3a and 3b. A class 3 laser may be hazardous under direct and specular reflection viewing conditions, but the diffuse reflection is usually not a hazard. A Class 3 laser is normally not a fire hazard. A class 4 laser (high-power) is a hazard to the eye and skin from the direct beam and sometimes from a diffuse reflection and also can be a fire hazard.

Although these guidelines relate specifically to the laser product and its potential hazard, the conditions under which the laser is used, the level of safety training of individuals using the laser, and other environmental and personnel factors are important considerations in determining the full extent of safety control measures. Because such situations require informed judgments by responsible persons, major responsibility for such judgment has been assigned to a person with requisite training authority and responsibility, namely, the Laser Safety Officer (LSO).

Lasers or laser systems certified for a specific class by a manufacturer in accordance with the Federal Laser Product Performance Standard may be considered as fulfilling all classification requirements of this standard. In cases where the laser or laser system classification is not provided or where the class level may change because of a change of engineering control measures, the laser or laser system shall be classified by the LSO.

The recommended procedure contained in the ANSI standard for evaluating the safety of lasers is as follows: (1) Determine the appropriate class of laser or

Table 39.13. Accessible Emission Limits for Selected Continuous-Wave[a] Lasers and Laser Systems

Range (μm)	Emission Duration (sec)	Class 1[b] (W)	Class 2 (W)	Class 3 (W)	Class 4 (W)
Ultraviolet (0.2–0.4)	3×10^4	$\leq 0.8 \times 10^{-9}$ to $\leq 8 \times 10^{-6}$ depending on wavelength (see Tables 39.10, 39.11, and 39.12)	—	$>$ Class 1 but ≤ 0.5 depending on wavelength (see Tables 39.10, 39.11, and 39.12)	>0.5
Visible (0.4–0.7)	3×10^4	$\leq 0.4\, C_B \times 10^{-6c}$	$>$ Class 1 but $\leq 1 \times 10^{-3}$	$>$ Class 2 but ≤ 0.5	>0.5
Near infrared (0.7–1.06)	3×10^4	$\leq 0.4 \times 10^{-6}$ to $\leq 200 \times 10^{-6}$ depending on wavelength	—	$>$ Class 1 but ≤ 0.5 depending on wavelength	>0.5
Near infrared (1.06–1.4)	3×10^4	$\leq 200 \times 10^{-6}$	—	$>$ Class 1 but ≤ 0.5	>0.5
Far infrared (1.4×10^2)	>10	$\leq 0.8 \times 10^{-3}$	—	$>$ Class 1 but ≤ 0.5	>0.5
Submillimeter (10^2 to 10^3)	>10	≤ 0.1	—	$>$ Class 1 but ≤ 0.5	>0.5

[a]Emission duration ≥ 0.25 sec.
[b]When the design or intended use of the laser or laser system ensures personnel exposures of less than 10^4 sec in any 24-hr period, the limiting exposure duration may establish a higher exempt power level.
[c]For C_B see notes in Tables 39.10 and 39.11.

Table 39.14. Summary of Levels (Energy and Radiant Exposure Emissions) for Single-Pulsed Laser and Laser System Classification[a]

Wavelength Range (μm)	Emission Duration (sec)	Class 1 (J)	Class 3 (J/cm²)	Class 4 (J/cm²)
Ultraviolet[b] (0.2–0.4)	$>10^{-2}$	$\leq 24 \times 10^{-6}$ to 7.8×10^{-3}	>Class 1 but ≤ 10	>10
Visible (0.4–0.7)	10^{-9} to 0.25	$\leq 0.2 \times 10^{-6}$	>Class 1 but $\leq 31 \times 10^{-3}$	$>31 \times 10^{-3}$
		$\leq 0.25 \times 10^{-3}$	>Class 1 but ≤ 10	>10
Near infrared 0.7–1.06	10^{-9} to 0.25	$\leq 0.2 \times 10^{-6}$ to 2×10^{-6}	>Class 1 but $\leq 31 \times 10^{-3}$	$>31 \times 10^{-3}$
		$\leq 0.25 \times 10^{-3}$ to 1.25×10^{-3}	>Class 1 but ≤ 10	>10
1.06–1.4	10^{-9} to 0.25	$\leq 2 \times 10^{-6}$	>Class 1 but $\leq 31 \times 10^{-3}$	>10
		$\leq 1.25 \times 10^{-3}$	>Class 1 but ≤ 10	>10
Far infrared (1.4–10^2)	10^{-9} to 0.25	$\leq 80 \times 10^{-6}$	>Class 1 but ≤ 10	>10
		$\leq 3.2 \times 10^{-3}$	>Class 1 but ≤ 10	>10
Submillimeter (10^2–10^3)	10^{-9} to 0.25	$\leq 10 \times 10^{-3}$	>Class 1 but ≤ 10	>10
		≤ 0.4	>Class 1 but ≤ 10	>10

[a]There are no Class 2 single-pulsed lasers.
[b]Wavelength dependent (see Tables 39.10, 39.11 and 39.12).

laser system. (2) Comply with the measures specified for that class of laser or laser system, using the following table as a guide. This procedure will in most cases eliminate the need for measurement of laser radiation, quantitative analysis of hazard potential, or use of the MPE values.

Class	Control Measures	Medical Surveillance
1	Not applicable	Not applicable
2	Applicable	Not applicable
2a	Applicable	Not applicable
3a	Applicable	Not applicable
3b	Applicable	Applicable
4	Applicable	Applicable

Typical laser classifications for continuous wave and single pulse lasers are given in Table 39.15 and 39.16, respectively.

5.5 Measurement of Laser Radiation

No single instrument can cover the required dynamic range, sensitivity, accuracy, precision, wavelength specificity, and stability of operation for proper measurement of the output from the wide variety of laser systems now in use. The wavelength range available in laser devices extends from the submillimeter to the VUV; power ranges from microwatts to terawatts; and operational modes encompass both the CW and pulse durations from 10^{-1} to 10^{-13} sec. The legislation promulgated by the Department of Health and Human Services (52) requiring the manufacturer to indicate certain operating parameters and classification of each laser device on a properly affixed label tends to reduce the need for an elaborate measurement system. If such equipment is purchased, however, it is encouraging to note that high-quality performance is readily obtainable. The output power of laser devices as specified by the manufacturer has rarely been found to differ from precision calibration data by more than a factor of 2 (3).

In choosing the appropriate detection and readout device for a specific measurement problem, one should strive to obtain (1) linear response over the wavelength the range of interest, (2) uniform response over the area of the receiving aperture, (3) response time that is suitable for the measured signal (e.g., time constants that are much shorter than the rise and decay times of Q-switched or mode-locked pulses), and (4) detector materials whose saturation or damage thresholds are much higher than the irradiances (W/cm^2) or radiant exposures (J/cm^2) of incident laser beams. See Table 39.17 for recommended maximum aperture diameters.

High current vacuum photodiodes are useful for measuring the output of Q-switched systems and can operate with a linear response over a wide range of wavelengths (e.g., from the UV to the near infrared). The response time is limited by the transit time of electrons from the photocathode to collector, but is typically

Table 39.15. Typical Laser Classification—Continuous-Wave (CW) Lasers

Wavelength Range	Laser	Wavelengths	Exempt–Class 1[a] (W)	Class 2 (W)	Class 3 (W)	Class 4 (W)
Ultraviolet (0.1–0.28 μm)	CW neodymium: YAG (quadrupled)	266 nm only	$\leq 0.8 \times 10^{-9}$	—	> Class 1 but ≤ 0.5	> 0.5
Ultraviolet (0.315–0.4 μm)	Helium–cadmium Argon Krypton	325 nm only 351.1, 363.8 nm only 350.7, 356.4 nm only	$\leq 8 \times 10^{-6}$	—	> Class 1 but ≤ 0.5	> 0.5
Visible	Helium–cadmium Argon (visible)	441.6 nm only 457.9, 476.5, 488, 514.5 nm, etc.	$\leq 0.4 \times 10^{-6}$	> Class 1 but $\leq 1 \times 10^{-3}$	> Class 1 but ≤ 0.5	> 0.5
	Helium–selenium	460.4–700 nm only (30 lines)				
	CW Neodymium: YAG (doubles) Helium–neon Krypton	532 nm 632.8 nm 647.1, 530.9, 676.4 nm, etc.	$\leq 0.4 \times C_B \times 10^{-6}$	> Class 1 but $\leq 1 \times 10^{-3}$	> Class 1 but ≤ 0.5	> 0.5
Near infrared (0.7–1.4 μm)	CW Gallium–aluminum arsenide	0.85 μm (20°C)	$\leq 80 \times 10^{-6}$	—	> Class 1 but ≤ 0.5	> 0.5
	CW Gallium arsenide	0.905 μm (20°C)	$\leq 0.1 \times 10^{-3}$	—	> Class 1 but ≤ 0.5	> 0.5
	CW neodymium YAG	1.064 μm	$\leq 0.2 \times 10^{-3}$	—	> Class 1 but ≤ 0.5	> 0.5
	Helium–neon	1.08, 1.152 μm only	$\leq 0.2 \times 10^{-3}$	—	> Class 1 but ≤ 0.5	> 0.5
Far infrared (1.4–100 μm)	Hydrogen fluoride Carbon monoxide Carbon dioxide Helium–Neon	4–6 μm 5.0–5.5 μm 10.6 μm 3.39 μm only	$\leq 0.8 \times 10^{-3}$	—	> Class 1 but ≤ 0.5	> 0.5
Far infrared (0.1–1 mm)	Water vapor Hydrogen cyanide	118 μm 337 μm	≤ 0.1	—	> Class 1 but ≤ 0.5	> 0.5

[a]Assumes no mechanical or electrical design incorporated into laser system to prevent exposures from lasting to λ_{min} = 8 hr (one workday); otherwise the Class 1 AEL could be larger than tabulated.

Table 39.16. Typical Laser Classification—Single-Pulse Lasers

Wavelength Range	Laser	Wavelengths	Pulse Duration (sec)	Exempt-Class 1 (J)	Class 3 (J/cm²)	High-Power-Class 4 (J/cm²)
Ultraviolet	Neodymium: YAG Q-SW (quadrupled) Ruby (doubled)	266 nm	10 to 30 × 10^{-9} (Q-sw)	No criteria	≤ 10	> 10
		347.1 nm	10 to 30 × 10^{-9} (Q-sw)	No criteria	≤ 10	> 10
Visible (0.4–0.7 μm)	Neodymium: YAG (doubled)	532 nm	~20 × 10^{-9} (Q-sw)	≤ 0.2 × 10^{-6}	> Class 1 but ≤ 74 × 10^{-3}	> 75 × 10^{-3}
	Ruby	694.3 nm				
	Ruby (long pulse)	694.3 nm	~1 × 10^{-3}	≤ 0.2 × 10^{-6}	> Class 1 but ≤ 3.1	> 3.1
	Rhodamine 6G (dye laser)	450–650 nm	~20 × 10^{-9}	≤ 0.2 × 10^{-6}	> Class 1 but ≤ 0.31	> 0.31
Infrared	Neodymium: YAG	1.064 μm	~20 × 10^{-9} (Qa-sw)	≤ 2 × 10^{-6}	> Class 1 but ≤ 0.16	> 0.16
	Erbium: glass	1.54 μm	~10 to 100 × 10^{-9} (Q-sw)	≤ 8 × 10^{-3}	> Class 1 but ≤ 10 J	> 10
	Carbon dioxide	10.6 μm	1 to 100 × 10^{-9} (Q-sw)	≤ 80 × 10^{-6}	> Class 1 but ≤ 10	> 10

Table 39.17. Maximum Aperture Diameters (Limiting Aperture) for Measurement Averaging

Measurement	Exposure Duration, t (sec)	Aperture Diameter (mm) vs. Wavelength			
		Ultraviolet $(0.2\text{–}0.4\ \mu m)$	Visible and Near Infrared $(0.4\text{–}1.4\ \mu m)$	Medium and Far Infrared $(1.4\text{–}10^2\ \mu m)$	Submillimeter $(0.1\text{–}1\ mm)$
Eye MPE	10^{-9} to 3×10^4	1	7	1	11
Skin MPE	10^{-9} to 3×10^4	1	1	1	11
Laser classification[a]	10^{-9} to 3×10^4	50	50	50	50

[a]The apertures are used for the measurement of total output power or output energy for laser classification purposes, to distinguish between all classes of CW lasers or between Class 1 and Class 3 pulsed lasers. The use of the 50-mm apertures as shown in the horizontal line labeled "Laser Classification" applies only to those cases where the laser output is intended to be viewed with optical instruments (excluding ordinary eyeglass lenses) or where the Laser Safety Officer determines that there is some probability that the output will be accidentally viewed with optical instruments and that such radiation will be viewed for a sufficient time duration so as to constitute a hazard. Otherwise the apertures listed for eye MPE and skin MPE are to be used.

For the specific case of optical viewing (beam collecting) instruments, the apertures listed for eye MPE and skin MPE apply to the exit beam of such devices.

0.5 nsec or less. Although the dark current is somewhat high (i.e., of the order of 5 nA at room temperature), the operation of photodiodes is very stable if they are used within the specified current limitations, and if the voltage of the power supply is well maintained. Because the sensitivity of the photocathode is usually nonuniform, a large fraction of the photocathode surface area must be well illuminated; otherwise measurement errors will be induced. The nonuniformity problem might be improved through the use of a planar photocell. Because the limiting incident power level of many vacuum photodiodes is of the order of 1 W, it may become necessary to attenuate the beam. This can be accomplished by inserting optical interference filters, neutral density filters, or diffuse reflectors into the system, or by beam splitting.

Average power measurements of CW laser systems may be made with conventional thermopiles or photovoltaic cells. A typical thermopile can detect signals in the power range from 10 μW to about 100 mW. Because thermopiles are composed of many junctions, the response may be nonuniform unless the entire surface is fully illuminated by the laser beam. Bismuth–silver junctions are recommended for high sensitivity; copper–constantan junctions are better for rugged stability. Table 39.4 lists the sensitivity, impedance, and response time of typical junction detectors. Gold-black or platinum-black junctions provide the fastest response times. The use of the thermopile is not recommended for irradiance levels exceeding approximately 100 to 125 mW/cm^2. For the measurement of super high CW power (e.g., 1 to 5 kW), the use of a cone flow thermopile is recommended.

For high-energy pulsed radiation, the ballistic thermopile should be considered. In this device the incident radiation travels down the inner surfaces of an active carbon cone, where absorption takes place. The active cone is balanced against an identical reference cone, to measure the difference in temperature and to minimize drift due to ambient temperature fluctuations. These devices employ a rather long response time characteristic (e.g., 1 to 1.5 min) to integrate the pulsed signals properly.

Photovoltaic cells consist of a junction of two materials that are specially selected to produce a contact potential. The two materials can be a semiconductor and a metal, or a type n and p semiconductor such as a silicon solar cell. The photovoltaic cells offer the following advantages: (1) the cell output is very stable with time, (2) external power supplies are not needed, and (3) the detector unit and associated circuitry is small enough to permit the construction of a portable device. Materials that are suitable as detector elements include indium arsenide and indium antimonide. In certain devices these detector elements are installed in glass Dewar flasks to cool them with liquid nitrogen. When this is done, the detectable wavelength band is expanded to include an upper limit of approximately 6 μm. Selenium cells are good for power measurements in the near ultraviolet; silicon solar cells are preferred for near-infrared measurements, and Schottky surface barrier silicon photodiodes are preferentially used for radiation in the blue wavelengths. Many calorimeters measure total energy, but they can also be used to measure total power if the time history of the radiation is known. Other calorimeters do not measure the temperature rise in the absorber; rather, they measure small changes

Table 39.18. Devices for Energy and Power Measurements

Device	Range of Operation	Typical Response Time	Surface Damage[a]
Energy			
Cone calorimeters	10^{-2} to 2×10^3 J	1–20 sec	10-20 mW/cm² in 50 nsec
Meter disk calorimeter	10^{-2} to 10 J	10 sec	50 mW/cm² in 50 nsec
Rat's nest calorimeter	10^{-3} to 10 J	10^{-4} sec	—
Wire calorimeter	10^{-2} to 0.5 J	10 sec	10 mW/cm² in 50 nsec
Liquid calorimeter	1–500 J	10–60 sec	—[b]
Torsion pendulum	0.5–500 J	60 sec	—
Integrating photocurrent	10^{-6} to 10^{-3} J	1 sec	1 mW/cm²
Thermopile	10^{-6} to 1 J	10^{-1} sec	300 mW/cm²
Copper sphere	5×10^{-4} to 10 J	180 sec	—
Power			
Phototube	10^{-8} to 10^{-3} W/cm²	3–10 nsec	1 mW/cm²
Photodiode	10^{-4} to 6 W/cm²	0.3–4 nsec	10 W/cm²[c]
Nonlinear crystal	10^3 to 10^{12} W/cm²	10^{-5} sec	10^{12} W/cm²[d]

Source: S. S. Charschan, Ed., *Lasers in Industry*, Van Nostrand Reinhold, New York, 1972, p. 521.
[a]Surface damage may occur at the indicated power density.
[b]Local boiling of the liquid should be avoided.
[c]Above this power, response of device is nonlinear.
[d]Breakdown of quartz at this power density.

in polarization, which in turn produce a potential difference across a detector. This "pyroelectric effect" occurs in ferroelectrics because such materials possess a permanent polarization that is temperature dependent. The pyroelectric detectors are especially suitable for the measurement of far-infrared radiation of the carbon dioxide and hydrogen cyanide lasers; however, the input signals are usually chopped to prevent degradation of the sensing element. Table 39.18 summarizes certain characteristics of energy and power measuring devices.

For measurement throughout the infrared region, many quantum detectors are available. For example, germanium junction photodiodes are used for the near infrared up to approximately 1.8 μm; PbS and InAs photodiodes may be used in the 1.8 to 3 μm range; PbTe, PbSe, and InSb are suitable for the 3 to 7 μm range, whereas germanium combined with a wide variety of metals (mercury, cobalt, zinc, copper, and gold) is preferred for the 7 to 40 μm range. Many of these detectors need liquid nitrogen cooling for proper performance.

Thermal detectors such as bolometers are still used where broad-band spectral response is desired. A bolometer using a platinum ribbon sensor element gives a flat response over a wavelength range of 1 to 26 μm. The change in electrical resistance due to heating by the absorbed radiation is proportional to the absorbed radiant power in these devices. Because the time constants are usually of the order to 10^{-4} sec however, they are not recommended for the measurement of Q-switched pulses.

The type of readout device for the various detector systems varies with individual requirements. Microammeters and microvoltmeters are often used with CW systems; microvoltmeters or electrometers are frequently employed in conjunction with oscilloscopes to measure pulsed laser system parameters. All these devices may be coupled to recorders or panel displays. For purposes of calibration, tungsten ribbon filament lamps are available from the National Institute for Standards and Technology as secondary standards of spectral radiance over the wavelength region of 0.2 to 2.6 μm. The calibration procedures for these devices permit comparisons within about 1 percent in the near UV and about a 0.5 percent in the visible. All radiometric standards are based on the Stefan–Boltzmann and Planck laws of blackbody radiation. The spectral response of measurement devices should always be specified, because the ultimate use of the measurements is a correlation with the spectral response (action level) of the biological tissue receiving the radiation result.

The effective aperture (aperture stop) of any device that is used for measuring the irradiance (W/cm^2) or the radiant exposure (J/cm^2) in the visible wavelengths should closely approximate 0.7 cm (7 mm), the diameter of the fully dilated or dark-adapted pupil. Apertures other than 7 mm may be used for radiation at other wavelengths (see Table 39.17).

The ANSI reasoning behind the recommended use of a 1 mm aperture for IR and UV wavelengths is that a practical minimum aperture diameter should be employed to measure the radiant exposure or irradiance received by a relatively small area of tissue; yet the diameter should be large enough to impinge upon acceptable sensing areas of transducing elements currently available in measuring instruments. Furthermore, there is no need to specify such a relatively large aperture size (7 mm), a value that is based on the magnification of visible light levels (not UV or far infrared) on their passage from the cornea to the retina. The absorption of UV and far infrared is, after all, a surface absorption phenomena. Finally, whereas a larger area of irradiation may be necessary for proper operation of thermal detectors (e.g., calorimeters), an impingement area of 1 mm (aperture) and even smaller is well within the acceptable operating requirements of state-of-the-art quantum detectors.

5.6 Special Precautions

Probably the most significant hazard associated with lasers is electrical shock rather than exposure to optical radiation. Therefore all pertinent provisions of the National Electrical Code should be followed, including first aid training for all persons working with lasers. Particular attention must be paid to adequate grounding, proper discharge of energized circuits, well-designed interlocks and warning signals, and periodic educational programs.

6 MICROWAVE RADIATION

6.1 Physical Characteristics of Microwave Radiation

Microwave wavelengths vary from about 10 m to 1 mm; the respective frequencies range from approximately 3 MHz to 300 GHz. Certain reference documents, how-

Table 39.19. Letter Designation of Microwave
Frequency Bands

Band	Frequency (MHz)
L	1,100–1,700
LS	1,700–2,600
S	2,600–3,950
C	3,950–5,850
XN	5,850–8,200
X	8,200–12,400
Ku	12,400–18,000
K	18,000–26,500
Ka	26,500–40,000

ever, define the microwave frequency range as beginning as low as 10 MHz and
as high as 300 GHz. The region between about 30 KHz and the infrared is generally
referred to as the radio-frequency (RF) region. Reference may be made to Figure
39.1 to determine the position in the electromagnetic spectrum occupied by microwaves relative to other electromagnetic radiations. Certain bands of microwave
frequencies have been arbitrarily assigned letter designations by industry (see Table
39.19). However, there is no universal agreement on these letter designations.
Other discrete frequencies have been assigned by the Federal Communications
Commission (FCC) for industrial, scientific, and medical applications (labeled ISM
bands) as shown in Table 39.20.

6.2 Sources of Microwave Radiation

Although microwave radiation has important applications in communications and
navigational technology such as satellite communications systems, acquisition and
tracking radar, air and traffic control radar, weather radar, and UHF TV transmitters, there is a large number of commercial applications in other fields—microwave ovens, diathermy equipment, and industrial drying equipment. Some typical primary sources of microwave energy are klystrons, magnetrons, backward

Table 39.20. Industrial, Scientific, and Medical
(ISM) Frequencies Assigned by the FCC

13.56 MHz ± 6.78 kHz
27.12 MHz ± 160 kHz
40.68 MHz ± 20 kHz
915 MHz ± 25 MHz
2450 MHz ± 50 MHz
5800 MHz ± 75 MHz
22,125 MHz ± 125 MHz

wave oscillators, and semiconductor transit time devices such as IMPATT diodes. Such sources may operate in a continuous mode (CW), as in the case of some communications systems, in an intermittent mode, as with microwave ovens, induction heating equipment, and diathermy equipment; or in a pulsed mode, as is the case with radar and digital communication systems. Various frequency designations and uses of microwave energy are given in Table 39.21. Of particular interest are the different designations used by the Eastern European countries, the USSR, and the Western countries.

Natural sources of RF and microwave energy can produce peak electric field strengths exceeding 100 V/m at ground level when cold weather fronts move through a given geographical area. Solar radiation intensities range from 10^{-18} to 10^{-17} W/$(m^2)(Hz)$, but the integrated intensities at the earth's surface for the frequency range 0.2 to 10 GHz are approximately 10^{-8} mW/cm^2. This value is to be compared with an average value of 10^2 mW/cm^2 on the earth's surface attributable to the entire (UV, visible, IR, and microwave) solar spectrum.

Given the formation of our solar system, the natural background of nonionizing electromagnetic radiation in the terrestrial biosphere arises primarily from our sun and from reradiation from the earth. The maximum solar spectral irradiance (E_λ) is found at a wavelength of 0.5084 μm; the maximum for terrestrial black-body reradiation is found at 10.06 μm. For all practical purposes the RF/microwave levels of any significance at the earth's surface arise from man-made sources. Also noteworthy is the fact that cosmic radiation and natural terrestrial radioactivity can create nonionizing electric fields of the order of 100 to 200 V/m, in the earth's atmosphere, and the presence of lightning may produce fields as high as 10^3 V/m in the ELF region of 3 to 300 Hz and in the VLF region of 1 to 10 kHz (53).

6.3 Biological Effects of Microwave Radiation

When the human body is exposed to microwave radiation, the usual process of absorption, reflection, transmission, and scattering take place, depending on the wavelength and wave front characteristics. Microwave wavelengths less than 3 cm are absorbed mostly in the outer skin surface. Wavelengths between 3 and 10 cm penetrate more deeply (1 mm to 1 cm), and at wavelengths of 10 to 20 cm, penetration and absorption are sufficiently great that the potential for causing damage to internal body organs must be considered. The human body is thought to be essentially transparent to wavelengths greater than about 500 cm. At about 300 MHz (1 m wavelength) the depth of penetration changes rapidly with frequency, declining to millimeter depths at frequencies of about 3000 MHz. For purposes of comparison, the skin penetration at 3000 MHz is approximately 16 mm, whereas that for 10,000 MHz it is approximately 4 mm. One of the difficulties in establishing the relationships between the characteristics of microwave fields and the bioeffects produced by those fields is our limited ability to measure and describe the incident fields as well as the special patterns of absorbed power within the tissue. Furthermore, when attempting to make environmental measurements one often encounters distortion of the field pattern when a measurement probe is inserted into that field.

Table 39.21. Radio-frequency and Microwave Band Designations

United States	U.S.S.R.	Wavelengths	Frequencies	Typical Uses
Radio-frequency bands				
Low frequency (LF)	Long (VCh)	10^4–10^3 m	30–300 kHz	Radio navigation, radio beacon
Medium frequency (MF)	Medium (HF)	10^3–10^2 m	0.3–3 MHz	Marine radiotelephone, Loran, AM broadcast
High frequency (HF)	Short (UHF)	10^2–10 m	3–30 MHz	Amateur radio, worldwide broadcasting, medical diathermy, radio astronomy, citizen bands
Microwave bands				
Very high frequency (VHF)	Ultra-short (meter)	10–1 m	30–300 MHz	FM broadcast, television, air traffic control, radio navigation
Ultra high frequency (UHF)	Decimeter	1–0.1 m	0.3–3 GHz	Television, microwave point-to-point, microwave ovens, telemetry, tropo scatter and meteorological radar, mobile telephone
Super high frequency (SHF)	Centimeter (SHF)	10–1 cm	3–30 GHz	Satellite communication, airborne weather radar, altimeters, shipborne navigational radar, microwave point-to-point radio
Extra high frequency (EHF)	Millimeter	1–0.1 cm	30–300 GHz	Radio astronomy, cloud detection radar, space research, hydrogen cyanide emission, millimeter wave communication

See Section 6.5, "Electromagnetic Fields and Absorption in Tissue," and Section 6.7, "Measurement of Microwave Radiation," for a more detailed description of measurement techniques and special problems in quantifying the deposition of microwave energy in tissue.

A review of the evidence assembled by Western investigators concerning the biological effects of microwave radiation leads to the overall conclusion that most of the reported effects are explainable on the basis of heat deposition in tissue. The photon energy of electromagnetic radiation at microwave frequencies is considered to be too low to cause the removal of electrons from molecules regardless of the number of quanta absorbed; hence ionization effects are excluded. One of the continuing sources of controversy is whether "nonthermal" effects can be induced in tissue as a result of microwave irradiation. The term "nonthermal" effect usually designates a biological change produced in the absence of any detectable rise in temperature in the test system. The phenomenon of "pearl chain" formation, an alignment of particles with the lines of force of an electromagnetic field, has been offered by some investigators as an example of a nonthermal effect. However, Sher and Schwan (57) concluded that "the implications for pearl chain formation are that on no account can biological pearl chains occur for particles smaller than 3 μm in diameter without risking overheating of tissues. Particles smaller than about 30 μm would not form pearl chains; freely moveable particles of this size are not available in the body."

The current consensus is that pearl chain formation does not occur in the human body as a result of microwave irradiation. Frey (58) reported "hearing" microwave pulses at an average power density as low as 100 μW/cm². The nature of the perception was described as a buzz or a ticking, depending on the pulse length and the pulse repetition rate. The subjects reported the sensation of sound within or behind the head. Selective shielding revealed that the most sensitive area was the region directly over the temporal lobe of the brain and that the greatest sensitivity was to the frequency range from 300 to 1200 MHz.

The current explanation for such microwave audition is that the microwave energy does not exert direct action on the auditory nerve or brain centers; instead it is due to induction of thermoelastic waves in the head. It is generally conceded that certain RF pulses can produce thermal gradients that in turn generate elastic shock waves at the boundaries of tissue or bone with dissimilar dielectric properties, and that the shock waves are transmitted to the cochlea and ossicular chain of the middle ear, where they are perceived as clicks. To date there is no evidence that the auditory clicks constitute a risk of injury.

Research performed in this country has generally concluded that microwave energy is capable of producing cataracts in exposed persons. This conclusion is based almost entirely on animal experimentation. One experienced ophthalmologist has openly expressed his doubts about the possibility of producing cataracts in humans as a result of microwave exposure unless a massive, extremely high level of exposure occurs. Despite this, most knowledgeable persons concede the point that microwave radiation can cause cataracts under special exposure conditions.

One of the notable early animal cataract studies was that of Carpenter and Van

Ummersen, who investigated the effects of microwave radiation on the production of cataracts in rabbit eyes. Exposures to 2.45 GHz radiation were made at power densities ranging from 80 to 400 mW/cm² for different exposure durations. They found that repeated doses of 67 J/cm² spaced a day, a week, or 2 weeks apart produced lens opacities even though the single threshold cataractogenic dose (at a power density 280 mW/cm²) was 84 J/cm². When the single dose was reduced to 50 J/cm², opacities were produced when the doses were administered 1 or 4 days apart, but when the interval between exposures was increased to 7 days, no opacification was noted even after five such weekly exposures. At a power density of 80 mW/cm² (dose of 29 J/cm²), the lowest exposure level used in the study, no effect developed; but when this dose was administered daily for 10 to 15 days, cataracts were produced. These experiments indicate a possible cumulative effect on the lens of the eye if exposures are repeated at a rate such that the lens does not have sufficient time to recover between exposures, even at levels below the single dose cataractogenic threshold (78).

The published allegation (61) that microwave radiation caused the death of a man has been challenged by medical authorities who claimed that the case in question was one of acute appendicitis that led to profound shock and death on the tenth postoperative day. The case was discounted as being due to microwave exposure in an Armed Forces Institute of Pathology Memorandum of July 25, 1957 (62).

A number of epidemiologic studies have been conducted on human populations that had at any time been exposed to microwave radiation. Many of these studies were designed to elicit information on the incidence of cataracts and general physiological effects. The studies of Zaret et al. (63, 64) have shown a small but statistically significant increase in the number of changes at the posterior polar surface of the lens; however, cause and effect relationships in each individual case have not been subsequently elucidated. Barron and Baraff (65) compared the environmental and clinical data on 335 microwave workers to those of a controlled, nonexposed population. No significant differences were found in mortality, disease, sick leave, subjective complaints, or the results of clinical analyses. Czerski (66) has made a similar study of some 800 Polish workers who were exposed to microwaves during the course of their work. The incidence of functional disturbances could not be correlated with exposure level (0.2 to 6 mW/cm²) or duration of exposure. The study by Robinette and Silverman (67) of 20,000 former Navy personnel each in the exposed and control groups did not yield significant correlations between exposure to radio frequencies and long-term mortality or between exposure and various clinical manifestations that required hospitalization. The epidemiology study by Lilienfeld and co-workers of personnel who resided in the U.S. embassy in Moscow during periods when the embassy was irradiated with low levels of radio frequencies was unable to find any statistically significant differences associated with radio frequencies in total mortality or mortality from specific causes or in other illnesses (including cancer) between personnel in the Moscow embassy and control groups comprised of personnel in other Eastern European U.S. embassies (68).

The investigations in the Soviet Union and Eastern European countries have largely stressed nonthermal microwave effects at the central nervous system and cellular levels: cataract formation in rabbits has been reported after a 60-min exposure to 10 mW/cm^2 (74), a result that has not been produced in this country under the conditions described by the Soviets. It is the consensus of Soviet research that exposure to a microwave power density of the order of 10 mW/cm^2 for long periods of time constitutes a pathogenic factor (morphologic lesions in the nervous system, changes in the reproductive function, and other borderline conditions). In response to American suggestions that some of the reported results may reflect adaptive or protective reactions to microwave radiation, the Soviets claim that their results are unequivocal and cannot be regarded as harmless regulatory, adaptive, or compensatory reactions. Effects reported by the Soviets for exposure levels below 10 mW/cm^2 include lowered endurance, retarded weight gain, inhibition of conditioned reflexes, and neurosecretion and neurophysiological disturbances. Even at intensities as low as 250 to 500 μW/cm^2, biological effects such as "change in the activity of the brain" and "immunobiologic resistance" are reported to occur. An interesting insight into Soviet research philosophy is contained in the statement (69) of a Soviet investigator:

When speaking about criteria of significance of reactions, once you keep in mind that not all reactions are of importance but some, while lacking pathologic traits under certain conditions, should nevertheless be given attention when possible harmful consequences are considered; it is therefore necessary to introduce the term "potentially harmful reaction" as a main criterion of importance of a symptom, as opposed to either a threshold biologic (regulatory) reaction in general or a threshold pathologic reaction.

Assuming this philosophy is used for the development of standards of permissible exposure, it is easy to understand why the present Soviet permissible exposure levels are so much more stringent than those in Western countries.

Soviet in vitro studies of 10 cm irradiation of catalase and cholinesterase have been entirely negative; however, in vivo investigations have shown changes in the activity of a whole range of enzymes including cholinesterase, at exposure levels significantly lower than 10 mW/cm^2 (i.e., down to a level of 1 mW/cm^2). According to the investigators, the difference between the in vivo and in vitro experiments "results not from a direct action on molecular structures but from changes in enzyme concentrations in the tissue which are apparently related to disturbances in the neurohormonal regulation of metabolic processes" (70, 71).

Another series of Soviet studies (72) on the effects of microwave radiation on cell membranes, particularly on the selective permeability of the membranes to potassium and sodium ions when subjected to a power density of 1 mW/cm^2, has revealed statistically significant changes in the rates of transport of potassium and sodium ions across all membranes. The transport rates were apparently unrelated to a rise in the temperature of the suspension or to the radiation frequency.

The effect of 2.45 GHz continuous wave radiation on rabbit erythrocytes and rat lymphocytes has been studied in two separate cell systems. Rabbit erythrocytes were exposed at a power density of 5 mW/cm^2 for 3 hr, thus duplicating the experimental conditions of a study by the Polish investigator Baranski, who reported an increase of some 1520 times in potassium efflux and a decrease in osmotic resistance. No changes were found in potassium efflux or osmotic resistance if the exposed and controlled cells were maintained at the same temperature. Although it was later learned that the cells in the Polish experiment had been washed and suspended in unbuffered saline, whereas buffered saline was used in the experiments at the U.S. National Institute of Environmental Health Sciences (NIEHS), the repeated experiments conducted at NIEHS using unbuffered saline again showed no changes in potassium efflux or osmotic resistance as long as the exposed and controlled cells were maintained at the same temperature (73). The importance of this study is highlighted by the Soviet hypothesis that the influence of microwaves on excitable cells such as erythrocytes is due to alterations in the permeability of the cell membranes and that these alterations in permeability are brought about by the transport of ions across the membranes, which in turn influences the membrane potential.

One of the experimental tools used by Soviet investigators to describe biological effects is the conditioned reflex. This fact accounts in part for the difference in threshold effects obtained by Western and Eastern European investigators. The Soviet investigator Livshits comments on the dilemma:

> In conditioned reflex investigations it has been discovered that irradiation induces various disturbances of the higher nervous activity. It must be mentioned that such an effect on radiation is observed only in experiments conducted according to the schemes used by I. P. Pavlov and his followers. Among the authors that have investigated the effects of radiation upon conditioned reflexes in different animals, there is no complete unanimity in the evaluation of the phenomena that they observed and the understanding of their mechanism. The opinion of certain foreign researchers that the change in the latent period and values of the conditioned reflex expresses only or primarily a change in the excitability of the unconditioned reflex centers, in general contradicts the considerable material obtained in the laboratories of I. P. Pavlov and his followers. In radiation pathology, in particular, this question is complex and will require a special investigation (74).

In commenting on Soviet research, Dodge has stated:

> An often disappointing facet of the Soviet and East European literature on the subject of clinical manifestations of microwave exposure is the lack of pertinent data presented on the circumstances of irradiation, frequency, effective area or irradiation, orientation of the body with respect to the source, wave form (continuous or pulsed; modulation factors), exposure schedule and duration, natural shielding factors; and a whole plethora of important environmental factors (heat, humidity, light, etc.) are often omitted from clinical and hygienic reports. In addition, the physiological and psychological status of human subjects such as health, previous or concomitant medication, and mental status is also more often than not omitted. These variables, both individually and combined, affect the human response to microwave radiation (75, 2).

Whereas on the one hand Soviet researchers claim that low level ($\ll 10$ mW/cm^2) nonthermal radiation produces behavioral effects in humans, McAfee indicates that it is thermal stimulation of the peripheral nervous system that produces the neurophysiological and behavioral changes (76, 77) and that the interaction between the peripheral nervous system and the central nervous system accounts for the heart rhythm and blood chemistry effects reported by the Soviets.

The debate over whether microwave radiation can produce cumulative effects continues unabated. As already indicated, the possibility of cumulative effects on the lens from repeated subthreshold cataractogenic exposures of rabbit eyes has been suggested by Carpenter and Van Ummersen (78). The reported cumulative effect may well be the accumulation of damage resulting from repeated exposures, each one capable of producing some degree of damage. If the repetitive exposures take place at time intervals shorter than those needed for the repair of damage, tissue damage is produced (79). Michaelson (2) claims that because this is the process that actually takes place, the suggestion of cumulative effects from microwave exposure is of questionable validity.

There is no strong support for the proposition that the blood–brain barrier (BBB) is adversely affected by brief exposures to the RF fields tested to date. The question about possible effects on the BBB with long-term exposure has not been addressed in experimental work.

With respect to possible RF interactions with the nervous system, there is evidence that some tissue interactions that produce minuscule increases in tissue temperature, that is, less than 0.1°C, can produce physiological changes that are not directly attributable to increased temperature. In such cases, RF fields may couple to tissue through amplification of incident energy at certain frequencies and power levels, that is, "frequency and power windows." Weak oscillating electric gradients in tissue have reportedly been effective in modifying cell functions by mechanisms that appear to amplify the triggering field. Such "cooperative" processes are already recognized in immunological and endocrine reactions and they appear to play a role in neurobiological excitation, especially at the surface of the cell membrane. It is unlikely that weak extracellular fields could couple across the plasma membrane with its extremely high electric gradient; therefore some amplification of fields would almost certainly have to occur along the length and area of the membrane, and the amplification process itself may be calcium-ion dependent in a way that has yet to be elucidated (59).

6.4 Skin Sensation

Questions are often raised about whether a person can "feel" microwaves or the heat that is produced by the incident radiation. Although it is difficult to quantify spurious reports of persons who have detected increases in skin temperature, some quantitative information is available. Schwan et al. (80) found that if a person's forehead is exposed to a power density of 74 mW/cm^2 at 3000 MHz, the reaction time for experiencing a sensation of warmth varies between 15 and 73 sec. The perception of warmth for a power density of 56 mW/cm^2 varies from 50 sec to 3 min of exposure. Hendler's data (81, 82) show that when an area of 37 cm² of the forehead is exposed to microwaves, thermal sensation can be elicited within 1 sec at power densities of 21 mW/cm² when using 10 GHz (10,000 MHz) radiation and 58.6 mW/cm² when using 3 GHz radiation. These data further indicate that if the entire face were exposed to 10 GHz radiation, the threshold for thermal sensation would be 4 to 6 mW/cm² within 5 sec, or approximately 10 mW/cm² within 0.5 sec (2). The sensation of warmth or pain from microwave heating apparently differs little from those felt from infrared heating. In the experiments of Cook (83) an average skin temperature rise of 15°C is required to achieve pain when the initial skin temperature is approximately 32.5°C.

6.5 Electromagnetic Fields and Absorption in Tissue

Prior to the time NCRP Report 67 was published in 1981 (84) there was very little standardization of quantities and units. Most investigators were tolerating the use of terms such as "dose," "dose rate," and "absorbed power density" in the very low frequency ranges. The System Internationale (SI) was being ignored. The proposed use of the term "absorbed power density" produced sufficient revulsion in the rank and file that everyone agreed that the quantities and units had to be improved. The use of such loose terminology reflected poorly on the quality of work being performed and it caused difficulty when trying to compare the results of different studies. Report 67 was a key forerunner in the NCRP entry into the nonionizing radiation arena in that it laid down meaningful, consistent definitions after an exhaustive review and resolution of the many disparate definitions found in electrical engineering dictionaries. Report 67 covered such fundamentals as field quantities, electrostatics, magnetostatics, electrodynamics, analogies with circuit quantities, wave propagation, polarization, modulation, reflection/refraction, standing waves, spherical waves, transmission line fields, antennas, energy transfer processes, and field and SAR measurements. The appendixes of this report contain excellent information on the natural background of nonionizing radiation in the biosphere, RF and microwave absorption in biopolymers, and molecular dynamics in the presence of electromagnetic field perturbation, to name just a few. A major contribution of the report was the introduction of the term specific absorption rate (SAR) into the RF/microwave radiation vocabulary.

SAR was defined as the rate at which electromagnetic energy is absorbed at a point in a medium, per unit mass of the medium, expressed in W/kg. Because energy absorption is a continuous and differentiable function of space and time one may speak of its gradient and rate. Therefore the time derivative of the incremental energy (dW) absorbed in an incremental mass (dm) contained in a volume element (dV) of a given density (p) is expressed as

$$\text{SAR} = \frac{d}{dt}\left(\frac{dW}{dm}\right) = \frac{d}{dt}\left(\frac{dW}{\rho\, dV}\right)$$

The term SAR should not be confused with a measure of the rate of heating of a tissue, although measuring the rate of temperature rise is one of many methods for measuring SAR. SAR is strictly a dosimetric quantity that provides no information as to interaction mechanism, that is, thermal or athermal.

It is most important to remember that the amount of microwave energy deposited and absorbed in biological tissue depends on frequency and on body size, configuration, and homogeneity. Furthermore, it is not possible to classify any of the field quantities measured exterior to an exposed subject as being hazardous, nonhazardous, thermal, or nonthermal. One must first know something about the exposure conditions, frequency, subject size, and subject geometry, to permit determination of the specific absorption rate (SAR), formerly called the absorbed power density, in the subject. For subjects who are exposed to microwave fields, the equation for the time rate of change of temperature T (°C per second) per unit volume of subcutaneous tissue is

$$\frac{dT}{dt} = \frac{0.239 \times 10^{-3}}{c(W_a + W_m - W_c - W_b)}$$

where W_a is the specific absorption rate, W_m is the metabolic heating rate, W_c is the heat loss due to thermal conduction, W_b is the power dissipated by blood flow, all expressed in watts per kilogram, and c is the thermal conductivity, expressed in kilocalories per kilogram per degree celsius (85).

The specific absorption rate for tissue exposed to an electromagnetic field is

$$W_a = \frac{10^{-3}\,\sigma}{\rho\,E^2}$$

where σ is the electrical conductivity in mhos per meter, ρ is the tissue density in grams per cubic centimeter, and E is the root mean square value of the electric field in the tissue, in volts per meter. Before the tissue is exposed to electromagnetic fields, it is assumed that a steady-state condition exists where $W_a = dT/dt = 0$, requiring $W_m = W_c + W_b$. Under normal conditions the metabolic rate W_m averages 1.3 W/kg of the whole body, 11 W/kg for brain tissue, and 33 W/kg for heart tissue (85).

According to the energy equation we would expect to see some change in tissue temperature due to applied electromagnetic fields if the specific absorption rate W_a were of the same order of magnitude as W_m or more. The former nonwavelength-dependent safety standard of 10 mW/cm² of incident power was at least partially based on limiting the average value of W_a to the average resting value of W_m. Therefore specific absorption rates that are much greater than W_m would be expected to produce marked thermal effects, whereas specific absorption rates that are much less than W_m would not be expected to produce any significant thermal effects (85).

Peak values of specific absorption rate (50 W/kg $\leq W_a \leq$ 170 W/kg) have been used to provide vigorous local therapeutic heating of deep vasculated tissues in man treated with diathermy (86). Also, values of W_a that are equal to or much

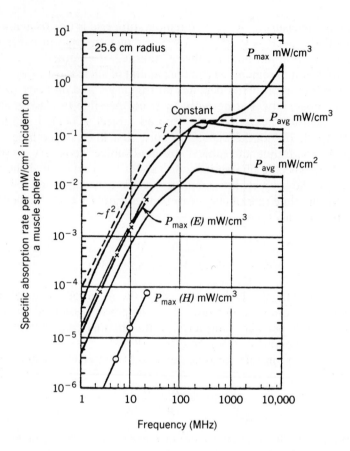

Figure 39.6. Specific absorption rate patterns versus frequency in spherical muscle model of 70-kg man exposed to plane wave 1 mW/cm² source. After Guy (85).

greater than 4 W/kg in the brain tissue of cats yielded measurable increases in brain temperature and decreases in the latency times of evoked potentials (87). Body temperature increases, accompanied by behavioral changes, have been observed in rats by Justesen and King (88) for values of W_a equal or greater than 3.1 W/kg and by Hunt et al. (89) and Phillips et al. (90) for W_a values equal to or greater than 6.3 W/kg averaged over the body of the animal (85). Cataracts have been induced in rabbits by acute exposures where the specific absorption rate in the eye was greater than 138 W/kg (91).

Figure 39.6 shows the variation of specific absorption rate with frequency in a spherical muscle model with the same mass as a 70-kg man. In the frequency range from 1 to 20 MHz the absorption varies as the square of the frequency and is due primarily to the magnetically induced electric fields. The maximum specific absorption rate induced by the incident E field is denoted by the curve marked with

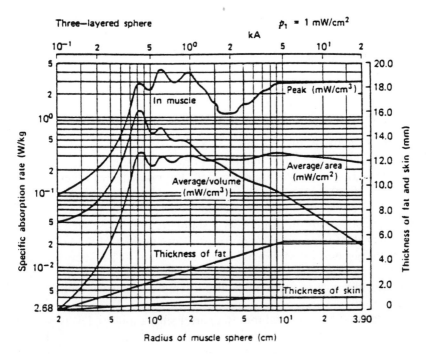

Figure 39.7. Specific absorption rate versus outside radius in spherical tissue layer model of animal exposed to 2.45 GHz 1 mW/cm² plane wave.

crosses, and that due to the incident H field is denoted by the curve with circles. In this range the maximum specific absorption rate is only 10^{-5} to 10^{-2} W/(kg)(mW)(cm²) of incident power. In the frequency range of 100 to 1000 MHz internal reflections are significant for the man-sized sphere, and the average absorption attains a maximum of 2×10^{-2} W/(kg)(mW)(cm²) of incident power at 200 MHz which remains relatively constant with frequency up to 10 GHz. The maximum specific absorption rate increases with frequency above 1000 MHz, approaching that produced by nonpenetrating radiation. The dashed lines illustrate roughly the frequency dependence of the total or average absorbed power and indicate how safety standards might be relaxed as a function of frequency if the absorption characteristics in humans were the same as that for the sphere (85).

Figure 39.7 illustrates absorption patterns as a function of body size (radius of muscle sphere) for a sphere consisting of a central muscle core surrounded by concentric layers of subcutaneous fat and skin exposed to 2.45 GHz plane wave radiation at 1 mW/cm². Assuming that the spherical models have some relevance to the human body, Guy (85) makes the point that the peak specific absorption rate could be as high as 4.2 W/kg within the body or head of a small bird or animal but as low as 0.27 W/kg at the surface and 0.05 W/kg at a point 2.5 cm deep in the human body, when all are exposed to a 1 mW/cm², 918 MHz source. Therefore exposure to 10 mW/cm² could be of extreme significance, and 0.5 mW/cm² could be of mild thermal significance to the smaller animals in comparison with their metabolic rate. For the human model, on the other hand, 10 mW/cm^2 would appear to be minimally significant in terms of thermal insult; a power density of 0.5 mW/cm^2 would have negligible thermal significance.

6.6   Exposure Criteria

Schwan and Li (92) in 1953 examined the thresholds for cataractogenesis and thermal damage to tissue. The minimum power density necessary for producing such changes was found to be approximately 100 mW/cm^2, to which they applied a safety factor of 10 to obtain a maximum permissible exposure level of 10 mW/cm^2. This number was then incorporated into many official standards. The American National Standards Institute C95.1 standard (93) subsequently required a limiting power density of 10 mW/cm^2 for exposure periods of 0.1 hr or more; and a limiting energy density of 1 mWhr/cm^2 during any 0.1-hr period. The latter criterion permitted intermittent exposure at levels above 10 mW/cm^2 on the basis that such intermittency would not produce a temperature rise greater than 1°C in human tissue.

Later, Schwan (94) suggested that the permissible exposure levels be expressed in terms of current density rather than power density, especially when dealing with measurements in the near or reactive field, where the concept of power density loses its meaning. He proposed a permissible current density of approximately 3 mA/cm^2, because this value was considered to be comparable to a far-field value of 10 mW/cm^2. At frequencies below 100 KHz it was believed that this value could be somewhat lower, and for frequencies above 1 GHz, somewhat higher.

The recommendations of the ANSI C95.1 committee were qualified as follows:

Body temperature depends in part on sources of heat input such as electromagnetic radiation, physical labor, high ambient temperature and on heat dissipation capability as affected by clothing, humidity, etc. People who suffer from circulatory difficulties and certain other ailments are vulnerable. The power levels established by radiation guide numbers are related in a complicated way to power levels at which damage may occur. The guide numbers are appropriate for moderate environments. Under conditions of moderate to severe heat stress, the guide number given should be appropriately reduced. Under conditions of intense cold, higher guide numbers may also be appropriate after careful consideration is given to the individual situation. These formulated recommendations pertain to both whole body irradiation and partial body irradiation. Partial body irradiation must be included since it has been shown that some parts of the body, for example, the eyes, and testes, may be harmed if exposed to incident radiation levels significantly in excess of the recommended levels.

Mumford (95) proposed a means for adjusting the 10 mW/cm^2 exposure criterion on the basis of heat stress. According to his recommendation the exposure limit would be 10 mW/cm^2 when the temperature–humidity index (THI) is equal to or less than 70. When the THI is between 70 and 79, the limit is (80 − THI) mW/

Table 39.22. USSR 1984 RF Occupational
Standard

Frequency (MHz)	E-Field EL (V/m)	H-Field EL (A/m)
0.06–1.5	50	5
1.5–3	50	—
3–30	20	—
30–50	10	0.3
50–300	5	—
300–300000	—	—

cm^2. Thus when the THI is equal to or greater than 79, the exposure limit is 1 mW/cm^2.

The original Soviet criterion (96) for exposure to microwave radiation stated that the intensity of the radiation within the frequency range of 300 MHz to 300 GHz should not exceed a value of 10 μW/cm^2 during the working day. For exposures not longer than 2 hr per working day, the level should not exceed 100 μW/cm^2, and for exposures not longer than 15 to 20 min per working day, a power density of 1 mW/cm^2, provided the exposure level for the remainder of the working day did not exceed 10 μW/cm^2. Also, protective goggles had to be worn at the higher (1 mW/cm^2) level. From the public health standpoint, the Soviet standard stated: "In the microwave range the intensity of radiation should not exceed 1 microwatt per square centimeter (1 μW/cm^2) for places where human occupancy occurs and for persons not occupationally exposed." According to Gordon et al. (69), one of the main reasons for the relatively low permissible exposure levels was that clinical observations of microwave workers exposed for 6 to 8 years to power densities of the order of tenths of a milliwatt per square centimeter had, in their opinion, demonstrated functional disturbances, such as neural, vegetative, and asthenic syndromes.

The 1984 U.S.S.R. occupational standard lists the following maximum permissible exposure limits for electric field strengths. For a frequency range of 60 kHz to 3 MHz, a value of 50 V/m is acceptable. For the range of 3 to 30 MHz, a value of 20 V/m is used. For a frequency range of 50 to 300 MHz, a value of 5 V/m is permitted. For magnetic field strengths in the frequency range of 60 kHz to 1.5 MHz, a value of 5 A/m is considered to be acceptable. See Table 39.22. Values for public exposure are approximately half those for occupational exposure. For additional reading on the development of the U.S.S.R. standards, refer to References 98 and 99.

The original Bell System radiation protection guide (100) called for a three-step exposure criterion. First, power density levels in excess of 10 mW/cm^2 were considered to be potentially hazardous; hence personnel were not permitted to enter areas where major parts of the body might be exposed to such levels. Second, power density levels between 1 and 10 mW/cm^2 were permitted only for incidental,

Table 39.23. ANSI (1982) Radio-frequency Radiation Protection Guides

(1) Frequency (f) range (MHz)	(2) E-squared (V²/m²)	(3) H-squared (A²/m²)	(4) Power Density (mW/cm²)
0.3–3	400,000	2.5	100
3–30	4,000 × (900/f^2)	0.025 × (900/f^2)	900/f^2
30–300	4,000	0.025	1.0
300–1,500	4,000 × (f/300)	0.025 × (f/300)	f/300
1,500–100,000	20,000	0.125	5.0

occasional, or casual exposure. And third, power density levels below 1 mW/cm² were considered to be acceptable for indefinite or prolonged exposure. The Bell System standard has since been changed to conform fairly closely to the present ANSI criterion.

In 1982, the Subcommittee IV of the ANSI C95 Committee adopted a frequency-dependent standard for both occupational and general-public exposure to RF radiation to replace the ANSI Radiation Protection Guide published in 1974, which was 10 mW/cm² or the equivalent electric and magnetic field strengths over the entire frequency range from 10 MHz to 100 GHz. The newer limits, like the older ones, are not to be exceeded for exposures averaged over any 0.1-hr period. For the first time in the development of Radiofrequency Protection Guides (RFPGs), due cognizance was taken of such factors as body size, mass, and orientation, the polarization of the incident wave, the frequency and intensity of the radiation, the presence of reflective surfaces, and whether the irradiated subject was in conductive contact with a ground plane.

Because of sparse quantitative human data on which to model an exposure criterion, Subcommittee IV of the ANSI C95 Committee conducted a review of exposure data to determine the most significant, reliable, and independently replicated biological/physiological endpoints produced at the lowest specific absorption rate. Thresholds of behavioral impairment (behavioral disruption) were found within a narrow range of whole body averaged SARs, ranging from 4 to 8 W/kg (101). The corresponding range of power densities was 8 to 140 mW/cm². Because the thresholds of behavioral disruption in primates were found to be approximately 4 W/kg this value was chosen as the working threshold for untoward effects in humans in the frequency range of 3 MHz to 100 GHz. (At frequencies below 3 MHz body surface interactions and electric shock are the relevant criteria, not SAR.) The safety margin is a factor of 10 applied to a SAR value of 4 W/kg; hence compliance with the values in the tabular representation of the ANSI C95.1-1982 RFPG shown in Table 39.23 will ensure that no person will be exposed to a specific absorption rate in excess of 0.4 W/kg, averaged over any 6-min interval in the frequency range of 3 MHz to 100 GHz. Although the frequency range of primary interest with respect to SAR is approximately 3 to 300 MHz, the so-called body resonance range, it is believed that limiting the whole body averaged SAR to 0.4 W/kg will auto-

Table 39.24. ACGIH (1984) Radio-frequency/microwave Threshold Limit Values

Frequency (f) Range (MHz)	Power Density (mW/cm^2)	E-squared (V^2/m^2)	H-squared (A^2/m^2)
0.01–3	100	377,000	2.65
3–30	$900/f^2$	$3,770 \times (900/f^2)$	$900/(37.7 \times f^2)$
30–100	1	3770	0.027
100–1000	$f/100$	$3,770 \times f/100$	$f/(37.7 \times 100)$
1000–300,000	10	37,700	0.265

matically provide an additional safety factor for exposure to frequencies above this range.

The 1982 ANSI standard contains the following qualifications:

(1) At frequencies between 300 kHz and 100 GHz, the protection guides may be exceeded if the exposure conditions can be shown to produce SARs below 0.4 W/kg as averaged over the whole body, with spatial peak values not exceeding 8 W/kg as averaged over any one gram of tissue.

(2) At frequencies between 300 kHz and 1 GHz, the protection guides may be exceeded if the radiofrequency input power to the radiating device is 7 W or less. [This exclusion was provided in recognition of the fact that many low-power devices in common use by the general population may produce fields that appear to exceed the exposure guides in local regions of the body close to the devices but which would yield whole-body SARs much lower than those in the exposure guides.]

The American Conference of Governmental Industrial Hygienists published threshold limit values (ACGIH, 1984) also based on 0.4 W/kg but intended for occupational exposures only. These TLVs, to be averaged over any 6-min period, are shown in Table 39.24 (55).

One major difference between the 1982 ANSI and 1984 ACGIH standards is that the 1-mW/cm^2 value on the latter standard extends only from 30 to 100 MHz and rises with a slope $f/100$ at 100 MHz to 10 mW/cm^2 at 1 GHz. This difference is based on the premise that children, who have higher whole body resonant frequencies than adults, are not likely to be occupationally exposed to radio frequencies. Also different in the 1984 ACGIH standard is the lower frequency limit of 10 kHz instead of 300 kHz.

The standard proposed by Scientific Committee 53 of the National Council for Radiation Protection and Measurements (NCRP) is shown in Figure 39.8. The main difference between the 1982 ANSI criteria (102) and those proposed by NCRP Scientific Committee 53 is the inclusion of a separate, more stringent general population exposure criterion in the NCRP document. ANSI proposed that its criteria apply to persons occupationally exposed in the workplace as well as persons

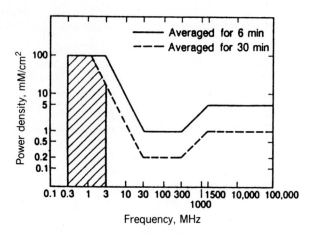

Figure 39.8.   Criteria for exposure to RFEM fields. Exposure, expressed in equivalent far-field power density (mW/cm²) for a whole body averaged SAR of 0.4 W/kg, is shown in the solid line, taken to be the occupational exposure criterion. The dashed line, one-fifth that of the occupational criterion, is the criterion for the general population. Note the time-averaging period allowed for each criterion. The cross-hatched area represents a frequency range in which whole body SAR has limited significance. The overall frequency range for the criteria is 0.3 MHz to 100 GHz. Depending on the circumstances, use of these criteria is constrained by a number of conditions and the criteria cannot be applied without reference to these conditions. From NCRP Report 86.

in the general population on the basis that RF exposures produce threshold effects and that a proper exposure envelope had been included to protect all persons. NCRP countered with the rationale that occupationally exposed persons are usually well informed about potential hazards, and are free to determine which risks are acceptable to them, but persons in the general population are generally unaware of RF exposure risks, have little control over exposure levels, and represent a much larger population comprised of people who are possibly more vulnerable to exposure, for example, the aged, pregnant females, infants, children, and the chronically ill. Scientific Committee 53 (101) has recommended that the exposure criterion for the general public be set at a level equal to one-fifth that for occupational groups. That is, the whole body averaged SAR for the general population should not exceed 0.08 W/kg with an exposure average averaging period of 30 min. This reduction is based on the relative exposure periods of the two groups: 168 hr versus 40 hr/week, that is, 40/168, which equals approximately 0.2.

A few observations about these latest exposure criteria are in order:

1. The limits of exposure below 30 MHz apply to free space conditions where a person is not in contact with any object, including the ground. For other conditions, such as standing on the ground with insulated shoes and making contact with a grounded object, or being grounded and touching an insulated metal object, the limits must be lowered and a case-by-case evaluation must

be made. The standard does not tell you exactly how to do this, other than to determine the applicable exposure limits through the use of three criteria: whole body SAR (0.4 W/kg); maximal local SAR (8 W/kg); and RF burns at the point of contact (200 mA). A great deal of caution is in order in applying the exposure criterion to situations where persons are exposed to high fields at frequencies below 3 MHz. Also, at higher frequencies the local or spatial peak SAR can be incredibly high while still conforming to the whole body averaged value of 0.4 W/kg.

2. Although the whole body average SAR may not exceed a value of 0.4 W/kg, data obtained with simulated human models indicate that the localized SAR values ("hot spots") may reach 20 times this value. Because this is also assumed to be the case for the animal species in the studies used to develop the RFPG, it was concluded that 8 W/kg in any gram of tissue is acceptable as long as the whole body SAR does not exceed 0.4 W/kg averaged over 0.1 hr (the occupational limit). We need more biological underpinning of the data before we can feel comfortable about permitting such high localized SARs. We also need this information in order to make a more quantitative assessment of the following apparently conflicting statements contained in NCRP Report No 86 (101). The first statement is: "In those cases where there are highly intense local concentrations of RF energy ("hot spots"), this knowledge should supersede whole body values and lead to a corresponding reduction in permissible exposure level." (In other words, this is a special case and permissible exposure level should be reduced.) The second statement is: "The deposition of energy in all or certain parts of the human body is a specific condition that lends itself uniquely to an evaluation through simulation modeling. Results of such procedures indicate that efficient thermal averaging over all body tissues of locally deposited RF energy will probably occur through convective heat transfer via blood flow." (In other words, this is not a special case and average values of energy diposition may be assumed.)

3. If a given carrier frequency is modulated at a depth of 50 percent or greater at frequencies between 3 and 100 Hz, the NCRP recommendation is to limit the exposure criteria for occupational exposure to those that apply to exposure of the general public. Such a practice may prove to be acceptable in the long run, but one cannot know with certainty whether this action is justified or whether the general population standard will guarantee the absence of biological effects when exposures occur to modulation frequencies between 3 and 100 Hz.

4. One can still conform to a whole body averaged SAR of 0.4 W/kg (averaged over 6 or 30 min) while exposing subjects to extremely high peak powers. There is a need to develop a limit on peak power. This is a very important research project.

5. The use of mobile and hand-held transceivers by members of the general public allows exposures that are in excess of the general population standard, but not above the occupational guidelines, as long as the user does not expose

others to levels exceeding the general population standard. It is difficult to understand how the average citizen will be able to determine whether or not he or she is exceeding the general population standard.

6. With the curtailment of research funds by the federal government, opportunities for further investigation of mechanisms of interaction between nonionizing electromagnetic energy and biological systems will undoubtedly be limited. There is a practical need to progress from our present macroscopic understanding of the basis on which exposure limits have been set to a better understanding of the effects of fields internal to the irradiated organism. There is a need to understand these matters on a microscopic scale.

7. The ANSI-C95.1-1982 Standard and the NCRP Report No. 86 represent a significant improvement over what we had in earlier times. But given the many qualifying remarks contained in the documents, the many conditions that require professional assistance in their interpretation and application, and conditions that cannot be evaluated for the lack of data, we need improved understanding through research.

A brief chronology of events associated with the development of RF exposure limits in the United States is given in Table 39.25.

6.7 Measurement of Microwave Radiation

One of the key factors underlying the controversy over the reported biological effects of RF radiation is the lack of standardization of (or in many cases even a means of comparing) the measurement techniques used to quantify the exposure of experimental animals or humans to electromagnetic fields. For these reasons, it is heartening to see that the National Council on Radiation Protection and Measurements (NCRP) plans to release a very useful report entitled: "A Practical Guide to the Determination of Human Exposures to Radiofrequency Fields."

An idealized conception of microwave propagation consists of a plane wave moving in an unbounded isotropic medium where the electric $|E|$ and magnetic field $|H|$ vectors are mutually perpendicular and both are perpendicular to the direction of wave propagation. Unfortunately the simple proportionality between the E and H fields in free space is valid only in the so-called far field (Fraunhofer region) of the radiating device. The far-field region is sufficiently removed from the source that the angular field distribution is essentially independent of the distance to the source. The power density in the far field is inversely proportional to the square of the distance from the source and directly proportional to the product of $|E|$ and $|H|$. Therefore the measurement of either the $|E|$ and $|H|$ vector in the far field is all that is needed to establish the exposure conditions. Plane wave detection in the far field is well understood and easily obtained with equipment that has been calibrated for use in the frequency range of interest. Most hazard survey instruments have been calibrated in the far field to read in units of milliwatts per square centimeter. A typical device for far-field measurements is a suitable antenna coupled to a power meter.

To estimate the power density levels in the near field (Fresnel region) of large-aperture circular antennas, one has traditionally used the following simplified relationships:

$$W = \frac{16P}{\pi D^2} = \frac{4P}{A} \quad \text{(near field)}$$

where P is the average power output, D is the diameter of the antenna, A is the effective area of the antenna, and W is power density. If this computation reveals a power density that is less than a specific exposure limit, no further calculation is necessary because the equation gives the maximum power density on the beam axis. If the computed value exceeds the exposure criterion, one then usually assumes that the calculated power density exists throughout the near field. The far-field power densities are then computed from the Friis free space transmission formula:

$$W = \frac{GP}{4\pi r^2} = \frac{AP}{\lambda^2 r^2} \quad \text{(far field)}$$

where λ is the wavelength, r is the distance from the antenna, G is the far-field antenna gain, and W, P, and A are as in the near-field equation.

The distance r from the circular antenna to the intersection of the near and far fields is approximately:

$$r_1 = \frac{\pi D^2}{8\lambda} = \frac{A}{2\lambda}$$

These simplified equations do not account for reflections from ground structures or surfaces where the power density could be as much as four times greater than the assumed free space value. If conditions are such that focusing effects can be produced, factors even greater than four can be achieved.

A number of uncertainties and errors accompany all measurements, but the uncertainties are particularly present when the measurements are made in the near field. The region closest to the radiating source is called the reactive near field where energy is stored. Between the reactive near field and the far field is the radiating near field (Fresnel region), where the phase and amplitude relationships between the E and H fields vary with distance from the source. The configuration of the source can also affect the pattern of radiation to a considerable extent. Multipath $|E|$ and $|H|$ vectors that exist in the reactive and radiative near fields may become so complex that the concept of power density is almost meaningless. For this reason there is considerable merit in measuring $|E|^2$ or $|E|$ or $|H|^2$ or $|H|$. Because $|E|^2$ and $|H|^2$ in plane waves are simply related to W, there is no problem comparable to that in the near field. Other potential measurement errors may be brought about by the characteristics of the monitoring probe.

To make measurements in the near or reactive field, the measurement device

Table 39.25. Chronology of Selected Events Associated with RF Exposure
Limits in the U.S.A.

Prior to 1950	Fragmented reports on electrical excitation of tissue
1952	Dr. Fred Hirsch reported eye damage in the *AMA Archives of Industrial Health* (103). Estimated "power density" to which humans were exposed: 100 mW/cm²
1953	Naval Medical Research Meeting; concluded that 100 mW/cm² was detrimental. Schwan estimated 10 mW/cm² as possible exposure criterion. Bell Laboratories estimated safe level at 0.1 mW/cm² (105). (Private speculation that Russians may have miscopied this value as 0.01 mW/cm²)
1955	General Electric established a level of 1 mW/cm² (106). Although many power density values were proposed there was little scientific basis for them because of the paucity of definitive biological data.
	Mayo Clinic attendees reviewed data on RF-induced cataracts in animals (107). Schwan and Li proposed 10 mW/cm² at all RF frequencies (104). Basis: metabolism of food generates energy at a rate of 5 mW/cm²; doubling this should not cause harm. No reported detrimental effects at levels of approximately 10 mW/cm² (108). The need for additional biological substantiation was recommended by Schwan
1956	Microwave "hearing" reported (RADAR) (109)
1957–58	The 10 mW/cm² value adopted by Bell Laboratories, GE, U.S. Army, U.S. Navy, U.S. Air Force (110). Consensus was thermal basis although nonthermal possibility discussed
1957–1961	Tri-Service Conferences; Col. G. M. Knauf, U.S.A.F. reported international concurrence with 10 mW/cm² value (111)
1961	Soviet standard 0.01 mW/cm² maximum in 1 day (based on reported low-level CNS effects). Soviet regulations required eye protection at 1 mW/cm² (112). Translations of Soviet work often inaccurate. Also, few technical details available on experimental design, measurements, and statistical treatment
1962	Documentation of low-intensity microwave irradiation to the U.S. embassy in Moscow (113). Security issues
1966	ANSI C95 Radiofrequency Standard adopted (frequency independent from 10 MHz to 100 GHz) (114). Johns Hopkins Applied Physics Laboratory attempted modeling of U.S. exposure criteria
1968	Senate Commerce Committee Meetings (115)
1968	Radiation Control for Health and Safety Act (PL 90-602) Technical Electronic Products Radiation Safety Standards Committee (TEPRSSC) established (116)
1969	Bureau of Radiological Health (BRH) standards adopted under Public Law 90-602 (117): TV, microwave ovens, electron tubes, diagnostic X-ray equipment. Symposium on Biological Effects and Health Implications of Microwave Radiation, Richmond, Virginia. Renewed interest in research. First Symposium since the Tri-Service meetings (118)
1970	Hirsch reevaluated 1952 report (119). Claimed exposure level was much greater than 100 mW/cm²

Table 39.25. (Continued)

1971	Electromagnetic Radiation Management Advisory Council (ERMAC) established in the President's Office of Telecommunications Policy (OTC). Five-year program, multi-agency, $63 million proposed budget. One objective was the determination of the long-term effects of low level microwave radiation
1973	Senate Committee on Commerce. Hearings on Radiation Control for Health and Safety Act of 1968. (Public Law 90-602) (120). Consumer Union recommends against purchase of microwave ovens, claiming unknown biological effects (121). TEPRSSC recommends the 1 and 5 mW/cm² microwave oven emission limits to BRH. American Home Appliances Manufacturers (AHAM) make presentations in support of the manufacture of microwave ovens (121)
1974	National Academy of Science Committee on the Biosphere Effects of Extremely Low Frequency Radiation (ELF) (122). SEAFARER, formerly SANGUINE project
1974	ANSI C95 1966 Standard reconfirmed with minor revisions (123)
1975	U.S. Air Force proposes 50 mW/cm² (10 kHz–10 MHz) and 10 mW/cm² (10 MHz–300 GHz) (124). Early U.S. evidence of a frequency-dependent exposure criterion
1976	Irradiation of U.S. embassy continues in Moscow (125). Lilienfeld et al. epidemiologic study (126)
1981	NCRP Report No. 67—"Radiofrequency Electromagnetic Fields: Properties, Quantities and Units, Biophysical Interaction and Measurements." The report develops and proposes the quantity specific absorption rate (SAR) (84)
1981	WHO classifies East European Standards: Group I (USSR); 10 μW/cm² not to exceed 1 mW/cm²; Group II (GDR, Poland, Czechoslovakia): general population standard 10–100 μW/cm² (127)
1982	New ANSI C95 Standard. Frequency dependent and based on a whole body averaged SAR of 0.4 W/kg (102)
1986	NCRP Report No. 86, "Biological Effects and Exposure Criteria for Radiofrequency Electromagnetic Fields" (101)

should be able to measure $|E|^2$ and $|H|^2$ fields separately. The probe sensor itself must be sufficiently small compared with wavelengths in the field, to minimize perturbation of those fields. And the probe should be essentially nondirectional and capable of responding to all polarizations. The polarization characteristics of the device are particularly important in the near field because vertical linear, horizontal linear, elliptical, and circular polarizations may be present at any point in space regardless of source configuration. Circular polarization exists when two equal-amplitude, linearly polarized waves (e.g., one in the x-direction and the other in the y-direction) are superimposed but 90° out of phase. An elliptically polarized wave consists of two unequal-amplitude, orthogonal, linearly polarized waves that are 90° out of phase. The polarization of a wave largely determines its transmission, reflection, and scattering characteristics when impinging upon materials, including human tissue.

Many probes measure electric field strength in volts per meter by means of a thermocouple whose dc output is proportional to volts per square meter. This output may be calibrated and translated into a power density reading, usually milliwatts per square centimeter, taking into account the space impedance Z_o that relates the $|E|$ to the $|H|$ component of the propagated wave. This proportionality constant or impedance for the far-field region has a value of 377Ω. If the space impedance is not 377Ω (the value at which the survey meter has been calibrated) at the point of measurement, a calibration error results.

In general, all devices that are designed to measure microwave radiation for the purpose of assessing hazards to personnel should possess certain minimum characteristics: the instrument impedance should be matched to the field to prevent backscatter from the probe to the source, and the probe should behave like an isotropic receiver, that is, be sensitive to all polarizations, possess a response time that is adequate for handling peak to average power, and exhibit a flat response over the frequency bands of interest.

Most microwave measuring devices are based on bolometry, calorimetry, voltage and resistance changes in detectors, rectification, and radiation pressure on a reflecting surface. The latter three methods are self-explanatory. Bolometry measurements are based on the absorption of power in a temperature-sensitive resistive element, the change in resistance being proportional to the absorbed power. For direct measurement of peak power, a bolometer is the element of choice because the typical time constant is of the order of 300 μsec. Bolometers have a positive temperature coefficient (i.e., the resistance increases with temperature); thermistors have a negative temperature coefficient. Thermistors are more rugged than bolometers and have a greater power-handling capacity. They are preferred for the measurement of average power of pulsed signals because of their log time constant (0.1 to 1 sec).

Crystal diode detectors are useful for the measurement of microwave power, but care must be taken to avoid burnout of the diode in the vicinity of high powered equipment.

Through the use of three orthogonal dipoles, it is possible to design a probe that is both independent of polarization and isotropic in its response. The probe consists of resistive thin-film elements that are folded back in parallel on themselves with the output terminals at one end (Figure 39.9a).

In the example shown in Figure 39.9, the probe elements may consist of thin films of overlapping antimony and bismuth deposited on a plastic substrate. Thermal energy is dissipated primarily at the narrow, high-resistance strips, thereby producing an increase in temperature at these points (hot junctions). The low-resistance, wider sections of the thermocouple act as cold junctions. The resultant dc voltage is directly proportional to the energy dissipated in the narrow, high-resistance strips of the thermocouple. Variation in the sensitivity of the device due to changes in ambient temperature is reportedly less than 0.05 percent because of close spacing of the cold junctions (131). The placement of the three mutually

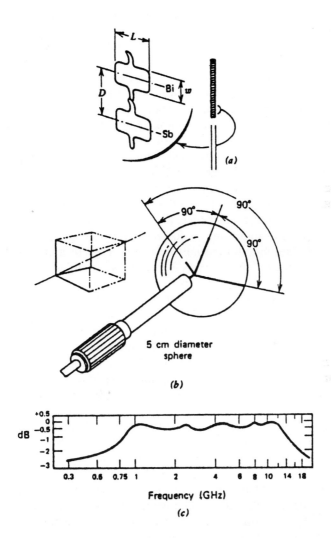

Figure 39.9. (*a*) Antenna probe element with distributed thermocouple film elements; tapered film leads reduce interaction between leads and elements. (*b*) Relationship of the three mutually orthogonal probe elements; probe elements are contained within a 5-cm sphere. (*c*) Typical frequency response characteristics. Reproduced with permission of the Narda Microwave Corporation.

orthogonal sensor elements within a polystyrene sphere to obtain a 5 cm spacing between the sensor elements and the outer surface of the sphere is illustrated in Figure 39.9b. A representative frequency response of the probe appears in Figure 39.9c. Note that the midfrequency bandwidth of interest (1 dB bandwidth) is 850 MHz to 15 GHz, thereby incorporating a flat response feature at the two microwave oven frequencies of 915 and 2450 MHz (2.45 GHz). This type of broad-band radiation monitor may be used for the measurement of radiation emitted from a wide variety of sources, but the original design was fashioned to provide a useful probe for measuring the leakage radiation from microwave ovens. The 5-cm electrically transparent Styrofoam spacing device was included to ensure compliance with the microwave standard under Public Law 90-602 (Radiation Control for Health and Safety Act). Under the provision of the regulations adopted under this Act, newly manufactured microwave ovens must not emit levels in excess of 1 mW/cm^2; the permissible level throughout the useful life of the product is 5 mW/cm^2.

Certain broad-band devices in the frequency range below 300 MHz have been designed to measure the $|H|$ field. The development of such instruments has been encouraged by some recent evidence (84) that the $|H|$ fields of low-frequency radiation produce more significant biological effects than do $|E|$ fields. The sensing probes of the $|H|$ field monitors usually contain a series of mutually perpendicular loops. Low-frequency radiation below 300 MHz is usually measured with loop or short whip antennas, tuned voltmeters, or field intensity meters.

When measurements must be made of electric and magnetic fields, particularly by a person who is not skilled in making such measurements, the following format is suggested: (1) become thoroughly familiar with the data that served as the technical basis from which the RF Protection Guideline was derived. Pay particular attention to the limitations in the data on biological effects and the field exposure measurement data, as well as the assumptions that have been made (because of the lack of data) in constructing the guideline; (2) become familiar with the RF equipment and exposure time patterns at the site to be assessed; and (3) become familiar with available measurement instrumentation, then choose those that most closely match requirements. If one understands the strengths and limitations of all three parameters one is in a much better position to make a realistic determination of risk.

Some examples of the usefulness of understanding the three aforementioned parameters include the following: (1) knowing something about how the Protective Guideline was derived will increase one's awareness of estimates, assumptions, lack of information, and degrees of precision in the data used to develop an exposure criterion; (2) determining whether measurements must be made in the near field, or whether a simple comparison with a Protective Guideline will suffice, or whether very precise measurements must be made in order to calculate the SAR for specific individuals, can make a big difference in approach; and (3) being aware of the fact that although many instruments have readout displays in units one wishes to measure and record, the device may actually respond to another parameter. For example, readouts may indicate electric and magnetic field strength but the instrument actually responds to the square of the electric and magnetic fields, or a device calibrated for the measurement of AM broadcast E fields may actually respond to the magnetic field. This is not to say that the readout is necessarily fallacious; it just means that some engineering assumption about field conditions has been made in the design of the instrument, an assumption that may not hold true for all conditions. Another fact is that many instruments exhibit a significant out-of-band response, that is, a resonance above or below the frequency range being measured. The out-of-band resonance may cause serious errors in measurements.

6.8 Control Measures

Engineering control measures should be given consideration over any use of personal protective equipment. Engineering control measures may range from the restriction of azimuth and elevation settings on radar antennas (to reduce ground level power density levels), to the complete enclosures found in microwave ovens. Because leaking radiation adversely affects the proper performance of a system, high system performance and safety are mutually reinforcing goals. Personnel who work with microwave energy should have the benefit of orientation and training sessions on the potential hazards of microwaves and the proper means for controlling undue exposures. Safe work practices should be formally prepared and instituted. Emphasis must be placed on the safe design of equipment to include shielding, enclosures, interlocks to prevent accidental energizing of circuits, dummy load terminations, warning signals, and control of electric shock potentials. Ionizing radiation (X-rays) may be produced by certain microwave generators because of the existence of accelerating voltages in excess of 16 kV.

Certain protective suits have been constructed of nylon mesh on which has been deposited a layer of metallic silver. The mesh permits a fair degree of ventilation for the body. Because the silver deposits bestow good electrical conductivity properties to the mesh, special care must be taken to avoid direct contact with exposed electrical terminals. Before using the protection garment, one should have a thorough understanding of its attenuation characteristics as a function of radiation frequency, to be certain that the external field intensity does not exceed the safe level by an amount greater than the attenuation provided by the garment. In this connection, Figure 39.10 showing the transmission loss through a wire grid may prove useful (132). Similarly, the attenuation provided by various types of material as given in Table 39.26 may be of use in designing shields or enclosures (133).

More than a decade ago, it was demonstrated that the function of certain cardiac pacemakers, particularly those of the demand type, could be seriously compromised by pulsed or intermittent microwave radiation. The most effective method of reducing the susceptibility of these devices to microwave interference was improved shielding, and the manufacturers of cardiac pacemakers have since instituted major programs to minimize such interference.

The judicious use of appropriate signs and labels may prove useful in alerting people to the presence of microwave sources. Figure 39.11 illustrates the RF and

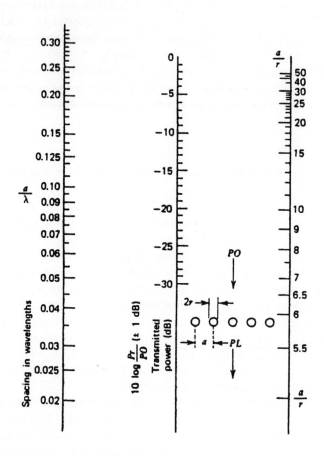

Figure 39.10.   Transmission through a grid of wires of radius r and spacing a.

microwave warning signs adopted by the American National Standards Institute C95 Committee (93).

7 EXTREMELY LOW FREQUENT (ELF) RADIATION

Until recently, the potential biological hazard associated with radio frequencies and extremely low frequencies were generally regarded as not as great as that posed by microwaves in the 800 to 3000 MHz frequency range. With the increasing concern and uncertainty about ELF magnetic fields, particularly at 60 Hz, major attention is now devoted to these portions of the electromagnetic spectrum, rather than the electric fields produced by the radio frequencies.

Table 39.26. Attenuation Factors (dB) (Shielding)

Material	Attenuation Factor (dB)			
	f = 1–3 GHz	3–5 GHz	5–7 GHz	7–10 GHz
60 × 60 mesh screening	20	25	22	20
32 × 32 mesh screening	18	22	22	18
16 × 16 window screen	18	20	20	22
0.25-in. mesh (hardware cloth)	18	15	12	10
Window glass	2	2	3	3.5
0.75-in. pine sheathing	2	2	2	3.5
8-in. concrete block	20	22	26	30

7.1 Biological Effects

Most of the information on the biological effects of extremely low frequencies (ELF) is derived from studies of the effects of radiation produced by low-power frequencies (50 to 60 Hz) and from the U.S. Navy's Project Sanguine (also called Seafarer) (143). Novitskiy et al. (139) have reported a wide range of physiological effects resulting from exposure to ELF, including changes in heartbeat and respiration, anaphylactic shock, and even death. Field strengths that produced the effects varied from approximately 50 to 5000 V/cm, with most of the effects being reported between 200 and 2000 V/cm. On the other hand, Knickerbocker et al. (140) failed to show any effects on general health, behavior, or reproductive ability of mice exposed to a 60-Hz field of 4 kV/in. (1.5 kV/cm) for 1500 hr during the course of a 10.5-month study period. Soviet studies of occupations where employees have been exposed to ELF fields often report complaints such as listlessness, excitability, drowsiness, fatigue, headache, and the so-called neurasthenic syndrome. Ulrich and Ferin (141) have reviewed these studies and have concluded that the irregular shifts such as night work, dry air, and other factors were the probably causes of the signs and symptoms. Kouwenhoven et al. (142) have studied linemen in the United States who have worked on high-voltage transmission lines for many years. Observations over a 3-year period of physical, mental, and emotional health parameters failed to reveal any detrimental effects.

Project Sanguine/Seafarer (143) was begun in the 1960s to investigate the design and operation of a large antenna system that would radiate in the ELF range as a means of communicating with nuclear submarines at depths of approximately 200 m at any location in the world. Because the original proposal envisioned an antenna whose area was in the tens of thousands of square miles, the residents of the areas in which the antenna was likely to be installed became concerned about environmental and ecological effects. Later designs specified an overall antenna area of approximately 3000 square miles. In 1969 the U.S. Navy initiated what is generally regarded as the most comprehensive research program ever attempted to determine whether biological or ecological effects could be expected from exposure to ELF radiation in the immediate vicinity of the antenna, and at various distances from

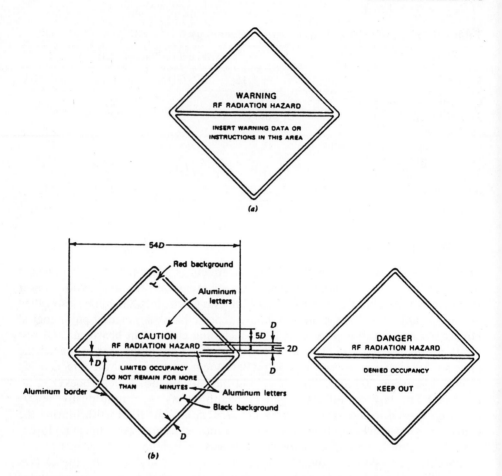

Figure 39.11. Radio-frequency signs. (*a*) General warning. (*b*) Limited occupancy. (*c*) Denied occupancy. ANSI Instructions as follows: (1) Place handling and routing instructions on reverse side. (2) Use *D*-scaling unit. (3) Dimensions in *b* indicate lettering ratio of letter height to thickness of letter lines. (4) Symbol is square, triangle, or right-angle isosceles. Derived from ANSI C95.2 Standard, American National Standards Institute, New York.

the antenna. The electric and magnetic field levels associated with the Seafarer system operating at a frequency of 75 Hz were approximately 0.07 V/m and 0.2 G, respectively, at the surface of the earth directly above a buried antenna.

Several studies have been conducted to determine whether exposure of fruit flies to ELF electric and magnetic fields would induce mutations in the X chromosome. A comprehensive fruit fly study performed by Mittler (144) at field strengths of 10 V/m and 1.0 G at 45, 60, and 75 Hz yielded the conclusion that there was no induction of sex-linked recessive lethal genes, no translocations between II and III chromosomes, no loss of the X or Y chromosomes, and no nondisjunctions or dominant lethal genes. The lethal genes reported in the initial study of Coate and Negherbon (145) were discounted. Exposure of *E. coli* cells to field strengths of up to 20 V/m and 2.0 G at 45 and 75 Hz has failed to produce any mutagenic effect in these bacteria (146). Onion roots exposed to 20 V/m and 2 G at 45 and 75 Hz did not yield any evidence of chromosomal damage (147).

Exposure of rats through the second generation at levels up to 20 V/m and 2 G at 45 and 75 Hz failed to show any significant differences in fertility between exposed controlled populations (148). Factors measured during the study were mortality, body weight, and anatomical defects. A study (149) of blood pressure, body weight, and anatomical and physiological changes among canines was made under the same exposure conditions (i.e., 20 V/m, 2 G, at 45 and 75 Hz) with negative results except for a spurious increase in hypertension. The finding of hypertension was considered to be inconclusive.

Results were inconclusive in a study (150) of the thresholds of perception of ELF electric fields by marine animals at levels that were roughly 1000 times greater than levels proposed for Seafarer. A comprehensive study (151) of the behavior in pigeons and rats at field levels of up to 2 G and 100 V/m at 45, 60, and 75 Hz indicated no effect. Studies of reaction time in primates exposed to ELF fields showed the central nervous system to be relatively insensitive to a magnetic field of 10 G at 45 Hz (152, 153). Tests of immediate memory and general motor activity in primates exposed to 1.4 V/m and 10 G at 75 Hz had no influence on performance (154). Studies have demonstrated the exquisite sensing apparatus of sharks to ELF at a frequency of 10 Hz (155). The detection threshold for ELF radiation at 75 Hz is being investigated, but it is expected to be higher than that at 10 Hz.

Because many biological functions exhibit a 24-hr periodicity, they are often referred to as circadian (*circa* = about; *dies* = day) rhythms. Although their mechanism remains obscure, circadian rhythms are endogenous to living organisms rather than the result of external influences. Also, there is some evidence that alteration of these internal timing systems may be associated with deterioration in psychological and physiological function.

When environmental time cues such as the normal day–night cycle have been eliminated, Wever (156) has demonstrated in humans a "free-running" rhythmic period of approximately 24 hr for activity, rest, and body temperature. When all known and recognized environmental cues have been eliminated from the test environment, the rhythm appears to be "free running"; that is, it is not synchronized by environmental cues. Wever reports that under such conditions there is an increased tendency for the circadian rhythms to desynchronize from each other, producing what he terms "internal desynchronization." Such desynchronization apparently increases the normal circadian periods from approximately 25 hr to time periods that may be twice as great. The presence of natural or artificial fields such as ELF apparently strengthens the internal coupling of biological timing mechanisms, a presumably desirable effect. Therefore there is some implication in Wever's work that it is the absence of natural or artificial fields such as ELF, rather than their presence, that may produce undesirable physiological effects.

Halberg (157) undertook a series of investigations of ELF radiation effects on silk tree leaflets, flour beetles, and mice. Field conditions ranged from 45 to 75

Hz, 0.4 to 2 G, and 1 to 180 V/m, with the duration of exposure varying from a few days to several months. The effects of magnetic fields (60 Hz, 1 G) and electromagnetic fields (60 Hz, 0.4 G, 100 V/m; and 60 Hz, 1 G, 100 V/m) on the circadian rhythms of silk tree leaflets appeared to be negative or at best inconclusive. There was no effect of exposure to 60 to 70 Hz, 1 G fields on the circadian toxicity susceptibility of the flour beetle when exposed to an insecticide (Dichlorvos), nor did continuous exposure of mice to a 45 Hz, 136 V/m electric field for 1 week before and after ouabain injection increase susceptibility to the drug. No significant difference in mice mortality appeared to exist between control and experimental groups exposed to 75 Hz, 25 V/m electric fields. Body weight was found to be unchanged even when mice were continuously exposed to 75 Hz, 1 G fields for several months. The effects of ELF fields ranging from 1 to 10 V/m and 0.5 to 2 G, singly and in combination, on the circadian temperature rhythm, food consumption, and survival rate, were negative. Even in cases of ELF fields reportedly increasing the synchronization of biorhythms, such results must likewise be considered to be suggestive rather than conclusive.

Although most investigations into the possible biological effects of ELF radiation have yielded negative data, a few findings bear careful follow-up investigation. These include (1) the need to elucidate the reasons for an apparent increase in serum triglycerides in exposed Navy personnel, (2) the need to follow up on the exquisite sensitivity of the shark (and possibly other marine life) to ELF radiation, and (3) the need to clarify the effect of ELF on bird migration patterns and avian biorhythms.

7.2 ELF Magnetic Fields

7.2.1 Static and Time-Varying Magnetic Fields

Today's intense interest in the biological effects of electromagnetic fields has materialized because of a recent concern that relatively low-level fields produced by everyday electrical appliances, wiring in the home, and power transmission lines may be causally related to a number of detrimental health effects. For example, certain epidemiologic studies have claimed a statistical association between exposures to power distribution lines in the vicinity of residences and cancer in children, but an approximately equal number of studies have not shown such an association. A report by the Congressional Office of Technology Assessment stated: "The emerging evidence no longer allows one to categorically assert that there are no risks, but it does not provide a basis for asserting that there is a significant risk." In summary, the implications for public health remain unclear.

Before examining the full range of events surrounding the concern about electric and, particularly, magnetic fields, it is well to review some facts about the existence and the prevalence of magnetic fields, beginning with the field surrounding the earth. Although the earth's magnetic field varies in magnitude across the United States, its average value is approximately 0.6 G. The inclination of the earth's magnetic field ranges from 36° from horizontal in Hawaii to 81° in Alaska and from 55° in southern Texas to 76.5° in northern Minnesota. Inclination angles vary annually from 0′ in Hawaii to 7.5′ in Nantucket, Massachusetts.

The naturally occurring magnetic field has a continuous frequency spectrum, ranging from the low end of the extremely low frequency (ELF) region to radio frequencies from lightning discharges. Weak 50 to 60 Hz fields exist naturally at approximately 10^{-5} mG, orders of magnitude below those created by man-made power systems. Relatively strong natural electromagnetic fields may exist for short periods during a local thunderstorm. The magnetic fields produced by lightning may be of the same order of magnitude as those produced by power systems although the magnetic fields generated by power systems persist for much longer periods of time. On the other hand, magnetic fields around the home and on the edge of rights of way of transmission lines may be lower than the earth's natural magnetic field. The predominant sources of residential magnetic fields appear to be external to private residences, the principal contributors appearing to be the electric power distribution lines in the immediate vicinity of the home (159).

7.2.2 Applications and Sources

Electromagnetic fields are generated by a wide variety of commonly used appliances, including radio-frequency induction heaters, microwave ovens, hair dryers, and electric ranges, and mixers. The average magnetic fields in residences in various geographic areas range from about 0.5 to 1.0 mG, but a small number of homes exceed this by an order of magnitude or more. Nearby transmission lines may be an external source. Also, ground currents in proximity to the house may generate magnetic fields. Localized 60-Hz magnetic flux densities measured near (30 cm from surface) household appliances such as refrigerators, ranges, electric drills, food mixers, and shavers range from 0.3 to 270 mG, with approximately 95 percent below 100 mG. Measurements of highly localized fields 1.5 m from the surfaces showed that fields of 95 percent of appliances were below 1 mG.

7.2.3 Medical Magnetic Resonance Imaging Systems

Medical magnetic resonance imaging (MRI) systems are based on the nuclear spin resonance of the hydrogen atom. The systems generate three different fields: (1) a high-intensity static magnetic field (0.04 to 1.5 T) that aligns the magnetic dipoles of hydrogen nuclei, producing longitudinal magnetization that determines the intensity of the MRI signal; (2) transverse magnetization produced from the longitudinal magnetization by a RF signal at a frequency determined by the magnitude of the magnetic field and the magnetic moment of the nuclei (for hydrogen nuclei, a 0.35-T magnetic field requires a pulsed RF signal at 15 MHz with a pulse repetition frequency of approximately one per second); and (3) a ramp-like magnetic field placed across the exposed tissue (at millitesla levels) and oscillating at a few hertz to produce a magnetic gradient that shifts the spin frequency slightly, thereby providing position information as a function of spin frequency.

The RF fields used in MRI interact with tissue primarily through the deposition of thermal energy. The specific absorption rate of the fields is proportional to the electrical conductivity and E^2 in tissue, where E is the electric field intensity. When energy deposition exceeds 4 W/kg, the thermoregulatory capacity of the body is exceeded, and tissue temperature rises. However, the specific absorption rates that can produce such effects are not exceeded in the present generation of MRI equipment.

In several laboratory and epidemiologic studies of the health effects of static and time-varying fields, exposure to high-intensity fields for the brief periods typical of MRI and in vivo spectroscopy has not produced adverse physiological effects. However, special attention must be paid to certain specific hazards, such as the torques exerted on ferromagnetic materials such as aneurysm clips and prostheses, the heating of prosthetic devices by radio-frequency fields, pacemaker malfunctions associated with closure of the reed relay switch, and electromagnetic interference caused by time-varying fields.

Nearly all current MRI scanners employ static magnetic fields with flux densities less than 2.5 T. The time variation of the flux density in these devices is less than 3 T per second, and the whole body specific absorption rates are less than 0.4 W/kg.

7.2.4 Cancer Risk and ELF Field Exposure

A number of epidemiologic studies have suggested that chronic exposure to ELF magnetic fields may increase the risk of cancers such as leukemia and CNS tumors. Since 1979, 27 publications have appeared on residential or occupational exposure to power frequency fields. The Wertheimer and Leeper studies showed a correlation of cancer (leukemia) in children with the presence of 60-Hz magnetic fields from high-current primary and secondary wiring configurations in the vicinity of their residences (160). Some investigators found similar results in adults in the same geographic area (161). In contrast, five other epidemiologic studies showed no clear relation between childhood or adult cancer incidence and the presence of power lines (161–165). The Savitz et al. study of childhood cancer and exposure to 60-Hz magnetic fields (166) attempted a replication of the original Wertheimer and Leeper study using a similar case-control method. However, in addition to estimating exposure to residential 60-Hz fields (based on the high-current or low-current configuration of power distribution lines) actual measurements were made of 60-Hz electric and magnetic fields in the residences of the cases and controls. No statistically significant correlation between 60-Hz fields and the incidence of childhood cancer was found. The cancer risk was found to be correlated with power distribution line configurations, but this correlation was marginally significant. It is difficult to see how the use of codes to estimate the density of power distribution lines could be superior to the measurement of actual magnetic fields in and around the residences.

Many confounding factors, such as smoking habits and exposure to industrial pollutants of known carcinogenic potential, were ignored in nearly all the aforementioned studies. Seven new epidemiologic studies designed to overcome many of these deficiencies are being undertaken in North America and Europe.

7.2.5 Exposure Guidelines

No national standards exist for magnetic field exposures, although guidelines have been generated within certain government agencies. Most guidelines were established for limiting exposures to static magnetic fields, but the more widespread use of MRI devices has caused increasing attention to time-varying fields.

In terms of whole body 8-hr exposure to static magnetic fields, it is suggested that the field be limited to 0.01 T. For exposures of less than 1 hr, the field should be limited to 0.1 T, and for exposures of the order of a few minutes perhaps 0.2 T. For exposures confined to the arms or hands, the respective limits may be in the range of 0.1 to 2.0 T: the lower limit might be permitted for an 8-hr exposure, whereas the upper limit could be permitted for an exposure time not to exceed 10 min. Until additional definitive data are collected on the biological effects of magnetic fields, it is suggested that peak exposures not exceed 2.0 T.

The increasing use of MRI led the National Radiological Protection Board in England and the FDA in the United States to establish guidelines for medical exposures to magnetic fields. These guidelines limit static magnetic fields to 2.5 T and time-varying magnetic fields to 20 T/second for pulses of 10 msec or longer, or a root mean square rate of change of magnetic flux density of $2/t^{1/2}$ T for shorter times, where t is the pulse duration in seconds. The average associated whole body RF specific absorption rate should not exceed 0.4 W/kg. Finally, consideration should be given to the installation of warning signs in areas near MRI devices that exceed 0.5 mT. Tentative suggested guidelines for the general public are 0.5 mT and for occupational exposures, 150 mT limited to an 8-hr day.

REFERENCES

1. Public Law 90-602, Radiation Control for Health and Safety Act of 1968, Washington, D.C.

2. S. M. Michaelson, "Human Exposure to Radiant Energy—Potential Hazards and Safety Standards," *Proc. IEEE*, **60** (April 1972).

3. G. M. Wilkening, in *The Industrial Environment, Its Evaluation and Control*, Department of Health Education and Welfare, NIOSH, Washington, DC, 1973, chapter on "Non-Ionizing Radiations."

4. M. Kleinfeld, C. Giel, and I. R. Tabershaw, "Health Hazards Associated with Inert-Gas-Shielded Metal Arc Welding," *Am. Med. Assoc. Arch. Ind. Health*, **15**, 27–31 (1957).

5. K. G. Hansen, "On the Transmission Through Skin of Visible and Ultraviolet Radiation," *Acta Radiol.*, Suppl. 71 (1948).

6. S. Lerman, "Radiation Cataractogenesis," *N.Y. State J. Med.*, **62**, 3075–3085 (1962).

7. J. A. Zuclich and J. S. Connolly, "Ocular Damage Induced by Near Ultraviolet," Technology Inc., San Antonio, TX.

8. T. M. Murphy, "Nucleic Acids: Interaction With Solar UV Radiation," in *Current Topics Radiat. Res.*, **10**, 199 (1975).

9. I. Willis, A. Kligman, and J. Epstein, "Effects of Long Ultraviolet Rays on Human Skin: Photoprotective or Photoaugmentative?" *J. Invest. Dermatol.*, **59**, 419 (1973).

10. C. E. Keeler, *Albinism, Xeroderma Pigmentosum, and Skin Cancer*, National Cancer Institute, Monograph 10, 1963, p. 349.

11. R. L. Elder, "Phototherapy Warning," *J. Pediatr.*, **84**, 145 (1974).

12. L. E. Bockstahler, C. D. Lytle, and K. B. Hellman, "A Review of Photodynamic Therapy for Herpes Simplex: Benefits and Potential Risks," Publication No. (FDA) 75-8013, U.S. Public Health Service, Bureau of Radiological Health, Rockville, MD, 1975.

13. L. F. Mills, C. D. Lytle, F. A. Andersen, K. B. Hellman, and L. E. Bochstahler, "A Review of Biological Effects and Potential Risks Associated with Ultraviolet Radiation as Used in Dentistry," Publication No. (FDA) 76-8021, U.S. Public Health Service, Bureau of Radiological Health, Rockville, MD, 1976.

14. F. Daniels, *Ultraviolet Carcinogenesis in Man*, National Cancer Institute, Monograph 10, p. 407, 1963.

15. M. Faber, personal communication, 1975.

16. G. Funding, O. M. Henriques, and E. Rekling, *Über Lichtkanzer*, Verh 3., Int. Kongress für Lichtforsch, Wiesbaden, 1936, p. 166.

17. O. A. Jensen, "Malignant Melanomas of the Uvea in Denmark, 1943–1952," Thesis, Copenhagen, 1963.

18. D. G. Pitts, "The Ocular Ultraviolet Action Spectrum and Protective Criteria," *Health Phys. J.*, **25**, 559–566 (December 1973).

19. *Threshold Limit Values for Physical Agents*, published by American Conference of Governmental Industrial Hygienists, Cincinnati, OH, 1989.

20. J. R. Richardson and R. D. Baertsch, "Zinc Sulfide Schottky Barrier Ultraviolet Detectors," in *Solid State Electronics*, Vol. 12, Pergamon Press, New York, 1969, pp. 393–397.

21. P. Bassi, G. Busuoli, L. Lembo, and O. Rimonde, G. *Fis. Sanit. Prot. Contro Radiaz.*, **18**(4), 137–142 (October–December 1974).

22. I. Matelsky, "Non-Ionizing Radiations," in *Industrial Hygiene Highlights*, Vol. 1, L. V. Cralley and G. D. Clayton, Eds., Pittsburgh, PA.

23. H. C. Weston, "Illumination and the Variation of Visual Performance with Age," *Proceedings of the International Commission on Illumination*, Vol. 2, Stockholm, 1951.

24. C. M. Edbrooke and C. Edwards, "Industrial Radiation Cataracts: The Hazards and the Protective Measures," *Arch. Occup. Hyg.*, **10**, 293–304 (1976).

25. K. L. Dunn, "Cataract from Infrared Rays. Class Workers' Cataract—A Preliminary Study on Exposures," *Arch. Inc. Hyg. Occup. Med.*, **1**, 166–180 (1950).

26. K. L. Dunn, "A Preliminary Study on Glass Workers' Cataract Exposures," *Trans. Am. Acad. Ophthalmol. Otolaryngol.*, **54**, 597–605 (1950).

27. G. F. Keatinge, J. Pearson, J. P. Simons, and E. E. White, "Radiation Cataract in Industry," *Arch. Ind. Health*, **11**, 305–315 (1955).

28. D. G. Cogan, D. D. Donaldson, and A. B. Reese, "Clinical–Pathological Characteristics of Radiation Cataract," *Arch. Ophthalmol.*, **47**, 55–70 (1952).

29. H. Goldmann, H. Koenig, and F. Maeder, "The Permeability of the Eye Lens to Infrared," *Ophthalmologica*, **120**, 193–205 (1950).

30. H. Goldmann, "The Genesis of the Cataract of the Glass Blower," *Ann. Ocul.*, **172**, 13–41 (1935); also in *Am. J. Ophthalmol.*, **18**, 590–591 (1935).

31. R. K. Langley, C. B. Mortimer, and C. McCullock, "The Experimental Production of Cataracts by Exposure to Health and Light," *Am. Med. Assoc. Arch. Ophthalmol.*, **63**, 473–488 (1960).

32. M. Wertheimer and W. D. Ward, "Influence of Skin Temperature upon Pain Threshold as Evoked by Thermal Radiation—A Confirmation," *Science* **115**, 499–500 (1952).

33. J. D. Hardy, "Thermal Radiation, Pain and Injury," in *Therapeutic Heat*, Vol. 2, S. Licht, Ed., New Haven, CT, 1958, pp. 157–178.

34. *IES Lighting Handbook*, 4th ed., Waverly Press, Baltimore, 1966.

35. American National Standards Institute Standard for the Safe Use of Lasers, Z136.1, New York, 1986.

36. M. L. Wolbarsht, Ed., *Laser Applications in Medicine and Biology*, Vol. 1, Plenum Press, New York, 1971.

37. A. Javan, W. R. Bennett, and D. R. Herriott, "Population Inversion and Continuous Optical Laser Oscillation in a Gas Discharge Containing a He-Ne Mixture," *Phys. Rev. Lett.*, **6**, 106 (1961).

38. R. C. Miller and W. A. Nordlung, "Tunable Lithium Niobate Optical Oscillator with External Mirrors," *Appl. Phys. Lett.*, **10**, 53 (1967).

39. G. M. Wilkening, "The Potential Hazards of Laser Radiation," *Proceedings of Symposium on Ergonomics and Physical Environmental Factors*, Rome, September 1968, International Labor Office, Geneva, Switzerland, pp. 16–21.

40. Peter P. Sorokin, James J. Wynne, John A. Armstrong, and Rodney T. Hodgson, "Resonantly Enhanced, Nonlinear Generation of Tunable Coherent, Vacuum Ultraviolet (VUV) Light in Atomic Vapors," Third Conference on the Laser, *Ann. N.Y. Acad. Sci.*, **267**, 36 (1976).

41. G. M. Wilkening, "A Commentary on Laser Induced Biological Effects and Protective Measures," *Ann. N.Y. Acad. Sci.*, **68**, part 3, 621–6236 (1970).

42. W. T. Ham, Jr., R. C. Williams, H. A. Mueller, D. Guerry, A. M. Clarke, and W. J. Geeraets, "Effects of Laser Radiation on the Mammalian Eye," *Trans. N.Y. Acad. Sci.*, **28**(2), 517 (1965).

43. W. K. Noell, V. S. Walker, B. S. Kang, and S. Berman, "Retinal Damage by Light in Rats," *Invest. Ophthalmol.* **5**, 450 (1966).

44. W. T. Ham, Jr., H. A. Mueller, and D. H. Sliney, "Retinal Sensitivity to Damage from Short Wavelength Light," *Nature*, **260** (March 11, 1976).

45. A. I. Goldman, W. T. Ham, Jr., and H. A. Mueller, "Mechanisms of Retinal Damage Resulting from Exposure of Rhesus Monkeys to Ultrashort Laser Pulses," *Exp. Eye Res.*, **21**, 457–469 (1975).

46. J. Taboada and D. W. Ebbers, "Ocular Tissue Damage due to Ultrashort 1060 μm Light Pulses from a Mode-Locked Nd:Glass Laser," *Appl. Opt.*, **14**, 1759 (August 1975).

47. A. I. Goldman, report accepted for publication in *Experimental Eye Research*, 1976.

48. K. W. McNeer, M. Ghosh, W. J. Geeraets, and D. Guerry, "ERG After Light Coagulation," *Acta Ophthalmol.*, Suppl. 76, 94 (1963).

49. T. P. Davis and W. J. Mautner, "Helium-Neon Laser Effects on the Eye," Annual Report, Contract No. DADA 17-69-C 9013, U.S. Army Medical Research and Development Command, Washington, DC, 1969.

50. L. Goldmann, W. Kapuscinska-Czerska, S. Michaelson, R. J. Rockwell, D. H. Sliney, B. Tengroth, and M. Wolbarsht, "Health Aspects of Optical Radiation with Particular Reference to Lasers," World Health Organization, 1978.

51. Techical Committee on Laser Products, International Electrotechnical Commission, Geneva, Switzerland, 1984.

52. U. S. Department of Health, Education and Welfare, Bureau of Radiological Health, Laser Products Performance Standard, *Fed. Reg.*, **40**, 148 32252–32265 (July 31, 1975).

53. *Radiofrequency Electromagnetic Fields: Properties, Quantities, and Units.* Biophysical Interacting and Measurements, Report No. 67, National Council on Radiation Protection and Measurements, Washington, DC, 1981.

54. G. M. Wilkening, "Laser Hazard Control Procedures," in *Electronic Product Radiation and the Health Physicist*, U.S. Department of Health, Education and Welfare REPORT BRH/DEP 70-26, October 1970, pp. 275–290.

55. American Conference of Governmental Industrial Hygienists, Threshold Limit Values ACGIH, Cincinnati, OH.

56. Occupational Safety and Health Administration Title 29, Code of Federal Regulations 1910, OSHA Standards 1972, under revision.

57. L. D. Sher and H. P. Schwan, "Mechanical Effects of AC Fields on Particles Dispersed in a Liquid, Biological Implications," Ph.D. dissertation (L. D. Sher), University of Pennsylvania, under Contract AF30(602) ONR Technical Report Dissertation No. 37, 1963.

58. A. H. Frey, "Human Auditory System Response to Modulated Electromagnetic Energy," *J. Appl. Physiol.*, **17**, 689–692 (1962).

59. W. R. Adey, "Electromagnetic Fields, Cell Membrane Amplification and Cancer Promotion," *Proceedings of 1986.* Annual Meeting Devoted to the Nonionizing Radiations, National Council on Radiation Protection and Measurements, Washington, DC.

60. A. W. Guy et al., "Microwave Interaction with the Auditory Systems of Humans and Cats," IEEE G-MTT International Symposium Digest of Technical Papers, 1973, p. 321.

61. J. T. McLaughlin, "Tissue Destruction and Death From Microwave Radiation (Radar)," *Calif. Med.*, **86**, 336–3339 (1957).

62. T. S. Ely, "Microwave Death," letter to the editor, *J. Am. Med. Assoc.*, **217**, 1394 (1971).

63. M. M. Zaret, S. Cleary, B. Pasternack, M. Eisenbud, and H. Schmidt, "A Study of Lenticular Imperfections in the Eyes of A Sample of Microwave Workers and a Control Population," New York University, Final Report No. RADC-TDR-63-125, 1963; ASTIA Doc. AD 413294.

64. M. M. Zaret and M. Eisenbud, "Preliminary Results of Studies of the Lenticular Effects of Microwaves Among Exposed Personnel," in *Proceedings of the Fourth Annual Tri-Service Conference on Biological Effects of Microwave Radiating Equipment and Biological Effects of Microwave Radiations*, M. F. Peytong, Ed., Plenum Press, New York, 1961, pp. 293–308 (Technical Report No. RADC-TR-60-180).

65. C. I. Barron and A. A. Baroff, "Medical Considerations of Exposure to Microwaves (Radar)," *J. Am. Med. Assoc.*, **168**, 1194–1199 (1958).

66. P. Czerski, M. Siekierzynski, and J. Gidynski, "Health Surveillance of Personnel Occupationally Exposed to Microwaves," Parts I, II, III, *Aerosp. Med.*, 1137–1148, October 1974.

67. C. D. Robinette and C. Silverman, "Causes of Death Following Occupational Exposure

to Microwave Radiation (Radar) 950-1974," in D. G. Hazzard, Ed., Symposium on Biological Effects and Measurement of Radiofrequency/Microwaves, Department of Health, Education and Welfare, Washington, DC, HEW Publication No. (FDA) 79-8026 (1977).

68. A. M. Lilienfeld, J. Tonascia, S. Tonascia, C. H. Libauer, G. M. Cauthen, J. A. Markowitz, and S. Weida, "Foreign Service Health Status Study: Evaluation of Status of Foreign Service and other Employees From Selected East European Posts. Ernal Report, July 31, 1978, Contract No. 6025-619073 Dept. of Epidemiology, School of Hygiene and Public Health, The Johns Hopkins University, Baltimore, MD, 1978.

69. Z. V. Gordon, A. V. Roscin, and M. S. Byckov, "Main Directions and Results of Research Conducted in the U.S.S.R. on the Biologic Effects of Microwaves," Proceedings of the International Symposium on Biological Effects and Health Hazards of Microwave Radiation, Polish Medical Publishers, Warsaw, 1973.

70. S. V. Nikogosjan, in *O biologiceskom deistvii EMP radiocastot.*, Vol. 3, Moscow, 1968, p. 97.

71. V. M. Stemler, in *Gigiena truda i biologiceskoe deistvii elektromagnitnyk voln radiocastot.*, Moscow, 1968, p. 175.

72. V. M. Stemler, in *Sbornik Bionika, 1973, Materialy 4 Vsisojuznoj konferenci po bionike*, Moscow, 1973, pp. 3, 87.

73. Fourth report on "Program for Control of Electromagnetic Pollution of the Environment: The Assessment of Biological Hazards of Nonionizing Electromagnetic Radiation," Office of Telecommunications Policy, Executive Office of the President, Washington, DC, June 1976, p. 21.

74. N. N. Livshits, "On the Causes of the Disagreements in Evaluating the Radiosensitivity of the Central Nervous System Among Researchers Using Conditioned Reflex and Maze Methods," *Radiobiology* 7, 238–261 (1967).

75. C. H. Dodge, "Clinical and Hygienic Aspects of Exposure to Electromagnetic Fields," in *Biological Effects and Health Implications of Microwave Radiation*, Symposium Proceedings, S. F. Cleary, Ed., U. S. Department of Health, Education and Welfare, 1970, pp. 140–149.

76. R. D. McAfee, "Physiological Effects of Thermode and Microwave Stimulation of Peripheral Nerves," *Am. J. Physiol.*, **203**, 374–378 (1962).

77. R. D. McAfee, "The Neural and Hormonal Response to Microwave Stimulation of Peripheral Nerves," in *Biological Effects and Health Implications of Microwave Radiation*, Symposium Proceedings, S. F. Cleary, Ed., U.S. Department of Health, Education and Welfare, 1970, pp. 150–153.

78. R. L. Carepenter and C. A. Van Ummersen, "The Action of Microwave Radiation on the Eye," *J. Microwave Power*, **3**, 3–19 (1968).

79. S. M. Michaelson, "Biomedical Aspects of Microwave Exposure," *Am. Ind. Hyg. Assoc. J.*, **32**, 338–345 (1971).

80. H. P. Schwan, A. Anne, and L. Sher, "Heating of Living Tissues," U.S. Naval Air Engineering Center, Philadelphia, Report No. NAEC-ACEL-534, 1966.

81. E. Hendler, "Cutaneous Receptor Response to Microwave Irradiation," in *Thermal Problems in Aerospace Medicine*, J. D. Hardy, Ed., Surrey, England, 1968, pp. 149–161.

82. E. Hendler, J. D. Hardy, and D. Murgatroyd, "Skin Heating and Temperature Sen-

sation Produced by Infrared and Microwave Irradiation," in *Temperature Measurement and Control in Science and Industry*, Part 3, Biology and Medicine, J. D. Hardy, Ed., Reinhold, New York, 1963, pp. 221–230.

83. H. F. Cook, "The Pain Threshold for Microwave and Infrared Radiations," *J. Physiol.*, **118**, 1–11 (1952).

84. NCRP (1981), National Council on Radiation Protection and Measurements: Report No. 67: "Radiofrequency Magnetic Fields: Properties, Quantities and Units, Biophysical Interaction and Measurements," Washington, DC.

85. A. W. Guy, "Quantitation of Induced Electromagnetic Field Patterns in Tissue and Associated Biologic Effects," *Proceedings of the International Symposium on Biological Effects and Health Hazards of Microwave Radiation*, Polish Medical Publishers, Warsaw, 1973.

86. A. W. Guy, J. F. Lehmann, and J. S. Stonebridge, "Therapeutic Applications of Electromagnetic Power," *IEEE Proc. Ind., Sci., Med. Appl. Microwaves*, special issue, January 1974.

87. C. C. Johnson and A. W. Guy, *Proc. IEEE*, **60**, 692 (1972).

88. D. R. Justesen and N. W. King, "Behavioral Effects of Low Level Microwave Irradiation in the Closed Space Situation," in *Biological Effects and Health Implications of Microwave Radiation*, S. F. Cleary, Ed., U.S. Department of Health, Education and Welfare, Report No. BRH/DBE 70-2 (PB 193 8 58), Rockville, MD, 1970, p. 154.

89. E. I. Hunt, "General Activity of Rats Immediately Following Exposure to 2450 MHz Microwaves," in *Digest*, 1972 IMPI Symposium, Ottawa, Canada, May 1972.

90. R. D. Phillips, N. W. King, and E. L. Hunt, "Thermoregulatory Cardiovascular and Metabolic Response of Rats to Single or Repeated Exposures to 2450 MHz Microwaves," in *1973 Symposium on Microwave Power Institute*, Ottawa, Canada, September 1973.

91. P. Kramar, A. F. Emery, A. W. Guy, and J. C. Lim, "Theoretical and Experimental Studies of Microwave Induced Cataracts in Rabbits," in *Applications in the 70s*, 1973 IEEE G-MTT International Microwave Symposium, University of Colorado, Boulder, 1973, p. 265.

92. H. P. Schwan and K. Li, *Proc. IRE*, **41**, 1735 (1953).

93. C95.1 Committee of the American National Standards Institute, "Safety Level of Electromagnetic Radiation with Respect to Personnel," ANSI, New York.

94. H. P. Schwan, in *Biological Effects and Health Implications of Microwave Radiation*, S. F. Cleary, Ed., U.S. Department of Health, Education and Welfare, Report No. BRH/DBE 70-2 (PB 193 8 58), Rockville, MD, 1970.

95. W. W. Mumford, "Heat Stress Due to RF Radiation," *Proc. IEEE*, **57**, 171–178 (1969).

96. "Sanitarnyje normy i pravila pri raboti s istocuikami electromagnitnyck polei vysokih, ultravipokih i sverhvysokih castot," Ministerstvo Zdravohranenija, SSSR 30, 03 NS 848-70, 1970.

97. "Performance Standard for Microwave Ovens," Title 42, Part 78, Suppl. C Sec. 78.212, Code of Federal Regulation; *Fed. Reg.*, **35** (194), 15642 (October 6, 1970).

98. "Jednotna metodika stanoveni intensity pole a ozareni elektromagneticRymi vinami v pasma vysokych fredvenci a velmi vysokych frekvenci k hygienickym ucelum." Vynos hlavniho hygienika CsRcj. HE 344.5-3.2.70/Priloha c. 3k Informacnim zpravam z oboru hygieny prace a nemoci z povolani, Praha, kveten, 1970.

99. "Rozpovzadzenie Rady Ministrow z dnia 25.02.1972 w. spraure bezpieczenstwa i higieny pracy przy stosowaniu urzadzen wytroarzajacych pola elektromagnetyczne w zakresie mikrofalowym," Dziennik Ustaw PRL No. 21, II, Poz. 153, 1972.

100. M. M. Weiss and W. W. Mumford, "Microwave Radiation Hazards," *J. Health Phys.*, **5**, 160 (1961).

101. National Council on Radiation Protection and Measurements (NCRP) Report No. 86, "Biological Effects and Exposure Criteria for Radiofrequency Electromagnetic Fields," NCRP, Bethesda, MD, 1986.

102. American National Standards Institute "Safety Levels with Respect to Human Exposure to Radio Frequency Electromagnetic Fields, 300 kHz to 100 GHz." Report No. ANSI C95.1 (The Institute of Electrical and Electronics Engineers Inc., New York, 1982.

103. F. G. Hirsch, "Microwave Cataracts" presented at NSAE Conference on Industrial Health, Cincinnati, Ohio, April 25, 1952. Also F. G. Hirsch and J. T. Parker, "Bilateral Lenticular Opacities Occurring in a Technician Operating a Microwave Generator," *AMA Arch. Ind. Health*, **6**, 512–517 (Dec. 1952).

104. H. P. Schwan and K. Li, "The Mechanism of Absorption of Ultrahigh Frequency Electromagnetic Energy in Tissues, as Related to the Problem of Tolerance Dosage," *IRE Trans of Medical Electronics*, Vol. ME-4, pp. 45–59, February 1956.

105. Central Safety Committee, Bell Telephone Laboratories, Nov. 1953.

106. B. L. Vosburgh, "Problems Which are Challenging Investigators in Industry," *IRE Trans. Medical Electronics*, Vol. ME-4, pp. 5–7, February, 1956. Also: B. L. Vosburgh, "Recommended Tolerance Levels of MW Energy; Current Views of the General Electric Company's Health and Hygiene Service," *Proc. 2nd Annual Tri-Service Conference on Biological Effects of Microwave Energy*, RADC, Griffiss AFB NY, pp. 188–125; July 8–10, 1958.

107. "Symposium of Physiologic and Pathologic Effects of Microwaves," Mayo Foundation House, Rochester, MN, Sept. 23, 24, 1955.

108. H. P. Schwan, "The Physiological Basis of Injury," *Proc. 1st Annual Tri-Service Conference on Biological Hazards of Microwave Radiation*, RDAC Griffiss AFB, NY, pp. 60–63, July 15–16, 1957.

109. A. H. Frey, "Auditory System Response to Modulated Electromagnetic Energy," *J. Appl. Physiol.* **17**, 689–692, 1962.

110. U.S. Air Force, "Microwave Radiation Hazards," Rome A.F. Depot, Griffiss AFB, NY, Urgent Action Tech. Order, 31-1-511, June 17, 1957.

111. "Biological Effects of Radio Frequency Energies," 1940–1957. *Proc. 1st Annual Tri-Service Conf. on Biological Hazards of Microwave Radiation*, RADC, Griffiss AFB, NY, July 15–16, 1957. Bibliography. Appendix B, pp. 94–103.

112. Temporary Sanitary Rules for Working with Centimeter Waves, Ministry of Health Protection of the USSR, 1958, 1961.

113. The irradiation of the U.S. embassy in Moscow was reportedly known in the 1950s, but documented much later. Refer to References No. 27 and 28.

114. American National Standards Institute (ANSI) Committee 95.1, Safety Level of Electromagnetic Radiation With Respect to Personnel, C95.1, New York 1966.

115. Hearings Before the Committee on the Commerce, United States Senate, Ninetieth

Congress, Serial No. 90-49, US Government Printing Office, Washington, DC, May 1968.

116. Technical Electronic Products Radiation Safety Standards Committee (TEPRSSC) established under the Radiation Control for Health and Safety Set of 1968 (Public Law 90-602).

117. Radiation Control for Health and Safety Act of 1968. (Public Law 90-602). See Annual Reports to the Congress on the Administration of DL 90-602. Bureau of Radiological Health, Washington, DC.

118. Proceedings of the Symposium on the Biological Effects and Health Implications of Microwave Radiation, Richmond, VA, Sept. 1969, S. F. Cleary, Ed., U.S. Dept. of HEW, June 1970.

119. F. G. Hirsch, "Microwave Cataracts—A Case Report Reevaluated," *Electronic Product Radiation and the Health Physicist*, Proceedings of the 4th Annual Symposium of the Health Physics Society, Louisville, Kentucky, Jan. 28–30, 1970 HEW Publication BRH/DEP 70-26.

120. Radiation Control for Health and Safety Hearings before the Committee on Commerce, U.S. Senate, Ninety-Third Congress, Serial No. 93-24, U.S. Government Printing Office, Washington, DC, 1973.

121. Presentations of Consumers Union before the Technical Electronic Products Radiation Safety Standards Committee (TEPRSSC), Washington, DC, 1973. Presentations of American Home Appliances Manufacturers before the Technical Electronic Products Radiation Safety Standards Committee (TEPRSSC), Washington, DC, 1973.

122. National Academy of Sciences, "Biosphere Effects of Extremely Low Frequency Radiation," (Sanguine/Seafarer) Washington, DC, 1974–1977.

123. American National Standards Institute (ANSI), "Safety Levels of Electromagnetic Radiation with Respect to Personnel," C95.1, New York, 1974.

124. U.S. Air Force, "Radiofrequency Radiation Health Hazard Control," AFR 161-42, Washington, DC, 1975.

125. U.S. Senate, Microwave Irradiation of the U.S. Embassy in Moscow. Hearings before the Committee on Commerce, Science, and Transportation Committee, Print 43-949, U.S. Government Printing Office, Washington, DC, 1979.

126. A. M. Lilienfeld, S. Tonascia, C. H. Libauer et al., "Foreign Service and Other Employees From Selected Eastern European Posts," Final Report Contract No. 6025-619073 Department of State, Washington, DC, 1978.

127. World Health Organization (WHO), Environmental Health Criteria for Radiowaves in the Frequency Range from 100 kHz to 300 GHZ (Radiofrequency and Microwaves). WHO Environmental Criteria Program, United Nations, New York, 1981.

128. G. M. Wilkening, "Protection Against Nonionizing Electromagnetic Radiation—An Evolutionary Process," *Proceedings of the Twenty-fifth Annual Meeting of the National Council on Radiation Protection and Measurements*. Also entitled *Radiation Protection Today, the NCRP at Sixty Years*, April 5–6, 1989, p. 235.

129. W. A. Palmisano and A. Pexczenik, "Some Considerations of Microwave Hazards Exposure Criteria," *Mil. Med.*, **131**, 611 (1966).

130. C95.3 Committee of the American National Standards Institute, "Techniques and Instrumentation for the Measurement of Potentially Hazardous Electromagnetic Radiation at Microwave Frequencies," ANSI, New York, 1973.

131. Narda Microwave Corporation, Catalog No. 20, Plainview, NY.

132. W. W. Mumford, "Some Technical Aspects of Microwave Radiation Hazards," *Proc. IRE*, **49**, 427 (1961).

133. W. A. Palmisano, U.S. Army Environmental Hygiene Agency, presented at the American Industrial Hygiene Conference, Akron, OH, 1967.

134. J. H. Heller and A. A. Teixerira-Pinto, "A New Physical Method of Creating Chromosomal Aberrations," *Nature*, **183**, 905–906 (1959).

135. A. A. Teixeira-Pinto, L. L. Nejelski, J. L. Cuttler, and J. H. Heller, "The Behavior of Unicellular Organisms in an Electromagnetic Field," *Exp. Cell Res.*, **20**, 548–564 (1960).

136. J. N. Bollinger, "Detection and Evaluation of Radiofrequency and Microwave Energy in Macaca Mulatta," Southwest Research Institute, San Antonio, TX, Final Report under Contract No. F41609-70-C-0025, SWR1 05-2808-01, February 1971.

137. S. A. Bach, "Biological Sensitivity to Radiofrequency and Microwave Energy," *Fed. Proc.*, **24**, Suppl. 14, 22–26 (1965).

138. S. A. Bach, A. J. Luzzio, and A. S. Brownell, *Effects of Radiofrequency and Microwave Radiation*, M. F. Peyton, Ed., Plenum Press, New York, 1961, pp. 117–133.

139. Y. I. Novitskiy, Z. V. Gordon, A. S. Pressman, and Y. A. Kholodov, "Radio Frequencies and Microwaves, Magnetic and Electric Fields," Nassau Technical Translation No. Nassau TTF-14.021, 1971.

140. G. G. Knickerbocker, W. B. Kouwenhoven, and H. C. Barnes, "Exposure of Mice to a Strong AC Electric Field-An Experimental Study," *IEEE Trans. Power Appar. Syst.*, PAS-96, 498 (1967).

141. L. Ulrich and G. Ferin, "The Effect of Working on Power Transmitting Stations upon Certain Functions of the Organisms (in Czechoslovakian)," *Pracovni Lek.* (Prague), **11**, 500 (1959).

142. W. B. Kouwenhoven, O. R. Langworthy, M. L. Singleweld, and G. G. Knickerbocker, "Medical Evaluation of Man Working in AC Electric Fields," *IEEE Trans. Power Appar. Syst.*, PSD-86, 506 (1967).

143. Project Seafarer, U.S. Department of the Navy, Washington, DC.

144. S. Mittler, "Low Frequency Electromagnetic Radiation and Genetic Aberrations," Northern Illinois University, DeKalb, Final report, September 15, 1972 (AD 749959).

145. Project Sanguine Biological Effects Test Program Pilot Studies, Hazelton Laboratories, Final Report Contract No. N00039-69-C-1572, November 1970 (AD 717408).

146. R. A. Pledger and W. B. Coate, "Bacteria Mutagenis Study," in Reference 145, Chapter F.

147. W. B. Coate and S. S. Ho, "Plant Cytogenic Study," in Reference 145, Chapter G.

148. W. B. Coate and F. E. Reno, "Rat Fertility Studies," in Reference 145, Chapter C.

149. W. R. Teeters and W. B. Coate, "Canine Physiological Study," in Reference 151, Chapter D.

150. "Effects of Low Frequency Electrical Current on Various Marine Animals," Naval Air Systems Command Air Task Report No. 0410801, Work Unit 0100, June 1972 (AD 749335).

151. M. J. Marr et al., "The Effects of Low Energy, Extremely Low Frequency (ELF) Electromagnetic Radiation on Operant Behavior in the Pigeon and the Rat," Georgia Institute of Technology, Atlanta, February 28, 1973 (AD 759415).

152. J. D. Grissett et al., "Central Nervous System Effects as Measured by Reaction Time in Squirrel Monkeys Exposed for Short Periods to Extremely Low Frequency Magnetic Fields," Naval Aerospace Medical Research Laboratory Report No. NAMRL-1137, August 1971 (AD 731994).

153. J. D. Grissett, "Exposure of Squirrel Monkeys for Long Periods to Extremely Low Frequency Magnetic Fields: Central Nervous System Effects as Measured by Reaction Time," Naval Aerospace Medical Research Laboratory Report No. NAMRL-1146, October 1971 (AD 735456).

154. J. deLorge, "Operant Behavior of Rhesus Monkeys in the Pressure of Extremely Low Frequency-Low Intensity Magnetic and Electric Fields," Experiment 1, November 1972 (AD 754058).

155. A. J. Kalmijn, Woods Hole Marine Biological Laboratory, Woods Hole, MA, personal communication; also A. J. Kalmijn, "The Electric Sense of Sharks and Rays," J. Exp. Biol., 55, 371–383 (1971), and A. J. Kalmijn, "Electro-Orientation in Sharks and Rays: Theory and Experimental Evidence," Scripps Institution of Oceanography, University of California, San Diego, Report Contract No. N00014-69-A-0200-6030, Office of Naval Research, November 1973.

156. R. Wever, "Einfluss schwacher elektromagnetische Felder auf die circadiane Periodik des Menschen," Naturwissenschaften, 55(1), 29–32 (1968).

157. F. Halberg, "Circadian Rhythms in Plants, Insects and Mammals Exposed to ELF Magnetic and/or Electric Fields and Currents," University of Minnesota, Final Report to the Office of Naval Research, August 1975.

158. Committee on Biosphere Effects of Extremely Low Frequency (ELF) Radiation, National Academy of Sciences-National Research Council, Washington, DC.

159. W. T. Kaune, R. G. Stevens, N. J. Callahan, et al. Residential Magnetic and Electric Fields, Bioelectromagnetics, 8, 315–335 (1987).

160. N. Wertheimer and E. Leeper, "Electrical Wiring Configurations and Childhood Cancer," Am J. Epidemiol., 109, 273–284 (1979).

161. N. Wertheimer and E. Leeper, "Adult Cancer Related to Electrical Wires Near the Home," Int. J. Epidemiol., 11, 345–355 (1982).

162. J. P. Fulton, S. Cobb, L. Preble, et al., "Electrical Wiring Configurations and Childhood Leukemia in Rhode Island," Am. J. Epidemiol., 111, 292–296 (1980).

163. M. E. McDowell, "Mortality of Persons Resident in the Vicinity of Electricity Transmission Facilities," Br. J. Cancer, 53, 271–279 (1986).

164. A. Myers, B. A. Cartwright, J. A. Bonnell, et al., "Overhead Power Lines and Childhood Cancer," International Conference on Electric and Magnetic Fields in Medicine and Biology, London, December 1985.

165. R. G. Stevens, R. K. Severson, W. T. Kaune, et al., "Epidemiological Study of Residential Exposure to ELF Electric and Magnetic Fields and Risk of Acute Non-lymphocytic Leukemia," DOE/EPRN/NY State Power Lines Project Contractors Review, Denver, November 1986.

166. D. A. Savitz, H. Wachtel, F. A. Barnes, et al., "Case Control Study of Childhood Cancer and Exposure to 60 Hz Magnetic Fields," Am. J. Epidemiol. 128, 21–38 (1988).

CHAPTER FORTY

Ergonomics

Erwin R. Tichauer, Sc.D.

In recent years OSHA has intensified its emphasis of the ergonomics facet of occupational health, and has recently stated that ergonomics will receive top priority during the 1990s. In 1989 the citations and the fines assessed industry increased, especially in the meatpacking industry. On November 1, 1989 OSHA reached an agreement with the United Auto Workers and the Chrysler Corporation regarding violations that had occurred in May of 1987.

OSHA has developed ergonomic guidelines for some industries; the four parts of the ergonomic program outlined in the guides are as follows:

1. *Work site analysis* deals with identification of ergonomic stressors.
2. *Hazard prevention and control* refer to methods for controlling or eliminating the stressors.
3. *Medical management* deals with early detection and return to work policies.
4. *Training and education* provide employees and management with sufficient information so they may actively participate in the ergonomic program.

It is anticipated OSHA will go beyond the guidelines and issue an ergonomics standard in the 1990s. It is expected this will be a complicated procedure, because the hazards are so diverse, ranging from repetitive motion in some industries to those related to the use of VDTs. Another problem being encountered is that the definition of familiar terms is changing.

Because no industry can escape the potential of an ergonomic problem, it is readily apparent that ergonomics will have an increasingly prominent role in the

Patty's Industrial Hygiene and Toxicology, Fourth Edition, Volume 1, Part B, Edited by George D. Clayton and Florence E. Clayton.
ISBN 0-471-50196-4 © 1991 John Wiley & Sons, Inc.

total occupational health picture. It is recommended that every major industry have a minimum of one professional on the staff who is well versed in ergonomics. Contact OSHA for the most current information on the standard and guidelines issued.

This chapter gives basic data on this important subject.

The Editors

1 HISTORICAL BACKGROUND

The systematic study of the ill effects on humans of poorly designed work situations is by no means of recent origin. Ramazzini (1), around 1700, elaborated on the disastrous effects of work stress and its disabling consequences for those engaged in physical labor. He wrote: "Manifold is the harvest of diseases reaped by crafts-men. . . . As the . . . cause I assign certain violent and irregular motions and unnatural postures . . . by which . . . the natural structure of the living machine is so impaired that serious diseases gradually develop." At that time, however, and for centuries to come, there existed no technique for the study of the anatomy of function of the living body. Also laborers were considered to be expendable, and occupational disease was ordinarily rewarded by dismissal.

Thus with few noteworthy exceptions, such as Thackrah (2), industrialists, physicians, and even the precursors of today's social scientists remained insensitive to the exposure of the work force to ergogenic disease produced by mechanical noxae until World War I, when labor became a scarce resource essential to the very survival of the warring nations. This stimulated physiologists (3), as well as psychologists, to embark on an intensive study of the effects of working conditions on human performance and well-being.

Experimental methods for these disciplines were already well established, and their application to problems of humans at work was quite easy. One of the human-oriented life sciences lagged sadly behind, however: anatomy, which had remained a cadaver-based, geographic discipline. Thus physiological responses and behavioral reactions to the demand of work situations were investigated for several decades before the exploration of the structural and mechanical basis of musculoskeletal performance.

Change was pioneered by Adrian (4), who recorded electromyograms during movement, relating them to kinesiological events. The Great Depression slowed down progress in these avenues of scientific inquiry, but during World War II biological and behavioral scientists again had the opportunity to lead the quest for an improved utilization of human resources.

During the postwar period work physiology, industrial psychology, and other specialties consolidated into a broad discipline dedicated to the study of man at work: ergonomics. The Ergonomics Research Society was organized in 1949 to serve the need of those professionals in a variety of disciplines concerned with the effects of work on man.

The founders of the new society argued about the most suitable name, suggesting (5) "Society for Human Ecology" and "Society for the Study of Human Environment." Finally, "ergonomics" was adopted as a term neutral with respect to the relative importance of the behavioral sciences, physiology, and anatomy. An anatomical methodology for the study of work did not yet exist. Thus the thrust of ergonomics had to stem initially from the other two disciplines. However, a speedy development of experimental biomechanics and novel instrumentation available added "live body anatomy" to the spectrum of research resources available to ergonomists.

Gradually, as interest grew in the structural basis of human performance, the electrogoniometer was perfected (6). This made it possible to use simultaneously goniometry and Adrian's myography when investigating the relationships between muscular activity and ensuing movements. Thus information about the functional anatomy of the living body could be applied by ergonomists to the problems germane to both occupational health and industrial productivity. One of the numerous pioneers was Lundervold (7), who related myoelectric signals obtained from typists to posture and hand usage (this was perhaps the first comprehensive biomechanical analysis of a common work situation). Thus electrophysiological kinesiology was developed.

Utilizing procedures of electrophysiological kinesiology, Tichauer (8) added the biomechanical profile to the techniques available for the study of interaction between worker and industrial environment. Meanwhile, behavioral scientists such as Lukiesh and Moss (9) pioneered research into the effects of light and illumination on human performance. Work physiologists such as Belding and Hatch (10) described and explained the effects of climate on working efficiency and studied noise and its effects on workers, not only with respect to deafness, but also in relation to many other physiological parameters (11). The study of performance decrements due to adverse working conditions became the object of an entire new school of students of human fatigue (12).

Thus today, knowledge gathered from many tributary sciences has been blended into a unified discipline dealing with the effects of work on humans.

1.1 Ergonomic Stress Vectors

The basic philosophy of ergonomics (Figure 40.1) considers humans to be organisms subject to two different sets of laws: the laws of Newtonian mechanics and the biological laws of life. It is part of this philosophy to postulate that humans in work situations are surrounded by the external physical working environment, and inside the human body the "internal biomechanical environment," an array of levers and springs also known as the musculoskeletal system, responds to the demands of the task. The stress vectors commonly acting upon humans in the industrial environment have been set out in flow diagram form (Figure 40.2 and 40.3).

1.2 Definition of "Ergonomist"

An ergonomist is a professional trained in the health, behavioral, and technological sciences and competent to apply them within the industrial environment for the

Figure 40.1. A worker surrounded by external physiological and mechanical environment, which must be matched to his internal physiological and biomechanical environments, symbolizes the concept of modern ergonomics (biomechanics). From Reference 79.

purpose of reducing stress vectors sufficiently to prevent the ensuing work strain from rising to pathological levels or producing such undesirable by-products as fatigue, careless workmanship, and high labor turnover. Ergonomics as a discipline aims to help the individual members of the work force to produce at levels economically acceptable to the employer while enjoying, at the same time, a high standard of physiological and emotional well-being.

2   THE ANATOMY OF FUNCTION

Anatomy is concerned with the description and classification of biological structures. Systematic anatomy describes the physical arrangement of the various physiological systems (e.g., anatomy of the cardiovascular system); topographic anatomy describes the arrangement of the various organs, muscular, bony, and neural features with respect to each other (e.g., anatomy of the abdominal cavity); and functional anatomy focuses on the structural basis of biological functions (e.g., the description of the heart valves and ancillary operating structures; the description of the anatomy of joints). As distinct and different from the aforementioned categories, the anatomy of function is concerned with the analysis of the operating

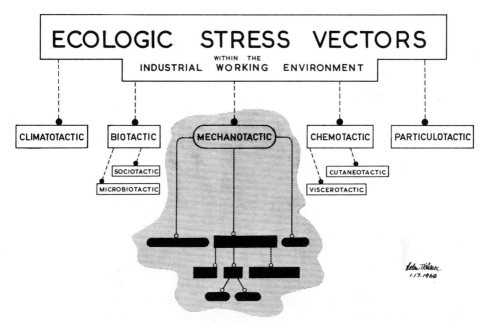

Figure 40.2. The scheme of ecologic stress vectors common to all working environments. Work stress is derived from contact with climate, contact with living organisms such as fellow man or microbe, contact with chemical elements or compounds, contact with hostile particles such as silica, or asbestos and finally, contact with mechanical devices. From Reference 13.

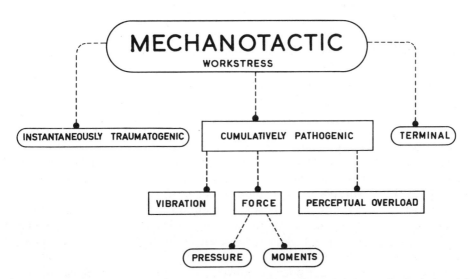

Figure 40.3. The mechanotactic stress vectors leading to hazard exposure in the industrial environment include instantaneous traumatogenic (e.g., an arm is torn off); terminal (i.e., death occurs immediately); and most frequently, cumulatively pathogenic. The latter term describes the gradual development of disability or disease through repeated exposure to mechanical stress vectors over extended periods of time. From Reference 13.

characteristics of anatomical structures and systems when these interact with physical features of the environment, as is the case in the performance of an industrial task. Whenever the "motions and reactions inventory" demanded by the external environment is not compatible with the one available from the internal biomechanical environment, discomfort, trauma, and inefficiency may arise (3).

The anatomy of function is the structural basis of human performance; thus it provides much of the rationale by which the output measurements derived from work physiology and engineering psychology can be explained.

2.1 Anatomical Lever Systems

The neuromuscular system is, in effect, an array of bony levers connected by joints and actuated by muscles that are stimulated by nerves. Muscles act like lineal springs. The velocity of muscular contraction varies inversely as the tension within the muscles. With very few exceptions, lever classifications and taxonomy in both anatomy and applied mechanics are identical. Each class of anatomical levers is specifically suited to perform certain types of movement and postural adjustment efficiently without undue risk of accidents or injury, but may be less suited to perform others. Therefore a good working knowledge of location, function, and limitation of anatomical levers involved in specific occupational maneuvers is a prerequisite for the ergonomic analysis and evaluation of most human–task systems.

First-class levers have force and load located on either side of the fulcrum acting in the same direction but opposed to any force supporting the fulcrum (Figure 40.4). This is exemplified by the arrangement of musculoskeletal structures involved in head movement when looking up and down. Then the atloido-occipital joint acts as a fulcrum of a first-class lever. The muscles of the neck provide the force necessary to extend the head. This is counteracted by gravity acting on the center of mass of the head, which is located on the other side of the joint, and hence constitutes the opposing flexing weight.

First-class levers are found often where fine positional adjustments take place. When a person stands or holds a bulky load, static head movement in the midsagittal plane produces the fine adjustment of the position of the center of mass of the whole body, necessary to maintain upright posture. Individuals suffering from impaired head movement (e.g., arthritis of the neck), should not be exposed to tasks in which inability to maintain postural equilibrium constitutes a potential hazard. Likewise, workplaces where unrestricted head movement is difficult should be provided with chairs or other means of postural stabilization. Special attention should be paid to any feature within the working environment that may cause head fixation (e.g., glaring lights). Sometimes even an unexpected acoustic stimulus, such as a friendly "hello" directed at an individual carrying a heavy and bulky load, may cause inadvertent sideways movement of the head, which can interfere with postural integrity and result in a fall.

Second-class levers have the fulcrum located at one end and the force acting at the other end, but in the same direction as the supporting part of the fulcrum. The

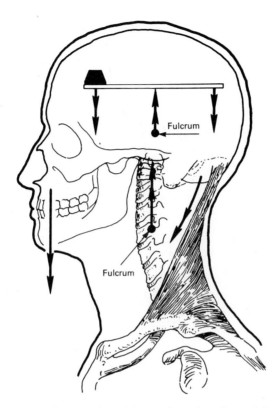

Figure 40.4. The action of the muscles of the neck against the weight of the head is an example of a first-class lever formed by anatomical structures. The atlantooccipital joint acts as a fulcrum. Adapted from Reference 48.

weight acts on any point between fulcrum and force in a direction opposed to both of them. Second-class levers are optimally associated with ballistic movements requiring some force and resulting in modifications of stance, posture, or limb configuration. The muscles inserted into the heel by way of the Achilles tendon (i.e., force) and the weight of the body transmitted through the ankle joint and the weight of the big toe (i.e., fulcrum) are a good example of a second-class lever system used in locomotion (Figure 40.5). The movements of this type of lever are never very precise, therefore foot pedals should have adequately large surfaces and their movement should be terminated by a positive stop rather than by relying on voluntary muscular control. Another example of a second-class lever is provided by the structural arrangement of the shoulder joint. Here the head of the humerus acts as a fulcrum, and the anterior and posterior heads of the deltoid provide the force; the "weight" is provided by the inertia of the mass of the arm. Hence it follows that a shoulder swing moves the hand to a rather indeterminate location; thus such a movement should be terminated by a positive stop, or alternatively, when an object has to be dropped at the end of a motion, the receptacle should

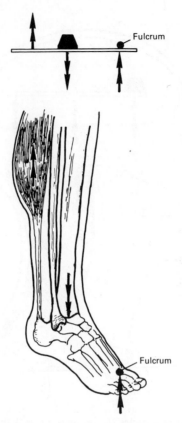

Figure 40.5. The ankle joint, as an example of an anatomical second-class lever system. The fulcrum is located at the base of the big toe. Adapted from Reference 48.

be sufficiently large, or have a flared inlet, to overcome the kinesiological deficiency.

Third-class levers have the fulcrum at one end, and the weight acts on the other end in the same direction as the supporting force of the fulcrum. The "force" acts on any point between weight and fulcrum, but in a direction opposed to them both. Tasks that require the application of strong but voluntarily graded force are best performed by this type of anatomical lever system. Holding a load with forearm and hand when the brachialis muscle acts on the ulna, with the elbow joint constituting a pivot, is a typical example (Figure 40.6).

Torsional levers are a specialized case of third-class lever. Here the axis of rotation of a limb or long bone constitutes a fulcrum. The force generating muscle of the system is inserted into a bony prominence and produces rotation of the limb whenever the muscle contracts. The "weight" is represented by the inertia of the limb plus any external torque opposing rotation. An example is the supination of the flexed forearm (Figure 40.7). Here the fulcrum is the longitudinal axis of the radius; the force is exerted by the biceps muscle inserted into the bicipital tuberosity

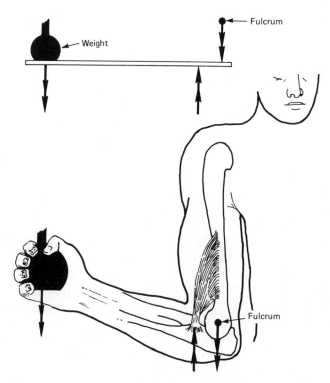

Figure 40.6. An anatomical third-class lever is formed between ulna and humerus. The brachialis muscle provides the activating force; the fulcrum is forced by the trochlea of the humerus. Adapted from Reference 48.

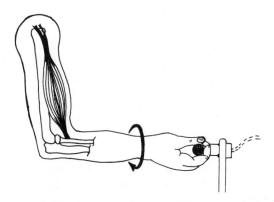

Figure 40.7. The torsional lever system involving radiohumeral joint, biceps, and a resistance against supination of the forearm can be employed with advantage in such operations as closing of valves.

of the radius. The opposing load is made up by the inertia of the forearm and hand, plus the resistance of, for example, a screw being driven home. Tasks to be performed for strength and precision and at variable rates of speed are best assigned to torsional lever systems.

For identification and classification of other lever systems involved in specific occupational maneuvers, standard text books on kinesiology (14–16) should be consulted.

2.2 Occupational Kinesiology

Occupational kinesiology is the discipline concerned with the basic study of human movement and its limitations in work situations. Unfortunately, with the exception of brief monographs (17), all texts and reference books in the field of kinesiology relate to either athletics or rehabilitation medicine.

Kinesiology describes the laws and quantitative relationships essential for the understanding of the mechanisms involved in human performance, either of individuals or of groups of individuals interacting with one another (i.e., a working population). Its basic tributaries are anatomy, physiology, and Newtonian mechanics. It describes and explains the behavior of the whole body, its segments, or individual anatomical structures in response to intrinsic or extrinsic forces.

The student of kinesiology should be thoroughly familiar with the nomenclature and organization of mechanics. Here statics is concerned with the generation and maintenance of equilibrium of bodies and particles. In the context of kinesiology, "bodies" are generally synonymous with anatomical structures and "particles" become anatomical reference points. Likewise, the biodynamic aspects of kinesiology are explained through kinematics, which is concerned with the geometry and patterns of movements, but not with causative forces producing motion. Kinetics, on the other hand, deals with the relation between vectors and forces producing motion and also with the output from body segments in terms of force, work, and power, including the resulting changes in temporal and spatial coordinates of anatomical reference points.

Before the kinematics basic to a specific kinesiological maneuver can be explored, the kinematic element involved must be identified. A kinematic element consists of bones and fibrous and ligamentous structures pertaining to a single joint inasmuch as they affect the geometry of motion. Because kinematics is not concerned with forces, muscles do not normally form part of a kinematic element. As an example, consider the kinematic element of forearm flexion, which consists of the humerus, the ulna, the joint capsule of the humeroulnar joint, and associated ligaments.

Kinematic elements can have several degrees of freedom of motion; the higher this number is, the greater the variety of movements that can be produced. However, accurate movements produced by elements possessing a high degree of freedom require a proportionally higher level of skill: for example, it is easier to position the hand accurately by means of humeroulnar flexion than by a shoulder swing. Likewise, the higher degree of freedom, the greater the influence of musculoskeletal configuration on the effectiveness of movement. As an example, the following are mentioned:

1. Humeroulnar joint: one degree of freedom (i.e., flexion); effectiveness quite independent of general musculoskeletal configuration.

2. Whole elbow joint: two degrees of freedom (i.e., flexion, pro/supination); effectiveness somewhat dependent on musculoskeletal configuration.

3. Hip joint: three degrees of freedom (i.e., flexion, ad/abduction, circumduction); effectiveness of this chain actually, in a number of situations, such as walking, quite dependent on musculoskeletal configuration.

Often in workplace layout, kinematic elements are considered in the initial planning of the geometry of the work situation; but when activity tolerance and other work and effort relationships are important, kinetic elements must be considered. These include constituents of kinematic elements, but in addition they incorporate muscular structures as well as the stimulating nerves and nutrient blood vessels because these affect the immediacy, strength, and endurance aspects of a specific kinesiological maneuver (Figure 40.8).

2.3 Application of Kinesiology to Workplace Layout

Kinesiological concepts can be applied with advantage to the design of work situations when it is essential to minimize physical stress and fatigue. When analyzing motion patterns incidental to the performance of industrial tasks, it is desirable to start with the preparation of a length–tension diagram (Figure 40.9). This is produced by making the protagonist muscle of the kinetic element performing the motion contract isometrically against measured resistance at different points of the motion's pathway. We plot on the x-axis the included angle between the major

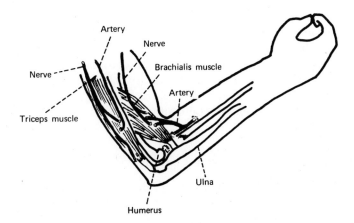

Figure 40.8. The kinetic element of forearm flexion and extension. From Reference 48.

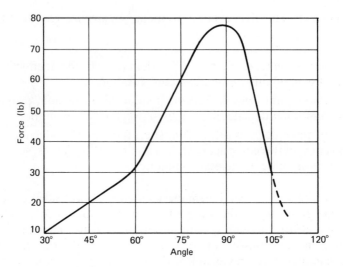

Figure 40.9. Length–tension diagram produced by flexion of the forearm in pronation. "Angle" refers to included angle between the longitudinal axes of forearm and upper arm. The highest parts of the curve indicate the configurations where the biomechanical lever system is most effective.

bony elements involved. On the y-axis the maximal force exerted during contraction in each position is recorded. Generally, the only range of joint movement to be utilized in workplace layout is that which coincides with the highest portion of this curve. The length–tension diagram will show that most kinetic chains can be utilized effectively only throughout a very narrow angle of joint movement, and it will identify the limits of this range.

The force–velocity diagram is another useful graphic representation of the effectiveness of joint movement (Figure 40.10). The rate of change in joint configuration is plotted against maximum forces developed by the muscle, forming roughly a negative exponential curve. The zero velocity value corresponds to the maximum of force. At maximum velocity the force exerted by a muscle approaches zero as a limit. Hence high velocity and high muscular forces are mutually exclusive. The plotting of a force–velocity curve is a complex undertaking not always feasible under field conditions or even in the laboratory. However, an awareness of its general configuration often serves to protect the practicing ergonomist against the selection of ineffective musculoskeletal configurations in workplace layout.

2.4 Optimal Placement of Equipment Controls

In the operation of many equipment controls and other industrially used devices there is no noticeable displacement of anatomical reference points while muscles contract. Under such conditions, isometricity of movement may be assumed. This implies that forces exerted by protagonist and antagonist muscles acting on a limb are in equilibrium with each other, even when exerting maximal force. Occa-

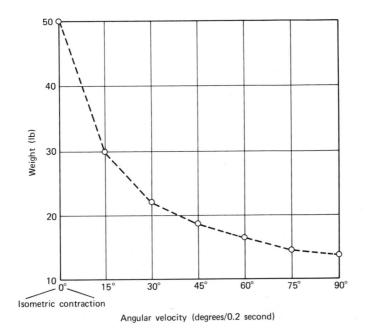

Figure 40.10. Force–velocity curve of elbow flexion; forearm in supination. Excursion of the limb through the narrow range between 75° and 110° of included angle between forearm and upper arm. The diagram shows that high strength and high velocity of movement are mutually exclusive conditions. Furthermore, the highest strength is developed under conditions of zero velocity (i.e., isometricity).

sionally—for example, in the raising of a leg—the force of gravity also acts on a joint; in this case, the sum of all three forces, protagonist, antagonist, and gravity, must be zero.

Kinesiological analysis of anatomical lever systems is of special usefulness whenever it becomes necessary to optimize the position of apparently innocuous but nevertheless potentially traumatogenic equipment controls.

The analysis and computation of some of the forces generated within the kinetic element of forearm flexion during the operation of a pushbutton are described. This may serve as example of the step by step procedure to be followed under similar circumstances.

Step 1. Those anatomical structures absolutely essential to the performance of the task are identified and all others are eliminated from consideration. These are (Figure 40.8) the humerus, ulna, and brachialis muscle.

Step 2. Those forces that, if excessive, may lead to anatomical failure are identified. In this example these are tension in the brachialis muscle and thrust on the elbow joint.

Step 3. A force diagram (such as is used in mechanics) is drawn true to scale (Figure 40.11).

Step 4. The mechanical assumptions essential to the solution of the problem are made. As the condition of isometricity exists, the sum of torques acting clockwise on the elbow joint must be equal to the sum of those acting counterclockwise. (The symbols used in the following computations are defined in the legend to Figure 40.11.)

The following clockwise torques act on the elbow joint:

$$(B + C)F_1 \sin \Omega \tag{1}$$

$$(B + C + D)F_2 \sin \Omega \tag{2}$$

$$(B + C + D)F_3 \tag{3}$$

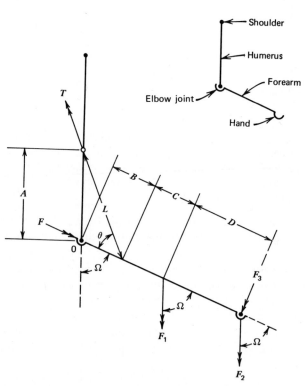

Figure 40.11. Vector diagram used in the computation of forces acting on elbow joint and tension generated in brachialis muscle when operating a push button. A, distance from the average origin of the brachialis to the center of rotation of the humero-ulnar joint; B, distance from the center of rotation to the average insertion of the brachialis on the ulna; C, distance from the insertion to the center of gravity of the forearm; D, distance from the center of gravity to the application of the load (either F_2 or F_3); F_1, weight of the forearm and hand; F_2, a weight in the hand; F_3, a force normal to the hand at all angles; T, tension in the brachialis; F, compressive force exerted on the elbow joint; L, distance between origin and insertion of brachialis; Ω, angle of flexion of the forearm; θ, angle of insertion of the brachialis.

The counterclockwise torque is

$$(BT \sin \theta) \tag{4}$$

where T is the unknown tension.

Equating clockwise and counterclockwise torques and regrouping terms, we have

$$T = \frac{[(F_1 + F_2)(B + C) + DF_2] \sin \Omega + (B + C + F)F_3}{B \sin \theta} \tag{5}$$

The force thrusting the ulna against the humerus is expressed as

$$F = \frac{T(B + A \cos \Omega)}{(A^2 + B^2 + 2AB \cos \Omega)^{1/2}} - (F_1 + F_2) \cos \Omega \tag{6}$$

If the weight of forearm and hand is assumed to be approximately 10 lb and to act in a vertical direction on the center of mass of the limb, and when a push button control, such as is used on a crane—also estimated at 10 lb—acts normally on the palm of the hand, then depending on the included angle between forearm and upper arm, the tension in the brachialis muscle will vary from 2150 lb to as low as 170 lb. The thrust exerted on the elbow joint will vary from a high of 2140 lb to a low positive value of 5 lb to a strong negative (joint separation) force of 700 lb. It can be seen (Figure 40.12) that unless the push button is located so that the included angle between upper arm and the forearm is between 80° and 120°, the combined effects of tension in the muscle and thrust acting on the joint surfaces can create conditions conducive to joint injury.

If the safe range of joint movement has been extended, tensions in the muscles increase rapidly. Likewise, under the same circumstances, dangerous thrust forces develop within the joint. Alternatively, poor workplace layout may also produce "separation" forces of substantial magnitude. These may not injure the surfaces of the synovial linings but can create conditions conducive to severe luxations. Quantitative biomechanical and kinesiological analysis of human–task systems is essential to the protection of the work force from the deleterious influences of mechanical noxae.

Standard references (18) permit the rapid quantitative assessment of forces and stresses generated in anatomical lever systems by work. Once the analysis has been made, standard reference tables (19) should be consulted to ascertain whether muscles are stressed to excess or are too weak to accomplish the task as planned.

The term "muscular strength" is defined as the maximum tension per unit area that can be developed within a muscle. A maximal effort resulting from the strongest of motivations will yield a tension of approximately 142 psi. However, many authors (20–22) agree that under normal working conditions, heavy work would generate approximately 50 to 60 psi tension, while light to medium work would produce tension values of the order of 20 to 30 psi. All these values presume maximal

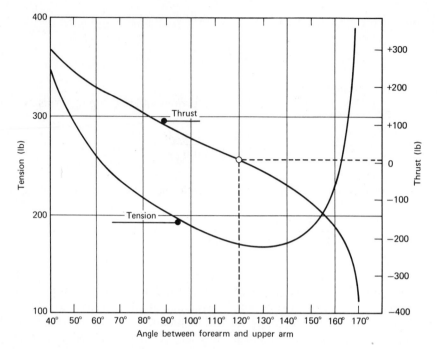

Figure 40.12. Tension in the brachialis muscle and thrust on the elbow joint generated by a push button requiring 10 lb of pressure, applied normal to the longitudinal axis of the forearm, for its operation. The combined weight of forearm and hand is assumed to be 10 lb. ($X° = 180°$ minus angle.)

shortening of the muscle and are applicable only to continued "ordinary" nonathletic work situations (Table 40.1).

Occasionally the force available from muscles involved in a kinetic chain is adequate to perform a given task but is unavailable because of a condition known as "muscular insufficiency." A state of active insufficiency exists whenever a muscle passing over two or more joints is shortened to such a degree that no further increase of tension is possible and the full range of joint movement cannot be completed. For example, when a person is seated too low, if the knee joint is hyperflexed it becomes impossible to plantar-flex the foot and operate a pedal. On the other hand, passive insufficiency exists when, in a particular limb configuration, the antagonists passing over one joint are extended to such a degree that it is impossible for the protagonist to contract further. A good example is the inability to close a fist and hold a rod while the wrist is hyperflexed (Figure 40.13).

3 PHYSIOLOGICAL MEASUREMENTS

Physiology is the discipline that deals with the qualitative and quantitative aspects of physical and chemical processes intrinsic to the function of the living body. As

Table 40.1. Maximal Work Capacities of Flexors of Elbow (19–22)

Muscle	Forearm in Supination		Forearm in Pronation	
	Cross Section (in.²)	Maximal Work Capacity (lb-ft)	Cross Section (in.²)	Maximal Work Capacity (lb-ft)
Brachialis	1.0	28	1.0	28
Biceps	1.1	35	0.8	25
Pronator teres	0.5	9.11	0.65	12
Whole flexor forearm as lumped muscle mass (e.g., grip strength)	3.2	90		

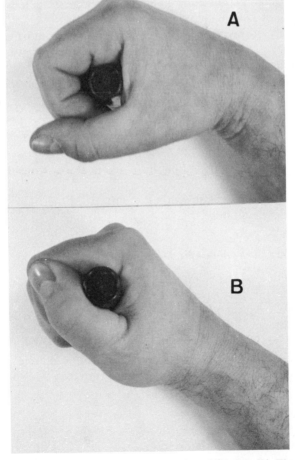

Figure 40.13. The flexed wrist (*a*) cannot grasp a rod firmly. (*b*) The straight wrist can grip and hold firmly. From Reference 48.

far as the study of humans is concerned, the term "physiology" relates by general agreement to the function of the healthy body. The mechanisms of body functions specific to disease form the field of pathophysiology. Both these disciplines are extremely wide and cover almost all aspects of life or death.

Work physiology is a much narrower field. It is restricted to the effects of work and exercise on physiological function. In industrial practice, work physiologists tend to limit themselves to the study of two narrow aspects of this discipline: the assessments of humans' capacity to perform physical work, and the study, as well as description, of the effects of fatigue. Most quantitative results obtained from work physiological evaluation are "output measurements." They constitute physical behavior resulting from the combined effects of a variety of inputs. Oxygen metabolism is not purely a function of severity of exercise or of the lean muscle mass involved; it is also determined, to some degree, by the obesity of the subject, postprandial status, and several emotional factors. The result, however, is a single number: oxygen uptake expressed in milliliters per minute. Solely on the basis of this result, it will not be possible to make any meaningful statement about the magnitude or quality of the contributing vectors. Therefore most work physiology procedures measure effect but do not permit one to establish cause, except in rare cases of near-perfect technique and experimental design.

The following procedures are commonly employed in industry:

1. Metabolic and quasi-metabolic measurements.
2. Electromyography.
3. The measurement of cardiac performance.
4. The measurement of body temperature and heat loss from the body.

Procedures 3 and 4, however, fall within the purview of the trained specialist and are therefore not discussed here.

3.1 Metabolic and Quasi-Metabolic Measurements

It is common practice in some branches of industry to determine physiological energy expended in the performance of certain tasks by direct measurement or indirect estimation of oxygen consumption. During respiration, oxygen passes through the walls of the pulmonary alveoli and surrounding capillaries into the blood vessels. It is then taken up by the red blood corpuscles, which are pumped by the heart to body tissues, such as muscles. Oxygen is unloaded in the muscles to take part in physiological combustion processes. The bioenergetics may be represented as follows (23):

At rest

$$\text{chemical energy in nutrients contained in body tissues} + \text{oxygen} \longrightarrow \text{heat} \tag{7}$$

At work

chemical energy in nutrients   heat (approximately 78%)
contained in body tissues   + oxygen

                                        energy available for   (8)
                                        work (approximately
                                        22%)

1. One liter of oxygen consumed releases approximately 5 kcal.
2. 5 kcal ≈ 21,000 J.
3. 4600 J (22 percent of 21,000 J) is available for the performance of work.

Thus energy available per liter of oxygen consumed per minute is approximately 0.1 hp. However, ambient fresh air contains approximately 20 percent oxygen, of which, roughly a quarter, or 0.05 liter, becomes available for metabolic activities.

Therefore, as a bench mark figure, it may be assumed that 20 liters of air per minute passing through the lungs over and above normal rest levels of pulmonary ventilation corresponds roughly to 0.1 hp expended in the pursuit of physiological work. This constitutes, however, an extremely crude and not always accurate approximation. The true level of net energy expenditure per liter of air passing through the lungs depends on a wide variety of physiological variables. Yet this has not deterred some industrial enterprises from basing computations of physiological effort, often under conditions of heavy work, on pulmonary ventilation (24). Most commonly, the measurement is performed by a knapsack gasometer, and various models are commercially available (Figure 40.14). To obtain readings that are of any use at all, it is necessary to measure total airflow through the lungs from the onset of the task to be investigated until the time after termination of work when pulmonary ventilation has returned to resting level. To obtain the net airflow ascribable to the demands of the task, an air volume corresponding to pulmonary ventilation at rest is then subtracted from the total.

More accurate is a procedure developed by Weir (25). In its application, both the knapsack gasometer and an oxygen analyzer are employed. The volume of air passing per minute through the lungs is measured and reduced to the volume corresponding to standard temperature, pressure, and dryness. To compensate for the inaccuracies of the equipment used, this value *must* be further corrected by multiplication with a "calibration constant" specific for the actual individual instrument used. The result is obtained as follows:

$$\text{kcal/min} = \frac{1.0548 - 0.0504V}{1 + 0.082d} \times \text{liters vent/min} \tag{9}$$

where V = percentage (vol) oxygen in expired air
      d = decimal fraction of total dietary kilocalories from protein

Figure 40.14. Metabolic measurement with the knapsack gasometer. Adapted from Reference 80.

Under most circumstances it is possible to ascertain the protein content of a worker's diet with reasonable accuracy. There is general consensus that this value is relatively constant for each individual.

Methods of metabolic measurement more accurate than those just outlined are complex and sophisticated and should be attempted only by experienced work physiologists. Therefore they are not discussed here. Practitioners active in enterprises where metabolic measurements are taken routinely should familiarize themselves with the basic theory and appropriate techniques through reliable references (26).

3.2 Electromyographic Work Measurement

The aforementioned changes in pulmonary or metabolic activity when measured and quantified by suitable instrumentation can indicate the level of effort demanded for the performance of a specific task. Changes in metabolism and pulmonary ventilation represent both dynamic and isometric work performed by muscles; therefore the accuracy of measurement is greatest whenever the musculature of the entire body, or at least large muscle masses such as those of the thighs or the back, must be applied to the successful completion of a job. However, in light work, where only mild muscular activity takes place or only small muscle groups or single muscles are utilized, the percentage change in metabolic activity is proportional to the relationship

$$\frac{m}{M} \times 100 = \Delta u$$

where M = total lean muscle mass of body
 m = mass of lean muscle applied to task
 Δu = percentage change of metabolic activity due to work

In many instances this percentage may be too small to permit a meaningful statement to be made about the level of effort expended, and under such circumstances experimental and computational error may render the result meaningless.

Whenever light work or effort expended by small kinetic chains is to be determined, electromyography can serve as a useful estimator of effort and fatigue.

The term "electromyography," as well as the purpose of the procedure, means different things to ergonomists, anatomists, physiologists, and physicians. Electrodes, signal conditioning, and display instrumentation vary widely between professions.

In ergonomics, myoelectricity may be assumed to be the "by-product" of muscular contraction which makes it possible to estimate strength and sequencing of muscular activity through techniques noninvasive to the human body. The myogram is an analogue recording of this bioelectric activity.

Each individual muscle fiber maintains in its resting state a negative potential within its membrane wall. This is termed the "resting potential." Excitation produces a transient reversal of the resting potential, causing a characteristic "depolarization" pattern to appear. A discussion of natural and quantitative events related to changes in membrane potential is beyond the scope of this chapter, and specialized literature should be consulted (27).

Modern electrophysiological thinking assumes (28) that muscle fibers probably never contract as individuals. Instead, small groups contract at the same moment. It has been established that all the fibers in each contracting group are stimulated by the terminal branches of one single nerve fiber, the axon of a motor cell whose body is located in the grey matter of the spinal cord. The nerve cell body proper plus the axon, its branches, and the muscle fibers supplied by them, has been named a "motor unit" (Figure 40.15). Because an impulse from the nerve cell causes all muscle fibers connected to the axon to contract almost simultaneously, the action potential resulting from such contraction constitutes the elemental event basic to all electromyographic work. This signal can be picked up by a variety of electrodes and amplified and recorded.

By insertion of very fine needle or wire electrodes into the muscle, it is possible to display action potentials generated by a single motor unit. These individual action potentials, or a sequence of them, are mainly used for electrodiagnostic purposes in medicine. The transducers of choice for electromyographic work measurement are surface electrodes, either permanent or disposable disks of silver

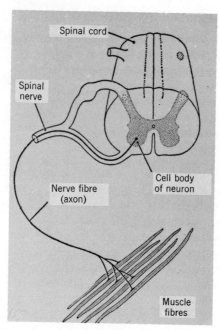

Figure 40.15. Scheme of motor unit. From Reference 28, by permission. © Williams and Wilkins, Baltimore, MD.

coated with silver chloride. More recently, conductive adhesive tape has become commercially available and is often preferred because of economy, flexibility, and ease of application (Figure 40.16).

Needle electrodes are best avoided in ergonomics. They are invasive, produce pain, and call for great skill in application; moreover, their use entails always a certain risk of infection, particularly under the circumstances prevailing on the shop floor or in the industrial laboratory. Even if the aforementioned difficulties could be overcome, however, the signal generated by needle electrodes is only of limited use unless the ergonomist specializes in the more basic aspect of work physiology or electrophysiological kinesiology. The study of single action potentials does not generally permit us to arrive at conclusions relating to the total effort expended by a muscle, and this is why adhesive surface electrodes are preferred in ergonomics. The principal argument against the use of surface electrodes is that unlike needles, they do not permit the study of single action potentials; rather, the signal gathered by them is merely a representation of the level of contractile activity within a relatively large volume of muscle considered to be a "lumped muscle mass" (29). This, however, is the specific advantage of surface myography in the study of the activity of whole muscles as opposed to single motor units.

Thus the surface electromyogram may be considered to represent of the sum of electrical activity generated simultaneously by a large number of motor units. To understand in more detail the theoretical basis of the procedure, the reader should consult specialized reference works (30).

Figure 40.16. A disposable electrode kit, which is inexpensive and noninvasive and avoids danger of cross-infection between subjects: *A*, electrode; *B*, adhesive collar; *C*, conductive jelly. The elimination of substantial amounts of time spent on the cleaning of permanent electrodes makes use of this kit very economical.

3.3 Electromyographic Technique

In usual practice the myoelectric signal, which is in the microvolt or millivolt range, is gathered by suitably placed electrodes and amplified by a factor of approximately 1000 prior to display by oscilloscope or oscillographic pen recorder.

Myographic apparatus embodying high-gain operational amplifiers is generally inexpensive and requires only two electrodes for the production of a myogram. These characteristics make its use more economical, especially where a large number of readings are to be taken. However, the operational amplifier, at the levels of magnification involved in myography, produces a noisy signal, often difficult to interpret. Furthermore, a two-electrode system does not permit the investigator to produce repeatable results easily.

Where simplicity of operation and noise-free signals under field conditions are desired, apparatus embodying differential amplifiers is definitely the equipment of choice. A differential amplifier uses three electrodes, one reference and two active ones. However, it augments only the difference between the two active electrodes based on the potential difference between each of them and the reference. Because any interference from external causes will produce identical changes in the potential of all three electrodes, the display signal will not change. This "common mode" rejection makes it imperative to use differential amplifiers in such settings, where electrical interference from fluorescent tubes, radio transmitters, and other equipment is abundant.

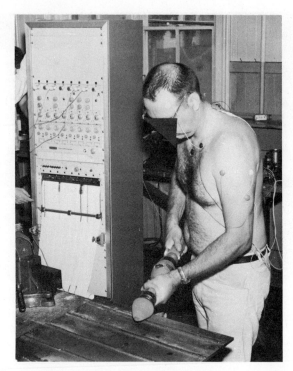

Figure 40.17. The recording of surface electromyograms by means of a commercially available oscillographic recorder for the purpose of physiological work measurement.

In most instances it is desirable (31) to record the myogram by means of one of the numerous commercially available oscillographic recorders, equipped with modular amplifiers and couplers provided by the manufacturer (Figure 40.17).

Most standard recording equipment has been specifically designed for short-term myography. Whenever data gathering exceeds 1 hr of operational time, large machines can become uneconomical, and it is recommended that modern miniaturized magnetic tape recorders be employed. These can be easily attached to the belt of a subject, and they record myograms continuously for up to 24 hr on miniature magnetic tape cassettes. In ergonomics the myogram serves three main purposes:

1. To determine the level of effort expended by a specific muscle mass
2. To determine the nature of sequencing of protagonist and antagonist muscles involved in specific kinesiological maneuvers
3. To identify or predict localized muscular fatigue

Correct electroding technique is a prerequisite to successful myography. It starts with muscle testing to determine the location and surface relationships of the individual muscle to be investigated (Figure 40.18).

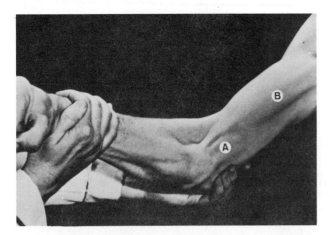

Figure 40.18. Correct procedure for muscle testing produces the contours and relationships of the biceps *B* as well as the brachialis muscle *A*. Adapted from Reference 32, by permission. © Williams and Wilkins, Baltimore, MD.

By way of example, we discuss the relationship between the biceps and the brachialis muscles. The brachialis is a short and stout muscle that originates from the lower third of the humerus and inserts into the ulna, just distal to the coronoid process (Figure 40.6). It is a powerful and precise flexor of the forearm (see Section 2.1). More superficially situated, and covering the brachialis, is the biceps. This muscle originates from the scapula and inserts medially into the proximal end of the shaft of the radius (Figure 40.19).

The biceps is a powerful supinator of the forearm but a comparatively weak flexor. Quite often, especially in materials handling, it is desirable to ascertain the relative magnitude of involvement of either the biceps or the brachialis in the maneuver under study.

These two muscles are located so close together that it is not possible to obtain separate surface myograms for each one unless special precautions are taken. First the brachialis is palpated. This is done by asking the subject to flex the forearm to form an angle of 90° with the upper arm. Then the biceps is pronated strongly against resistance, while simultaneously the experimenter causes the brachialis to flex the forearm against powerful opposition (Figure 40.20). The outlines of the muscles are then palpated and the electrodes applied so that a maximum of muscle mass is triangulated by them. The reference electrode is placed conveniently over the triceps tendon. The subject is asked to supinate the flexed forearm strongly against external resistance, and the biceps is palpated and triangulated with an additional set of electrodes (31).

It is essential to provide a separate reference electrode for each muscle investigated. To verify the electrode placement, the subject is first asked to flex the forearm isometrically while the limb is being actively and strongly pronated. This yields a strong brachialis signal but little activity in the biceps (Figure 40.21). Subsequently the limb is supinated against resistance while being flexed strongly

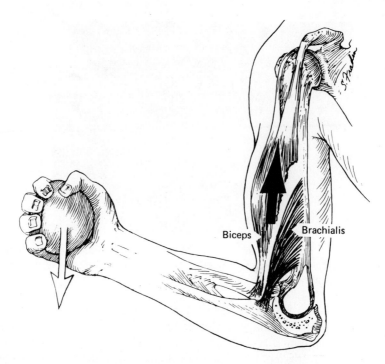

Figure 40.19. The topographic relationships between two muscles: the biceps and the brachialis. The biceps is superficial to the latter, inserts into the radius, is a powerful supinator but a less effective flexor of the forearm. The brachialis is a short but powerful muscle connecting humerus and ulna; because of the character of the humeroulnar joint as a hinge, this muscle is a very powerful flexor of the forearm. Separate electroding of the two muscles is desirable in electrophysiological work measurement. Adapted from Reference 81.

against an opposing force. This produces a strong biceps myogram, concurrent with light to moderate myoelectric activity in the brachialis (Figure 40.21). Standard reference works (32) should be consulted if it is necessary to acquire proficiency in electroding the muscles contracted during the performance of common industrial elements of work.

3.4 Interpretation of Myograms

Once it has been established through muscle testing that the electrodes indeed produce a specific signal, representative of the level of activity in the single muscle under investigation, the real task is performed and the myogram interpreted.

An indispensable prerequisite to the correct interpretation of the signal is an understanding of the circuitry that produces the display and of the operating characteristics of the recording apparatus. Owing to the nature of the procedure, myograms produced by surface electrodes are records of the summed signals from a number of action potentials generated simultaneously, near-simultaneously, or con-

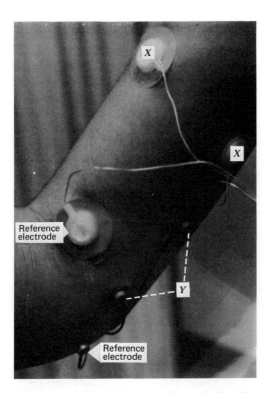

Figure 40.20. Differential electroding of biceps and brachialis: X, active electrodes of biceps; Y, active electrodes of brachialis.

secutively, within the muscle mass triangulated by the electrodes. Because the interval of time elapsing between the peaking of individual action potentials may be as little as a few microseconds, readout devices and pen recorders may be "overdriven." The signal is then not representative of the action potentials but is conditioned and distorted as a function of the quality of the recording device (Figure 40.22). Even a change in the viscosity of the recording ink may produce a drastic change in the pattern of a tracing from the same amplifier reproduced by the same recorder. Likewise, pen inertia and the quality of maintenance the instrument has received may cause a badly distorted signal, obfuscating the physiological status of the muscles studied.

To obtain a satisfactory resolution of the direct surface myogram, it is necessary in most instances to run paper recorders at speeds of 12.5 cm/sec. This procedure is uneconomical, and because of friction between pen and paper it gives a completely distorted signal. Therefore in ergonomics a conditioned type of myogram is employed, called the "integrated myogram." However, no true integration has taken place. Integrators are circuits that produce a signal that essentially records the sum total of the action potentials counted over a sampling interval of time. This type of myogram is therefore representative of the total number of muscle fibers contracting at any instant. Because of the physiological "all or none" law,

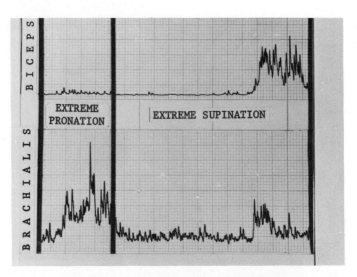

Figure 40.21. Differential myograms from both biceps and brachialis obtained while hand held 20 lb of weight; the included angle between forearm and upper arm was approximately 100°.

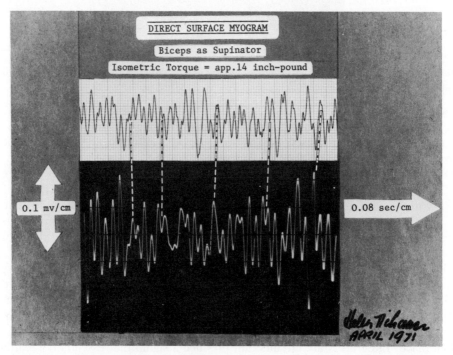

Figure 40.22. A simultaneous recording of the biceps muscle firing pattern displayed on a chart recorder (upper half) and an oscilloscope (lower half) at exactly the same sensitivity and speed. Only five points of similarity are evident because the signal speed of the myogram exceeds the rise time and slew rate of commercial chart recorders. From Reference 82, by permission. © American Institute of Industrial Engineers, Inc., Norcross, Ga.

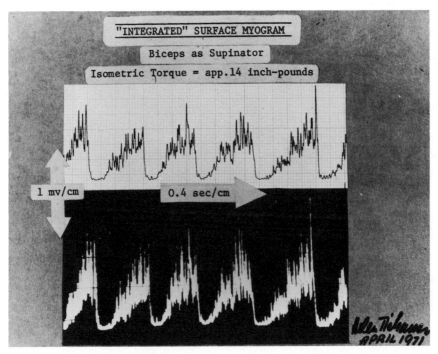

Figure 40.23. The same biceps contraction pattern shown in Figure 40.22: chart recorder (upper half), oscilloscope (lower half). However, the signals here have been conditioned by summing all action potentials over a time constant so that the trace now represents the analogue of the firing rate, which is indicative of the total activity of the muscle mass at any instant during the sampling period. The signals are fully compatible with the frequency response of the chart recorder. The "integrated" myogram produces repeatable and very reliable measurements of muscular activity levels. From Reference 82, by permission. © American Institute of Industrial Engineers, Inc., Norcross, Ga.

it is also representative of the effort expended at any instant. Therefore the area under the integrated myogram is proportional to the total effort made during the time interval under consideration. The generation of this signal requires a much slower recording speed, as low as 1 cm/sec, and therefore it can be reliably reproduced by almost any recording device available (Figure 40.23).

A myogram must not only be read, it must be "interpreted." This entails recourse to a judgmental process that takes into consideration magnitude of pen excursion, general pattern of the tracing, and some features hard to define, such as fuzziness of recording.

The ability to interpret a myogram is an acquired skill, which can be developed within a short time provided the ergonomist is restricted, at least in the beginning, to working with tracings obtained from the same make of recorder. This has a twofold advantage. First, it is easy to develop an appreciation about the "soundness" of the tracing. Second, the integrated surface myogram, the only type of myoelectric readout considered in this chapter, constitutes a signal conditioned and transformed into shapes and patterns, which may vary for different designs of recording equipment.

Recording speed should be kept constant to facilitate recognition of patterns and to keep slopes and tracings uniform. Many practitioners prefer to work at a recording speed of 1 cm/sec. Whenever an exchange of myographic information with other workers is planned, it is advisable for all to adopt the same paper speed. Recorders designed to produce integrated myograms take into consideration both the frequency of peaking of the raw signal and the amplitude of the action potential, but circuits developed by different manufacturers weigh each of these features differently; thus the appearance of myograms from the same muscles, tasks, and individuals, taken at the same occasion, differs considerably according to the make of recorder employed. It is therefore essential not to vary equipment from study to study.

The shape and quality of myograms may also be substantially affected by the following factors:

1. Loose electrodes.
2. Dry electrodes.
3. Loose or broken wiring.
4. Electrical interference from light fixtures or other machinery being used nearby.

It is essential to label myograms with the sensitivity settings and speed of the recording device; otherwise tracings obtained at different occasions cannot be compared. Likewise, the base line should be clearly indicated, or amplitude of pen excursion cannot be quantified. When an isometric task is performed (Figure 40.21), it is relatively easy to ascertain the degree to which each of the muscles investigated participates in the performance of the task, and how changes in musculoskeletal configuration produce a different distribution of work stress acting on the individual members of a kinetic chain. It is also quite simple, when the precondition of isometricity exists, as in static holding (Figure 40.24), to determine when a critical work stress level has been reached and when a relatively light increase in the severity of the task will produce an undesirably violent myoelectric response.

The area under the integrated myogram has the dimensions of force (volts), multiplied by time (paper speed), which is identical to the dimensions of "linear impulse" in physics and also to the "tension–time" concept used to work physiologists to quantify isometric work (33).

The interpretation of myograms of dynamic tasks, however, is a far more complex matter. The shape of the tracing, its slopes and troughs and amplitudes, are affected by a multitude of factors. These include force and velocity of contraction, tension within the muscle, and whether the contraction is eccentric or concentric.

Thus it is useless in most dynamic situations even to attempt the numeric quantification of the signal. The qualitative discussion, however, can yield information of considerable usefulness. This is the case, for instance, in the analysis of se-

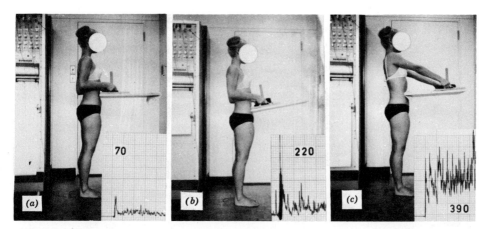

Figure 40.24. An isometric holding task. Numbers below the myograms represent incremental inch-pounds of torque, applied to the lumbosacral joint, which have elicited the electrophysiological signal. It can be seen that once a certain "critical" level of stress has been exceeded, the electrophysiological signal increases disproportionately as compared with the increment of stress. Under such conditions the subject is at risk. Adapted from Reference 31.

quencing of different muscles during the performance of a kinesiological maneuver (Figure 40.25). In wrist rotation the biceps acts as protagonist during supination, while the pronator teres is the antagonist that reverses the movement. Therefore the integrated myograms of both muscles show peak and valley phasing in a non-fatigued, efficiently working individual. When the protagonist fires, the antagonist should be relaxed, and vice versa. In a state of fatigue, however, this clear phasing of muscular activity becomes blurred. A fatigued muscle has lost the ability to relax quickly; therefore the weaker of the two muscles, pronator teres, may fire simultaneously with the biceps, slowing down movement and bringing about undue exertion by the antagonist, increasing further the level of fatigue.

Whenever it is desired to ascertain whether a given musculoskeletal configuration is conducive to an undue expenditure of effort resulting in fatigue or potential trauma, the integrated myogram is the method of investigation of choice. A wire-cutting operation may be deemed to be a quasi-isometric work situation (Figure 40.26). Very often, when a tool such as a side-cutter is employed while the wrist is in ulnar deviation, the tendons of the flexor muscles bunch against each other inside the carpal tunnel. This produces friction and necessitates a disproportionately large effort by the muscles of the flexors of the fingers and thumb for effective performance. This excessive effort becomes immediately apparent on inspection of the myogram.

It is impossible to discuss in detail all the potential uses of electromyographic kinesiology; specialized papers should be consulted (34–36). However, the proper and imaginative use of electromyography constitutes one of the most elegant and useful techniques of ergonomic work measurement.

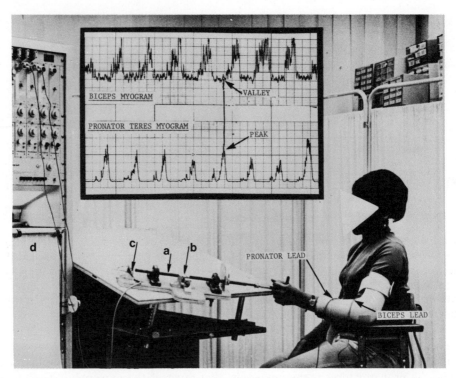

Figure 40.25. Kinesiometer and subject wired for two surface electromyograms; insert shows how antagonist myograms peak in proper sequence with one another. The kinesiometer consists of *A*, rotatable shaft; *B*, friction brake; *C*, potentiometer; *D*, recorder. Adapted from Reference 83.

4 WORK TOLERANCE

Within the context of ergonomics, any action on the living body by any vector intrinsic to the industrial environment is termed "work stress." It is irrelevant whether these vectors are forces and produce movement, whether they merely cause sensory perception, or whether, like heat, they increase metabolic activity. All physiological responses to work stress are identified as "work strain." Frequently work stress and the resulting strain occur in anatomical structures quite distant from each other, or the two effects may even be observed in separate physiological systems. High environmental temperatures are referred to as "heat stress," and the resulting increase in sweating rate is then correctly termed "heat strain." Likewise, in heavy physical work, the forces exerted on the musculoskeletal system are correctly identified as "work stress" and the resulting increase of metabolic activity is an example of "work strain."

The performance of any task, no matter how light, imposes some work stress, and consequently produces work strain in terms of physiological responses. Neither

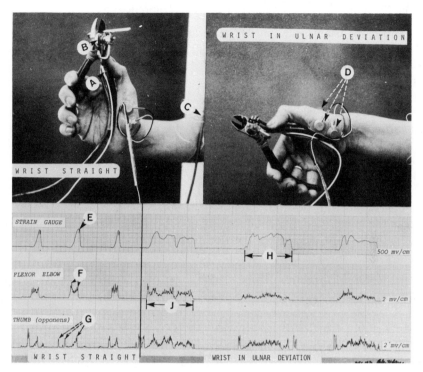

Figure 40.26. The profile of wire cutting (84). For explanation, see text. *A*, strain gauge; *B*, potentiometer; *C*, electrodes for common flexor myograms; *D*, electrodes for thenar myograms; *E*, "notched" strain signal; *F*, dicrotic flexor myogram; *G*, multicrotic thumb myogram; *H*, increased duration of strain signal; *J*, increased "tension time" of muscle. From Reference 84.

stress nor strain per se is undesirable unless it becomes excessive and diminishes work tolerance.

Work tolerance is defined as a state in which the individual worker performs at economically acceptable rates, while enjoying high levels of emotional and physiological well-being (13). It is common in industrial engineering practice to employ incentive schemes as inducements to increase production rates, thus reducing labor costs. Job enrichment as well as other procedures applied by behavioral and managerial scientists have their place in increasing job satisfaction and in enhancing the social well-being of the work force. However, no management technique available has been found to be successful in overcoming the results of physical discomfort and occupational disease resulting from a poorly designed work situation, ill-matched to the physical operating characteristics of man.

In a room illuminated by a defective spectrum, everyone is color blind. On a job where the motions and reactions inventory demanded by a task is not available from the musculoskeletal system of the worker, everybody is disabled (37), the physically impaired more so. The institution and maintenance of work tolerance has a high priority in the practice of industrial hygiene and ergonomics.

4.1 The Prerequisites of Biomechanical Work Tolerance

The 15 most important prerequisites of biomechanical work tolerance (38) have been arranged in the form of a table (Table 40.2) and can be employed as a checklist in industrial surveys. Proper use of the table can prevent workplace design from imposing physical demands that cannot be met by a wide range of individual workers. The use of this checklist may also help to avoid the generation of anatomical failure points, which may develop over a number of months, or years, as a result of cumulative work stress. Not all "prerequisites" are applicable to all work situations. However, a correctly designed working environment will not violate many of them because this will, beyond doubt, lead to low productivity, poor morale, feelings of ill health and, sometimes, real occupational disease (13).

The "prerequisites" have been arranged in three sets of five statements. The first is concerned with postural integrity; the second relates to the proper engineering of the human–equipment interface; and the third set may be used to ensure that the motions demanded from the workforce are kinesiologically effective.

4.2 The Postural Correlates of Work Tolerance

P1   Keep the Elbows Down

Abduction of the unsupported arm for long intervals may produce fatigue, severe emotional reactions, and also decrements in production rates. The need to keep the unsupported elbow elevated is often the result of poor workplace layout. For example, a seat positioned only 3 in. too low with respect to the workbench will produce an angle of abduction of the upper arm of approximately 45° (Figure 40.27). When this is the case, wrist movement in the horizontal plane, normally performed by rotation of the humerus, could require a physically demanding shoulder swing. The resulting fatigue over several hours may reduce the efficiency rating by as much as 50 percent. Also, when the seat is too low—especially in assembly operations—the left arm is frequently used as a vise, while the right hand is employed to manipulate objects. This may result in the left arm being held in abduction for several hours. After the elapse of an hour or two, particularly under

Table 40.2. Prerequisites of Biomechanical Work Tolerance

Postural	Engineering	Kinesiological
P1 Keep elbows down	**E1** Avoid compression ischemia	**K1** Keep forward reaches short
P2 Minimize moments on spine	**E2** Avoid critical vibrations	**K2** Avoid muscular insufficiency
P3 Consider sex differences	**E3** Individualize chair design	**K3** Avoid straight-line motions
P4 Optimize skeletal configuration	**E4** Avoid stress concentration	**K4** Consider working gloves
P5 Avoid head movement	**E5** Keep wrist straight	**K5** Avoid antagonist fatigue

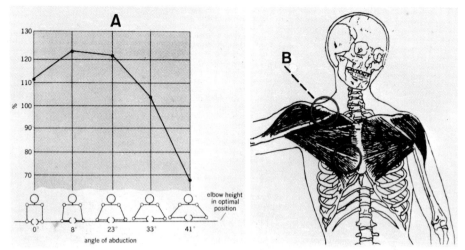

Figure 40.27. (*a*) Chair height determines the angle between upper arm and torso, also the moment of inertia of the moving limb. For example, a chair 3 in. too low may raise this moment to a level at which increased effort causes performance to drop to 70 percent of "standard." (*b*) Also, continued tension in the muscles stabilizing the arm in the raised posture may cause great discomfort at the shoulder over the breast bone. From Reference 37, by permission. © American Institute of Industrial Engineers, Inc., Norcross, Ga.

incentive conditions, some vague sense of discomfort may be felt in the general region of the origin of the left pectoralis major and deltoid muscles, which stabilize the abducted arm. Elderly and overweight workers especially, or individuals with a history of cardiac disease, may develop an unjustified fear of an impending heart attack, and they themselves, as well as all those around them, may suffer from the ensuing undesirable emotional difficulties (13).

P2 Minimize Moments Acting Upon the Vertebral Column

Lifting stress is not solely the result of the weight of any object handled. Its magnitude must be expressed in terms of a "biomechanical lifting equivalent" in the form of a "moment."

The location of the center of mass of the body proper causes a bending moment to be exerted on the axial skeleton even when no object is handled (Figure 40.28). The muscles erecting the trunk counteract the moment and thus help to maintain upright posture. Thus even simple variations in posture and trunk–limb configuration may modify and substantially increase or decrease, according to the circumstances, the forces exerted on the lumbar spine according to the contribution of the individual body segments to the total moment sum (39). The ergonomic effects of these posture-generated forces are discussed in Section 5. The present checklist is merely concerned with the additional moments imposed on the back by an external load. Very often a light but bulky object (Figure 40.29) imposes a heavier lifting stress than a heavy load of great density. It should be remembered that the

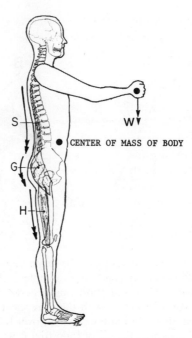

Figure 40.28. Even when no object is handled, very often a bending moment acts on the vertebral column because of the location of the center of mass of the body. The erector muscles of the trunk counteract this: *S*, sacrospinalis; *G*, glutei; *H*, hamstrings; *W*, weight. From Reference 51.

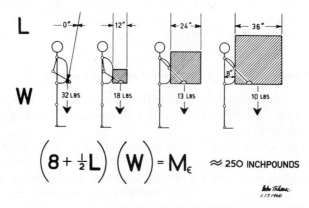

$$\left(8 + \tfrac{1}{2}L\right)\left(W\right) = M_\epsilon \quad \approx 250 \text{ INCHPOUNDS}$$

Figure 40.29. The "moment concept" applied to the derivation of biomechanical lifting equivalents. All loads represented produce approximately equal bending moments on the sacrolumbar joint, approximately 250 lb-in. Moments exerted by body segments are neglected (13). In the equation, 8 is the approximate distance (inches) from the joints of the lumbar spine to the front of the abdomen, a constant for each individual; *L* is the length (inches) of one side of the object; *W* is the weight (pounds) of the object; M_ϵ is the biomechanical lifting equivalent, approximately 250 lb-in. in this example. From Reference 13.

only way to reduce the lifting stress exerted by an object resides in devising a handling method that will bring the center of mass of the article as close to the lumbar spine as possible.

P3   Consider Sex Differences

If employment opportunities for both sexes are to be equal, work environments must be engineered in a manner that takes cognizance of, and compensates for, any sex-dependent differences in anatomy that may affect work tolerance. In the context of lifting tasks, it must be appreciated that male hip sockets are located directly below the bodies of the lumbar vertebrae; in the female they are situated more forward (Figure 40.30). A line through the center of the sockets of both hip joints in a woman is located several inches in front of a vertical line passing through the center of gravity of the female body. This activates a force couple. Therefore any object handled by a woman exerts a moment on the back approximately 15 percent larger than if it were handled by a male of identical size or strength.

P4   Optimize Skeletal Configuration

Through faulty workplace design, musculoskeletal configuration, especially angular relationships of long bones and muscles, may impose great stress on joints and produce physical impairment (Figure 40.31). Variations of a few inches horizontally

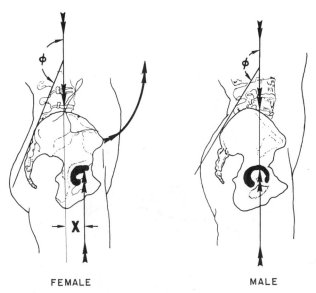

FEMALE   MALE

Figure 40.30.   Hip sockets in the male are located directly under the bodies of the lumbar vertebrae in the same plane as the center of mass of the body. In the female, the sockets are located further forward, represented by the distance X. This produces a force couple so that the lifting stress in the back muscles in women, for the same object, can be as much as 15 percent higher than in males. From Reference 85, by permission, American Industrial Hygiene Association, Akron, Ohio.

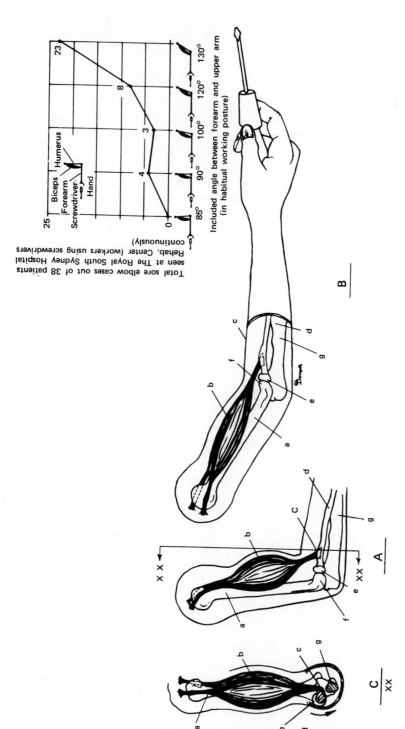

Figure 40.31. The mechanical advantage of the biceps depends on the angle of flexion of the forearm. This muscle is not only a flexor, but also, because of the mode of attachment, the most powerful outward rotator of the limb. The worker who sits too far away from his workplace has to overexert himself when using a screwdriver because the biceps operates at mechanical disadvantage. Sore muscles and excessive friction between the bony structures of the elbow joint are the results (80). (*A*) The forearm flexed at 90°: *a*, humerus; *b*, biceps; *c*, attachment of biceps; *d*, radius; *e*, head of radius; *f*, capitulum of humerus; *g*, ulna. (*B*) The angle of the forearm extended when the efficiency of the biceps as an outward rotator is reduced. Here the muscles will pull the radius strongly against the humerus, causing friction and heat in the joint. (*C*) Cross section X-X through *A*, showing why the biceps is an outward rotator of the forearm. From Reference 80.

in the distance between the chair and the work space may make the difference between a productive working population and one that must perform under medical restriction because of great mechanical stress imposed on joint surfaces.

P5 Avoid the Need for Excessive Head Movement During Visual Scanning of the Workplace

It is not possible to estimate correctly, and/or easily, the true sizes or the relative distance of objects except under conditions of binocular vision, which can take place without head movement only within a visual cone of 60° of included angle. The axis of this cone originates from the root of the nose and is located in the midsagittal plane of the head (Figure 40.32). Head movement at the workplace is often invoked as a "protective reaction" (17) whenever it is necessary to reestablish binocular sight if the visual target is located outside of the cone. Simultaneous eye and head movements take much time, and this may produce a hazard whenever fast-moving equipment—such as motor vehicles, airplanes, or conveyors—are operated. Binocular vision without head movement can be instituted either by dimensioning the workplace appropriately or by changing the position of the operator or the adjustment of the working chair.

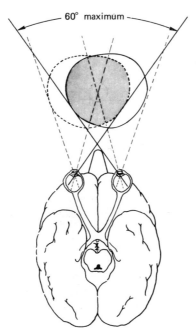

Figure 40.32. Eye travel and binocular vision. Whenever an object is located outside a binocular field of vision (shaded area) depth perception is impossible; head movement will automatically ensue to correct the deficiency. Heavy dotted lines indicate convergence at 16 inches. From Reference 48.

4.3   The Engineering of the Human–Equipment Interface

E1   Avoid Compression Ischemia

The term "ischemia" describes a situation in which blood flow to the tissues is obstructed. It is essential that the designer as well as the evaluator of tools and equipment be familiar with the location of blood vessels vulnerable to compression. Improperly designed or misused, a piece of equipment may have the effect of a tourniquet. Of special importance is a knowledge of the location of blood vessels and other pressure-sensitive anatomical structures in the hand. A poorly designed or improperly held hand tool may squeeze the ulnar artery (Figure 40.33). This may lead to numbness and tingling of the fingers. The afflicted worker will put down his tools and may make use of any reasonable excuse to absent himself temporarily from the workplace as the only means of relief open to him. Apart from the resulting drop in productivity, the health of the working population under such circumstances is in serious jeopardy inasmuch as cases of thrombosis of the ulnar artery and other instances of permanent damage have been reported. Unless the engineering and medical departments are alerted to a possible mismatch between hand and tool, the complaint of numb and tingling fingers could be erroneously attributed to one of the numerous other causes of these symptoms, and the sufferer improperly diagnosed or treated (41). Generally speaking, handles of implements should be designed to make it impossible for the tools to dig into the palm of the hand or to exert pressure on danger zones.

E2   Avoid Vibrations in Critical Frequency Bands

Vibrations transmitted at the human–equipment interface can easily lead to somatic resonance reactions. "White finger syndrome" or intermittent blanching and numbness of the fingers, sometimes accompanied by lesions of the skin, has been identified for many years as an occupational disease associated with the operation of pneumatic hammers and other vibrating tools (42, 43). It is cited as only one example of numerous diseases and injuries caused by exposure to vibrations. When exposed to critical ranges of vibration, various viscera, muscle masses, and bones may react in an undesirable manner. This can simulate a wide range of diseases that are commonly associated with musculoskeletal discomfort, including back pain, respiratory difficulties, and visual disturbances. The critical ranges of vibration that may produce such undesirable side effects are fortunately very narrow, and it is often easy, through recourse to such normal engineering procedures as construction of vibration-absorbing tool handles, to avoid exposure to noxious frequencies. An "epidemic" of otherwise inexplicable afflictions of the musculoskeletal system should always alert the ergonomist to consult a reliable reference work on industrial vibration (44).

E3   Individualize Chair Design

The design of any seating device should match the need of individual work situations. The anthropometric and biomechanical basis of chair design and adjustment

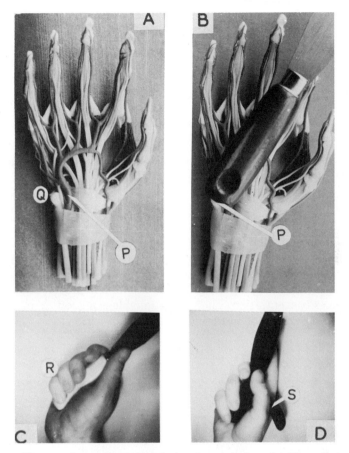

Figure 40.33. Ergonomic considerations in hand tool design. (*a*) The relations of bones, blood vessels, and nerves in the dissected hand. (*b*) A paint scraper is often held so that it presses on a major blood vessel *P* and directs a pressure vector against the hook of the hamate bone *Q*. (*c*) In the live hand, this results in a reduction of blood flow to, among others, the ring and little fingers, which shows as light areas on infrared film *R*. (*d*) A modification of the handle of the paint scraper causes it to rest on the robust tissues between thumb and index finger *S*, thus preventing pressures on the critical areas of the hand. From Reference 40, by permission. American Industrial Hygiene Association, Akron, Ohio.

are discussed in section 7. When evaluating work situations on the shop floor, it should be remembered that universally useful "ergonomic work chairs" do not exist. Likewise, diagrams that show "standard dimensions" of the components of a work chair, or of seating height, are highly suspect unless they list a wide range of tolerances for each dimension. By way of example, it is mentioned that the pilot seat in an aircraft should support the trunk while the pilot is sitting still. On the other hand, the chair on an assembly line should give adequate lumbar support but at the same time permit the body to perform all necessary productive movements (45) (Figure 40.55). Working chairs should have an adjustable-height backrest that swivels about a horizontal axis, to be able to adapt to the demands of the contours of the back. It should be small enough not to interfere with the free movement of the elbows during work. The seat should be adjustable in height, and it should be complemented by an adjustable footrest to relieve pressure exerted by the edge of the seat on the back of the thigh.

E4   Avoid Stress Concentration on Vulnerable Bones and Joints

Sometimes features in tool and equipment design look deceptively advantageous but are in reality most dangerous to the integrity of the skeletal system. Finger-grooved tools are an example (Figure 40.34). They fit perfectly one hand—the hand of the designer. If gripped firmly by a hand too large or too small, the metal ridges of the handle may exert undue pressure on the delicate structures of the interphalangeal joints (46). This makes it painful to grip the tool firmly. Sometimes worse, the working population may show signs of discomfort and absenteeism and will be exposed to medical restriction of performance levels. Finally, under the worst of circumstances, permanent and disabling bone and joint disease may result. Sometimes simple shielding devices are quite effective to protect anatomical structures from stress concentration. The tailor's thimble is a good example of this—it is perhaps the oldest device in history protecting tissues against stress concentration.

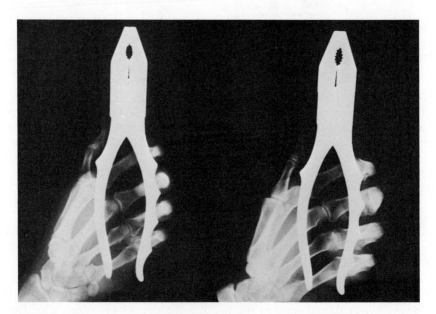

Figure 40.34. Form-fitting grips on hand tools may cause severe pressures on the finger joints of a person with a larger hand than the hand size for which the tool was designed. From Reference 46, by permission. © Journal of Occupational Medicine, Chicago, Ill.

E5   Keep the Wrist Straight while Rotating Forearm and Hand

Four wrist configurations, particularly when they approach the extremes of their range, are conducive to fatigue, discomfort, and sometimes disease. These are (*a*) ulnar deviation, (*b*) radial deviation, (*c*) dorsiflexion, and (*d*) palmar flexion.

Especially unhygienic situations are those in which these positions alternate fairly rapidly during the work cycle, or occur in combination with each other. Unfortunately tools and workplaces are often so designed that the aforementioned movements are demanded as part of the normal work cycle. This affects both health and efficiency. The principal flexor and extensor muscles of the fingers originate in the elbow region, or from the forearm, and are connected to the phalanges by way of long tendons. The extensor tendons are held in place by the confining transverse ligament on the dorsum of the wrist, and the flexor tendons on the palmar side of the hand pass through the narrow carpal tunnel, which contains also the median nerve.

Failure to maintain the wrist straight causes these tendons to bend, to become subject to mechanical stress, and to traumatize such ancillary structures as the tendon sheaths and some ligaments. Ulnar deviation, combined with supination, favors the development of tenosynovitis. Radial deflexion, particularly if combined with pronation, increases pressure between the head of the radius and the capitulum of the humerus. This is conducive to epicondylitis or epicondylar bursitis. These conditions are frequently observed in those who operate hand tools. If hand tools such as screwdrivers or pliers are pronated and supinated against resistance for only a few minutes during the day, no harm results. However, continuous production jobs may constitute a hazard to the working population. Generally speaking, it is safer to bend the implement than to bend the wrist. Furthermore, both ulnar and radial deviations of the wrist, when combined with simultaneous pronation or supination, reduce the range of rotation of the forearm by more than 50 percent (46). Under these circumstances, the afflicted individuals are obliged to go through double the number of motions to perform a given task, such as looping a wire around a peg (i.e., they have to work twice as hard for half the output). This may dramatically increase personnel turnover and lead to massive dropouts of new workers during training (13) (Figure 40.35).

4.4   The Development of Effective Kinesiology

K1   Keep Forward Reaches Short

Numerous industrial engineering texts describe and define "normal" and "extended" reach areas, specifying that the motion elements "reach" and "transport" may safely be included in repetitive tasks for continuous work provided they do not exceed 25 in. (47). This assumption is fallacious. The protagonist muscles of forward flexion of the upper arm operate at biomechanical disadvantage. One of their principal antagonists is the large and powerful latissimus dorsi (Figure 40.36). Whenever an extended and fast reach movement in the sagittal plane away from the body is performed, a strong stretch reflex is produced in the latissimus dorsi.

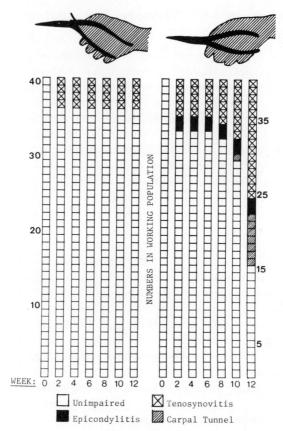

Figure 40.35. Comparison of two groups of trainees in electronics assembly shows that it is better to bend pliers than to bend the wrist (37). With bent pliers and wrists straight, workers become stabilized during second week of training. With wrist in ulnar deviation, a gradual increase in disease is observed. Note the sharp increase of losses in tenth and twelfth weeks of training: only 15 of 40 workers in the sample remained unimpaired. From Reference 37, by permission. © American Institute of Industrial Engineers, Inc., Norcross, Ga.

This in turn subjects the vertebral column to compressive and banding stresses. Frequent and rapid forward reach movements, especially when associated with the disposal of objects, require that the length of this motion be kept to somewhat less than 16 in. The operative word here is "frequent." Extended reaches per se, especially when performed slowly and relatively rarely during the working day, are quite harmless.

K2 Avoid Muscular Insufficiency

Sometimes a specific muscle is unable to produce an expected full range of joint movement. Such a situation is defined by the term "muscular insufficiency" (15). It occurs when the protagonist is contracted to such a degree that it cannot further

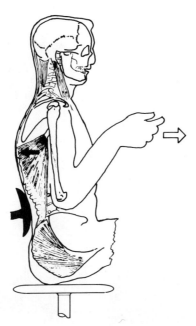

Figure 40.36. The disposal of objects in a direction away from the body and in the mid-sagittal plane is conducive to discomfort and early fatigue because of the strong antagonist activity in the latissimus dorsi muscle. From Reference 48.

shorten, or an antagonist has become hyperextended and impedes further joint movement. By way of example, when the wrist is fully flexed, the hand cannot grasp a rod firmly because the extensors of the fingers are overextended, and the flexors are overcontracted (Figure 40.37).

K3   Movements Along a Straight Line Should be Avoided

All joints involved in productive movements at the workplace are hinges. Therefore the pathway of an anatomical reference point is always curved when it results from the simple movement of a single joint. Such curved movements can be produced by a contraction of one single muscle. On the other hand, motion along a straight line requires higher skills not always available from a subclinical impaired working population (e.g., the aged). Straight line movements generally require longer learning, produce early fatigue, and are less precise than simple ballistic motions (48).

K4   Working Gloves Should be Correctly Designed

Often occupational hazards and operational inefficiencies are erroneously ascribed to improper tool or usage when, instead, the working glove is implicated. The end organs of the sensory nerves of the hand are, among others, also distributed along the interdigital surfaces of the fingers (49) (Figure 40.38). Because it is impossible to close the fist firmly while the fingers exert pressure against each other, it is

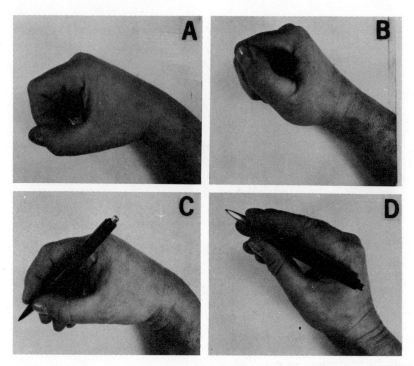

Figure 40.37.   Flexed wrist (*A*) cannot grasp a rod firmly; the straight wrist (*B*) can grip and hold it firmly. Conversely, the flexed wrist (*C*) is well positioned for fine manipulation, but when extended (*D*), freedom of finger movement is severely limited. From Reference 48.

difficult to maintain grip strength under such circumstances. If a working glove is too thick between the fingers, high pressure on the interdigital surfaces may be generated before the hand is firmly closed about a tool handle or equipment control. This results in an insecure grasp. Lack of awareness of this lowered capability may cause heavy objects to slip out of the hands of workers, resulting in accidents. It may also lead to inadequate control over cutting tools or dials operated in a cold climate.

K5 Antagonist Fatigue Should be Considered in Task Design

In the performance of many simple movements, the muscles that reestablish the initial condition after a specific maneuver has been performed are often weaker than those which bring about the primary movement. For example, when inserting screws by means of a ratchet screwdriver, the biceps is the outward rotator of the forearm, and this strong muscle will not easily fatigue even when operating against strong resistance. Even when operating against a resistance too small to be measurable, however, the opposing inward rotator of the forearm, the pronator teres, fatigues easily because of its small size (19). Physiological work stress should never exceed the capacity of the smallest muscle involved in a kinetic chain. Reliable

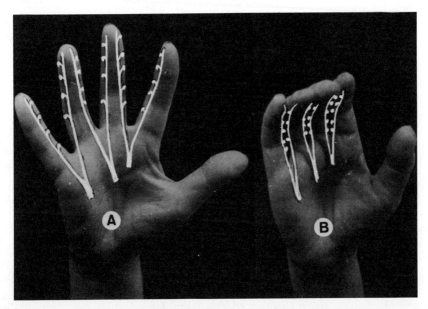

Figure 40.38. Nerve endings that provide feedback information about the degree of closure of the hand are located between the fingers (*A*). When the hand executes a gripping motion (*B*), the fingers abut and the nerve endings press against each other. Work gloves that are too thick may produce pressure against the interdigital surfaces too early and provide misleading information about the firmness of grip when holding a heavy or slippery object. From Reference 37, by permission. © American Institute of Industrial Engineers, Inc., Norcross, Ga.

tables relating name of muscle, function, cross section, and working capacity should be consulted by the practicing ergonomist (15).

The 15 prerequisites of biomechanical work tolerance just given are no substitute for a comprehensive knowledge of the theory and practice of ergonomics. They are merely a convenient expedient for the rapid and gross evaluation of work situations during discussions on the shop floor. Properly employed in discussions with first-line supervisors and engineering personnel engaged in the planning and design of work situations, these prerequisites can help to reduce, without access to complex esoteric and costly laboratory facilities, the incidence of occupational accident and disease right where it occurs: on the shop floor.

5 MANUAL MATERIALS HANDLING AND LIFTING

Almost one-third of all disabling injuries at work—temporary or permanent—are related to manual handling of objects (48). Many of these incidents are avoidable and are the consequence of inadequate or simplistic biomechanical task analysis. The relative severity of materials handling operations, and differences in lifting methods, can be evaluated only when all elements of a lifting task are considered together as an integral set (Table 40.3). All these elements have different dimensional characteristics, but nevertheless have one basic property in common: any major change in magnitude of any element of a lifting task produces a change in the level of metabolic activity.

No matter what the dimensions of mechanical stress imposed on the human body during materials handling, the physiological response will always result in changes of energy demand and release, customarily expressed in kilocalories. Thus physiological response to lifting and materials handling stress has always the dimensions of work. Hence when the task is heavy enough, the measurement of metabolic activity provides a convenient experimental method for the objective comparison of the relative severity of materials handling chores.

Current consensus (26) assumes on the basis of an 8-hr working day that the limit for heavy continuous work has been reached when the oxygen uptake over and above resting levels approaches 8 kcal/min. The upper limit for medium-heavy continuous work seems to be 6 kcal/min, and an increment of 2 kcal/min appears to be the dividing line between light and medium-heavy work. However, in many instances the application of metabolic measurement is not feasible. The complex procedure may be too difficult to perform on the shop floor. Furthermore, the assessment of work stress through the analysis of respiratory gas exchange is, by definition, an ex post facto procedure. The job exists already, and the energy demands are merely computed to decide whether corrective action is indicated. It is of course much better to analyze a task objectively while both job and workplace layout are still in the design stage. Recourse must then be taken to "elemental analysis."

5.1 Elemental Analysis of Lifting Tasks

Any activity producing a moment that acts—no matter in which direction—on the vertebral column, must be classified as a "lifting task." A "moment" is defined as the magnitude of a force multiplied by the distance from the points of its application. In most instances ergonomic analysis of lifting tasks is concerned with moments acting on the lumbar spine or, when a specific task involves head fixation, with moments acting on the cervical spine.

There are three static moments to be considered (Table 40.3). In many instances they are easily computed by direct measurement or with the aid of drawings or photographs. Sometimes, however, it is more convenient to estimate them by

Table 40.3. The Elements of a Lifting Task (48)

Static Moments	Gravitational Components	Inertial Forces
Sagittal	Isometric	Acceleration
Lateral	Dynamic	Aggregation
Torsional	Negative	Segregation
	Frequency of task	

speculative analysis or visual inspection of the work situation. They are conveniently expressed in pound-inches (lb-in), or kilogram-centimeters (kg-cm). This value is obtained by multiplying the force acting on an anatomical structure with the distance from the point of maximal stress concentration.

The heaviest article normally handled by man at work is his own body, or its subsegments. Only rarely do workers handle objects weighing 150 lb and, in most instances, the mass of an object moved is quite insignificant when compared with the weight of the body segment involved in the operation. For example, the majority of hand tools or mechanical components in industry weigh considerably less than 0.5 lb, but an arm, taken as an isolated body segment, weighs 11 lb (50).

The *sagittal lifting moment* is the one most frequently encountered and easiest to compute. It is most conveniently derived by graphical methods. First the weights of the body segments involved in a specific task are obtained from reliable tables (18). Then a "stick figure" of proper anthropometric dimensions (Figure 40.39) is drawn, and the location of the center of mass for each body segment is marked, as well as the center of mass for the load handled. Finally, the sum of all moments acting on the selected anatomical reference structure (in this case, the lumbosacral joint) is computed and becomes the sagittal biomechanical lifting equivalent of the specific task under consideration. Whenever the vector representing the moment sum of all gravitational moments is directed at a point on the floor located in front of or behind the soles of the feet, a prima facie hazard exists because of inherent postural instability and the ensuing likelihood of falls.

The estimation of sagittal lifting equivalents is of great practical usefulness in the comparison of work methods or in the assessment of the relative magnitude of lifting stress due to sex differences (Figure 40.40).

Males and females of approximately the same height and weight may be subject to different stresses when handling the same object. This is because of sex-dependent differences of the relative proportions of body segments. Moments acting on the lumbosacral joint during lifting depend largely on work surface height; therefore females are often at a disadvantage when picking up objects from the floor, but a pallet 12 to 14 in. high reduces sex differences in lifting stress during load acquisition (51).

Sometimes it is necessary to decide whether a task would be better performed sitting, as opposed to standing (Figure 40.41). Then a sketch, true to scale, or a photograph, becomes a convenient aid to decision making. If the pictorial representation of the work situation is supplemented by the estimated weight of the body segments involved and the location of the respective centers of gravity, the relative lifting stress for both postures can be easily estimated. Because the procedure does not compare lifting stresses between different individuals but merely establishes how the same person is affected by changes in posture, it may be assumed that the body segments involved (i.e., the torso above the lumbosacral joint, plus neck, head, upper limb, and the object manipulated) are identical in both the seated and standing postures.

According to standard data (52) the body mass in the case of a 110-lb female would be 45 lb. To this must be added the weight of the object handled—20 lb in the example under consideration. For this example, the distance from the lum-

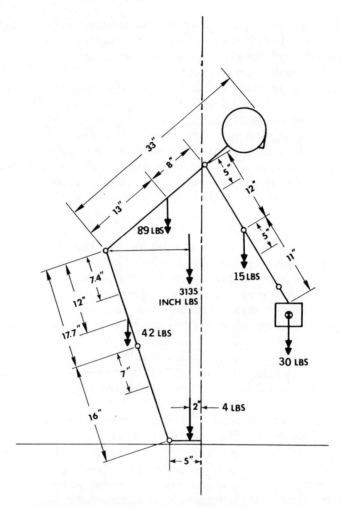

Figure 40.39. Graphic computation of the location of center of mass of whole body and body segments as well as of the sagittal moment acting on the lumbosacral joint can be conveniently accomplished through the use of stick figures. This example shows that in improper working posture, a load weighing only 30 lb combined with the mass of the various body segments involving in a lifting task, may produce a torque exceeding 300 in.-lb, which is the lifting equivalent of a very severe task. Illustration courtesy of Dr. C. H. Saran; from Reference 48.

bosacral joint to the center of mass of body segments and load combined may be estimated to be 1½ ft when standing. This exerts a torque of approximately 98 ft-lb. However, if the individual is seated, this value increases to approximately 2½ ft because of the forward leaning posture of the trunk and the outstretched arm. Therefore the torque now exerted on the lumbar spine amounts to 146 ft-lb, or an increase of nearly 50 percent compared with the standing position. This explains

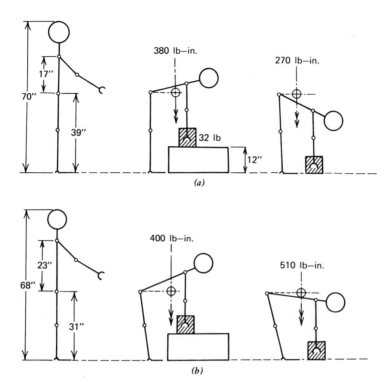

Figure 40.40. (*a*) Males and (*b*) females of approximately the same height and weight may be subject to quite different stresses when handling the same object, owing to sex-dependent differences of the relative proportions of body segments. Because moments acting on the lumbosacral joint when lifting depend largely on work surface height, females are at a disadvantage in certain postures during the load acquisition, whereas in others sex differences are minimal. Adapted from Reference 51.

why, in so many instances, employees complain—and rightly so—about much increased work stress when chairs are introduced unnecessarily into a work situation.

Analyzing lifting tasks routinely in terms of moments tends to develop in supervisors not only a "clinical eye" for the magnitude of a task but also a healthy and critical attitude toward "cookbook" rules of lifting. The principle of "knees bent—back straight—head up" is well enough known. However, in many work situations concessions must be made to the influence of body measurements. In Figure 40.42 the male, long legged and short torsoed, does not benefit at all from the application of the standard lifting rule. A female, however, having a differently proportional body, can get under the load and close to it.

Thus the sagittal lifting moment acting on the lumbosacral joint becomes much less, and work stress is approximately halved. Under such circumstances, provided the height of the work bench cannot be changed, the standard lifting rule may be applied to the female, whereas in the case of the male no benefit will be derived. Working according to the "approved" lifting posture could lull the male worker

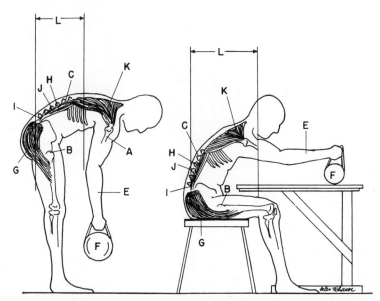

Figure 40.41. Awareness of the "hidden" lifting task should exist. Because of increase of dimension L, hence higher torques exerted on the lumbar spine, a seated job, instead of being "light work" may be the physiological equivalent of a severe lifting task. In seated work, the rule is: "get the job close to the worker." A, humerus; B, socket of hip joint; C, vertebral column; D, shoulder joint; E, arm; F, load; G, muscles of the buttocks (gluteus maximus); H, muscles of the back (sacrospinalis); I, lumbosacral joint; J, spinous process of a vertebra; K, trapezius muscle; L, distance from the center of mass of combined body–load aggregate to the joints of the lumbar spine. From Reference 58, by permission. © American Institute of Industrial Engineers, Inc., Norcross, Ga.

into a sense of false security. Therefore it is always advisable to temper the categorical instruction of "knees bent—back straight—head up" with the additional explanation "provided it helps to get the load closer to your body."

It has already been described (Figure 40.29) how a light but bulky object often imposes a lifting stress much greater than the one exerted by a heavier article of greater density. A metal ingot held close to the body exerts a lesser sagittal bending moment on the spine than a box of equal weight containing small miniaturized components, such as transistors packed for shipping in a Styrofoam container. This age of miniaturization and containerization has added a serious and sinister overtone to the age-old joke: "Which is heavier, a pound of lead or a pound of feathers?" The feathers, of course; they are so much bulkier.

In all instances, however, the two other moments in addition to the sagittal one, must also be considered. *Lateral bending moments* are all-important whenever a job calls for "sidestepping" (Figure 40.43). This often occurs when, for example, the serving of food at lunch counters is involved, or when components have to be transferred to a tray from a jig mounted on a machine.

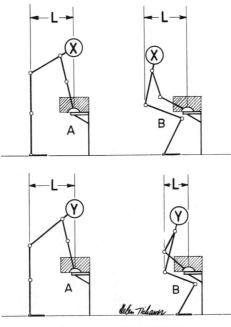

A "WRONG" POSTURE
B "APPROVED" POSTURE

Figure 40.42. Postural corrections in training for lifting should be aimed at reducing torques acting on the spine: X, an anthropometric male, does not benefit materially from the "approved" lifting posture because L, the distance from the center of mass of load to the fourth lumbar vertebra, does not shorten materially; Y, an anthropometric female, does benefit from the "bent knees, straight back" rule because she can get under the load. When matching worker and task, the measurements of the individual worker as well as the dimensions of the workplace should be considered. From Reference 58, by permission. © American Institute of Industrial Engineers, Inc., Norcross, Ga.

Generally speaking, bench work involves unnecessary sidestepping. Lateral bending moments can be of considerable magnitude, and in special cases (e.g., when an individual suffers from a mild nerve root entrapment syndrome), they may impose considerable hazard and suffering. A workplace designed for sidestepping is a workplace designed for trouble.

Consideration of *torsional moments* acting on the vertebral column becomes necessary when materials are transferred from one service or work bench to another (Figure 40.44). The L-shaped work surfaces appeal to both architect and industrial engineer because they combine aesthetic considerations with opportunities for performance efficiency. Whenever a lifting task requiring rotation of the torso about the vertical axis of the body is to be performed by a standing person, a serious hazard is not presented because of the interaction between ankle, knee, and hip joints. However, when such a task is performed seated, the pelvis is securely

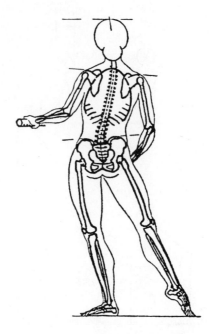

Figure 40.43. Sidestepping induces heavy lateral bending moments acting on the spine (86).

anchored and the entire torsional moment must be absorbed by the lumbar and thoracic spine. This condition is conducive to great mechanical stress on the vertebral column; it may aggravate preexisting back pain caused by pathology in the lumbar or lumbosacral region, and it can cause distress, sometimes of a respiratory nature, to scoliotic or kyphotic individuals.

It is easy to avoid excessive torsional moments in seated work situations by providing well-designed swivel chairs. It is possible to determine numerically the severity of moment action on the human body during a lifting task by the use of computerized models (53). These have been developed to a fairly high degree of perfection, and their use constitutes the procedure of choice in mass production operations under conditions of rigidly controlled work methods. One of the principal advantages of the computerized lifting models resides in their capability to permit accurate estimation of moments exerted by the mass of various body segments on a wide variety of anatomical reference points. Sometimes, however, this accuracy and convenience must be sacrificed when the need occurs to make rapid ad hoc decisions under field conditions or when the small size of the work force makes recourse to a large computational facility uneconomical.

It is for this reason that ergonomists should aim to develop the knack of "guesstimating" the magnitude of all three moments by looking at the worker, at motion pictures of an operation, or even at a drawing of the workplace layout. Then these guessed moments should be added, not algebraically, but vectorially.

By way of bench mark, it may be assumed that when the vector sum of all three moments is 350 in.-lb or less, the work is light and can be performed with ease by

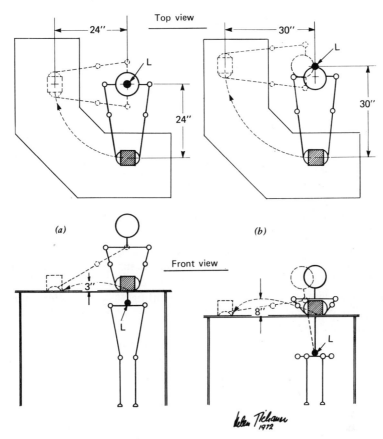

Figure 40.44. The schematic drawing shows that because of relative fixation of the pelvis, a seated lifting task may produce very heavy torsional bending moments acting on the lumbar spine; L, is the lumbosacral joint. From Reference 48.

untrained individuals, male as well as female, irrespective of body build. Moments above this level, but below 750 in.-lb, classify a task as "medium-heavy," requiring good body structure as well as some training. Tasks above this moment but below 1200 in.-lb may be considered to be heavy, requiring selective recruitment of labor, careful training, and attention to rest pauses. Whenever the vector sum of moments exceeds those stated before, the work is very heavy, cannot always be performed on a continuing basis for the entire working day, and requires great care in recruitment and training.

It is my personal experience that the ability to guess the magnitude of moments exerted by the load on the body is easily acquired. The same does not hold true, however, for moments exerted by segments of the living body. Therefore the validity of the aforementioned bench mark values is somewhat limited because they do not take into consideration individual differences in lifting stress caused by body segments, but refer to moment increases caused by external loads only.

The *gravitational components* are elements of a lifting task closely related to the concept of "work" in the sense of physics. In mechanics, work is defined as the product of force multiplied by the distance through which it acts. Thus lifting 10 lb against gravity to a height of 5 ft will constitute 50 ft-lb of work. Likewise, pushing an object horizontally for a distance of 5 ft when 10 lb of pushing force is required will also result in 50 ft-lb of work. This definition, however, is not always applicable to a situation involving the human body at work. For example, if an individual pushes with all his force against a wall and moves neither his body nor the wall, he has not accomplished any work according to the rigid definitions of physics. Nevertheless, during the entire time muscles have been under tension, metabolic activities have increased, and the added energy demands of the living organism have manifested themselves in the expenditure of additional calories that, in physics, are assigned the dimensions of work.

The effort expended on an activity that requires the application of force for a period of time without concurrent displacement of an object called "isometric activity." Sometimes the term "tension time" is also applied to this kind of activity. "Isometric work" is assigned the physical dimensions of "impulse," which equals force multiplied by time. This makes mathematical processing somewhat difficult inasmuch as gravitational components of a lifting task have the dimensions of "work," which equals force multiplied by distance. Mathematical transformations to overcome this difficulty have been developed (54) and are useful whenever recourse to computerized models must be taken. For all practical purposes, it is often adequate to estimate *isometric work* by taking the weight of the object handled plus the estimated weight of the body segment involved in this task, and to multiply these by the time the muscles are under tension.

Dynamic work is defined as the product of the weight of an object handled, multiplied by the vertical distance through which it is lifted upward against gravity. It has the dimensions of work as defined in physics and can be generally computed with ease.

Negative work is performed whenever an object is lowered at velocities and accelerations of less than gravity so that work against the "*g*" vector is performed.

To avoid complex computations, it is practical to assume, under industrial working conditions (55), that one-third of the work that would have been expended when lifting the same object over the same distance in an upward direction is approximately equal to the negative work performed.

Finally, in the evaluation of a lifting task, the *inertial forces* must be considered. When an object handled is in motion, acceleration is generally insignificant as far as work stress is concerned. However, the forces involved in *aggregation* and *segregation* of man and load may impose severe stresses on the human body.

To maintain the unstable equilibrium of upright posture, it is necessary that the center of mass of the body be located over a line connecting the sesamoid bones of the big toes. Whenever a load is lifted, object and human body become one single aggregate, and as soon as the load has left the ground, the body, through changes in postural configuration, must place the center of mass of this body–load aggregate over the area of support. This requires displacement of the center of mass of the body proper which, during normal acquisition of the load from the ground, takes place over a time interval of roughly 0.75 sec. During this brief time interval "stress spikes" will be observable if myograms of the muscles of the back are taken (56). Acquisition stress can best be minimized by having the point of pickup of the load as close to the worker's body as possible.

A stress far more severe is experienced on segregation of the load. Because release of an object is normally fairly rapid, segregation may take place over as small an interval of time as 40 msec, or $\frac{1}{20}$ of the time involved in acquisition. This requires that postural adjustments be far more rapid during release than during pickup, which, in turn, produces great stress on the musculoskeletal structures involved. Electromyographic studies (31) have shown that stress experienced during release may be a multiple of the physical work stress generated during the rest of the work cycle. This has been confirmed by other investigators (56). Segregation stress can best be reduced by having the point of release of the load as high above floor level as possible. It is also essential that workers be made aware of the postural adjustments occurring during load release, including pelvic rotation, and be trained to perform these slowly.

Finally, *frequency of lift* is often one of the elements deserving considerable attention. The number of times a lifting task is performed during the day is equivalent to "productivity," and therefore it is determined by economic needs. Where bulky loads, such as television tubes, are handled, it may be quite impossible to control the severity of a lifting operation by attention to the frequency of lift, because this will not be amenable to modification. Yet often a number of relatively small units, several at a time, have to be handled, and the frequency of the operation then can be controlled by optimizing or by changing the amount to be handled at one time.

Gilbreth, at the beginning of the century (57), stated ". . . lifting 90 lb of brick on a packet [sic] to the wall will fatigue a bricklayer less than handling the same number of bricks one or two at a time. . . ." This remark was, of course, based on purely empirical and subjective observations because work physiological instrumentation available during the lifetime of Gilbreth simply did not permit the objective substantiation of such hypotheses. The bricklaying task was repeated by a group of volunteers during the 1950s (57).

In the Gilbrethian example, "lightness" of the task when handling 90 lb of bricks (18 bricks) one at a time is illusory. Each time a 5-lb brick is handled, the worker must move, bend, erect, rotate, and so on, approximately 100 lb of body mass. Thus the task load imposed by the handling of the body itself becomes much more severe than the work stress caused by the material being handled. Approximately 1800 lb of lifting body mass must be maneuvered to shift 90 lb of brick.

When 90 lb of brick was handled at one time, according to the Gilbrethian intuitive prescription, the ratio between body mass moved and inanimate material handled was roughly 1 : 1. As was to be expected, the physiological response to the task, under the circumstances described, was quite moderate and was essentially a function of the rate of productive work, not of unnecessary and physiologically expensive body movements. However, when the task load was increased to 120 lb

Table 40.4. Results of a Replication of the
Gilbrethian Lifting Experiment (57)

Frank Bunker Gilbreth, the father of motion study,
was intuitively right when he stated that ". . . to lift
90 pounds of brick at a time is most advantageous
physiologically as well as economically . . ." (57).

	I	II	III
Bricks per lift	1	18	24
Weight per lift (lb)	5	90	120
Work per hour (kcal)	520	285	450
Bricks per hour	250	600	300

(i.e., 24 bricks) handled at the same time, the metabolic cost of the job rose out
of proportion to the increment in the dynamic component of the task (Table 40.4).
This apparent paradox was rationally explained at the time of the experiment as
a result of oxygen debt. The significance of the element "frequency of lift," as
shown previously, can easily cause the handling of numerous small loads to become
the equivalent of a very stressful task. It is therefore essential to be aware of the
concept of "optimal load" per lift.

5.2 Queueing Situations

In materials handling situations, loads are frequently passed from one worker to
the other, or they arrive at the lifting station by way of conveyor belts and are
then handled manually. Much useless materials handling takes place when conveyor
belts, device pallets, trolleys, and other devices, or areas for temporary holding or
storage of products, are too small. Bottlenecks occur, and emergency measures
may have to be taken to clear the congested areas. The dimensions of all temporary
storage areas and devices should be computed on the basis of queueing theory,
assuming that the arrival at the handling station follows the Poisson distribution
and service times are exponentially distributed. Frequently the following formula
can be used to advantage (58):

$$N = \frac{\log P}{\log R} \tag{11}$$

where N = required capacity of the area

P = greatest acceptable probability that the area will become temporarily
       overloaded; this principle is normally determined on the basis of a
       subjective management decision

R = mean arrival rate of units per time divided by the mean processing
       rate; these values are normally available from industrial engineers

It should not be forgotten that in addition to readily discernible or "overt" lifting tasks, workplace design may embody hidden or "covert" situations. A covert lifting task exists whenever a moment is exerted on the axial skeleton without an extraneous object being handled. This may be the case when a typist's head is bent over the typewriter, or an arm has to be held out for extended periods of time because of the poor location of storage bins, or working shoes have heels that are too high, so that they increase the concavity of the lordotic curve of the lumbar spine (31). To assess the true severity of such a situation, textbooks of classical biomechanics (18) or computerized mathematical models should be consulted (39).

In conclusion, it is reemphasized that the elements of a lifting task are heterogeneous in their physical dimensions and their physiological effects; therefore, although all contribute to the level of work stress, the individual effects cannot be "summed." The reality of most materials handling situations demand that each element be considered separately and that those amenable to control be reduced in magnitude and severity as far as possible.

Unfortunately, at the present state of the art, there are only partial and no total solutions available to eliminate hazards from manual lifting and materials handling tasks.

6   HAND TOOLS

Toolmaking is probably as old as the human race itself. Almost all the basic tools used today, as well as the "basic machines," were invented in the dawn of the prehistoric ages, evolving gradually and parallel with the development of technology until modern times.

However, the current technology explosion proceeds too rapidly to permit the gradual development of tools to suit the new industrial processes. Now "instant" creation of new and specialized implements often becomes an acute economic necessity, if an industry wishes to accommodate itself to the rapid changes of workforce and technology. Hand tools are more often than not standardized to be "fairly acceptable" to the broadest possible spectrum of populations and activities. Only rarely are they designed to fit perfectly the specialized needs of a particular manufacturing pursuit or the anthropometric attributes of the workforce in a given plant.

Thus considering their widespread use, the number of varieties of hand tools is quite limited, and each species of tool is normally employed by vast numbers of users. Therefore hand-tool-generated work stress, trauma, and ergogenic disease may at times reach epidemic proportions, disabling numerous individuals, seriously impairing the productive capacity of a manufacturing plant, and having very detrimental effects on labor–management relations. All this is quite apart from the medical costs and the human suffering involved. Therefore the occupational health specialist should make the proper selection, evaluation, and usage of hand tools one of his major concerns.

6.1 Basic Considerations in Tool Evaluation

Though the shapes of tools are varied and the functions of the different classes of implement diverse, there are nevertheless many principles of biomechanics and ergonomics that are applicable to the prevention and solution of problems created by hand tools, no matter how different their fields of application.

Many of the principles enunciated in this section extend in application beyond the narrow field of hand tools proper. They are equally useful in the analysis of equipment controls, because the latter are merely the "tools" that permit the operation of machinery. The initial purpose of primeval tools, such as stone axes and scrapers, was to transmit forces generated within the human body onto inanimate materials, food, or live animals of prey. As the spectrum of artisanal and industrial pursuits widened, the basic purpose of tools became more varied, and today these implements are designed to extend, reinforce, and make more precise range, strength, and effectiveness of limb movement engaged in the performance of a given task.

In this context, the word "extend" does not solely imply a magnification of limb function. Often tools such as tweezers and screwdrivers also make possible far smaller and finer movements than the unarmed hands would be capable of performing. An even better example is the micromanipulator of the "master–slave" type, which serves as attenuator rather than amplifier of human force in motion. A third example is a suction tool that makes it possible to transport small, fragile, and soft workpieces without injury either to their dimensions or their surface finish.

Selection and evaluation of all hand tools, whether manual or power operated, should be based on all the following considerations: technical, anatomical, kinesiological, anthropometric, physiological, and hygienic (59).

In the course of technological development, tool designers were conditioned to focus their attention on a real or imagined need to maximize the tool force output obtainable from a minimal muscular force input from the hand. This, however, should not be overdone.

The operation of a tool should always require sufficient force to provide adequate sensory feedback to the musculoskeletal system in general, and the tactile surfaces of the hand in particular. This is frequently a process of optimization. For example, if a fine screw thread is tapped by hand and the handle of the threading tool is too large, the force acting on the tool becomes excessive, resulting in stripped threads, broken taps, or bruised knuckles. If, on the other hand, the ratio of force output to force input is too small, an unduly large number of work elements must be repeated, and this makes the job fatiguing. An example would be the pounding of a large nail with a very small hammer.

A tool should provide a precise and optimal stress concentration at a specific location on the workpiece. Thus, up to a certain limit, an ax should be as finely honed as possible to fell a tree with a minimum number of strokes, but the edge should not be so keen that it requires frequent resharpening or is fragile. Preferably the tool should be shaped so that it will be automatically guided into a position of optimal advantage where it will do its job best without bruising either hand or workpiece. The Phillips screwdriver, as compared with the ordinary, flat, blade tool, illustrates the latter point.

Hand tool usage causes a variety of stress vectors to act on the human–equipment interface. These may be mechanical, thermal, circulatory, or vibratory, and they are often propagated to quite distant points within the body, far from the actual locus of application of force. The cause of severe pain in the neck muscles or numbness and tingling in the fingers of the left hand may quite conceivably be due to the transmission of somatic resonance vibrations triggered by a vibrating hand tool held in an unergonomic configuration in the right hand. Numerous other examples could be cited. Whenever a single specific anatomical region becomes the locus of repeated manifestations of signs and symptoms of trauma, no matter how far away from the tool operating hand, the work situation should be carefully analyzed for the possible implication of hand tool design or usage as a traumatogenic vector.

Contact surfaces between the tool and the hand should be kept large enough to avoid concentration of high compressive stresses (Figure 40.45). Pressure and impact acting on the hand may be transmitted either directly or by rheological propagation on vulnerable structures. A poorly designed or improperly held scraping tool may squeeze the ulnar artery and sometimes the ulnar nerve between the timber of the handle and the bones of the wrist. This may deprive the ring and little fingers of proper blood supply (Figure 40.33), which may cause numbness and tingling of the fingers. Under such circumstances, the afflicted worker will devise an excuse to leave the workplace temporarily, because this is the only means of relief open to him. Apart from the resulting drop in productivity, the health of the working population is in serious jeopardy. Literature (60) suggests that compressive stresses applied against the medial side of the hook of the hamate can traumatize the ulnar artery and may result in thrombosis or other irreversible injury. The ulnar artery and nerve may also be injured indirectly by stress propagation whenever the palm of the hand is used as a hammer or repeatedly pushes a tool against strong resistance. The resulting lesion, be it vascular or nervous, is known as "hypothenar hammer syndrome" (61).

6.2 The Anatomy of Function of Forearm and Hand

Further chances of injury exist when the motions inventory demanded by specific features in the design of the tool is not readily available from arm and hand. The kinesiology of the upper extremity is basically determined by the structure and arrangement of the skeleton of arm and hand (Figure 40.46). There are two bones in the forearm, the ulna and the radius. The ulna is stout at its joint surface of contact with the humerus and forms a hinge bearing there. It is, however, slender at the distal end, where it articulates well with the radius but only poorly with the carpus formed by the bones of the root of the hand. The radius, on the other hand, is slender at its point of contact with the humerus and stout at its distal end. At the proximal end it forms a thrust bearing with the humerus and a journal bearing with the ulna. The distal end forms a joint with the carpal bones, the primary articulation between forearm and hand. The unique configuration of the mating surfaces of this joint permits movements in only two planes, each one at an angle of approximately 90° to each other. The first of these maneuvers is palmar flexion, or when performed in the opposite direction, dorsiflexion (Figure 40.47). The

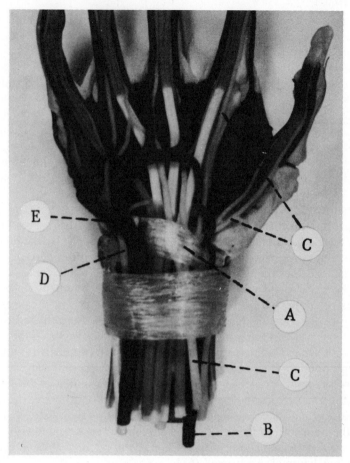

Figure 40.45. Through the carpal tunnel *A* pass many vulnerable anatomical structures: blood vessels *B* and the median nerve *C*. Outside the tunnel, but vulnerable to pressure, are the ulnar nerve *D* and the ulnar artery *E*. From Reference 48.

second set of wrist movements possible consists of either ulnar or radial deviation of the hand (Figure 40.48). The wrist joint does not allow rotation of the carpus about the longitudinal axis of the forearm. Swiveling the wrist without forearm rotation is not possible (Figure 40.49).

This geometry of joint movement causes the axis of longitudinal rotation of the forearm–hand aggregate to run roughly from the lateral side of the elbow joint through a point at the base of the ulnar side of the middle finger (Figure 40.46). The palmar aspect of the carpal bones forms a concave surface roofed by the transverse carpal ligament. The resulting channel is known as the "carpal tunnel." Through this conduit pass the tendons of the flexor muscles of the fingers, which originate from the medial side of the elbow. Some blood vessels and nerves also pass through this tunnel. A similar, albeit much shallower, passage for the extensor

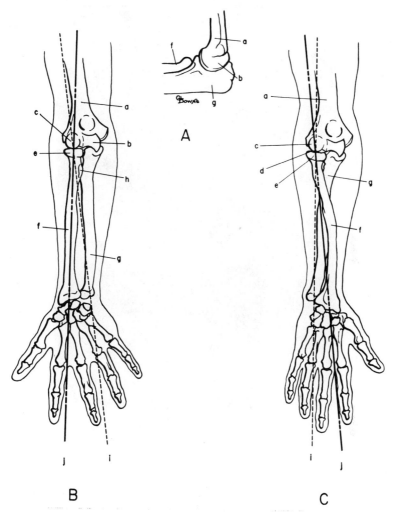

Figure 40.46. The construction of the skeleton of the forearm. (*A*) The hinge joint between ulna and humerus from the medial side. (*B*) The right forearm outwardly rotated. (*C*) The right forearm medially rotated: *a*, humerus; *b*, trochlea of humerus; *c*, capitulum of humerus; *d*, thrust bearing formed by capitulum and head of radius; *e*, head of radius; *f*, radius; *g*, ulna; *h*, attachment of biceps; *i*, axis of rotation of forearm; *j*, optimal axis for thrust transmission. From Reference 48.

tendons and some nerves is formed on the dorsal surface by another ligament and the dorsal aspects of parts of the carpus, the radius, and the ulna (Figure 40.50). Each of the two passages is further divided into several longitudinal compartments, and this produces considerable friction and lateral pressures between tendons, tendon sheaths, adjacent nerves, and vascular structures. If the fist is forcefully opened, closed, or rotated, these forces, as well as the friction between anatomical structures, can become unduly high whenever the carpal tunnel and its homologue

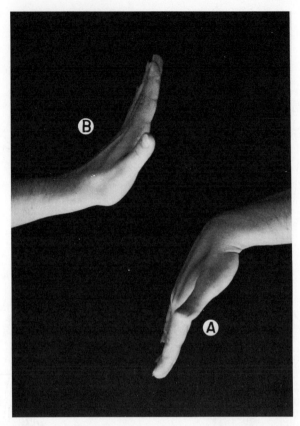

Figure 40.47. (*A*) Hand in palmar flexion. (*B*) Hand in dorsiflexion.

compartment on the dorsum of the hand are not properly aligned with the longitudinal axis of the forearm. Such alignment exists only if the wrist is kept perfectly straight, so that the metacarpal bone of the ring finger is reasonably parallel with the distal end of the ulna. A potentially pathogenic situation may exist whenever manipulative maneuvers, especially forceful ones that require ulnar deviation, radial deviation, palmar flexion, or dorsiflexion, either singly or in combination, are performed. The misalignments of tendons at the wrist under such circumstances and their bunching up against each other multiply drastically the already high interstructural forces and frictions produced by the muscles operating the hand; early fatigue may result, among other undesirable manifestations (Figure 40.51). The long interval of time (several weeks, months, even years) of repetitive multiple work stress elapsing before occupational disease of forearm or hand becomes clinically evident has militated against the definitive establishment of direct cause and effect relationships between the forearm–wrist configuration and specific pathology. Nevertheless, certain motion elements are almost inevitably associated with fairly narrow spectra of occupational disease of the hand (2).

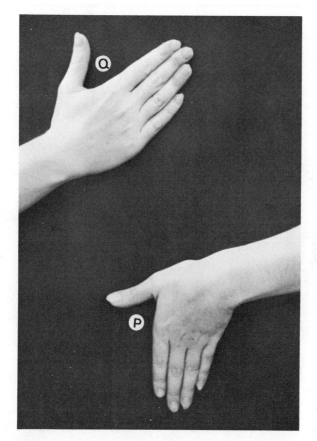

Figure 40.48. *P*, ulnar deviation; *Q*, radial deviation.

6.3 Elemental Analysis of Hand Movements

The design of tool and handle determines, limits, and defines the motion elements that are necessary for the purpose of the productive process. This intimate relationship between humans and hand tools affects occupational health and safety most directly.

It is almost impossible to describe innocuous as well as potentially pathogenic manipulative maneuvers, either verbally or by way of two-dimensional pictorial representation. As a somewhat simplistic yet practical alternative, recourse to the analysis of manipulative maneuvers in terms of their pertinence to "clothes wringing" (Figure 40.52) is recommended. "Clothes wringing" has been associated for more than a century with tenosynovitis and other undesirable conditions of the hand (2). Here we assume that the wringing is done by a clockwise movement of the right fist and counterclockwise action of the left.

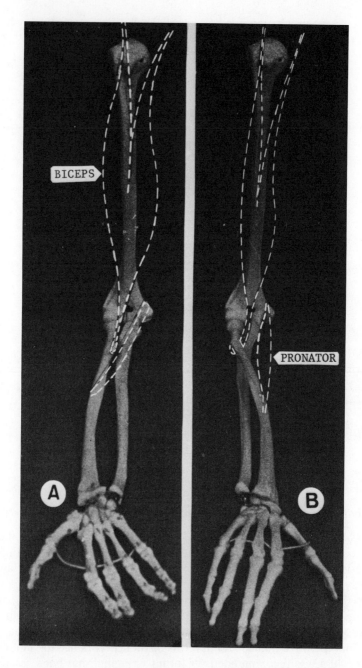

Figure 40.49. The bones of the wrist articulate with only one of the two long bones of the forearm, the radius, with which they form a firm aggregate. Therefore swiveling the wrist without rotating the forearm is impossible. (*A*) Forearm in supination. (*B*) Forearm in pronation.

Whenever a detailed elemental analysis of hand motions is performed, it should be remembered that stress injurious to the hands is produced by four basic conditions:

1. Excessive use against resistance
2. Hand use while in a potentially pathogenic configuration
3. Repetitive maneuvers and cumulative work stress rather than "single episode" overexertion
4. Use of the hand in an unaccustomed manner, as in training, for example (Figure 40.35)

Figure 40.52 shows that in "clothes wringing" the right hand is engaged simultaneously in supination, ulnar deviation, and palmar flexion. This motion pattern is frequently associated with occurrences of tenosynovitis of the extensor tendons

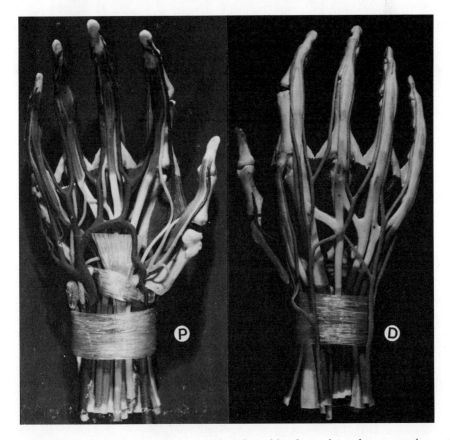

Figure 40.50.   The complex arrangements of tendons, blood vessels, and nerves underneath the ligaments of the wrist: *P*, palmar aspect; *D*, dorsal aspect. Whenever the wrist is deviated, these bunch up against each other. The ensuing friction may lead to trauma and disease.

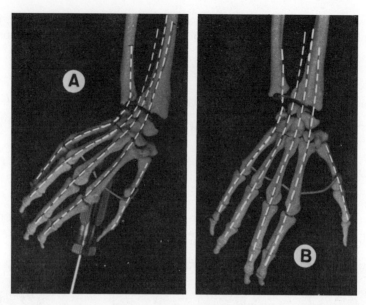

Figure 40.51. The need to align a tool with the axis of the forearm often forces the hand to deflect toward the ulna (*A*). The tendons operating the fingers get "kinked" and bunch up. This causes friction between these delicate anatomical structures which, in turn, produces discomfort and, occasionally, diseases of the wrist. When the wrist can be maintained in a straight configuration (*B*) by good tool design, the tendons are well separated, run straight, and can operate efficiently. From Reference 37, by permission. © American Institute of Industrial Engineers, Inc., Norcross, Ga.

Figure 40.52. Any manipulative motion element involving the wrist that may be considered to be part of a "clothes wringing" operation puts the worker at risk.

of the wrist or the abductors of the thumb, the latter affliction also known as De Quervain's disease. It is an inflammation of the synovial lining of the tendon sheaths or associated structures. It frequently becomes manifest under conditions of unaccustomed hand usage in new employees (Figure 40.35) and is therefore occasionally considered quite without justification to be "a training disease." However, the basic cause of much exposure to excessive work stress resides not in the learning process but in the misuse of the hand, which has been forced by poor workplace or tool design into unergonomic configuration.

The similarity between the movement of the right hand motion in "clothes wringing" and the use of pliers when looping wires around pegs is quite evident. Because three factors are implicated in the generation of the undesirable work stress, elimination of one of them may successfully reduce the stress to a nonpathogenic level. For example, ulnar deviation can be eliminated by bending the tool handle (Figure 40.35).

In some other work situations involving the same wrist configuration, it may be more desirable, or easier, to reduce excessive palmar flexion by small modifications in workplace layout. A simple change of the distance between worker and work bench, location of the work chair, work surface height, or the degree of tilt of the work space, may constitute adequate remedial action (40). The experienced practitioner will apply the same principles to the improvement of other pathogenic work situations, such as the manual insertion of screws, the manipulation of rotating switches, or the operation of electrical or air-powered nut setters suspended over the workplace.

In Figure 40.52 the left hand is engaged in pronation, radial deviation, and dorsiflexion of the wrist. This configuration is conducive to pressures between the head of the radius and the mating joint surface of the humerus (Figure 40.46). This posture should be strongly discouraged through proper tool design because it may produce a high incidence of the group of diseases known in the vernacular as "tennis elbow." The condition may occur in such tasks as overhead use of wire brushes in maintenance work. This is a good example of the basic principle that work strain may affect a site distant from the location of work stress. The wrist is being stressed, but it is the elbow that gets sore (62). Strong and repeated dorsiflexion of the wrist, especially in combination with some other hand or forearm movement, is conducive to carpal tunnel syndrome—a disease that may also be provoked by direct trauma to the region of the hand over the carpal tunnel, compression of the median nerve in the tunnel through tenosynovitis and swelling of the flexor tendons, as well as several other causes that are not occupationally related (61). Implicated in unergonomic imposed patterns conducive to carpal tunnel syndrome are tossing motions, the operation of valves located overhead or on vertical walls, and a number of poor handle designs of power tools. This list, however, is by no means exhaustive.

Finally, the wrists of both hands can be severely stressed when a two-handled tool is designed so that the longitudinal axes of both handles coincide. Under such circumstances, the included angle between the axes of the handles should approximate 120°.

6.4   Trigger-Operated Tools

Occasionally the condition of "trigger finger" is encountered. The afflicted person typically can flex the finger but cannot extend it actively. It must be righted passively by external force. When snapped back in such a way, an audible click may be heard. This has been attributed to the generation of a groove in the flexor tendon which snaps into a constriction produced by a fibrous tunnel guiding the tendon along the palmar side of the finger. Small ganglia arising in the tendon sheath have also been implicated (63). This affliction has been observed under conditions of overusage of the index finger as an equipment control, such as the trigger of a tool embodying a pistol grip. The association between overusage of the index finger and the lesion seems to occur most frequently if the tool handle is so large that the distal phalanx of the finger has to be flexed while the middle phalanx must be kept straight. This can easily be the case when females, with relatively small hands, operate tools designed for males. A tool handle a bit too large may put a female working population to distress, but the tool designer who errs on the side of smallness will produce an implement that can be operated with equal effectiveness by both sexes (48). As a rule, frequent use of the index finger should be avoided, and thumb-operated controls should be put into tools and implements wherever possible, because for all practical considerations, the thumb is the only finger that is flexed, abducted, and in addition to muscles crossing the wrist, opposed by strong, short muscles located entirely within the palm of the hand. It can therefore actuate push buttons and triggers repeatedly, strongly, without fatigue, and without exposure to undue hazard.

6.5   Miscellaneous Considerations

Numerous hand-held tools, especially of the power-driven type, are advertised as "light." The lightest tool is not always the best tool. Optimal tool weight depends on a number of considerations. If a power tool houses vibrating components, it should be heavy enough to possess adequate inertia. Otherwise it may transmit vibrations of pathogenic frequency onto the body of the operator (see Section 4.3) (42, 43).

Heavy tools should be designed so that the center of mass of the implement is located as close as possible to the body of the person holding it. When this type of facilitation is inadequate, recourse should be taken to suspension mechanisms and counterweights. Strength of handgrip is the most important single factor limiting the weight of a tool that must be held without external assistance from supporting mechanisms. Under such circumstances, a reliable and specialized reference work should be consulted (64).

The design of the tool–hand interface should be based on carefully selected anthropometric considerations. There are a number of specialized anthropometric reference works available (65–67). However, not all the hand dimensions identified in the literature are necessarily representative of a working population of specialized age, sex, or ethnic origin. Furthermore, the geometry of manipulative movements imposed on the hand by the specifics of tool design or usage may force the user population to employ a geometry of hand movement based on special anthropometric parameters quite different from those available from the majority of reference works. Likewise, the use of working gloves changes the dimensions of the hand. Gloves vary in design and thickness of material used. This does affect grip strength and sensory feedback from the tool (see Section 4.4).

It is recommended that producers as well as consumers of hand-held tools conduct their own anthropometric testing based on the hand and other body dimensions of the specialized working population under study, as well as of the tools and gloves to be used. Of course basic reference charts can be employed with advantage as the point of departure for anthropometric hand studies designed to suit specialized needs (68).

Gloves may affect the working hand in a number of additional ways. Mild trauma due to pressure may be aggravated through contacts with irritants entrapped unintentionally in working gloves. In the case of abrasives, such as metal chips or mineral particles, the foreign substances may be worked into the skin, producing in some cases benign tumors, such as talcum granuloma. Certain fat-soluble chemicals, such as many common solvents and detergents, may be soaked up in the material of the glove and transferred onto the skin, causing maceration, dermatoses, or other tissue damage. Sometimes it may be advisable to wear thin cotton gloves under these working gloves to absorb perspiration and improve hygiene.

Any study of work stress, occupational disease, or safety involving hand tools is incomplete unless the potential effects of working gloves are fully considered, inasmuch as they could modify fit between tool and hand, change the distribution of pressure, or become detrimental to the integrity of the skin.

7 CHAIRS AND SITTING POSTURE

Many jobs require performance in the seated posture, and chairs are among the most important devices used in industry. They determine postural configuration at the workplace as well as basic motion patterns. A well-constructed chair may add as much as 40 productive minutes to the working day of each productive individual (48). Furthermore, poorly designed seating and inadequate supervision of seating posture constitute frequent and definitive occupational hazards. Properly designed working chairs are a prerequisite to the maintenance of occupational health and safety of many working populations, including individuals with preexisting disabilities of the back.

A number of reference works (69, 70) provide dimensions for working chairs and benches considered by many as optimal. However, such optimality can never be general; it applies only to the restrictive parameters of a population defined by age, sex, ethnic origin, cultural background, and specific working conditions. Slide rule-like devices (71) are available that permit the adaptation of anthropometric data to a wide variety of body dimensions, and in some countries the anthropometric basis for industrial seating has been subject to national standardization (72, 73). However, all dimensional aids to chair design must be personalized and revalidated for each individual application on the basis of certain aspects of functional and surface anatomy.

7.1   Anatomical, Anthropometric, and Biomechanical Considerations

It was realized almost a century ago that ". . . our chairs almost without exception are constructed more for the eye than for the back . . ." (74). This statement was made in an article appearing in a journal of preventive medicine and stating the urgency of providing better lumbar support to seated workers. Unfortunately, in many instances, the same plea can be made today with respect to some of the most modern office and factory furniture. Somewhat later Strasser (75) quantified the forces exerted by the backrest and the seat on the lumbar region and the buttocks. He also showed how these changed drastically as a function of the slope of these features. The first comprehensive study of the biomechanics of seating was not conducted until 1948 (45). This existing body of knowledge should be put to good use in the design of working chairs.

To facilitate a description of seated work situations, a nomenclature of planes of reference suitable for the definition of relationships between postural configuration and the position of equipment controls, as agreed by convention, is employed throughout this section (Figure 40.53). The "axis of support" of the seated torso is a line in a coronal plane passing through the projection of the lowest point of the ischial tuberosities on the surface of the seat. This is, in essence, a "two-point support" (Figure 40.54). As a result, the compressive stresses exerted on the areas of the buttocks underlying the tuberosities is quite high and has been estimated as 85 to 100 psi. This, of course, varies with body weight and posture. Stress can be nearly double when a person is sitting cross-legged. These high pressures make it necessary to vary sitting posture and position on a seat frequently to provide the necessary stress relief for the body tissues. Therefore the seat of the chair should be approximately 25 percent wider than the total breadth of the buttocks. To further facilitate change of position, all coronal sections of the seating surface of the chair should be straight lines. A coronally contoured seating surface tends to restrict postural freedom at the workplace and, especially when poorly matched to the curvatures of the buttocks, may cause severe discomfort. Improper contouring may also interact with sanitary napkins and other devices worn by women during the menstrual period, with the ensuing further reduction in physical well-being during an already trying time.

The height of the seat has been the subject of much argument, and literature abounds with numerical data, most stating the relevant dimension as the vertical distance of the highest point of the seat from the floor. This is ergonomically wrong. The back of the thigh is ill-equipped to withstand pressure (Figure 40.55)—compression applied there may deform the limb severely, irritate important nerves and blood vessels, and interfere with the circulation in the lower extremity. Many people have had the experience that a chair too high causes the leg "to go to sleep." This is an especially undesirable condition when the circulation is already impaired by preexisting disease, such as diabetes or varicose veins.

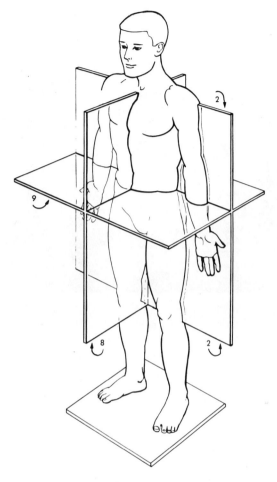

2 = CORONAL PLANE

8 = MID-SAGITTAL PLANE

9 = TRANSVERSE PLANE

Figure 40.53. The basic planes of reference for biomechanical and anatomical description. From Reference 76, by permission. © W. B. Saunders Company, Philadelphia, PA.

The maximum elevation of the seat above the surface supporting the feet—be this the floor or a footrest—should be 2 in. less than the crease at the back of the hollow of the knee, known as the popliteal crease. For this reason, the seat height of the chair should be adjustable in a limited number of discrete steps (Figure 40.55). The frontal end of the seating surface should terminate in a "scroll" edge, which does not cut into the back of the thigh. For further protection, there should be a distance of at least 5 in. between the scroll and the popliteal crease. The seating surface should have a backward slant of approximately 8° in a sagittal direction. This encourages the use of the backrest and prevents forward sliding of

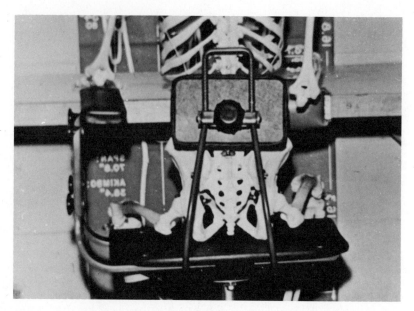

Figure 40.54. The structure of the pelvis and the location of the ischial tuberosities demand that the coronal sections of a chair be not contoured.

the buttocks. Because the thighs are tapered, any inadvertent forward sliding may produce compression of the limb between the lower edge of the work bench and the top of the chair. The seating surface should be slightly padded and covered with a porous, rough, fabric that "breathes" and facilitates adequate conduction of heat away from the contact area between buttocks and chair.

The backrest of a chair employed in manufacturing operations should provide lumbar support. It should be small enough not to exert pressure against the bony structures of the pelvis or the rib cage. At least the top of the backrest should be below all but the "false" ribs. Many work situations require continuous and rhythmic movement of the torso in the sagittal plane, and a backrest that is too high will produce bruises on the backs of a considerable number of the working population.

The best designed backrests are so small that they do not interfere with elbow movement and can swivel freely about a horizontal axis located in the coronal plane (Figure 40.55). Thus they fit well into the hollow of the lumbar region and provide the needed support for the lower spine without detrimental interference with soft tissues. When much materials handling and twisting of the torso in the seated position takes place a backrest that is overly wide will make frequent and repeated contact with the breasts of female workers (Figure 40.56). This can be most uncomfortable, especially during the premenstrual period, when the breasts of many women are quite tender.

Whenever backrests produce either bruising or even only slight discomfort, workers protect themselves. The painful effects of excessive interaction between chair and human body are reduced under such circumstances by ad hoc protective devices, such as pillows brought from home and strapped to the backrest. Work

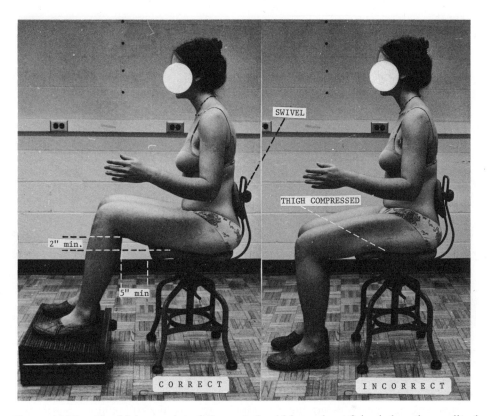

Figure 40.55. The highest point of the seat should be at least 2 in. below the popliteal crease of the worker. If necessary, this must be accomplished by a footrest. The backrest should swivel about the horizontal axis to align with the lumbar curve.

sampling studies show that several hours per worker per week may be wasted by the effort needed to keep such improvised cushioning devices in place. Properly designed backrests are much cheaper than improvisation, both in the long run and in the short run.

Some working chairs are equipped with casters. These make possible limited mobility of the worker and permit materials handling without abandoning the seated posture. In many situations casters reduce unnecessary torsional moments acting on the lumbar spine. Also, in some cases of circulatory disturbance in the lower extremity, a chair that can be pushed around by leg movement, while the person maintains seated posture, can activate the "muscle pump," consequently improving circulation. The disadvantage of any chair mounted on rollers resides in the risk that the chair may roll accidentally away or be inadvertently removed while the user gets up for a brief time. If the same person afterward tries to sit down without being aware of the changed situation, a dangerous fall can result. Whenever possible under such circumstances, therefore, a cheap restraining device, such as a nylon rope or chain or, sometimes, a more rigid linkage between chair and work bench, should be considered.

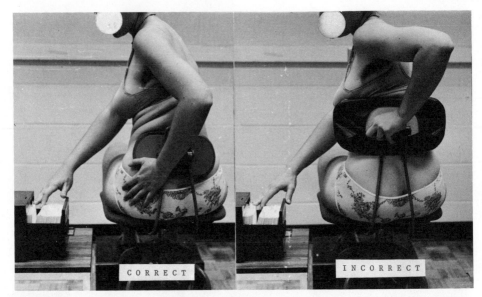

Figure 40.56. A poorly located backrest that is overly wide may severely traumatize the breasts of female workers during torsional movements (77).

7.2 Adjustment of Chairs on the Shop Floor

A frequent but sometimes dangerous response to chair-generated discomfort is temporary absenteeism from the workplace. When the level of personal tolerance has been exceeded, many individuals simply "take a walk." It is found occasionally that individuals involved in accidents are at the place of injury without authorization. Because no accident is possible unless victim and injury-producing agent meet at the same spot and at the same time, temporary absenteeism from the workplace may result in unnecessary exposure to potentially hazardous situations (40).

To guard against this, especially in new plants or when existing working chairs are being replaced by different models, the work force should be polled and a subjective assessment of chair comfort established through attitudinal measurement techniques specifically designed for the evaluation of seating accommodations (78). Although subjective comfort evaluations decrease the longer a chair has been in use, it should also be remembered that a comfortable chair is not necessarily the "best chair," regardless of whether high levels of productivity or optimal physiological compatibility with the anatomy of the worker is the guiding considerations. Therefore it is essential that the necessary initial attitudinal measurement be either preceded or immediately followed by rational adjustment of the seating accommodations as related to the work bench on the basis of biomechanical and ergonomic considerations.

Supervisors as well as workers should receive formal instruction in the proper adjustment of working chairs. First, the height between seating surface and top of the work bench is adjusted so that an angle of abduction of approximately 10°

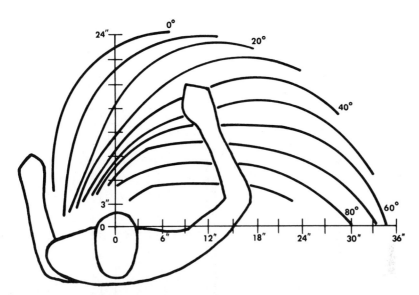

Figure 40.57. The natural motions pathway of the wrist changes with the angle of abduction. From Reference 48.

between the upper arms and the torso can be maintained during activity (Section 4.2). In specialized work situations, adjustment for an angle of abduction other than optimal may be necessary. It should be taken into consideration that the natural motions pathway of the wrist changes with the angle of abduction (Figure 40.57), and the implements of work—components, tools, jigs, and fixtures—should be arranged accordingly on the work bench; otherwise imprecise movements, poor eye–hand coordination, and early fatigue will result. Second, correct seat height with respect to the floor as well as the popliteal crease is established by means of an adjustable footrest. Footrails, because they cannot be easily adjusted, are less desirable. Occasionally drawers located underneath work benches may interfere with proper seat height adjustment, and these should be removed if necessary. Third, the height and position of the backrest should be arranged so that the minimal distance of 5 in. between the front edge of the seat and the popliteal crease is maintained. The adjustment of the backrest should enable the worker to have it low enough to permit freedom of trunk movement as demanded by the work situation, but not so low that it interferes with the bony structures of the pelvis.

Sometimes it is desirable to provide the opportunity to perform a job from either the seated or standing posture, or to alternate both postures in the course of the working day, according to the personal preference of the worker. Then the basic angle of abduction of the upper arm should be attained by a suitable height of the work bench. To maintain this for both postures, the chair and the footrest should be of the appropriate dimensions (Figure 40.58).

Finally, the distance of the backrest of the chair from the near edge of the work bench should be standardized to optimize skeletal configuration (Sections 4.2 and 4.4), to facilitate the development of the most effective kinesiology.

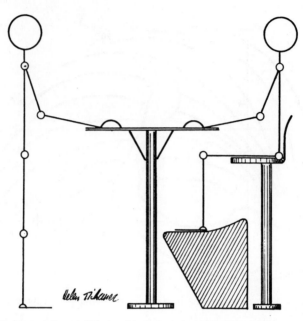

Figure 40.58. When work is possible in either seated or standing position, work bench and seating design should permit change of posture without change of musculoskeletal configuration. From Reference 48.

7.3 Ancillary Considerations

No single chair design can possibly be optimal for all work situations. Seating analysis should always be conducted to meet the needs of specific, not generalized situations, giving due weight to all relevant features of the task under consideration. It is highly desirable that standard reference works (68–70) be available for perusal.

In addition to the analysis of the seated posture with respect to physical comfort and biomechanical correctness, it is necessary to consider the changes in kinesiology of the lower extremity resulting from seated posture. The seated leg and foot can rotate only with difficulty, and unless the height of the chair is extremely low, such as is the case in motor vehicles, operation of foot pedals may become cumbersome and fatiguing. On the other hand, the seated "leg and thigh aggregate" can abduct and adduct voluntarily, precisely, and strongly, without fatigue for long intervals of time. It is therefore frequently advantageous to make use of this kind of movement in the design of machine controls. In the seated posture, knee switches, such as are often employed in the operation of industrial sewing machines, are generally superior to foot pedals.

8 ERGONOMIC EVALUATION OF WORK SITUATIONS

Historically, all industrial and technological development in the United States has been triggered by the need to overcome and solve problems in the design, production, distribution, and use of manufactured articles. Traditionally, industry has subscribed to the "improvement approach" as the principal avenue toward economic efficiency and viability of American enterprise.

Before the development of industrial and human engineering, it was common practice to conceive products hastily, determine an adequate manufacturing process intuitively, and gradually remove deficiencies in product design or manufacturing methods during actual production. Such improvement extended over a period of several months or even years. This approach was to some extent acceptable in past decades, and the model T Ford remained in production for approximately 30 years. In today's fast-changing marketplace, this kind of policy often leads to economic disaster. By the time a product has been improved, it may have become redundant. Unfortunately the "improvement" approach to product design and manufacture is still maintained in many industries under the guise of the term "continuous cost reduction." This often conveniently excuses failure to design the product and process correctly before production.

In some industries "cost reduction" is expected to be practiced routinely by supervisory and engineering personnel during the first months or years of production of a new article. This tempts those who are responsible for "efficiency" to design new products and work methods initially imperfect, so that easy opportunities for later cost reduction are not lost.

Unfortunately, a similar attitude often prevails when the occupational health and safety of the working population is at stake. Too often industry waits for evidence of work-induced occupational disability before commissioning an occupational health specialist to identify the causes, which are then removed by a process of not always satisfactory gradualism.

Only too often the practitioner of ergonomics is challenged to justify the removal of pathogenic vectors from the working environment by a prediction of potential savings in medical costs, or an increase of productivity likely to accrue from his activities. It is therefore important that those who are engaged in the practice of ergonomics be ready to prove that the maintenance of occupational health and high levels of productivity are inseparable. Three main areas of ergonomic evaluation are of prime interest to the practitioner in industry: (*a*) historical evaluation, (*b*) analytical evaluation, and (*c*) projective evaluation.

8.1   Historical Evaluation

An active interest of management in ergonomics is often initially triggered by noticeable breakdowns in occupational health occurring in the performance of a well-defined, essential, and easily identifiable productive operation. This then results in increased manufacturing expense, high medical cost, potential retribution by a regulatory agency, and other undesirable side effects. Under such circumstances, a request for historical evaluation of past activities and events may be made. Such study is best conducted keeping in mind the "four big C's" of occupational health investigation:

1. Cause
2. Consequence

3. Cost
4. Cure

Frequently consequence and cost are known, and the cause and the cure remain to be discovered. Here theoretical analysis is the most expedient tool of research. Experimental methods are generally an unnecessary expense and quite superfluous in developing a critique of past events. Furthermore, a theoretical analysis offers a degree of confidentiality not available in experimentation with humans.

It must be reemphasized that this book concentrates only on those narrower aspects of ergonomics (79) that are related to its subspecialty, biomechanics (80). To treat all the disciplines tributary to ergonomics and their applications within the confines of the space available would have led to a superficial discussion or sometimes mere mention of numerous important aspects of the industrial environment, its evaluation, and control.

It is important that no evaluation of work situations be undertaken unless the investigator is familiar with the ergonomic aspects of climate control, noise, vibration, illumination, circadian rhythms, and related topics. All these should be included in an environmental workup, which should always precede the general study of physical interaction between humans and the implements of the workplace. Figure 40.2 should provide helpful guidance in the systematic conduct of such a preliminary workup. The next step would be to ascertain whether the prerequisites of biomechanical work tolerance (Section 4.1) were to some extent disregarded in the design of the product manufactured or the work method. A step-by-step checkout of each situation with Table 40.2 in hand is the best approach; afterward it should be possible to suggest some potential causes for the observed anatomical, physiological, or behavioral failure. Sometimes it may then be possible to suggest remedial action immediately. However, the use of tables and "cookbooks" is no substitute for speculative analysis based on sound professional knowledge.

Superficial study of occupational accident or disease vectors based on mere guidelines may very often obfuscate cause and effect relationships. What may appear to be the cause of an accident could be, in fact, an effect produced by a less obvious mechanism. Discussion of an actual case may illustrate the need to probe energetically for the primary cause of occupational injury.

In a chemical factory the number of workers hit by forklift trucks while crossing aisles increased suddenly and dramatically without any apparent cause. Whenever such a "lost time" accident report was filed, either pedestrian error or driver error was listed as the cause. Subsequently, common human factors engineering approaches likely to eliminate the problem were explored. The trucks were made more visible by painting them in conspicuous colors, and illumination in the aisles was improved. At some of the places of greatest accident frequency, automatic warning horns were installed which signaled whenever a vehicle approached. When none of these measures proved to be successful, investigators began to try to discover why so many individuals were walking around the factory instead of remaining seated safely at their workplaces. Accident frequency was proportional to the number of pedestrians present in the aisles at any given time. Brief periods of absenteeism from the workplace suddenly increased dramatically, leading in turn to an increase in pedestrian traffic density.

The time of this change coincided with the introduction of a new tool (see Figure 40.33). An electrical brush used to clean trays was replaced by a much less expensive but equally effective paint scraper that produced insults to the ulnar artery (Section 4.3). This reduced blood supply to the ring and little fingers. The resulting numbness and tingling caused the individuals afflicted to lay down their tools occasionally and seek relief by exercising their hands. To avoid ensuing arguments with supervisors, workers were tempted to make use of every opportunity of brief absences from the job. Trips to the washroom, the toolroom, and so on, became much more frequent, and this was the true cause of increased exposure of the factory population to the risk of traffic accidents. Thus the first of the four big C's was identified. The cure: the handle of the paint scraper was redesigned. The result: the workers spent more time per day in productive activity; thus the output and economy of the operation increased, while at the same time, because of diminished risk exposure, the accident rate returned to normal.

Whenever the frequency of incidence of occupational ill health or accident increases after a manufacturing process has been in safe operation for some time, the following question should be asked: *what change in equipment, product design, tools used, working population employed, or work method applied has taken place immediately before the breakdown of occupational health?*

8.2 Analytical Evaluation

The procedures of analytical evaluation are called for whenever an existing manufacturing operation is generally satisfactory but has to be improved to make it more competitive, to reduce training time, or to eliminate operator discomfort and ill health.

Theoretical analysis (Section 8.1) is the initial step in all analytical evaluation. However, this should often be followed by some additional procedures. The simplest and perhaps most effective aid in this kind of study is cinematography and subsequent frame-by-frame analysis. This permits a detailed evaluation of the workers' reaction to each event at the workplace and to each contact with tool, machine, or manufactured article. Slow-motion viewing of the operation not only reveals biomechanical or ergonomic defects but is instrumental in discovering reflex reactions to mild localized repetitive trauma that cannot be detected by the naked eye because of the brief duration of many such events. Stage magicians know that the hand is quicker than the eye. The current trend toward the use of video tape for work analysis should be discouraged because the tapes are inadequate for the detection of fine details of expression (81) or blanching of the skin, or frame-by-frame analysis. Furthermore, color video taping is exorbitantly expensive, whereas color film, especially in the Super-8 size, is economical and tells much more than a black and white picture. Finally, manufacture of a video tape from movie film is very inexpensive, whereas the converse (i.e., the manufacture of a movie film from video tape) is a very costly operation. Furthermore, being magnetic, video tape requires more careful storage, is sensitive to magnetic fields, and often is erased accidentally.

It is important for motion picture analysis of a work situation to allow the viewing of the workplace in at least two different planes, if necessary, with the aid of mirrors. When motion picture analysis alone is not adequate for process evaluation, recourse to other experimental technologies must be taken. Some investigative procedures, such as metabolic measurement, electromyography, and electromyographic kinesiology, have already been described (Section 3). This work can often be complemented by the production of biomechanical profiles, which are of special usefulness when the potential pathogenic effects of individual and brief motion elements of work are under discussion (37); therefore their main field of application resides in projective evaluation.

8.3   Projective Evaluation

Whenever possible, a job should be ergonomically evaluated while it is still in the planning phase. This makes it possible to "design out" of a task features, equipment, and maneuvers that are potentially traumatogenic. Projective evaluation should include reliable predictions with respect to the work tolerance of a specific population, estimated duration of training, and counseling procedures useful in overcoming difficulties in the training process.

All projective evaluation should include, when necessary, biomechanical profiles (82, 83) as a means both to establish reliable effort input–production work output relationships and to predict potential anatomical failure points.

The profile is a polygraphic recording produced during the performance of a standard element of work that includes displacement, velocity, and acceleration of at least one anatomical reference point. Such measurement is performed by kinesiometers—apparatus that permits the performance of a specific motion element against a known resistance. *Displacement* indicates of range and pattern of motion; *velocity* serves as an index of speed as well as strength (slow joint movement is often associated with muscular weakness) (37).

Finally, *acceleration* reflects control over precision and quality of motion (84). Abnormal acceleration and deceleration signatures are invariably associated with imprecise and unsafe movements due to the inability to terminate a motion at the correct place and time. These biomechanical parameters are recorded simultaneously with integrated myograms of selected muscles involved in the performance of the task under study. The usefulness of the biomechanical profile becomes evident when a tossing motion is investigated. Many individuals have difficulty with wrist extension, but this does not interfere with their competence to perform assembly tasks in an entirely satisfactory manner. Once the task is finished, however, the individual may not be able to dispose of the article by tossing it into a bin (Figure 40.59). Inability to toss is more frequent than is commonly assumed. If caused by common industrial disorders, such as tenosynovitis, it will persist throughout the working day. Alternatively, when caused by fatigue, it may become evident only after several hours of work have elapsed.

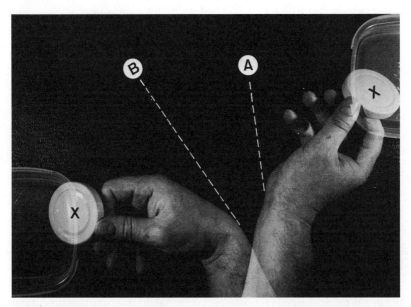

Figure 40.59. Hyperextension of the wrist is not a recognized element in motion study, but it is often essential for disposal of items at workplaces *X*. Hand and wrist disease are frequent occurrences, preventing the successful execution of this biomechanical motion, especially in women and aged workers. A simple change of location of the disposal bin may make the difference between occupational disability *A* or ability *B*. From Reference 37, by permission. © American Institute of Industrial Engineers, Inc., Norcross, Ga.

To establish a cause and effect relationship between the muscular effort involved in wrist extension and the pattern and quality of the tossing motion produced, a biomechanical profile is constructed. The kinesiometer employed for this purpose appears in Figure 40.60. The subjects are tested before the start of the working day, then retested after several hours of normal productive work. The manner in which the task affects the performing individual can be established by comparison of the "prework" and "postwork" biomechanical profiles (Figure 40.61).

In the case of a tendency toward or a history of tenosynovitis, characteristic changes in the profile can be observed with ease. Displacement shows no change. However, the myograms are slightly stronger, indicating a greater effort necessary to produce movement. The velocity curve displays notches at peak speed. This demonstrates that multiple, repetitive efforts must be made during each movement to achieve peak performance. The biomechanical profile cannot be used to make a definitive diagnosis such as tenosynovitis. The results obtained permit only the statement that some anatomico-mechanical obstruction interferes with the tossing motion. This could be caused by a large number of conditions—orthopedic, arthritic, or others. Medical diagnosis, of course, is under the jurisdiction of a physician.

Fatigue changes the profile in quite a different fashion. No meaningful change whatsoever can be observed in the type of myogram used here. However, the displacement tracing shows that it is not possible to maintain wrist extension for

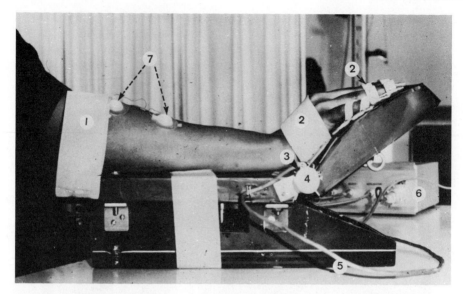

Figure 40.60. Wrist extension kinesiometer: 1, arm stabilizer; 2, wrist and finger stabilizers; 3, hinge under radiocarpal joint producing 10 lb-in.; 4, electrogoniometer; 5, cable to computation module; 6, analogue computation module generating biomechanical profile; 7, electrodes for extensor myogram. From Reference 84.

the necessary interval of time, and the acceleration signature displays evidence of muscular rigidity and fine tremors (Figure 40.61). This establishes that the task is fatiguing. Either the rhythm of the work cycle should be changed or the length of work periods and rest pauses should be significantly modified. A less expensive, less complex, and more feasible way to deal with both classes of disability may be a simple change in the position of the disposal bin, which would simply eliminate the necessity for wrist extension at the end of the work cycle (Figure 40.59). This demonstrates that the biomechanical profile, as distinct from other methods of performance measurement, not only indicates that quality and/or magnitude of performance are defective, but also makes it possible to pinpoint the physical cause of the deficiency so that successful remedial action may be taken.

The biomechanical profile, one of the newest procedures in ergonomic work measurement, bridges the gap that has existed since the beginning of scientific management, industrial psychology, and work physiology, plaguing most practitioners in industry. Workers as early as the Gilbreths had fully established a scientific rationale on which the disciplines and biomechanics as practiced today are based. Nevertheless, they lacked instrumentation adequate to conduct experimental investigations into the physical effort expended by individual muscles in the performance of a specific task. The analytical thinking of these pioneers was simply 50 years ahead of the technology available at that time. The second industrial revolution has introduced productive processes that constrain workers to a relatively rigid posture, which forces them to maintain repetitive motion patterns throughout a long working day. This has produced or aggravated numerous known, as well as

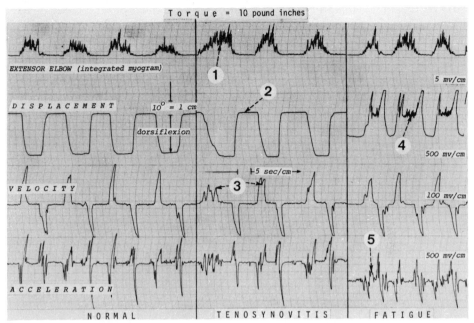

Figure 40.61. Biomechanical profile produced by the wrist extension kinesiometer in Figure 22.60: 1, increased extensor myogram; 2, normal displacement signature; 3, notched velocity peaks; 4, inability to maintain wrist extension; 5, signs of rigidity in acceleration signature. From Reference 84.

previously unknown, elements and complaints. However, the new technologies (85, 86), as a by-product, have produced instrumentation now available for effective ergonomic work measurement. These make possible the prevention of occupational disability, increasing levels of physiological and emotional well-being of the working population, as well as the productive capacity and the competitive posture of American enterprise.

9 PURPOSE OF CHAPTER

This chapter does not purport to be a comprehensive treatise on the practice of the broad field of ergonomics. The discipline draws on so many aspects of the biological, behavioral, medical, and technological sciences that a complete coverage of the field within the confines of a chapter would have been impossible. Instead, the subject matter presented relates to the aspects of ergonomics pertinent to the biomechanics of work situations, very often requiring immediate decisions in the field. The analysis of problems created by the general industrial environment— including those stemming from noise, light, climate, and fatigue—has been omitted. Separate descriptive chapters are contained elsewhere in this volume.

ACKNOWLEDGMENTS

This chapter is based on three main references: 37, 38, and 48; it was developed with the partial support of the National Institute for Occupational Safety and Health and, to some extent, the Social Rehabilitation Administration of the Department of Health, Education and Welfare. Additional information has been derived from my lecture notes distributed to my students at the University of New South Wales (Australia), Texas Technological University, and New York University.

Helen Tichauer contributed most of the graphic work and illustrations. Dr. H. Gage contributed to the section on "Handtools." Special thanks are due to Audrey Lane for the painstaking editorial efforts taken with this manuscript.

REFERENCES

1. B. Ramazzini, *Essai sur les Maladies de Artisans* (translated from the Latin text *De Morbis Artificum* by M. de Fourcroy), Chapters 1 and 52, 1777.

2. D. Hunter, *The Diseases of Occupations*, 4th ed., Little, Brown, Boston, 1969, p. 120.

3. J. Amar, *Organization Physiologique du Travail*, H. Dunod, Paris, 1917.

4. E. D. Adrian, "Interpretation of the Electromyogram," *Lancet*, **2**, 1229–1233; 1283–1286 (1925).

5. O. G. Edholm and K. F. H. Murrell, *The Ergonomics Research Society, A History*, 1949–1970, Wykeham Press, Winchester, 1973.

6. P. V. Karpovich and G. P. Karpovich, "Electrogoniometer: A New Device for the Study of Joints in Action," *Fed. Proc.*, **18**, 311 (1959).

7. A. J. S. Lundervold, *Electromyographic Investigations of Position and Manner of Working in Typewriting*, A. W. Broggers Boktrykkeri A/S, Oslo, 1951.

8. E. R. Tichauer, "Electromyographic Kinesiology in the Analysis of Work Situations and Hand Tools," *Proceedings of the First International Conference of Electromyographic Kinesiology, Electromyography*, Supplement 1 to Vol. 8, 1968, pp. 197–212.

9. M. Lukiesh and F. K. Moss, "The New Science of Seeing," in *Interpreting the Science of Seeing into Lighting Practice*, Vol. 1, General Electric Company, Cleveland, 1932, pp. 1927–1932.

10. H. S. Belding and T. F. Hatch, "Index for Evaluating Heat Stress in Terms of Resulting Physiological Stress," *Heat. Pip. Air Cond.*, **27**, 129 (1955).

11. B. L. Welch and A. S. Welch, Ed., *Physiological Effects of Noise*, Plenum Press, New York, 1970.

12. W. F. Floyd and A. T. Welford, Eds., *Symposium on Fatigue*, H. K. Lewis, London, 1953.

13. E. R. Tichauer, "Potential of Biomechanics for Solving Specific Hazard Problems," in *Proceedings 1968 Professional Conference*, American Society of Safety Engineers, Park Ridge, IL, 1968, pp. 149–187.

14. P. J. Rasch and R. K. Burke, *Kinesiology and Applied Anatomy*, 3rd ed., Lea & Febiger, Philadelphia, 1967.

15. S. Brunnstrom, *Clinical Kinesiology*, 3rd ed., revised by R. Dickinson, F. A. Davis, Philadelphia, 1972.

16. D. L. Kelley, *Kinesiology: Fundamentals of Motion Description*, Prentice-Hall, Englewood Cliffs, N.J., 1971.

17. E. R. Tichauer, *Occupational Biomechanics (The Anatomical Basis of Workplace Design)*, Rehabilitation Monograph No. 51, Institute of Rehabilitation Medicine, New York University Medical Center, New York, 1975.

18. M. Williams and H. R. Lisner, *Biomechanics of Human Motion*, Saunders, Philadelphia, 1962.

19. R. Fick, *Anatomie und Mechanik der Gelenke*, Vol. 3, Spezielle Gelenk und Muskelmechanik Fisher, Jena, 1911, pp. 318–389.

20. N. Recklinghauser, *Gliedermechanik und Lähmungsprothesen*, Springer, Berlin, 1920.

21. H. A. Haxton, "Absolute Muscle Force in Ankle Flexors of Man," *J. Physiol.*, **103**, 267–273 (1944).

22. R. W. Ramsey and S. F. Street, "Isometric Length-Tension Diagram of Isolated Skeletal Muscle Fibers in Frog," *J. Cell. Comp. Physiol.*, **15**, 11–34 (1940).

23. B. A. Houssay, *Human Physiology*, translated by J. T. Lewis and O. T. Lewis, McGraw-Hill, New York, 1955, p. 385.

24. L. Brouha, *Physiology in Industry*, 2nd ed., Pergamon Press, Oxford, 1960.

25. J. B. de V. Weir, "New Methods for Calculating Metabolic Rate with Special Reference to Protein Metabolism," *J. Physiol.*, **109**, 1–9 (1949).

26. C. F. Consolazio, R. E. Johnson, and L. J. Pecora, *Physiological Measurements of Metabolic Functions in Man*, McGraw-Hill, New York, 1963.

27. F. H. Norris, Jr., *The EMG: A Guide and Atlas for Practical Electromyography*, Grune & Stratton, New York, 1963.

28. J. V. Basmajian, *Muscles Alive*, 3rd ed., Williams & Wilkins, Baltimore, MD, 1974, p. 7.

29. H. H. Ju, "A Statistical Multi-Variable Approach to the Measurement of Performance Effectiveness of a Lumped System of Human Muscle," Master's thesis, New York University, New York, 1970.

30. J. F. Davis, *Manual of Surface Electromyography*, Wright Air Development Center Technical Report No. 59-184, Wright-Patterson Air Force Base, Ohio, 1959.

31. E. R. Tichauer, *Biomechanics of Lifting*, Report No. RD-3130-MPO-69, prepared for Social and Rehabilitation Service, U.S. Department of Health, Education and Welfare, Washington, D.C., 1970.

32. H. O. Kendall et al., *Muscles: Testing and Function*, 2nd ed., Williams & Wilkins, Baltimore, MD, 1971.

33. A. V. Hill and J. V. Howarth, "The Reversal of Chemical Reactions of Contracting Muscle During an Applied Stretch," *Proc. Roy. Soc., S.B.*, **151**, 169 (1959).

34. B. Jonsson, "Electromyographic Kinesiology, Aims and Fields of Use," in *New Developments in Electromyographic and Clinical Neurophysiology*, Vol. 1, J. E. Desmedt, Ed., Karger, Basel, 1973, pp. 498–501.

35. B. Jonsson and M. Bagberg, "The Effect of Different Working Heights on the Deltoid Muscle," *Scand. J. Rehab. Med.*, Suppl. 3, 26–32 (1974).

36. E. Asmussen et al., *Quantitative Evaluation of the Activity of the Back Muscles in Lifting*, Communication No. 21, Danish National Association for Infantile Paralysis, Hellerup, 1965.

37. E. R. Tichauer, "Biomechanics Sustains Occupational Safety and Health," *Ind. Eng.*, February 1976.

38. E. R. Tichauer, "Occupational Biomechanics and the Development of Work Tolerance," in *Biomechanics V-A*, P. V. Komi, Ed., University Park Press, Baltimore, MD, 1976, pp. 493–505.

39. D. B. Chaffin and W. H. Baker, "A Biomechanical Model for Analysis of Symmetric Sagittal Plane Lifting," *AIIE Trans.*, **2**, 1, 16–27 (1970).

40. E. R. Tichauer, "Ergonomics: The State of the Art," *Am. Ind. Hyg. Assoc. J.*, **28**, 105–116 (1967).

41. J. Hasan, "Biomedical Aspects of Low Frequency Vibration," *Work-Environment-Health*, **6**, 1, 19–45 (1970).

42. G. Loriga, in *Occupation and Health, Encyclopedia of Hygiene, Pathology and Social Welfare*, ILO, Geneva, 1934.

43. A. Hamilton, J. P. Leake, et al., Bureau of Labor Statistics Bulletin No. 236, Department of Labor, Washington, D.C., 1918.

44. D. E. Wasserman and D. W. Badger, *Vibration and the Worker's Health and Safety*, Technical Report No. 77, National Institute for Occupational Safety and Health, Government Printing Office, Washington, D.C., 1973.

45. B. Akerblom, *Standing and Sitting Posture*, A.-B. Nordiska Bokhandelns, Stockholm, 1948.

46. E. R. Tichauer, "Some Aspects of Stress on Forearm and Hand in Industry," *J. Occup. Med.*, **8**, 2, 63–71 (1966).

47. B. W. Niebel, *Motion and Time Study*, 4th ed., Irwin, Homewood, IL, 1967, p. 169.

48. E. R. Tichauer, in *The Industrial Environment—Its Evaluation and Control*, National Institute for Occupational Safety and Health, Department of Health, Education and Welfare, Washington, DC, 1973, pp. 138–139.

49. M. Arnold, *Reconstructive Anatomy*, Saunders, Philadelphia, 1968, p. 391.

50. W. T. Dempster, "The Anthropometry of Body Action," in *Dynamic Anthropometry*, R. W. Miner, Ed., *Ann. NY Acad. Sci.*, **63**, 4, 559–585 (1955).

51. E. R. Tichauer et al., *The Biomechanics of Lifting and Materials Handling*, Report No. HSM 99-72-13, submitted to the National Institute of Occupational Safety and Health, New York, 1974.

52. L. E. Abt, "Anthropometric Data in the Design of Anthropometric Test Dummies," in *Dynamic Anthropometry*, R. W. Miner, Ed., *Ann. NY Acad. Sci.*, **63**, 4, 433–636 (1955).

53. J. B. Martin and D. B. Chaffin, "Biomechanical Computerized Simulation of Human Strength in Sagittal-Plane Activities," *AIIE Trans.*, **4**(1), 19–28 (1972).

54. I. Starr, "Units for the Expression of Both Static and Dynamic Work in Similar Terms, and Their Application to Weight-Lifting Experiments," *J. Appl. Physiol.*, **4**, 21 (1951).

55. P. V. Karpovich, *Physiology of Muscular Activity*, Saunders, Philadelphia, 1959.

56. I. J. Schorr, *Changes in Myoelectric Activity of the Erector Spinae, Gluteus Maximus and Hamstring Muscles During Pick-Up and Release of Loads for Various Workplace Geometrics*, Master's thesis, New York University, New York, 1974.

57. F. B. Gilbreth, *Motion Study*, Van Nostrand, New York, 1911.

58. E. R. Tichauer, "Industrial Engineering in the Rehabilitation of the Handicapped," *J. Ind. Eng.*, **19**, 2, 96–104 (1968).

59. R. Drillis, D. Schneck, and H. Gage, "The Theory of Striking Tools," *Hum. Factors*, **5**, 5 (October 1963).

60. J. M. Little and A. F. Grand, "Hypothenar Hammer Syndrome," *Med. J. Aust.*, **1**, 49–53 (1972).

61. D. Briggs, "Trauma," in *Occupational Medicine*, C. Zenz, Ed., Year Book Medical Publishers, Chicago, 1975, pp. 254 ff.

62. E. Grandjean, *Fitting the Task to the Man*, Taylor and Francis, London, 1969.

63. H. Bailey et al., *A Short Practice of Surgery*, H. K. Lewis, London, 1956.

64. F. Fitzhugh, *Gripstrength Performance in Dynamic Tasks*, Technical Report, University of Michigan, Ann Arbor, 1973.

65. C. E. Clauser et al., *Anthropometry of Air Force Women*, AMRL-TR-70-5, Aerospace Medical Research Laboratory, Wright-Patterson Air Force Base, Ohio, 1972.

66. J. W. Garrett, "The Adult Human Hand: Some Anthropometric and Biomechanical Considerations," *Hum. Factors*, **13**, 2 (1971).

67. J. W. Garrett et al., *A Collation of Anthropometry*, AMRL-TR-68-1, 2, 2 vols., Wright Air Development Center, Wright-Patterson Air Force Base, OH, 1971.

68. H. Dreyfuss, *The Measure of Man*, 2nd ed., Whitney Library of Design, New York, 1967.

69. K. H. E. Kroemer, *Seating in Plant and Office*, AMRL-TR-71-52, Aerospace Medical Research Laboratory, Wright-Patterson Air Force Base, OH, 1971.

70. E. Grandjean, *Fitting the Task to the Man—An Ergonomic Approach*, Taylor and Francis, London, 1967.

71. N. Diffrient et al., *Humanscale 1/2/3*, Henry Dreyfuss Associates, MIT Press, Cambridge, Mass., 1974.

72. *Specification for Office Desks, Tables and Seating*, B. S. No. 3893, British Standards Institution, 1965.

73. *Anthropometric Recommendations for Dimensions of Non-Adjustable Office Chairs, Desks and Tables*, B. S. No. 3079, British Standards Institution, 1959.

74. F. Staffel, "Zur Hygiene des Sitzens," *Z. Allg. Gesundheitspflege*, **3**, 403–421 (1884).

75. H. Strasser, *Lehrbuch der Muskel- und Gesundheitspflege*, Vol. 2, Springer, Berlin, 1913.

76. S. W. Jacob and C. A. Francone, *Structure and Function in Man*, Saunders, Philadelphia, 1970, p. 8.

77. S. Slesin, "Biomechanics," *Ind. Design*, **18**, 3, 36–41 (1971).

78. B. Shackel, K. D. Chidsey, and Pat Shipley, "The Assessment of Chair Comfort," in *Sitting Posture*, E. Grandjean, Ed., Taylor and Francis, London, pp. 155–192.

79. *The Origin of Ergonomics*, Ergonomics Research Society, Echo Press, Loughborough, England, 1964.

80. E. R. Tichauer, in: *Biomechanics Monograph*, E. F. Byars, R. Contini, and V. L. Roberts, Eds., American Society of Mechanical Engineers, New York, 1967, p. 155.

81. R. J. Nagoe and V. H. Sears, *Dental Prosthetics*, Mosby, St. Louis, 1958.

82. E. R. Tichauer, H. Gage, and L. B. Harrison, "The Use of Biomechanical Profiles in Objective Work Measurement," *J. Ind. Eng.*, **4**, 20–27 (1972).

83. E. R. Tichauer et al., "Clinical Application of the Biomechanical Profile of Pronation and Supination," *Bull. NY Acad. Med.*, 2nd ser., **50**, 4, 480–495 (1974).

84. E. R. Tichauer, in: *Rehabilitation After Central Nervous System Trauma*, H. Bostrom, T. Larsson, and M. Ljungstedt, Eds., Nordiska Bokhandelns Förlag, Stockholm, 1974.

85. E. R. Tichauer, M. Miller, and I. M. Nathan, "Lordosimetry: A New Technique for the Measurement of Postural Response to Materials Handling," *Am. Ind. Hyg. Assoc. J.*, **34**, 1–12 (1973).

86. C. Sparger, *Anatomy and Ballet*, A. & C. Black, London, 1960.

CHAPTER FORTY-ONE

ODOR

A LEGAL OVERVIEW

Ralph E. Allan, C.I.H., J.D.

1 INTRODUCTION

The practice of industrial hygiene has changed tremendously in the past 20 years, since the passage of the Occupational Safety and Health Act of 1970. With the issues of indoor air pollution generally and tight building syndrome specifically becoming a part of an industrial hygienist's daily involvement, odors are and will continue to be more and more an issue of contention. Is odor an index of hazard potential? Can it be such an index? How should odors be handled in reference to industrial hygiene activity? What are the legal issues considering an odor present without a quantitative determination of hazard potential? Currently we do not have all the answers available to answer fully all questions posed. However, history sometimes reveals indications of the future.

2 HISTORICAL REVIEW

Since ancient times, it has been indicated that pleasing odors preserve health and unpleasant odors are injurious. In the early days this thinking provided the basis for the application of aromatic eau de cologne and of pomanders stuffed with balsam for health preservation. Alternatively, diseases have been attributed to atmospheric "miasmas." In fact, the word malaria is related to the Italian translation of "bad air," that is, *malaria*.

Patty's Industrial Hygiene and Toxicology, Fourth Edition, Volume 1, Part B, Edited by George D. Clayton and Florence E. Clayton.
ISBN 0-471-50196-4 © 1991 John Wiley & Sons, Inc.

Because of this thinking, prior to the nineteenth century people avoided crowded places because it was generally believed that crowded locations, with the accompanying odors, were responsible for the breeding of disease, and the odors specifically were responsible for the spread of infection. Even as new information developed into the twentieth century, it was difficult for people to accept the premise that contagion is primarily a fingerborne phenomenon and not airborne. In fact, it was a convenient notion to associate foul smelling air with unhealthy situations. What better way to provide a basis for freshening the air! Church activities that included the burning of incense supported the idea that a pleasant odor represents healthy air. This concept was also supported by pre-nineteenth century medical practice. As the twentieth century approached, scientists' experiments tended to conclude that there was not consistent hazard associated with occupancy odor (1).

In 1923, a New York State Commission focused on ventilation considerations, including such functions and indexes as comfort, body temperature, intellectual performance, motivation, respiration, metabolism, condition of the nasal mucosa, frequency of colds, blood pressure, hematocrit, appetite, and rate of physical work. There was no cause for medical concern found under normal conditions of occupancy. The study determined that control of occupancy odor had to stand on the basis of comfort rather than ill health (2).

As understanding of odors progressed, and more and more substances became part of everyday living through the industrial revolution and into the twentieth century, it became important further to organize scientific thinking concerning odors and categorize them according to known properties.

One such classification follows (3):

- Odorous substances that have been well established as toxic to humans
- Odorous substances that have produced well-defined pathological changes in animals that have not been identified in humans
- Odorous substances that have not been identified as toxic to humans, but that evoke violent and alarming physical symptoms in a substantial fraction of an exposed population whenever odor intensity is high and exposure more than fleeting
- Odorous substances that have not been identified as toxic to humans, but that are capable of evoking violent and alarming physical symptoms in a small number of people even when exposure is moderate and fleeting
- Odorous substances that have not been identified as toxic to humans (or are present at concentrations substantially below a well-established toxic threshold), but that produce more than passing vexation by the continuing or frequent presence of their unpleasant odor
- Odorous substances that have no known toxic properties and are universally recognized as pleasant or neutral, but that produce vexation in a substantial fraction of an exposed population because of unusual intensity or persistence
- Odorous substances of no known toxicity that are sensed to the point of conscious recognition, but that evoke only pleasant or indifferent sensations

The above classification is indeed important to nurture and develop especially considering the need for unambiguous classification, identification and definition of responses to odors, and their effect on human populations. A clear and unequivocal as possible assessment is important in order to minimize litigation and provide guidance for the development of laws, regulations, and standards. It therefore becomes desirable to attempt to deal with the effect of odorous substances on humans by means of a series of precisely defined categories, rather than with a continuum of effects that stretches from highly pleasing (however, what may be highly pleasing to one may be antagonistic to another!) to violently repugnant (perhaps to all?) and then ultimately merges into the clearly toxic category. The sole unifying factor through the continuum remains the ability of each substance to stimulate or act on the olfactory system in some fashion.

3 HUMAN RESPONSE

Odors affect the well-being of an individual by precipitating unpleasant feelings, by initiating harmful responses and other physiological interactions, and by adjusting olfactory activity. Some objective responses to unpleasant odors include vomiting, nausea, headache, shallow breathing and cough, sleeplessness, stomach problems, interference with appetite, and eye-, nose-, and throat irritation; emotional upset resulting in lack of sense of well-being and interference with the enjoyment of home, food, and general environment; and irritability and depression. Physiological effects of unpleasant odor include decreased heart rate, constriction of the blood vessels of the skin and muscles, release of epinephrine, and changes in the size and condition of the cells in the olfactory bulbs of the brain. Science, however, has not provided us an understanding of the relationships between the symptoms and the intensity or duration of the exposure to the odor. Indication, therefore, of changes of olfactory function or sensitivity in populations exposed under controlled conditions are not available to provide a clear insight into the phenomenon of attributable incidence. There is no doubt that certain physiological phenomena are related; that is, respiratory and cardiovascular responses are elicited by stimulation of receptors in the nasal mucosa. These effects have been documented in various animal species and include sneezing, bronchodilation, decrease in breathing rate, decrease in heart rate, increase in arterial blood pressure, decrease in cardiac output, and vasoconstriction in various parts of the body. All information, regarding adverse reactions occurring in humans as a result of environmental odors, however, has come from complaints and surveys, which are difficult to verify and evaluate scientifically (3).

4 LEGAL ASPECTS

The law, meanwhile, considering this continuum, does indeed have difficulty in dealing with odor problems, somewhat as it does with the definition of noise as unwanted sound: what is noise to some is music to others' ears; similarly what is odiferous to one (or even toxic to one) is not to another. The cattleman's "pleasant" animal odor is foul smelling to the uninterested, or perhaps to those who are unacclimatized to the economic aspects of the cattle industry. Differently from noise, however, where there is a definite measurement of hearing loss accruing from high-intensity sound (whether or not it is classified as noise), there are no easy identifying biological markers for odor intensity unless it tends to trip into a frankly toxic effect. Odors therefore do indeed present a difficult issue to handle from the legal viewpoint because of differences of perspective, insufficiency of data, or ambiguities in data. Despite the difficulty, there is no doubt that malodor should be regulated by one means or another, as a matter of public policy. There are many cases in the legal archives relating to odor. Whether it be a case concerning the odors emanating from a rendering plant (4) or a chicken processing plant (5), the issue of public or private nuisance becomes involved. Cases that arise under private or public nuisance have developed through a long line of judicial precedents and are codified in some situations by local ordinances. Although more scientific approaches to odor control are becoming more involved, nuisance law is the oldest and strongest source of law for controlling odors in our society.

4.1 Nuisance

Nuisance law is divided into public nuisance and private nuisance. A public nuisance is created when an act invades a right common to all members of the public. *Black's Law Dictionary* (6) defines nuisance as follows:

That which annoys and disturbs one in possession of his property, rendering its ordinary use or occupation physically uncomfortable to him. Yaffe v. City of Ft. Smith, 178 Ark,406, 10 S.W.2d 886, 890, 61 A.L.R. 1138. Everything that endangers life or health, gives offense to senses, violates the laws of decency, or obstructs reasonable and comfortable use of property. Hall v. Putney, 291 Ill. App. 508, 10 N.E. 2d 204, 207. Annoyance; anything which essentially interferes with enjoyment of life or property. Holton v. Northwestern Oil Co., 201 N.C. 744, 161 S.E. 391, 393. That class of wrongs that arise from the unreasonable, unwarrantable, or unlawful use by a person of his own property, either real or personal, or from his own improper, indecent, or unlawful personal conduct, working an obstruction of or injury to the right of another or of the public and producing such material annoyance, inconvenience, discomfort, or hurt, that the law will presume resulting damage. City of Phoenix v. Johnson, 51 Ariz. 115, 75 P.2d 30; Wood, Nuis § 1; District of Columbia v. Totten, 55 App. D.C. 312, 5F.2d 374, 380, 40 A.L.R. 1461. Anything that unlawfully worketh hurt, inconvenience, or damage. 3 Bl. Comm. 216; City of Birmingham v. Hood-McPherson Realty Co., 233 Ala. 352, 172 So. 114, 120, 108 A.L.R. 1140. Anything which is injurious to health, or is indecent or offensive to the senses, or an obstruction to the free use of property, so as to interfere with the conferrable enjoyment of life or property, or which unlawfully obstructs the free passage or use, in the customary manner, of any navigable lake or river, bay, stream, canal, or basin, or any public park, square, street, or highway, is a nuisance. Civ. Code Cal. § 3479; Veazie v. Dwinel, 50 Me. 479; Bohan v. Port Jervis Gaslight Co., 122 N.Y. 18, 25 N.E. 246, 9 L.R.A. 711; Baltimore & P.R. Co. v. Fifth Baptist Church, 137 U.S. 568 11 S.Ct. 185, 34 L.Ed. 784; Ex parte Foote, 70 Ark. 12, 65 S.W. 706, 91 Am. St. Rep. 63.

In determining what constitutes a "nuisance," the question is whether the nuisance will or does produce such a condition of things as in the judgment of reasonable men is naturally productive of actual physical discomfort to persons or ordinary sensibility and ordinary tastes and habits. Meeks v. Wood 66 Ind. App. 594, 118 N.E. 591, 592.

Nuisances are commonly classed as public and private, and mixed. A public nuisance is one which affects an indefinite number of persons, or all the residents of a particular locality, or all people coming within the extent of its range or operation, although the extent of the annoyance or damage inflicted upon individuals may be unequal. Burnham v. Hotchkiss, 14 Conn. 317; Chesbrough v. Com'rs, 37 Ohio St. 508; Lansing v. Smith, 4 Wend., N.Y., 30, 21 Am. Dec. 89. A private nuisance was originally defined as anything done to the hurt or annoyance of the lands, tenements, or hereditaments of another. 3 Bl. Comm. 216; Whittenmore v. Baxter Laundry Co., 181 Mich. 564, 148 N.W. 437, 52 L.R.A., N.S., 930, Ann. Cas. 1916C, 818. As distinguished from public nuisance, it includes any wrongful act which destroys or deteriorates the property of an individual or of a few persons or interferes with their lawful use or enjoyment thereof, or any act which unlawfully hinders them in the enjoyment of a common or public right and causes them a special injury different from that sustained by the general public. Therefore, although the ground of distinction between public and private nuisances is still the injury to the community at large or, on the other hand to a single individual, is evident that the same thing or act may constitute a public nuisance and at the same time a private nuisance. Heeg v. Licht, 80 N.Y. 582, 36 Am. Rep. 654; Baltzeger v. Carolina Midland R. Co., 54 S.C. 242, 32 S.E. 358, 71 Am. St. Rep. 789; Wilcox v. Hines, 100 Tenn. 538, 46 S.W. 297, 41 L.R.A. 278; Harris v. Poulton, 99 W. Va. 20, 127 S.E. 647, 650, 651, 40 A.L.R. 334. A mixed nuisance is of the kind last described; that is, it is one which is both public and private in its effects—public because it injures many persons or all the community, and private in that it also produces special injuries to private rights. Kelley v. New York, 27 N.Y.S. 164, 6 Misc. 516.

A private nuisance, therefore, involves an invasion of a private party's interest in the use and enjoyment of his property.

To result in fulfilling the elements of a private nuisance cause of action, odors complained of must be judged a substantial annoyance based upon standards established by the ordinary reasonable person living in that specific locality. If the odor were located in an industrial community, a highly sensitive person may find it impossible to establish a cause of action on the basis of odor pollution because (a) the odor is a common characteristic of the industrial community and (b) the odor is considered harmless by most residents of the area. Contrarily, if a foundry were located within a residential community where the average homeowner is not used to the smell of amines and other decomposition products from the pouring process, it would be extremely difficult to defend against a complaint based upon private nuisance. It must, however, be shown by the plaintiff that the odors produced were unreasonable—the liability is not automatic. The plaintiff will have to show that the harm to him is greater than he should be required to bear without compensation. Elements of consideration by the trier of fact can include the economic viability consideration of the community. If the suit is successful at all, the usual remedy is an award of damages, rather than an injunction that would force the defendant to abate the odor.

Other legal aspects in addition to the difficulty of obtaining injunctions also limit the role of private litigation as a technique for regulating odorants. If one literally moves into the vicinity of an odor-emitting source, a defense for the defendant can arise. Called appropriately "coming to the nuisance," this defense can prevent recovery by a plaintiff. In *McLung v. Louisville* (7) plaintiffs purchased land near an old railroad and were denied injunction after the railroad recommended active operation. Sometimes delay in presenting legal rights can cause a detrimental change in the defendant's position and ultimately bar the nuisance action altogether. Further, from the legal standpoint, the private nuisance is tied to interests in land; therefore an action in nuisance cannot be maintained by an employee or by any person who has no property right in the affected land. Finally, because private litigation is a costly and uncertain route, plaintiffs only rarely have the resources available to pursue remedies available to them under the law of nuisance.

Despite any regulation or standard implementation, whether based upon federal legislation or regulation and subsequent compliance, these basic nuisance actions would not be explicitly preempted and would remain valid avenues for seeking control of undesirable odors (1).

4.2 Government Control

As presented earlier, whether or not odor presents a public health hazard is a difficult question to answer in many situations. Thus federal government intervention concerning odor control occurs only indirectly, as air pollution regulation of basic air pollutants provides control requirements for gas, vapor, and particulate emissions. There is therefore no specific nationwide effort directed at controlling odors per se.

Many states and local authorities, however, have implemented odor control regulations as part of a total program for air quality maintenance. In attempts to answer the question what constitutes an "acceptable" level of odor, local jurisdictions have used the basic public nuisance criteria that include community consensus as an important factor in defining acceptable limits for odor perception. As indicated earlier, pleasantness or unpleasantness of odor can be determined by one's interest in the source of emission; that is, a farmer finds odors of the farm at least unobjectionable (if not pleasant) because the odors reflect and relate to a vital personal economic interest. On the other hand, a nonfarmer, commuter resident living near the farm may indeed associate the odors with unpleasantness. An example of a local regulation relating to odor control is as follows:

> No person shall discharge . . . one or more air contaminants (including odors) or combinations thereof in such concentrations and of such duration as are . . . injurious to . . . human health or welfare, animal life, vegetation or property . . . (1).

The violation of the regulation is proved primarily on the basis of testimony from affected residents of the community. This type of regulation essentially codifies the traditional public nuisance cause of action. More specific regulatory direction has been developed in some jurisdictions for designated industries. The regulations are directed toward reducing highly odorous emissions from sources such as rendering plants by requiring incineration control methods.

Another approach at establishing regulatory control is based on sensory evaluation of odors in the general environment. Using a dilution system by mixing portions of the subject-contaminated air with portions of clean air, a violation is determined by assessing whether or not an odor exists after continued dilution of the odorous air with the clean air. There are, however, obvious drawbacks to this direction of regulatory control including the following possible criticisms: a questionable link between dilution factors and community annoyance, the problem associated with obtaining ambient air samples, and questions concerning the reliability of the odor panel(s) involved in the assessment.

Effort by First (8) and by Copley International Corporation (9) resulted in presentation of model ordinances with accentuation on a recognition that a number of variables relate to the acceptability of an odor condition.

First's proposal considers factors such as odor intensity and quality, duration, and frequency of the odor and includes the time and the day of the week as well as the wind direction. By appropriately considering the weighting of these factors, an "odor perception index" is used to assess the magnitude of the annoyance condition and thereby determine whether or not a nuisance requiring control is present.

Problems are associated with this approach, however, in that the numerical values assigned can be open to challenge based on arbitrary and capricious establishment of the index system. In addition, given the subjective nature of the establishment of the standard, a procedure is necessary in order to maintain the rights of offenders to due process of law, yet at the same time provide the moving party (agency) the opportunity to present a violation even though the specific index may not have been exceeded.

The Copley International Corporation system is based on a public attitude survey that includes the elements of establishment of the presence of a community odor problem and prescribes odor control requirements as appropriate. A subsequent public attitude survey assesses the sufficiency of abatement. The legal system could have difficulty dealing with the Copley approach because the surveys may be considered hearsay unless additional information supports the public attitude survey as evidence is presented to the court.

4.3 Audit

Schroeder, in "Industrial Odor Technology Assessment" (10), has developed a checklist for evaluation of an odor situation that can provide an insight into the status of legal significance for interested parties. The checklist follows:

1. What is the nature of the problem?
2. Who in the company has authority to control it?

3. Is the situation covered by any law or ordinance covering odors or air pollution, such as the Occupational Safety and Health Act or general nuisance law? Exactly what do pertinent provisions require?

4. Does the problem involve an accident, periodic occurrence, or constant emission? If a violation has been charged, is it substantiated or capable of being substantiated? Could there have been mistakes in measurement, procedures, interpretation, or conclusions?

5. Background history:
   a. Have there been previous complaints, by private persons or a public agency, by workers, by the community? (Has the scope of the community changed since then?)
   b. Company responses.
   c. Resolution of problem.

6. Present company measures or equipment to curb odors, if any. Can this be documented?

7. What new measures or equipment would be needed to prevent such an episode or abate the odor?
   a. Cost.
   b. Lead time to get equipment.

8. If a complaint has been filed, how bad is the effect on complainant or complainants?
   a. Mere annoyance.
   b. Health effects.
   c. So severe as to support a preliminary injunction to cease operations pending trial, or compliance with statute or ordinance.

9. Consequences of shutdown cost in
   a. Jobs.
   b. Production.
   c. Start-up costs.
   d. Inventory spoilage.
   e. Financial stability of company.

10. What court or government agency has enforcement jurisdiction?

11. What is the purpose of the action?
    a. Close down the plant.
    b. Have company control the odor.
    c. Money damages for actual injuries, punitive damages, exemplary damages?

12. What are the steps of the legal procedure?
    a. Who has the burden of proof?
    b. Avenues of appeal.
    c. Possibilities of delaying or expediting procedure.

13. Estimate of costs in alternative courses of action: paying damages, settlement, legal fees, control equipment costs, production interruption.

14. Is there any relationship between this action and other court cases or enforcement actions in this industry or political subdivision that would affect the outcome?
    a. Court decisions on what constitutes a nuisance.

b. Decisions on availability of abatement technology.

c. Admissions on economically feasible control measures contained in consent decrees.

15. Is legislation pending that will require compliance anyway? Will defiance only assure passage of the legislation? Or will good faith efforts to control the odor give legislators an example of what is economically and technically feasible and thereby possibly help achieve reasonable legislation?

5 SUMMARY

Based upon the historical and scientific basis for odor evaluation and control, the legal system continues to struggle with appropriate legal basis for objective assessment of alleged odor problems. It is the nature of the issue, as with most toxic substance issues, that an objective, convenient judicial basis for noncontroversial problem resolution is difficult to achieve. Resolution of odor problems through the science-law mechanism will therefore be slow to develop into efficient, rapid processes; however, continued progress from the scientific standpoint will lead the way to a more objective legal basis of resolution in this most difficult area of community interest.

REFERENCES

1. National Research Council—Committee on Odors from Stationary and Mobile Sources, National Academy of Sciences, Washington, DC, 1979.
2. Committee on Indoor Pollutants, National Research Council, National Academy Press, Washington, DC, 1981.
3. Committee on Odors from Stationary and Mobile Sources, Odors from Stationary and Mobile Sources, National Academy of Sciences, Washington, DC, 1979.
4. Cox v. Schlachter, 147 Ind. App. 530, 262 N.E. 2d 550, 1 ERC 1681 (1970).
5. Ozark Poultry v. Garman, 251 Ark. 389, 472 S.W. 2d 714, 3 ERC 1545 (1971).
6. H. C. Black; *Black's Law Dictionary*, West Publishing Company, St. Paul, MN, 1968.
7. McLung v. Louisville and N.R. Co., 255 Ala. 302, 51 Sr.2d 371 (1951).
8. M. W. First, "A Model Odor Control Ordinance" in H. M. Englund and W. T. Perry, eds., *2d International Clean Air Congress Proceedings*, New York, Academic Press, 1971, pgs. 1255–1259.
9. Copley International Corporation, "A Study of the Social and Economic Impact of Odors. Phase III. Development and Evaluation of a Model Odor Control Ordinance," A report to U.S.E.P.A. EPA Publication #650/5-73-001. Contract #68-02-0095. Washington, D.C., Feb. 1973.
10. Paul N. Cheremisinoff, P.E. and Richard A. Young, *Industrial Odor Technology Assessment*, Ann Arbor Science, Ann Arbor, MI, 1975.

B   ODOR-MEASUREMENT AND CONTROL

Amos Turk, PH.D., and Angela M. Hyman

1   INTRODUCTION

The scope and applications of odor measurement and control are frequently confused by various uses of the word *odor*. In the past, "odor" has been used to mean either (*a*) the perception of smell, referring to the sensation, or (*b*) that which is smelled, referring to the stimulus. To eliminate confusion, *odor* should be used only for the former meaning, and *odorant* should be defined as any odorous substance. *Odor intensity*, then, is the magnitude of the olfactory sensation produced on exposure to an odorant. *Odor control* is a term that can be used to describe any process that makes olfactory experiences more acceptable to people. The perceptual route to this objective is usually, but not always, the reduction of odor intensity. An alternative route is the change of odor quality in some way that is considered to be an improvement.

When the reduction of odor intensity is accomplished by removal of odorant from the atmosphere, the process is equivalent to gas and vapor abatement, or to air cleaning, but with some special considerations. These include (*a*) problems related to the need to attain very low concentrations, often approaching threshold levels, (*b*) uncertainties with regard to the reliability of sensory or chemical analyses (see Section 4), and (*c*) difficulties associated with diffuse or sporadic sources.

When an odorous atmosphere is improved by the addition of another (usually pleasant) odorant under conditions in which chemical reactions are not involved, the process is called *odor modification*, referring to modification of the odor perception, not of the odorant. These and other methods of odor control are considered in some detail in Section 6.

2   OLFACTORY PERCEPTION

2.1   Anatomy of the Olfactory System

The olfactory system (Figure 41.1) is unique among the sensory systems in that the cells responsible for the receipt and transduction of odorous stimuli also transmit the encoded information to the brain. In humans, olfactory receptor cells can number in the hundreds of millions. The olfactory receptor cells are interspersed among supporting and basal cells in a yellow pigmented epithelium called the *olfactory mucosa*. The term *mucosa* refers to a mucus-secreting membrane; a layer of mucus 10 to 40 μm thick coats the olfactory mucosa, which lie in the upper part of the nasal cavity above the path of the main air currents that enter the nose with normal inspiration. To reach the olfactory mucosa, therefore, odorous molecules must either diffuse up to the olfactory receptor cells or be drawn up by sniffing.

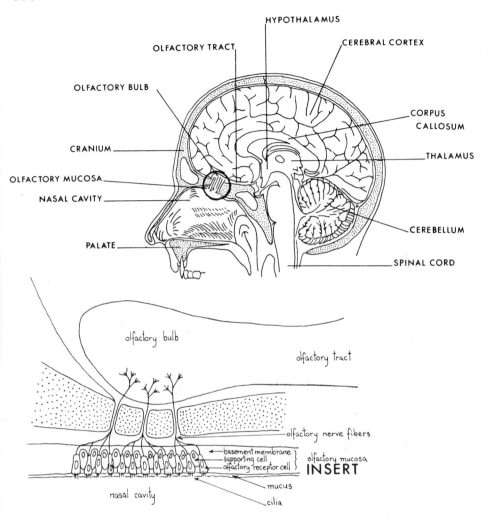

Figure 41.1. Schematic of human olfactory system in sagittal view showing location of olfactory receptor cells.

Olfactory receptor cells are bipolar neurons; that is, they are nerve cells with two main processes. The *apical* or *dendritic* process of each cell reaches the surface of the mucosa, and several hairlike projections called *cilia* extend into the mucus. The cilia are typically 0.3 μm in diameter and 50 to 150 μm long. The basal process of each olfactory receptor cell forms an unmyelinated axon about 0.2 μm in diameter. This diameter is small, and propagation of nerve impulses along the length of these axons consequently is slow. The axons bundle together as they pass out of the olfactory mucosa and form the *olfactory nerve* or *cranial nerve I*. The human olfactory nerve is only a few millimeters long. The olfactory nerve courses centrally, enters the cranium through a series of perforations called the *cribiform plate*, and terminates in the surface layers of a region of cortex called the *olfactory bulb*. Thus olfactory bulb is the first synaptic relay center in the olfactory pathway. Within the olfactory bulb there is a complex pattern of connections and interconnections among various types of neuron. The synaptic organization of the bulb is thought to be related to the processing of olfactory information. The principal relay neurons of the bulb send their axons out from the bulb through the *olfactory tract* to project to several other parts of the brain. Olfactory information is believed to participate in complex patterns of behavior such as feeding and reproduction, and in the control of emotional responses such as fear, pleasure, and excitement. The olfactory bulb is merely two or three relays from nearly every other brain area. Information from other brain areas is likewise relayed to the olfactory bulb; thus there exists the potential for modulation of activity in the olfactory system.

Cellular replication has recently been observed in basal cells located within the olfactory mucosa of the adults of some nonhuman species. Some investigators believe that the olfactory receptor cell population turns over and suggest that basal cells divide to form precursors of new olfactory receptor cells.

2.2 Sensory Characteristics of Odors

An odor can be characterized by its absolute threshold, its intensity, its quality, and its affective tone or pleasantness–unpleasantness dimension. Determinations of thresholds of odor perception and measurements of odor intensity are considered in Sections 2.4 and 2.5, respectively. Discussion of measurements of odor quality and odor acceptability are reserved for Section 4.4. It is important to recognize throughout that any program of odor measurement should be related to realistic analytical objectives. For example, if the odor from a landfill operation is to be controlled by masking agents, it is irrelevant to appraise odor intensities or threshold levels; a more appropriate measurement would describe the changed-odor quality or character and would assay its acceptability to the people in the community. On the other hand, a nonselective method of odor reduction, like ventilation or activated carbon adsorption, may well be monitored by measurements of intensity or by threshold dilution techniques.

2.3 Psychophysics and the Measurement of Odor Perception

G. T. Fechner is generally considered to be the founder of psychophysics, which he defined as the exact theory of the functionally dependent relations of body and soul, or more generally, of the physical and the psychological worlds (1). Fechner developed the concept of estimating or measuring the magnitude of sensations by assigning numerical values, reasoning that the relative magnitudes of sensations would be mathematically related to the magnitudes of the corresponding physical stimuli. In the years since Fechner's work, psychophysicists have endeavored to determine the laws relating physical stimuli to the resulting conscious sensations of the human observer. These efforts have concerned the responses of subjects to a number of basic questions about physical stimuli. Referring specifically to odors, these questions and the types of sensory evaluation needed to answer them can be tabulated:

Questions About an Odor Stimulus	Relevant Category of Sensory Evaluation
Is an odor present?	Determination of sensitivity to the detection of odor
Is a particular odor (e.g., that of phenol) present?	Determination of sensitivity to the recognition of odor
Is this odor different from that?	Determination of sensitivity to the discrimination between odors
How strong (intense) is this odor?	The scaling of odor intensity
What does this odor smell like, or what is the quality of this odor?	The scaling of the degree of similarity or dissimilarity between a given odor and each of a set of different odors
How pleasant (or unpleasant) is this odor?	The scaling of the hedonic, or like–dislike, response to an odor

Each except the last of the problems named has a counterpart in quantitative neurophysiology. Indeed, applications of psychophysical techniques in some fields of research have aided in the investigation of underlying neural mechanisms. However, electrophysiological signals, even if it were convenient to monitor them routinely from human subjects, are not direct expressions of subjective experience. The magnitudes of sensations are therefore measured along psychophysical scales, and for olfactory sensations the reference points on such scales consist of the perceptions experienced on exposure to standard odorants.

2.4   Decision Processes and Thresholds of Odor Perception

The word *threshold* means a boundary value, a point on a continuum that separates values that produce a physiological or psychological effect from those that do not. The *upper threshold* is the odorant concentration above which further increases do not produce increases in perceived intensity. The *detection* or *absolute threshold* is the minimum odorant concentration that can be distinguished from an environment free of that odor. Correspondingly, the *recognition threshold* is the minimum concentration at which an odorant can be individually identified. The recognition threshold of a particular odorant is never lower than its absolute threshold.

Thresholds are not firmly fixed values. Sensitivity fluctuates irregularly, and a certain odorant concentration will elicit a response at one time but not at another. Classical psychophysicists recognized the instability inherent in sensitivity, and they defined the absolute threshold statistically as that odorant concentration that is detected as often as not over a series of presentations. The probability of detection of such a stimulus is 50 percent. This definition of threshold has prevailed in the literature of air pollution and industrial hygiene. The theory of signal detection, originally engineered for telephone and radio communication systems by Shannon and Weaver (2) and later translated into a more general theory by Swets and others (3), contributed significantly to the development of modern psychophysics by specifying the dependence of threshold determinations on certain experimental variables. Detection theory has revealed the criteria employed by subjects in making perceptual judgments.

To illustrate how odor thresholds differ from other sensory evaluations of odor, refer to the questions about odor tabulated in Section 2.3 and note that the first three questions, which deal with thresholds, can be answered only by "yes" or "no." Because either answer can be right or wrong, four possibilities emerge when a subject is asked one of these questions. These possibilities are set out in a "response matrix" as follows:

	Response	
Odorant	Yes	No
Present	Hit	Miss
Absent	False alarm	Correct rejection

In such an experimental situation, the subject must decide whether in an observation period of set duration there is only *noise* (i.e., background interference either introduced by the experimenter or inherent in the sensory process), or whether a *signal* (i.e., the designated odor stimulus) is present as well. Two kinds of error can be made by the subject in such a situation: a *miss* (i.e., the failure to detect an odor stimulus when one is present) and a *false alarm* (i.e., the report of perception of an odor stimulus when one is absent). Detection theory emphasizes the relation between the occurrence of these two types of error. According to detection theory (4), a positive response by a subject is favored (*a*) by positive expectations that are enhanced by a high rate of stimulus presentation or suggested by the experimenter's instructions, and (*b*) by reluctance to miss the presence of an odorant in accordance with rewards obtained for hits and/or punishments incurred for misses. Conversely, a negative response is favored (*a*) by an expectation, perhaps resulting from experience, that no odor stimulus is present, and (*b*) by reluctance to score a false alarm, perhaps resulting from rewards obtained for correct rejections and/or punishments incurred for false alarms. These personal biases can be so strong that it is possible, in laboratory situations, to manipulate a subject's responses by manipulation of the variables influencing the decision process. Thus the experimental paradigm can drive the value determined for the "odor threshold" to a particular target. For example, a subject can avoid false alarms by not saying "yes"; this behavior generates a high value for the threshold. Conversely, a subject will not miss by not saying "no"; the consequence of such a decision is the generation of a low value for the threshold. Young and Adams (5), Steinmetz et al. (6), and Johansson et al. (7) describe such effects of the manipulation of experimental variables on odorant threshold determinations.

It is against this background of psychological biases that determinations of the limits of sensory perception must be evaluated. The absolute thresholds of different odorants determined in the same investigation can vary over six orders of magnitude, for example, 0.00021 ppm for trimethylamine and 214 ppm for methylene chloride (8). Such wide ranges may reflect real differences among the odors of different substances, yet since some reported threshold concentrations are very low, we have the implication that analytical and calibration errors are large and that the accuracy of the threshold values is suspect. Furthermore, the signal detection model of threshold determinations suggests that the variance can reflect decision as well as sensory processes. This suggestion could account for the wide range of scatter found in values of the absolute threshold generated for the same odorant in different investigations, for example, 3.2×10^{-4} to 10 ppm for pyridine (9). Another reasonable explanation for such scatter is the possibility that different samples of nominally identical odorants may really be very dissimilar, owing to different contents of odorous impurities. For example, phosphine, for which values of the detection threshold reported in the literature range from 0.2 to 3 ppm, has been shown to be odorless when pure (10), the reported odors being due to impurities in the form of organic phosphine derivatives. Similarly, it has been demonstrated (11) that ultrapurification engenders radical changes in the odor of nominally "pure" samples. Interindividual differences in sensitivity are yet another possible cause of large differences among values of the absolute threshold generated for the same odorants in different investigations. Interindividual variation can be high, but typically this factor is ignored in data analyses for the determination of odorant thresholds. There is also the problem of diversity in experimental procedures, for example, the mode of odorant presentation and sampling, which could introduce adaptation effects (Section 2.7), odor masking (Section 6.7), and errors in values reported for odorant concentration (12).

2.5 Factors Relating Odor Intensity to Odorant Concentration

A major objective of psychophysics is to measure the dynamic properties of sensory systems. Odorant concentration is the one property of the olfactory stimulus that can be varied somewhat systematically, although its effective control at the level of the olfactory receptor cell or even the olfactory mucosa has yet to be demonstrated. In the past, odorant concentration has been estimated or measured at some distance from the olfactory receptor cells. Psychophysicists have strived to characterize the functional dependence of odor intensity on such values of stimulus magnitude.

Psychophysical scaling attempts to determine the mathematical form of the relation between odor intensity and odorant concentration. *Direct scaling* refers to methods in which direct assessments of psychological quantities are made on an equal-interval scale (i.e., a graduated series with a constant unit) or on a ratio scale (i.e., a graduated series with a constant unit and a true zero). Direct scaling procedures include *estimation methods* in which stimuli are manipulated by the experimenter and judged by subjects, and *production methods* in which stimuli are manipulated by the subject to achieve a defined relation. In *category estimation* a given segment of a sensation continuum (usually predefined by stimuli supplied as anchors by the experimenter) is partitioned by subjects into a predetermined number of perceptually equal intervals, and subsequently presented stimuli are distributed by subjects among these categories. In *ratio estimation* the apparent ratio corresponding to a pair of stimulus magnitudes is numerically estimated. *Magnitude estimation* is a method in which various magnitudes of a stimulus are individually presented, and subjects assign a number to each signal in proportion to perceived intensity. For a complete discussion of psychophysical scaling methods, refer to Engen (13). There are numerous important variations in these procedures, and psychophysical scales generated with different experimental methods often do not correspond (14).

Stevens (15) proposed the psychophysical relation in which perceived intensity is a power function of stimulus magnitude, specifically the relation $R = cS^n$, where R is the perceived intensity, S is the stimulus magnitude, and c and n are constants referring to the intercept and the slope, respectively, when R and S are plotted on log-log coordinates. This yields a linear relation between the logarithm of the magnitude of the stimulus and the logarithm of the magnitude of the sensation. This relation has been verified for a wide range of sensory continua (14, 16) and has been proposed to be the fundamental psychophysical law.

Not all olfactory psychophysical scaling data conform to the power function proposed by Stevens. Magnitude estimations of odor intensities reported by Engen (17) for diacetone alcohol, *n*-heptane, and phenylethyl alcohol diluted in benzyl benzoate, and by Cain (18) for air dilutions of *n*-butyl acetate and 1-propanol, depart from the psychophysical law at low odorant concentrations. One reasonable explanation for these departures is that control and manipulation of concentration become increasingly difficult at greater dilutions. Another consideration for work with benzyl benzoate is that the diluent itself is odorous, and contributions of the diluent to odor intensity are greatest at the lowest concentrations of test odorant. Yet another possibility is that dilution does not produce a universal, unidimensional psychophysical scale for all odorants. Changes in odorant concentration can produce changes in odor quality as well as in odor intensity, and these transitions may not be identical for all subjects for all or even any odorants. In the experiments cited, shifts in odor quality might not have been ignored by subjects in their assessments of odor intensity. Indeed, it is questionable whether human subjects in any experimental situation can reliably distinguish odor intensity from other attributes of the stimulus. There is also the problem of psychological impurity, which is discussed in Section 2.6.

For most odorants, however, perceived intensity is reported to be a power function of stimulus magnitude. Hyman (12) has tabulated values of the exponent of the psychophysical function for odor intensity for 33 odorants generated in different investigations. The values are typically less than 1.00, with some exceptions for some subjects. They range from 0.03 for 1-decanol (19) to 0.82 for 1-hexanol (20). Some of the results of the scaling of perceived intensity for the same and even for different odorants are quite similar, whereas considerable disagreement exists among the results of many other investigations of the intensity of different and even of the same odorants. Hyman (12) has discussed the factors capable of influencing the outcome of olfactory psychophysical scaling in the hope of elucidating possible sources of discrepancies. In brief, the factors concerning stimulus magnitude are (a) the use in different investigations of different physical scales that are not linearly related; (b) the use in some investigations of excessively strong stimuli or intertrial intervals so short that adaptation effects are introduced (see Section 2.7); (c) the use in some investigations of stimuli in concentrations so low that analytical and calibration errors are large; (d) the adsorption of vapors on the walls of gas-sampling vessels, which introduces errors in reported odorant concentrations; (e) the evaporation of volatiles from mixtures, which alters their composition during a given study; and (f) deviations from ideal behavior as given by Raoult's law, which is used to predict the vapor pressure of a dissolved substance on the basis of proportionality to solute concentration.

Deviations from Raoult's law readily explain differences among values of the exponent of the psychophysical function for odor intensity generated for the same as well as for different substances under different methods of dilution (21). Furthermore, because the extent of these deviations depends on the solvent and on concentration, the possible error in values of stimulus magnitude used in the determination of odor intensity function parameters also applies for odorants diluted in different liquids and for odorants diluted in the same liquid but at different ranges of concentration. There are also problems of psychological impurity (Section 2.6) and factors concerning chemical impurities and contaminants of stimuli that can introduce adaptation effects (Section 2.7) and odor interactions (Section 6.7) into the test situation.

With the exception of a few investigations (22–24) in which reported values of the exponent of the psychophysical function for odor intensity are the mean of values of individual subjects, published values of the exponent seem to be based on odor intensity assessments collected from a group of subjects, then pooled to form one regression plot from which the parameters of the psychophysical function are found. The assumption implicit in such a pooling procedure is that subjects are interchangeable psychophysical transducers, or that they represent a normal distribution of such transducers and that the distribution is the same for all odorants. But variation across individuals can be high (22, 25), and individual subjects can be consistent in the values of the exponent they generate for a particular odorant over repeated test sessions (25). It therefore seems improper to regard subjects as interchangeable for the purpose of pooling data in the indiscriminate manner traditionally employed for the calculation of odor intensity function parameters. Indeed, interindividual variation can be as large as differences between values of the exponent of the psychophysical function for different odorants derived from odor intensity assessments pooled from groups of subjects. Group measures may be obscuring important attributes of the olfactory system which these investigations are endeavoring to elucidate. Not only can the value of the exponent of the psychophysical function for odor intensity for any particular odorant vary greatly between subjects, but large intraindividual differences can exist between values of the exponent for different odorants (22, 25). Thus it seems appropriate to direct investigations to individual behavior over a variety of stimulus and response modes within the olfactory sensory continuum while concurrently striving to minimize variation in experimental procedures.

2.6 "Pure" and "Impure" Odor Perceptions

The experience of smell can be taken either to mean any perception that results from nasal inspiration or to refer only to sensations perceived by way of the olfactory receptor cells. An odorant that is sensed only by the olfactory receptor cells (e.g., air containing 0.01 ppm vanillin) is said to be *psychologically pure*. An odorant that stimulates both olfactory receptor cells and other sensitive cells (the so-called common chemical sense) is said to be *psychologically impure*; an example is air containing 50 ppm propionic acid. The common chemical sense includes sensations of heat, cold, pain, irritation, and dimensions of pungency described by our language inadequately. The trigeminal nerve, or cranial nerve III, is the primary mediator of common chemical sensations. The receptors responsible for the transduction of these perceptions remain incompletely characterized. The concept of psychological purity implies nothing about the chemical composition of the stimulus.

Differences between psychologically pure and impure odors are neglected in many odor measurements, especially in industrial hygiene and community air pollution applications, where irritants and stenches are grouped together as objectionable atmospheric contaminants that ought to be removed. It is not always operationally feasible to make distinctions among odors based on degrees of psychological purity. In establishing sensory measurement scales, however, it is important to recognize that "irritating odor" is not necessarily an extension in magnitude of "strong odor" but refers to a different type of sensation.

The transitions in perceived quality that can accompany changes in the perceived intensity of an odorant may be purely olfactory in origin, or they may be due to a change from purely olfactory stimulation at low odorant concentrations to olfactory plus trigeminal stimulations at high odorant concentrations. Evidence has accumulated in support of the notion that the trigeminal nerve contributes to the overall intensity of odorants. For human subjects with unilateral destruction of the trigeminal nerve, the magnitudes of perceived intensities for certain odorants are consistently lower through the deficient nostril, with the most dramatic differences at high odorant concentrations (26). Doty (27) reported the detection of odorous vapors by what he considered to be anosmic human observers; such detection presumably occurs by way of the trigeminal system. The olfactory and trigeminal systems can have different psychophysical functions for a particular odorant (28), and psychophysical scaling data reported for odor intensity could actually be some combination of the two. There is usually no effort to eliminate the trigeminal component from assessments of odor intensity. It is therefore plausible that excitation of trigeminal receptors is responsible, at least in part, for differences among values of the exponent of the psychophysical function for odor intensity and for departures from that relation reported for some odorants in some investigations.

2.7 Olfactory Adaptation

Olfactory adaptation is the decrement in sensitivity to an odorant following exposure to what is termed an "adapting" odorous stimulus. The rate and degree of loss in sensitivity and subsequent recovery depend on the adapting stimulus and on its concentration (29). Self-adaptation to an odorant affects the perceived intensity of the same odorant, whereas cross-adaptation affects the perceived intensity of other odorants. Olfactory adaptation is common, yet its physiological mechanism has not yet been elucidated.

Olfactory adaptation has been studied by measuring the time required for the odor of a continuously presented stimulus to disappear, and by determining absolute threshold values before and after exposure to an odorant for a given period. An increase in odorant concentration usually increases the time required for disappearance of an odor, and for a particular experimental design, absolute threshold values increase with an increase in the magnitude of the adapting odorous stimulus.

Self-adaptation affects the magnitude of the olfactory sensation as a function of stimulus magnitude. Values of the exponent of the psychophysical function for odor intensity are significantly greater after self-adaptation (20, 30). High concentrations of self-adapting stimuli generate steeper psychophysical functions for odor intensity than do self-adapting stimuli of low concentrations (30). The use of weak stimuli closely spaced in magnitude or long periods of exposure to an adapting odorous stimulus can permit resolution of the influence of duration of an adapting stimulation on the psychophysical function for odor intensity (31). However, without consideration of these factors, increasing the duration of self-adaptation produces only minor effects on the odor intensity function (32).

Cross-adaptation has approximately the same effect as self-adaptation on the form of the psychophysical function for odor intensity. It is important to recognize, however, that the effects of cross-adaptation within pairs of odorants can be asymmetric; for example, concentrations of 1-pentanol and 1-propanol closely matched for odor intensity have unequal cross-adapting effectiveness (30).

Olfactory adaptation may be responsible for some of the differences among values of the exponent of the psychophysical function for odor intensity reported for the same and even for different odorants. Adaptation effects can be introduced in an experiment by (*a*) the use of excessively strong stimuli, (*b*) short intertrial intervals, (*c*) the presence of odors apart from those generated by stimuli in intertrial intervals, and (*d*) the presentation of a standard reference stimulus immediately before test concentrations (32).

In the context of industrial hygiene and air pollution situations, two important questions arise. Does prolonged exposure to odors produce a permanent loss of ability to smell? And to what extent does temporary adaptation interfere with the sensory measurements of odor intensity carried out in the workspace or in the outside community?

With regard to the first question, there is simply no evidence on which a reliable answer can be based. In view of the many difficulties involved in quantifying sensory odor measurements under controlled test conditions, it would seem hopelessly unreliable to attempt any studies that depended on retrospective estimates of exposures to odors.

To answer the second question, we must consider patterns of recovery from adaptation and the opportunities for such recovery during typical odor survey programs. Köster (33) studied recovery times from adaptation to the odors of benzene and various alkylbenzenes, as well as to isopropyl alcohol, dioxane, cyclopentanone, and β-ionone. Typically, 60 to 70 percent recovery of sensory response to the odors was realized in about 2 min. The implication of these findings, which is in accord with practical experience, is that if a subject is downwind from a point source such as a stack and is exposed to the odor intermittently depending on changes in wind direction, he can make perfectly good sensory evaluations when the plume comes his way again after a lapse of 5 min or more. However, a person in an odorous enclosure, who is subjected to unrelieved exposure to odor, will indeed become adapted to a point that can invalidate any sensory judgment.

2.8   Theories of Olfactory Mechanism

Theories of olfaction attempt (*a*) to delineate the mechanism by which odorant stimuli elicit responses in olfactory receptor cells, and (*b*) to correlate the sensory characteristics of odors with physicochemical properties of the odorants. The objective is to understand the physiological basis of olfactory sensory discriminations.

The perception of smell begins with the impingement of molecules upon the olfactory receptor. The molecular property responsible for the initiation of excitation in olfactory receptor cells has yet to be identified. Cross-adaptation has been applied as a basis for the classifying of odorants with the aim of isolating the physical correlate of olfactory reception, the idea being that odorants with a common characteristic for olfactory stimulation will cross-adapt. Almost every physicochemical property has been implicated (34). Theories based exclusively on molecular size and shape have been found to be inadequate. There have been attempts to combine several physicochemical properties into complex functions in the hope that a certain combination will correlate with the available data on odor quality and thresholds. Intermolecular interactions, electron affinities, and even bulk properties such as vapor pressure and molar volume have been considered. Such attempts to formulate an appropriate composite function are more realistic than attempts to discern a unitary determinant of olfactory reception, but the correct balance of molecular properties that will predict all odors has not yet been produced. Indeed, even these attempts may be futile in that the parameters of a composite function may be correlated with molecular properties not considered in the function.

Much attention has been given to intramolecular vibrations as the physical correlate of odors. Wright (35, 36) has proposed that the energy state of olfactory receptor molecules becomes altered when an odorant molecule of appropriate vibrational frequencies is encountered, and the change in state of the receptor molecule induces excitation in the olfactory receptor cell. This theory is rather vague, but as with other theories of olfaction, vibrational measures do successfully predict some odors. However, the pertinent spectral features have not been identified, and many odorants that produce different odors have similar vibrational properties. Furthermore, optical enantiomers, which have identical spectra, can have different odors (37).

Although direct evidence is lacking, it is thought that the cilia that project from each olfactory receptor cell contain the sites at which chemical stimuli are received. It is generally believed that odorant molecules are adsorbed at specific receptor sites, where they interact with proteins or other macromolecular constituents of the cilia surface, causing conformational changes that somehow initiate the transduction of molecular energy into electrical energy. Davies (38, 39) has proposed that odorant molecules cross the interface between the surrounding medium and the nerve cell membrane, creating temporary defects in the nerve cell membrane through which ions can pass; the result is a collapse of the nerve membrane potential, which provides the drive for the development of action potentials. Many other theories on the generation of neural signals by odorant stimuli have been proposed, but each is supported by circumstantial evidence only.

Not only is the sequence of events leading to excitation of olfactory receptor cells not understood, but the encoding of sensory information in olfactory nerve signals remains a mystery. On the microscopic level, there are no discernible differences in olfactory receptor cells to account for sensory discrimination. However, electrophysiological studies have revealed that individual olfactory receptor cells differ in their responses to odor stimuli with regard to the extent of their excitation and the odorant concentrations at which they become excited. Such a system could provide for sensory discrimination by the pattern of activity of many receptors. However, the code has yet to be deciphered.

In sum, the functional parameters of the olfactory stimulus remain obscure, and knowledge of the physiological basis of olfactory stimulation and sensory discrimination is scanty. Even investigations of the relation between odor intensity and odorant concentration have not provided insight into the operating characteristics of the olfactory sensory system.

3 CHARACTERISTICS OF ODOROUS SUBSTANCES

3.1 Odor Sources

A vented storage tank being filled with liquid ethyl acrylate from a delivery truck discharges to the atmosphere a volume of air saturated with the acrylate vapor at the prevailing temperature of the liquid. A system of this type is a simple example of a confined odor source: the location, molecular aggregation, composition, concentration, and volumetric discharge of the source can be quantitatively specified. Such definite characterizations facilitate the establishment of relationships between the odor source and the odor measurements made in the work space or the community. In general, an odor source may be said to be confined when its rate of discharge to the atmosphere can be measured and when the atmospheric discharge is amenable to representative sampling and to physical or chemical processing for purposes of odor abatement. For meteorological diffusion calculations, the location at which the odor is discharged into the atmosphere is assumed to be a point in space.

The characterization of a confined odor source should include (*a*) the volumetric rate of gas discharge (gas volume/time), (*b*) the temperature at the point of discharge, (*c*) the moisture content, (*d*) the location, elevation, and the area and shape of the stack or vent from which the discharge is emitted, and (*e*) a description of the state of aggregation (gas, mist, etc., including particle size distribution) of the discharge. Such information is useful when meteorological factors relating to odor reduction by dispersion and dilution are being evaluated. The characterization of an odor source in terms of chemical composition will depend on the relationships to be established between sensory and chemical measurements. If there are no technical or legal uncertainties regarding the source of a given community odor, a detailed chemical analysis is not needed; instead, a comparative method of appraising the effect of control procedures will suffice. On the other hand, chemical characterization may be helpful in tracing a community odor to one of several alleged sources, in relating variations in odor from a given source to changes in process conditions, or in appraising the effectiveness of chemical procedures that are designed for odor abatement.

The direct sensory characterization of a confined odor source is frequently impossible because extremes of temperature or of concentration of noxious components make the source intolerable for human exposure. Methods suitable for the dilution and cooling for sensory evaluation of odor sources are described in Section 4.2.

A drainage ditch discharges odorous vapors to the atmosphere along its length. The composition of the contents of the ditch changes from place to place, and the rate of discharge of odorants to the atmosphere is affected by wind and terrain. No single air sample is likely to be representative of the odor source at the time the sample is taken. Such a configuration is an unconfined odor source. Other examples are garbage dumps, settling lagoons, and chemical storage areas. An unconfined source may be represented by an imaginary emission point for the purpose of meteorological diffusion calculations. Such assignment may be made on the basis of supposing that if all the odor from the unconfined area were being discharged from the "emission point," the dispersion pattern would just include the unconfined source. Figure 41.2 schematically illustrates this procedure.

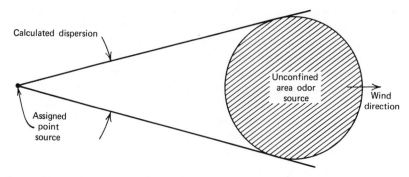

Figure 41.2.   Assignment of emission point to an undefined area odor source.

3.2 Odorous Gases and Vapors

In general, gases and vapors are odorous. The relatively few odorless or practically odorless exceptions include oxygen, nitrogen, hydrogen, steam, hydrogen peroxide, carbon monoxide, carbon dioxide, methane, and the noble gases. No precise relationship has been established between the olfactory quality of a substance and its physicochemical properties, although attempts have been made for many years. This topic is treated in more detail in Section 4.1. Most odorants encountered in the work space or in the outside air are complex mixtures, however, and it cannot be assumed that the odor of the mixture is that of its major component. Thus a "phenolic" odor from the curing of a phenolic resin is not the same as the odor of pure phenol, and the pungent odor of burning fat is not the same as that of acrolein, though it is often so characterized. Even more dramatic are instances in which the ultrapurification of a supposedly "pure" substance produces a considerable change in odor (11).

3.3 Odors Associated with Airborne Particulate Matter

It is often assumed that all odorants are gases or vapors. However, there is evidence that some particles can stimulate the sense of smell because the particles themselves are volatile or because they are desorbing a volatile odorant (40). There is also speculation that some particulate matter is capable of stimulating the sense of smell. Regardless of the mechanism by which particles may stimulate the olfactory sense, they definitely appear to be involved.

The idea that odors are associated with particles is supported by observations that filtration of particles from an odorous airstream can reduce the odor level. It has been shown (41) that the removal of particulate matter from diesel exhaust by thermal precipitation effects a marked reduction in odor intensity. The precipitation method was selected because it provides minimal contact between the collected particles and the gaseous components of the diesel exhaust stream. Thus effects that could be produced by a filter bed, such as removal of odorous gases by adsorption in the filter cake, are eliminated. Therefore the observed odor reduction must have resulted directly from the removal of particulate matter. Other more or less causal observations on the role of particulate matter in community odor nuisance problems appear occasionally in the literature (42). Several hypothetical mechanisms have been developed for the association of odor with particles.

3.3.1 Volatile Particles

Liquid or even solid aerosols may be sufficiently volatile that their vaporization on entering the nasal cavity produces enough gaseous material to be detected by smell. Such aerosols may be relatively pure substances such as particles of camphor, or they may be mixtures that release volatile components. The retention of the odorous properties of volatile aerosols, of course, depends on the prevailing temperature and on the length of time they are dispersed in air. In a cold atmosphere, the relatively greater temperature rise accompanying inhalation accelerates the production of gaseous odorant.

3.3.2 Desorption of Odorous Matter by Particles

The interaction between gas molecules and the surface of airborne particles has been discussed by Goetz (43). The theoretical considerations were directed to the question of transfer of toxicants by particles, but they are also applicable to odors. Even if a given aerosol is intrinsically odorless, it could act as an odor intensifier (a) if the sorptive capacity of the aerosol particles for the odorant were smaller than the affinity of the odorant for the nasal receptor, and (b) if at the same time, the sorptive capacity of the aerosol particles were large enough to produce an accumulation of odorant on the particle surface. Such aerosol particles would concentrate odorous molecules on their surfaces, but the odorous matter would be transferred to olfactory receptors when the aerosol entered the nasal cavity. The odorous matter would then be present at the receptor sites in concentrations higher than in the absence of the aerosol. The resulting effect would be synergistic. If an odorant is more strongly adsorbed by the aerosol particles than by the olfactory receptors, transfer of the odorant to the receptors would be impeded and the particles would actually attenuate the odor.

3.3.3 Odorous Particles

No study has rigorously defined the upper limit of particle size for airborne odorous matter. Particles up to about 8 or 10×10^{-4} μm in diameter are considered to be molecules that can exist in equilibrium with a solid or liquid phase from which they escape by vaporization. The vapor pressure decreases as the molecular weight increases, and particles above about 10^{-3} μm do not generally exist in significant concentrations in equilibrium with a bulk phase; hence we do not consider them to be "vapors." Nonetheless it is possible that odorant properties do not disappear when particle sizes exceed those of vapor molecules. Our knowledge about particles in the size range of 1 to 5×10^{-3} μm (up to about the size of small viruses) is relatively meager, and we do not know whether they can be odorous, or what effect an electrical charge might have on their odorous properties. Larger particles may also be intrinsically odorous, although their more significant role may be to contribute to odor by absorbing and desorbing odorous gases and vapors.

4 ODOR MEASUREMENT

4.1 Relation Between Sensory and Physicochemical Measurements

No one can predict the odor of a compound from its molecular structure, nor is it possible to specify the molecular structure of a substance that will yield a predicted odor. In the context of odorous air pollutants, this means that attempts to predict odor nuisance from a knowledge of sources and processes can serve as a guide to signal possible problems, but not as a substitute for direct sensory evaluation. Recent statistical attempts at multidimensional scaling of such attributes (44) lead to conclusions that explain the degree of similarity between different odors in terms of physicochemical variables. The information generated by such methods, however, seems to be of the type that is already common knowledge—for example, that compounds differing greatly in molecular weight are likely to have very different odors, or that sulfides do not smell like aldehydes. Because most organic chemists believe they can identify the functional group (alcohol, amine, ester, etc.) in a compound by smell, it is interesting to determine the degree to which such attempts are successful. Brower and Schafer (45) conducted such a study and found that for most representative compounds the functional group was correctly identified in 45 percent of the cases. The performance was poor for alcohols, ethers, and halides, and excellent for amines, sulfur compounds, esters, phenols, and carboxylic acids. When the subjects missed the functional group, they used the labels alcohol, ester, and ketone twice as often as average. The label "sulfur compound" was misapplied in only 1 percent of all cases. Bulky hydrocarbon groups near the functional group can weaken or obliterate the odor quality, but aliphatic amines and sulfur compounds are very resistant to steric hindrance. On the other hand, they are greatly weakened by electron-withdrawing groups. Aliphatic compounds bearing a multiplicity of methyl groups have the odor of camphor or menthol.

Of course, most odor problems are produced by complex mixtures of odorants, which severely complicates the task of predicting sensory effects. The literature from 1958 through March 1969 was surveyed for instances of instrumental–sensory correlation studies under the sponsorship of ASTM Committee E-18 on Sensory Evaluation of Materials and Products (46). Of the several thousand articles reviewed in 65 major technical journals, including journals devoted to air pollution, only 45 were judged to have any direct value toward progress in establishing standards for instrumental–sensory correlations. Of these, some 40 were concerned with foods, and the remainder with pure chemicals and other topics. None dealt with odorous air pollutants per se.

There have been significant advances, however, in the analyses of various organic substances that are known to be odorous. For example, the short-term analysis of malodorous sulfur-containing gases can be performed in the ambient air (47, 48). For most other odorous air pollutants, however, direct analytical methods that are demonstrably related to odor are not available.

4.2 Sampling for Odor Measurements

In the case of nuisance odors in the work space or in the community, sensory judgments are usually made from evidence gained on direct exposure of human subjects to the odorous atmosphere in question. Where odor intensities are low enough to be tolerable and where they vary from time to time in accordance with atmospheric turbulence, changes in indoor air currents or outdoor wind direction, and changes in the odor source, it is much better for the observers to expose themselves directly to the odorous atmosphere rather than to a previously collected sample. Furthermore, when odor levels are variable, there are often enough intervals of low odor intensity to provide adequate recovery from odor fatigue. Under such circumstances, it would require considerable effort to collect samples that were representative of all the experiences a roving observer could accumulate in a single session of an hour or so.

In some instances, however, it is advantageous to collect a sample of odorant before presenting it for sensory measurements. Such instances occur (a) when the odorant must be diluted, concentrated, warmed, cooled, or otherwise modified before people can be exposed to it; (b) when it is necessary to have a uniform sample large enough to be presented to a number of judges; or (c) when the samples must be transported to another location at which sensory evaluations are made.

4.2.1 Grab Sampling

A grab sample places a volume of odorant at barometric pressure into a container from which it can subsequently be presented to judges for evaluation. The general principles of sampling for gases and vapors are described in Chapter 27. The subject has also been reviewed by Weurman (49). However, the objective of preserving the integrity of a sample for odor measurement may present special problems because mass concentrations of odorants are often very low. As a result, adsorption or absorption by the container walls may alter the odor properties of the sample. It is therefore important to use containers that are known to be inert to the odorant and are large enough to minimize wall effects. A number of container materials have been used, including glass, stainless steel, and various inert plastics (50, 51).

When the grab sample is to be evaluated by human subjects, it must be expelled from the container into the space near the judge's nose. This expulsion may be effected by displacement of the sample with another fluid (usually water) or by collapsing the container. It must be recognized that when a person smells a jet of air from a small orifice, the aspiration of ambient air around the jet creates some additional dilution before the odorant reaches the subject. Springer (52) has designed a conical funnel that is positioned in front of the subject's nose to prevent this effect.

4.2.2 Sampling with Dilution

Dilution procedures for sampling odor sources such as oven exhausts can reduce concentrations and temperatures to levels suitable for human exposure without permitting condensation of the odorant material. The dilution ratio must be specified; for this purpose it is convenient to define a dilution ratio Z such that $Z = C/C_t$, where C is the concentration of odorant at the source and C_t is the target concentration to be reached by dilution (53), and both concentrations are expressed in the same units.

The material of choice for sampling of hot odor sources with dilution is stainless steel. A suitable device is a 1000-in.^{3} (16-liter) stainless steel tank, fitted with valves, a vacuum gauge, and a pressure gauge. The tank is first evacuated, then filled through a metal probe to a desired pressure P_1, which must be low enough to prevent condensation when the gas warms to room temperature. This dilution is $Z_1 = 1$ atm/P_1 (atm). The tank is then pressurized with odor-free air to a new pressure P_2, which is above atmospheric pressure. The diluted, pressurized sample may then be sniffed by one or more judges for odor evaluation. The dilution that occurs when pressurized gas is released to the atmosphere is $Z_2 = P_2(\text{atm})/1$ atm. The overall dilution ratio is

$$Z = Z_1 \times Z_2$$

$$= \frac{1 \text{ atm}}{P_1 \text{ (atm)}} \times \frac{P_2(\text{atm})}{1 \text{ atm}} = \frac{P_2}{P_1}$$

For example, if the evacuated tank is first filled to 0.1 atm, then pressurized to 2 atm (gauge), $P_1 = 0.1$ atm, and $P_2 = (2 + 1)$ atm (absolute), and $Z = 3/0.1 = 30$.

4.2.3 Sampling with Concentration

A dilute odorant may have to be concentrated before it can be adequately characterized by chemical analysis. Such circumstances are likely to arise when a community malodor is to be traced to one or more possible sources. The ratio of concentrations between the odor source and the odorant outdoors where it constitutes a nuisance may be in the range of 10^3 to 10^5. Under such circumstances the comparison of the odorant with the alleged source is greatly facilitated if the concentration is increased to approximate that of the source.

The most widely used methods of concentration are freezeout trapping and adsorptive sampling. Chapter 27 elaborates on both these methods.

When the concentration ratio must be high, the accumulation of water in a cold trap is a serious drawback, and adsorptive sampling is usually preferred. Activated carbon has been used, but it is so retentive to organic vapors that recovery of the sample often requires special methods (54, 55). More recently, various porous polymers (56) have been found to be both effective and convenient, especially when the concentrated sample is to be injected into a gas chromatograph.

4.3 Sensory Evaluation

4.3.1 General Conditions

Sensory testing requires concentration and high motivation on the part of human judges. Therefore outside interferences, such as noise, extraneous odors, and any other environmental distractions or discomforts, must be kept to a minimum (57). Odor control is generally achieved by use of air conditioning combined with activated carbon adsorbers. The testing room should be under slight positive pressure to prevent infiltration of odors. All materials and equipment inside the room should be either odor free or have a low odor level. Transite partitions have proved to be very effective as wall and ceiling material. If highly odorous products are to be examined or high humidities are anticipated, these partitions may be sprayed with an odorless, strippable, soft-colored coating that can be replaced if it becomes contaminated. Low odor asphalt tile has proved effective as floor material.

The panel moderator must exert every precaution to avoid bias with respect to any of the tests. Ideally, the moderator, like the judges, should be unaware of the identity of the samples, so that the test is "double blind."

4.3.2 Selection and Training of Judges

The general criteria for selection of judges, as outlined in an ASTM manual (57), involve the individuals' natural sensitivity to odors, their motivation, and their ability to work in a test situation. Wittes and Turk (58) have pointed out that three variables characterize the efficacy of any screening procedure: (*a*) the cost, as determined by the number of sensory tests, (*b*) the proportion of potentially suitable candidates rejected by the screen, and (*c*) the proportion of potentially unsuitable candidates accepted by the screen. These variables are functionally dependent; that is, specification of any two determines the value of the third. Therefore a screening procedure can be designed to favor any two of these variables, but not all three.

It is generally best, especially with novices, to start with familiar substances such as food flavors. The transition to malodors can come later. Screening tests are based on the ability of a judge to identify the odd sample in a group of three odors of which two are the same ("triangle" test), or to order samples in a sequence of increasing intensity, or to identify the components of a mixture. Details are given by Wittes and Turk (58).

After a group of judges has been selected, they must be trained. The screening tests should be repeated, and this time any errors should be discussed and analyzed. The judges are then introduced to the types of sensory tests they will carry out. If possible, the original measurements should be made on known standards.

4.4 Odor Surveys and Inventories

Because the ASTM "odor unit" is expressed in cubic feet and is thus an additive value, it is tempting to use it to quantify odorous emissions, as suggested by Hemeon (59). Moreover, because the ASTM "odor emission rate" is the number of odor units discharging per minute from a stack or vent, this value also is additive, and the total emission from a given industrial area can be taken to be the sum of all the odor emission rates (60). For a nonconfined source such as a lagoon or drainage ditch, some estimate of emission rate can be made on the basis of the odor concentration of the air in equilibrium with the source, and an assumed velocity of transfer of odor to the atmosphere by wind action. Then the total odor emission rate is the emission rate per square foot of surface times the total area in square feet. Of course, such procedures suffer from all the limitations of threshold measurements: they ignore considerations of odor quality or objectionability, as well as differences in the odor intensity exponent among different odorants; the threshold level itself depends strongly on the response criterion; and problems in preserving the integrity of odorant samples become severe when the dilutions approach threshold levels, as evidenced by the large differences obtained by different dilution devices.

An approach to odor surveying that distinguishes among sources of different qualities was used as early as the 1950s in Louisville, Kentucky (61). In that study a kit of 14 reference odorants was chosen to "represent the principal odors expected to occur" in the area being surveyed. Some of the reference standards were actual samples of presumed odorant sources, such as a creosote mixture used as a wood preservative in an operation that was known to generate an odor nuisance. In the actual survey, the observers reported daily the nature (by identification with the reference samples), the intensity (category scale), and the location of the odors they smelled. In later studies odor reference standards have been used in conjunction with improved methods of intensity scaling to yield quality–intensity inventories of odor sources of various kinds (60, 62).

Assessments of odor intensity and quality, however, do not serve as reliable predictors of human affective responses to unpleasant odors. There have been two approaches to direct surveying of such reactions. In one approach, developed by Springer and co-workers (52) for diesel exhaust odors, the odor source to be surveyed is set up in a transportable facility, under controlled conditions of dilution and presentation to judges. Responses are elicited from a large number of randomly selected, untrained people. For the diesel odor, the exhaust from an engine was diluted and piped into compartments in a trailer for sensory evaluations. The source–trailer combination was then used in several sites in various cities in the United States, to constitute a national survey. The questionnaire submitted to each judge was designed around the cartoon scale of Figure 41.3. Note that the facial expressions, the bodily actions, and the descriptive adjectives are all mutually reinforcing.

The second approach to measuring human affective responses to odors is a survey of annoyance reactions. Jönsson (63) has reviewed human reactions studied in different odor surveys and concludes that responses to questions about annoyance

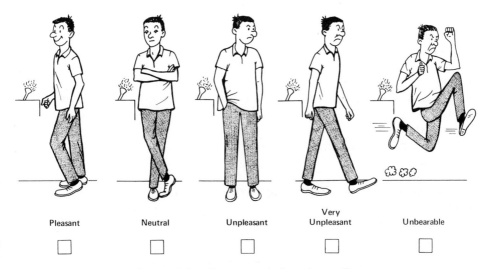

			Very	
Pleasant	Neutral	Unpleasant	Unpleasant	Unbearable
□	□	□	□	□

Figure 41.3. Cartoon scale for odor testing.

are more reliable than such other indexes as willingness to sign a petition or to take direct action to modify the environment (e.g., by using household deodorants or installing air conditioning). A detailed survey procedure is described in "A Study of the Social and Economic Impact of Odors. Phase II," also known as the Copley Report (64). An important feature of this procedure is the use of an odor-free area as a control to account for the fact that in some instances annoyance to odor is expressed when no odor source exists. Even in an odorous area, complainants may call attention to the odor problem, but their opinions are not generally typical of the majority of the community. Consumer and social research studies have found repeatedly that the likes and dislikes of persons who volunteer their opinions are different from the likes and dislikes of their neighbors who must be solicited for their opinions. The purpose of the survey, then, is to compare the attitudes of people residing in a community thought to be an odor problem area, with attitudes of similar people residing in an odor-free area. Attitudes of both groups are determined by conducting interviews by telephone with residents of both areas. The survey proceeds in the following sequence:

1. The first task is to define the possible odor problem area.
2. Next, a matching odor-free area is located.
3. Utilizing a street address (reverse order) telephone directory, a list of telephone numbers in each area is made, and a sample of these telephone numbers is selected at random.
4. Utilizing the questionnaire provided, telephone interviews are conducted with the adult occupant of the house for each telephone number included in the sample.
5. The total number of responses to key questions asked in both areas is tabulated and compared for problem identification.
6. If an odor problem is found, an odor problem index number is calculated.

The questionnaire itself explores the respondent's length of residence in the community, attitude to various categories of complaints about environmental problems, personal experiences of odor, degree of annoyance by odors, and opinion with regard to the origin of the odors; place of employment is also ascertained for all members of the household. The final calculated odor index expresses the degree to which it can be confidently stated that there is a statistically significant difference between the test area and the control area; such a difference is said to constitute an "odor problem identification."

5 SOCIAL AND ECONOMIC EFFECTS OF ODORS

5.1 Social Effects

For convenience in definition, social effects can be said to differ from economic effects in that the former cannot be measured directly in monetary terms (65). The social effects of odors include (a) interference with the everyday activities of the exposed individuals, (b) feelings of annoyance caused by offensive smell, (c) physical symptoms of physiological changes, (d) actual complaints to an authority, and (e) various forms of direct individual action to modify the environment other than through complaints.

All these effects are very difficult to quantify; for example, the average homeowner cannot say with assurance how many times an unpleasant odor has prevented him from using his yard to entertain friends. As described in the preceding section, the principal tool used to assess social effects has been the attitude survey, and the results yield only a statistical measure of the confidence that an odor problem is correctly identified, not a rating of the intensity of human reactions.

5.2 Economic Effects

Assuming that unpleasant odors annoy people, can we say that they also cause economic loss? If so, how can this loss be determined? Again the most direct approach is to survey the affected area by asking a representative sample of persons in the community how much they would be willing to pay to get rid of the odors. Though theoretically valid, this approach runs into the practical difficulty of separating what people say they would pay from what they actually will pay if asked to comply with their own responses.

A more fruitful approach lies in determining what people actually have paid to obtain an odor-free environment. This requires recognition that the economic impact of odor pollution is most likely to manifest itself in reduced property values, reduced productivity of industrial companies, and reduced sales in commercial areas.

Economic theory states that if odors are bothersome, people should be willing to pay more to live in an odor-free area. Thus two similar properties in similar neighborhoods should sell for different prices if one area is affected by odors and the other is not. Thus we are assuming that all economic losses due to the presence of odors are capitalized negatively into property values and that buyers need only know that they prefer some properties to others and be willing to pay more for them.

Odors may affect industrial property values in somewhat the same way residential areas are affected. However, another form of loss to commercial and industrial establishments is possible, namely, odors may reduce the productivity of employees because of induced illness or distraction from work assignments. Productivity losses are likely to be particularly noticeable if such odors are intermittent as well as strong. Such a situation would be offensive, while also tending to inhibit persons from becoming adapted to their work environment.

Commercial areas may suffer economic losses from odors in the form of a general loss of customers and reduced sales per customer.

The one serious attempt to measure such economic effects was carried out in Los Angeles in 1969 as part of the Copley Report (64). Unfortunately, the attempt was not successful, either because the methods were not sensitive enough to isolate economic effects caused by odors, or because transitory odors are not capitalized into property values.

6   ODOR CONTROL METHODS

6.1   Process Change and Product Modification

The modification of malodorous chemical processes merits first consideration because of the possibility that slight changes may yield significant results. Certainly a review of existing facilities and practices to increase process efficiency, separation efficiency, and collection efficiency—in other words, to reduce waste—can only be helpful. The same may be said of upgrading the quality of valves, pumps, drainlines, and other potential sources of leaks. Changes of process conditions, however, yield less easily predictable results. For example, an increase in process temperature may promote more complete oxidation to odorless products, or it may bring about more volatilization and cracking, to release more odorants. When such changes are contemplated, their possible effects must be carefully assessed.

It is also attractive to substitute chemicals with low odors for more highly odorous ones. Frequently such substitution can be made for solvent mixtures when different types are interchangeable in function, if not in cost. Differences may be less easy to accommodate with regard to change of product, but product substitution should be considered before making a major investment in odor abatement equipment.

6.2   Dispersal and Dilution Techniques

6.2.1   Ventilation Systems in Enclosed Spaces

A time-honored approach to controlling odors in enclosed spaces has been to exhaust the malodorous air to the outdoors. There are several fundamental limitations to such a remedy. First, the exhausted air may transfer its nuisance to the outdoors. Such instances are particularly troublesome in congested areas where people may be exposed to the exhaust, or where one building's exhaust becomes another's intake. Under some conditions such atmospheric short-circuiting may even occur between the vents of the same structure. Second, the exhausted air must be replaced, and if the makeup air requires heating or cooling, large consumptions of energy may be required. Moreover, the odorant concentration is not reduced to zero; rather, it approaches an equilibrium level in which generation and removal rates are equal, and $C_\infty = G/Q$, where C_∞ is the concentration of odorant at equilibrium (mg/m^3), G is the rate of generation of odor (mg/min), and Q is the ventilation rate (m^3/min). This approach occurs at an exponentially *decreasing* rate (66, 67). In addition, depending on the mixing characteristics of the space, there may be areas in which people are exposed to greater than average odorant concentrations. Finally, the perceived odor intensity does not decrease linearly with the decrease of odorant concentration (see Section 2), but more slowly, because the intensity exponents are typically less than unity.

6.2.2   Validity of Outdoor Dispersion Models

Some of the problems of ventilation described in the preceding section have their counterparts in outdoor dispersal methods. Thus exhausted ventilation air must be replaced, often with heating or cooling; and the perceived odor reduction does not match the physical dilution. In addition, the results of efforts to predict by conventional diffusion models the extent of odor travel from a source have not been very successful. Early efforts (68) revealed extreme discrepancies in which odors were experienced at far greater distances than were predicted. More recent approaches by Högström (69) have established a model to predict the frequency of occurrence, as a function of distance from the source, of instantaneous concentrations equal to or above an absolute odor threshold level. Högström reports fairly good agreements between observed and predicted frequencies up to several kilometers from the source, but at distances between 5 and 20 km the observed frequencies are larger than those calculated by a factor of 2 or 3. The reasons for these discrepancies are not well understood, but they may involve (*a*) failure of the dispersion model to account for peak concentrations of short duration, (*b*) irregularities in threshold responses, or (*c*) participation of particulate matter.

6.3 Adsorption Systems

6.3.1 General Principles

Any gas or vapor will adhere to some degree to any solid surface. This phenomenon is called *adsorption*. When adsorbed matter condenses in the submicroscopic pores of an adsorbent, the phenomenon is called *capillary condensation*. Adsorption is useful in odor control because it is a means of concentrating gaseous odorants, thus facilitating their disposal, their recovery, or their conversion to innocuous or valuable products. When an odorous airstream is passed through a fresh adsorbent bed, almost all the odorant molecules that reach the surface are adsorbed, and desorption is very slow. Furthermore, if the bed consists of closely packed granules, the distance the molecules must travel to reach some point on the surface is small, and the transfer rate is therefore high. In practice, the half-life of airborne molecules streaming through a packed adsorbent bed is of the order of 0.01 sec, which means that a 95 percent removal can occur in about 4 half-lives, or around 0.04 sec (70). This means that the very high efficiencies required to deodorize a highly odorous airstream may be achieved with a bed of moderate depth at reasonable airflow rates.

Disposal of adsorbed odorants may be effected in any of the following ways. (*a*) The adsorbent with its adsorbate may be discarded. Because even the saturated adsorbent is relatively nonvolatile, this step seldom entails difficult problems. (*b*) The adsorbate may be desorbed and recovered, if it is valuable, or discarded; the adsorbent is recovered in either case. (*c*) The adsorbate may be chemically converted to a more easily disposable product, preferably with preservation and recovery of the adsorbent.

6.3.2 Adsorbents

Adsorbents are most significantly characterized by their chemical natures, by the extent of their surfaces, and by the volume and diameter of their pores. The most important chemical differences among adsorbents are those of electrical polarity.

Activated carbon, consisting largely of neutral atoms of a single element, presents a surface with a relatively homogeneous distribution of electrical charge. As a result, carbon exhibits less preference for highly polar molecules such as water than for most organic substances; it is therefore suitable for the overall decontamination of an airstream that contains odorous organic matter.

Table 41.1 gives ranges of surface areas and pore volumes for several different adsorbents. Among these, activated carbon is generally highest in surface area and pore volume, and these are the properties that primarily determine overall adsorptive capacity.

The pore size distributions of activated carbons are important determinants of their adsorptive properties. Pores less than about 25 Å in diameter are generally designated as micropores; larger ones are called macropores. The distinction is

Table 41.1. Surface Areas and Pore Sizes of Adsorbents

	Activated Carbon	Activated Alumina	Silica Gel	Molecular Sieve
Surface area (m^2/g)	1100–1600	210–360	750	—
Surface area (m^2/cm^3)	300–560	210–320	520	—
Pore volume (cm^3/g)	0.80–1.20	0.29–0.37	0.40	0.27–0.38
Pore volume (cm^3/cm^3)	0.40–0.42	0.29–0.33	0.28	0.22–0.30
Mean pore diameter (Å)	15–20[a]	18–20	22	3–9

[a]Refers to micropore volume (<25 Å diameter); macropores (>25 Å) not included.

Figure 41.4. Aggregated flat cell thin-bed adsorber. The small test element located on the upstream side of the cell contains carbon that is to be analyzed after some period of service for degree of saturation, thereby to predict the remaining capacity of the cell. Photograph courtesy of Connor, Inc., Danbury, CT.

important because most molecules of concern in air pollution range in diameter from about 4 to about 8.5 or 9 Å. If the pores are not much larger than twice the molecular diameter, opposite-wall effects play an important role in the adsorption process by facilitating capillary condensation. Maximum adsorption capacity is determined by the liquid packing that can occur in such small pores.

6.3.3 Equipment and Systems

When odorant concentrations are low, thin-bed adsorbers often provide a useful service life while offering the advantage of low resistance to airflow. Flat, cylindrical, or pleated bed shapes are retained by perforated metal sheets (Figures 41.4 to 41.6). Commercially available cylindrical canisters are designed for about 25 cubic feet of air per minute (cfm); the larger pleated cells handle 750 to 1000 cfm, and cells comprising aggregates of flat bed components handle 2000 cfm.

Thick-bed adsorbers are used when large adsorbing capacity is needed and when on-site regeneration is used. Bed depths are in the range of 1 to 6 ft (0.3 to 1.8 m). Design airflow capacities are up to 40,000 cfm (67,960 m^3/hr). The ratio of weight of carbon to design airflow capacity is typically about 0.5 lb/cfm [0.27 kg/(m^3)(hr)]. Typical thick-bed adsorbers, such as are used in solvent recovery systems are shown in Figure 41.7. Other systems include fluidized, rotating, and falling bed adsorbers.

The service life of the adsorbent is limited by its capacity and by the contaminating load. Provisions must therefore be made for determining when the adsorbent is saturated and for renewing it. The weight of an adsorbent is not a valid measure of its saturation because its moisture content, which depends on the relative humidity of the gas streaming through it, is likely to be variable. If mechanically

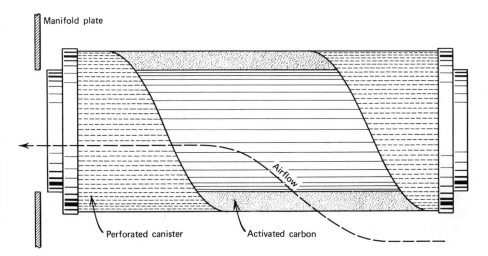

Figure 41.5. Cylindrical thin-bed canister adsorber. Schematic courtesy of Connor, Inc., Danbury, CT.

Figure 41.6.   Pleated cell, thin-bed adsorber. Photograph courtesy of Barneby-Cheney Co., Columbus, OH.

feasible, a representative element or portion of the adsorbent bed may be removed and chemically analyzed to determine the degree of saturation of the entire bed (71). In many cases a schedule for renewal of adsorbent is determined by actual deterioration of performance (odor breakthrough), or it may be based on a time schedule calculated from previous performance history.

When the adsorbent is saturated it must be replaced or regenerated. Thin-bed adsorbers, which are used for light odorant loads and thus are expected to have long service lives, are normally replaced when they are exhausted. For thick-bed adsorbers and heavy contaminant loads, it is generally economical to regenerate the adsorbent by on-site stripping with superheated steam. The adsorbate is thereby also removed and may be recovered if it is valuable. When the adsorbate is not worth recovering, either because its intrinsic value is low or because the recovery procedure is too difficult or expensive, it may nonetheless pay to regenerate the adsorbent at the site. The desorbed matter is then disposed of or destroyed. In such applications the preferred regenerating gas is hot air, at either atmospheric or reduced pressure. The desorbate may then be removed from the effluent steam by incineration or scrubbing. In effect, the adsorber serves as a vapor concentrating medium. For example, benzene at a concentration of 150 ppm in air can be effec-

Figure 41.7. Thick-bed adsorbers used in a solvent recovery system. Photograph courtesy of Union Carbide Corp., New York.

tively stripped by a carbon bed and returned to a regenerating airstream at concentrations up to about 3 percent, or 30,000 ppm (70). This ratio represents a 200-fold magnification, which greatly reduces the cost of any subsequent treatment.

The oxidation of the adsorbate by air may also occur on the adsorbent surface, preferably in the presence of a catalyst. It has been shown (72, 73) that various oxide and noble metal catalysts are effective for such applications, that hydrocarbons and oxygenates can be completely oxidized before the carbon bed itself starts to oxidize, and that repeated cycles of adsorption and catalytic oxidation can be carried out without impairing the function of the carbon.

In general, activated carbon adsorption is the method of choice for deodorizing at ambient temperature an odorous airstream whose vapor concentrations are low (ppm range or below). At higher temperatures and concentrations other methods, as described in the following sections, become progressively more attractive, and the choice of activated carbon usually must be justified by some additional benefit such as recovery of a valuable solvent. When a less efficient but cheaper method can serve to remove the bulk of contaminant organic matter from an airstream, an activated carbon adsorbent may be used as a final stage to advance the cleanup to a condition of complete deodorization.

For low emission rates of malodorous air streams, earth filters are sometimes suitable and have been used in several European sewage plants (74).

6.4   Oxidation by Air

The complete oxidation of most odorants in air results in deodorization. Some final products are odorless (H_2O, CO_2), but others (SO_2, SO_3, NO, NO_2) have higher odor thresholds than their precursors. When malodorous waste gases containing halogens are oxidized, however, the products may include the free halogens, the halogen acids, or other toxic halogen compounds such as phosgene. All such substances must be removed by scrubbing before the gas stream is discharged to the atmosphere.

In addition, partial oxidation of hydrocarbons and oxygenates often yields intermediate products that are more highly odorous than their precursors. Unsaturated aldehydes, unsaturated ketones, and unsaturated carboxylic acids, all having highly pungent odors, are frequently encountered.

The system to be used depends on the reactivity of the contaminants with oxygen, their heat content, and the concentration of oxygen in the gas stream. Many malodorous vapors, especially those formed in decomposition reactions, are relatively easy to oxidize. These include such odorants as rendering plant emissions, cooking vapors, and coffee roasting effluents. On the other hand, many hydrocarbons, such as toluene, represent products of considerable chemical evolution and are much more resistant to oxidation.

The heat content of the oxidizable vapors determines the temperature rise of the gas stream during oxidation. It has been found (75) that this rise, ΔT, can be expressed in terms of the lower explosive limit (LEL) of the vapor. Such a relationship is convenient because combustible gas meters are scaled directly in "percentage of the LEL." The expression is

$$\Delta T(°F) = 29 \times (\text{percentage of the LEL})$$

$$\Delta T(°C) = 16 \times (\text{percentage of the LEL})$$

The choice of mode of operation depends on the various factors outlined previously. In flame incineration air and fuel are used to sustain a flame (Figure 41.8). This is in effect an enclosed flare, and it operates at temperatures of 2500°F (1371°C) or higher. The fuel cost is so high that the method is practical only when the combustible gas concentration exceeds 50 percent of the LEL, thus contributing at least about 1500°F (816°C) temperature rise. Such high vapor loadings often suggest that activated carbon solvent recovery may be the better choice.

In thermal incineration the operating temperature is about 1200 to 1400°F (649 to 760°C); the half-life of reactive odorants is about 0.1 sec, and that of the more stable hydrocarbons is slightly longer. Consequently, a detention time of about 0.5 sec is sufficient for most odor control objectives. Figure 41.9 schematically represents a thermal incinerator.

Catalytic incineration is designed to give performance like that of a thermal incinerator but at a lower temperature and faster detention time. The advantages result from the action of a solid catalyst that consists of a noble metal alloy or in some cases a metallic oxide mixture. Operating temperatures are typically in the 600 to 900°F (316 to 482°C) range.

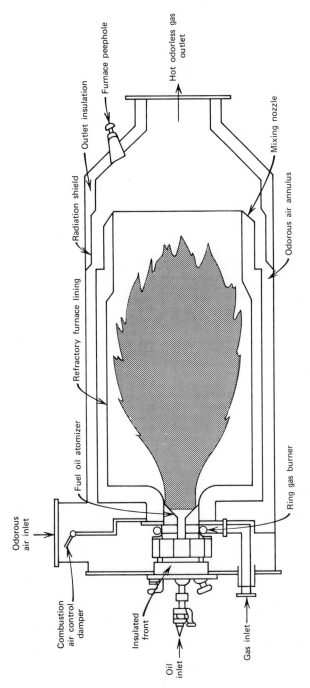

Odorous air inlet

Combustion air control damper

Insulated front

Oil inlet

Gas inlet

Fuel oil atomizer

Ring gas burner

Refractory furnace lining

Radiation shield

Outlet insulation

Furnace peephole

Hot odorless gas outlet

Mixing nozzle

Odorous air annulus

Figure 41.8. Direct-fired air heater. Schematic courtesy of Peabody Engineering Co.

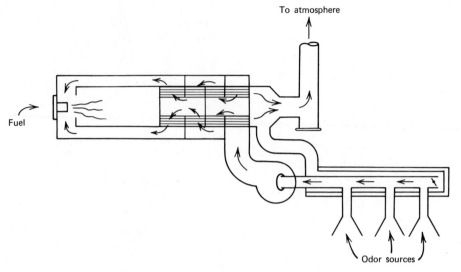

Figure 41.9. Thermal incinerator system.

The loss of catalyst activity, which determines catalyst life, hence equipment maintenance costs, is related to three major factors: (*a*) the presence of catalyst poisons (such as metallic or organometallic vapors) in the odorous air, (*b*) the obstruction of catalyst surface by deposit of inorganic materials such as silicates from silicone resins, and (*c*) the mechanical loss of catalyst through abrasion by solid particles in the air stream.

For air free from particles and metal-containing vapors, a long catalyst life may be realized, and some installations are reported (76) to have given more than 23,000 hr of service without catalyst regeneration. In other cases, however, loss of activity may be quite rapid. A pilot run before installing full-scale equipment is generally advisable.

6.5 Liquid Scrubbing

Liquid scrubbing is widely used for odor control. The mechanisms for its action include (*a*) solution of the odorous vapors into the scrubber liquid, (*b*) condensation of odorous vapors by the cooling action of the liquid, (*c*) chemical reaction of the odorants with the scrubber liquid to yield an innocuous product, and in some cases (*d*) adsorption of odorant onto particles suspended in the scrubber liquid.

The physical actions of solution, condensation, and absorption generally approach equilibrium conditions that still involve a significant partial pressure of the odorant vapors. Therefore such actions are only partially effective in deodorizing gas streams. For example, the water scrubbing of a gas stream containing ammonia and other nitrogenous odorants will remove much of the ammonia but only a small portion of some of the organic nitrogen compounds, which may be extremely odorous.

Chemical conversions in scrubbers are generally oxidations or acid–base neutralizations. Because the latter category involves very rapid proton exchange, the important determinant of effective action is the equilibrium condition. Thus soluble acidic odorants such as hydrogen sulfide and phenol are effectively scrubbed by basic solutions, and basic odorants such as ammonia and soluble amines can be neutralized by acids.

Reagents for chemical oxidation include potassium permanganate, sodium hypochlorite, chlorine dioxide, and hydrogen peroxide. In general, such oxidations are much slower than flame reactions and require considerably more residence time for effective odor control. Furthermore, the absolute rate of vapor removal is virtually independent of the chemical nature of the odorants in the case of physical adsorption, moderately dependent in the case of incineration, and extremely dependent in the case of ambient temperature oxidation. These differences are significant because a ratio of, say, 5 : 1 in the rates of removal of two vapors has only marginal significance when dealing with hundredths of seconds in a granular bed; it has more significance in the ranges of tenths of seconds in a flame, and it can well be of overriding importance in the much slower, ambient temperature reactive systems.

The various oxidants cited earlier differ in their reaction pathways for different odorants, and it is meaningless to specify which one is "best." Potassium permanganate (77) is generally used under mildly alkaline conditions (pH $\approx$ 10), where its reduction product is the insoluble manganese dioxide (MnO_2), which poses a waste disposal problem. However the MnO_2 slurry acts as an oxidation catalyst (78), and thus makes it possible for air to participate in the oxidation of some very reactive odorants, such as mercaptans and some amines. Sodium hypochlorite has been found to react more rapidly with rendering plant emissions and has been recommended for such applications (79).

Figure 41.10 represents a typical scrubber installation for control of rendering plant odors. The malodorous gases enter the first stage of the system, where they pass through water in a venturi scrubber. This step removes particulate matter and cools and saturates the gas stream. The gases are then passed through a packed bed, where they contact a countercurrent stream of scrubbing liquid containing a permanganate or hypochlorite oxidant. Malodorous gases are absorbed and oxidized. The scrubbed gas stream leaves the packed bed, flows through a mist elimination section, and is exhausted to the atmosphere. The depleted scrubbing liquid is collected and recycled to the scrubber. A portion of the depleted scrubbing solution is continuously removed from the recycle stream and replaced with makeup water and chemicals. The bleed stream is combined with the wastewater from the venturi scrubber and sent to a sewage treatment facility.

6.6   Ozonation

Ozone is a reactive ambient temperature oxidizing agent that has been used for gas phase conversion of malodorants to less offensive products. The toxicity of ozone renders it unfit for use in occupied spaces, however. Ozone-producing devices

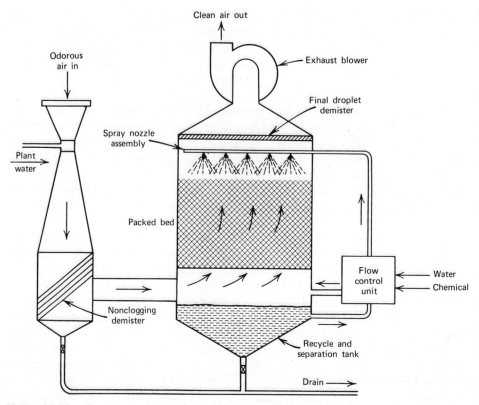

Figure 41.10. Two-stage chemical scrubbing system. Schematic courtesy Environmental Research Corp., St. Paul, MN.

have been offered for indoor use, but they generate such low concentrations that their effect in controlling malodorants is nil and the injury they cause is probably too little to be evident.

For controlling odorants before they are discharged to the outdoors, ozone is introduced into the odorous airstream in concentrations of 10 to 30 ppm, and a reaction time of 5 sec or more is provided during passage of the stream through a stack or special detection chamber. Ozonation is chemically selective and is not equivalent to thermal incineration or catalytic oxidation, which converts malodorants to their ultimate oxidation products. Ozone reacts with mercaptans to produce sulfones and sulfoxides, with amines to produce amine oxides, and with unsaturated hydrocarbons to produce aldehydes, ketones, and carboxylic acids. The reactions with mercaptans and amines are often so rapid that the available detention times are sufficient to provide considerable conversion, and any degree of oxidation of these malodorous compounds is a great improvement. Consequently ozonation has been reported (80) to be effective in odor control from sources such as sewage treatment plants. For unsaturated hydrocarbons or oxygenates, however, odor control performance is less reliable. For example, ozonation of a mixture of styrene and vinyl toluene from resin operations yields a mixture of aromatic aldehydes with a distinct cherrylike odor. Attempts to deodorize acrylic esters by ozonation during the detention time available in typical industrial stacks have been unsuccessful.

6.7   Odor Masking and Counteraction

When a mixture of odorants is smelled, the odor qualities of the components may be perceived separately or may blend into one quality such that the individual components cannot be recognized. The odor mixture may be perceived as stronger than, equal to, or less than the sum of the odor intensities of the components. Likewise, the odor intensity of any single component of such a mixture may be stronger than, equal to, or less than the odor intensity of that component smelled alone. When referring to the mixture as a whole, these effects are designated hyperaddition, complete addition, and hypoaddition, respectively. When referring to any individual component, the effects are called synergism, independence, and antagonism, respectively.

Interaction effects on odor intensity have been studied for some two-component mixtures (81). For example, the perceived intensity of vapor phase mixtures of various concentrations of pyridine and a second component such as linalyl acetate, linalool, or lavandin oil, is less than the sum of the perceived intensities of the two components smelled alone. The addition of the second component to a relatively weak stimulus of pyridine causes an increase in overall odor intensity, but the addition of the same amount of the second component to a relatively intense stimulus of pyridine causes a reduction in overall odor intensity. Mixtures of 1-propanol and n-amyl butyrate have been reported to interact similarly (82). These data suggest the existence of complex interactions in the perceived intensity of odorous mixtures.

When mixtures of odor components are perceived as a single blend, a vector summation model of odor interaction has been suggested as a means of predicting the odor intensity of mixtures of malodorants such as dimethyl disulfide, dimethyl monosulfide, hydrogen sulfide, methyl mercaptan, and pyridine (83–85). For components equal in perceived intensity when smelled alone, a direct proportionality has been reported (83, 84) between odor intensity of the mixtures and the arithmetic sum of the odor intensities of the components.

The interpretation of the application of these phenomena to practical odor control objectives presents difficulties, and the common industrial terminology does not make matters easier. *Counteraction* has been used to connote reduction of intensity, although it is not always clear whether this refers to the blend or to the malodorant alone. *Cancellation* means reduction to zero intensity, a phenomenon that has never been convincingly documented. *Masking* refers to a change in odor quality that makes the malodorant unrecognizable; the connotation of concealment has made the term unpopular. In spite of this variety of terminology, the odor control practices to which the words refer are operationally indistinguishable. The materials used are selected from industrially available high intensity odorants, often from by-product sources. They may be applied in undiluted form or as an aqueous emulsion. They may be incorporated into the process or product that constitutes the malodorous source, sprayed into a stack or over a stack exit, or vaporized over a large outdoor area.

The general method has the important practical advantages of low initial equipment costs, negligible space requirements, and greater freedom from the necessity of confining the atmosphere into a closed space for treatment. It is not applicable when irritation or toxicity accompany odor.

Clearly it is very difficult to estimate the effectiveness of this category of odor control methods. Not the least of the problems is that of choosing a criterion for evaluation. Furthermore, industrial or commercial installations are not designed to be controlled experiments. Instead, they are generally combined with other beneficial actions, such as improvements in sanitation and general housekeeping, to maximize the opportunities for odor reduction. As a result, information concerning the performance of such systems consists entirely of descriptions of actual operations and other anecdotal reports.

6.8   Epilogue

Odor control has long been considered to be more or less equivalent to the reduction of gaseous emissions that happen to be odorous. Furthermore, if the criterion of odor control is the reduction of odorant concentration to the threshold level, the efficiency required of a control device can readily be calculated. It is true that the ratio of high source concentrations to some very low odor threshold levels led to rather unprecedented requirements for the performance of gas-cleaning devices, but the usual remedy was to count on atmospheric dispersal to help solve the problem. Then, in the schematic diagram of Figure 41.11, we assume an odorant concentration C_s at a source that is treated by an abatement device (dashed lines) of efficiency E_a to discharge the abated concentration C_a to the atmosphere. Atmospheric dispersal of efficiency E_d further reduces the odorant concentration to the target or threshold value C_t before it reaches ground level. Then,

$$E_a = \frac{C_s - C_a}{C_s}$$

$$E_d = \frac{C_a - C_t}{C_a}$$

and the overall efficiency is

$$E = \frac{C_s - C_t}{C_s}$$

$$= E_a + E_d - E_a E_d$$

Now, if we assume that the required overall efficiency can be measured by human judges exposed to source samples of progressively greater dilution, and that

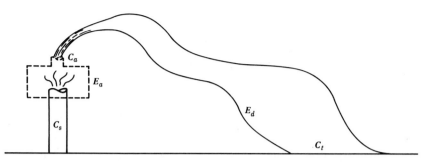

Figure 41.11. Overall efficiency of control of gaseous emission is $E = E_a + E_d - E_a E_d$, where E_a is the efficiency of a vapor control device and E_d is the efficiency of vapor dilution by atmospheric dispersal.

the atmospheric dilutions can be predicted by dispersion models, we can readily calculate the required design efficiency of the abatement equipment to be applied to the source.

We have pointed out the many deficiencies in this sequence, such as the inadequacies of dispersion models, the inconstancy of the odor threshold, the variability of the exponent of odor intensity functions and the influence of odor quality on human reactions to odor. The technical and anecdotal literature offers many instances (68) in which this strategy fails to predict human responses to unpleasant odors, in terms of either the distance at which odors can be detected or the affective reactions or overt social initiatives they elicit.

The recognition that odor control effectiveness is most truly manifested by its reduction of human annoyance has also, unfortunately, been of little help. Although such recognition does allow for the consideration of odor masking and counteraction methods that other strategies do not accommodate, it fails to offer a basis for establishing design criteria for odor abatement systems. Furthermore, the measurement of economic effects of odors, such as the depression of property values or the reduction in work efficiency, has not been successfully accomplished in any precise manner.

On the other hand, methods of controlling gaseous emissions are being improved continually by means such as the design of more efficient scrubbers, the development of new catalysts, and advances in the effectiveness of systems for heat recovery and the regeneration of adsorbent beds. In addition, the continued refinement of instrumented monitoring systems serves all these abatement methods. Thus although there has been no spectacular breakthrough such as the discovery of a universal ambient temperature oxidation catalyst, the options available for odor control are better than ever.

The overall result has been that odor control strategies are determined in large measure by the available technology, which is often very good. Attempts to predict how the control of gaseous emissions reduces odors are described by two terms currently in vogue. The first is *dose–response relationships*, which purport to relate odorous emissions at the source to odor problems in the community or the workspace. However, we have seen that such relationships have not been quantitatively established. Instead, they are generally approximated in the give-and-take fashion that is typical of political and social processes in an open society—by persuasion, by negotiation, and sometimes by legal adjudication. The second popular term is *cost–benefit analysis*, which purports to predict how much odor control each dollar will buy. It should be obvious that human "benefit" from odor control is no less difficult to quantify than human "response." However, it is much easier to compare costs of alternative abatement methods that yield the same reductions of odorant concentrations, and such comparisons are valid and important. The main findings of the latter approaches have been that as the costs of energy continue to increase, the search for more sophisticated approaches to odor control, starting with a full review of the role of the process itself, becomes ever more urgent.

REFERENCES

1. G. T. Fechner, *Elemente der Psychophysik*, Breitkopf and Härterl, Leipzig, 1860. English translation of Vol. 1 by H. E. Adler (D. H. Howes and E. G. Boring, Eds.), Holt, Rinehart and Winston, New York, 1966.

2. C. E. Shannon and W. Weaver, *The Mathematical Theory of Communication*, University of Illinois Press, Urbana, 1949.

3. J. A. Swets, W. P. Tanner, Jr., and T. G. Birdsall, *Psychol. Rev.*, **68**, 301 (1961).

4. T. Engen, "Psychophysics. 1. Discrimination and Detection," in *Woodworth and Schlosberg's Experimental Psychology*, 3rd ed., J. W. Kling and L. A. Riggs, Eds., Holt, Rinehart and Winston, New York, 1971, pp. 11–46.

5. F. A. Young and D. F. Adams, *Proceedings of the 74th Annual Convention of the American Psychological Association*, New York, 1966, 75.

6. G. Steinmetz, G. T. Pryor, and H. Stone, *Percept. Psychophys.*, **6**, 142 (1969).

7. B. Johansson, B. Drake, B. Berggren, and K. Vallentin, *Lebensm.-Wiss. Technol.*, **6**, 115 (1973).

8. G. Leonardos, D. Kendall, and N. Barnard, *J. Air Pollut. Control Assoc.*, **19**, 91 (1969).

9. E. M. Adams, "Physiological Effects," in: *Air Pollution Abatement Manual*, Manufacturing Chemists Association, Washington, D.C., 1951, Chapter 5.

10. E. Fluck, *J. Air Pollut. Control Assoc.*, **26**, 795 (1976).

11. A. Turk and J. Turk, "The Purity of Odorant Substances," in *Methods in Olfactory Research*, D. G. Moulton, A. Turk, and J. W. Johnston, Eds., Academic Press, New York, 1975, pp. 63–73.

12. A. M. Hyman, *Sensory Processes*, **1**, 273 (1977).

13. T. Engen, "Psychophysics. II. Direct Scaling Methods," in *Woodworth and Schlosberg's Experimental Psychology*, 3rd ed., J. W. Kling and L. A. Riggs, Eds., Holt, Rinehart and Winston, New York, 1971, pp. 47–86.

14. S. S. Stevens, in: *Psychophysics: Introduction to Its Perceptual, Neural and Social Prospects*, G. Stevens, Ed., Wiley, New York, 1975.

15. S. S. Stevens, *Psychol. Rev.*, **64**, 153 (1957).

16. G. Ekman and L. Sjöberg, *Ann. Rev. Psychol.*, **16**, 451 (1965).

17. T. Engen, "Report from the Psychological Laboratory," University of Stockholm, 1961, p. 106.

18. W. S. Cain, *ASHRAE Trans.*, **80**, 53 (1974).

19. K. E. Henion, *Psychon. Sci.*, **22**, 213 (1971).

20. H. Stone, G. T. Pryor, and G. Steinmetz, *Percept. Psychophys.*, **12**, 501 (1972).

21. H. G. Haring, "Vapor Pressures and Raoult's Law Deviations in Relation to Odor Enhancement and Suppression," in *Human Responses to Environmental Odors*, A. Turk, J. W. Johnston, Jr., and D. G. Moulton, Eds., Academic Press, New York, 1974, pp. 199–226.

22. B. Berglund, U. Berglund, G. Ekman, and T. Engen, *Percept. Psychophys.*, **9**, 379 (1971).

23. M. J. Mitchell and R. A. M. Gregson, *Percept. Mot. Skills*, **26**, 720 (1968).

24. M. J. Mitchell and R. A. M. Gregson, *Quart. J. Exp. Psychol.*, **22**, 301 (1970).

25. M. J. Mitchell and R. A. M. Gregson, *J. Exp. Psychol.*, **89**, 314 (1971).

26. W. S. Cain, *Ann. NY Acad. Sci.*, **237**, 28 (1974).

27. R. L. Doty, *Physiol. Behav.*, **14**, 855 (1975).

28. W. S. Cain, *Sens. Processes*, **1**, 57 (1976).

29. G. Steinmetz, G. T. Pryor, and H. Stone, *Percept. Psychophys.*, **8**, 327 (1970).

30. W. S. Cain, *Percept. Psychophys.*, **7**, 271 (1970).

31. G. T. Pryor, G. Steinmetz, and H. Stone, *Percept. Psychophys.*, **8**, 331 (1970).

32. W. S. Cain and T. Engen, "Olfactory Adaptation and the Scaling of Odor Intensity," in *Olfaction and Taste*, Vol. 3, C. Pfaffmann, Ed., Rockefeller University Press, New York, 1969, pp. 127–141.

33. E. P. Köster, *Adaptation and Cross-Adaptation in Olfaction*, Bronder-Offset, Rotterdam, 1971.

34. A. Dravnieks, "Theories of Olfaction," in *Chemistry and Physiology of Flavors*, H. W. Schultz, E. A. Day, and L. M. Libbey, Eds., Avi Publishing Co., Westport, CT, 1967, pp. 95–118.

35. R. H. Wright, *J. Appl. Chem.*, **4**, 611 (1954).

36. R. H. Wright, *Ann. NY Acad. Sci.*, **116**, 552 (1964).

37. E. E. Langenau, "Correlation of Objective-Subjective Methods as Applied to the Perfumery and Cosmetics Industry," in: *Correlation of Subjective-Objective Methods in The Study of Odors and Taste*, American Society for Testing and Materials Special Publication No. 440, ASTM, Philadelphia, 1968, pp. 71–86.

38. J. T. Davies, *Symp. Soc. Exp. Biol.*, **16**, 170 (1962).

39. J. T. Davies, *J. Theor. Biol.*, **8**, 1 (1965).

40. W. R. Roderick, *J. Chem. Educ.*, **43**, 510 (1966).

41. A. T. Rossano and R. R. Ott, "The Relationship Between Odor and Particulate Matter in Diesel Exhaust," paper presented at the Pacific Northwest Section Meeting of the Air Pollution Control Association, Portland, Ore., November 5–6, 1964.

42. W. A. Quebedeaux, *Air Repair*, **4**, 141 (1954).

43. A. Goetz, *Int. J. Air Water Pollut.*, **4**, 168 (1961).

44. S. S. Schiffman, *Science*, **185**, 112 (1974).

45. K. R. Brower and R. Schafer, *J. Chem. Educ.*, **52**, 538 (1975).

46. American Society for Testing and Materials, *Reviews of Correlations of Objective-Subjective Methods in the Study of Odors and Taste*, Special Technical Publication No. 451, ASTM, Philadelphia, 1969.

47. R. A. Schmall, *Atmospheric Quality Protection Literature Review—1972*, Technical Bulletin No. 65, National Council of the Paper Industry for Air and Stream Improvement, New York, 1972.

48. R. A. Rasmussen, *Am. Lab.*, **4**(12), 55 (1972).

49. C. Weurman, "Sampling in Airborne Odorant Analysis," in *Human Responses to Environmental Odors*, A. Turk, J. W. Johnston, and D. G. Moulton, Eds., Academic Press, New York, 1974, pp. 263–328.

50. C. A. Clemons and A. P. Altshuller, *J. Air Pollut. Control Assoc.*, **14**, 407 (1964).

51. W. D. Connor and J. S. Nader, *Am. Ind. Hyg. Assoc. J.*, **25**, 291 (1964).

52. K. Springer, "Combustion Odors," in: *Human Responses to Environmental Odors*, A. Turk, J. W. Johnston, and D. G. Moulton, Eds., Academic Press, New York, 1974, pp. 227–262.

53. A. Turk, *Atmosph. Environ.*, **7**, 967 (1973).

54. A. Turk, J. I. Morrow, and B. E. Kaplan, *Anal. Chem.*, **34**, 561 (1962).

55. A. Turk, J. I. Morrow, S. H. Stoldt, and W. Baecht, *J. Air Pollut. Control Assoc.*, **16**, 383 (1966).

56. A. Dravnieks and B. K. Krotoszynski, *J. Gas Chromatogr.*, **6**, 144 (1968).

57. American Society for Testing and Materials, *Manual on Sensory Testing Methods*, Special Publication No. 434, ASTM, Philadelphia, 1968.

58. J. Wittes and A. Turk, "The Selection of Judges for Odor Discrimination Panels," in *Correlation of Subjective–Objective Methods in the Study of Odors and Taste*, American Society for Testing and Materials Special Publication No. 440, ASTM Philadelphia, 1968, pp. 49–70.

59. W. C. L. Hemeon, *J. Air Pollut. Control Assoc.*, **18**, 166 (1968).

60. A. Turk, *Pollut. Eng.*, **4**, 22 (1972).

61. *The Air Over Louisville*, Technical Report of the Public Health Service, U.S. Department of Health, Education and Welfare, 1956–1957.

62. A. Turk, J. T. Wittes, L. R. Reckner, and R. E. Squires, *Sensory Evaluation of Diesel Exhaust Odors*, National Air Pollution Control Administration, Publication No. AP-60, 1970.

63. E. Jonsson, "Annoyance Reactions to Environmental Odors," in *Human Responses to Environmental Odors*, A. Turk, J. W. Johnston, and D. G. Moulton, Eds., Academic Press, New York, 1974, pp. 330–333.

64. Copley International Corp., "A Study of the Social and Economic Impact of Odors, Phase I, 1970; Phase II, 1971, Phase III, 1973," U.S. Environmental Protection Agency, Report of Contract No. 68-02-0095.

65. R. D. Flesh and A. Turk, "Social and Economic Effects of Odors," in *Industrial Odor Technology Assessment*, P. N. Cheremisinoff and R. A. Young, Eds., Ann Arbor Science Publishers, Ann Arbor, MI, 1975, pp. 57–74.

66. A. Turk, *J. ASHRAE*, October 1963.

67. A. Turk, "Concentrations of Odorous Vapors in Test Chambers," in *Basic Principles of Sensory Evaluation*, American Society for Testing and Materials Special Publication No. 433, ASTM, Philadelphia, 1968, pp. 79–83.

68. H. C. Wohlers, *Int. J. Air Water Pollut.*, **7**, 71 (1963).

69. U. Högstrom, "Transport and Dispersal of Odors," in *Human Responses to Environmental Odors*, A. Turk, J. W. Johnston, and D. G. Moulton, Eds., Academic Press, New York, 1974, pp. 164–198.

70. A. Turk, "Adsorption," in *Air Pollution*, Vol. 4, 3rd ed., A. C. Stern, Ed., Academic Press, New York, 1977.

71. A. Turk, H. Mark, and S. Mehlman, *Mater. Res. Stand.*, **9**, 24 (1969).

72. J. Nwanko and A. Turk, *Ann. NY Acad. Sci.*, **237**, 397 (1974).

73. J. Nwanko and A. Turk, *Environ. Sci. Technol.*, **9**, 846 (1975).

74. H. L. Bohn, *J. Air Pollut. Control Assoc.*, **25**, 953 (1975).

75. R. J. Ruff, "Catalytic Method of Measuring Hydrocarbon Concentrations in Industrial Exhaust Fumes," in: American Society for Testing and Materials Special Publication No. 164, ASTM, Philadelphia, 1954, p. 13.

76. R. J. Ruff, *Am. Ind. Hyg. Assoc. Quart.*, **14**, 183 (1953).

77. H. S. Posselt and A. H. Reidies, *Ind. Eng. Chem. Prod. Res. Develop.*, **4**, 48 (1965).

78. D. F. S. Natusch and J. R. Sewell, "A New Solid Filter for Industrial Odors," in *Proceedings of the Second International Clean Air Congress*, Academic Press, New York, 1971, p. 948.

79. T. R. Osag and G. B. Crane, *Control of Odors from Inedibles—Rendering Plants*, U.S. Environmental Protection Agency Publication No. 450/1-74-006, 1974.

80. C. Nebel, W. J. Lehr, H. J. O'Neill, and T. C. Manley, *Plant Eng.*, March 21, 1974.

81. W. S. Cain and M. Drexler, *Ann. NY Acad. Sci.*, **237**, 427 (1974).

82. W. S. Cain, *Chem. Sens. Flav.*, **1**, 339 (1975).

83. B. Berglund, U. Berglund, and T. Lindvall, *Acta Psychol.*, **35**, 255 (1971).

84. B. Berglund, U. Berglund, T. Lindvall, and L. T. Svensson, *J. Exp. Psychol.*, **100**, 29 (1973).

85. B. Berglund, *Ann. NY Acad. Sci.*, **237**, 35 (1974).

CHAPTER FORTY-TWO

Fire and Explosion Hazards of Combustible Gases, Vapors, and Dusts

Joseph Grumer*

1 GENERAL CONSIDERATIONS

Explosions and ensuing fires of combustible gases, vapors, and dusts are among the major hazards in homes, commerce, and industry. Understanding of these hazards and awareness of means of combating and preventing them are obvious necessities. The earlier literature is voluminous, and a total extract of it is outside the scope of this chapter. Here I want to update three prior reviews, which have summarized the literature into useful texts for those concerned with combating hazards due to the possible pressures of combustible gases, vapors, and dusts mixed with air; some consideration is also given to hazards due to these combustibles mixed with oxygen. The reader is referred to these reviews (1–3) for original references to the literature prior to 1962. To conserve space, this overview omits references to original sources of quoted data, provided the origins are cited in References 1, 2, or 3. Furthermore the present writing undertakes particularly to update Reference 2—namely, Chapter XVI, Section 1, by G. W. Jones, and Section 2 by I. Hartmann, of the second revised edition of Patty's *Industrial Hygiene and Toxicology*—and closely follows the organization of that earlier work. Other recommended references are those of the National Fire Protection Association, such as References 4 and 5.

*Deceased.

Patty's Industrial Hygiene and Toxicology, Fourth Edition, Volume 1, Part B, Edited by George D. Clayton and Florence E. Clayton.
ISBN 0-471-50196-4 © 1991 John Wiley & Sons, Inc.

Certain discussions of terminology are in order. Fires generally involve diffusion flames of a liquid or solid fuel bed in which there is no premixing of fuel and air before combustion (e.g., a wood fire or an oil tank fire). Explosion constitutes a sudden release of pressure, not necessarily as a consequence of combustion. For example, an explosion occurs when compressed air bursts its container suddenly. As a consequence, "explosive limits" is a poor term when applied to limits involving combustion; "flammable limits" or "limits of flammability" is preferable.

1.1 Limits of Flammability

Limits of flammability are the extremes in concentration between which homogeneous gas–air mixtures, vapor–air, or dust–air mixtures can be burned when subjected to an ignition source of adequate temperature and energy. For example, trace amounts of methane in air can be readily oxidized on a heated surface, but a flame will propagate from an ignition source at ambient temperatures and pressures only if the surrounding mixture contains at least 5 but less than 15 volume percent methane. The more dilute mixture is known as the lower limit, or the combustible-lean limit mixture; the more concentrated mixture is known as the upper limit, or combustible-rich limit mixture. In practice, the limits of flammability of a particular premixed system of gaseous combustibles and oxidant are affected by temperature, pressure, direction of flame propagation, and prefential diffusion due to gravity. The flammability limits of premixed combustible dusts and oxidants are affected by additional factors such as particle size.

It is impractical to differentiate between mixtures on the basis of the amount of violence produced, because more than mixture composition is involved. If confined in a long tube, open at one end and ignited at the open end, mixtures just within the limits of flammability will propagate flame quietly and slowly through the tube. (Propagation is usually at a uniform speed, and the speed for a given concentration of combustible in oxidant is faster for upward than for downward flame propagation.) If this mixture is confined in a large enough closed vessel and ignited, however, flame propagation will accelerate and propagate at speed several times that in the open tube, particularly if the mixture is initially in gentle or turbulent motion; pressures up to 30 psi or more may be attained. This is only to say that comfort should not be taken in expectations that at worst only a limit composition can be encountered in a specific operation. Any unwanted flammation is inherently disastrous.

Admittedly the rate of flame propagation through a flammable mixture and the maximum attainable pressure depend on a number of factors, including temperature, pressure, and mixture composition. For a given fuel–oxidant system, flame speed and peak pressure are relatively low at the limits of flammability and high near stoichiometric. However, destruction can be fearful even from explosions due to relatively mild flames. Consider that at the lean limit of deflagration of 5 percent methane in air, peak pressures attainable in an adiabatic closed container are theoretically about 5 atmospheres or about 75 psi. This is enough to cause failure of structural elements, as evidenced by the critical peak overpressures given in Table 42.1.

Table 42.1. Conditions of Failure of Peak Overpressure-sensitive Elements (3)

Structural Element	Failure	Approximate Incident Blast Overpressure (psi)
Glass windows, large and small	Usually shattering, occasional frame failure	0.5-1.0
Corrugated asbestos siding	Shattering	1.0-2.0
Corrugated steel or aluminum paneling	Connection failure, followed by buckling	1.0-2.0
Wood siding panels, standard house construction	Usually failure occurs at main connections, allowing a whole panel to be blown in	1.0-2.0
Concrete or cinderblock wall panels, 8 or 12 in thick (not reinforced)	Shattering of the wall	2.0-3.0
Brick wall panel, 8 or 12 in. thick (not reinforced)	Shearing and flexure failures	7.0-8.0

The data in this survey are based on well-mixed, flammable, homogeneous mixtures. In homes, commerce, industry, and nature, heterogeneous single-phase gas and multi-phase (gas, liquid, and dust) incompletely mixed, locally flammable mixtures probably occur more frequently and are more important than are homogeneous mixtures. The occurrence and accounting of the hazardous characteristics of such heterogeneous mixtures is unfortunately still unpredictable in adequate detail. It is important to recognize, for example, that heterogeneous mixtures can ignite at overall concentrations (as if completely mixed) that would normally be nonflammable. For example, 1 liter of methane can form a flammable mixture with air near the top of a 100-liter container, although a 1.0 volume percent, nonflammable mixture would result if complete mixing occurred; only about 6 liters of the 100 liters of available air need mix with the methane in this hypothetical situation to form a flammable mixture. Once ignition occurs in a flammable zone in a given volume, unpredictable signatures of pressure, temperature, and composition rule, with possibly disastrous results. Not only lighter than air, roof-hugging flammable gases such as hydrogen and methane present layering hazards. Flammable gases that are heavier than air, such as propane, butane, and gasoline vapors, can produce disasters by way of ground-hugging layers. This is an important consideration, because layering can occur in both stationary and flowing mixtures. Consequently, good safety practice prohibits closely skirting flammability limits of homogeneous mixtures. Under the Mine Safety and Health Act of 1969 (6), precautionary measures such as shutting down machinery are mandated at methane gas concentrations in exhaust air of 20 percent of the lower flammability limits, and evacuation of the mine is required when gas concentration in exhaust air reaches 30 percent of this limit. Such margins avoiding hazardous concentrations (7) are well advised, though not necessarily sufficient in all instances.

Combustion may propagate through a mixture by deflagration or by detonation. If the propagation rate relative to the unburned gas is subsonic, the process is deflagration, and if the rate is supersonic, the process is detonation. If it is subsonic, the pressure caused by the deflagration process will equalize at the speed of sound throughout the enclosure in which combustion is taking place. The pressure drop across the flame front will be relatively small, generally on the order of millimeters of water pressure or less. If the flame propagation rate is supersonic, the pressure caused by the detonation process will equalize less rapidly than the flame propagation rate, and there will be an appreciable pressure drop across the flame front. Moreover, with most of air and combustibles at ordinary temperatures, the ratio of peak to initial pressure due to a deflagration within a closed vessel seldom exceeds 8:1 but that due to a detonation may be more than 40:1 (3). The pressure buildup is especially great when detonation follows a large pressure rise due to deflagration. The distance required for a deflagration to transit to a detonation varies with many factors, but the most important is the strength of the ignition; with a sufficiently powerful ignition, detonation may occur immediately on ignition, even in open air (3). Detonation limits are generally not as lean as the lean limit and not as rich as the rich limit for upward flame propagation. Available data suggest that the limits of detonability are within the limits of downward flame propagation (8).

The peak pressure that may be expected theoretically in an adiabatic situation in a closed vessel with central ignition is given by:

$$P_b = P_1 \frac{\overline{M_i} T_b}{M_b T_i} \tag{1}$$

where P is pressure, M is molecular weight, and T is absolute temperature. The overbar indicates an average value, and the subscripts i and b indicate initial and burned gas parameters, respectively. The minimum elapsed time (in milliseconds) to reach peak pressure has been estimated (3) for paraffin hydrocarbons and gasoline vapors to equal 75 times the cube root of the volume (in cubic feet) of the enclosure.

2 COMBUSTIBLE GASES AND VAPORS

2.1 Factors Affecting the Limits of Flammability

Some years ago the U.S. Bureau of Mines adopted a standard apparatus for limit-of-flammability determinations (1). This apparatus (Figure 42.1) was originally designed for determining limits of gases and vapors of liquids that are volatile enough at room temperatures to give flammable mixtures. In Figure 42.1, a is the 5-cm-i.d. glass tube in which the mixture is tested. The lower end of the glass tube is closed by a lightly lubricated ground-glass plate b sealed with mercury c. After evacuation by pumping through tube j, gas is introduced from a gas holder through stopcock, r, or if a vapor is to be tested, it is drawn from its liquid in container p.

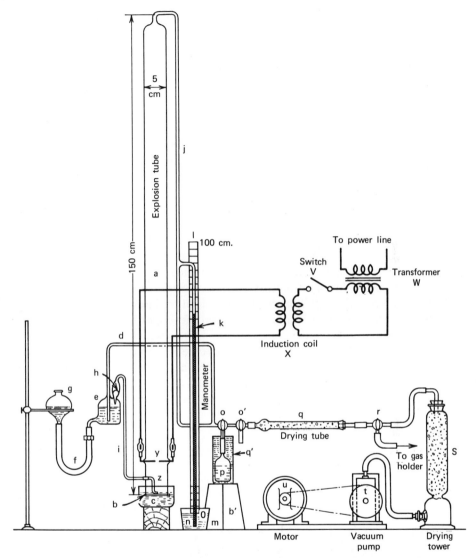

Figure 42.1. Apparatus for determining the limits of flammability of gases and vapors (1). Schematic courtesy of Bureau of Mines, U.S. Department of the Interior.

The amount of gas or vapor introduced is measured by the manometer k. Air or "other atmosphere" is then admitted through the drying tube q until atmospheric pressure is reached. Gaseous oxidant and fuel are thoroughly mixed by raising and lowering the mercury vessel g repeatedly for 10 to 30 min, depending on the density differences of the components in tube a. The mercury seal is then removed, glass plate b is slid off the tube, and the flammability of the mixture is tested by sparking at y or by passing a small flame across the open end of the tube.

If ignition is by sparking, the energy in the spark must be adequate. Figure 42.2 illustrates the effect of mixture composition on the electrical spark energy requirements for ignition of methane–air mixtures (3, 9). A 0.2-mJ spark cannot ignite any methane–air mixture at atmospheric pressure and room temperature. A 1-mJ spark can ignite mixtures containing between 6 and 11.5 volume percent methane; stronger sparks are needed to ignite leaner than 6 volume percent and richer than 11.5 volume percent flammable methane–air mixtures. Limit mixture compositions that depend on the ignition source strength are not limits of flammability; they are limits of ignitability, or ignitability limits. Limit mixtures, compositions that are essentially independent of the ignition source strength and support flames capable of propagating beyond the region of influence of the ignition source, are limits of flammability.

Flammability limit determinations must be made in apparatus large enough that flame quenching by vessel walls is practically eliminated. A 5-cm-i.d. vertical tube has been found to be large enough for determinations of flammability limits of paraffin hydrocarbons (methane, ethane, etc.) at atmospheric pressure and room temperature. However, a 5-cm tube is not large enough for many halogenated and other compounds, or for paraffin hydrocarbons at very low temperatures and pressures. For example, the lower flammability limit of trichloroethylene–air mixtures had to be measured in a 17.8-cm-i.d. tube and the upper limit in a 20-cm sphere (3) to attain apparatus-independent flammability limits. Recently the Bureau of Mines apparatus for flammability limit measurements was changed to incorporate 10- and 30-cm-i.d. tubes (10). Furthermore, the apparatus must be long enough to ensure continued propagation of flame beyond the zone wherein heat from the ignition source contributes significantly to flame propagation. An apparatus about

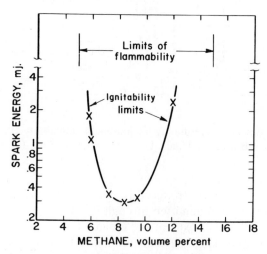

Figure 42.2. Ignitability curve and limits of flammability for methane-air mixtures at atmospheric pressure and 26°C (3). Graph courtesy of Bureau of Mines, U.S. Department of the Interior.

1.25 m long or more is generally long enough to allow observation of flammability limits free of influence of the ignition source. If limits are observed within very short distances close by the ignition source, the apparent limits can be forced leaner of the lower flammability limit and richer of the upper flammability limit. Therefore in some practical applications, the composite hazard due to ignition source and combustibles should be considered. Vessel shape (e.g., spherical vs. cylindrical) as well as size can affect observed limits if the vessel's smallest dimension is less than needed to yield apparatus-free flammability limits; spheres about 10 cm in diameter should generally be large enough for attaining flammability limits.

Minor variations of temperature such as those in a laboratory do not significantly affect flammability limits, but major elevations of temperature will widen the composition range of fuel-oxidant between these limits. Consider, for example, the shift of point A in Figure 42.3 from a nonflammable composition–temperature point to point B, which is in the flammable zone. Likewise, normal variations in barometric pressure have no appreciable effect, but limits—particularly those of paraffin hydrocarbon—at high pressure are generally wider than at atmospheric pressure.

The quantity of water vapor in air at room temperatures affects the lower limit of flammability very little because the mixture contains excess oxygen. Such is not the case at the upper limit of flammability, which is lowered because the mixture is oxygen short for stoichiometric combustion and the oxygen content of the mixture is reduced by the water vapor content.

Wider flammability limits are obtained for upward propagation of flame than for horizontal or downward propagation. From the safety point of view, flammability hazards should be evaluated on the basis of limits for upward propagation.

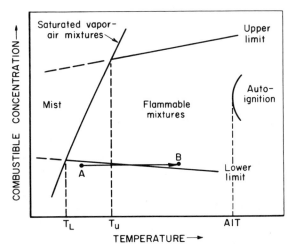

Figure 42.3. Effect of temperature on the limits of flammability of a combustible vapor in air at a constant initial pressure (3). Graph courtesy of Bureau of Mines, U.S. Department of the Interior.

2.2 Limits of Flammability, Lower Temperature Limits, and Minimum Autoignition Temperatures

Most of the values of flammability limits of mixtures of air and combustibles given in Table 42.2 were obtained by using an apparatus such as described in Figure 42.1 and are believed to have been determined in adequately large vessels with an adequately strong ignition source. These limits are for upward propagation at room temperature and pressure, except as noted in footnotes. Table 42.2 also contains data on lower temperature limits and minimum temperatures for autoignition of combustible–air mixture. The three types of limit can be explained with reference to Figure 42.3.

Table 42.2. Summary of Limits of Flammability, Lower Temperature Limits (T_L), and Minimum Autoignition Temperatures (AIT) of Individual Gases and Vapors in Air at Atmospheric Pressure (3)

Combustible	Limits of Flammability (volume percent)		T_L (°C)	AIT (°C)
	L_{25}	U_{25}		
Acetal	1.6	10	37	230
Acetaldehyde	4.0	60		175
Acetic acid	5.4[a]		40	465
Acetic anhydride	2.7[b]	10[a]	47	390
Acetanilide	1.0[d]			545
Acetone	2.6	13		465
Acetone cyanohydrin	2.2	12		
Acetophenone	1.1[d]			570
Acetylacetone	1.7[d]			340
Acetyl chloride	5.0[d]			390
Acetylene	2.5	100		305
Acrolein	2.8	31		235
Acrylonitrile	3.0		−6	
Adipic acid	1.6[d]			420
Aldol	2.0[d]			250
Allyl alcohol	2.5	18	22	
Allyl amine	2.2	22		375
Allyl bromide	2.7[d]			295
Allyl chloride	2.9		−32	485
o-Aminodiphenyl	.66	4.1		450
Ammonia	15	28		
n-Amyl acetate	1.0[a]	7.1[a]	25	360
n-Amyl alcohol	1.4[a]	10[a]	38	300
tert-Amyl alcohol	1.4[d]			435
n-Amyl chloride	1.6[e]	8.6[a]		260
tert-Amyl chloride	1.5[f]		−12	345
n-Amyl ether	0.7[d]			170
Amyl nitrite	1.0[d]			210

Table 42.2. (*Continued*)

Combustible	Limits of Flammability (volume percent)		T_L (°C)	AIT (°C)
	L_{25}	U_{25}		
n-Amyl propionate	1.0[d]			380
Amylene	1.4	8.7		275
n-Amyl nitrate	1.1			195
Aniline	1.2[g]	8.3[g]		615
Anthracene	0.65[d]			540
Benzene	1.3[a]	7.9[a]		560
Benzyl benzoate	0.7[d]			480
Benzyl chloride	1.2[d]			585
Bicyclohexyl	0.65[a]	5.1[h]	74	245
Biphenyl	0.70[i]		110	540
2-Biphenylamine	0.8[d]			450
Bromobenzene	1.6[d]			565
Butadiene (1,3)	2.0	12		420
n-Butane	1.8	8.4	−72	405
1,3-Butanediol	1.9[d]			395
Butene-1	1.6	10		385
Butene-2	1.7	9.7		325
n-Butyl acetate	1.4[e]	8.0[a]		425
n-Butyl alcohol	1.7[a]	12[a]		
sec-Butyl alcohol	1.7[a]	9.8[a]	21	405
tert-Butyl alcohol	1.9[a]	9.0[a]	11	480
tert-Butyl amine	1.7[a]	8.9[a]		380
n-Butylbenzene	0.82[a]	5.8[a]		410
sec-Butylbenzene	0.77[a]	5.8[a]		420
tert-Butylbenzene	0.77[a]	5.8[a]		450
n-Butylbromide	2.5[a]			265
Butyl cellosolve	1.1[h]	11[j]		245
n-Butyl chloride	1.8	10[a]		
n-Butyl formate	1.7	8.2		
n-Butyl stearate	0.3[d]			355
Butyric acid	2.1[d]			450
α-Butryolactone	2.0[h]			
Carbon disulfide	1.3	50		90
Carbon monoxide	12.5	74		
Chlorobenzene	1.4		21	640
m-Cresol	1.1[h]			
Crotonaldehyde	2.1	16[k]		
Cumene	0.88[a]	6.5[a]		425
Cyanogen	6.6			
Cycloheptane	1.1	6.7		
Cyclohexane	1.3	7.8		245
Cyclohexanol	1.2[d]			300

continued on next page

Table 42.2. (*Continued*)

Combustible	Limits of Flammability (volume percent)		T_L (°C)	AIT (°C)
	L_{25}	U_{25}		
Cyclohexene	1.2[a]			
Cyclohexyl acetate	1.0[d]			335
Cyclopropane	2.4	10.4		500
Cymene	0.85[a]	6.5[a]		435
Decaborane	0.2			
Decalin	0.74[a]	4.9[a]	57	250
n-Decane	0.75[t]	5.6[m]	46	210
Deuterium	4.9	75		
Diborane	0.8	88		
Diesel fuel (60 cetane)				225
Diethylamine	1.8	10		
Diethylaniline	0.8[d]		80	630
1,4-Diethyl benzene	0.8[a]			430
Diethylcyclohexane	0.75			240
Diethyl ether	1.9	36		160
3,3-Diethylpentane	0.7[a]			290
Diethyl ketone	1.6			450
Diisobutyl carbinol	0.82[a]	6.1[j]		
Diisobutyl ketone	0.79[a]	6.2[a]		
2-4-Diisocyanate			120	
Diisopropyl ether	1.4	7.9		
Dimethylamine	2.8			400
2,2-Dimethylbutane	1.2	7.0		
2,3-Dimethylbutane	1.2	7.0		
Dimethyldecalin	0.69[a]	5.3[i]		235
Dimethyldichlorosilane	3.4			
Dimethyl ether	3.4	27		350
n,n-Dimethylformamide	1.8[a]	14[a]	57	435
2,3-Dimethylpentane	1.1	6.8		335
2,2-Dimethylpropane	1.4	7.5		450
Dimethyl sulfide	2.2	20		205
Dimethyl sulfoxide			84	
Dioxane	2.0	22		265
Dipentene	0.75[h]	6.1[h]	45	237
Diphenylamine	0.7[d]			635
Diphenyl ether	0.8[d]			620
Diphenylmethane	0.7[d]			485
Divinyl ether	1.7	27		
n-Dodecane	0.60[d]		74	205
Ethane	3.0	12.4	−130	515
Ethyl acetate	2.2	11		
Ethyl alcohol	3.3	19[k]		365
Ethylamine	3.5			385

Table 42.2. (*Continued*)

Combustible	Limits of Flammability (volume percent)		T_L (°C)	AIT (°C)
	L_{25}	U_{25}		
Ethylbenzene	1.0[a]	6.7[a]		430
Ethyl chloride	3.8			
Ethylcyclobutane	1.2	7.7		210
Ethylcyclohexane	2.0[n]	6.6[n]		260
Ethylcyclopentane	1.1	6.7		260
Ethyl formate	2.8	16		455
Ethyl lactate	1.5			400
Ethyl mercaptan	2.8	18		300
Ethyl nitrate	4.0			
Ethyl nitrite	3.0	50		
Ethyl propionate	1.8	11		440
Ethyl propyl ether	1.7	9		
Ethylene	2.7	36		490
Ethylenimine	3.6	46		320
Ethylene glycol	3.5[d]			400
Ethylene oxide	3.6	100		
Furfural alcohol	1.8[o]	16[p]	72	390
Gasoline				
100/130	1.3	7.1		440
115/145	1.2	7.1		470
Glycerine				370
n-Heptane	1.05	6.7	−4	215
n-Hexadecane	0.43[d]		126	205
n-Hexane	1.2	7.4	−26	225
n-Hexyl alcohol	1.2[a]			185
n-Hexyl ether	0.6[d]			
Hydrazine	4.7	100		400
Hydrogen	4.0	75		
Hydrogen cyanide	5.6	40		
Hydrogen sulfide	4.0	44		360
Isoamyl acetate	1.1[a]	7.0[a]	25	350
Isoamyl alcohol	1.4[a]	9.0[a]		460
Isobutane	1.8	8.4	−81	
Isobutyl alcohol	1.7[a]	11[a]		430
Isobutylbenzene	0.82[a]	6.0[j]		
Isobutyl formate	2.0	8.9		465
Isobutylene	1.8	9.6		
Isopentane	1.4			460
Isophorone	0.84			
Isopropyl acetate	1.7[d]			
Isopropyl alcohol	2.2			440
Isopropyl biphenyl	0.6[d]			

continued on next page

Table 42.2. (*Continued*)

Combustible	Limits of Flammability (volume percent)		T_L (°C)	AIT (°C)
	L_{25}	U_{25}		
Jet fuel				240
JP-4	1.3	8		230
JP-6				210
Kerosene			-187	540
Methane	5.0	15.0		
Methyl acetate	3.2	16		
Methyl acetylene	1.7			385
Methyl alcohol	6.7	36^k		430
Methylamine	4.2^d			
Methyl Bromide	10	15		
3-Methyl-1-butene	1.5	9.1		
Methyl butyl ketone	1.2^e	8.0^a		380
Methyl cellosolve	2.5	20^g	46	
Methyl cellosolve acetate	1.7^q			
Methyl chloride	7^d			
Methyl cyclohexane	1.1	6.7		250
Methyl cyclohexanol	1.0^d			295
Methyl cyclopentadiene	1.3^a	7.6^a	49	445
Methyl ethyl ether	2.2^d			
Methyl ethyl ketone	1.9	10		
Methyl ethyl ketone peroxide			40	390
Methyl formate	5.0	23		465
Methyl isobutyl carbinol	1.2^d		40	
Methyl isopropenyl ketone	1.8^e	9.0^e		
Methyl lactate	2.2^a			
α-Methylnaphthalene	0.8^d			530
2-Methylpentane	1.2^d			
Methyl propionate	2.4	13		
Methyl propyl ketone	1.6	8.2		
Methyl styrene	1.0^d		49	495
Methyl vinyl ether	2.6	39		
Methylene chloride				615
Monoisopropyl bicyclohexyl	0.52	4.1^r	124	230
2-Monoisopropyl biphenyl	0.53^j	3.2^r	141	435
Monomethylhydrazine	4			
Naphthalene	0.88^s	5.9^t		526
Nicotine	0.75^a			
Nitroethane	3.4			
Nitromethane	7.3			
1-Nitropropane	2.2			
2-Nitropropane	2.5		27	
n-Nonane	0.85^t		31	205
n-Octane	0.95		13	220

Table 42.2. (*Continued*)

Combustible	Limits of Flammability (volume percent)		T_L (°C)	AIT (°C)
	L_{25}	U_{25}		
Paraldehyde	1.3			
Pentaborane	0.42			
n-Pentane	1.4	7.8	−48	260
Pentamethylene glycol				335
Phthalic anhydride	1.2[g]	9.2[v]	140	570
3-Picoline	1.4[d]			500
Pinane	0.74[w]	7.2[w]		
Propadiene	2.16			
Propane	2.1	9.5	−102	450
1,2-Propanediol	2.5[d]			410
β-Propiolactone	2.9[c]			
Propionaldehyde	2.9	17		
n-Propyl acetate	1.8	8		
n-Propyl alcohol	2.2[t]	14[a]		440
Propylamine	2.0			
Propyl chloride	2.4[d]			
n-Propyl nitrate	1.8[p]	100[p]	21	175
Propylene	2.4	11		460
Propylene dichloride	3.1[d]			
Propylene glycol	2.6[x]			
Proplyene oxide	2.8	37		
Pyridine	1.8[k]	12[y]		
Propargyl alcohol	2.4[e]			
Quinoline	1.0[d]			
Styrene	1.1[z]			
Sulfur	2.0[aa]		247	
p-Terphenyl	0.96[d]			535
n-Tetradecane	0.5[d]			200
Tetrahydrofuran	2.0			
Tetralin	0.84[a]	5[h]	71	385
2,2,3,3-Tetramethylpentane	0.8			430
Tetramethylene glycol				390
Toluene	1.2[a]	7.1[a]		480
Trichloroethane				500
Trichloroethylene	12[bb]	40[y]	30	420
Triethylene amine	1.2	8.0		
Triethylene glycol	0.9[h]	9.2[cc]		
2,2,3-Trimethylbutane	1.0			420
Trimethylamine	2.0	12		
2,2,4-Trimethylpentane	0.95			415
Trimethylene glycol	1.7[d]			400
Trioxane	3.2[d]			

continued on next page

Table 42.2. (*Continued*)

Combustible	Limits of Flammability (volume percent)		T_L (°C)	AIT (°C)
	L_{25}	U_{25}		
Turpentine	0.7^a			
Unsymmetrical dimethylhydrazine	2.0	95		
Vinyl acetate	2.6			
Vinyl chloride	3.6	33		
m-Xylene	1.1^a	6.4^a		530
o-Xylene	1.1^a	6.4^a		465
p-Xylene	1.1^a	6.6^a		530

$^aT = 100°C.$	$^kT = 60°C.$	$^uT = 43°C.$
$^bT = 47°C.$	$^lT = 53°C.$	$^vT = 145°C.$
$^cT = 75°C.$	$^mT = 86°C.$	$^wT = 160°C.$
^{d}Calculated.	$^nT = 130°C.$	$^xT = 96°C.$
$^eT = 50°C.$	$^oT = 72°C.$	$^yT = 70°C.$
$^fT = 85°C.$	$^pT = 117°C.$	$^zT = 29°C.$
$^gT = 140°C.$	$^qT = 125°C.$	$^{aa}T = 247°C.$
$^hT = 150°C.$	$^rT = 200°C.$	$^{bb}T = 30°C.$
$^iT = 110°C.$	$^sT = 78°C.$	$^{cc}T = 203°C.$
$^jT = 175°C.$	$^tT = 122°C.$	

Flammable mixtures considered in Figure 42.3 fall in one of three regions, defined by the parameters of initial temperatures and concentration of the combustible. To the left of the line labeled "saturated vapor–air mixtures" exists the region of flammable mists or flammable mixtures of droplets in air. A flammable mixture can form at temperatures below the flash point of a liquid combustible either if such a combustible is sprayed into air or any suitable oxidant, or if a mist or foam forms. Some details of the complex behavior of mists and sprays are given in Reference 3 and in the literature mentioned therein. The second region lies along the curve for saturated vapor–air mixtures. The intercept of this curve and the one for the lower limit of flammability defines the lowest temperature at which a homogeneous flammable vapor–air mixture can be formed from a liquid; this computed temperature is given in Table 42.2 as T_L. At a given pressure, the most common hazards are due to temperature–compositions in the region labeled "flammable mixtures." The autoignition curve defines the limit temperature–composition point to the right of which mixtures can ignite spontaneously. The minimum temperature along this curve is the minimum autoignition temperature (AIT) and is also listed in Table 42.2. It is the lowest temperature at which ignition can occur when a flammable mixture is heated to an elevated temperature. The time that elapses between the instant a mixture is raised to a given temperature and the formation of a flame is called the ignition lag or time delay before ignition. If the ignition lag is relatively short—certainly if it is less than a second, as may be the case in a flowing system—the ignition temperature increases above the minimum AIT as heating time decreases. From the standpoint of safety, the minimum temperature at which ignition can occur spontaneously is needed, and this temperature must be determined as a function of heating time lasting long enough (many minutes) that the temperatures for the appearance of flame are independent of increasing heating time. Again, more detail can be obtained in Reference 3 and also in Reference 2.

2.3 Prediction of Limits of Flammability

2.3.1 One Combustible

Although the limits of flammability of about 250 single-component combustibles are presented in Table 42.2, an estimate may be needed of the limits of a combustible not yet tested. No hard and fast calculations of limits are possible, but past experience provides guidelines, which may be used to obtain rough estimates (1–3) of unmeasured lower limits in air. For paraffin hydrocarbons containing four or more carbons, the lower limit, in volume percent, is about (107 ÷ molecular weight) and about 48 mg per liter of air at 25°C. Another approximation for paraffinic hydrocarbons in air at 25°C is that the lower limit equals 0.55 times the volumetric percentage of the combustible for stoichiometric combustion. Table 42.3 contains a list by type of combustible of ratios of the volumetric percentage of the lower limit to the stoichiometric volumetric percentage and of the upper limit to the stoichiometric percentage.

Other rule-of-thumb techniques are available for estimating whether a mixture is flammable. The following three criteria (11, 12) pertain to carbon–hydrogren–oxygen–nitrogen systems with an average molecular weight of likely combustion products of about 29. An enthalpy release of 300 to 350 cal/g, an exothermicity of about 10 kcal per mole of product, or an adiabatic flame temperature of 1500 to 1600°K are marginal conditions for potential sustained combustion in most C–H–O–N systems. Among the exceptions are systems with unusually low combustion initiation temperatures, which manifest cool flames and smoldering combustion.

2.3.2 Binary and Multiple-Component Combustibles

Limits of flammability of combustible mixtures can be calculated using the measured or estimated (Section 2.3.1) limits of each combustible and the percentage of each combustible in the mixture by means of the Chatelier mixture rule. By no means are these estimated values for mixtures a substitute for actual measurement, but they are better than no appraisal of hazard limits. This mixture rule is based on the premise that if separate limit combustible mixtures are mixed, the resultant mixture will also be a limit mixture. The Le Chatelier mixture rule is expressed as follows:

$$L = \frac{100}{C_1/N_1 + C_2/N_2 + \cdots} \tag{2}$$

where C is the volumetric percentage of the combustible gas in the air-free and inert gas-free mixture (so that $C_1 + C_2 + \cdots = 100$), N is the respective lower or upper limit of flammability of the combustible in air, and L is the lower or upper limit of flammability of the mixture in air. For example, a mixture containing 80 volume percent methane, 15 volume percent ethane, and 5 volume percent propane has a lower limit in air of

$$\frac{100}{80/5.0 \ + \ 15/3.0 \ + \ 5/2.1} = 4.3$$

Flammable liquid mixtures obeying Raoult's law (i.e., the partial pressure of each component equals the vapor pressure of that component when alone, multiplied by its mole fraction in the mixture) can also be treated by the Le Chatelier mixture rule. Exceptions to the rule include mixtures containing considerable excess nitrogen over that in air, mixtures of hydrogen–ethylene, hydrogen–acetylene, hydrogen sulfide–methane, methane–dichloroethylene, methyl-ethyl chlorides, and mix-

Table 42.3. Average Ratios of Volumetric Percentages of Lower and Upper Limits of Flammability to Stoichiometric Composition with Air at 25°C (1, 3)

Combustible	Ratios	
	Lower/Stoichiometric	Upper/Stoichiometric
Hydrocarbons		
Paraffinic	0.55	2.9
Olefinic	0.52	3.3
Acetylenic	0.33	—[a]
Aromatic	0.53[b]	3.4[b]
Cyclic	0.55	3.3
Other combustibles		
Alcohols	0.51[b]	3.0[b]
Aldehydes and ketones	0.54	3.2[b]
Ethers	0.54	5.7
Acids and esters	0.54	2.7
Hydrogen and deuterium	0.15	2.5
Organic oxides	0.51	5.2
Nitrogen compounds	0.56	3.6
Sulfur compounds	0.53	4.4
Halogen compounds	0.67	2.3
Selected exceptions		
Ethylene	0.42	4.4
Acetylene	0.32	10.4
Acetaldehyde	0.51	7.4
Diethylether	0.55	10.8
Carbon monoxide	0.42	2.5
Ethylene oxide	0.39	13.0
Hydrazine	0.27[b]	5.8[b,c]
Carbon disulfide	0.19	7.7

[a]Pure acetylene can propagate flame at atmospheric pressure in tubes with diameters over 12 cm.
[b]Initial temperature in the range of 50 to 100°C.
[c]Corresponds to 100 percent hydrazine.

tures containing carbon disulfide (2). Generally, the Chatelier mixture rule does well with mixtures containing components having about the same (within 10 percent) ratio of the limit to the stoichiometric volume percentages (Table 42.3). More information about using the mixture rule, particularly with respect to complex mixtures containing a little air or air vitiated with nitrogen or carbon dioxide can be found in Reference 1. It is apparent from these exceptions that measured limits of flammability are preferred to those estimated by calculation.

2.4   Limits of Flammability and Minimum Autoignition Temperature of Gases and Vapors in Oxygen

The flammability of combustible gases and vapors in oxygen or in atmospheres of oxygen-enriched air is a vital consideration in many instances—for example, in space flight vehicles, in the compressed gas industry, and in hospitals. The limits of flammability and minimum autoignition temperatures in Table 42.4 are compiled from References 2, 3, and 5.

Oxygen-enriched atmospheres pose not only an explosion problem but also a fire problem, as in hospital and hyperbaric oxygen chambers (13–15). The partial pressure of oxygen, better than the percentage of oxygen, is directly related to the increased hazard with respect to fire spread rate, ignition temperature, and minimum ignition energies of sparks.

Table 42.4. Limits of Flammability and Minimum Autoignition Temperatures of Gases and Vapors in Oxygen (2, 3, 5)

Combustible	Limits of Flammability (volume percent)		Minimum Autoignition Temperatures (°C)
	Lower	Upper	
Acetaldehyde	4	93	159
Acetic acid	≤5.4	—	490
Acetone	≤2.6	60[a]	485
Acetylene	≤2.5	100	296
Ammonia	15.0	79	—
n-Amyl acetate	≤1.0	—	234
Aniline	≤1.2	—	—
Benzene	≤1.3	30	—
Bromochloromethane	10.0	85	368
Bromodifluoromethane	29.0	80	453
Butane	1.8	49	278
Isobutane	1.8	48	319
Butene-1	1.7	58	310
Butene-2	1.7	55	—
Butyl chloride	1.7	52[a]	235
Carbon disulfide	≤1.3	—	107
Carbon monoxide	12.5	94	588
2-Chloropropene	4.5	54	—

continued on next page

Table 42.4. (*Continued*)

Combustible	Limits of Flammability (volume percent)		Minimum Autoignition Temperatures (°C)
	Lower	Upper	
Isocrotyl bromide	6.4	50	—
Isocrotyl chloride	2.8	66	
Cyclopropane	2.5	60	454
Deuterium	4.9	94	—
Dichloroethylene	10.0	26	—
Diethyl ether	2.0	82	182
Dimethyl ether	3.9	61	252
Divinyl ether	1.8	85	166
Ethane	3.0	66	506
Ethyl acetate	≤2.2	—	—
Ethyl alcohol	≤3.3	—	329
Ethyl bromide	6.7	44	—
Ethyl chloride	4.0	67	468
Ethylene	2.9	80	485
Ethylene chloride	4.0	68	470
Ethylene oxide	≤3.6	100	—
Ethyl-*n*-propyl ether	2.0	78	—
Gasoline (100/130)	≤1.3	—	316
Glycol	≤3.5	—	—
n-Hexane	1.2	52[a]	218
Hydrogen	4.0	95	542
Hydrogen sulfide	≤4.0	—	220
Kerosene	0.7	—	216
Methane	5.1	61	556
Methyl alcohol	≤6.7	93	461
Methyl bromide	14.0	19	—
Methyl chloride	8.2	66	—
Methylene chloride	11.7[a]	68[a]	606
Naphtha (Stoddard)	≤1.0		216
n-Octane	≤0.8	—	208
Propane	2.2	52	468
Propylene	2.1	53	423
n-Propyl alcohol	≤2.2	—	328
Propylene oxide	≤2.8	—	—
Isopropyl ether	2.2	69	—
Toluene	1.2	—	—
Trichloroethane	5.5[a]	57[a]	418
Trichloroethylene	7.5[a]	91[a]	396
Vinyl chloride	4.0	70	396

[a]Determinations made at about 100°C.

2.5 Flash Points of Liquids and Temperature Range of Flammability

The concepts of minimum temperatures (T_L) for ignition by a source hot enough and energetic enough and the minimum autoignition temperature (AIT) have been introduced and tabulated in Table 42.2 when air is the oxidant, and in Table 42.4 when oxygen is the oxidant: T_L is the intercept of the curve of the lean flammability limit of the vapor in air as a function of temperature and the curve of the vapor pressure of the liquid as a function of temperature. The T_L involves ignition sources such as flames and sparks, with ignition occurring very rapidly, in much less than a second. Combustion can also be started by slow heating, resulting in autoignition at temperatures corresponding to the AIT values in Tables 42.2 and 42.4; the time scale is longer, of the order of seconds and many minutes. Flash points are related to T_L in that they are an attempt to approximate T_L values. Closed-cup and open-cup techniques are used to obtain these flash points. It must be realized that these two types of flash point in most instances are probably somewhat too high to be absolutely the minimum temperature at which a liquid fuel will ignite. In both the open-cup and closed-cup methods, a given volume of liquid is heated in a prescribed container at a definite rate, and periodically a test flame or spark is passed across the surface of the liquid. The temperature at which a flame passes across the surface of the liquid is taken as the flash point. The container is open to the air in the open-cup method and closed to the air in the closed-cup method. A small opening is uncovered in the latter instance as the test flame is introduced. Lower values are obtained by means of the closed-cup method than by the open-cup method. Values listed in Table 42.5 were obtained by the closed-cup method. A far more

Table 42.5. Flash Point of Combustible Liquids and Vapors in Air (2)

Name	Flash Point (°C)
Acetal	37
Acetaldehyde	−38
Acetanilide	174
Acetic acid	40
Acetic anhydride	53
Acetone	−18
Acetophenone	105
Acetyl acetone	79
Acetyl chloride	4
Acrolein	< −18
Acrylonitrile	−5
Adipic acid	196
Aldol	83

continued on next page

Table 42.5. (*Continued*)

Name	Flash Point (°C)
Allyl alcohol	22
Allyl chloride	−32
Amyl acetate	25
Isoamyl acetate	25
Amyl alcohol	38
tri-Isoamyl alcohol	42
tert-Amyl alcohol	19
Amylbenzene	66
Amyl chloride	13
tert-Amyl chloride	−12
Amylene (*n* and β)	−18
Amyl ether	57
Amyl propionate	41
Aniline	76
Anthracene	121
Benzaldehyde	64
Benzene	−11
Benzoic acid	121
Benzyl acetate	102
Benzyl alcohol	101
Benzyl benzoate	148
Benzyl chloride	60
Bromobenzene	65
Butyl acetate	22
Isobutyl acetate	18
Butyl alcohol	29
Isobutyl alcohol	28
sec-Butyl alcohol	21
Dioxane	12
Dipentene	45
Diphenyl	113
Diphenylamine	153
Diphenylmethane	130
Dodecane	74
Ethyl acetate	−4
Ethyl alcohol	13
Ethylamine	< −18
Ethylaniline	85
Ethylbenzene	18
Ethyl butyrate	26
Ethyl chloride	−50
Ethylene chlorohydrin	60
Ethylene dichloride	13
Ethylene glycol	111
Ethylene oxide	< −18

Table 42.5. (*Continued*)

Name	Flash Point (°C)
Ethyl ether	−45
Ethyl formate	−20
Ethyl lactate	46
Ethyl propionate	12
Formaldehyde	54
Furfural	75
Furfural alcohol	75
Gasoline, regular	−44
Glycerine	160
Heptane	−4
Hexane	−26
Isohexane	<−29
Hexyl alcohol	58
Hexyl ether	77
Hydrogen cyanide	−18
Isophorone	84
Isoprene	−54
Kerosene	38–74
Methyl acetate	−10
Methylal	−18
Methyl alcohol	12
Methylamine	−18
Methylbutyl ketone	35
Methyl cellosolve	41
Methyl cyclohexane	−4
Tung	289
Turkey-red	247
Whale	230
Paraffin	199
Paraformaldehyde	70
Paraldehyde	17
Pentane	−40
Isopentane	<−51
Petroleum ether	−56
Phenol	79
Phthalic anhydride	152
Pinene	33
Propyl acetate	14
Isopropyl acetate	4
Propyl alcohol	15
Isopropyl alcohol	12
Isopropylamine	−26
Propylbenzene	30
Isopropylbenzene	39
Propyl chloride	<−18

continued on next page

Table 42.5. (*Continued*)

Name	Flash Point (°C)
Propylenediamine	22
Propylene dichloride	16
Propylene glycol	97
Propylene oxide	−37
Isopropyl ether	−28
Propyl formate	−3
tert-Butyl alcohol	11
Butylamine	8
Isobutylamine	−9
Butyl bromide	13
Butyl carbitol	78
Butyl cellosolve	61
Butyl chloride	−28
Butyl formate	18
Butyl propionate	32
Butyl stearate	160
Butyraldehyde	−18
Butyric acid	77
Carbon disulfide	−30
Cellosolve	40
Cellosolve acetate	51
Chlorobenzene	29
o-Cresol	81
m-Cresol	86
p-Cresol	86
Crotonaldehyde	13
Cumene	39
Cyanamide	141
Cyclohexane	−17
Cyclohexanol	68
Cyclohexanone	64
p-Cymene	47
Decalin	58
Decane	46
Diisobutyl carbinol	—
Diisobutyl ketone	48
o-Dichlorobenzene	66
1,2-Dichloro-*n*-butane	52
sec-Dichloroethylene	14
Dichloroethyl ether	55
Diethanolamine	138
Diethylamine	< −18
Diethylene glycol	124
Dimethyl aniline	63
2,3-Dimethylbutane	< −29

Table 42.5. (*Continued*)

Name	Flash Point (°C)
Dimethyl ether	−41
Dimethylformamide	57
o-Methylcyclohexanol	68
Methyl ethyl ketone	−1
Methyl formate	−19
Methyl lactate	49
Methyl propionate	−2
Methyl propyl ketone	16
Methyl salicylate	101
Naphtha	−7–42
Naphthalene	80
Nitrobenzene	88
Nonane	31
Octane	13
Isooctane	−12
Oil	
Castor	229
Coconut	216
Corn	254
Cottonseed	252
Creosote	74
Fish	216
Gas	66
Lard	184
Linseed, raw	222
Linseed, boiled	206
Lubricating spindle	
Lubricating, turbine	204
Menhaden	224
Mineral seal	77
Neat's-foot	243
Olive	225
Palm	162
Paraffin	229
Peanut	282
Pine	78
Pine tar	62
Rape	163
Rosin	130
Soybean	282
Sperm	220
Straw	157
Tallow	256
Transformer	146

continued on next page

Table 42.5. (*Continued*)

Name	Flash Point (°C)
Isopropyl formate	−6
Pyridine	23
Stoddard solvent	38–43
Stearic acid	196
Styrene	31
Tetradecane	100
Tetrahydrofurfural alcohol	75
2,2,3,3-Tetramethylpentane	—
Toluene	4
o-Toluidine	85
p-Toluidine	87
Trichloroethylene	—
Triethylamine	−7
Triethylene glycol	177
Trimethylamine	—
Trimethylbenzene	—
2,2,4-Trimethylpentane	−12
Trioxane	45
Turpentine	35
Valeric acid	—
Valeric aldehyde	—
Vinyl acetate	−8
o-Xylene	17
m-Xylene	25
p-Xylene	25
o-Xylidine	97

extensive compendium, listing materials by trade names, has been published by the National Fire Protection Association (16), and by other organizations for lubricants (17) and jet aircraft fluids (18). The great value of flash point data lies in the ease of measurement, the standardization of open-cup and closed-cup methodology, and the comparative classification of the hazard of flammability of each listed liquid with respect to the others; the higher the flash point of a liquid, the safer it is with respect to fires and explosions. However, flash points are not the absolute minimum temperature at which flammable mixtures may be produced by the vapors from a liquid; the flash point may be a few degrees to about 12°C above the minimum flammable temperature (2). As the temperature of some liquids is raised toward the boiling point, the liquid's vapors remain flammable until an upper temperature limit corresponding to the upper concentration limit of flammability is reached, at which point the percentage of vapor in saturated air is so high that the mixture is incapable of propagating flame. If this mixture becomes diluted with additional air, the resultant mixture will be flammable.

2.6   Methods of Minimizing Explosions and Fires

2.6.1   *Reducing the Oxygen Content of the Atmosphere*

Oxygen present in a flammable atmosphere or in air feeding a fire may be reduced by dilution with inert gases such as nitrogen, carbon dioxide, steam, or combinations of these inerts in the form of exhaust gases from flues, automobile engines, or jet engines. Carbon dioxide and compressed nitrogen fire extinguishers are frequently used to put out fires and to smother initial kernels of explosions. If the fuel bed is hot, it must be cooled generally by large quantities of water, if reignition is to be prevented on dissipation of the applied inert gases. Halons (Freons) may also be used to lower the oxygen content to levels not supporting combustion. In some applications, special reagents and catalysts can reduce the oxygen content of a potentially explosive atmosphere. More generally, the explosion has started and there is very little time to add inert material fast enough to quench the progressing explosion before structural damage occurs; fast-acting valves, operated often by safely contained explosive activators, are often used to release very rapidly the explosion-quenching agent. Halons are frequently employed as quenching agents because they can chemically inhibit combustion as well as lower the available oxygen concentration. It must be remembered that no toxic gas or vapor is suitable for use as a diluent in situations where it may be inhaled; likewise, atmospheres deficient in oxygen may not be respirable.

  Table 42.6 gives the minimum oxygen concentrations required to support flames of combustible gases and vapors, when the reduction of oxygen concentration is brought about by the addition of nitrogen or carbon dioxide. The critical oxygen concentrations for any combustible vary with the concentration of the combustible present; the oxygen concentrations in Table 42.6 are minimal, therefore covering any concentration of the combustible that may be present.

Table 42.6. Oxygen Percentages Below Which Flames of Combustible Gases and Vapors are Extinguished (2)

	Oxygen Percentage Below Which No Mixture Is Flammable	
Compound	Nitrogen as Diluent in Air	Carbon Dioxide as Diluent in Air
Methane	12.1	14.6
Ethane	11.0	13.4
Propane	11.4	14.3
Butane	12.1	14.5
Isobutane	12.0	14.8
Pentane	12.1	14.4

continued on next page

Table 42.6. (*Continued*)

	Oxygen Percentage Below Which No Mixture Is Flammable	
Compound	Nitrogen as Diluent in Air	Carbon Dioxide as Diluent in Air
Isopentane	12.1	14.6
Hexane	11.9	14.5
2,2-Dimethylbutane	12.1	14.7
Heptane	11.6	14.2
Natural gas	12.0	14.4
Gasoline	11.6	14.4
Ethylene	10.0	11.7
Propylene	11.5	14.1
Butene-1	11.4	13.9
Butene-2	11.7	14.0
Isobutylene	12.1	14.8
3-Methylbutene-1	11.4	13.9
1,3-Butadiene	10.4	13.1
Benzene	11.2	13.9
Cyclopropane	11.7	13.9
Methyl alcohol	9.7	11.9
Ethyl alcohol	10.6	13.0
Dimethyl ether	10.3	13.1
Diethyl ether	10.3	13.2
Acetone	11.6	14.3
Methylisobutyl ketone	11.4	—
Methyl formate	10.1	12.5
Ethyl formate	10.4	12.8
Methyl acetate	10.9	13.6
Isobutyl formate	12.4	14.8
Methylamine	10.7	—
Allyl chloride	12.6	15.1
Hydrogen	5.0	5.9
Carbon disulfide	5.4	7.6
Carbon monoxide	5.6	5.9

Figure 42.4 illustrates the relation between the concentrations of combustible and the minimum oxygen percentage for flammability when nitrogen or carbon dioxide is the diluent. The graph shows the flammable areas of possible relevant mixtures of butadiene, air, and added nitrogen or carbon dioxide. The line a–d represents mixtures of butadiene and pure air. Point b on this line is the lower limit of flammability (2.0 percent) and point c is the upper limit of flammability

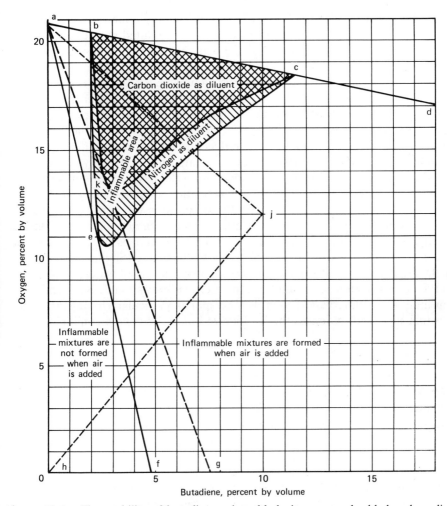

Figure 42.4. Flammability of butadiene, air, added nitrogen, and added carbon dioxide (2). Graph courtesy of Bureau of Mines, U.S. Department of the Interior.

(11.5 percent); mixtures between *b* and *c* are flammable. As the oxygen concentration is lowered by the dilution with carbon dioxide (double-cross-hatched area) or with nitrogen (single-hatch area), the spread in composition between lower and upper limits narrows until the close or nose of the respective hatched area is reached, namely, at 10.4 percent oxygen when nitrogen is the diluent (point *e*) and at 13.1 percent oxygen when carbon dioxide is so employed (point *k*). The mixture that will propagate flame with a minimum of oxygen present (added nitrogen being the diluent) contains 2.6 percent butadiene. If the mixture contains 5.0 percent of butadiene, Figure 42.4 indicates that the oxygen concentration has to be reduced with added nitrogen to 13.2 percent. A caution is in order; Figure 42.4 shows that if this mixture containing 5.0 percent butadiene and 13.2 percent oxygen is diluted with air, the nonflammable mixture becomes flammable. Only mixtures diluted with excess nitrogen falling along the line $a-e-f$, or diluted with excess carbon dioxide falling along $a-k-g$, can be diluted with air and remain nonflammable.

A frequent problem is that of flushing out equipment containing a nonflammable gaseous mixture without running through the flammable range. This operation can be accomplished successfully by using information such as that contained in Figure 42.4. Consider point j, which is a nonflammable mixture containing 12 percent oxygen, 10 percent butadiene, and 78 percent nitrogen. To stay out of the flammable area $b-c-e$, the oxygen content must be reduced to a value below 10.4 percent to pass safely from the composition of point j to compositions given by any point to the left of line $a-e-f$. Nitrogen can be added until the oxygen content is reduced to less than 4.6 percent, thereby shifting composition along the line $j-h$. When a mixture thus obtained is diluted with air, the composition will move along the line $a-e-f$, or to the left of it, and the mixture always will be nonflammable. At this state, air can be added to the container and the combustibles swept out without danger of explosion. Figure 42.4 applies only to butadiene. Similar diagrams for other flammable gases and vapors and dealing with the use of water vapor as the diluent appear in References 1 and 3. An extensive discussion in Reference 3 of the flushing of gasoline vapors with water vapor is particularly noteworthy; the logic is that just used to discuss the flushing of butadiene from a container.

2.6.2 Operating Outside the Range of Flammability

If possible, the concentration of combustibles should be kept below their lower limits of flammability, because any ingress of air into the mixture will not make the mixture flammable. If the concentrations of combustible must be above the lower limit of flammability, explosion hazards can be avoided if the combustibles are above the upper concentration limit of flammability; in this situation danger does arise when additional air finds its way into the mixture. With liquid combustibles, temperature control is another means of seeking safety. As Figure 42.3 indicates, temperature can be held low enough (T_L) that the corresponding low vapor pressure of the liquid does not produce a saturated vapor–air mixture with enough concentration of the combustible vapor to match the percentage at the lower limit of flammability; or more crudely, the temperature of the liquid's flash point is never reached for safety purposes. Nonflammability will be gained again when the temperature is raised to correspond to a vapor pressures equal to and above that corresponding to the concentration of the combustible at its upper limit of flammability $(T_u,$ in Figure 42.3). It is noted again that combustion can occur at and above the upper temperature limit, if the air space is not saturated or if fresh air is admitted.

2.6.3 Use of Less Flammable Materials and Chemical Flame Inhibitors

Various halogenated hydrocarbons called Halons or Freons by their manufacturers can be used to advantage where noninflammable liquids are required. Not only are they nonflammable, but in many applications relatively small concentrations of these materials added to flammable mixtures can render the overall mixture nonflammable. Thus, for example, the National Fire Protection Association publication NFPA No. 12A (19) lists 2.0 volume percent of Halon 1301 (trifluorobromomethane) in the mixture as enough to inhibit the combustion of any methane–air mixture; other sources cite about 3.2 percent (20) or 4.7 percent (21). Whichever of the three values is accepted, the effectiveness of this agent exceeds that of carbon dioxide, because about 24 percent carbon dioxide is required to render all methane–air mixtures nonflammable (1).

2.6.4 Elimination of Ignition Sources

Combustible gases and vapors, mixed with an oxidant such as air or pure oxygen in proportions to give flammable mixtures and at temperatures below their AIT, will ignite if exposed to an ignition source. The atmospheres in the gasoline tanks of millions of automobiles and trucks on our road contain flammable mixtures, or mixtures that are not flammable because they are above the upper limit for gasoline but become flammable when air is admixed. Yet ignitions are rare because sources of ignition are rarely present in the tank or around the filler cap. Another way of describing ignition hazard is to say that ignitions of a fuel–gas or vapor–air mixture can occur only if the mixture composition is within the flammability range and a critical volume of that mixture is heated sufficiently to produce an exothermic chemical reaction capable of propagating flame beyond the initiation zone. Ignition sources are therefore of prime interest in safety, and representative ignition sources are as follows:

1. Electrical sparks and arcs, generated by static electricity, electrical shorts, or lightning
2. Flames such as open pilot or burner flames, matches and cigarette lighters, flames in space heaters, cooking stoves, hot water heaters, and so on, burning materials, and incinerators
3. Hot surfaces such as frictional sparks, incendiary particles, heated wires, rods or fragments, hot vessels or pipes, glowing metals, hot cinders, overheated bearings, and breaking electric light bulbs to expose hot filaments
4. Hot gases brought about by shock compression, adiabatic compression, and hot gas jets
5. Lasers
6. Pyrophoricity or catalytic reaction
7. Self-heating or spontaneous combustion

Often temporal and spatial characteristics of heat sources are useful in evaluating ignition sources. In the case of an electrical spark, the ignition source is highly localized in space, and the duration of heating can be as little as a few microseconds; here temperatures are very high. However, ignition is determined by the amount and rate of energy input into the flammable mixture. Lasers fall into this category when the medium readily absorbs the laser radiation. In contrast, a heated vessel represents a much more distributed ignition source spatially, and the duration of heating can extend to several minutes. Here temperature, rather than energy, is the critical factor and it is tied to the heating duration. Temperature for ignition is minimal when heating duration or ignition lag is so long that the temperature for ignition does not decrease with further increase in heating time; these are the AITs in Table 42.2. Both temperature and heating rate tend to be important with small hot surface sources as heated wires, hot metal fragments, and frictional sparks. Generally, ignition temperatures increase with a decrease in size or surface area of the heat source. Incendiary ignitions are a case of hot surface particle ignition augmented by the chemical reactivity of the heat sources. Hot gas ignitions also differ in their temporal and spatial characteristics, with shock waves being most localized in space, and heating time being less than milliseconds; adiabatic compression heating is more spatially distributed, extends over a few seconds, and can produce much lower temperature ignition than is possible by shock compression. Self-heating or spontaneous combustion involves slow oxidation or decomposition of the combustibles and requires hours or even days for ignition to occur. Pyrophoric or catalytic ignitions involve the combustible reacting with moisture or oxygen in the air; ignition lags are of the order of seconds.

Sources of ignitions just discussed can be controlled by proper safety practices, such as by using flameproof and vaporproof electrical equipment, and building industrial plants so that boilers, water heaters, incinerators, and other equipment with open flames and incandescent materials are isolated at safe distances from operations involving flammable materials. Static electricity is one of the most difficult ignition sources to control. Static electricity is a more serious hazard in dry atmospheres where the relative humidity is below 50 percent; where practical, it is advisable to maintain room humidity above 50 percent to reduce static sparks. Friction due to such situations as slipping belts or pulleys, revolving machinery, and passage of dust, pellets, liquids, or gases at high velocity through small openings can generate static electricity, and so may processes of impact, cleavage, or induction. Proper grounding of machinery, pipes, and even personnel by means of semiconducting floors (9) is advisable, if charges may accumulate.

2.6.5 Segregation of Hazardous Operations

Operations involving large volumes of flammable gases or vapors should be separated from other operations. This requires installation of hazardous operations in a building at a safe distance from others and suitably diked to avoid flow of hazardous liquids and vapors toward other structures in the event of spillage. Safe distance criteria are hard to define. Some guidance in this respect can be gotten from quantity–distance tables for liquid hydrogen developed by the Bureau of Mines (22), the Compressed Gas Association (23), and the Armed Services Explosives Safety Board (24). More definitive work is needed in this highly important area.

2.6.6   Adequate Ventilation

Adequate ventilation is necessary where flammable gases and vapors are used and transported—in the spaces around equipment in which the combustibles may be present, for example, and in conduits, trenches, and tunnels carrying pipelines. Facile inspection for leaks should always be feasible. Whenever possible, operations should be carried out in the open air, as is done in certain processes in the petroleum industry, with closed structures being used to house instruments and equipment that must be protected from the weather.

2.6.7   Combustible Gas Indicators

Combustible gas indicators or recorders should be installed in all locations where hazardous concentrations of combustibles may occur, and if large locations are involved, sampling should be from different sites in the volume at risk. Flammables must not be lost to the sampling tubing walls because of improperly selected or maintained sampling lines, and temperature should be high enough to avoid condensation if heavy combustible vapors are being tested. Considerable progress has been made in combustible gas indicators, for example, in those for hydrogen (25), methane (26), and several other gases (27).

2.6.8   Venting Explosions and Flame Arresting

In addition to procedures discussed under Sections 2.6.1 to 2.6.7, an installation in which an explosive mixture could conceivably be present should be protected by adequate release diaphragms for venting a possible explosion before destructive pressures develop. To protect an installation properly against such damage or destruction, determination must be made of the type and concentration range of combustibles that may be present, and the maximum pressure the enclosure can withstand versus the maximum pressure an explosion could develop. Then the engineer must determine the kind of material, area, thickness, and location of a release diaphragm or device that is capable of rupturing at a pressure safely below the maximum pressure the enclosure can withstand. For specific information on release or vent devices, the reader is referred to References 2 and 28 to 30. Such data are largely empirical and were obtained by means of experiments in relatively small vessels. Maisey (28) reviews simple theoretical relationships for the venting of closed vessel explosions of gas–air mixtures; these relationships are between flame speed, pressure and time, and the volume in which the explosion is occurring. Maisey (29) also discusses the similarities and differences between gas–air and dust–air explosions. He treats dust–air explosion venting by analogy to gas–air explosions and provides relationships predicting the characteristics of dust explosions and their venting. His starting points are measurements in equipment such

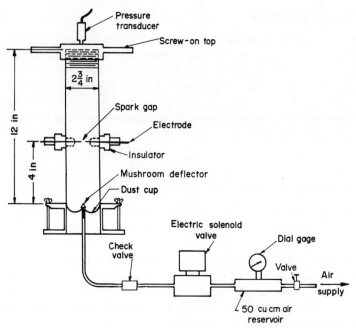

Figure 42.5. Hartmann tube, No. 2 (32). Schematic courtesy of Bureau of Mines, U.S. Department of the Interior.

as the Hartmann apparatus (Figure 42.5), the Godbert–Greenwald furnace, and related apparatus (31, 32); some 2500 materials, including dusts of plastics, chemicals, agricultural materials, carbonaceous materials, and metals, have been evaluated for their explosibility using such apparatus (33, 34).

The spread of explosion between physically connected equipment such as from flare stacks disposing of waste gases, ducting systems carrying solvent vapors to a recovery unit, pipes carrying gases to burners or furnaces, and exhaust pipes of internal combustion engines operating in flammable atmospheres, can be prevented by means of flame arrestors or flame traps. These consist of an assembly of narrow passages or apertures through which gases or vapors can flow, yet small enough to quench any flame attempting to pass through. Such flame traps are generally made of wire gauze, crimped metal, sintered metal, perforated metal plates, or pebble beds. They are often used in conjunction with pressure release vents. The flame trap restricts the explosion from passing from one vessel to another but does not reduce the pressure developed by the explosion. The release vent can keep the pressure down but cannot prevent the explosion from propagating throughout a plant. A combination of both can therefore be used to limit the spread and effects of an explosion. Data on flame arrestors are given in References 34 and 35.

In spite of an extensive literature on explosion venting and flame arresting, the behavior of an explosion, once started, is not yet sufficiently predictable to make measures to combat an explosion entirely reliable; prevention is better than the cure (29).

3 COMBUSTIBLE DUSTS

3.1 Factors Affecting the Limits of Flammability

Dust explosions are similar in some respects to gas explosions, particulary for dust particles approaching the respirable dust range of less than 5 μm in diameter. To explode, a combustible dust must be mixed with an oxidant, such as air or oxygen, and must be in flammable concentration when in contact with an ignition source or when spontaneously ignited. Lower and upper flammability limits can be measured for most dusts. There are dissimilarities, too. The flammability limits of dusts are less meaningful by themselves than is the case with combustible gases because dust clouds are unstable, spatially and temporally, and the dust particle size distribution and dust volatility are factors. Maximum pressures in a given volume can be greater for a dust explosion than for a gas explosion in the same volume, because more moles of reactants can be contained in a solid particle–air mixture than in a homogeneous gas mixture. The rate of pressure rise is usually higher in gas explosions than in dust explosions. The longer duration and greater impulse of dust explosions frequently make these more destructive than gas explosions. Dust is more likely to be accumulated than is gas because dust can become widely distributed by an industrial activity. An explosion hazard can exist with even seemingly little dust being spread about. For example, a layer of coal dust about 0.002 cm thick, on the roof, floor, and walls of a passageway about 3 m wide × 2 m high is enough dust to form a stoichiometric mixture with air in the passageway. The existence of a dust cloud may be an inherent feature of a manufacturing process, such as within a grinding mill or a pneumatic transport system; flammable concentrations may occur, as in equipment handling flammable gases and air.

Any process handling combustible dust-producing stock or dusts directly should be considered as a possible source of a destructive explosion or fire, or both. Parameters strongly affecting the violence of a dust explosion are briefly discussed next.

3.1.1 Chemical Composition of Dust

The chemical composition of a combustible dust determines its oxygen requirement for complete or stoichiometric combustion and its ease of oxidations as reflected by its ignition temperature and burning rate. The chemical composition also determines the heat released on combustion, and the volumes of gaseous products of combustion have a direct bearing on the maximum pressure developed during an explosion. The gases remaining after the combustion of metal powders are chiefly nitrogenous, the oxygen of the air involved in the combustion being in the solid oxides formed; these remaining gases occupy less space at a given temperature than the initial air. For the combustion of metal powders resulting in solid oxides, therefore, the pressure developed is due to the expansion of nitrogen by the heat of combustion. On the other hand, when sugar or starch dust burns in air, the volume of gaseous combustion products, carbon dioxide, nitrogen, and water vapor exceeds that of the initial air. Accordingly, the maximum pressure developed is due to a combination of increased volume of gases and temperature developed by the combustion.

Many dusts contain volatile matter, which can be evolved below the ignition temperature of the dust. This generally increases the combustion tendencies of the dust, as for example with coal dust (2). Figure 42.6 plots the explosibility of various coals, increasing with increasing volatile matter in the coal. (In Figure 42.6, note that pulverized coal contains dust particles whose size is −100 mesh, and 75 to 80 percent is −200 mesh, and mine size coal is all −20 mesh, with 20 percent being −200 mesh.) The data in Figure 42.6 indicate that coal dust containing less than 10 percent volatiles will not propagate explosions. Indeed, anthracite coal with about 12 percent volatiles shows little if any explosibility (2, 36). Early investigators of coal dust explosions in mines were well aware that the explosibility of coal is directly related to the coal's content of volatile matter, and more recent studies (37, 38) continue to demonstrate that this is true. Speculation leads to the postulate that in mine scale coal dust explosions, volatiles are pyrolyzed out of the dust in the preheat zone of the explosions, and homogeneous gas phase combustion follows after the volatiles interdiffuse with air to form a more or less continuous unburned combustible gas–air zone: the residual coked coal particles burn at a later stage. This is the so-called predistillation theory of the ignition of pyrolytic dusts. However, evidence exists to support contrary speculation that in large-scale explosions of pyrolytic dusts, combustion proceeds heterogeneously, with the volatiles of each

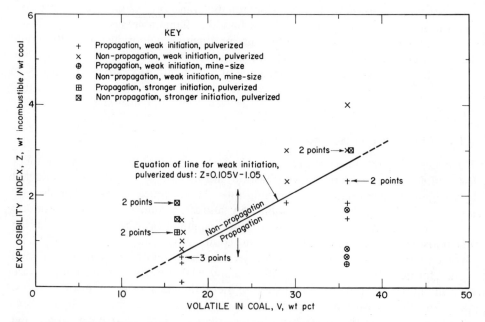

Figure 42.6. Explosibility index versus volatile in coal (36). Graph courtesy of Bureau of Mines, U.S. Department of the Interior.

particle burning while still associated chemically and physically with the burning parent particle (39, 40).

Some combustible dusts vaporize on heating, and their explosions involve essentially homogeneous combustion. The boiling points of these materials, such as aluminum and magnesium, are lower than their flame temperatures, and their heats of vaporization are smaller than the energy needed to initiate rapid surface oxidation. Under certain conditions, aluminum, magnesium, iron, lead, thorium, titanium, zirconium, and other metal powders can combine with nitrogen in the air to form nitrides, thereby increasing the consumption of the dust during an explosion and the maximum pressure that is developed.

3.1.2 Fineness and Physical Structure of Dust

The explosibility of dusts generally increases with decreasing particle size because small dust particles can be thrown into suspension more readily; moreover, they disperse more uniformly and remain in suspension longer than coarse dust. Also, because smaller particles have a greater specific surface, they can absorb more oxygen per unit weight and can oxidize more rapidly than larger ones. Greater electric charges can develop on finer particles because of greater electric capacitance. The combined effect of these factors is a lower ignition temperature and a lower ignition energy (Figure 42.7). A lower minimum explosive concentration, faster flame speed, and as shown in Figure 42.8, higher maximum pressures, are

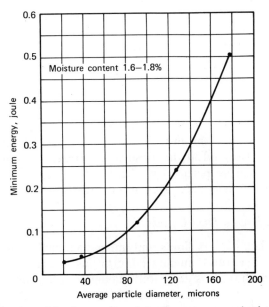

Figure 42.7.   Effect of particle size on the minimum energy required to ignite dust clouds of cornstarch in air by electric sparks (2). Graph courtesy of Bureau of Mines, U.S. Department of the Interior.

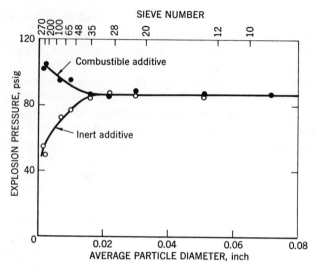

Figure 42.8. Effect of particle diameter of additive at 1.0 oz/ft³ concentration on explosion pressure of cornstarch at 0.5 oz/ft³ concentration (41). Graph courtesy of Bureau of Mines, U.S. Department of the Interior.

also observed. An extreme example of the effect of fineness on flammability is the pyrophoric combustion of very finely divided powders such as those of magnesium, lead, iron, and uranium. In addition to fineness, particle shape and surface characteristics resulting in differences in surface area, density, adsorbed gas, or protective surface layers can affect flammability. For example, flat, thin aluminum and magnesium particles produced by stamping are more flammable than nearly spherical particles of the same mesh produced by atomization or milling. These effects have been reported by Hartmann and co-workers (2), as well as by others.

Fineness and physical structure of dust enters into the aerodynamics of formation of dust clouds. These characteristics as they pertain to coal dust explosion in mines have been studied by Singer and co-workers (42, 43) and by Rae (44). Minimum velocities of about 5 to 30 m/sec were reported for coal and rock dust particles in the range of −74 to −14 μm dust. For very thin layers, the lowest required air velocity was observed around 30 μm; for smaller particles the minimum air velocities increased with decreasing particle size because of increasing immersion of the particles in the viscous boundary layer of the airflow (42).

3.1.3 Concentrations of Dust Clouds

Measured lower flammability limits and upper flammability limits of dusts exist (2). The lower limit is the minimum uniform dust concentration in an oxidant, generally air, in which there is just barely enough dust to propagate flame through a cloud after ignition at a localized point. The upper limit is the maximum uniform dust concentration through which flame can barely propagate. Between these two limits there is a concentration or a small concentration range at which the maximum pressure developed by that mixture, or the maximum pressure rise rate, or a related

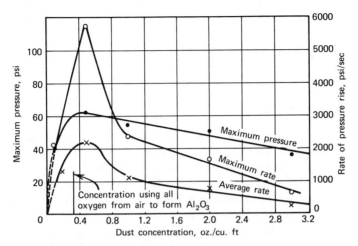

Figure 42.9. Maximum pressure and rates of pressure rise developed during explosions of aluminum powder at various concentrations (2). Graph courtesy of Bureau of Mines, U.S. Department of the Interior.

parameter, maximizes for the fuel–oxidant system. Figure 42.9 presents such data for the combustion of aluminum powder in air. Perhaps the lower limit is indicated by these data, but certainly the upper limit is poorly defined, even though the measurements extend to eightfold stoichiometric. Similar findings have been obtained with other metal powders, metallic hydrides, and bituminous coal (2). The difficulty in establishing flammability limits of dusts stems from several sources. The role of fineness and physical structure of dust has been discussed in Section 3.1.2; such complications could be handled by developing a family of lower and upper flammability limits for each combustible dust. But the uncertainty level of flammability limits of combustible dusts stems from the transient nature of dust clouds, particularly with respect to local concentrations in a cloud as a function of time and space. Consider the case of a coal dust explosion in a coal mine. Immediately before, dust is at rest on the floor, roof, and walls of the passageways. The initial ignition may be due to an electric or frictional spark or to an explosive shot that ignites a relatively small pocket of flammable methane–air mixture (firedamp), which may explode with relatively little violence. If this quantity of methane were the only fuel available, the damage caused by such an explosion would be relatively slight and localized. However, with coal dust present, such a small explosion can generate an airflow capable of lifting coal dust into the air and can ignite the resulting cloud. Then the coupled processes of lifting dust ahead of the propagating explosion and propagation into the freshly formed cloud can continue as long as coal dust is available. Dust concentration ahead of the combustion front builds until the dust supply is nearly consumed, and dust concentration is not uniform during a large-scale dust explosion. Even if the nominal or average concentration is below a lower flammability limit of the dust, the real concentration in part of the passageway can be above this limit, making flame propagation feasible. Likewise if the nominal or average concentration is above an upper flammability limit of the dust, the real concentration in part of the passageway can be low enough to support flame propagation. Such observations have been made in the Experimental Mine of the U.S. Bureau of Mines during explosions, with flame filling the passageway in strong explosions and not reaching the walls in weak explosions (40).

3.1.4 Composition of Atmosphere

The partial pressure of oxygen or the percentage of oxygen in the atmosphere carrying a dust cloud influences the flammability of the mixture; usually if the initial pressure is atmospheric, the percentage of oxygen is the parameter followed. Dust clouds ignite at lower temperatures, their minimum ignition energy is less, and the concentration at the lower flammability limit is less in oxygen than in air. Pressure rise rates and final pressures reached by explosions in closed vessels are generally higher in oxygen than in air. Vitiated air, with less oxygen than in normal air, generally leads to decreased ease of ignition as evidenced by the increase in minimum ignition energy of aluminum dust clouds (Figure 42.10). Such trends are also encountered in fires of solid fuels and are a crucial factor in the safe operation of hyperbaric chambers (14, 15). Furthermore, with some dusts a critical maximum oxygen concentration exists below which flame will not propagate. For example, the limiting oxygen concentration for flame propagation of bituminous coal dust clouds is between 16 and 17 percent oxygen (45). It must be remembered that

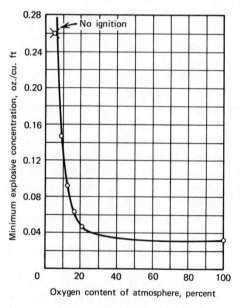

Figure 42.10. Effect of oxygen content of atmosphere on the minimum explosive concentration of atomized aluminum dust clouds (2). Graph courtesy of Bureau of Mines, U.S. Department of the Interior.

elimination of oxygen does not prevent flame propagation of all dusts, because a number of metal powders including zirconium, magnesium, titanium, thorium, and some magnesium–aluminum alloys can be ignited by electric sparks and will propagate flame when dispersed in carbon dioxide. Diluents such as nitrogen, carbon dioxide, water vapor, argon, and helium vary in their effectiveness because of differences in specific heat, thermal conductivity, and thermal molecular dissociation. Atmospheric humidity introduces relatively minor concentrations of water vapor into a dust cloud, thus appearing to have little direct effect on the flammability of dust clouds. However, high humidity favors agglomeration of particles and reduces their dispersibility. More important, high humidity also promotes leakage of static electricity and thereby reduces the hazard of ignition by static electrical sparks.

3.1.5 Effects of Ignition Source and Explosion Chamber

Dust clouds in contact with hot surfaces for long periods of time can often be ignited at lower temperatures than when exposed to nearly instantaneous contact. Large ignition sources are generally more effective than small ones. Electric sparks and arcs are less effective than open flames and hot surfaces; data in Tables 42.7 and 42.8 indicate that ignition in a furnace at 700°C is more readily achieved than by sparking and that layers of dust are ignited at lower temperatures than clouds of dust. However, the temperatures in Table 42.8 cannot be regarded as fundamental properties of the fuel–oxidant mixture to the degree that minimum autoignition temperatures (AIT) and minimum ignition temperatures T_L in Table 42.2 may be regarded as fundamental properties of the mixture; the data in Table 42.8 are apparatus dependent and mixture dependent. Nevertheless, because the equipment and procedures were fairly constant throughout the many years of the investigations, the data are internally sound and constitute important guides on the relative explosibility of dusts examined. Ignition sources can strongly influence the possibility of ignition, the amount of inerting agent needed to prevent flame propagation, and the intensity of an explosion, particularly in large-scale chambers and passageways such as in coal mines (36, 44, 46). Turbulence in fast-flowing dust clouds leads to more violent explosions, as in a grinding mill, than in a slow-moving dust cloud of the same composition (2).

3.2 Laboratory Studies of Dust Explosibility

Potential explosion hazards of about 2500 dusts have been measured by means of laboratory tests and reported in a series of Bureau of Mines publications, including reports on the explosibility of agricultural (47), carbonaceous (48), chemical (49), metallic (50), plastic (51), and miscellaneous dusts (33). In addition, similar data have been published by other organizations such as the National Fire Protection Association (4). Some of these data are in Tables 42.7 to 42.9, based on Reference 33. Because the initial purpose of the Bureau of Mines investigators was to obtain laboratory results comparable to those obtained in tests in their Experimental Mine at Bruceton, Pennsylvania, it is likely that the equipment and procedures are tuned

Table 42.7. Description, Moisture Content, Particle Size, and Explosibility Index of Dusts (33)

Line No.	Moisture Content (percentage)	Particle size, (percentage finer than micrometer size)	Explosibility Index	Materials
		Agricultural Dusts		
1	—	80/74	—	Coconut shell charcoal (2% moisture, 18% volatile matter, 78% fixed carbon, 2% ash)
2	—	98/74	—	Coconut shell dust, raw
3	0	95/74	>10	Cork dust, drom factory dust collector
4	—	100/297	—	Cornstarch, 10%; stable metal oxide, 90%
5	—	100/297	—	Cornstarch, 20%; stable metal oxide, 80%
6	—	100/297	—	Cornstarch, 30%; stable metal oxide, 70%
7	—	100/297	—	Cornstarch, 50%; stable metal oxide, 50%
8	—	100/297	0.6	Cornstarch, 70%; stable metal oxide, 30%
9	0.6	100/44	6.3	Dextrin, U.S. Pharmacopoiea grade
10	4.9	42/74	—	Grain, mixed (oats, barley, wheat), with addition of linseed oil, cake meal, soybean oil, fish meal, bone meal, limestone, and salt
11	0	77/44	0.2	Grass seed
12	2.0	100/74	—	Horseradish
13	—	100/250	—	Rockweed, airdried
14	—	53/74	—	Wheat bran, with organic reducing material
		Cellulosics		
15	3.0	100/840	—	Cellucotton, coarse broke waste
16	4.5	100/74	—	Cellucotton, fluff, normal product
17	4.2	100/840	—	Cellucotton, from suction duct
18	4.0	100/840	—	Cellucotton, from beams and pipes in storeroom
19	—	70/74	—	Cellucotton, from dryer discharge
20	—	98/74	—	Cellucotton, contains paper, clay, starch, wood pulp, alum, casein, and color dyes, from balcony
21	—	100/74	0.1	Cellucotton wadding, treated with zinc peroxide, from cyclone collector

Table 42.7. (*Continued*)

Line No.	Moisture Content (percentage)	Particle size, (percentage finer than micrometer size)	Explosibility Index	Materials
		Cellulosics		
22	—	100/74	2.0	Cellucotton wadding, treated with zinc peroxide, from air ducts of ventilation system
23	—	100/840	—	Celotex, fiberboard
24	2.3	100/74	6.0	Ethyl hydroxyethyl cellulose
25	2.2	100/74	6.9	Hydroxyethyl cellulose
26	2.3	100/74	>10	Hydroxypropl cellulose
27	2.1	1/74	—	Hydroxypropyl methyl cellulose (29% methoxy, 6% hydroxy pro-poxy, 65% cellulose)
28	0	100/74	<0.1	Ligninsulfate paper waste
29	0	100/74	0.8	Ligninsulfate paper waste
30	0	100/74	—	Ligninsulfonate, desulfonated and demethylated
31	0	100/74	<0.1	Ligninsulfonate, sodium, Marasperse N, 76%, with 10% organic and 14% inorganic ingredients
32	0	100/74	<0.1	Ligninsulfonate, sodium, Maracell A, 44%, with 21% organic and 35% inorganic ingredients
33	0	100/74	<0.1	Ligninsulfonate, sodium, Maracell A, heat-treated
34	2.9	100/74	0.2	Ligninsulfonate mixture, RDA 49-116
35	0	100/74	0.2	Ligninsulfonic acid, calcium, spray-dried
36	0	100/74	0.1	Ligninsulfonic acid, calcium, standard
37	0	100/74	0.2	Ligninsulfonic acid, calcium, spray-dried
38	1.2	61/74	>10	Paper coating solids, from drying oven of paper coating operation
39	0	100/840	0.8	Paper dust, from dust collector
		Chemicals, Gums, and Resins		
40	0.1	100/74	—	Acetal copolymer, Celcon, molding material
41	0	100/74	0.3	Acrylamide and methyl methacrylate copolymer
42	1.1	100/74	>10	Acrylonitrile–butadiene–styrene (22-18–60 copolymer)

continued on next page

Table 42.7. (*Continued*)

Line No.	Moisture Content (percentage)	Particle size, (percentage finer than micrometer size)	Explosibility Index	Materials
				Chemicals, Gums, and Resins
43	—	23/74	—	Batu gum regin
44	0	100/74	0.2	Bone glue, steer bones
45	—	—	—	Cream of tartar (potassium acid tartrate)
46	0	100/74	4.9	1,4-Cyclohexylene dimethylene isophthalate and 1,4-cyclohexylene dimethylene terephthalate copolymer
47	2.1	100/74	—	Dodecyl diphenyl oxide and disodium sulfonate
48	0	100/74	—	Formaldehyde-naphthalene sulfonic acid copolymer, sodium salt of, from cyclone dust collector
49	0.3	53/74	—	Latex mastic material (Neoprene, cement, marble chips, accelerator), for deck covering
50	0.5	100/44	—	Molybdenum disulfide, <5 μm
51	0.9	—	—	Polystyrene, rubber modified
52	0.7	100/44	>10	Rosin residue, extracted from pinewood
53	—	100/149	1.8	Santowax R, mixture of o-, m-, and p-terphenyls
54	0	100/74	—	Tanning extract, vegetable, spray-dried
				Detergents and Soaps
				*Detergents*
55	4.6	100/149	—	Detergent (approximately 30–40% alkyl aromatic sulfonates combined with inorganic builders, including sodium sulfate, sodium tripolyphosphate, and sodium silicate)
56	3.9	100/149	—	Detergent (approximately 30–35% alkyl aromatic sulfonates combined with builders consisting chiefly of sodium sulfonate and small amounts of pyrophosphate)
57	1.6	100/74	—	Detergent, NS-NR, S
58	0.6	96/74	—	Detergent, low active grade (38% sodium alkylaryl sulfonate, 62% sodium sulfate)

Table 42.7. (*Continued*)

Line No.	Moisture Content (percentage)	Particle size, (percentage finer than micrometer size)	Explosibility Index	Materials
			Detergents and Soaps	
59	0	100/74	—	Detergent, (70% sodium alkylaryl sulfonate, 27% sodium sulfate, 3% unsulfonated oil)
60	2.8	100/74	—	Detergent fines (85% alkyl sulfate plus inorganic chlorides and sulfates)
61	5.7	100/74	—	Detergent powder, S, derived from Alkane
62	3.8	72/74	—	Disinfectant rinse (33% monosodium phosphate, 15% sodium dodecyl-benzene sulfonate, 13% trichloro-melamine, 24% citric acid, 13% sodium bicarbonate, 2% polyethylene glycol)
			Soap Powder	
63	—	100/44	—	Soap powder (96% soap, 4% preservatives and additives)
64	—	100/44	—	Soap powder
65	2.8	100/74	—	Soap powder
66	3.1	100/74	—	Soap powder
67	1.2	100/74	—	Soap powder, fines (2% anhydrous soap)
68	1.0	100/74	0.7	Soap powder, fines (96% anhydrous soap)
69	3.6	100/74	0.2	Soap powder, from dust collectors
70	0.7	100/74	0.5	Soap powder, from dust collectors
71	1.1	100/74	1.0	Soap powder, from dust collectors
72	0.6	98/74	0.6	Soap powder, from filters
73	0	100/74	—	Soap powder aggregate (58% anhydrous soap)
74	1.7	100/74	0.4	Soap powder mixed with cadmium and barium silicates
75	0	100/149	<0.1	Soap product (65% anhydrous soap plus inorganic silicates and phosphates)
76	—	100/44	0.2	Aluminum, atomized, 35% (average 10 μm); barium nitrate, 65% (average 15 μm)

continued on next page

Table 42.7. (*Continued*)

Line No.	Moisture Content (percentage)	Particle size, (percentage finer than micrometer size)	Explosibility Index	Materials
		Detergents and Soaps		
77	—	100/74	<0.1	Aluminum, grade B, 40% (average 16 μm); barium nitrate, 30% (average 50 μm); potassium perchlorate, 30% (average 18 μm)
78	—	100/74	—	Aluminum–magnesium alloy, 45.0% (60% finer than 74 μm); barium nitrate, 55.0% (95% finer than 74 μm)
79	—	100/840	—	Ammonium nitrate
80	—	100/840	—	Ammonium nitrate, 78%; potassium nitrate, 4%; charcoal, 16%; No. 2 fuel oil, 2%
81	0	100/30	—	Ammonium nitrate, 94%; carbon, 6%
82	4.0	100/840	—	Ammonium nitrate-fuel oil mixture (96% ammonium nitrate, 4% No. 2 fuel oil)
83	5.4	100/840	—	Ammonium nitrate-fuel oil mixture (94% ammonium nitrate, 10–200 μm, 6% No. 2 fuel oil)
84	8.1	100/840	—	Ammonium nitrate-fuel oil mixture (90% ammonium nitrate, from 10–200 μm; 10% No. 2 fuel oil)
85	—	100/74	—	Ammonium nitrate No. 3, 73%; sodium nitrate, 10%; starch 5%; wood pulp, 12%
86	0	100/74	—	Ammonium perchlorate, 3–50 μm
87	—	100/74	—	Ammonium picrate, 50%; sodium nitrate, 45%; resin binder, 5%
88	—	—	—	Anthracene, 45%; potassium perchlorate, 55%
89	—	100/74	—	Barium nitrate, average 15 μm
90	—	100/149	—	Black powder mixture (91% No. 6 grade A black powder, 9% Acrawax)
91	0.2	100/74	7.2	Dinitrobenzamide
92	0.1	100/74	4.0	Dinitrobenzoic acid
93	0.3	100/74	0.7	Dinitro-*sym*-diphenylurea (dinitrocarbanilide)
94	0.1	100/74	>10	Dinitrotoluamide (3,5-dinitro-*o*-toluamide)
95	—	100/149	—	Guanidine nitrate

Table 42.7. (*Continued*)

Line No.	Moisture Content (percentage)	Particle size, (percentage finer than micrometer size)	Explosibility Index	Materials
			Detergents and Soaps	
96	0.8	100/74	—	Napalm (78% ferronapalm, 16% Pfister napalm, 6% caustic calcine magnesia)
97	—	100/840	—	Nitroguanidine
98	—	100/840	—	Nitroguanidine
99	—	100/840	—	Nitroguanidine
100	—	100/840	—	Nitroguanidine
101	—	100/840	—	Nitrostarch
102	—	100/840	—	Potassium chlorate, 50%; antimony sulfide, 30%; dextrin, 20%
103	—	100/44	—	Potassium perchlorate, <18 μm
104	—	100/840	—	Silicon, 20%; no binder; magnesium, grade B, 20%; lead chromate, 60%
105	—	100/840	—	Silicon, 20%; with binder; magnesium, grade B, 20%; lead chromate, 60%
106	—	100/840	<0.1	Silicon, 10%; lead oxide (red lead), 90%
107	—	100/74	—	Silicon, 6%; manganese, 40%; lead oxide (red lead), 54%
108	—	100/840	—	Silicon, 26%; aluminum, 13%; ferric oxide, 22%; potassium nitrate, 35%; carbon, 4%
109	—	100/149	—	Sodium nitrate
110	—	100/840	—	Thermite (28% aluminum, 72% iron oxide), 58%; plus igniter—barium 28%; nitrate
			Feeds and Fertilizers	
112	0	100/74	<0.1	Animal meal, defatted (bones, meat)
113	0	100/74	0.2	Animal meal, defatted (bones, meat), from machinery and ledges, processing plant
114	2.0	98/74	—	Bone meal, from machinery and ledges in the grinding room
115	2.0	94/74	—	Bone meal, from packaging department
116	2.3	95/74	1.0	Fermentation mash, spray-dried
117	—	100/74	—	Fertilizer (sodium nitrate, tankage, fish meal, potassium chloride, calcium triorthophosphate)

continued on next page

Table 42.7. (*Continued*)

Line No.	Moisture Content (percentage)	Particle size, (percentage finer than micrometer size)	Explosibility Index	Materials
			Feeds and Fertilizers	
118	4.0	100/74	—	Guano, bat, from cave in Grand Canyon
119	—	100/74	—	Guano, Gaviota, containing free sulfur
120	—	100/74	—	Hoof and horn meal
121	2.6	100/74	—	Molasses fermentation residue (49% nitrogen free extract, 4% moisture, 28% crude protein, 4% crude fiber, 15% ash)
122	3.2	100/44	0.1	Organisms, 65%; dextrin, 20%; thiourea, 5%; ammonium chloride, 5%; ascorbic acid, 5%
123	—	100/44	0.8	*Serratia marcescens* (protein) cells, 90%; miscellaneous material, 8%; moisture, 2%
124	0	100/74	1.0	*Serratia marcescens* (protein) cells, 47%; sucrose, 34%; skim milk, 14%; thiourea, 5%
125	1.8	100/74	1.2	*Serratia marcescens* (protein) cells, 66%; dextrin, 19%; ammonium chloride, 5%; ascorbic acid, 5%; thiourea, 5%
126	5.7	100/840	<0.1	Sewage, from ventilator openings in storage building of disposal plant
127	0	100/74	0.2	Sewage sludge, heat-dried, consisting of mixture of brown dust and vari-colored chaff
128	1.3	100/74	—	Sewage sludge, black, dust deposit in breeching
129	0	100/74	—	Sewage sludge, dried, fertilizer product
130	0	100/74	—	Sheep manure
131	8.0	80/74	—	Water-soluble proteins and carbohydrates
132	7.6	87/74	—	Water-soluble proteins and carbohydrates
133	5.6	90/74	—	Water-soluble proteins and carbohydrates
134	5.2	43/74	—	Wood flour, 55%, plus wheat bran and organic reducing materials
135	4.6	23/74	—	Wood flour mixed with desiccated animal tissue

Table 42.7. (*Continued*)

Line No.	Moisture Content (percentage)	Particle size, (percentage finer than micrometer size)	Explosibility Index	Materials
		Plant Dusts		
		Agricultural		
136	7.5	39/74	—	Corn dust, from cyclone collector in cleaning operation, prior to separation, cereal manufacturer
		Carbonaceous		
137	0.8	43/74	—	Boilerhouse dust, from vicinity of a porthole in a duct leading to a 135-ft stack, steel manufacturer
138	0	32/74	—	Boilerhouse dust (22 percent moisture), from bottom of stack, slurry-like, where dust had accumulated to a depth of 25 ft, steel manufacturer
139	—	—	—	Carbonaceous dust, particle size from 15 to over 100 µm, from pitch conveyor in paste plant, aluminum manufacturer
140	—	—	—	Carbonaceous dust, particle size from 15 to over 100 µm, from I-beams of paste plant, third floor, carbon department, aluminum manufacturer
141	—	—	—	Carbonaceous dust, particle size from 15 to over 100 µm, from I-beams of paste plant, fourth floor, carbon department, aluminum manufacturer
142	—	—	—	Carbonaceous dust, particle size from less than 5 to about 70 µm, from screw-conveyor, paste plant, sixth floor, carbon department, aluminum manufacturer
143	0.8	79/74	—	Coal dust (1% moisture, 21% volatile matter, 51% fixed carbon, 27% ash (contains 1.6% sulfur)), from I-beam in electric power station, boiler room
144	1.0	40/74	—	Coal dust (54 percent ash), from fourth floor level above boilers,

continued on next page

Table 42.7. (*Continued*)

Line No.	Moisture Content (percentage)	Particle size, (percentage finer than micrometer size)	Explosibility Index	Materials
			Carbonaceous	
				near head of coal dump, at forced draft fan level, steam generating plant, steel manufacturer
145	0.5	85/74	—	Coal dust–coke breeze mixture (52% ash), from above boiler, steam generating plant, steel manufacturer
146	0.8	52/74	—	Flue dust (55% ash), from blast furnace gas, collected from rafters on fourth floor level above boiler at forced draft fan level, steam generating plant, steel manufacturer
147	0.7	73/74	—	Flue dust (88% ash), from above is burned, steam generating plant, steel manufacturer
			Metal Dusts	
148	0.5	71/74	—	Aluminum (42% aluminum, 7.5% oil, plus lint and fibers), from filters above cyclone collector used in connection with grinding aluminum castings, tool manufacturer
149	0	71/74	<0.1	Aluminum with oil removed
150	1.5	95/74	—	Aluminum, grained, plus dust from crane rail, metal alloy manufacturer
151	0.8	90/74	—	Aluminum, grained, plus dust on roof of small enclosure in main building, metal alloy manufacturer
152	0.4	91/74	—	Aluminum, from collector on grinding machines, metal alloy manufacturer
153	0.8	90/74	—	Aluminum, from bag in portable vacuum cleaner used around air conditioning machinery, metal alloy manufacturer
154	0.8	94/74	—	Iron (89% ash), from building trusses, malleable iron manufacturer
155	1.6	81/74	—	Iron (70% ash), from building trusses in coreroom, malleable iron manufacturer

Table 42.7. (*Continued*)

Line No.	Moisture Content (percentage)	Particle size, (percentage finer than micrometer size)	Explosibility Index	Materials
		Metal Dusts		
156	0.9	99/74	—	Iron (83% ash), from building trusses in sand handling room, malleable iron manufacturer
157	0.5	95/74	—	Iron (43% ash), from carburizing department, malleable iron manufacturer
158	4.1	90/74	—	Iron (23% combustible), suspensions of dust settled on overhead surfaces above molding machine, iron foundry
159	0.1	77/74	—	Iron dust (21% iron plus lint and wool), tool manufacturer
160	—	100/840	—	Smelter dust, variety of metals, smelter and refining company
161	0.8	80/74	—	Titanium, contains wood, sand, paper, coal, concrete, slag, and flakes of paint, from around a collector, titanium plant
162	5.9	100/840	—	Titanium, contains wood, sand, paper, coal, concrete, slag, and flakes of paint, from outside bag collector, titanium plant
163	0.5	100/74	—	Titanium, residual dust from abrasive grinding of titanium billets, titanium plant
164	0.2	100/74	—	Titanium, residue from grinding titanium sheets, titanium plant
165	1.4	100/74	—	Titanium, from area above crushers in sponge preparation room, titanium plant
166	0.8	100/74	—	Titanium, from area above blender in sponge preparation room, titanium plant
167	2.9	100/74	—	Titanium, from around feeders in melt shop, titanium plant
168	2.6	100/74	—	Titanium (24% titanium, 30% carbonaceous material, 13% silicon dioxide, 4% manganese, 13% iron, 16% metal oxides), from structural members in furnace building, titanium plant

continued on next page

Table 42.7. (*Continued*)

Line No.	Moisture Content (percentage)	Particle size, (percentage finer than micrometer size)	Explosibility Index	Materials
		Metal Dusts		
169	0.6	100/74	<0.1	Titanium, mixed with iron, from grit blasting, titanium plant
170	0.8	91/74	0.2	Zinc (5% zinc, 4% silicon dioxide, 55% carbon, 36% ash), from beams and platform near hammer-mill coal crusher, zinc smelter plant
171	0.9	97/74	—	Zinc (32% zinc, 10% silicon dioxide, 26% carbon, 32% ash), from roof beams, zinc smelter plant
172	1.1	98/74	—	Zinc (22% zinc, 8% silicon dioxide, 38% carbon, 32% ash), from roof beams, zinc smelter plant
		Plastics		
173	1.3	29/74	—	Buffring machine dust, a mixture of Liquabrade 4787 (Lea Liquid buffing compound), cotton buffing wheel lint, and cellulose acetate butyrate molding compound, electrical manufacturer
174	0.5	36/44	0.9	Residue, including polyvinyl chloride resins, compound fines, dioctyl-phthalate (plasticizer), calcium stearate, coloring pigments, clays, and titanium oxide, from structural framework in millroom, chemical company
175	0.7	7/44	—	Residue, including polyvinyl chloride resins, compound fines, dioctyl-phthalate (plasticizer), calcium stearate, coloring pigments, clays, and titanium oxide, from structural framework above baggers and cubers, chemical company
176	0.8	15/74	—	Residue, including polyvinyl chloride resins, compound fines, dioctyl-phthalate (plasticizer), calcium stearate, coloring pigments, clays, and titanium oxide, from structural framework above mills, chemical company

Table 42.7. (*Continued*)

Line No.	Moisture Content (percentage)	Particle size, (percentage finer than micrometer size)	Explosibility Index	Materials
		Plant Dusts		
		Rubber		
177	—	100/840	—	Residue from rafters of tire recapping plant
178	—	100/840	—	Residue from under rafters of tire recapping plant
		Miscellaneous		
179	0.9	100/74	—	Blasting machine dust (quartz, hematite, magnetite, alpha iron) from collector used in cleaning airplane parts, aircraft maintenance depot
180	3.6	100/74	>10	Wood dust, from refinishing bowling alleys, first cut, with lacquer
181	5.6	100/74	>10	Wood dust, from refinishing bowling alleys, second or finishing cut

to simulate large-scale explosions, but one cannot be certain that such is the case for every measurement. The equipment used and procedures followed are given in Reference 32. The investigators (32, 33, 47–51), following the lead largely set by I. Hartmann at the Bureau of Mines, developed three empirical indexes: the index of sensitivity of ignition, the index of explosion severity, and the overall explosibility index (47). Values of selected relative parameters are compared to those of Pittsburgh coal dust, which by definition has an explosibility index of unity; a dust with greater explosion hazard has an explosibility index greater than unity. The ignition sensitivity and explosion severity indexes of a dust and the explosibility index are defined as follows:

$$\text{ignition sensitivity} = \frac{\left(\begin{array}{c}\text{ignition temperature} \times \\ \text{minimum ignition energy} \times \\ \text{minimum flammable concentration}\end{array}\right) \text{Pittsburgh coal dust}}{\text{(same parameters) other dust}}$$

$$\text{explosion severity} = \frac{\left(\begin{array}{c}\text{maximum explosion pressure} \times \\ \text{maximum rate of pressure rise}\end{array}\right) \text{Pittsburgh coal dust}}{\text{(same parameters) other dust}}$$

$$\text{explosibility index} = (\text{ignition sensitivity})(\text{explosion severity})$$

Table 42.8. Parameters Affecting Ignition (33)

Line No.[a]	Ignition Sensitivity	Minimum Ignition Temperature (°C)		Minimum Ignition Energy (J)		Minimum Explosive Concentration (oz/ft³)	Relative Flammability (percentage inert)		Limiting Oxygen Concentration,[b] (percentage)	
		Cloud	Layer	Cloud	Layer		In Spark Apparatus	Furnace, 700°C	Spark	Furnace, 850°C
						Agricultural Dusts				
1	—	730	—	—	—	—	—[c]	—	—	—
2	—	470	—	—	—	—	80	90+	—	—
3	3.6	460	210	0.035	—	0.035	85	90+	—	—
4	—	500	—	—	—	—	—[b]	—	—	—
5	—	495	—	—	—	—	Arc[d]	—	—	—
6	—	495	—	—	—	—	Arc[d]	—	—	—
7	<0.1	480	—	0.080	—	0.700	—[d]	—	—	—
8	0.7	485	—	0.060	—	0.105	—[d]	—	—	—
9	2.5	410	440	0.040	—	0.050	—[d]	—	N, 10; C, 14	—
10	—	460	—	—	—	0.120	75	90+	—	—
11	0.5	530	—	0.060	—	0.140	—[d]	—	—	—
12	—	—[c]	—	—	—	<0.100	—[c]	90+	—	—
13	—	520	—	—	—	—	—[c]	90+	—	—
14	—	550	—	—	—	—	60	90+	—	—
						Cellulosics				
15	—	610	300	—	—	0.160	—[d]	—	—	—
16	—	650	—	—	—	<0.200	—[d]	—	—	—
17	—	440	—	—	—	0.075	—[d]	—	—	—
18	—	510	—	—	—	0.060	—[c]	60	—	—
19	—	590	—	—	—	—	—[c]	75	—	—
20	—	550	—	—	—	—	80	90+	—	—
21	0.2	480	280	0.320	—	0.055	85	90+	—	—
22	1.4	470	270	0.060	—	0.050	85	—	—	—
23	—	—[c]	—	—	—	—	—[f]	—	—	—
24	8.6	390	—	0.030	—	0.020	—[d]	—	C, 16	—
25	4.9	410	—	0.040	—	0.025	—[d]	—	—	—
26	8.4	400	—	0.030	—	0.020	—[d]	—	—	—
27	—	430	—	—	—	0.800	5	85	—	—
28	<0.1	380	350	0.390	—	0.200	20	85	C, 18	C, 5

#											
29	0.6	530	260	—	0.080	—	0.085	50	90+	—	C, 4
30	—	390	—	—	—	—	0.230	—d	90+	C, 17	C, 4
31	0.2	490	230	—	0.140	—	0.200	45	90+	C, 17	C, 4
32	<0.1	400	340	—	0.260	—	0.350	30	90+	C, 17	C, 4
33	0.2	390	350	—	0.140	—	0.250	45	85	—	—
34	0.2	650	330	—	0.140	—	0.150	10	60	—	—
35	0.2	560	410	—	0.160	—	0.095	65	55	—	—
36	0.1	670	410	—	0.240	—	0.120	40	70	—	—
37	0.1	590	470	—	0.160	—	0.150	50	—	—	—
38	3.7	390	170	—	0.020	—	0.070	—d	—	—	—
39	1.4	440	270	—	0.060	—	0.055	—d	—	—	—

Chemicals, Gums, and Resins

#											
40	—	470	—	—	—	—	0.060	—d	—	—	—
41	1.5	510	—	—	0.060	—	0.045	—d	—	—	—
42	7.1	470	—	—	0.030	—	0.020	90+	90+	—	—
43	—	420	—	—	—	—	—	75	90	—	—
44	0.3	550	—	—	0.140	—	0.030	70	90+	—	—
45	—	520	—	—	—	—	2.000	—	60	—	—
46	5.4	500	360	—	0.025	—	0.165	70	—	C, 13	—
47	<0.1	540	420	—	8.320	—	—b	hcd	—	—	—
48	—	620	440	—	—b	—	—	20	80	—	—
49	—	640	290	—	—b	—	0.800	—c	45	—	—
50	—	570	—	—	—	—	—	—c	75	—	—
51	—	460	—	—	0.015	—	0.016	50	90+	C, 17	C, 8
52	>10	470	—	—	—	—	.035	90	90+	—	—
53	1.2	620	—	—	.080	—	.200	90+	90+	—	—
54	—	650	340	—	—	—	—	—d	90	—	—

Detergents and Soaps

#											
55	—	510	260	—	—	—	—	gcd	90	—	—
56	—	520	250	—	—	—	—	gcd	90+	—	—
57	—	530	280	—	—	—	—	gcd	90+	—	—
58	—	540	300	—	—	—	—	—d	90+	—	—
59	—	300	570	—	—	—	0.130	30	90+	—	—
60	—	530	—	—	—	—	—	70	90	—	—
61	—	660	390	—	—	—	—	gcd	90	—	—
62	—	—	—	—	—	—	—	—b	90+	—	—

continued on next page

Table 42.8. (Continued)

Line No.[a]	Ignition Sensitivity	Minimum Ignition Temperature (°C)		Minimum Ignition Energy (J)		Minimum Explosive Concentration (oz/ft³)	Relative Flammability (percentage inert)		Limiting Oxygen Concentration[b] (percentage)	
		Cloud	Layer	Cloud	Layer		In Spark Apparatus	Furnace, 700°C	Spark	Furnace, 850°C
Detergents and Soaps										
Soap Powder										
63	—	635	—	—	—	—	—[c]	60	—	—
64	—	580	—	—	—	—	—[c]	80	—	—
65	—	—	—	—	—	<0.045	—[d]	—	—	—
66	—	—	480	—	—	—	—[d]	90+	—	—
67	0.3	560	500	—	—	—	gc[d]	90+	—	—
68	0.4	640	380	0.120	—	0.085	90	90+	—	—
69	0.6	600	600	0.120	—	0.075	55	85	—	—
70	0.9	430	460	0.100	—	0.085	70	90+	—	—
71	0.8	600	450	0.060	—	0.060	85	90+	—	—
72	—	540	310	0.100	—	0.045	80	90+	—	—
73	—	630	260	0.240	—	—	25	90+	—	—
74	0.4	380	430	0.120	—	0.125	80	90+	—	—
75	<0.1	650	—	0.960	—	0.070	70	90+	—	—
Explosives and Related Compounds										
76	0.1	700	—	0.120	0.032	0.200	65	5	N. 5	C. 3
77	0.1	700	—	0.375	—	0.270	55	10	N. 11	C. 3
78	—	440	—	—	0.056	0.430	25	55	N. 11	C. 3
79	—	400	190	—	—	—	—[c]	30	—	—
80	—	490	200	—	—	—	gc[d]	—	—	—
81	—	360	—	—	—	—	hc[d]	80	—	—
82	—	390	—	—	—	—	20	65	—	—
83	—	380	—	1.600	—	2.000	45	75	—	—
84	—	370	160	0.104	—	0.370	—[d]	75	—	—
85	—	310	—	—	—	—	—[c]	65	—	—
86	—	—	260	—	—	—	—[d]	—	—	—
87	—	—	250	0.160	—	0.200	—[c]	—	—	—
88	—	530	700	—	—	0.160	65	90+	—	—
89	—	—	—	—	—	—	—[c]	—	—	—

continued on next page

No.									Ref.
90	0.2	340	—	0.320	—	0.120	—d	—	
91	2.2	500	—	0.045	—	0.040	85	90+	
92	1.9	460	—	0.045	—	0.050	80	90+	
93	0.6	550	—	0.060	—	0.095	85	90+	C. 13
94	5.4	500	—	0.015	0.024	0.050	90	90+	N. 8
95		850	—	—	—	—	—c	—c	
96	5.6	450	—	0.040	—	0.020	90	90+	N. 12
97	—	400	—	<7.200	—	—	—d	80	
98	—	670	—	<7.200	—	—	—d	55	
99	—	680	—	—	—	—	—c	70	
100	—	850	—	—	—	—	—d	(²)	
101	3.8	190	165	0.040	—	0.070	25	80	
102	—	280	300	—d	—	0.070	—c	90	
103	—	—	—	8.000	0.004	0.265	15	90+	
104	0.1	620	520	—c	0.320	<3.000	—d	25	
105	—	650	520	0.360	0.0016	<3.000	10	48	
106	0.1	540	540	0.350	—	0.650	10	15	
107	0.1	540	450	—d	—	—	—c	—	
108	—	<700	400	—	—	—	—	(²)	
109	—	—	—	0.240	0.400	0.760	15	—	
110	0.1	720	—	0.075	—	0.070	—d	—	
111	—	—	—						

Feeds and Fertilizers

No.									Ref.
112	0.3	530	350	0.180	—	0.065	75	90+	
113	0.5	530	310	0.120	—	0.060	80	90+	
114	—	490	230	—	—	—	gc^d	85	
115	0.7	560	250	0.080	—	0.070	gc^d	75	
116	—	500	310	—	—	—	80	90+	
117	—	—c	380	—	—	—	arc, gc^d	90+	
118	—	460	240	—	—	—d	30	90+	
119	—	330	—	—	—	—	45	85	
120	—	660	240	—	—	0.155	—d	90+	
121	0.3	660	410	0.100	—	0.080	65	90+	
122	0.6	500	200	0.080	—	0.130	—d	—	
123	0.4	490	180	0.080	—	0.080	—d	—	N. 14
124	0.6	470	200	0.080	—	0.165	—d	—	N. 14
125	0.3	490	190	0.120	—	0.095	—	90+	N. 14
126	0.7	390	160	0.080	—		60		
127		390							

Table 42.8. (Continued)

Line No.[a]	Ignition Sensitivity	Minimum Ignition Temperature, (°C)		Minimum Ignition Energy (J)		Minimum Explosive Concentration (oz/ft³)	Relative Flammability (percentage inert)		Limiting Oxygen Concentration.[b] (percentage)	
		Cloud	Layer	Cloud	Layer		In Spark Apparatus	Furnace, 700°C	Spark	Furnace, 850°C
Feeds and Fertilizers										
128	—	530	180	0.960	—	—	—[d]	90	—	—
129	—	420	—	—	—	—	—[d]	90+	C, 16	C, 5
130	—	730	—	—	—	—	—[c]	—	—	—
131	—	550	—	—	—	—	45	90+	—	—
132	—	620	—	—	—	—	—[d]	90+	—	—
133	—	520	—	—	—	—	—[d]	90+	—	—
134	—	480	—	—	—	—	80	90+	—	—
135	—	470	—	—	—	—	80	90+	—	—
Plant Dusts										
Agricultural										
136	—	430	290	—	—	—	75	90+	—	—
Carbonaceous										
137	—	—	510	—	—	—	—[c]	—	—	—
138	—	—	510	—	—	—	—[c]	—	—	—
139	—	690	—	—	—	—	—[c]	15	—	—
140	—	650	—	—	—	—	—[c]	45	—	—
141	—	630	—	—	—	—	60	80	—	—
142	—	660	—	—	—	—		40	—	—
143	—	560	180	—	—	—	arc, gc[d]	85	—	—
144	—	800	520	—	—	—	—[c]	—	—	—
145	—	740	510	—	—	—	—[c]	—	—	—
146	—	800	550	—	—	—	—[c]	—	—	—
147	—	—	590	—	—	—	—[c]	—	—	—
Metal Dusts										
148	—	470	580	—	—	0.200	15	85	—	—
149	<0.1	550	660	0.280	—	0.140	55	35	—	—
150	—	—	460	—	—	—	—[c]	—[c]	—	—

No.	Min. conc.	Ign. temp. cloud (°C)	Ign. temp. layer (°C)	Min. energy	Min. energy	Ignition source[d]	Explosibility (%)
151	—	770	450	—	—	—[c]	—[c]
152	—	—	510	—	—	—[c]	—[c]
153	—	—	440	—	—	—[c]	—[c]
154	—	750	410	—	—	gc[d]	85
155	—	470	220	—	—	—[c]	70
156	—	560	320	—	—	arc[d]	—
157	—	770	470	—	—	gc[d]	—
158	—	780	430	—	—	gc[d]	10
159	—	—	520	—	—	—[c]	30
160	—	700	—	—	—	—[c]	—
161	—	650	460	—	—	arc, gc[d]	70
162	—	740	480	—	—	20	70
163	—	950	380	—	—	arc[d]	60
164	—	740	690	—	—	30	50
165	<0.1	490	220	6.400	0.800	50	55
166	—	500	260	—	—	gc[d]	90
167	—	640	240	—	—	gc[d]	50
168	<0.1	480	250	—	—		80
169	<0.1	600	210	0.180	0.370		—
170	0.4	640	230	0.100	0.080		—
171	—	640	270	—	—		—
172	—	600	240	—	—		—
					Plastics		
173	<0.1	530	230	0.180	1.000	45	80
174	1.7	540	—	0.050	0.045	—[d]	—
175	—	580	—	—	—	—[d]	—
176	—	530	—	—	—	—[d]	—
					Rubber		
177	—	530	—	—	—	50	80
178	—	540	—	—	—	40	90
					Miscellaneous		
179	—	600	—	—	—	—[c]	—
180	4.6	360	180	0.035	0.035	90	90+
181	3.5	360	260	0.040	0.040	90	90+

[a] See Table 42.7 for identity of material.
[b] Prefix letter denotes diluent gas: C, carbon dioxide; N, nitrogen.
[c] No ignition.
[d] Dust ignited by spark except as noted: arc, carbon arc; gc, guncotton; hc, heated coil.

Table 42.9. Explosion Severity, Pressures, and Rates of Pressure Rise of Dust Explosions (33)

Line No.[a]	Explosion Severity	Maximum Pressure (psig)	Concentration 0.50 oz/ft^3	
			Rate of Pressure Rise (psi/sec)	
			Average	Maximum
Agricultural Dusts				
3	3.3	96	1700	6500
8	0.9	69	1800	2500
9	2.5	81	1800	6000
10	0.5	83	450	1200
11	0.3	52	300	900
12	0.4	78	600	900
Cellulosics				
15	0.6	80	400	1500
16	<0.1	43	100	250
17	2.0	88	1300	4300
18	2.3	98	1600	4500
21	0.7	78	1200	1800
22	1.4	87	1500	3100
24	0.7	84	800	1500
25	1.4	106	1200	2600
26	1.3	84	1200	2900
28	0.1	42	300	500
29	1.4	105	1100	2500
30	<0.1	38	150	300
31	0.2	75	400	600
32	0.3	62	500	800
33	0.6	77	600	1400
34	1.2	99	1000	2300
35	1.0	79	1000	2300
36	1.0	84	1000	2400
37	1.8	94	1100	3700
38	3.7	80	900	2700
39	1.8	96	1300	3600
Chemicals, Gums, and Resins				
40	2.4	71	2100	6500
41	0.2	57	400	600
42	1.8	71	1700	4700
44	0.6	73	500	1500
46	0.9	79	800	2200
48	0.2	68	200	500

Table 42.9. (*Continued*)

Line No.[a]	Explosion Severity	Concentration 0.50 oz/ft³		
		Maximum Pressure (psig)	Rate of Pressure Rise (psi/sec)	
			Average	Maximum
Chemicals, Gums, and Resins				
52	2.0	75	1600	5000
53	1.5	67	1200	4200
54	<0.1	15	<100	150
Detergents and Soaps				
Detergents				
55	<0.1	5	<100	100
57	<0.1	5	<100	100
59	1.4	108	800	2400
60	1.1	90	700	2300
61	<0.1	10	150	350
Soap Powder				
65	0.9	97	900	1700
66	<0.1	30	100	200
67	<0.1	39	200	400
68	2.4	116	1200	4000
69	0.5	74	800	1200
70	0.9	67	1200	2800
71	1.1	84	1300	2600
72	0.7	78	900	1800
73	0.2	61	250	500
74	0.9	69	900	2400
75	0.2	76	300	500
Explosives and Related Compounds				
76	2.0	77	2400	5000
77	0.3	56	450	900
78	1.1	68	1200	3000
87	0.9	74	1400	2400
91	3.2	94	2600	6500
92	2.1	92	1800	4300
93	1.1	87	1000	2500
94	5.6	106	3200	10000+
101	6.1	116	6600	10000+
111	0.7	63	600	2100

continued on next page

Table 42.9. (*Continued*

Line No.[a]	Explosion Severity	Concentration 0.50 oz/ft^3		
		Maximum Pressure (psig)	Rate of Pressure Rise (psi/sec)	
			Average	Maximum
Feeds and Fertilizers				
112	0.2	58	300	700
113	0.4	61	400	1200
116	1.4	81	1200	3200
118	<0.1	6	100	200
121	<0.1	23	100	200
122	0.4	63	500	1200
123	1.3	76	1300	3200
124	2.5	87	1600	5500
125	2.0	88	1200	4400
126	<0.1	35	150	400
127	0.3	58	450	900
129	0.1	49	200	400
Metal Dusts				
148	0.2	30	100	150
149	0.1	52	300	400
157	<0.1	12	100	150
169	<0.1	25	100	200
170	0.6	72	700	1700
171	<0.1	14	100	200
172	<0.1	19	100	150
Plastics				
173	—	—	—	—
174	0.5	64	700	1500
175	<0.1	10	<100	100
176	0.1	45	400	600
Rubber				
177	0.3	69	500	900
178	0.2	67	300	600
Miscellaneous				
180	4.0	115	3100	6700
181	3.0	99	1600	5700

[a]See Table 42.7 for identity of material.

The characteristics of the Pittsburgh coal dust (-74 μm) used as the reference dust are as follows:

Characteristic	Value
Explosibility index	1.0
Ignition sensitivity	1.0
Explosion severity	1.0
Minimum ignition temperature of dust cloud	610°C
Minimum ignition energy of dust cloud	0.06 J
Minimum explosive concentration	0.055 oz/ft^3
Maximum explosion pressure*	83 psig
Maximum rate of pressure rise*	2300 psi/sec

Explosion hazards were then classified in terms of these indexes as follows. [Carbonaceous materials having a dust cloud ignition temperature of 730°C or more and not ignitable by an electric spark were rated primarily as fire hazards (33).] Data in Tables 42.7 to 42.9 pertain to some of the tested dusts that evidenced an explosion hazard.

Explosion Hazard	Ignition Sensitivity	Explosion Severity	Explosibility Index
Weak	<0.2	<0.5	<0.1
Moderate	0.2–1.0	0.5–1.0	0.1–1.0
Strong	1.0–5.0	1.0–2.0	1.0–10
Severe	>5.0	>2.0	>10

These dusts are grouped as agricultural dusts, cellulosics, chemicals, gums, and resins; detergents and soaps; explosives and related compounds; feeds and fertilizers; and industrial plant dusts. A line number identifying the dust, a description of the dust, explosibility index (when applicable), moisture content, and particle size are given in Table 42.7. Particle size is designated by two numbers; the number to the left of the diagonal stroke indicates the percentage passing, and the number to the right the sieve opening, in micrometers. The dusts were passed through a No. 20 U.S. Standard sieve (840 μm). Table 42.8 contains the calculated ignition sensitivity and the related parameters of minimum ignition temperature, minimum ignition energy, and minimum explosive concentration. Data are given also on layer ignition temperature, the relative flammability (which is the percent by weight of calcined fuller's earth required in admixture to prevent flame propagation), and the limiting atmospheric oxygen required to prevent ignition by electric spark and furnace at 850°C. Relative flammability data are given for both the spark and furnace (700°C) ignition sources. The column of relative flammability (spark) is used also to indicate those dusts not igniting by spark but by stronger sources. Table 42.9 contains the calculated explosion severities and the explosion pressures and rates of pressure rise at a dust concentration of 0.50 oz/ft^3 (500 mg/liter). For a few dusts that did not ignite by spark, data are included for ignition by flame from guncotton and heated coil.

Additional data are in References 33 and 47 to 51; Reference 33 contains information on materials that presented primarily a fire hazard and those which did not present a dust explosion hazard.

3.3 Fire Hazard

In addition to the laboratory data in Reference 33 on fire hazards due to dusts, it must be emphasized that the hazard of dust layers catching fire is greater than that of the same material in bulk. In general, the dusts ignite at lower temperatures and burn more rapidly, and the fires are more difficult to extinguish than is the case for larger pieces. Should the fire-fighting efforts generate a flammable dust cloud, an explosion may follow. The ignition temperature of undispersed layers and minimum spark energies for ignition of dust layers, such as those in Table 42.8, are indexes of the potential fire hazard. Dust layers can smolder even in thin layers, presenting at times a hidden and prolonged hazard (2). Smoldering is possible at oxygen concentrations below that required for propagation of an explosion of that dust (45, 52); some indirect evidence of this hazard can be noted in Table 42.8, where the limiting oxygen concentrations are generally lower for furnace ignition of dust clouds than for their spark ignition. The minimum spark energies listed in Table 42.8 for ignition of dust layers are so low that fires obviously can be started in dusts by weak electrical sparks or arcs, including static sparks, as well as by hot surfaces, glowing particles, frictional sparks, open flames, or other ignition sources. Most dusts oxidize so slowly that no significant temperature rise occurs, but some dusts, including those of some agricultural products, activated carbons, and metal powders, can oxidize rapidly enough to combust spontaneously. In addition, some metal powders react with moisture, especially at elevated temperatures, and thereby facilitate their spontaneous ignition.

3.4 Prevention and Control of Dust Explosions and Fires

Dust explosions and fires constitute an ever-present hazard in industry whenever materials that engender combustible dusts are processed or stored. Too often a small dust explosion or a small dust fire starts a highly destructive chain of events. A small dust explosion disperses additional dust, which explodes more violently and in turn generates a still larger dust cloud, which explodes with greater force than before; the process continues till all the dust is exhausted or scattered fires in undispersed dust end the event. Alternatively, if a fire starts first in dust layers, enough dust may be stirred up to lead to an explosion, followed by more explosions. The problem of preventing dust explosions and fires has been investigated by several research organizations and has lead to codes such as that of the National Fire Protection Association (4) and related NFPA listings to be found therein. The

---

*Dust concentration of 0.50 oz/ft³.

reader is referred to these codes as a helpful guide in the design and construction of building and equipment and in regard to safe operating procedures.

Measures recommended to prevent fires and explosion of combustible dust include the following:

1. Ignition sources should be eliminated. Open flames, open lights, or smoking should be prohibited; electric or gas cutting or welding equipment is to be avoided, unless the vicinity is dustfree; all equipment that may produce electric static sparks should be grounded; the National Electrical Code, especially as it pertains to hazardous locations, is to be followed in electric installations and operations, and nonsparking fans, shafts, and belts are to be used wherever possible; magnetic separators should be used to prevent ignition of dusts by frictional sparks produced by tramp metal particles passing through equipment.

2. Buildings should be constructed to avoid collection of dust on beams, ledges, and other surfaces. Good housekeeping is necessary. In removal of dust, the use of compressed air is highly dangerous; vacuum or brushing is advisable.

3. Equipment in which dust clouds may be generated should be as dust tight as possible, but strong enough to contain explosion pressures; or it should be provided with venting adequate to prevent an explosion from being disastrous. Unrestricted openings, hinged windows, panels, light wall construction in rooms, and release diaphragms can be designed to release explosion pressures without damage (2). Recommendations on the proper area of vents and suitability of particular vent closures are in the National Fire Protection Guide for Explosion Venting (30) and in Reference 35. Recommended unrestricted venting areas range from 1 ft^2 for each 10 to 30 ft^3 of enclosure for small equipment and light construction, to 1 ft^2 per 80 ft^3 for large rooms and heavy reinforced concrete buildings (2).

4. Preferably dust collectors should be located outside buildings or in detached rooms with explosion vents. Ducts leading to and from collectors should be as short and straight as possible, and the blowers should be on the clean air side of collectors.

5. Grinding, conveying, and other equipment frequently can be protected by maintaining an inert atmosphere in the equipment, sufficient to reduce the oxygen content below that at which the dust will explode. Some examples of the maximum oxygen percentage for flame propagation in dust clouds are given in Table 42.8, and more data are in References 48 to 52. Methods of producing and using inert gas are in the NFPA standard for fire and explosion prevention by inerting (53).

6. The explosion hazard can sometimes be combated by adding inert dusts to the combustible dust, as is done in coal mines to prevent the propagation of coal dust explosions (36), or by using flame inhibitors such as alkali bicarbonates or Halons (20, 45). Devices triggered by the early stages of an explosion to eject inert dust or water or flame inhibitors into the propagating explosion wave are presently being developed (39).

7. Equipment for fighting fires of flammable dusts should not be likely to produce a dust cloud, with a consequent explosion. Water pails, soda–acid extinguishers, hand-operated water pump tanks, hoses with spray-type or fog nozzles,

foam nozzles, and automatic sprinklers are satisfactory for use with most dusts. Small fires in magnesium, aluminum, or other metal powders are best extinguished by sand, talc, limestone, soapstone, or other dry inert powders.

REFERENCES

1. H. F. Coward and G. W. Jones, "Limits of Flammability of Gases and Vapors," U.S. Bureau of Mines Bulletin No. 503, 1952, 155 pp.; NTIS [National Technical Information Service] AD701575.

2. F. A. Patty, Ed., *Industrial Hygiene and Toxicology*, 2nd rev. ed., Wiley-Interscience, New York, 1958, Chapter XVI, pp. 511–578.

3. M. G. Zabetakis, "Flammability Characteristics of Combustible Gases and Vapors," U.S. Bureau of Mines Bulletin No. 627, 1965, 121 pp.; NTIS AD701576.

4. National Fire Protection Association, "Fire Hazard Properties of Flammable Liquids, Gases and Volatile Solids," NFPA No. 325M, 1969, 139 pp.

5. National Fire Protection Association, "Fire Hazards in Oxygen-Enriched Atmospheres," NFPA No. 53M, 1974, 89 pp.

6. Federal Coal Mine Health and Safety Act of 1969, PL 91-173, December 30, 1969.

7. J. Grumer, A. Strasser, and R. A. Van Meter, "Safe Handling of Liquid Hydrogen," *Cryogen. Eng. News*, August 1967, pp. 60–63.

8. A. L. Furno, E. B. Cook, J. M. Kuchta, and D. S. Burgess, "Some Observations on Near-Limit Flames," Paper presented at the 13th Symposium (International) on Combustion, the Combustion Institute, Pittsburgh, 1971, pp. 593–599.

9. P. G. Guest, V. W. Sikora, and B. Lewis, "Static Electricity in Hospital Operating Suites: Direct and Related Hazards and Pertinent Remedies," U.S. Bureau of Mines, Report of Investigation No. 4833, 1952, 64 pp.

10. J. M. Kuchta, A. L. Furno, A. Bartkowiak, and G. H. Martindell, "Effect of Pressure and Temperature on Flammability Limits of Chlorinated Hydrocarbons in Oxygen–Nitrogen and Nitrogen Tetroxide–Nitrogen Atmospheres," *J. Chem. Eng. Data*, **13**(3), 421–428 (1968).

11. D. Burgess, "Thermochemical Criteria for Explosion Hazards," Paper No. 10b, presented at the 62nd Annual Meeting of the American Institute of Chemical Engineers, Washington, DC, 1969, 44 pp.

12. D. Burgess and M. Hertzberg, "The Flammability Limits of Lean Full–Air Mixtures: Thermochemical and Kinetic Criteria for Explosion Hazards," *ISA Trans.*, **14**(2), 129–136 (1975).

13. P. G. Guest, "Oily Fibers May Increase Oxygen Tent Fire Hazard," *Mod. Hosp.*, May 1965, pp. 180–182.

14. J. M. Kuchta, A. L. Furno, and G. H. Martindill, "Flammability of Fabrics and Other Materials in Oxygen-Enriched Atmospheres. Part I. Ignition Temperatures and Flame Spread Rates," *Fire Technol.*, **5**(3), 203–215 (August 1969).

15. E. L. Litchfield and T. A. Kubala, Flammability of Fabrics and Other Materials in Oxygen-Enriched Atmospheres. Part II. Minimum Ignition Energies," *Fire Technol.*, **5**(4), 341–345 (November 1969).

16. National Fire Protection Association, "Flash Point Index of Trade Name Liquids," NFPA No. 325A, 1972, 258 pp.

17. J. M. Kuchta and R. J. Cato, "Ignition and Flammability Properties of Lubricants, *Trans. SAE*, **77**, 1008–1020 (1968).

18. J. M. Kuchta, "Summary of Ignitions Properties of Jet Fuels and Other Aircraft Combustible Fluids," Report No. AFAPL-TR-75-70, 1975, 54 pp.

19. National Fire Protection Association, "Halogenated Fire Extinguishing Agent Systems—Halon 1301," NFPA No. 12A, 1973, 80 pp.

20. J. Grumer and A. E. Bruszak, "Inhibition of Coal Dust–Air Flames," U.S. Bureau of Mines, Report of Investigations No. 7552, 1971, 14 pp.

21. J. M. Kuchta, "Fire and Explosion Manual for Aircraft Accident Investigators," U.S. Bureau of Mines Report No. 4193, 1973, 117 pp.

22. M. G. Zabetakis and D. Burgess, Research on the Hazards Associated with the Production and Handling of Liquid Hydrogen," U.S. Bureau of Mines, Report of Investigations No. 5707, 1961, 50 pp.

23. Compressed Gas Association, "Standards for Liquefied Hydrogen Systems at Consumer Sites," Pamphlet No. G-5.2, CGA, 1966.

24. Armed Services Explosions Safety Board, "Quantity-Distance Criteria for Liquid Propellants," March 26, 1964.

25. A. Strasser, I. Liebman, and S. R. Harris, "Hydrogen Detectors," *Cryogen. Eng. News*, **2**(12), 16–20 (1967).

26. M. C. Irani, A. Tall, B. M. Bench, and P. W. Heran, "A Continuous-Recording Methanometer for Exhaust Fan Monitoring," U.S. Bureau of Mines, Report of Investigations No. 7951, 1974, 18 pp.

27. H. B. Carroll, Jr., and F. E. Armstrong, "Accuracy and Precision of Several Portable Gas Indicators," U.S. Bureau of Mines, Report of Investigations No. 7811, 1973, 42 pp.

28. H. R. Maisey, "Gaseous and Dust Explosion Venting. Part I." *Chem. Process Eng.*, 527–535, 563 (October 1965).

29. H. R. Maisey, "Gaseous and Dust Explosion Venting. Part 2." *Chem. Process Eng.*, 662–672 (December 1965).

30. National Fire Protection Association, "Explosion Venting Guide," NFPA NO. 68, 1974, 84 pp.

31. J. Nagy and W. M. Portmann, "Explosibility of Coal Dust in an Atmosphere Containing a Low Percentage of Methane," U.S. Bureau of Mines, Report of Investigation No. 5815, 1961, 16 pp.

32. H. G. Dorsett, Jr., M. Jacobson, J. Nagy, and A. P. Williams, "Laboratory Equipment and Test Procedures for Evaluating Explosibility of Dusts," U.S. Bureau of Mines, Report of Investigation No. 5624, 1960, 21 pp.

33. J. Nagy, A. R. Cooper, and H. G. Dorsett, Jr., "Explosibility of Miscellaneous Dusts," U.S. Bureau of Mines, Report of Investigation No. 7208, 1968, 31 pp.

34. K. N. Palmer and Z. W. Rogowski, "The Use of Flame Arrestors for Protection of Enclosed Equipment in Propane–Air Atmospheres, I." Chemical Engineering Symposium Series No. 25, Institution of Chemical Engineers, London, 1968, pp. 76–85.

35. Ministry of Labour, "Guide to the Use of Flame Arresters and Explosion Reliefs," Safety, Health and Welfare Booklets New Series No. 34, Her Majesty's Stationery Office, London, 1965, 55 pp.

36. J. K. Richmond, I. Liebman, and L. F. Miller, "Effect of Rock Dust on Explosibility of Coal Dust," U.S. Bureau of Mines, Report of Investigation No. 8077, 1975, 34 pp.

37. J. K. Richmond and I. Liebman, "A Physical Description of Coal Mine Explosions," paper presented at the 15th Symposium (International) on Combustion, the Combustion Institute, Pittsburgh, 1975, pp. 115–126.

38. W. Cybulski, "Researches on the Relationship Between Coal Dust Explosibility and the Kind of Coal," Restricted Conference of Directors of Safety in Mines Research, Paper No. 1, 1961, 52 pp.

39. J. Grumer, "Recent Research Concerning Extinguishment of Coal Dust Explosions," paper presented at the 15th Symposium (International) on Combustion, the Combustion Institute, Pittsburgh, 1975, pp. 103–114.

40. I. Hartmann, "Studies on the Development and Control of Coal Dust Explosions in Mines," U.S. Bureau of Mines, Information Circular No. 7785, 1957, 27 pp.

41. J. Nagy, A. R. Cooper, and J. M. Stupar, "Pressure Development in Laboratory Explosions," U.S. Bureau of Mines, Report of Investigation No. 6561, 1964, 19 pp.

42. J. M. Singer, N. B. Greninger, and J. Grumer, "Some Aspects of the Aerodynamics of Formation of Float Coal Dust Clouds," U.S. Bureau of Mines, Report of Investigation No. 7252, 1969, 26 pp.

43. J. M. Singer, E. B. Cook, and J. Grumer, "Dispersal of Coal- and Rock-Dust Deposits," U.S. Bureau of Mines, Report of Investigations No. 7642, 1972, 32 pp.

44. D. Rae, "The Initiation of Weak Coal Dust Explosions in Long Galleries and the Importance of the Time Dependence of the Explosion Pressure," paper presented at the 14th Symposium (International) on Combustion, the Combustion Institute, Pittsburgh, 1973, pp. 1225–1234.

45. J. Grumer, L. F. Miller, A. E. Bruszak, and L. E. Dalverny, "Minimum Extinguishant and Maximum Oxygen Concentrations for Extinguishing Coal Dust–Air Explosions," U.S. Bureau of Mines, Report of Investigation No. 7782, 1973, 6 pp.

46. D. Rae, "Experimental Coal-Dust Explosions in the Buxton Full-Scale Surface Gallery. IV. The Influence of the Dust Deposit and Form of Initiation of Explosions in a Smooth Gallery," SMRE Research Report No. 277, 1972, 57 pp.

47. M. Jacobson, J. Nagy, A. R. Cooper, and F. J. Ball, "Explosibility of Agricultural Dusts," U.S. Bureau of Mines, Report of Investigation No. 5753, 1961, 23 pp.

48. J. Nagy, H. G. Dorsett, Jr., and A. R. Cooper, "Explosibility of Carbonaceous Dusts," U.S. Bureau of Mines, Report of Investigation No. 6597, 1965, 30 pp.

49. H. G. Dorsett, Jr., and J. Nagy, "Dust Explosibility of Chemicals, Drugs, Dyes and Pesticides," U.S. Bureau of Mines, Report of Investigation No. 7132, 1968, 23 pp.

50. M. Jacobson, A. R. Cooper, and J. Nagy, "Explosibility of Metal Powders," U.S. Bureau of Mines, Report of Investigation No. 6515, 1964, 25 pp.

51. M. Jacobson, J. Nagy, and A. R. Cooper, "Explosibility of Dusts Used in the Plastics Industry," U.S. Bureau of Mines, Report of Investigations, No. 5971, 1962, 30 pp.

52. D. Burgess and J. Murphy, "Some Experiments with the Application of Bromotrifluoromethane to Coal Fires," Paper No. 34, presented at the International Conference of Safety in Mines Research, Tokyo, 1969, 17 pp.

53. National Fire Protection Association, Explosion Prevention Systems, NFPA No. 69, 1973, 53 pp.

CHAPTER FORTY-THREE

Industrial Hygiene Aspects of Hazardous Material Emergencies and Cleanup Operations

Gary R. Rosenblum and Lawrence R. Birkner, C.I.H.

1 INTRODUCTION

Industrial hygiene issues have a significant impact on all aspects of responding to hazardous material emergencies and on the process of cleaning up a hazardous material spill. The health and safety of all response and cleanup personnel depend on effective evaluation of potential exposures and expert determination of the appropriate means to protect against those exposures.

This chapter covers how the professional judgment and expertise of the industrial hygienist are counted on during all phases of a hazardous material emergency. The industrial hygienist is involved with contingency planning for potential incidents and recognizing, evaluating, and controlling health and safety risks during the incident. The industrial hygienist also helps develop the cleanup site safety plan and contributes to the incident termination process, where a critique of the response and cleanup is conducted, and final documentation is made.

Hazardous material emergencies were etched into the public's consciousness by two major catastrophes in the 1980s, the Bhopal, India methyl isocyanate release, and the Exxon Valdez oil spill. Both disasters resulted in major changes in U.S. industry practices and have put a new emphasis on the industrial hygiene profession for all aspects of hazardous materials management.

Patty's Industrial Hygiene and Toxicology, Fourth Edition, Volume 1, Part B, Edited by George D. Clayton and Florence E. Clayton.
ISBN 0-471-50196-4 © 1991 John Wiley & Sons, Inc.

In India, a catastrophic release of methyl isocyanate, a highly poisonous and irritating vapor, rapidly killed or injured thousands of people living near the Union Carbide facility. Images quickly flashed to the rest of the world showing the devastating results of an uncontrolled release from a chemical process that produced a deadly vapor cloud. As soon as it became clear how quickly so many lives could be lost, much of the industrialized world launched top-priority programs to examine its own chemical manufacturing processes to assess whether that kind of event could ever happen again, and to work to prevent it.

The Exxon Valdez oil spill took many weeks to develop fully, which gave the public time to develop a perception that the spill was of a magnitude that seemed beyond effective control, and the size of the area eventually touched by the spilled oil was astonishingly large. As a result of this incident, public attention became focused on the potential risks of hazardous materials transportation, from crude oil to just about any transported chemical.

The immediate result of focusing public attention on the potential risk of catastrophic release from industrial processes or hazardous material transportation was to spur the U.S. government into action. Federal, state, and local governments quickly created many important new laws and regulations to change the way hazardous materials are handled in industrial processes and how they are transported, in order to reduce the risk of a hazardous materials emergency. And if a release or spill were to occur now, many new requirements have also been added to help protect the health and safety of the emergency responders.

One important piece of this new legislation is the 1986 Superfund Amendment and Reauthorization Act (SARA), which contains a section, Title III, that covers emergencies arising from hazardous materials releases. SARA Title III established three major new requirements for all U.S. industry for emergency planning notification for hazardous materials, emergency release notification to the local community, and reporting on the hazards of the chemicals and releases to the local community. These requirements, described as community "right to know," are specifically designed to help inform the public of the potential risks of local industries handling hazardous materials.

A U.S. Department of Labor agency, the Occupational Safety and Health Administration, promulgated regulations designed specifically to protect the safety and health of personnel responding to hazardous materials emergencies. HAZWOPER is the euphonious acronym of the Hazardous Waste Operations and Emergency Response final rule. It specifies many safety and health protection aspects of the emergency response to hazardous materials. This regulation has also been adopted by the U.S. Environmental Protection Agency, so that it applies to government employees such as firefighters and hazardous materials response teams. HAZWOPER has dramatically changed the way emergency response to hazardous materials is performed; much of this chapter is devoted to covering industrial hygiene activities and decisions required by the rule.

A chart is provided to give a quick summary of some of the laws and regulations (Figure 43.1) that can apply.

COMPREHENSIVE ENVIRONMENTAL RESPONSE, COMPENSATION AND LIABILITY ACT (CERCLA)
Also known as "Superfund", addresses hazardous chemical releases into the environment, with those responsible for releases above "Reportable Quantities" required to notify the National Response Center.

RESOURCE CONSERVATION AND RECOVERY ACT (RCRA)
Covers management and disposal of hazardous wastes.

SUPERFUND AMENDMENTS AND REAUTHORIZATION ACT OF 1986
Contains Title III, which establishes major requirements relating to Emergency Planning Notification, Emergency Release Notification, and Reporting on Chemicals and Releases for Community Right To Know.

HAZARDOUS WASTE OPERATIONS AND EMERGENCY RESPONSE FINAL RULE (HAZWOPER)
An OSHA regulation designed to protect the health and safety of persons involved with hazardous waste site cleanups, and other hazardous waste activities, as well as all persons responding to a uncontrolled release of a hazardous substance, and all persons involved with cleaning up a hazardous waste spill.

OSHA GENERAL INDUSTRY SAFETY AND HEALTH STANDARDS
Found in the Code of Federal Regulations 29CFR 1910, these standards cover workplace activities concerning hazardous substances, environmental controls, fire protection, hand and portable tools, and walking/ working surfaces.

HAZARDOUS MATERIALS TRANSPORTATION ACT
DOT regulations cover the transportation of hazardous materials, which includes the design of transportation vehicles and vessels, and specifying shipping, packing, labeling, and placarding requirements for both carriers and shippers.

OSHA HAZARD COMMUNICATION STANDARD
Regulates the determination of workplace hazards, and requires communication of the hazards to the affected employees, and the training to mitigate the hazards.

Figure 43.1. Summary of federal laws and regulations pertaining to hazardous material incidents. (Not an all-inclusive list; other federal, state, and local laws and regulations may apply.)

2 ADVANCE PLANNING AND COORDINATION

2.1 Preparing for Emergency Response

The first priority for any group anticipating an emergency response to a hazardous materials incident, whether it is an industrial corporation, government agency, local planning committee, or volunteer response service, is planning and coordination in advance of any hazardous materials incident. This is a process of determining the range of anticipated events, and then realistically preparing a response plan to ensure that the group's efforts are timely, effective, and safe. Without advance planning and coordination, the emergency response to a hazardous materials incident is much more likely to result in confusion, delays, and possibly excess damage, unnecessary injuries, and even deaths.

Some of the key elements for advance planning and coordinating an emergency response plan are site management and control, anticipating the hazardous materials potentially involved, assuring that the appropriate exposure assessment equipment will be available in a timely manner, determining the different levels of personal protective equipment that could be needed, assessing the control and containment parameters for the various scenarios, determining the extent of decontamination that might be necessary, and anticipating the documentation needed during the response and after the incident has been terminated. Each one of these elements derives from predicting the type and extent of the range of possible incidents at a facility or during the transportation of a hazardous substance.

Planning site management and control combines coordination of the personnel and groups that will be involved in a potential incident and anticipating the nature of the hazardous materials emergency. Before response to a hazardous materials incident takes place, a charter should be developed for assigning key personnel roles and responsibilities. This includes who will staff the initial response, and who relief personnel will be. For example, the industrial hygienist needs to know in advance whom to report to, and his or her level of responsibility on the scene, which could perhaps include becoming the safety officer for the event. Also, it is important that the plan determine who will be the "on-scene commander," the person with final authority for tactical response decisions, with the ultimate responsibility for the safety and health of all response and cleanup personnel. This role and responsibility plan should be flexible to anticipate various staffing contingencies and potentially expanding roles, should the incident grow or change in scope.

The site management plan should include the coordination of communications equipment. The amount and types needed, as well as anticipating the equipment utilized by other responding organizations and agencies is important to ensure effective communication with all active response groups.

Another important element of the site management plan is determining the level of emergency that will require evacuation of any public nearby as well as evacuation of responding personnel. The conditions for evacuation do not have to be exactly laid out in advance. But it is possible to anticipate certain conditions, such as fire, involving certain materials or storage vessels that would lead the incident commander to call for evacuation. This element leads to another communication issue, which is the coordination of information with appropriate government agencies, responding teams, other authorities, and company management. Hazard potential and resources requirements must be clearly understood by all these groups, and the plan should be able to provide clear instructions to facilitate this process.

Other than site management, personnel coordination, and communications coordination, most of the other planning elements stem from anticipating the types of incidents and the materials involved. At first the sheer enormity of anticipating all possible hazardous materials incident scenarios can be daunting.

The hygienist should be aware of the importance of the identification of the materials likely to be involved in a hazardous materials incident, because all major safety and health decisions follow that determination. Preplanning of a hazardous materials incident can provide initial clues to the hazardous material identity, but it is also important to know and be able to utilize the labeling and placard systems in use. Later in this chapter (Section 3.1.2) site hazard identification covers these and other means for identifying the hazards of materials.

One key factor for industrial hygienists conducting hazardous materials incident advance planning is anticipating exposure monitoring needs. Having the appropriate monitoring equipment immediately on the incident site can be of critical importance for the entire emergency response action. One reason is that without good knowledge of the exposure potential and the actual exposures faced by responding personnel, the level of personal protective equipment must be maximized. As a result of poor planning, if the proper direct-reading instruments are not available when the "on-scene commander" determines that the hazardous materials team must immediately go in to the danger zone, the team must be protected with maximum personal protective equipment. This may require the responders to wear the cumbersome fully encapsulating suit and self-contained breathing apparatus, which would limit their effectiveness and perhaps increase their risk of heat stress. Hazardous materials incident planners and industrial hygienists should keep in mind that the federal regulations defined in the HAZWOPER rule require that emergency responders wear supplied air, unless air monitoring at the site documents that the exposure level is low enough to use either air-purifying respirators or no respiratory protection at all.

The advance planning of personal protective equipment (PPE) needs is also of great importance. Again, response efforts cannot be performed without appropriate protection from exposure. Preplanning can help ensure that the proper level of personal protective equipment is available when it is needed.

The quantity of PPE equipment needed must also be anticipated. For instance, fully encapsulating suits are always stocked in pairs, because they are always utilized in groups of at least two for safety. The buddy system provides the potential safety net for getting someone out of danger if there is a problem. Again, the logistics of properly storing the equipment and getting to the incident site should be covered in the plan to increase the chances that the PPE will be there when it is needed, functioning and ready to go without delay.

The amount and type of control and containment equipment and supplies should be determined in advance in the plan. Control and containment equipment includes all types of items, from granular absorbent to skimmers and booms to backhoes, bulldozers, and vacuum trucks. Again, the key is that the appropriate equipment and supplies needed to manage the incident must be on the scene at the right time, and be functional, which requires anticipation of the potential emergencies and thoughtful advance planning. Proper planning of the logistics of storing and transporting the equipment will help make the response proceed faster and smoother and will likely produce a more desirable result. Another planning element to consider for the control and containment equipment is proper disposal of contaminated material and decontamination of equipment and material that must be reused.

Decontamination is another important issue to consider during the advance planning stage. Based on the hazardous materials anticipated in the incident, a decontamination strategy can be developed. One primary issue to be considered in advance of an incident is whether to utilize reusable or disposable PPE and other equipment. Effectiveness, cost, and utility of the PPE are part of this consideration. If the plan calls for disposal as the decontamination procedure, steps must be taken to determine whether the material will be considered "hazardous waste" and thus require adherence to all applicable laws for transporting and disposing of hazardous waste from that location. Decontamination of reusable PPE, on the other hand, can be more labor intensive and require additional personnel for the multiple step cleaning and rinsing process.

Finally, the plan must cover the termination procedures that will be followed after the incident has ended. These procedures can consist of defining when the event has ended, when to demobilize the responders, and if possible, how clean the cleanup should be. An additional element of the termination procedure is to critique the response in a constructive manner. This also includes assessing the documentation of the event. The plan should therefore anticipate and provide many of the generic documentation sheets that can be helpful, such as an incident commander log, material hazard summary sheet, lists of personnel involved, and monitoring records. An example of this type of documentation is provided in Figure 43.2.

2.2 Emergency Response Site Safety and Health Plan

The site safety and health plan is a critical element to the effective operation of a response and cleanup operation; it may be required for compliance with HAZ-WOPER. This plan is different from emergency response planning in that it addresses specifics of health and safety issues for the particular incident. Therefore it can often become part of the emergency response planning process.

Whereas the emergency response plan is utilized to guide such items as training, resource allocation, stockpiling of supplies, personnel requirements, and compliance issues, the site safety plan is created specifically to address the safety and health needs for the personnel responding to and cleaning up that one particular incident. Because this is quite a large task, and it is done during an incident, it is highly recommended that a generic site safety and health plan be produced that simply contains blanks to be filled in as the information is developed. This fill-in-the-blanks approach has the advantage of reducing the chance that an important element of personnel safety is inadvertently forgotten in the heat of action, and it provides good documentation of the safety and health decision-making process when it is completed. An example of a model site safety plan blank is provided in the Appendix.

The minimum elements of a site safety plan include risk assessments for all site tasks and operations, frequency and type of air monitoring, personnel monitoring, environmental sampling techniques, specific personal protective equipment to be used by all employees for each task and operation, medical surveillance requirements, site control measures, decontamination procedures, confined space entry procedures if applicable, task-oriented employee training requirements, and a spill containment program.

The risk assessments are based on an evaluation of the tasks to be performed by the emergency responders and cleanup crews. The specific hazards of the materials to which the responders may be exposed are reviewed, and a task analysis is performed that includes characterizing exposure. A qualitative risk assessment is produced from these factors that enables the hygienist to consider the need for monitoring and possible personal protective equipment strategies.

The site safety plan then progresses to consideration of the type and frequency of air monitoring, as well as developing task-specific personal protection equipment strategies. Details of developing a monitoring strategy are covered in Section 3, Hazard Evaluation, but one key issue is identifying the type of direct-reading instruments for rapid airborne contaminant evaluation. The HAZWOPER final rule mandates exposure assessment (including air monitoring) before any reduction in respiratory protection below supplied air is considered. The site safety plan must provide for such exposure assessments.

```
COMMON NAME:_____CHEMICAL NAME:_____

I. PHYSICAL/CHEMICAL PROPERTIES
                                                          SOURCE
       Natural physical state: Gas____ Liquid____ Solid____  _____
       (at ambient temps of 20°C-25°C)
       Molecular weight              _____g/g-mole_____
       Densitya                      _____g/ml_____
       Specific gravitya             _____@_____°F/°C   _____
       Solubility: water             _____@_____°F/°C   _____
       Solubilityb:_____         _____@_____°F/°C   _____
       Boiling point                 _____°F/°C   _____
       Melting point                 _____°F/°C   _____
       Vapor pressure                ___mmHg @_____°F/°C   _____
       Vapor density                 _____@_____°F/°C   _____
       Flash point                   _____°F/°C   _____
       (open cup_____; closed cup_____)
       Other:_____        _____       _____

II. HAZARDOUS CHARACTERISTICS

A. TOXICOLOGICAL HAZARD  HAZARD?    CONCENTRATIONS         SOURCE
                                    (PEL, TLV, other)

   Inhalation            Yes  No    _____       _____
   Ingestion             Yes  No    _____       _____
   Skin/eye absorption   Yes  No    _____       _____
   Skin/eye contact      Yes  No    _____       _____
   Carcinogenic          Yes  No    _____       _____
   Teratogenic           Yes  No    _____       _____
   Mutagenic             Yes  No    _____       _____
   Aquatic               Yes  No    _____       _____
   Other:_____       Yes  No    _____       _____

B. TOXICOLOGICAL HAZARD  HAZARD?    CONCENTRATIONS         SOURCE
   Combustibility        Yes  No    _____       _____
   Toxic byproduct(s):   Yes  No    _____       _____

   _____                  _____       _____

   Flammability          Yes  No    _____       _____
     LFL                             _____       _____
     UFL
   Explosivity           Yes  No    _____       _____
     LEL                             _____       _____
     UEL                             _____       _____

aOnly one is necessary.
bFor organic compounds, recovery of spilled material by solvent extraction may
 require solubility data.
```

Figure 43.2.   Hazard summary sheet.

C. REACTIVITY HAZARD   HAZARD?   CONCENTRATIONS   SOURCE
                            Yes No
   Reactivities:

   _____   _____   _____
   _____   _____   _____

D. CORROSIVITY HAZARD   HAZARD?   CONCENTRATIONS   SOURCE
   ph _____   Yes No
   Neutralizing agent:

   _____   _____   _____
   _____   _____   _____

E. RADIOACTIVE HAZARD   HAZARD?   EXPOSURE RATE   SOURCE
   Background   Yes No   _____   _____
   Alpha particles   Yes No   _____   _____
   Beta particles   Yes No   _____   _____
   Gamma radiation   Yes No   _____   _____

III. DESCRIPTION OF INCIDENT:

   Quantity involved _____
   Release information _____
   _____

   Monitoring/sampling recommended _____
   _____

IV. RECOMMENDED PROTECTION:

   Worker _____
   _____

   Public _____
   _____

V. RECOMMENDED SITE CONTROL:

   Hotline _____
   _____

   Decontamination line _____
   _____

   Command Post location _____
   _____

VI. REFERENCES FOR SOURCES:

   _____
   _____
   _____
   _____
   _____
   _____
   _____

Figure 43.2. *(Continued)*

The collection of air contaminant exposure level data plays a large role in determining the level of protection to be assigned to individuals who will be entering the danger zone in a hazardous material emergency response and to those personnel providing direct assistance to those in the danger zone. In some cases where quantitative assessments of exposures are not possible, the hygienist must make a determination of the appropriate PPE based on qualitative factors. In almost all these cases, higher levels of protection are utilized to protect personnel against unknown risks. The model site safety plan should provide a format that allows the hygienist to review all the exposure data that is available in the situation, and provide areas to document the decision-making process for selecting the personal protection for the various tasks needed in the emergency response and cleanup.

Other site safety plan elements follow as the risk assessments, hazard determinations, and exposure characterizations are filled in on the generic form. Issues such as training and certification needed for particular emergency response and cleanup tasks are determined by the personnel roles and the potential exposure to the hazardous material. If confined space entry procedures are needed owing to site-specific factors, this is documented in the plan. A medical surveillance program based on the potential adverse effects that could be observed if overexposure occurs is also included in the plan. The need for decontamination or use of disposable equipment is also documented.

Site control procedures (also known as zone control system), which allow only protected trained individuals near the danger zone and keep the media and other members of the public at a safe distance, are always set up on a case-by-case basis. In fact, during the course of the incident the site control plan could change quite a few times as the material or wind moves, or the hazard is contained, and each change is accounted for and documented in the site safety plan.

Putting together the generic site safety plan for a particular location is done as part of the advance planning process. As far as the type of hazardous materials incident can be predicted, parameters of the site safety plan can be completed in advance. Thus if the possible hazardous materials in the incidents can be anticipated, then the hazards of that material, and the types of PPE that are most effective for that material, can be incorporated into the site safety and health plan in advance. Whenever the hazardous material involved in an incident can't be anticipated, the generic site safety and health plan is utilized, with blanks to be filled in on the site.

2.3 Emergency Response Training

In order to respond effectively to an incident, industrial hygienists who will be emergency responders must, in addition to their knowledge and experience in industrial hygiene, be trained in emergency response procedures, the incident command system and its management, the use of PPE, and decontamination procedures. More specific training requirements that include and go beyond these training requirements are covered in the HAZWOPER rule. Emergency response training mandated by the HAZWOPER rule is covered in Section q(6), and is outlined in Figure 43.3.

The hygienist who will respond to hazardous materials emergencies must be trained in emergency response procedures, which are based on a simple premise: an emergency response means that all actions must be done as rapidly as possible, but always in an orderly and safe fashion. Training in emergency response procedures emphasizes that if an action cannot be done safely, it must not be done. At the same time, the training emphasizes that the determination of risk is being done in a situation that is far from optimum for risk data gathering. The best available information is gathered to reach the best available assessment, which is not the way an industrial hygienist normally gathers data.

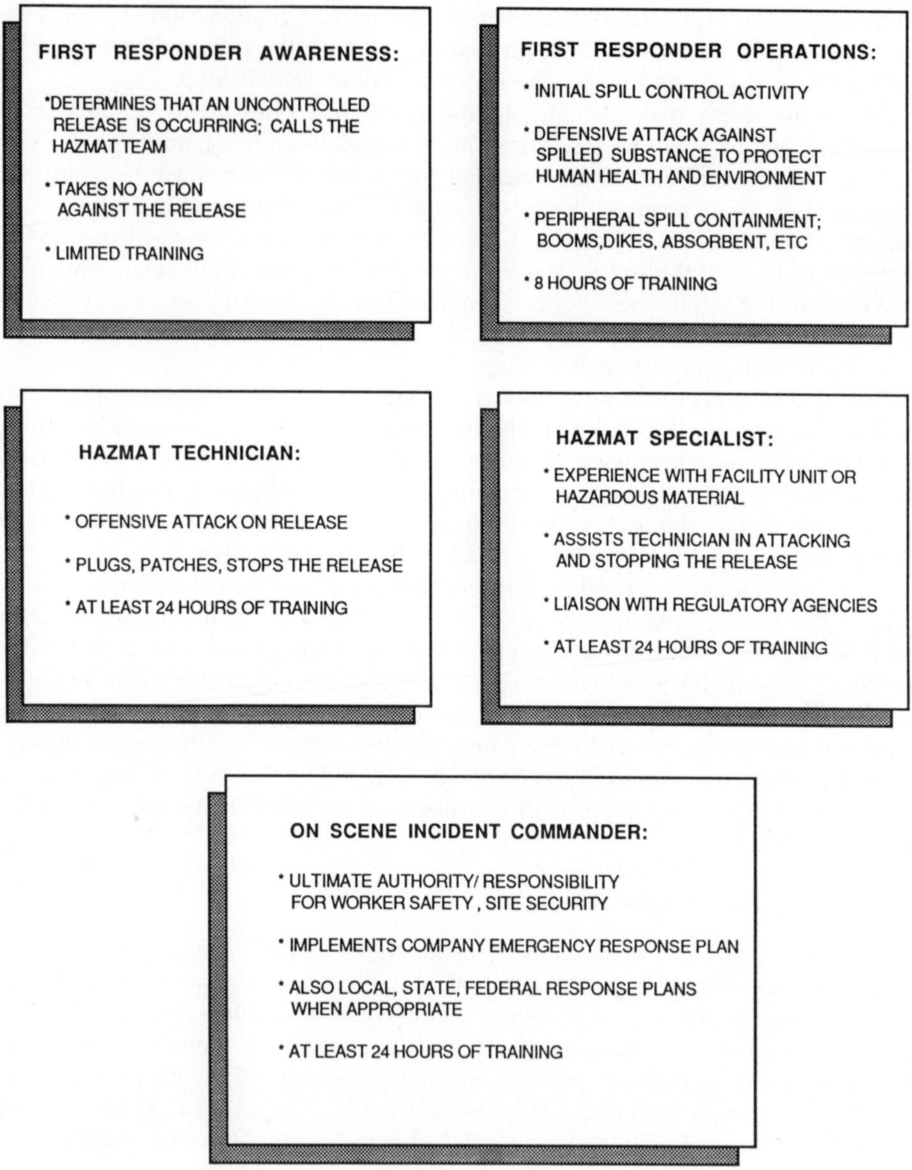

FIRST RESPONDER AWARENESS:

*DETERMINES THAT AN UNCONTROLLED
 RELEASE IS OCCURRING; CALLS THE
 HAZMAT TEAM

* TAKES NO ACTION
  AGAINST THE RELEASE

* LIMITED TRAINING

FIRST RESPONDER OPERATIONS:

* INITIAL SPILL CONTROL ACTIVITY

* DEFENSIVE ATTACK AGAINST
  SPILLED SUBSTANCE TO PROTECT
  HUMAN HEALTH AND ENVIRONMENT

* PERIPHERAL SPILL CONTAINMENT;
  BOOMS, DIKES, ABSORBENT, ETC

* 8 HOURS OF TRAINING

HAZMAT TECHNICIAN:

* OFFENSIVE ATTACK ON RELEASE

* PLUGS, PATCHES, STOPS THE RELEASE

* AT LEAST 24 HOURS OF TRAINING

HAZMAT SPECIALIST:

* EXPERIENCE WITH FACILITY UNIT OR
  HAZARDOUS MATERIAL

* ASSISTS TECHNICIAN IN ATTACKING
  AND STOPPING THE RELEASE

* LIAISON WITH REGULATORY AGENCIES

* AT LEAST 24 HOURS OF TRAINING

ON SCENE INCIDENT COMMANDER:

* ULTIMATE AUTHORITY/ RESPONSIBILITY
  FOR WORKER SAFETY , SITE SECURITY

* IMPLEMENTS COMPANY EMERGENCY RESPONSE PLAN

* ALSO LOCAL, STATE, FEDERAL RESPONSE PLANS
  WHEN APPROPRIATE

* AT LEAST 24 HOURS OF TRAINING

Figure 43.3.   HAZWOPER training summary.

The hygienist may get only a few moments to sample only once, perhaps while wearing a fully encapsulating suit, and must recognize that this kind of pressure is not unusual.

A critical part of emergency response procedures is training in the incident command system. This system was developed by the fire service to coordinate effectively the activities of many different responding groups at a single incident. For example, when many fire companies from different cities within a county are called to fight a large fire, each arrives with a fire chief. Without a preplanned system for coordinating each chief's orders to his or her personnel, through the selection of one incident commander, each chief would not know what the other is doing and resources are sure to be wasted. Use of the incident command system at a hazardous materials emergency assures that a chain of command is established quickly, and the roles and responsibilities of all responding personnel are defined.

Without going into detail beyond the scope of this chapter, the incident commander is the single authority for making all strategic decisions concerning the activities of all responding personnel. This may be accomplished through direct management, or indirectly through specific officers chosen to lead certain major categories of tasks, such as logistics, operations, and support. The management system depends, of course, on the type of incident, and the number of personnel needed and available. The incident commander has the final responsibility for the health and safety of all responding and cleanup personnel.

However, some of the decision-making authority for technical health and safety decisions may be delegated to a safety officer. The industrial hygienist may either be the site safety officer or may report directly to the site safety officer. The industrial hygienist may even be the incident commander in certain circumstances. This safety and health evaluation may take place in the advance planning stage to be carried out on the site by a non-industrial hygienist, or may be done by the hygienist under the pressure of instant analysis during a rapidly evolving uncontrolled release of a hazardous material. In all these cases, the professional judgment needed to make decisions on PPE and other safety issues comes from the hygienist(s) or other safety professional(s), who must know how this evaluation fits into the hazardous materials incident response process.

Although it may be assumed that a hygienist knows how to select and use PPE, one element of training not to be overlooked by the emergency response industrial hygienist is developing the ability to train others on the proper use of that equipment. The responding hygienist must be familiar with all the PPE that will be utilized at an incident and be able to instruct others on its use.

The other side of knowing how to select and use PPE, and being able to train others in its use, is the knowledge of when and how to decontaminate the PPE. The procedures for decontaminating nondisposable equipment are frequently site-specific and therefore require on-site review. The hygienist may be called upon to design the decontamination facility for the incident either in the emergency response plan or on site, and may also be called upon to train others in the proper decontamination methods. Decontamination is an area of expertise that is generally not part of a standard safety and health training program, and is usually only part of hazardous materials response training. Industrial hygienists who will be emergency responders must be familiar with decontamination procedures. Section 6 of this chapter covers decontamination in more detail.

2.4 Emergency Response Drills

Classroom training is a large part of the process of preparing emergency responders for a hazardous materials incident. However, it must be followed up with hands-on practical drills in order for the training to be truly effective. The industrial hygienist and the rest of the response team should participate in tabletop simulations of hazardous material incidents and also actual real-time drills that require donning and doffing PPE, using air and other monitoring equipment, and going through anticipated decontamination procedures. The drills should also emphasize documentation of all decisions and activities for study after the drill has been terminated. The hygienist has a role in both planning and executing the drills, as well as participating as a responder. Without drills and related exercises, an emergency response team should not be considered ready for action. The drills are also an important part of annual refresher training to maintain readiness and competency.

3   HAZARD EVALUATION

After a hazardous material release has occurred, many decisions must be made based on the nature and severity of the incident. Accurate data determining the hazards faced by the responding personnel lead to effective personnel protection, which produces more efficient performance. It is therefore very important for the industrial hygienist to be in a position to assess the hazards quickly and accurately.

This section covers some of the issues and methods affecting the rapid and accurate collection of data. It should be pointed out, however, that the industrial hygienist is frequently not the first responder on the scene. As a result, other first responder personnel will need to be able to make initial assessments of the scene and be able to perform some preliminary monitoring. Therefore this section is applicable for either the industrial hygienist on the scene or the first responder.

3.1   Developing the On-Scene Safety Plan

There are four basic steps to this process, which can be described as initial site assessment, identification of site hazards, evaluation of site-specific safety plan, and communication of the plan. Because managing hazardous materials incidents consists of caution in the face of the unknown, combined with aggressive management of known hazards, the person or group developing the on-scene safety plan needs to concentrate on separating verifiable facts from speculation and must be able to work quickly and effectively under pressure.

3.1.1   Initial Site Assessment

The initial site characterization consists of observing the general geographic features of the incident from a safe distance. If closer views are needed, then binoculars are recommended. This first stage of site characterization is essentially making a map of the incident with many important initial decisions based on the observations. Wind speed and direction, as well as pending weather conditions, such as rain, should be automatically noted. Because approaches to a hazardous materials incident are from the upwind direction, the topography of the upwind side must be assessed. Roads, fields, hills, cliffs, streams, rivers, ditches, fences, buildings, and walls are the types of things that will determine many of the most critical initial site decisions such as setting up the command post (upwind and uphill if possible), access for heavy equipment and personnel, and determining zone control boundaries and potential escape routes. The topography is also a major factor in predicting where the spill material is moving, and where firefighting water or foam will collect or run off. This influences decisions on diking, booming, or otherwise impeding the flow of the spilled material.

Other atmospheric conditions besides wind are important. Temperature affects the vaporization of hazardous materials, but also has an impact on the responders, who may need to wear bulky PPE that could induce heat stress. High humidity can also play a role in inducing heat stress, whereas very low humidity can stress responders because of extra rapid depletion of fluids. If it is raining, or about to rain, many decisions may change concerning access to the site by crossing creek beds or streams, influence firefighting activities, and even create an increased hazard if the hazardous material is unstable or pyrophoric with water contact. Fog, of course, reduces visibility, and lightning might inhibit the use of cranes or other heavy equipment.

Site geology is also assessed. The movement of the spilled material through the soils can range from rapid to virtually nil, depending on the local geology. Spilled material can soak through soils at one level and reappear below a cliff or seep down to groundwater. Hazardous materials soaked into soils can continue to expose workers throughout the cleanup phase of the operation as they remove the soil.

3.1.2   On-Scene Hazard Identification

The second aspect of on-scene risk evaluation deals with the identification of the hazards. These can be hazards to the responding personnel, the public, or both. The most critical aspect is to identify the hazardous material or materials involved in the incident.

Failing to identify the hazardous materials in an incident can lead to disastrous consequences for the responding personnel and restrict the incident commander's ability to manage the response. Without knowing the identity of the materials, it becomes impossible to send response personnel into hazardous material incident situations without taking maximum personal protection precautions. Maximum PPE, such as the level A totally encapsulated suit with self-contained breathing apparatus is difficult to work in, is inefficient because it reduces the amount of time a person can work, and requires substantially increased levels of resources because it must be used in pairs, with two backup personnel standing by and with protected personnel assisting with donning, doffing, and decontamination. Therefore, when the hazards are not positively identified, it can make the emergency response effort much more dangerous, difficult, and more costly.

Materials are identified through either administrative means, such as examining shipping or transport information, including placards and labels, or through investigative means, such as observation of the location and function of the facility and of the appearance of the released material or its container, or use of direct-reading instruments that can identify the substances and evaluate the exposure.

The significant health and safety hazards, as well as the most significant routes of exposure to be faced by the emergency responders, can then be determined after a positive identification is made of the released substance(s). A review of the toxic properties is done using the best available information sources, such as Material Safety Data Sheets (MSDS), handbooks of properties of toxic materials, the Department of Transportation (DOT) Emergency Response Guidebook, computer data sources, or hotline phone sources such as CHEMTREC (Chemical Transportation Emergency Center). Identifying emergency data sources, listing appropriate phone numbers, and providing responders with handbooks with this information is also part of the advance planning process.

There are many regulatory and administrative identifiers of hazardous materials that can be seen from a safe distance which may aid in determining the identity of the substance and the risks faced by responders. The U.S. Department of Transportation requires placarding and labeling of bulk hazardous materials transported in the United States by rail or motor vehicle. Hazardous cargoes must be marked with a standard format placard that would enable an emergency responder to identify the type of hazard the cargo presents (see Figure 43.4). The placard provides a pictogram of the hazard, a code number corresponding to the U.N. hazard classification system (see Figure 43.5). The placard or label may also be used with a four digit material-specific identification number that is coded to the DOT Emergency Response Guidebook. The number and guidebook provide emergency responders with general hazard and emergency response information for the material type identified by that code number.

Other label systems may also be used, and should be looked for to assist in identifying the hazard. The National Fire Protection Association has created a placarding system for allowing the rapid identification of material hazards. This placard contains symbols for fire, health, reactivity, and other hazards in one diamond-shaped sign (see Figure 43.6). Each section of the diamond contains a hazard summary number, with "0" indicating the minimum hazard and "4" indicating the maximum hazard. Petroleum companies may utilize a color code system of identification developed by the American Petroleum Institute (API). Each petroleum or petroleum product has a color or color combination associated with it. For instance, high-grade unleaded gasoline would be marked with an orange circle containing a white cross, and kerosene would be marked with a brown hexagon.

Shipping papers, bills of lading, and similar paper work often can confirm the identity of the hazardous material. Unfortunately, these papers often are carried directly with the material so when there is a spill, they are within the contaminated zone and are not readily available. In addition, if there is a fire situation, they could be destroyed. Finally, the shipping papers may not have been completed properly or indicate a mixed load, which could complicate an already difficult situation. As a result, it is often not possible to rely on these papers completely to identify the hazardous material.

In addition to relying on administrative identifiers, observation and deduction are also useful and important tools when sizing up a hazardous materials incident. The importance of advance planning information cannot be overemphasized for

- PLACARD MOTOR VEHICLES, FREIGHT CONTAINERS AND RAIL CARS CONTAINING ANY QUANTITY OF HAZARDOUS MATERIALS LISTED IN TABLE 1.
- PLACARD MOTOR VEHICLES, FREIGHT CONTAINERS AND RAIL CARS CONTAINING 1,000 LBS. OR MORE GROSS WEIGHT OF HAZARDOUS MATERIALS LISTED IN TABLE 2.
- PLACARD FREIGHT CONTAINERS 640 CUBIC FEET OR MORE CONTAINING ANY QUANTITY OF HAZARDOUS MATERIAL CLASSES LISTED IN TABLES 1 AND/OR 2 WHEN OFFERED FOR TRANSPORTATION BY AIR OR WATER.

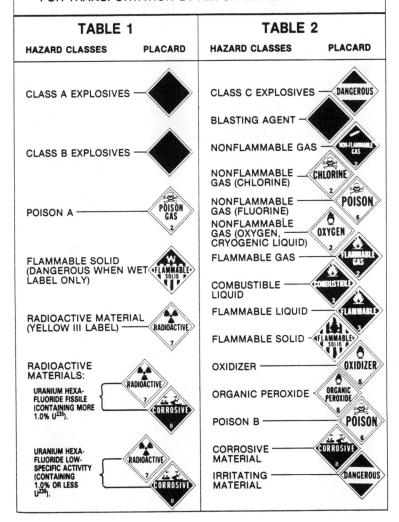

Figure 43.4. U.S. Department of Transportation hazardous materials placarding guidelines.

1. Explosives
2. Gases (Flammable and Nonflammable)
3. Flammable Liquids
4. Flammable Solids
5. Oxidizers and Organic Peroxides
6. Poisons
7. Radioactives
8. Corrosives
9. Miscellaneous Hazardous Materials

The Identification numbering system is based on the system adopted for worldwide use by the United Nations Committee of Experts on the Transportation of Dangerous Goods

These hazardous materials have been grouped by the U.N. task force from the most destructive to the least destructive during transportation.

EXAMPLE OF PLACARD AND PANEL WITH ID NUMBER

The ID number (ID No) for a material in transport may be displayed on placards. or on orange panels on tank trucks. Check the sides of the transport vehicle if the ID number is not displayed on the ends of the vehicle or tank truck

A Numbered Placard   Or

A Placard and an Orange Panel

Figure 43.5.   United Nations identification system.

helping to determine the identity of the hazardous material(s) through observation and deduction. An additional source of information is interviews with employees or other witnesses involved with the release or spill.

The type of facility or transport equipment involved is a key point to help determine the identity of an unknown hazardous material. The types of questions to answer would include whether the facility is a storage or a production facility, or whether or not the cargo is being shipped under pressure. For example, a round cylinder tank on a cargo truck would indicate it is pressurized, whereas an elliptical cylinder tank would indicate that the cargo is not pressurized. Quite different safety decisions would be made for some materials given a pressurized versus unpressurized situation. Besides shape, container color, and additional markings such as color, stripes can provide further clues. This is because some industries use standard colors and markings for transport of particular materials or material types. Observing the characteristics of the shipping container or the facility in combination with good advance planning data can go a long way toward identifying the hazardous material quickly and safely.

THE NFPA 704 MARKING SYSTEM DISTINCTIVELY INDICATES THE PROPERTIES AND POTENTIAL DANGERS OF HAZARDOUS MATERIALS. THE FOLLOWING IS AN EXPLANATION OF THE MEANINGS OF THE QUADRANT NUMERICAL CODES:

HEALTH (BLUE)

IN GENERAL, HEALTH HAZARD IN FIREFIGHTING IS THAT OF A SINGLE EXPOSURE WHICH MAY VARY FROM A FEW SECONDS UP TO AN HOUR. THE PHYSICAL EXERTION DEMANDED IN FIREFIGHTING OR OTHER EMERGENCY CONDITIONS MAY BE EXPECTED TO INTENSIFY THE EFFECTS OF ANY EXPOSURE. ONLY HAZARDS ARISING OUT OF AN INHERENT PROPERTY OF THE MATERIAL ARE CONSIDERED. THE FOLLOWING EXPLANATION IS BASED UPON PROTECTIVE EQUIPMENT NORMALLY USED BY FIREFIGHTERS:

4 MATERIALS TOO DANGEROUS TO HEALTH TO EXPOSE FIREFIGHTERS. A FEW WHIFFS OF THE VAPOR COULD CAUSE DEATH OR THE VAPOR OF LIQUID COULD BE FATAL ON PENETRATING THE FIREFIGHTER'S NORMAL FULL PROTECTIVE CLOTHING. THE NORMAL, FULL-PROTECTIVE CLOTHING AND BREATHING APPARATUS AVAILABLE TO THE AVERAGE FIRE DEPARTMENT WILL NOT PROVIDE ADEQUATE PROTECTION AGAINST INHALATION OR SKIN CONTACT WITH THESE MATERIALS.

3 MATERIALS EXTREMELY HAZARDOUS TO HEALTH, BUT AREAS MAY BE ENTERED WITH EXTREME CARE. FULL-PROTECTIVE CLOTHING INCLUDING SELF-CONTAINED BREATHING APPARATUS, COAT, PANTS, GLOVES, BOOTS AND BANDS AROUND LEGS, ARMS AND WAIST SHOULD BE PROVIDED. NO SKIN SURFACE SHOULD BE EXPOSED.

2 MATERIALS HAZARDOUS TO HEALTH, BUT AREAS MAY BE ENTERED FREELY WITH FULL-FACE MASK AND SELF-CONTAINED BREATHING APPARATUS WHICH PROVIDES EYE PROTECTION.

1 MATERIALS ONLY SLIGHTLY HAZARDOUS TO HEALTH. IT MAY BE DESIRABLE TO WEAR SELF-CONTAINED BREATHING APPARATUS.

0 MATERIALS WHICH WOULD OFFER NO HAZARD BEYOND THAT OF ORDINARY COMBUSTIBLE MATERIAL UPON EXPOSURE UNDER FIRE CONDITIONS.

FLAMMABILITY (RED)

SUSCEPTIBILITY TO BURNING IS THE BASIS FOR ASSIGNING DEGREES WITHIN THIS CATEGORY. THE METHOD OF ATTACKING THE FIRE IS INFLUENCED BY THIS SUSCEPTIBILITY FACTOR.

4 VERY FLAMMABLE GASES OR VERY VOLATILE FLAMMABLE LIQUIDS. SHUT OFF FLOW AND KEEP COOLING WATER STREAMS ON EXPOSED TANKS OR CONTAINERS.

3 MATERIALS WHICH CAN BE IGNITED UNDER ALMOST ALL NORMAL TEMPERATURE CONDITIONS. WATER MAY BE INEFFECTIVE BECAUSE OF THE LOW FLASH POINT.

2 MATERIALS WHICH MUST BE MODERATELY HEATED BEFORE IGNITION WILL OCCUR. WATER SPRAY MUST BE USED TO EXTINGUISH THE FIRE BECAUSE THE MATERIAL CAN BE COOLED BELOW ITS FLASH POINT.

1 MATERIALS THAT MUST BE PREHEATED BEFORE IGNITION CAN OCCUR. WATER MAY CAUSE FROTHING IF IT GETS BELOW THE SURFACE OF THE LIQUID AND TURNS TO STEAM. HOWEVER, WATER FOG GENTLY APPLIED TO THE SURFACE WILL CAUSE A FROTHING WHICH WILL EXTINGUISH THE FIRE.

0 MATERIALS THAT WILL NOT BURN.

REACTIVITY (STABILITY)

THE ASSIGNMENT OF DEGREES IN THE REACTIVITY CATEGORY IS BASED UPON THE SUSCEPTIBILITY OF MATERIALS TO RELEASE ENERGY EITHER BY THEMSELVES OR IN COMBINATION WITH WATER. FIRE EXPOSURE WAS ONE OF THE FACTORS CONSIDERED ALONG WITH CONDITIONS OF SHOCK AND PRESSURE.

4 MATERIALS WHICH (IN THEMSELVES) ARE READILY CAPABLE OF DETONATION OR OF EXPLOSIVE DECOMPOSITION OR EXPLOSIVE REACTION AT NORMAL TEMPERATURES AND PRESSURES. INCLUDES MATERIALS WHICH ARE SENSITIVE TO MECHANICAL OR LOCALIZED THERMAL SHOCK. IF A CHEMCIAL WITH THIS HAZARD RATING IS IN AN ADVANCED OR MASSIVE FIRE, THE AREA SHOULD BE EVACUATED.

3 MATERIALS WHICH (IN THEMSELVES) ARE CAPABLE OF DETONATION OR OF EXPLOSIVE DECOMPOSITION OR EXPLOSIVE REACTION BUT WHICH REQUIRE A STRONG INITIATING SOURCE OR WHICH MUST BE HEATED UNDER CONFINEMENT BEFORE INITIATION. INCLUDES MATERIALS WHICH ARE SENSITIVE TO THERMAL OR MECHANICAL SHOCK AT ELEVATED TEMPERATURES AND PRESSURES OF WHICH REACT EXPLOSIVELY WITH WATER WITHOUT REQUIRING HEAT OR CONFINEMENT. FIREFIGHTING SHOULD BE DONE FORM AN EXPLOSIVE-RESISTANT LOCATION.

2 MATERIALS WHICH (IN THEMSELVES) ARE NORMALLY UNSTABLE AND RAPIDLY UNDERGO VIOLENT CHEMICAL CHANGE BUT DO NOT DETONATE. INCLUDES MATERIALS WHICH CAN UNDERGO CHEMICAL CHANGE WITH RAPID RELEASE OF ENERGY AT NORMAL TEMPERATURES AND PRESSURES OR WHICH CAN UNDERGO VIOLENT CHEMICAL CHANGE AT ELEVATED TEMPERATURES AND PRESSURES. ALSO INCLUDES THOSE MATERIALS WHICH MAY REACT VIOLENTLY WITH WATER OR WHICH MAY FORM POTENTIALLY EXPLOSIVE MIXTURES WITH WATER. IN ADVANCE OR MASSIVE FIRES, FIREFIGHTING SHOULD BE DONE FROM A SAFE DISTANCE OR FROM A PROTECTED LOCATION.

1 MATERIALS WHICH (IN THEMSELVES) ARE NORMALLY STABLE BUT WHICH MAY BECOME UNSTABLE AT ELEVATED TEMPERATURES AND PRESSURES OR WHICH MAY REACT WITH WATER WITH SOME RELEASE OF ENERGY BUT NOT VIOLENTLY. CAUTION MUST BE USED IN APPROACHING THE FIRE AND APPLYING WATER.

0 MATERIALS WHICH (IN THEMSELVES) ARE NORMALLY STABLE EVEN UNDER FIRE EXPOSURE CONDITIONS AND WHICH ARE NOT REACTIVE WITH WATER. NORMAL FIREFIGHTING PROCEDURES MAY BE USED.

SPECIAL INFORMATION (WHITE)

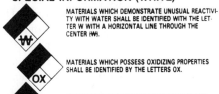

MATERIALS WHICH DEMONSTRATE UNUSUAL REACTIVITY WITH WATER SHALL BE IDENTIFIED WITH THE LETTER W WITH A HORIZONTAL LINE THROUGH THE CENTER (W̶).

MATERIALS WHICH POSSESS OXIDIZING PROPERTIES SHALL BE IDENTIFIED BY THE LETTERS OX.

MATERIALS POSSESSING RADIOACTIVITY HAZARDS SHALL BE IDENTIFIED BY THE STANDARD RADIOACTIVITY SYMBOL.

Figure 43.6. The NFPA 704 marking system.

Observations should not be limited to the containers. The spilled or released material itself can present many identifiers when observed from a safe distance. Color, physical state, viscosity, and vaporization rate are some examples of properties that can help identify a hazardous material. Of course, in addition to the senses, utilizing direct-reading instruments is important for many reasons. The direct-reading instrument can not only provide solid identification of a hazardous material, but can also provide very important exposure level data as well.

3.1.3   Site-Specific Safety Plan

Federal regulations require that the decision to send emergency responders into an area that could potentially have a hazardous atmosphere in anything less protective than self-contained breathing apparatus must be based on documented exposure monitoring data. This means that on-scene measurements of airborne contaminants could play a critical role in emergency response decision making. Because these data are needed immediately on the scene, direct-reading instruments are the apparent choice. However, the use of direct-reading instruments also has some drawbacks to overcome.

First, the material has to be identified to the extent that the appropriate instrumentation can be used. Of course the monitoring equipment must be available on the scene, which suggests sufficient advance planning to anticipate equipment needs or sufficient flexibility of the equipment.

Another factor to consider is that the person performing monitoring in a contaminated area may need the maximum protection during the sampling procedure and must also be competent in field calibration and sampling. If direct-reading instruments do not exist for a particular hazardous material or mixture, then other sampling strategies may have to be developed that include laboratory analysis of the samples. Again, advance planning is important so that the right equipment will be rapidly available, and so that an analytical laboratory is identified and available when the need arises. Even the best advance planning, however, will not enable monitoring data developed from an analytical laboratory to have an impact on the initial emergency response process. In these cases the hygienist will have to work within the constraints of these data gaps.

Ideally, once the identity of the hazardous material or material type has been established and the quantity of the released material has been sufficiently estimated, an evaluation of the situation follows. However, decisions are often made quickly under the time pressures inherent in an emergency response situation, and it is not always possible to wait until all the data are in. Thus worst-case exposure assumptions may be required if the situation just cannot wait for monitoring equipment or laboratory results to arrive back on scene. Evaluation of the situation must be done when it is needed, with the existing data in hand, so much of the evaluation process may be based on professional judgment, knowledge, and experience within the context of emergency response.

During a hazardous materials incident, the conditions are always changing. Weather fluctuations, involvement of additional substances, fire situations, and effects of containment and control procedures are just a few of the factors that can produce changes in the health and safety situation. Data must be continually collected, and constant vigilance be maintained so that the health and safety evaluation at the scene is maintained.

Environmental conditions can play a large role in the safety and health evaluation for the emergency response and the cleanup. Although the spilled material hazards determine the type and level of personal protective equipment, the relationship of that equipment, the personnel wearing the equipment, and the environmental conditions is complex. If the weather is hot, then heat stress is a major concern.

Personnel wearing totally encapsulated suits or other skin protection suits in hot weather may require reduced work time, careful maintenance of body fluids, additional rest time, and closer medical monitoring. In cold weather, or in marine situations where cold water is a concern, protecting personnel against hypothermia is important in the site evaluation. Additionally, the physical integrity of some protective suit materials could be compromised when the temperature drops below freezing.

Ergonomic factors also are a consideration for hazardous material incident activities. The tools utilized must match the dexterity levels of personnel wearing various glove types and layerings. The impact of wearing PPE must be assessed for those personnel operating heavy machinery. The proper handling methods for drums and other containers must be evaluated as well.

Noise levels also should be part of the hazard evaluation. If the type of equipment used and emergency response environment can produce excessive noise levels, then a hearing protection program to protect responders and cleanup personnel should be incorporated into the response plan.

Sometimes the hazardous material incident is combined with a confined space entry. A confined space is generally described as a location that is large enough so that a person can enter, has a limited or restricted means of entry or exit, is not designed for continuous personnel occupancy, and may contain a hazardous atmosphere or lack of oxygen. Special procedures to ensure the safety of personnel working with the additional hazards presented by the confined space are needed.

The evaluation of the hazardous materials incident combines an assessment of the chemical and physical hazards, the potential for hazardous exposures, the site and environmental conditions, and the risks to responding personnel. Many factors are weighed and balanced and interpreted. The industrial hygienist should be able to determine a site-specific health and safety strategy based on this evaluation. The interpretation of the hazard and exposure factors will lead to personal protective equipment recommendations that are appropriate for the level of involvement for each particular personnel group. Not all responders are performing the same task, and consequently each group of responders may be exposed to differing levels of hazardous exposures. The evaluations should therefore produce a PPE plan tailored to meet each group's objective.

The evaluation should also consider the time line of the incident. As the situation progresses, the release could be stabilized, and the risks could consequently become more defined. Conversely, it may become clear that the incident is progressing adversely and it will become worse as time progresses and the risks to responders increases. In some cases, the evaluation indicates that nothing can be done, and responders back off to wait for the appropriate time to reenter the scene.

3.1.4 On-Scene Communications

The final step within the process for establishing the on-scene safety plan is communication with the incident commander. The incident commander has the final authority for the safety of every single person on the hazardous material incident scene, and must be fully apprised of the complete extent of the risks from both the hazardous material and the use of the PPE. There is no time for ineffective communication techniques. The key assumptions and variables must be covered efficiently and effectively. The incident commander must be made aware of the reasons for key decisions concerning PPE selection and other key factors in the on-scene plan.

The main factors of immediate concern to the incident commander would include the nature of the hazard(s) faced by responding personnel, the levels of exposure, the PPE to be utilized to minimize exposure, recommended periods of exposure, decontamination procedures, and medical surveillance and/or biological monitoring requirements.

When the incident command system is operating as designed, all decision-critical information is brought to the incident commander, who may have received certain additional information that might change some of the evaluations made by the hygienist. This must be communicated in an effective give-and-take process as the on-scene plan is made final and implemented. In addition, as the emergency response progresses, communication channels providing feedback from the field must be established so that any necessary adjustments to health and safety decisions can be made and quickly implemented.

4 PERSONAL PROTECTIVE EQUIPMENT FOR EMERGENCY RESPONSE

One of the major tasks for the industrial hygienist who responds to hazardous material incidents is to select the appropriate level and type of personal protective clothing and equipment for all responding personnel. This decision is based on the evaluation discussed in the preceding section and also on an understanding of the PPE available, which includes knowing the specific protective features of the PPE and the limitations as well.

PPE is to be used whenever there is a probability of response personnel coming in contact with a hazardous material. This potential exposure can include gases, liquids, or solids, and can occur not only through direct exposure activities such as leak plugging, but also indirectly through such activities as helping the responders remove PPE during decontamination. If the potential exposure is unknown, then PPE must be used to protect the wearer from the maximum credible exposure scenario.

This section first covers the major aspects of an emergency response PPE program, personal protective clothing and its limitations, and finally the different levels of protection defined by the U.S. Environmental Protection Agency (EPA). Respiratory protection is covered in Chapter 19.

4.1 Elements of Hazardous Materials Incidents PPE Program

The key elements of the PPE program are to protect emergency response personnel from site-specific hazards and to protect the PPE wearer from injury resulting from incorrect use or malfunction of the PPE. The program should detail the hazard identification and the exposure characterization including any known exposure levels, the use, decontamination, and maintenance of the PPE, the medical monitoring requirements, and the training needs for successful program implementation.

Respiratory protection is of primary importance because inhalation is a significant route of exposure, and hazardous materials in a gaseous state not only have a great potential for dispersion but are in some cases not visible. Further details for selecting appropriate respiratory protection are found in Chapter 19. The principles for selecting respiratory protection for emergency responders are similar to those for the workplace, except that unless a site-specific exposure assessment including air monitoring has been performed, the maximum level of protection is warranted. Emergency responders should be prepared to wear supplied air respirators every time they arrive on scene. Only when the on-scene commander has reviewed on-scene exposure assessments, including monitoring data or other documented information, can lesser respiratory protection other than supplied air be considered.

Emergency responders must also always be prepared for the maximum level of skin and eye protection. Again, reduced levels of protective clothing may be worn when the on-scene commander deems it appropriate, based on site-specific exposure assessments including monitoring data or other documented information.

4.2 Personal Protective Clothing

The three main types of personal protective clothing are structural firefighting clothing, chemical protective clothing, and high-temperature protective clothing. The chemical protective clothing is of greatest importance for the hazardous material emergency responder. However, the structural firefighting clothing is usually worn to all calls by fire service responders, who may also be the hazardous materials responders in a particular locale, so a brief summary of this protective clothing and its limitations is provided here.

Structural firefighting clothing is often called "turnout gear," and is the standard type of PPE worn by firefighters. It is primarily designed to afford protection from burns, steam, hot particles, and falling debris resulting from structural fires. It generally consists of a helmet, a fire-resistant hood, positive pressure self-contained breathing apparatus, turnout coat, turnout pants, gloves, and boots. This protective clothing is worn by fire service personnel to fight structural fires, but is sometimes used when they are responding to hazardous materials incidents. When contact with hazardous chemical liquids or vapors is possible, structural firefighting clothing is not appropriate. Although sealing of the arms and legs is possible to a limited extent, chemical splashes or vapors can penetrate the suit with relative ease, and the materials commonly used, such as leather, may be difficult or impossible to decontaminate.

High-temperature protective clothing is designed to protect the wearer for short exposures to either heat or flame. A proximity suit protects the wearer from short exposures to close proximity to heat and flame, and is made of a highly reflective aluminized outer surface over an inner shell of flame-retardant fabric. The fire entry suit offers protection for short-duration entries into a flame environment; it is composed of multiple layers of flame-retardant material. Both types of suits are of limited use in firefighting and have no significant chemical protection properties.

When any chemicals are present, selection of proper protective clothing can be complex and should be performed by an industrial hygienist with knowledge and experience. Choosing the most appropriate clothing depends on the chemicals present and the tasks to be performed. In the selection process, the hygienist balances the performance of the protective clothing in protecting against exposure, the physical limitations created by using different types of protective clothing, and site-specific factors.

Performance against chemical exposure depends on how well the protective clothing material resists permeation, degradation, and penetration. Permeation is the process whereby the chemical moves through protective clothing material by dissolving through the molecular structure of the clothing material. Degradation occurs when chemical exposure, use, or environmental conditions actually break down the protective clothing material through a change in the fabric's chemical composition or structure. Penetration is direct passage of a chemical through openings such as zippers, seams, or imperfections in the protective clothing material. Physical limitations that may influence the selection of protective clothing for response personnel include, but are not limited to, heat stress, excessive mobility or vision restrictions, and incompatibility with appropriate respiratory protection.

In order to make decisions on appropriate PPE, the industrial hygienist can consult a variety of sources that present the wide range of materials in a matrix with the chemicals to be protected against. A sample matrix is presented in Figure 43.7. Because no single material is completely impermeable to all chemicals, one

Generic Class	Butyl Rubber	Polyvinyl Chloride	Neoprene	Natural Rubber
Alcohols	E	E	E	E
Aldehydes	E-G	G-F	E-G	E-F
Amines	E-F	G-F	E-G	G-F
Esters	G-F	P	G	F-P
Ethers	G-F	G	E-G	G-F
Fuels	F-P	G-P	E-G	F-P
Halogenated hydrocarbons	G-P	G-P	G-F	F-P
Hydrocarbons	F-P	F	G-F	F-P
Inorganic acids	G-F	E	E-G	F-P
Inorganic bases and salts	E	E	E	E
Ketones	E	P	G-F	E-F
Natural fats and oils	G-F	G	E-G	G-F
Organic acids	E	E	E	E

Key: E, excellent; G, good; F, fair; P, poor

Figure 43.7.   Chemical protection of clothing materials by generic class.

key factor in the selection process is determining which materials will allow slowest permeation, and/or last the longest under the conditions of use. In some cases the hygienist must estimate the maximum exposure times allowable before permeation or degradation becomes too great a risk and give the responders a time limit for exposure. Increasing thicknesses or doubling protection may be one remedy that is appropriate in some cases. Often in the case of mixtures or incidents involving multiple chemicals, a combination of protective clothing must be worn.

Chemicals in mixture sometimes present a special challenge for hygienists because determining appropriate protective clothing can be complex when one chemical can act as a carrier across the barrier, actually enhancing permeation of a second hazardous chemical. Environmental conditions of cold, rain, humidity, or sun and dryness may also change the rate of permeation or degradation, so the hygienist must carefully weigh many factors before making protective clothing selections. Material flexibility given the ambient conditions is another important factor, particularly for glove selection.

Materials used in personal protective clothing generally fall into three classes: elastomers, nonelastomers, and blends. Some examples of elastomers are rubber, butyl rubber, polyethylene, chlorinated polyethylene, polyvinyl chloride, nitrile, polyurethane, polyvinyl alcohol, and neoprene. Viton™ is a DuPont brand of fluoroelastomer. Nonelastomers include leather, Nomex™ and Tyvek™, DuPont brand products, and Gore-tex™. Blend materials often layer combinations of the different materials listed above to provide protection against a wider variety of hazards in a wider range of environmental conditions. The very wide range of different materials available for personal protective clothing highlights the need for proper advance planning of an incident.

4.3 PPE Ensembles for Hazardous Materials Incidents

Personal protective clothing when teamed with respiratory protection is often called an ensemble. The EPA has defined four levels of protection that combine the different protective values of respiratory PPE and protective clothing. These levels are termed Level A, B, C, and D. Each level is detailed below, but these groupings are considered the starting point for developing ensembles. Many site-specific circumstances will lead the hygienist to recommend modifications for each of the basic levels of protection.

Level A provides maximum protection against chemical exposure. It fully protects the eyes, skin, and respiratory system from exposure to hazardous materials. This level is selected when the chemical that has been identified has a high degree of skin, eye, or respiratory hazard, when the chemical is suspected to be a high skin hazard, when operations are conducted in a poorly ventilated or confined space, perhaps with less than 19.5 percent oxygen, when there are unknown vapors or gases in the air, or when potential skin exposures to an unknown chemical may occur. One task that often requires Level A PPE at a hazardous material incident includes the initial walk-through or monitoring of a site where vapor, gas, or particulate skin hazards are or may be present. The key elements of the Level A protection are the fully encapsulating suit, and the pressure-demand supplied-air breathing apparatus or pressure-demand airline respirator. Figure 43.8 provides a full list of the ensemble equipment.

If Level A protection is utilized, a minimum of at least four persons should be equipped. This is because entry into an area with potential exposures warranting Level A protection requires use of a "buddy system" so that an individual who might be injured or accidentally exposed, or has problems with heat stress, has a partner in visual contact to provide immediate assistance. It is a good practice to have two additional personnel suited up in a "ready" state, in case the first pair encounter unexpected trouble and can't escape. The backups are needed because the exposure situation is clearly hazardous, and all rescuers need the maximum level of protection as well.

Because Level A PPE is impermeable to many substances, it also is very hot and claustrophobic to wear. Heat stress is a genuine hazard to personnel wearing Level A protection even in temperate weather conditions, and appropriate medical surveillance precautions must be taken. Also, unless an airline respirator is used, the combination of limited air, heat stress, equipment weight, suit bulkiness, and limited visibility severely limit the useful work time the personnel may have in the exposure zone.

Level B protection is similar to Level A protection with the exception of slightly reduced skin protection. Level B is worn when the skin hazard of chemical exposure

-Pressure-demand Supplied air respirator approved by NIOSH/MSA:

* pressure-demand self-contained breathing apparatus, or

* pressure-demand airline respirator (with an escape bottle for IDLH conditions)

- Fully Encapsulated Chemical Resistant Suit

- Coveralls*

- Long Cotton Underwear*

- Chemical Resistant Gloves (inner)

- Chemical Resistant Boots with Steel Toe and Shank

- Hard Hat*

- Disposable Gloves and Boot Covers* (worn over fully encapsulated suit gloves and boots)

- Body Cooling Unit*

- 2-Way Radio Communicator (intrinsically safe)

(*) optional equipment

Figure 43.8.   Level A protection equipment list.

is known not to be significantly hazardous. Level B can be worn in Immediately Dangerous to Life and Health atmospheres, if those concentrations do not pose a skin hazard or if the chemicals present do not meet the selection criteria for use of air-purifying respirators. Essentially, supplied air is still needed, but the fully encapsulating suit is not. However, although the protective clothing is not fully encapsulated against gases, vapors, or particulates, it still affords a very substantial level of skin protection. The protective clothing is chemical resistant, hooded, and one piece, with taped ankles and wrists over chemical-resistant boots and gloves.

This level may also be used for initial site entry and reconnaissance, if a lack of skin hazard is verified. Level B is slightly less bulky and cumbersome than Level A, but can be just as hot and stressful. Also, the supplied air canisters are on the outside of the protective clothing, as opposed to Level A where they are often inside the encapsulated suit. When the air canisters are outside the protective clothing, they are subject to contamination and may therefore become additional equipment to decontaminate. The list of Level B equipment is in Figure 43.9.

Level C protection maintains the chemical-resistant suit protection of Level B but allows for reduced respiratory protection. Air-purifying respirators are utilized in Level C protection in combination with the chemical-resistant suit, with hood, gloves, and boots. The chemical-resistant suit may also be downgraded one step to a two-piece outfit depending on anticipated exposure conditions. In order to utilize air-purifying respirators, all criteria for selection must be met. This of course

- Pressure-demand, Supplied-air respirator approved
 by NIOSH/MSA

   * pressure-demand self contained breathing apparatus, or

   * pressure-demand, airline respirator (with escape bottle for
 IDLH conditions)

- Chemical Resistant Clothing (overalls and long sleeved jacket
 or hooded one or two piece chemical splash suit or disposable chemical resistant one piece splash suit)

- Long Cotton Underwear*

- Coveralls*

- Chemical Resistant Outer Gloves

- Chemical Resistant Inner Gloves

- Chemical Resistant Outer Boots with Steel Toe and Shank

- Disposable Chemical Resistant Outer Boot Covers*

- Hard Hat (face shield*)

- 2-way Radio Communicator (intrinsically safe)

(*) optional equipment

Figure 43.9. Level B protection equipment list.

requires identification not only of the chemicals but also their concentrations, and oxygen levels must be 19.5 percent. The site must not be likely to change exposure conditions or generate unknown compounds or excessive levels of known substances. Generally, the full-face air-purifying respirator is used. Level C has the advantages of providing a lighter, less bulky protection ensemble, and the personnel can usually work for longer periods of time than if they were wearing supplied air. The equipment list for Level C protection is given in Figure 43.10.

Level D protection is sometimes defined as primarily a work uniform providing minimal protection. There is no respiratory protection needed, and only limited skin protection, mainly through the use of a coverall. Wearing safety glasses is usually recommended as good safety practice in most situations, and splash goggles may be used in specific instances. Boots are also recommended. Gloves are optional depending on the chemicals involved and the task. It is not common for a hazardous materials incident to have responders outfitted in Level D protection, unless it is a very small, well-defined incident, such as a small container spill in a workplace, where the hazards of the material are well known and are not especially severe. Figure 43.11 lists Level D equipment.

Intermediate ensembles combining elements of different levels are also used. For hazardous materials incidents where monitoring has established that respiratory protection is not needed, and the primary hazard of concern is skin exposure, personnel are usually outfitted in what is called a "modified Level C," where the chemical-resistant clothing of Level C is used, but without the respiratory protection. In the place of the eye protection provided by the full-face respirator, glasses or goggles are substituted.

The EPA levels of protection provide guidelines for PPE ensembles. The specifics of equipment such as gloves or double gloves, boots, and varieties of chemical protective suits are determined on a site-specific basis. PPE is also in a constant state of improvement, as manufacturers are now regularly modifying and improving the materials, blends, and designs, and bringing out new products for the expanding hazardous materials response market. Manufacturers and local equipment suppliers should be contacted regularly to find out the latest specifications and ratings for new products.

5 EMERGENCY RESPONSE MONITORING

Monitoring the hazardous exposures at a hazardous materials incident is a critical function. Without accurate monitoring data, the on-scene incident commander will not have any basis to allow a reduction in PPE below maximum protection. Although it may be possible to determine that skin hazards may not be a factor, and reduce the level of protection to Level C, supplied air would still be required in a potentially hazardous atmosphere if exposure assessment and air monitoring did not show otherwise. Monitoring throughout the incident, even if not direct reading, will be valuable as documentation of exposures, which may be important if any injuries from chemical exposure are incurred, or if any litigation results. Sampling and analysis after the incident has stabilized or moved into the cleanup phase will also play an important role in the overall assessment of exposures and contamination caused by the hazardous materials spill.

- Air Purifying Respirator, Full Face, Canister-equipped, approved by NIOSH/MSA

- Chemical Resistant Clothing (coveralls or hooded, one piece or two piece chemical splash suit, or chemical-resistant hood and apron, or disposable chemical-resistant coveralls)

- Coveralls*, or

- Long Cotton Underwear*

- Chemical-resistant outer gloves

- Chemical-resistant inner gloves

- Chemical Resistant Outer Boots with Steel Toe and Shank

- Outer Disposable Chemical-resistant Boot Covers*

- Hard Hat (face shield*)

- Escape Mask*

- 2-Way Radio Communicator (intrinsically safe)

(*) optional equipment

Figure 43.10. Level C protection equipment list.

- Coveralls

- Gloves*

- Boots/Shoes, Leather or Chemical Resistant, Steel Toe and Shank

- Safety Glasses or Chemical Splash Goggles*

- Hard Hat (face shield*)

- Escape Mask

(*) optional equipment

Figure 43.11. Level D protection equipment list.

A detailed discussion of monitoring instrumentation can be found in Chapter 27.

6 DECONTAMINATION PRINCIPLES

6.1 Decontamination Planning and Implementation

Decontamination is a very critical process in the hazardous materials emergency. Decontamination is simply the process of removing or neutralizing contaminants from the personnel and their equipment. All the extensive efforts to ensure that emergency responders are properly protected from exposure to hazardous materials are wasted if decontamination is not effective. Materials transferred out of the hazardous material zone can produce adverse effects on the responders, support personnel, their families, and the next group of responders who use contaminated equipment. Very specific procedures must be set up and implemented in the decontamination plan. These procedures may range from a multi-step cleaning procedure to carefully disposing of all equipment and clothing in a hazardous waste disposal bin.

The plan for decontamination contains many elements that are site and material specific. However, based on advance planning, it may be possible to anticipate how the decontamination process will be handled. For instance, knowing that disposable protective clothing will be available and utilized enables the decontamination process to be structured one way, whereas knowing that expensive and high-maintenance reusable encapsulating suits will be utilized ensures that the decontamination setup will be handled differently.

Decontamination is one significant component of the site zone control system. When the PPE selection process and the site control zones are being established, the decontamination plan and setup must be established as well. Personnel must never enter a contamination zone without having the decontamination area fully prepared.

The zone control system helps ensure that only properly protected personnel enter a "hot zone" where hazardous materials are present. The responders always enter the "hot zone" through controlled entry points from the "warm zone," and always exit the "hot zone" through a decontamination area located in a "warm zone," and then exit the decontamination area into the "cold zone." This system helps ensure that responders who are in the area of major contamination are wearing the proper protective equipment, and that residual contamination from the decontamination area is isolated from personnel who are not wearing any protection in the cold zone. Also, personnel performing decontamination procedures on the exiting responders may also be required to wear substantial PPE. Determining the proper level of protection for decontamination personnel is a major part of the decontamination plan.

The plan consists of six basic elements which include determining the number and layout of the decontamination stations, determining the decontamination equipment needed, determining the appropriate decontamination methods, setting up procedures to prevent contamination of clean areas, establishing work practices to minimize personnel contact with contaminants in the hot zone and while removing contaminated equipment, and establishing the methods for properly disposing of the materials that are not fully decontaminated. The plan should be flexible to change with changing conditions or equipment needs, or if the site hazards are reassessed after new information is received.

6.2 Contamination Prevention

Preventing contamination is a major aspect of the process. Work practices such as walking around contamination or not directly touching contamination must be stressed. Whenever possible, remote handling equipment should be used. Monitoring and sampling instruments and equipment can be bagged. Sampling ports and sensors must of course be exposed through openings in the bag. Disposable outer garments and equipment are very effective in reducing the amount of decontamination to perform. Decontamination usually produces a certain amount of contaminated material such as waste water or sorbent wipes. The purpose of reducing decontamination is partially to reduce contaminated waste.

6.3   Decontamination Methods

There are two basic decontamination methods. These are to remove the contaminants physically or to inactivate, neutralize, or detoxify the contaminant. Physical removal of the contaminants generally consists of a variety of actions that include rinsing, scraping, and scrubbing, or using stream jets. The runoff from all these types of activities must be contained and disposed of as hazardous waste, or otherwise treated. It cannot be allowed to flow freely, spreading contamination.

Chemical removal can consist of activities such as dissolving contaminants with a solvent, utilizing specific surfactants, or solidifying the contamination through gelling or catalyzing agents, removal of water substrate with absorbents, or freezing. Again, the runoff or residue from these activities must be carefully collected and disposed or treated as a hazardous waste.

The effectiveness of the decontamination process must be monitored to demonstrate that the contaminants are indeed removed. If the decontamination methods prove ineffective, then they must be revised. For example, if the hazardous material has permeated the protective clothing to the extent that it can't be removed by scrubbing with the appropriate wash solution, then disposal should be considered. Visual observation, wipe sampling, analysis of cleaning solution runoff, and even permeation testing should each be considered as necessary. Another aspect of the decontamination process to consider is whether any of the materials or processes used could be hazardous themselves. Compatibility of the contaminants and cleaning agents, compatibility of the cleaning agents and the protective clothing materials, and the toxicity of the cleaning agents should be assessed.

6.4   Decontamination Facility Design

The general design for the decontamination facility or area should be based on the principle that the area closest to the exit of the hot zone will receive the most contaminated personnel and equipment, and the decontamination area exits into the cold zone where emerging personnel should be free of contamination. The actual design is highly site specific. The first stages handle the outer garments and equipment, and later stages handle the inner garments. Each removal or cleaning is done stepwise to prevent cross contamination, and is often set up in a line.

The design must also take into account potential emergency decontamination for injured personnel. One rule of thumb is if the injured or ill person will not be harmed by the delay from emergency decontamination, then it should be performed. It is highly desirable to avoid contaminating an ambulance, emergency room, and so on; some hospitals may not even accept contaminated personnel. If it is warranted, on-scene medical facilities can be coordinated with the emergency decontamination design. A typical decontamination facility is outlined in Figure 43.12.

Figure 43.13 provides a general list of some of the materials needed for decontamination facilities. Of course, if disposable materials are being heavily relied upon, the decontamination facility does not have to be as extensive as one where

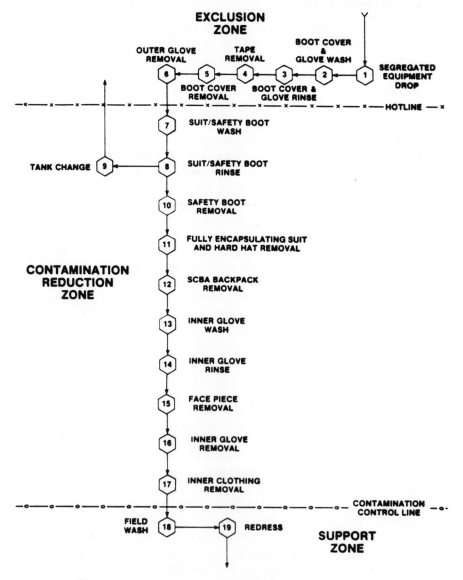

Figure 43.12. Decontamination facility—Level A protection.

- PLASTIC TARPS, HEAVY POLYETHYLENE SHEETING

- PORTABLE PLASTIC SWIMMING POOLS

- SOFT BRUSHES, LONG AND SHORT HANDLES

- BUCKETS

- WATER SUPPLY,

- HOSES, SPRAY APPLICATORS, WATER PRESSURE
  SPRAY UNITS

- SALVAGE DRUMS

- DRUM LINERS

- EQUIPMENT HANGERS (FOR SCBA, BOOTS ETC.)

- DUCT TAPE

- DISPOSABLE WIPING MATERIALS/ ABSORBENTS

- LIQUID SPILL CONTAINMENT/ABSORBENT MATERIALS

- CONTAMINANT NEUTRALIZATION CHEMICALS

ADDITIONAL EQUIPMENT:

- PERSONAL PROTECTIVE EQUIPMENT FOR DECON
  PERSONNEL

- CONTAMINANT TEST EQUIPMENT

Figure 43.13.   Typical decontamination equipment and materials.

multiple-step washes and rinses are utilized. When materials are collected for disposal, it must be done in a manner that is in compliance with local and all other applicable regulations governing the generation and disposal of hazardous waste. Advance planning must address this issue.

REFERENCES

U.S. E.P.A. Office of Emergency and Remedial Response, Emergency Response Division, *Standard Operating Safety Guides*, 1988.

NIOSH, OSHA, USCG, EPA, *Occupational Safety and Health Guidance Manual For Hazardous Waste Site Activities*, 1985.

G. G. Noll, M. S. Hildebrand, and J. G. Yvorra, *Hazardous Materials—Managing the Incident*, Fire Protection Publications, Stillwater, OK, 1988.

Hazardous Waste and Emergency Response, OSHA Publication # 3114.

NIOSH, *Hazardous Waste Sites and Hazardous Substance Emergencies*, Worker Bulletin, Publication #83-100.

OSHA Final Rule, *Hazardous Waste Operations and Emergency Response Final Rule*, 29 CFR Part 1910, 54 FR 9294, 1989.

"Department of Transportation Emergency Response Guidebook for Initial Response to Hazardous Materials Incidents," # DOT P5800.4, U.S. Department of Transportation, Washington, DC., 1987.

Appendix

MODEL SITE SAFETY PLAN

This appendix provides a generic plan based on a plan developed by the
U.S. Coast Guard for responding to hazardous chemical releases.[1] This
generic plan can be adapted for designing a Site Safety Plan for hazardous
waste site cleanup operations. It is not all inclusive and should only be
used as a guide, not a standard.

A. SITE DESCRIPTION
    Date_____   Location_____
    Hazards_____
    Area affected_____

    Surrounding population_____
    Topography_____
    Weather conditions_____
    _____

    Additional information_____
    _____
    _____
    _____
    _____

B. ENTRY OBJECTIVES - The objective of the initial entry to the contaminated
    area is to ___(describes actions, tasks to be accomplished; i.e., identify
    contaminated soil; monitor conditions, etc.)_____
    _____
    _____
    _____

C. ONSITE ORGANIZATION AND COORDINATION - The following personnel are
    designated to carry out the stated job functions on site. (Note: One
    person may carry out more than one job function.)

    PROJECT TEAM LEADER_____
    SCIENTIFIC ADVISOR_____
    SITE SAFETY OFFICER_____
    PUBLIC INFORMATION OFFICER_____
    SECURITY OFFICER_____
    RECORDKEEPER_____
    FINANCIAL OFFICER_____
    FIELD TEAM LEADER_____
    FIELD TEAM MEMBERS_____
    _____
    _____
    _____
    _____
    _____

---

[1]U.S. Coast Guard. Policy Guidance for Response to Hazardous Chemical
Releases. USCG Pollution Response COMDTINST-M16465.30.

Appendix (*Continued*)

FEDERAL AGENCY REPS __(i.e., EPA, NIOSH)_____
_____
_____

STATE AGENCY REPS _____
_____

LOCAL AGENCY REPS _____
_____
_____

CONTRACTOR(S) _____
_____
_____
_____

All personnel arriving or departing the site should log in and out with the Recordkeeper. All activities on site must be cleared through the Project Team Leader.

D. ONSITE CONTROL

__(Name of individual or agency_____ has been designated to coordinate access control and security on site. A safe perimeter has been established at __(distance or description of controlled area)_____
_____

No unauthorized person should be within this area.

The onsite Command Post and staging area have been established at _____
_____

The prevailing wind conditions are _____. This location is upwind from the Exclusion Zone.

Control boundaries have been established, and the Exclusion Zone (the contaminated area), hotline, Contamination Reduction Zone, and Support Zone (clean area) have been identified and designated as follows: __(describe_____ boundaries and/or attach map of controlled area)_____
_____
_____

These boundaries are identified by: __(marking of zones, i.e., red boundary___ tape - hotline; traffic cones - Support Zone; etc.)_____
_____
_____

continued on next page

Appendix (*Continued*)

E. HAZARD EVALUATION

The following substance(s) are known or suspected to be on site. The primary
hazards of each are identified.

Substances Involved	Concentrations (If Known)	Primary Hazards
(chemical name)		(e.g., toxic on inhalation)

The following additional hazards are expected on site: (i.e., slippery
ground, uneven terrain, etc.)

Hazardous substance information form(s) for the involved substance(s) have
been completed and are attached.

F. PERSONAL PROTECTIVE EQUIPMENT

Based on evaluation of potential hazards, the following levels of personal
protection have been designated for the applicable work areas or tasks:

Location	Job Function	Level of Protection
Exclusion Zone		A B C D Other
		A B C D Other
		A B C D Other
		A B C D Other
Contamination Reduction Zone		A B C D Other
		A B C D Other
		A B C D Other
		A B C D Other

Specific protective equipment for each level of protection is as follows:

Level A   Fully-encapsulating suit   Level C   Splash gear (type)
            SCBA   Full-face canister resp.
            (disposable coveralls)

Level B   Splash gear (type)   Level D
            SCBA

Other

Appendix (*Continued*)

The following protective clothing materials are required for the involved substances:

Substance	Material
(chemical name)	(material name, e.g., Viton)
_____	_____
_____	_____
_____	_____

If air-purifying respirators are authorized, __(filtering medium)__ is the appropriate canister for use with the involved substances and concentrations. A competent individual has determined that all criteria for using this type of respiratory protection have been met.

NO CHANGES TO THE SPECIFIED LEVELS OF PROTECTION SHALL BE MADE WITHOUT THE APPROVAL OF THE SITE SAFETY OFFICER AND THE PROJECT TEAM LEADER.

G. ONSITE WORK PLANS

Work party(s) consisting of _____ persons will perform the following tasks:

Project Team Leader ____(name)____   _____ (function) _____
                                              _____
                                              _____

Work Party #1   _____   _____
                                              _____
                                              _____
                                              _____

Work Party #2   _____   _____
                                              _____
                                              _____
                                              _____

Rescue Team   _____   _____
(required for   _____
entries to IDLH   _____
environments)   _____

Decontamination   _____
Team   _____   _____
                                              _____
                                              _____
                                              _____

The work party(s) were briefed on the contents of this plan at _____.

continued on next page

Appendix (*Continued*)

2. Emergency Medical Care

   __(names of qualified personnel)__ are the qualified EMTs on site.
   __(medical facility names)__, at __(address)__,
   phone _____ is located _____ minutes from this location.
   __(name of person)__ was contacted at __(time)__ and briefed on
   the situation, the potential hazards, and the substances involved. A map
   of alternative routes to this facility is available at __(normally Command
   Post)__.

   Local ambulance service is available from _____ at
   phone _____. Their response time is _____ minutes.
   Whenever possible, arrangements should be made for onsite standby.

   First-aid equipment is available on site at the following locations:

   First-aid kit   _____
   Emergency eye wash   _____
   Emergency shower   _____
   __(other)__   _____

   Emergency medical information for substances present:

Substance	Exposure Symptoms	First-Aid Instructions

   List of emergency phone numbers:

Agency/Facility	Phone #	Contact
Police		
Fire		
Hospital		
Airport		
Public Health Advisor		

3. Environmental Monitoring

   The following environmental monitoring instruments shall be used on site
   (cross out if not applicable) at the specified intervals.

   Combustible Gas Indicator - continuous/hourly/daily/other _____
   O_2 Monitor   - continuous/hourly/daily/other _____
   Colorimetric Tubes   - continuous/hourly/daily/other _____
   __(type)__   _____

   _____
   HNU/OVA   - continuous/hourly/daily/other _____
   Other _____   - continuous/hourly/daily/other _____
   _____   - continuous/hourly/daily/other _____

Appendix (*Continued*)

4. Emergency Procedures (should be modified as required for incident)

The following standard emergency procedures will be used by onsite personnel. The Site Safety Officer shall be notified of any onsite emergencies and be responsible for ensuring that the appropriate procedures are followed.

Personnel Injury in the Exclusion Zone: Upon notification of an injury in the Exclusion Zone, the designated emergency signal _____ shall be sounded. All site personnel shall assemble at the decontamination line. The rescue team will enter the Exclusion Zone (if required) to remove the injured person to the hotline. The Site Safety Officer and Project Team Leader should evaluate the nature of the injury, and the affected person should be decontaminated to the extent possible prior to movement to the Support Zone. The onsite EMT shall initiate the appropriate first aid, and contact should be made for an ambulance and with the designated medical facility (if required). No persons shall reenter the Exclusion Zone until the cause of the injury or symptoms is determined.

Personnel Injury in the Support Zone: Upon notification of an injury in the Support Zone, the Project Team Leader and Site Safety Officer will assess the nature of the injury. If the cause of the injury or loss of the injured person does not affect the performance of site personnel, operations may continue, with the onsite EMT initiating the appropriate first aid and necessary follow-up as stated above. If the injury increases the risk to others, the designated emergency signal _____ shall be sounded and all site personnel shall move to the decontamination line for further instructions. Activities on site will stop until the added risk is removed or minimized.

Fire/Explosion: Upon notification of a fire or explosion on site, the designated emergency signal _____ shall be sounded and all site personnel assembled at the decontamination line. The fire department shall be alerted and all personnel moved to a safe distance from the involved area.

Personal Protective Equipment Failure: If any site worker experiences a failure or alteration of protective equipment that affects the protection factor, that person and his/her buddy shall immediately leave the Exclusion Zone. Reentry shall not be permitted until the equipment has been repaired or replaced.

Other Equipment Failure: If any other equipment on site fails to operate properly, the Project Team Leader and Site Safety Officer shall be notified and then determine the effect of this failure on continuing operations on site. If the failure affects the safety of personnel or prevents completion of the Work Plan tasks, all personnel shall leave the Exclusion Zone until the situation is evaluated and appropriate actions taken.

continued on next page

Appendix (*Continued*)

H. COMMUNICATION PROCEDURES

Channel ____ has been designated as the radio frequency for personnel in the
Exclusion Zone. All other onsite communications will use channel ____.

Personnel in the Exclusion Zone should remain in constant radio communication
or within sight of the Project Team Leader. Any failure of radio
communication requires an evaluation of whether personnel should leave the
Exclusion Zone.

__(Horn blast, siren, etc.)__ is the emergency signal to indicate that all
personnel should leave the Exclusion Zone. In addition, a loud hailer is
available if required.

The following standard hand signals will be used in case of failure of radio
communications:

        Hand gripping throat --------------- Out of air, can't breathe
        Grip partner's wrist or ------------ Leave area immediately
          both hands around waist
        Hands on top of head --------------- Need assistance
        Thumbs up -------------------------- OK, I am all right, I understand
        Thumbs down ------------------------ No, negative

Telephone communication to the Command Post should be established as soon as
practicable. The phone number is _____.

I. DECONTAMINATION PROCEDURES

Personnel and equipment leaving the Exclusion Zone shall be thoroughly
decontaminated. The standard level _____ decontamination protocol shall be
used with the following decontamination stations: (1) _____
(2) _____ (3) _____ (4) _____ (5) _____
(6) _____ (7) _____ (8) _____ (9) _____
(10) _____ Other _____

Emergency decontamination will include the following stations: _____
_____
_____
_____

The following decontamination equipment is required: _____
_____
_____
_____

__(Normally detergent and water)__ will be used as the decontamination
solution.

J. SITE SAFETY AND HEALTH PLAN

1. _____(name)_____ is the designated Site Safety Officer and is
directly responsible to the Project Team Leader for safety recommendations on
site.

Appendix (*Continued*)

The following emergency escape routes are designated for use in those situations where egress from the Exclusion Zone cannot occur through the decontamination line: __(describe alternate routes to leave area in__ __emergencies)__

In all situations, when an onsite emergency results in evacuation of the Exclusion Zone, personnel shall not reenter until:

1. The conditions resulting in the emergency have been corrected.
2. The hazards have been reassessed.
3. The Site Safety Plan has been reviewed.
4. Site personnel have been briefed on any changes in the Site Safety Plan.

5. Personal Monitoring

The following personal monitoring will be in effect on site:

Personal exposure sampling: __(describe any personal sampling programs__ __being carried out on site personnel. This would include use of sampling__ __pumps, air monitors, etc.)__
Medical monitoring: The expected air temperature will be __(°F)__ . If it is determined that heat stress monitoring is required (mandatory if over 70°F) the following procedures shall be followed: __(describe procedures in effect, i.e., monitoring body temperature, body__ __weight, pulse rate)__

All site personnel have read the above plan and are familiar with its provisions.

Site Safety Oficer _____ (name) _____   _____ (signature) _____
Project Team Leader_____   _____
Other Site Personnel_____   _____

Index

989

Combined Index for Parts A and B